LQ7 86.10

SUMMARY OF CONTENTS

Table of Contents   v
Acknowledgments   xxi
List of Figures   xxv
List of Tables   xxix
Introduction   1
  1 — Octanol/Water Partition Coefficient   1-1
  2 — Solubility in Water   2-1
  3 — Solubility in Various Solvents   3-1
  4 — Adsorption Coefficient for Soils and Sediments   4-1
  5 — Bioconcentration Factor in Aquatic Organisms   5-1
  6 — Acid Dissociation Constant   6-1
  7 — Rate of Hydrolysis   7-1
  8 — Rate of Aqueous Photolysis   8-1
  9 — Rate of Biodegradation   9-1
 10 — Atmospheric Residence Time   10-1
 11 — Activity Coefficient   11-1
 12 — Boiling Point   12-1
 13 — Heat of Vaporization   13-1
 14 — Vapor Pressure   14-1
 15 — Volatilization from Water   15-1
 16 — Volatilization from Soil   16-1
 17 — Diffusion Coefficients in Air and Water   17-1
 18 — Flash Points of Pure Substances   18-1
 19 — Densities of Vapors, Liquids and Solids   19-1
 20 — Surface Tension   20-1
 21 — Interfacial Tension with Water   21-1
 22 — Liquid Viscosity   22-1
 23 — Heat Capacity   23-1
 24 — Thermal Conductivity   24-1
 25 — Dipole Moment   25-1
 26 — Index of Refraction   26-1

Appendices
  A — Bibliography of Standard Chemical Property
     Data Sources   A-1

  B — Simple Linear Regression   B-1

  C — Evaluating Propagated and Total Error in
     Chemical Property Estimates   C-1

Index follows Appendix C

HANDBOOK OF
CHEMICAL PROPERTY
ESTIMATION METHODS

ABOUT THE EDITORS

WARREN J. LYMAN, Ph.D., is a senior consultant at Arthur D. Little, Inc., Cambridge, Massachusetts, where he serves the needs of a variety of commercial and government clients in the area of environmental chemistry. Recent projects have focused on the fate and transport of chemicals in the environment, environmental fate modeling, exposure assessments, hazardous waste treatment, chemical spill response, and water quality problems.

WILLIAM F. REEHL has been a member of Arthur D. Little's Communication Services group for over 20 years and has edited hundreds of technical reports to government and industrial clients.

DAVID H. ROSENBLATT, Ph.D., is the Division Chemistry Advisor for the Environmental Protection Research Division, U.S. Army Medical Bioengineering Research and Development Laboratory at Fort Detrick, Frederick, Maryland. He holds several patents for chemical synthesis; has published numerous papers and reviews on chemical kinetics and mechanisms, analytical methods, synthesis, and environmentally significant data bases; and has developed methods for chemical risk assessment.

HANDBOOK OF
CHEMICAL PROPERTY
ESTIMATION METHODS
Environmental Behavior of Organic Compounds

Warren J. Lyman, Ph.D.
William F. Reehl

Arthur D. Little, Inc., Cambridge, Massachusetts

David H. Rosenblatt, Ph.D.

U.S. Army Medical Bioengineering Research and Development
Laboratory, Fort Detrick, Frederick, Maryland

McGraw-Hill Book Company

New York   St. Louis   San Francisco   Auckland
Bogotá   Hamburg   Johannesburg   London   Madrid   Mexico
Montreal   New Delhi   Panama   Paris
São Paulo   Singapore   Sydney   Tokyo   Toronto

Library of Congress Cataloging in Publication Data
Lyman, Warren J.
    Handbook of chemical property estimation methods.

    Includes index.
    1. Organic compounds—Analysis.   I. Reehl,
William F.   II. Rosenblatt, David Hirsh.   III. Title.
QD271.L95   547.3   81-23662
ISBN 0-07-039175-0   AACR2

1234567890   KPKP   898765432

ISBN 0-07-039175-0

TABLE OF CONTENTS

ACKNOWLEDGMENTS   xxi

LIST OF FIGURES   xxv

LIST OF TABLES   xxix

INTRODUCTION   1

    Overview   1
    Appendices   4
    Objectives   4
    Benefits of Estimation   6
    Contents of Each Chapter   7
    Limitations of this Handbook   8
    Errors Associated with Chemical Property Estimates   8
    Documentation and Reporting of Estimated Values   9

1   OCTANOL/WATER PARTITION COEFFICIENT

    1-1   Introduction   1-1

        Definition and Measurement   1-1
        Environmental Significance   1-2
        Estimation Methods Described in This Handbook   1-3

    1-2   Overview of Available Estimation Methods   1-5

    1-3   Leo's Fragment Constant Method   1-10

        Principles of Use   1-10
        Method Error   1-12
        Fragments and Factors   1-16
        Basic Steps   1-29

    1-4   Estimation with Solvent Regression Equations   1-39

        Principles of Use   1-39
        Solvent Regression Equations   1-39
        Method Errors   1-42
        Basic Steps   1-42

    1-5   Estimation from (Estimated) Activity Coefficients   1-47

        Introduction   1-47

		Relating K_{ow} to γ	1-47
		Estimating γ^w and γ^o	1-49
	1-6	Available Data	1-49
	1-7	Symbols Used	1-50
	1-8	References	1-51
2	**SOLUBILITY IN WATER**		
	2-1	Introduction	2-1
		Definition	2-1
		Units and Range of Values	2-2
		Estimation Methods Provided	2-2
	2-2	Overview of Available Estimation Methods	2-4
	2-3	Factors Influencing Solubility	2-11
		Method of Measurement	2-11
		Temperature	2-11
		Salinity	2-12
		Dissolved Organic Matter	2-12
		pH	2-13
	2-4	Estimation of S from K_{ow}	2-13
		Equations Available	2-13
		Basis for Estimation Method	2-24
		Method Errors	2-25
		Selection of Appropriate Equation(s)	2-33
		Basic Steps	2-35
	2-5	Estimation of S from Structure (Method of Irmann)	2-39
		Method Errors	2-39
		Basic Steps	2-41
	2-6	Available Data	2-46
	2-7	Symbols Used	2-46
	2-8	References	2-48
3	**SOLUBILITY IN VARIOUS SOLVENTS**		
	3-1	Introduction	3-1
		Liquid-Liquid Systems	3-1
		Solid-Liquid Systems	3-2
		Gas-Liquid Systems	3-2

3-2	Basic Approach	3-2
3-3	Other Estimation Methods Considered	3-3
	Use of Solvent/Water Partition Coefficients	3-3
	Use of Solubility Parameters	3-5
	Gas Solubilities in Polar and Non-Polar Solvents	3-5
3-4	Liquid-Liquid Binary Solutions	3-6
	Basis for Estimation Method	3-6
	Calculation of Vapor-Liquid Equilibria	3-10
	Method Errors	3-10
	Basic Steps	3-13
3-5	Solid-Liquid Binary Solutions	3-17
	Basis for Estimation Method	3-17
	Method Errors	3-20
	Basic Steps	3-21
3-6	Available Data	3-26
3-7	Symbols Used	3-27
3-8	References	3-28
4	**ADSORPTION COEFFICIENT FOR SOILS AND SEDIMENTS**	
4-1	Introduction	4-1
	The Adsorption Coefficient, K_{oc}	4-1
	Overview of Estimation Methods	4-3
	Factors Influencing the Values of K and K_{oc}	4-4
4-2	Available Estimation Methods	4-8
	Regression Equations	4-8
	Selection of the Most Appropriate Equation(s)	4-20
	Basic Steps for Estimating K_{oc}, K, and x/m	4-20
	Uncertainty in Estimated Values	4-23
4-3	Available Data	4-28
4-4	Symbols Used	4-29
4-5	References	4-30
5	**BIOCONCENTRATION FACTOR IN AQUATIC ORGANISMS**	
5-1	Introduction	5-1
5-2	Basic Approach	5-3
5-3	Methods of Estimation	5-4

	Estimation from Octanol-Water Partition Coefficient	5-4
	Estimation from Water Solubility	5-10
	Estimation from Soil Adsorption Coefficients	5-13
	Other Regression Equations	5-13

5-4   Uses and Limitations of Estimated Values   5-17

Sources of Discrepancies between BCF
Estimates and Laboratory Data   5-17
Application of BCF Estimates to Field Situations   5-21

**5-5   Other Approaches to Estimating the
Accumulation of Organic Compounds   5-23**

5-6   Available Data   5-26

5-7   Symbols Used   5-26

5-8   References   5-27

6   ACID DISSOCIATION CONSTANT

6-1   Introduction   6-1

6-2   Experimental Measurement of K_a 6-5

6-3   Overview of Estimation Method   6-6

**6-4   Estimation of K_a for Aromatic Acids —
Hammett Correlation   6-9**

Basic Steps for Substituted Benzoic Acids   6-10
Basic Steps for Other Aromatic Acids   6-16

**6-5   Estimation of K_a for Aliphatic Acids —
Taft Correlation   6-20**

Basic Steps   6-21

6-6   Uncertainty in Estimated Values   6-22

6-7   Available Data   6-24

6-8   Symbols Used   6-24

6-9   References   6-27

7   RATE OF HYDROLYSIS

7-1   Introduction   7-1

7-2   Characteristics of Hydrolysis   7-4

Hydrolysis Mechanism   7-4
Hydrolysis Rate Law   7-7

		Measurement of Hydrolysis Rate	7-11
		Temperature Dependence of k	7-14
		Effect of Reaction Medium	7-16
	7-3	Overview of Estimation Methods	7-18
	7-4	Uncertainty in Estimating Values	7-22
	7-5	Estimation of k_H from the Hammett Correlation	7-22
		Basic Steps	7-23
	7-6	Estimation of k_H from the Taft Correlation	7-25
		Basic Steps	7-25
	7-7	Estimation of k_0 from the Hammett Equation	7-26
		Basic Steps	7-26
	7-8	Estimation of k_{OH} from the Hammett Equation	7-28
		Basic Steps	7-28
	7-9	Estimation of k_{OH} from the Taft Equation	7-31
		Basic Steps	7-31
	7-10	Estimation of k_{OH} from the pK_a of the Leaving Group	7-32
		Basic Steps	7-32
	7-11	Available Data	7-36
	7-12	Symbols Used	7-36
	7-13	References	7-46
8		RATE OF AQUEOUS PHOTOLYSIS	
	8-1	Introduction	8-1
	8-2	Basic Principles of Excitation/Deactivation	8-2
		Excitation	8-2
		Deactivation: Internal Conversion and Intersystem Crossing	8-4
		Energy Transfer: Sensitization and Quenching	8-7
		Summary	8-8
	8-3	Absorption of Light	8-9
		Chromophores and Characteristic Absorption Bands	8-9
		Quantitative Calculation of Absorption of Solar Energy	8-16
		Compound-specific Inputs (ϵ_λ values)	8-19
		Ecosystem-specific Inputs (I_λ values)	8-22

8-4   Photochemical Reactions   8-29

    General Considerations   8-29
    Some Specific Examples   8-36
    Real-World Complications   8-39

8-5   Symbols Used   8-40
8-6   References   8-41

9   RATE OF BIODEGRADATION

9-1   Introduction   9-1
9-2   Principles of Biodegradation   9-2

    Definition   9-2
    Characterization of the Biological System   9-3
    Variables in Biodegradation   9-21

9-3   Standard Test Methods   9-33

    Principles of Use   9-33
    Characteristics of Typical Tests   9-37
    Effect of Method and Analytical
      Technique on Measured Rates   9-46

9-4   Biodegradation Rate Constants   9-47

    Derivation   9-47
    Rate Constants for Various Organic Compounds   9-51
    Extrapolation of Laboratory Results to Field Conditions   9-52

9-5   Estimation of Biodegradation Rates   9-57

    Solubility   9-62
    BOD/COD   9-62
    Hydrolysis   9-69

9-6   Available Data   9-70
9-7   Symbols and Abbreviations   9-74
9-8   References   9-75

10   ATMOSPHERIC RESIDENCE TIME

10-1   Introduction   10-1
10-2   Selection of Appropriate Method   10-3
10-3   Steady-State Model   10-13

    Principles of Use   10-13
    Basic Steps   10-15

10-4   Nonsteady-State, One-Compartment Model   10-16

    Principles of Use   10-16
    Basic Steps   10-16

10-5   Nonsteady-State, Two-Compartment Model   10-18

    Principles of Use   10-18
    Basic Steps   10-20

10-6   Use of Chemical Reactivity Data   10-21

    Principles of Use   10-21
    Basic Steps   10-26

10-7   Correlation with Mean Standard
      Deviation (Junge's Correlation)   10-27

    Principles of Use   10-27
    Basic Steps   10-29

10-8   Symbols Used   10-29

10-9   References   10-31

11   ACTIVITY COEFFICIENT

11-1   Introduction   11-1

11-2   Available Methods   11-2

11-3   Method Errors   11-7

11-4   Method 1 — Infinite Dilution Activity Coefficients   11-10

    Basic Steps   11-14

11-5   Method 2 — UNIFAC   11-20

    Basic Steps   11-25

11-6   Available Data   11-49

11-7   Symbols Used   11-49

11-8   References   11-51

12   BOILING POINT

12-1   Introduction   12-1

12-2   Selection of Appropriate Method   12-3

12-3   Meissner's Method   12-8

    Principles of Use   12-8
    Basic Steps   12-13

12-4   Lydersen-Forman-Thodos Method   12-16

    Principles of Use   12-16
    Calculation of Temperature Ratio, θ 12-17
    Estimation of Critical Temperature   12-17
    Basic Steps   12-28

12-5   Miller's Method   12-33

    Basic Steps   12-34

12-6   Method of Ogata and Tsuchida   12-39

    Principles of Use   12-39
    Basic Steps   12-40

12-7   Method of Somayajulu and Palit   12-42

    Principles of Use   12-42
    Basic Steps   12-42

12-8   Kinney's Method   12-44

    Principles of Use   12-44
    Basic Steps   12-46

12-9   Method of Stiel and Thodos   12-47

    Principles of Use   12-47
    Basic Steps   12-48

12-10 Factors Affecting Boiling Point   12-50

12-11 Available Data   12-52

12-12 Symbols Used   12-52

12-13 References   12-54

13 HEAT OF VAPORIZATION

13-1   Introduction   13-1

13-2   Available Estimation Methods   13-2

13-3   Selection of Appropriate Method   13-4

13-4   Estimation of ΔH_{vb} from Critical Constants   13-4

    Principles of Use   13-4
    Basic Steps   13-9

13-5   Estimation of ΔH_{vb} from Vapor Pressure Data   13-12

    Principles of Use   13-12
    Basic Steps   13-15

13-6   Estimation of ΔH_{vb} from Compound Structure   13-16

    Principles of Use   13-16
    Basic Steps   13-17

13-7   Estimation of ΔH_{vb} at Temperatures Other than
    the Boiling Point   13-23

    Basic Steps   13-24

13-8   Available Data   13-25

13-9   Symbols Used   13-25

13-10 References   13-26

14   VAPOR PRESSURE

14-1   Introduction   14-1

14-2   Selection of Appropriate Method   14-2

    Evaluation of Parameters   14-2
    General Characteristics   14-4

14-3   Method 1   14-7

    Derivation   14-7
    Basic Steps   14-8

14-4   Method 2   14-12

    Derivation   14-12
    Basic Steps   14-14

14-5   Estimation from Boiling Points at Reduced Pressure   14-15

    Basic Steps   14-16

14-6   Available Data   14-18

14-7   Symbols Used   14-18

14-8   References   14-19

15   VOLATILIZATION FROM WATER

15-1   Introduction   15-1

15-2   Modeling Volatilization in the Environment   15-2

15-3   Approaches to Estimation of the Volatilization Rate   15-4

    Method of Mackay and Wolkoff   15-4
    Method of Liss and Slater   15-5
    Method of Chiou and Freed   15-6
    Method of Smith *et al.*   15-6

15-4	Method Errors	15-7
15-5	Methods of Estimation	15-9
	Recommended General Method	15-9
	Method of Smith *et al.*	15-17
	Basic Steps of Calculation	15-27
	Basic Steps of Calculation via Reaeration Coefficient	15-30
15-6	Symbols Used	15-31
15-7	References	15-33

16   VOLATILIZATION FROM SOIL

16-1	Introduction	16-1
16-2	Factors Affecting the Volatilization Process	16-2
	Properties on Which Volatilization Is Dependent	16-2
	Compound Distribution and Equilibria	16-3
	Adsorption	16-3
	Vapor Density	16-4
	Water Content of the Soil	16-4
	Partitioning between Water, Soil, and Air	16-5
	Temperature	16-6
	Atmospheric Conditions	16-7
	Diffusion	16-8
	Mass Transport by the Wick Effect	16-10
16-3	Methods for Estimating Volatilization of Chemicals from Soil	16-11
	Hartley Method	16-11
	Hamaker Method	16-12
	Mayer, Letey, and Farmer Method	16-13
	Jury, Grover, Spencer, and Farmer Method	16-19
	Dow Method	16-25
16-4	Selection of Method	16-27
	Basic Steps	16-27
16-5	Evaluating the Error Function	16-43
16-6	Symbols Used	16-46
16-7	References	16-49

17   DIFFUSION COEFFICIENTS IN AIR AND WATER

17-1	Introduction	17-1

Air-Water Interfaces   17-2
Interstitial Waters   17-4
Groundwater   17-4

17-2   Diffusivity in Air   17-6

17-3   Available Methods of Estimating
Diffusion Coefficients in Air   17-7

17-4   Selected Methods of Estimating
Gaseous Diffusion Coefficients of Organics in Air   17-9

FSG Method   17-9
Basic Steps   17-10
WL Method   17-13
Basic Steps   17-14

17-5   Diffusivity in Water   17-17

17-6   Available Methods for Estimating
Diffusion Coefficients in Water   17-18

17-7   Recommended Methods of Estimating
$\mathscr{D}_{BW}$ for Organic Liquids and Vapors   17-20

Basic Steps   17-20

17-8   Available Data   17-22

17-9   Symbols Used   17-22

17-10   References   17-24

18   FLASH POINTS OF PURE SUBSTANCES

18-1   Introduction   18-1

18-2   Selection of Appropriate Method   18-3

18-3   Affens' Method   18-3

Principles of Use   18-3
Basic Steps   18-5

18-4   Prugh's Method   18-6

Principles of Use   18-6
Basic Steps   18-7

18-5   Butler's Method   18-9

Principles of Use   18-9
Basic Steps   18-10

18-6   Flash Points of Mixtures   18-12

18-7   Available Data   18-13

18-8   Symbols Used   18-13

18-9   References   18-14

19   DENSITIES OF VAPORS, LIQUIDS AND SOLIDS

19-1   Introduction   19-1

19-2   Vapor Density Estimation   19-2

        Basic Steps   19-3

19-3   Available Methods for Estimating Liquid Density   19-3

19-4   Selection of Appropriate Method for Liquid Density   19-5

19-5   Grain's Method (Liquid Density)   19-7

        Basic Steps   19-11

19-6   Bhirud's Method (Liquid Density)   19-12

        Basic Steps   19-13

19-7   Available Methods for Estimating Solid Density   19-14

19-8   Method of Immirzi and Perini   19-16

        Basic Steps   19-19

19-9   Available Data   19-21

19-10   Symbols Used   19-21

19-11   References   19-23

20   SURFACE TENSION

20-1   Introduction   20-1

20-2   Available Methods   20-2

20-3   Method 1 (MacLeod-Sugden)   20-3

        Basic Steps   20-8

20-4   Method 2 (Grain)   20-10

        Basic Steps   20-13

20-5   Available Data   20-15

20-6   Symbols Used   20-15

20-7   References   20-16

21   INTERFACIAL TENSION WITH WATER

21-1   Introduction   21-1

21-2   Available Methods   21-2

21-3   Method 1   21-4

     Basic Steps   21-4

21-4   Method 2   21-7

     Basic Steps   21-10

21-5   Available Data   21-14

21-6   Symbols Used   21-14

21-7   References   21-15

22   LIQUID VISCOSITY

22-1   Introduction   22-1

22-2   Available Estimation Methods   22-3

22-3   Selection of Appropriate Method   22-3

22-4   Method 1   22-5

     Derivation   22-5
     Basic Steps   22-8

22-5   Method 2   22-9

     Derivation   22-9
     Basic Steps   22-10

22-6   Method 3   22-15

     Derivation   22-15
     Basic Steps   22-16

22-7   Available Data   22-17

22-8   Symbols Used   22-18

22-9   References   22-19

23   HEAT CAPACITY

23-1   Introduction   23-1

23-2   Estimation Methods for Gases   23-2

23-3   Method of Rihani and Doraiswamy   23-4

     Principles of Use   23-4
     Basic Steps   23-4

23-4   Method of Benson, Cruickshank, *et al.*   23-9

     Principles of Use   23-9
     Basic Steps   23-15

23-5   Estimation Methods for Liquids   23-16

23-6   Method of Johnson and Huang   23-16

      Principles of Use   23-16
      Basic Steps   23-16

23-7   Method of Chueh and Swanson   23-19

      Principles of Use   23-19
      Basic Steps   23-19

23-8   Available Data   23-21

23-9   Symbols Used   23-21

23-10   References   23-21

24   THERMAL CONDUCTIVITY

24-1   Introduction   24-1

24-2   Method Errors   24-4

24-3   Method of Sato and Riedel   24-5

      Basic Steps   24-6

24-4   Method of Robbins and Kingrea   24-6

      Basic Steps   24-7

24-5   Available Data   24-8

24-6   Symbols Used   24-8

24-7   References   24-9

25   DIPOLE MOMENT

25-1   Introduction   25-1

25-2   Available Estimation Methods   25-3

25-3   Estimation of Dipole Moments for Aromatic Compounds   25-4

      Substituted Benzenes with No Hydrogen Bonding   25-10
         Symmetrical Functional Groups — Basic Steps   25-10
         One Asymmetrical Functional Group — Basic Steps   25-12
         More than One Asymmetrical Functional Group —
           Basic Steps   25-13
      Naphthalene Derivatives   25-15
         Basic Steps   25-15
      Heterocyclics   25-18
         Basic Steps   25-18
      Miscellaneous Compounds   25-20

25-4   Available Data   25-20
25-5   Symbols Used   25-21
25-6   References   25-21

26   INDEX OF REFRACTION
26-1   Introduction   26-1
26-2   Available Estimation Methods   26-2
       Group Contribution Methods   26-2
       Connectivity Method   26-3
26-3   Selection of Appropriate Method   26-3
       Recommended Methods   26-3
       Method Errors   26-4
26-4   Eisenlohr's Method   26-6
       Basic Steps   26-7
26-5   Vogel's Methods   26-10
       Basic Steps   26-10
26-6   Hansch's Method   26-14
       Basic Steps   26-14
26-7   Effects of Temperature on n   26-19
26-8   Available Data   26-19
26-9   Symbols Used   26-19
26-10 References   26-21

APPENDICES
A   BIBLIOGRAPHY OF STANDARD CHEMICAL PROPERTY
     DATA SOURCES
       Bibliography   A-5

B   SIMPLE LINEAR REGRESSION
B-1   Introduction   B-1
B-2   Examining the Data   B-3
B-3   Parameter Estimation   B-7
B-4   Evaluating the Regression   B-13
B-5   Predicting Future Values of Y   B-25
B-6   Symbols Used   B-29
B-7   References   B-30

C   EVALUATING PROPAGATED AND TOTAL ERROR IN
    CHEMICAL PROPERTY ESTIMATES

    C-1   Introduction   C-1
    C-2   Components of Error   C-2
    C-3   Theoretical Background   C-4
    C-4   Evaluating Propagated Error   C-7

        Single Estimated Input (Method 1)   C-7
        Basic Steps   C-7
        Multiple Input Values (Method 2)   C-9
        Basic Steps   C-9
        Simplified Method for Multiple Inputs (Method 3) C-11
        Basic Steps   C-11

    C-5   Additional Examples   C-12
    C-6   Symbols Used   C-15
    C-7   Reference   C-17

INDEX FOLLOWS APPENDIX C

ACKNOWLEDGMENTS

This book is the result of a project undertaken by Arthur D. Little, Inc., under contract to the U.S. Army Medical Research and Development Command. A project report exists which does not differ substantially from the material in this publication.[1]

The original idea for an environmentally oriented handbook of chemical property estimation methods was by our Project Officer, Dr. David H. Rosenblatt of the U.S. Army Medical Bioengineering Research and Development Laboratory (Fort Detrick, Frederick, MD). His initiative, guidance, and subsequent assistance in the preparation of the handbook — especially as an editor — were essential to the successful completion of this program.

The U.S. Army Medical Bioengineering Research and Development Laboratory funded both Phase I of this program, a preliminary problem definition study, and a major portion of Phase II, the writing of the handbook. Financial support for Phase II was also provided by the U.S. Army Toxic and Hazardous Materials Agency (Aberdeen Proving Ground, MD) and the U.S. Environmental Protection Agency's Office of Pesticides and Toxic Substances (Washington, DC).

The authors of individual chapters are listed on the first page of each chapter. Special credit should be given to William F. Reehl, who served as both a technical and style editor for each chapter.

All of the chapters in this handbook were rigorously reviewed by individuals in the U.S. Army, the U.S. Environmental Protection Agency, Arthur D. Little, Inc., and various universities and other organizations. These reviewers provided many helpful comments and pointed out several errors in our initial drafts. The authors are, however, responsible for any errors that may remain. The extramural reviewers who assisted us are listed below; special mention must be made of Dr. Robert Reid's contribution in the review of eleven chapters.

---

1. Lyman, W.J., W.F. Reehl and D.H. Rosenblatt (eds.), "Research and Development of Methods for Estimating Physicochemical Properties of Organic Chemicals of Environmental Concern, Final Report, Phase II," U.S. Army Medical Research and Development Command, Fort Detrick, Frederick, MD (June 1981). (Available from N.T.I.S.)

Dr. George Armstrong, U.S. EPA and National Bureau of
    Standards, Washington, DC

Mr. Sami Atallah, Gas Research Institute, Chicago, IL

Mr. George Baughman, U.S. EPA, Athens, GA

Dr. Howard Bausum, U.S. Army, Fort Detrick, Frederick, MD

Dr. Robert Boething, U.S. EPA, Washington, DC

Dr. Robert Brink, U.S. EPA, Washington, DC

Dr. David Brown, U.S. EPA, Athens, GA

Dr. Joseph Bufalini, U.S. EPA, Research Triangle Park, NC

Dr. William D. Burrows, U.S. Army, Fort Detrick, Frederick, MD

Dr. John Carey, National Water Research Institute,
    Burlington, Ontario

Dr. James Davidson, University of Florida, Gainesville, FL

Dr. James Dragun, U.S. EPA, Washington, DC

Dr. Walter Farmer, University of California, Riverside, CA

Dr. Lewis Gevantman, National Bureau of Standards,
    Washington, DC

Dr. Corwin Hansch, Pomona College, Claremont, CA

Dr. Albert Leo, Pomona College, Claremont, CA

Dr. Donald Mackay, University of Toronto, Toronto, Ontario

Dr. Doris Paris, U.S. EPA, Athens, GA

Dr. Kenneth Partymiller, U.S. EPA, Washington, DC

Dr. John Prausnitz, University of California, Berkeley, CA

Dr. Robert Reid, Massachusetts Institute of Technology,
    Cambridge, MA

Mr. David Renard, U.S. Army, Aberdeen Proving Ground, MD

Dr. Ivan Simon (independent consultant), Cambridge, MA

Dr. Arthur Stern, U.S. EPA, Washington, DC

Dr. Gilman Veith, U.S. EPA, Duluth, MN

Dr. John Walker, U.S. EPA, Washington, DC

Dr. N. Lee Wolfe, U.S. EPA, Athens, GA

Dr. Richard Zepp, U.S. EPA, Athens, GA

Dr. Gunter Zweig, U.S. EPA, Washington, DC

In addition to the above, special thanks are due to Dr. Joan Berkowitz, Dr. George Harris, Dr. John Ketteringham and Dr. Philip Levins, all of Arthur D. Little, Inc., who served as program reviewers.

Warren J. Lyman
Program Manager
Arthur D. Little, Inc.

Disclaimer

This handbook describes a variety of methods for obtaining estimates of chemical property values for organic chemicals. All estimation methods are subject to some limitations and involve some range of method errors (i.e., deviations of the estimates from the real values). It is incumbent upon the user to determine what estimation methods might be appropriate for each property and chemical, and to make a careful assessment of the possible errors in the predicted values. Neither the authors nor McGraw-Hill Book Co. provide any guarantee, express or implied, with regard to the general or specific applicability of any method, the range of errors that may be associated with any estimation routine, or the appropriateness of using an estimated property value in any subsequent calculation, design, or decision process. The authors and McGraw-Hill Book Co. accept no responsibility for damages, if any, suffered by any reader/user of this handbook as a result of decisions made or actions taken based on information contained herein.

LIST OF FIGURES

Figure No.		Page No.
	INTRODUCTION	
1	Sample Form for Reporting Estimated Chemical Properties	10
	OCTANOL/WATER PARTITION COEFFICIENT	
1-1	Pathways for the Estimation of Octanol/Water Partition Coefficients	1-6
	SOLUBILITY IN WATER	
2-1	Pathways for Estimating the Aqueous Solubility of Organic Chemicals	2-5
2-2	Correlation Between Solubility and Octanol/Water Partition Coefficient	2-27
	SOLUBILITY IN VARIOUS SOLVENTS	
3-1	Phase Stability as a Function of Temperature in Three Binary Liquid Mixtures	3-6
3-2	Free Energy of Mixing Curves for Binary Solutions	3-7
3-3	Plot of x_1 vs γ^∞ from Equations 3-14 and 3-15	3-9
3-4	Sample Plot for Heptane(1)-Acetonitrile(2) System	3-16
3-5	Sample Plot for Water(1)-Butanol(2) System	3-17
3-6	Phase Stability as a Function of Temperature in Two Binary Solid(1)-Solvent(2) Systems	3-18
	ADSORPTION COEFFICIENT FOR SOILS AND SEDIMENTS	
4-1	Correlation Between Adsorption Coefficient and Water Solubility	4-16
4-2	Correlation Between Adsorption Coefficient and Octanol/Water Partition Coefficient	4-17
4-3	Correlation Between Adsorption Coefficient and Bioconcentration Factors	4-18
	BIOCONCENTRATION FACTOR IN AQUATIC ORGANISMS	
5-1	Correlation Between Bioconcentration and Octanol-Water Partition Coefficient	5-20

RATE OF HYDROLYSIS

7-1 Examples of the Range of Hydrolysis Half-Lives for Various
Types of Organic Compounds in Water at pH 7 and 25°C 7-6

7-2 pH Dependence of k_T for Hydrolysis by Acid-, Water-,
and Base-Promoted Processes 7-10

7-3 Linear Free Energy Relationships for the Alkaline
Hydrolysis of O,O-Dimethyl and O,O-Diethyl-O-Alkyl and
Aryl Phosphorothioate in Water at 27°C 7-35

RATE OF AQUEOUS PHOTOLYSIS

8-1 Spectral Distribution of Extraterrestrial Solar
Radiation and of Solar Radiation at Sea Level
on a Clear Day 8-10

8-2 Seasonal Variation in Solar Radiant Energy
at the Earth's Surface 8-11

8-3 Attenuation of Solar Spectrum in Natural Waters 8-12

8-4 Seasonal Variations in log Z_λ and log W_λ 8-28

8-5 Relative Midday Half-life for Direct Photolysis of
Sevin® at Midseason for Northern Latitudes 8-30

8-6 Diurnal Variation of Direct Photolysis Rate of
Sevin® at Latitude 40°N 8-30

8-7 Computed Depth Dependence of Direct Photolysis of
Sevin® at Midday and Midsummer, Latitude 40°N 8-31

RATE OF BIODEGRADATION

9-1 Anaerobic and Aerobic Biodegradation of DDT 9-3

9-2 Microbial Microhabitats in a Generalized Freshwater System 9-10

9-3 Microbial Habitats in a Generalized Marine System 9-11

9-4 Microbial Microhabitats in a Generalized Soil System 9-14

9-5 A Conventional Activated Sludge System for
Secondary Biological Waste Treatment 9-15

9-6 Anaerobic Wastewater Treatment Process 9-16

9-7 Degradation of Organic Compounds 9-17

9-8 Lag Period in Biodegradation of m-Chlorobenzoic Acid 9-18

9-9 Population and Substrate Concentrations During
Biodegradation 9-20

9-10 Amount of ^{14}C-trichlorophenol Degraded with Time
at Two Temperatures 9-28

9-11 Arrhenius Temperature Plots for Biodegradation in Water 9-29

9-12 Effect of Temperature on Efficiency of Biological Processes
for an Activated Sludge System 9-30

9-13 First-order Disappearance Curve of a Chemical 9-48

9-14 Second-order Reaction Rate as a Function of Substrate
Concentration Following Equation 9-3 9-51

9-15 Correlation of Second-order Alkaline Hydrolysis
Rate Constants with Second-order Biodegradation
Rate Constants for Two Groups of Compounds 9-71

ATMOSPHERIC RESIDENCE TIME

10-1 Plot of Yearly Emissions for Example 10-3 10-18

ACTIVITY COEFFICIENT

11-1 Calculated Activity Coefficients in the System
n-Pentane-Acetone at 760 mm Hg 11-19

11-2 Calculated vs. Experimental Vapor-Liquid Equilibrium
in the System n-Pentane-Acetone at 760 mm Hg 11-20

BOILING POINT

12-1 Carbon Types Used in Forman-Thodos Estimation of
Critical Temperature 12-19

VOLATILIZATION FROM WATER

15-1 Two-Layer Model of Gas-Liquid Interface 15-10

15-2 Solubility, Vapor Pressure and Henry's Law Constant for
Selected Chemicals 15-12

15-3 Volatility Characteristics Associated with Various Ranges of
Henry's Law Constant 15-16

15-4 Effect of Molecular Weight and Environmental Characteristics
on Liquid-Phase Exchange Coefficient 15-22

15-5 Effect of Molecular Weight, Wind Speed and Current
on Gas-Phase Exchange Coefficient 15-23

VOLATILIZATION FROM SOIL

16-1 Desorption Isotherms for Lindane 16-6

16-2 Dieldrin Flux: Comparison of Measured Values with Those
Calculated by Model II 16-20

16-3 Lindane Flux from Treated Gila Silt Loam: Comparison of
Measured and Calculated Values 16-21

16-4 Dieldrin Volatilization Flux from Uniformly Treated
Gila Silt Loam at Two Air Velocities:
Comparison of Measured and Calculated Values   16-21

16-5 Decision Chart for Selecting Estimation Method   16-30

16-6 Error Function   16-45

FLASH POINTS OF PURE SUBSTANCES

18-1 Graphical Determination of Equilibrium Flash Point   18-5

18-2 Determination of Equilibrium Flash Point of n-Hexane by
Affens' Method   18-6

18-3 Nomograph for Estimation of Flash Point   18-8

18-4 Graphical Determination of Open- or Closed-Cup Flash Point   18-12

INTERFACIAL TENSION WITH WATER

21-1 Graphical Estimation of Interfacial Tension (Method 1)   21-5

21-2 Mixture Surface Tension of *n*-Butanol-Water   21-8

LIQUID VISCOSITY

22-1 Variation of Viscosity with Temperature   22-7

DIPOLE MOMENT

25-1 Examples of the Determination of θ 25-9

APPENDICES

SIMPLE LINEAR REGRESSION

B-1 Scatter Plot of Twenty-five Observations of Two Variables   B-2

B-2 Violation of Regression Assumptions   B-5

B-3 Least-Squares Regression Line   B-8

B-4 Histogram of Residuals   B-20

B-5 Normal Probability Plot of Residuals   B-22

B-6 Residual Plots   B-23

B-7 95% Prediction Invervals   B-27

B-8 Problems with Extrapolation   B-28

EVALUATING PROPAGATED AND TOTAL ERROR IN CHEMICAL PROPERTY ESTIMATES

C-1 Graph of η_L as a Function of T_b C-3

LIST OF TABLES

Table No.		Page No.
	INTRODUCTION	
1	Physicochemical Properties Covered in this Report	2
	OCTANOL/WATER PARTITION COEFFICIENT	
1-1	Overview of Estimation Methods for K_{ow} Provided in This Handbook	1-4
1-2	Pathways Leading to the Estimation of Octanol/Water Partition Coefficients	1-7
1-3	Percentage of Compounds for Which log K_{ow} Could be Calculated	1-11
1-4	Comparison of Observed Values of log K_{ow} with those Calculated from Fragment Constants	1-13
1-5	Fragment Constants	1-17
1-6	Summary of Rules for Calculating Factors	1-22
1-7	Aliphatic H/S-Polar Interactions: α-Halogen Factors ($F_{H/SP}$)	1-23
1-8	General Solute Classes	1-40
1-9	Solvent Regression Equations	1-41
1-10	Hydrogen Bonding Corrections, I_H, for Eq. 1-38	1-43
1-11	Comparison of Observed log K_{ow} with Values Estimated from Solvent/Water Partition Coefficients	1-44
	SOLUBILITY IN WATER	
2-1	Overview of Solubility Estimation Methods Provided in this Handbook	2-3
2-2	Pathways Leading to the Estimation of Aqueous Solubility for Organic Chemicals	2-6
2-3	Regression Equations for the Estimation of S	2-14
2-4	Additional Information on Equations for Estimating S	2-16
2-5	Compounds Used for Regression Equation 2-2	2-17
2-6	Compounds Used for Regression Equation 2-3	2-18
2-7	Compounds Used for Regression Equation 2-4	2-19

2-8   Compounds Used for Regression Equations 2-5 to 2-15   2-20

2-9   Compounds Used for Regression Equation 2-16   2-21

2-10   Compounds Used for Regression Equation 2-17   2-22

2-11   Compounds Used for Regression Equation 2-18   2-23

2-12   Compounds Used for Regression Equation 2-19   2-23

2-13   Compounds Used for Regression Equation 2-20   2-24

2-14   Comparison of Measured and Estimated Values of S for
       Selected Chemicals   2-28

2-15   Analysis of Errors Associated with Methods Using
       Correlations with K_{ow} 2-32

2-16   Parameters for the Calculation of Water Solubility   2-40

2-17   Deviations Between Measured and Calculated Solubilities
       Using Irmann's Method   2-42

2-18   Deviations Between Measured and Calculated Solubilities for
       Compounds with More Accurately Measured Solubilities   2-42

SOLUBILITY IN VARIOUS SOLVENTS

3-1   Comparison of Observed and Estimated Activity Coefficients
       and Solubilities for Liquid-Liquid Binaries   3-11

3-2   Observed and Estimated Solubility of Naphthalene in
       Various Solvents at 40°C   3-21

3-3   Observed and Estimated Solubility of Anthracene in
       Various Solvents at 20°C   3-22

3-4   Observed and Estimated Solubility of Phenanthrene in
       Various Solvents at 20°C   3-23

3-5   Observed and Calculated Eutectics in Binary Mixtures   3-24

3-6   Observed and Estimated Aqueous Solubilities for
       Solid-Liquid Binaries   3-25

ADSORPTION COEFFICIENT FOR SOILS AND SEDIMENTS

4-1   Regression Equations for the Estimation of K_{oc} 4-9

4-2   Information on Equations Given for Estimation of K_{oc} 4-10

4-3   Compounds Used by Kenaga and Goring for
       Regression Equations   4-11

4-4   Compounds Used by Karickhoff *et al.* for
       Regression Equations   4-13

4-5   Compounds Used by Chiou *et al.* for Regression Equation   4-14

4-6   Compounds Used by Brown *et al.* for Regression Equations   4-14

4-7   Compounds Used by Rao and Davidson for
Regression Equation   4-14

4-8   Compounds Used by Briggs for Regression Equation   4-15

4-9   Comparison of Measured and Estimated Values of K_{oc}
for Selected Chemicals   4-24

4-10   Deviation from Linearity for the Freundlich
Adsorption Isotherm   4-27

4-11   Errors Associated with Assumption of Reversible Adsorption   4-28

BIOCONCENTRATION FACTOR IN AQUATIC ORGANISMS

5-1   Recommended Regression Equations for Estimating log BCF,
Based on Flow-through Laboratory Studies   5-5

5-2   Additional Regression Equations for Estimating log BCF,
Based on Flow-through Laboratory Studies   5-6

5-3   Compounds Used to Derive Regression Equation 5-2   5-7

5-4   Compounds Used to Derive Regression Equations 5-3,
5-4 and 5-7   5-11

5-5   Compounds Used to Derive Regression Equation 5-5   5-14

5-6   Compounds Used to Derive Regression Equation 5-6   5-15

5-7   Compounds Used to Derive Regression Equation 5-8   5-16

5-8   Compounds Used to Derive Regression Equation 5-9   5-16

5-9   Compounds Used to Derive Regression Equation 5-10   5-16

5-10   Comparison of Estimated Values with Laboratory
Measurements of BCF   5-18

5-11   Comparison of Estimated BCF with Observed BCF from
Field Data   5-22

5-12   Regression Equations Derived from Model Ecosystem Tests
and Static Bioassays   5-24

ACID DISSOCIATION CONSTANT

6-1   Range of pK_a Values for Organic Acids   6-5

6-2   Commonly Encountered LFERs and Substituent Parameters   6-8

6-3   Hammett Substituent Constants   6-11

6-4   Values of σ for Multiple Substituents   6-13

6-5   Default Values of σ for Substituents in Generalized Categories
for Aromatic Acids   6-15

6-6   Hammett Reaction Constants (ρ) for Equilibrium Reactions in Water   6-17

6-7   Reaction Parameters for Acid Dissociation   6-21

6-8   Substituent Constants for Taft Equation   6-21

6-9   Some Correlation Data for Linear Free Energy Relationships of Hammett and Taft   6-23

6-10 Comparison of Measured and Estimated Values of Dissociation Constants   6-24

RATE OF HYDROLYSIS

7-1   Types of Organic Functional Groups That Are Generally Resistant to Hydrolysis   7-4

7-2   Types of Organic Functional Groups That Are Potentially Susceptible to Hydrolysis   7-5

7-3   pH Regimes in Which Specific Acid/Base Catalysis Is Significant for Organic Functional Groups   7-10

7-4   Characteristics of Estimation Methods Described   7-19

7-5   Data for Estimation of k_H from the Hammett Correlation   7-23

7-6   Taft, σ^*, and Steric, E_s, Substituent Constants for Taft Correlation   7-25

7-7   Correlation of Neutral Hydrolysis Rate Constant with Hammett Substituent Constant   7-27

7-8   Correlation of Alkaline Hydrolysis Rate Constant with Hammett Substituent Constant   7-29

7-9   Correlations of Alkaline Hydrolysis Rate Constant with pK_a of Leaving Group   7-33

7-10 Example Rate Data and Estimated Half-Lives (25°, pH 7) for Organic Compounds of Various Types in Aqueous Solution   7-37

7-11 Disappearance Rate Constants for Acid, Neutral and Alkaline Hydrolyses of Common Pesticides   7-42

RATE OF AQUEOUS PHOTOLYSIS

8-1   Types of Molecular Transition and Associated Energy Levels   8-3

8-2   Some Approximate Bond Dissociation Energies   8-3

8-3   Triplet Energies for Selected Compounds   8-8

8-4 Chromophoric Groups and Their Characteristic Absorption
     Maxima at $\lambda > 290$ nm   8-13

8-5 UV/Visible Absorption Maxima and Molar Absorptivities
     for Selected Organic Compounds   8-14

8-6 Comparison of Features of n $\rightarrow$ π and π $\rightarrow$ π^* Transitions   8-17

8-7 Molar Absorptivities, $\epsilon\lambda$, of Selected Compounds
     as a Function of Wavelength   8-20

8-8 Solar Irradiation Intensity, $W\lambda$, from Eq. 8-20   8-24

8-9 Solar Irradiation Intensity, $Z\lambda$, from Eq. 8-21   8-26

8-10 Primary Photochemical Reaction Modes Typical of
     Various Organic Compound Categories   8-32

8-11 Disappearance Quantum Yield, ϕ, for Photolysis in
     Aqueous Media   8-37

8-12 Half-Life for Disappearance via Direct Photolysis in
     Aqueous Media   8-38

RATE OF BIODEGRADATION

9-1 Examples of Biodegradation Reactions   9-4

9-2 Presence of Microbial Populations in Various Aquatic Systems   9-12

9-3 Distribution of Microorganisms in Various Soil Horizons   9-13

9-4 Representative Microbial and Protozoan Genera Found in
     Different Environments   9-16

9-5 Methods for Estimating Microbial Populations and Biomass   9-22

9-6 Microbial Population Density in Various Environments   9-23

9-7 Microbial Biomass in Various Environments   9-24

9-8 Variables Potentially Affecting Rate of Biodegradation   9-25

9-9 Effect of Soil Water Potential on Microorganisms   9-31

9-10 Analytical Techniques Commonly Used in
     Biodegradation Tests   9-35

9-11 Comparison of Biodegradation Test Methods   9-37

9-12 Biodegradation Tests Recommended for Screening
     Organic Compounds   9-38

9-13 Summary List of Standard Tests for Measuring Biodegradation   9-38

9-14 Standard Laboratory Test Methods for Measuring
     Biodegradation   9-40

9-15 Units for Biodegradation Rate Constants   9-52

9-16   Biodegradation Rate Constants for Organic Compounds in Aquatic Systems   9-53

9-17   Biodegradation Rate Constants for Organic Compounds in Soil   9-55

9-18   Biodegradation Rate Constants for Organic Compounds in Anaerobic Systems   9-56

9-19   Rate Constants for Biodegradation of Organic Compounds by Activated Sludge Cultures   9-57

9-20   Rules of Thumb for Biodegradability   9-58

9-21 BOD_5/COD Ratios for Various Organic Compounds   9-64

9-22   COD Removal and Rate of Removal for Various Compounds   9-66

9-23   Refractory Indices for Various Organic Compounds   9-70

ATMOSPHERIC RESIDENCE TIME

10-1   Time Scales for Atmospheric Phenomena   10-4

10-2   Estimation Methods Considered   10-5

10-3   Data Required for Estimation   10-7

10-4   Estimated Atmospheric Residence Times for Selected Chemicals   10-8

10-5   Rate Constants for Reaction of Organic Chemicals with OH Radical at 300K   10-23

10-6   Rate Constants for Reaction of Organic Chemicals with Ozone at 300K   10-25

10-7   Typical Concentrations of OH· and O_3 in the Atmosphere   10-27

10-8   Data Used in Junge's Correlation   10-28

ACTIVITY COEFFICIENT

11-1   Some Models for the Excess Gibbs Energy and Subsequent Activity Coefficients for Binary Systems   11-5

11-2   Characteristics of Methods 1 and 2   11-8

11-3   Comparison of Experimental and Calculated Activity Coefficients (Method 2)   11-9

11-4   Correlating Constants for Activity Coefficients at Infinite Dilution, Homologous Series of Solutes and Solvents   11-12

11-5   Correction Factors for log γ_i^∞, per Group   11-15

11-6   Group Volume and Surface-Area Parameters   11-23

11-7   UNIFAC Group Interaction Parameters, a_{mn} 11-26

11-8	Group Parameters for Example 11-5	11-34
11-9	Group Parameters for Example 11-6	11-42

BOILING POINT

12-1	Summary of Methods for Estimating Normal Boiling Points	12-4
12-2	Comparison of Boiling Point Estimation Methods for Selected Chemicals	12-5
12-3	Contributions to Molar Refraction and Parachor	12-10
12-4	McGowan Parachor Contributions	12-12
12-5	Constant B for Various Chemical Classes	12-12
12-6	Lydersen's Increments for Calculating $\Sigma \Delta T$	12-18
12-7	Group Contributions $\Delta a^{2/3}$ and $\Delta b^{3/4}$ for Saturated Aliphatic Hydrocarbons	12-21
12-8	Equations for $\Delta a^{2/3}$ and $\Delta b^{3/4}$ for Saturated Aliphatic Hydrocarbons Containing n Carbons ($2 \leq n \leq 15$)	12-22
12-9	Double- and Triple-Bond Contributions in Forman and Thodos Method	12-25
12-10	Naphthenic and Aromatic Contributions in Forman and Thodos Method	12-27
12-11	Constants for Equations that Establish Functional Group Contributions for Organic Compounds in Forman and Thodos Methods	12-29
12-12	Lydersen's Increments for Calculating $\Sigma \Delta P$ and $\Sigma \Delta V$	12-35
12-13	Values of y for Equation 12-20	12-40
12-14	Values of Parameters p and q for Equation 12-20	12-41
12-15	Atomic Numbers of Some Common Elements	12-43
12-16	Values of r, s, and t for Equation 12-21	12-43
12-17	Atomic and Group Boiling Point Numbers	12-45

HEAT OF VAPORIZATION

13-1	Summary of Recommended Methods for Estimating Heats of Vaporization	13-5
13-2	Accuracy of Estimation Methods for Heat of Vaporization	13-6
13-3	Values for Klein Constant	13-8
13-4	Lydersen's Critical Pressure Increments	13-10
13-5	Antoine's Constant C for Organic Compounds	13-13

13-6   Estimation of Compressibility-Factor Difference from
       Temperature   13-15

13-7   Values of Exponent n as a Function of T_b/T_c 13-17

13-8   Structural Group Increments for ΔH_{vo} 13-18

13-9   Rules for Using Type A or Type B Halogen Increments   13-20

13-10 Correction Increments   13-21

VAPOR PRESSURE

14-1   Recommended Methods   14-5

14-2   Calculated vs. Experimental Vapor Pressures   14-6

14-3   Average and Maximum Errors in Estimated Vapor Pressure   14-7

14-4 K_F Factors for Aliphatic and Alicyclic Organic Compounds   14-9

14-5   Values of K_F for Aromatic Hydrogen Bonded Systems   14-11

VOLATILIZATION FROM WATER

15-1   Typical Values of Aquatic Turbulent Diffusivities   15-14

15-2   Measured Reaeration Coefficient Ratios for
       High-Volatility Compounds   15-18

15-3   Oxygen Reaeration Coefficients, $(k_v^o)_{env}$, for Water Bodies   15-20

15-4   Volatilization Parameters for Selected Chemicals   15-24

VOLATILIZATION FROM SOIL

16-1   Chemical Properties and Volatilization Rate Constants for
       Chemicals Applied to the Soil Surface   16-26

16-2   Comparison of Measured Volatilization Half-Lives with
       Values Predicted by the Dow Method   16-28

16-3   Comparison of Field-Measured, Laboratory-Measured, and
       Predicted Half-Lives Using the Dow Method   16-28

16-4   Models Used to Compute Volatilization of Chemicals from Soil 16-29

16-5   Chemical and Environmental Data for Estimation of
       Trichloroethylene Volatilization   16-35

16-6   TCE Vapor Diffusion Coefficients through Soil, Calculated from
       Known Properties of Other Compounds Using Eq. 16-5   16-37

16-7   Sample Calculations of Soil Concentration by Eq. 16-24   16-39

16-8   Values of the Error Function for $x \leq 2.2$ 16-44

DIFFUSION COEFFICIENTS IN AIR AND WATER

17-1   Methods for Estimating Gaseous Diffusion Coefficients   17-8

17-2   Comparison of Absolute Average Errors by Chemical Class   17-10

17-3   Some Physical Properties of Air   17-11

17-4   Atomic and Structural Diffusion Volume Increments   17-11

17-5   Additive Volume Increments for Calculating LeBas Molar
        Volume, V'_B 17-11

17-6   Available Methods for Estimating Diffusivity into Water   17-19

17-7   Viscosity of Water at Various Temperatures   17-21

FLASH POINTS OF PURE SUBSTANCES

18-1   Overview of Flash Point Estimation Methods   18-4

18-2   Flash Point Estimation Equations   18-11

DENSITIES OF VAPORS, LIQUIDS AND SOLIDS

19-1   Accuracy of the Ideal Gas Law and Redlich-Kwong Methods
        for Estimation of Vapor Density   19-4

19-2   Some Available Methods for Estimating Liquid Density   19-6

19-3   Accuracy of Two Methods for Estimating Liquid Density   19-8

19-4   Percent Difference Obtained by Grain's Equation Using the
        Approximation $T_c = 1.5 T_b$ 19-10

19-5   Incremental Values for Estimating Molar Volume, V_b,
        by Schroeder's Method   19-11

19-6   Percent Error in Calculated Density Using the
        Method of Tarver   19-15

19-7   Absolute Average Error Calculated for Density
        Using the Method of Nielsen   19-15

19-8   Errors in Estimates of Crystal Volume for
        53 Compounds, by Method of Immirzi and Perini   19-17

19-9   Volume Increments (v_i) for Common Elements and Ions   19-18

SURFACE TENSION

20-1   Some Methods for Estimating Liquid Surface Tension   20-4

20-2   Parachor Increments   20-5

20-3   Error in Estimating Surface Tension of Pure Liquids by
        MacLeod-Sugden Method   20-6

20-4   Values of Constants k and n   20-11

20-5   Comparison of Measured Surface Tensions with
        Values Calculated by Method 2   20-12

20-6   Temperature Dependence of Measured and Calculated
Surface Tensions   20-13

INTERFACIAL TENSION WITH WATER

21-1   Overview of Recommended Estimation Methods   21-2

21-2   Estimated vs. Experimental Interfacial Tensions with Water 21-6

21-3   Values of q for Equation 21-7   21-10

21-4   Surface Tension of Water at Various Temperatures   21-11

LIQUID VISCOSITY

22-1   Methods for Estimating Liquid Viscosity   22-4

22-2   Comparison of Calculated and Experimental Liquid Viscosities 22-6

22-3   Values of η_{Lb} to be Used with Figure 22-1   22-8

22-4   Functions for ΔN and ΔB 22-11

22-5   Values of n (Equation 22-18)   22-16

HEAT CAPACITY

23-1   Recommended Methods for Estimating Heat Capacity   23-3

23-2   Measured vs. Estimated Heat Capacities for Gases at 300K   23-4

23-3   Group Contributions for Equation 23-1   23-5

23-4   Benson Group Contributions to Heat Capacities at 300K   23-10

23-5   Measured vs. Estimated Heat Capacities for Organic Liquids 23-17

23-6   Group Heat Capacities for Organic Liquids   23-18

23-7   Group Contributions for Molar Liquid Heat Capacity at 20°C 23-20

THERMAL CONDUCTIVITY

24-1   Recommended Methods for Estimating Thermal Conductivity   24-2

24-2   H Factors for Method of Robbins and Kingrea   24-3

24-3   Measured vs. Estimated Thermal Conductivities of
Organic Liquids   24-5

DIPOLE MOMENT

25-1   Typical Values of Experimentally Determined
Dipole Moments of Aliphatic and Alicyclic Chemicals   25-5

25-2   Magnitude and Direction of Electrical Moments of
Benzene Derivatives   25-6

25-3   Comparison of Observed and Estimated Dipole Moments   25-7

INDEX OF REFRACTION

26-1   Overview of Recommended Methods   26-4

26-2   Comparison of Observed Molar Refraction with Values
       Estimated by Methods of Eisenlohr and Vogel   26-5

26-3   Observed Molar Refraction versus Values Estimated by
       Hansch's Method   26-6

26-4   Atomic and Structural Contributions to Molar Refraction by
       Additive Method of Eisenlohr   26-7

26-5   Atomic, Structural, and Group Increments for Vogel's Method   26-11

26-6   Hansch's Group Contributions to R_D of Aromatic Compounds   26-15

26-7   Methods of Deriving n from R_D 26-20

APPENDICES
BIBLIOGRAPHY OF STANDARD CHEMICAL PROPERTY
DATA SOURCES

A-1   Compilations of Data for Common Properties of
       Organic Chemicals   A-2

SIMPLE LINEAR REGRESSION

B-1   Twenty-Five Observations of Two Variables   B-2

B-2   Upper Percentage Points of the t Distribution   B-11

B-3   Analysis of Variance (ANOVA) Table   B-16

B-4   Upper Percentage Points of the F Distribution   B-18

B-5   Predicted Values and Residuals   B-21

B-6   Analysis of Variance   B-24

B-7   95% Prediction Intervals   B-29

HANDBOOK OF
CHEMICAL PROPERTY
ESTIMATION METHODS

INTRODUCTION

OVERVIEW

This report contains selected estimation methods for several physicochemical properties of organic chemicals. The full list of properties covered is shown in Table 1. The general style of the report is that of a handbook with specific instructions for the use of each estimation method. It is hoped that the descriptions and examples will be useful to environmental chemists and environmental program managers, who must frequently deal with problem chemicals for which even the most basic physicochemical properties may be missing from the literature or data collections. The report should also be of use to process engineers when they must estimate properties of chemicals at or near ambient temperatures and pressures.

The "properties" covered by this handbook include a variety of conventional properties of pure materials (e.g., density, boiling point, refractive index), some properties that describe how a chemical behaves or interacts with a second substance (e.g., solubility in water, diffusion coefficient in air, interfacial tension with water), and a set that describe the fate of trace concentrations of the chemical in specific environmental situations (e.g., rate of hydrolysis in water, atmospheric residence time, and volatilization from soil). The latter group — in particular, Chapters 8, 10, 15 and 16 — are related more to environmental fate than to physicochemical properties; these models require input information on the environmental compartment of concern as well as chemical-specific properties.

TABLE 1

Physicochemical Properties Covered in this Report

Chapter	Property
1	Octanol/Water Partition Coefficient
2	Solubility in Water
3	Solubility in Various Solvents
4	Adsorption Coefficient for Soils and Sediments
5	Bioconcentration Factor in Aquatic Organisms
6	Acid Dissociation Constant
7	Rate of Hydrolysis
8	Rate of Aqueous Photolysis
9	Rate of Biodegradation
10	Atmospheric Residence Time
11	Activity Coefficient
12	Boiling Point
13	Heat of Vaporization
14	Vapor Pressure
15	Volatilization from Water
16	Volatilization from Soil
17	Diffusion Coefficients in Air and Water
18	Flash Points of Pure Substances
19	Densities of Vapors, Liquids and Solids
20	Surface Tension
21	Interfacial Tension with Water
22	Liquid Viscosity
23	Heat Capacity
24	Thermal Conductivity
25	Dipole Moment
26	Index of Refraction

Two important properties, rate of aqueous photolysis (Ch. 8) and rate of biodegradation (Ch. 9), are included in this handbook even though the current state of the art does not permit quantitative estimation. These two chapters stress the importance of photolysis and biodegradation in environmental fate and should allow a qualitative determination of the susceptibility of an organic chemical to these forms of degradation. Additional research is required before quantitative estimation methods can be developed.

A few additional properties that are not the subject of separate chapters may be estimated from instructions given in this handbook. These properties are used as input parameters for other estimation methods and frequently must be estimated themselves. Included are:

Property	See
Critical temperature	§12-4
Critical pressure	§13-4
Henry's law constant	§15-5 (see also §11-4, Example 11-2)
Mass transfer coefficients[1]	§15-5
Molar refractivity	Ch. 26 (also §12-3)
Molar volume at the boiling point	§12-5 (or §19-5)
Parachor	§20-3 (or §12-3)

Most of the estimation methods in this handbook (excluding those dealing with environmental fate models) are based upon one of the following:

(1) Theoretical equations, usually containing parameters that are empirically derived (e.g., via fragment constants),

(2) Group or atomic fragment constants derived by regression analysis of data sets, or

(3) Correlations (usually in the form of linear regression equations) between two properties.

Types 1 and 2 are most frequently encountered for properties of the pure chemical (e.g., boiling point, heat of vaporization, density, heat capacity), while type 3 is more commonly used for certain environmental properties (e.g., aqueous solubility, soil adsorption coefficient, bioconcentration factor). Combinations of the above approaches are also possible. Because of the increasing importance of linear regression equations, a detailed discussion of this subject is provided in Appendix B. Type 2 methods (i.e., those requiring only the use of fragment constants) are favored in many circumstances where little is known about the chemical, since they require only a knowledge of the chemical's structure. In most other methods, one or more different properties of the chemical must be known (or estimated) before the desired property can be estimated.

---

1. In air and water near the air/water interface.

Some aspects of the novelty and innovation associated with this work should be noted. It is, to our knowledge, the first attempt to review and evaluate available estimation methods for a group of environ-mentally important physicochemical properties (particularly those covered in Chapters 1—10, 15, 16 and 18). For many of the remaining properties we relied heavily on the excellent review of estimation methods by Reid et al.[2] Secondly, although the original objective of this report was to cover only available (i.e., previously published) estimation methods, several authors found ways to improve or expand them. This has permitted the inclusion of some new or modified methods with enhanced utility.

APPENDICES

The following three appendices to this report provide important supplemental information:

A. "Bibliography of Standard Chemical Property Data Sources" lists selected reference books and articles that contain compilations of measured physicochemical properties of organic chemicals.

B. "Simple Linear Regression" describes linear regression analysis and explains (with examples) how readers can use regression analysis to formulate new estimation equations.

C. "Evaluating Propagated and Total Error in Chemical Property Estimates" discusses propagated and total error in chemical property estimates to cover those cases when both propagated error (i.e., error associated with an estimated or otherwise uncertain input parameter) and method error (i.e., that associated with the method when all input parameters are exactly known) must be considered. For such cases a scheme is provided that allows the user to obtain an estimate of the total error in the property value estimate, where the total error is a simple function of method error and propagated error. Schemes for both single and multiple estimated inputs are provided. Detailed instructions and examples are given.

OBJECTIVES

Over the past decade, the chemical contamination of our environment has justifiably aroused growing concern. A proper assessment of

2. Reid, R.C., J.M. Prausnitz and T.K. Sherwood, *The Properties of Gases and Liquids,* 3rd ed., McGraw-Hill Book Co., New York (1977).

the risk — to man and the environment — created by exposure to these chemicals generally includes attempts to measure or predict the concentrations in various environmental compartments in conjunction with toxicological data. Frequently, however, neither the concentration data nor the toxicological data are adequate for any realistic assessment. In addition, basic physical and chemical data are often unavailable, especially for new organic chemicals being considered for bulk manufacture. If, however, the most important physical and chemical properties of these chemicals could be estimated, their transport and fate in the environment could be better understood — even modeled in some cases — and the eventual environmental concentrations might be estimated.

In September 1978, with support from the U.S. Army Medical Bioengineering Research and Development Laboratory (Fort Detrick, Frederick, MD), Arthur D. Little, Inc., undertook a preliminary problem definition study to answer the following questions: (1) What are the important properties of organic chemicals with regard to their transport and fate in the environment? (2) What methods are available for the estimation of these properties? (3) What limitations and/or uncertainties are associated with these methods? and (4) Can a comprehensive user's manual incorporating the basic elements of these methods be prepared so they become not only easy to comprehend and use, but hard to misuse?

The results of this preliminary study[3] included a recommendation that a property estimation handbook be prepared covering 26 specific properties. In this study, a review of environmental fate models, hazard ranking schemes, federal regulations, and other material first led to the identification of about 50 physicochemical properties of interest. Additional literature surveys indicated that estimation methods were available for only half of the properties.

For the estimable properties, it has been our goal to select (and recommend) two or more estimation methods for each property. Our selection of methods was based upon the following considerations:

---

3. Lyman, W.J., J.C. Harris, L.H. Nelken and D.H. Rosenblatt, "Research and Development of Methods for Estimating Physicochemical Properties of Organic Compounds of Environmental Concern, Final Report, Phase I," NTIS Report No. AD-A074829, U.S. Army Medical Bioengineering Research and Development Laboratory, Fort Detrick, Frederick, MD (February 1979).

- *Range of applicability* — Methods should be adaptable to a variety of chemical classes and structures and should be applicable over the range of values of interest to environmental chemists.

- *Ease of use* — The rules and equations should be relatively simple and capable of solution without a computer. We have presumed that the user would have at least one year of college-level organic chemistry and a reasonable facility with common mathematical functions.

- *Minimum input data requirements* — Very little property data (from actual measurements) is available for some chemicals. Thus, methods requiring minimum input data are desired. In several cases, estimation methods requiring only the chemical structure are available.

- *Accuracy* — While the highest possible accuracy is desirable, it should not overshadow the first three considerations. For environmental fate models, hazard assessments, etc., precise chemical property data are not always needed.

BENEFITS OF ESTIMATION

With such estimation methods at their disposal, environmental scientists and environmental program managers should find their tasks much simplified. In particular, we expect estimated values of organic chemicals to be used in lieu of measured values (temporarily, at least) to:

(1) Obtain a sufficient understanding of a chemical's fate and transport in the environment to allow decisions and actions (for environmental protection) to proceed in a timely manner;

(2) Run a variety of environmental fate models to predict concentrations in various environmental compartments (air, water, soil);

(3) Set research priorities, especially in cases where large numbers of chemicals must be considered;

(4) Check the reliability of reported measurements; and

(5) Design laboratory and/or field experiments.

Associated with the use of estimated values of physicochemical properties will be a significant saving of both time and funds. To obtain a measured value of many of the selected properties, several days to several weeks and hundreds to thousands of dollars (per property) are commonly required. For example, the cost of obtaining a measured value for all 26 properties of just one chemical could be in the range of $10,000 to $50,000, or perhaps higher. In some situations, estimation methods could be preferable to the usual literature search, which could cost $50-$500 for a single property of just one chemical. Estimation of a chemical property (excluding the models in Chapters 8, 10, 15 and 16) usually takes only 15 to 30 minutes if the user is reasonably familiar with the methods and there are no significant problems with input data requirements. This time will normally allow for some checking for calculational errors and for documentation of the method used.

CONTENTS OF EACH CHAPTER

A degree of uniformity of style and content has been imposed on each chapter in this handbook to facilitate its use. Each property estimation chapter (excepting Chapters 8 and 9) generally contains the following elements:

- *Introductory material* describing the property, its importance, range of values, and factors affecting the value.

- *Overview of available estimation methods,* including summary information on each recommended method (applicability, input requirements, method error) and a discussion of which methods the user should consider for a particular chemical.

- *Method description* (for each recommended method), including basis for method, necessary equations and tables, and an explicit set of instructions (labeled *Basic Steps)* for the use of the method. Additional information on method error is usually included.

- *Examples.* Two or more examples of each method are usually provided.

- *Available data.* References to major compilations of data on the property are provided. Supplemental information is given in Appendix A for the more common physicochemical properties.

- *Symbols used.* A listing, with definitions, of all symbols used.

- *References.* An alphabetical listing of the references cited in the chapter.

LIMITATIONS OF THIS HANDBOOK

As noted above, this is the first attempt to review, evaluate, and recommend estimation methods for a large group of environmentally important physicochemical properties. Several deficiencies and errors are likely to be present, which we hope to eliminate in future editions.

The basic limitation of this handbook is that only single-component (i.e., pure) organic chemicals are covered. Future editions may be able to cover organic mixtures (e.g., gasoline, fuel oil, or simple two-component mixtures) for some properties. Extensions to include polymers, salts, solutions, and inorganic chemicals are not presently contemplated.

For several estimation methods the reader will find that some chemical classes or structures cannot be handled, i.e., no estimate can be calculated. This may be due to a lack of appropriate fragment constants, to an unacceptably high method error for that class (or no information on method error at all for that class), or to the lack of an appropriate constant or equation. This is particularly true of organometallics. Improvisation is not recommended unless the reader is familiar with the method(s) and aware of the large errors that may result.

Another problem is the limited capability of many of the recommended methods to provide estimates either as a function of temperature (and pressure), or to provide estimates at temperatures (and pressures) outside of the normal range of ambient values. In addition, the value of many properties (e.g., solubility, adsorption coefficient, rate of hydrolysis) may be affected by other environmental factors in ways that are not understood.

Finally, the preparation of this handbook did not involve the compilation and evaluation of large sets of *measured* data on each property. As a result, we have not analyzed method applicability and method error for as many chemicals (and chemical classes) as would be desirable.

ERRORS ASSOCIATED WITH CHEMICAL PROPERTY ESTIMATES

Each chapter in this handbook contains some information on method error, i.e., the errors found when estimated values are compared with measured values for a set of chemicals. These method errors (the absolute average errors for the selected test sets) vary greatly. Some are as low as 1-2% (e.g., density, vapor pressure for values >10 mm Hg, heat capacity, and index of refraction); several are in the range of 3-20% (e.g., boiling point, heat of vaporization, diffusion coefficients, surface tension); and some have errors that are nearly one order of magnitude (e.g., aqueous solubility, soil adsorption coefficient, bioconcentration factors). Uncertainties of one order of magnitude for this last group are not a serious problem, when one considers the normal use to which these estimates will be put (risk assessments, fate modeling) and the fact that their values range over six orders of magnitude.

The error associated with a particular estimate obviously cannot be predicted until a measured value is obtained. We recommend that all reported estimates be listed with their associated uncertainty, taken from the information on method error given in each chapter. If method errors are given for several chemical classes, the value for the appropriate chemical class should be selected. If the information on method error for a particular class of chemicals is insufficient, the reader may wish to evaluate the likely method error by using the method to estimate values for several chemicals (of related structure) for which measured values are available.

In many instances the user will have to estimate one or more of the input parameters for a particular method. For example, method 3 for estimating liquid viscosity (§22-6) requires a value for the boiling point as input. If no measured value of the boiling point is available, it can be estimated by the methods described in Chapter 12. The uncertainty of this estimate should be combined with the method error for the viscosity estimation when one is calculating the likely total error, as described in Appendix C.

DOCUMENTATION AND REPORTING OF ESTIMATED VALUES

It is strongly recommended that all reported chemical property estimates be clearly labeled as estimates, and that the methods used to obtain them be explained in a footnote or reference. The uncertainty in the estimate (see above) should also be reported. It will often be desirable to prepare a more formal record of the procedures used to obtain an estimate; Figure 1 is a sample of one kind of form that could be used for internal documentation.

Chemical_____ Estimate of_____

Estimated Value _____ Uncertainty _____ Temperature_____

Estimated by _____ Date_____

Method Used _____ Ref. _____

Equations Used _____ Ref. _____

Values of Input Parameters:

	Parameter	Value	Comments	Ref.
1)	_____	_____	_____	_____
2)	_____	_____	_____	_____
3)	_____	_____	_____	_____
4)	_____	_____	_____	_____

List below: (1) important assumptions made, (2) values of key intermediate parameters,
(3) key equations, and (4) other pertinent comments on the method used and
the estimated value. (Attach details of complex calculations on separate sheet.)

Checked by _____ Date _____

References

1.
2.
3.

FIGURE 1. Sample Form for Reporting Estimated Chemical Properties

1

OCTANOL/WATER PARTITION COEFFICIENT

Warren J. Lyman

1-1 INTRODUCTION

Definition and Measurement. The octanol/water partition coefficient (K_{ow})[1] is defined as the ratio of a chemical's concentration in the octanol phase to its concentration in the aqueous phase of a two-phase octanol/water system.

$$K_{ow} = \frac{\text{Concentration in octanol phase}}{\text{Concentration in aqueous phase}} \qquad (1\text{-}1)$$

Values of K_{ow} are thus unitless. The parameter is measured using low solute concentrations, where K_{ow} is a very weak function of solute concentration. Values of K_{ow} are usually measured at room temperature (20 or 25°C). The effect of temperature on K_{ow} is not great — usually on the order of 0.001 to 0.01 log K_{ow} units per degree — and may be either positive or negative [28].

Measured values of K_{ow} for organic chemicals have been found as low as 10^{-3} and as high as 10^7, thus encompassing a range of ten orders of magnitude. In terms of log K_{ow}, this range is from -3 to 7. As noted later in this chapter, it is frequently possible to estimate log K_{ow} with an uncertainty (i.e., method error) of no more than ± 0.1-0.2 log K_{ow} units.

---

1. The symbol P is also commonly used.

The octanol/water partition coefficient is not the same as the ratio of a chemical's solubility in octanol to its solubility in water, because the organic and aqueous phases of the binary octanol/water system are not pure octanol and pure water. At equilibrium, the organic phase contains 2.3 mol/L of water, and the aqueous phase contains 4.5×10^{-3} mol/L of octanol [27]. Moreover, K_{ow} is often found to be a function of solute concentration for concentrations $\gg 0.01$ mol/L.

References 7, 20, and 28 describe various measurement techniques. The chemical in question is added to a mixture of octanol and water whose volume ratio is adjusted according to the expected value of K_{ow}. Very pure octanol and water must be used, and the concentration of the solute in the system should be less than 0.01 mol/L [7]. The system is shaken gently until equilibrium is achieved (15 min to 1 hr). Centrifugation is generally required to separate the two phases, especially if an emulsion has formed. An appropriate analytical technique is then used to determine the solute concentration in each phase.

A rapid laboratory estimate of K_{ow} may be obtained by measuring the retention time in a high-pressure liquid chromatography system; the logarithm of the retention time and the logarithm of K_{ow} have been found to be linearly related [1,7,20,30,31,43,44,45,48].

Environmental Significance. Interest in the K_{ow} parameter developed first with the study of structure-activity relationships, primarily with pharmaceuticals. Numerous studies showed that K_{ow} was useful for correlating structural changes of drug chemicals with the change observed in some biological, biochemical, or toxic effect. The observed correlations could then be used to predict the effect of new drugs for which a value of K_{ow} could be measured or estimated. References 14 and 28 contain interesting discussions of the history of this parameter.

In recent years the octanol/water partition coefficient has become a key parameter in studies of the environmental fate of organic chemicals. It has been found to be related to water solubility, soil/sediment adsorption coefficients, and bioconcentration factors for aquatic life. (Estimation of these three parameters solely on the basis of K_{ow} is described in Chapters 2, 4, and 5 respectively.) Because of its increasing use in the estimation of these other properties, K_{ow} is considered a required property in studies of new or problematic chemicals.

Values of K_{ow} can be considered to have some meaning in themselves, since they represent the tendency of the chemical to partition itself between an organic phase (e.g., a fish, a soil) and an aqueous phase. Chemicals with low K_{ow} values (e.g., less than 10) may be considered relatively hydrophilic; they tend to have high water solubilities, small soil/sediment adsorption coefficients, and small bioconcentration factors for aquatic life. Conversely, chemicals with high K_{ow} values (e.g., greater than 10^4) are very hydrophobic.

Estimation Methods Described in this Handbook. Table 1-1 summarizes the methods for estimating K_{ow} that are discussed in this handbook. (A more detailed review of available methods, including several not covered in this handbook, is provided in §1-2 of this chapter.) This chapter presents two different methods by which K_{ow} may be estimated:

(1) *From fragment constants* (see §1-3). This method requires only a knowledge of the chemical structure; for structurally complex molecules, however, it is helpful to have a measured value of K_{ow} for a structurally similar compound. If, for example, K_{ow} is sought for a complex compound R-OH and a measured value is available for the compound R-NH$_2$, the fragment constants for $-NH_2$ (f_{NH_2}) and $-OH$ (f_{OH}) would be used as follows:

$$\log K_{ow} \text{ for R-OH} = \log K_{ow} \text{ for R-NH}_2 - f_{NH_2} + f_{OH} \qquad (1\text{-}2)$$

When K_{ow} must be calculated "from scratch," a variety of fragment constants and structural factors must be combined. Users require some practice in order to become proficient in this method.

(2) *From other solvent/water partition coefficients* (K_{sw}) (see §1-4). If a measured value of the chemical's partition coefficient between an organic solvent and water (K_{sw}) is available, K_{ow} can be calculated from linear regression equations that relate $\log K_{sw}$ (for a particular solvent) and $\log K_{ow}$. The method is straightforward, and the calculations are simple.

These two methods should be given preference over the approach that uses regression equations with solubility (No. 3 in Table 1-1), which has larger method errors. However, the latter method provides a rough check on K_{ow} if the fragment-constant method is used and the user is unfamiliar with the procedure.

TABLE 1-1

Overview of Estimation Methods for K_{ow} Provided in This Handbook

No.	Location	Basis for Method	Information Required[a]	Comments
1	This chapter § 1-3	Fragment constants and structural factors	Structure (K_{ow} for structurally related compound)[b]	— Fairly accurate — Wide range of applicability — Requires some practice
2	This chapter § 1-4	Regression equations	K_{sw}	— Easy, rapid calculations — Limited applicability — Fairly accurate (less than No. 1)
3	Chapter 2	Regression equations	S	— Easy, rapid calculations — Wide range of applicability — Less accurate
The following two approaches are not recommended[c]				
4	Chapter 4	Regression equations	K_{oc}	— Relatively large method error
5	Chapter 5	Regression equations	BCF	— Relatively large method error
A method for the venturesome[d]				
6	Chapter 11 plus § 1-5 of this chapter	Uses estimated activity coefficients	Structure	— Calculations are lengthy and difficult — Limited applicability for functional groups — Fairly accurate

a. K_{ow} = octanol/water partition coefficient; K_{sw} = organic solvent/water partition coefficient; S = water solubility; K_{oc} = soil adsorption coefficient based on organic carbon; BCF = bioconcentration factors for aquatic life.

b. Helpful, but not required.

c. These methods involve relatively large method errors because values of K_{oc} and BCF are highly variable.

d. This method is only described in general terms; it has not been evaluated by this author. Detailed instructions for the calculation of activity coefficients are provided in Chapter 11.

Methods using regression equations with soil adsorption coefficients or bioconcentration factors (Nos. 4 and 5 in Table 1-1) are not recommended because of the relatively large method errors that would be involved.

A sixth possible estimation method, via the use of estimated activity coefficients, is outlined in §1-5. Activity coefficients are estimated via the methods described in Chapter 11. The calculations involved are relatively difficult, but the method promises to be fairly accurate.

1-2 OVERVIEW OF AVAILABLE ESTIMATION METHODS

Several pathways are available for the estimation of octanol/water partition coefficients (K_{ow}) for organic chemicals. These pathways are shown schematically in Figure 1-1 and described briefly in Table 1-2. (Each of the pathways in Figure 1-1 is numbered, and this number is used as the index for Table 1-2.) Some comments about these pathways are given below.

- Two pathways involve the use of substituent constants (#1) or fragment constants (#2) that, when summed for the molecule, yield values of K_{ow} directly. (Structural factors must also be considered for both #1 and #2.) The methods are closely related, but each has particular advantages in practical application. Method 1 requires a measured value of K_{ow} for a structurally related or base compound; this method is primarily of interest for aromatic compounds. A measured value for a structurally related compound is desirable for method 2 — especially if the compound in question has a complex structure — but is not required. When, as is often the case, no K_{ow} is available for a parent compound, one must start "from scratch" using the fragment-constant pathway (#2). Values of π and f are available for a large number of functional groups; in addition, measured values of K_{ow} for thousands of chemicals have been compiled [14,28]. Thus, these two pathways allow the (direct) estimation of K_{ow} for a broad range of chemicals.

- Two different sets of fragment constants, f, for pathway #2 have been reported, one by Hansch and Leo [14] (Leo's method) and one by Nys and Rekker [33,34,39]. A detailed analysis and comparison of the two methods have not been made, but the method of Hansch and Leo [14] appears to be

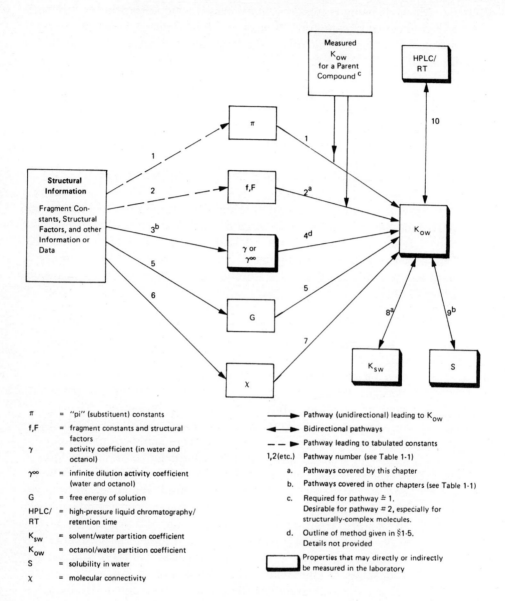

π   = "pi" (substituent) constants

f,F   = fragment constants and structural
          factors

γ   = activity coefficient (in water and
          octanol)

γ∞   = infinite dilution activity coefficient
          (water and octanol)

G   = free energy of solution

HPLC/   = high-pressure liquid chromatography/
RT   retention time

K_{sw} = solvent/water partition coefficient

K_{ow} = octanol/water partition coefficient

S   = solubility in water

χ   = molecular connectivity

⟶ Pathway (unidirectional) leading to K_{ow}

◄──► Bidirectional pathways

— — ► Pathway leading to tabulated constants

1,2(etc.) Pathway number (see Table 1-1)

a. Pathways covered by this chapter

b. Pathways covered in other chapters (see Table 1-1)

c. Required for pathway ≠ 1.
    Desirable for pathway ≠ 2, especially for
    structurally-complex molecules.

d. Outline of method given in §1-5.
    Details not provided

▭ Properties that may directly or indirectly
    be measured in the laboratory

FIGURE 1-1   Pathways for the Estimation of Octanol/Water Partition Coefficients

TABLE 1-2

Pathways Leading to the Estimation of Octanol/Water Partition Coefficients

Pathway	References	Approach[a]	Chemical Classes Covered	Comments
1	[4, 11, 14, 15, 28, 39]	G	Best for aromatic compounds, but works for aliphatic as well. π values for most functional groups are available. Few restrictions.	Need measured value of K_{ow} for structurally related compound.
2	[14, 27, 33, 34, 39]	G	Few restrictions. f values for most functional groups are available. Calculations may be difficult for complex structures.	Two different sets of f values are available, one in [14] and the other in [33,34,39].
3	[8, 9, 10, 29, 35, 36, 42]	G,R	Most of the more common classes/functional groups are covered. Structure must not be too complex.	Estimation methods are described in Chapter 11.
4	—	I	No restrictions beyond those for pathway 3.	Basic equations and approach are outlined in §1-5.
5	[19, 37]	I	Most chemical classes can apparently be handled. Fewer problems with complex structures.	Computer required for calculations.
6	[13, 23]	I	Organics with one or more of the following functional groups: $-NH_2$, $\geq NH$, $\geq N-$, $=N-$ (pyridine), $C\equiv N$, $-OH$, $-O-$, $=O$, $-Cl$, $-Br$, $-I$, $-F$, nitro N, furan O, and a few other special cases. Structure must not be too complex.	Computer program is available for the calculation of X values [12].
7	[23, 32]	R	Hydrocarbons, carboxylic acids, esters, ethers, alcohols, ketones, amines, mixed classes with functional groups.	Regression equations primarily for mono-functional chemical classes.

(continued)

TABLE 1-2 (Continued)

Pathway	References	Approach[a]	Chemical Classes Covered	Comments
8	[18, 27, 28, 39, 41]	R	Organic acids, alcohols, phenols, ketones, esters, ethers, amides, imines, imides, nitriles, aromatic amines, sulfonamides, barbiturates, aromatic hydrocarbons, other miscellaneous groups.	Regression equations for calculation of K_{ow} from K_{sw} in any of 20 different solvent/water systems are given. Equations primarily for mono-functional chemical classes.
9	[2, 5, 16, 22, 40, 46, 47]	R	Most of the common chemical classes are covered. A few equations cover mixed chemical classes.	Estimation methods are described in Chapter 2.
10	[1, 30, 31, 43, 44, 45, 48]	R	Many studies have used mixed chemical classes.	Requires a measured value of retention time (RT) on a calibrated HPLC system.

a. G = group, fragment, and/or structure-related constants; R = regression equation; I = other types of instructions or equations.

better suited for inclusion in this handbook, since more fragment constants are available and the rules for considering
structural factors are better explained. In addition, the Leo
method is more amenable to computerization; two computer
programs using this method have been reported [3,6]. For
these reasons, the method of Nys and Rekker is not included
here.

- The pathway via intermediate values of activity coefficients
  (#3-4) has not, to our knowledge, been described in the open
  literature. One industry group that has used this approach
  has reported fairly accurate estimates (errors typically
  <10%). The method requires the estimation of the of the
  activity coefficient for the chemical in both octanol and
  water; for compounds that are hydrophobic, it may suffice to
  estimate these values in pure water and octanol, but for
  hydrophilic compounds it may be necessary to estimate the
  values in octanol-saturated water and water-saturated octanol.

- The more complex parts of pathways #3, 5 and 6 have been
  computerized, but only pathway #5 actually requires a computer because of the difficult and lengthy calculations.

- Pathway #5 calculates the free energy of solvation (G) of the
  solute in both octanol (G_o) and water (G_w). This is done via a
  solvent-dependent conformational analysis procedure
  (SCAP) which allows G to be calculated as a function of
  molecular conformation; preferred minimum-energy conformations are then selected for subsequent use. In the computer program now being used, only the structure of the
  molecule (via a special numerical code) and output format
  instructions are necessary for a complete conformational
  analysis. The free-energy values (G_o and G_w) are then used to
  calculate activities which are, in turn, used to calculate K_{ow}.
  In one test of 20 compounds, SCAP was able to estimate log
  K_{ow} values with an average absolute error of 9% [19]. (Log
  K_{ow} values with an absolute average error of 5% were obtained for the same 20 compounds using the π-constant
  approach, pathway #1 [19].

- Pathway #6 requires the calculation of an intermediate, χ
  (the molecular connectivity index), which is a topological
  index. The regression equations that link χ to K_{ow} (pathway
  #7) cover only a relatively small number of monofunctional-
  group chemical classes.

- Pathways #8, 9 and 10 all involve two-parameter, linear regression equations using the log of each parameter. The utility of pathway #9 is enhanced by an available compilation of various solvent/water partition coefficients (K_{sw}) for thousands of chemicals [28]. The utility of pathway #9 is fairly well recognized; the regression equations are included in Chapter 2, which covers estimation methods for solubility (S). Pathway #10 is more of a laboratory estimation method than a computational method; it derives its main benefit from the fact that the measurement of retention time takes only about 25 minutes [44]. In a test of 18 compounds, the HPLC/RT method estimated values of log K_{ow} with average absolute error of ~23% [44].

- Only pathway #3-4 is intrinsically capable of estimating K_{ow} at any (reasonable) temperature. Essentially all of the other approaches use, or are derived from, data or fragment constants that come from measurements at room temperature. Quite frequently, data covering a range of temperatures (e.g., 15-30°C) have been used in the derivation of these approaches. Data from a number of solvent systems indicate that the effect of temperature on K_{ow} (actually, log K_{ow}) is on the order of 0.001 to 0.01 log units/deg and may be either positive or negative [28].

1-3   LEO'S FRAGMENT CONSTANT METHOD

Principles of Use. Leo's approach (Hansch and Leo [14]) to the estimation of octanol/water partition coefficients uses empirically derived atomic or group fragment constants (f) and structural factors (F). All calculations are carried out in terms of log K_{ow}:

$$\log K_{ow} = \text{sum of fragments (f)} + \text{factors (F)} \qquad (1\text{-}3)$$

The only input information required for this method is the structure of the chemical, since the fragment values and factors are known.

Fragment values (f) are provided in this chapter for over 100 atoms or atom groups. A fragment has different f values, depending on the type of structure (e.g., aliphatic or aromatic) it is bonded to. Thus, in total, about 200 f values are available. Fourteen different factors must be considered; these take into account molecular flexibility (e.g., possible rotation around bonds), unsaturation, multiple halogenation, branching, and interactions with H-polar fragments.

Because of the large number of f and F values available, this method is a fairly powerful one, and there are relatively few man-made chemicals for which a value of log K_{ow} cannot be calculated. One study [3], using a computerized version of this method, looked at several large data sets for organic compounds and found that, on average, log K_{ow} could be estimated for 84.5% of the compounds. Details are provided in Table 1-3.

TABLE 1-3

Percentage of Compounds for Which log K_{ow} Could Be Calculated

No. of Compounds	Class	Percentage Calculated
209	Heterogeneous set of mutagens and carcinogens	80.4
200	Polycyclic aromatic hydrocarbons	97.5
155	N-Nitroso compounds	75.5
90	Random selection from the 3052 compounds in Ref. 28	84.4
	Average	84.4

Source: Chou and Jurs [3]. *(Reprinted with permission from the American Chemical Society.)*

For complex molecules it is very desirable to have a measured value of log K_{ow} for a structurally similar compound.[2] This measured value can then be modified by adding or subtracting, as required, the appropriate f or F values:

$$\log K_{ow} \text{ (new chemical)} = \log K_{ow} \text{ (similar chemical)}$$

$$\pm \text{ fragments (f)} \pm \text{ factors (F)} \qquad (1\text{-}4)$$

If, for example, an estimate of log K_{ow} is desired for the compound R-Br (R = any organic base structure) and a measured value is available for R-Cl, then

$$\log K_{ow} \text{ (R-Br)} = \log K_{ow} \text{ (R-Cl)} - f_{Cl} + f_{Br} \qquad (1\text{-}5)$$

---

2. References 14 and 28 contain measured values of K_{ow} for thousands of chemicals.

This approach is recommended whenever a reliable measured value of K_{ow} is available for a base compound that differs from the compound of interest by the substitution of only one or just a few fragments. However, if different factors (F values) are involved in the two structures, the application of Eq. 1-4 can be rather difficult.

Because of the numerous f values and sometimes confusing F values involved in the use of this method, *the user is urged to study the procedure and the examples carefully and practice with a few compounds before tackling the first real estimate.* For complex structures it may be desirable to number each fragment and bond so that their f and F values can be considered in a systematic way. Use of an alternative method is advised if the user is unsure of the correct procedure here.

Method Error. This method assumes that log K_{ow} depends upon the structure of the solute in an additive-constitutive fashion, and that the more important structural effects are adequately described by the available factors (F values). Test calculations have shown that these assumptions are justified for most chemicals, but the user should keep in mind that some chemicals deviate seriously from the norm.

The results of one set of test calculations are shown in Table 1-4. For each chemical a value of log K_{ow} was calculated by a computer program from the fragment and factor values of Hansch and Leo [14]. The average absolute error for the 76 compounds was 0.14 log K_{ow} unit. Fifty chemicals (66%) had errors of less than 0.1 log K_{ow} unit, and 63 chemicals (83%) had errors of less than 0.2 log K_{ow} unit.[3] An average absolute error of 0.09 log K_{ow} unit is obtained from a comparison of the estimated and measured values for the chemicals used in Examples 1-1 to 1-37 (given later in this section). Three chemicals had no measured value, so the test set (chosen to exemplify the rules for the method) contined 34 chemicals. In this set, absolute errors were ≤ 0.1 log K_{ow} unit for 23 chemicals (68%) and ≤ 0.2 log K_{ow} unit for 31 chemicals (91%). The maximum error was +0.30 log K_{ow} unit.

Both of the test sets mentioned above contained relatively simple chemicals, and for these a method error ± 0.12 log K_{ow} unit appears valid. However, errors for more complex chemicals (including many pesticides and drugs) would probably be substantially larger. In addition, estimates

---

3. Not all of the fragment constants and factors of the Leo method were incorporated in this program. Their inclusion would have reduced the error for a few compounds and thus slightly lowered the average method error.

TABLE 1-4

Comparison of Observed Values of log K_{ow} with Those Calculated
from Fragment Constants[a]

Compound	Observed log K_{ow}	Δ log K_{ow} (Calculated —Observed)
Methylacetylene	0.94	−0.04
Fluoroform	0.64	0.00
Isobutylene	2.34	−0.16
Ethanol	−0.31	0.10
Dimethyl ether	0.10	0.02
Cyclohexane	3.44	0.07
Propane	2.36	−0.04
2-Propanol	0.05	0.06
tert-Butylamine	0.40	0.13
2-Phenylethylamine	1.41	0.03
N-Phenylacetamide	1.16	0.01
Halothane	2.30	0.16
Benzimidazole	1.34	0.17
p-Nitrophenol	1.91	0.06
Cyclohexene	2.86	0.10
1,2-Dichlorotetrafluoroethane	2.82	0.04
Hexachlorophene	3.93	−0.04
1,2-Methylenedioxybenzene	2.08	0.02
2-Phenyl-1,3-indandione	2.90	−0.08
Carbon tetrachloride	2.83	0.13
Dioxane	−0.42	0.43
2-Bromoacetic acid	0.41	0.07
2-Chloroethanol	0.03	0.00
Indene	2.92	0.07
Fluorene	4.12	−0.09
Anthracene	4.45	0.00
Pyrene	4.88	0.02
Quinoxaline	1.08	0.05
Carbozole	3.51	0.01
Menadione	2.20	−0.45
Chloramphenicol	1.14	−0.58
2-Hydroxy-1,4-naphthoquinone	1.46	−0.94
2-Methyl-3-hydroxyl-1,4-naphthoquinone	1.20	−0.02
2-Methoxy-1,4-naphthoquinone	1.35	−0.09
Benzothiazole	2.01	0.00
o-Phenanthroline	1.83	0.10
Thiazole	0.44	−0.02

(continued)

TABLE 1-4 (Continued)

Compound	Observed log K_{ow}	Δ log K_{ow} (Calculated —Observed)
Piperazine	−1.17	−0.08
Morpholine	−1.08	0.09
Salicylic acid	2.24	−0.27
Imidazole	−0.08	0.00
Cyclohexanol	1.23	0.19
o-Phenyleneurea	1.12	0.27
Tripropylamine	2.79	0.06
Di-n-propylamine	1.62	0.05
Coumarin	1.39	0.05
Trifluoromethylbenzene	2.90	−0.70
Trifluoromethylsulfonanilide	3.05	0.01
1,3-Indandione	0.61	0.66
9-Fluorenone	3.58	−0.71
Phenazine	2.84	−0.12
Morphine	0.83	0.35
2,2,2-Trifluoroethanol	0.41	0.00
2,2,2-Trifluoroacetamide	0.12	0.00
2,2,2-Trichloroethanol	1.35	0.04
2,2,2-Trichloroacetamide	1.04	0.00
Pyrimidine	−0.40	−0.06
Glucose	−3.24	−0.15
Cyclohexylamine	1.49	0.03
Neopentane	3.11	0.03
2-Methylpropane	2.76	−0.03
Crotonic acid	0.72	0.13
Cinnamonitrile	1.96	−0.04
Cinnamic acid	2.13	0.05
Cinnamamide	1.41	−0.41
Methyl cinnamate	2.62	−0.15
Phenyl vinyl ketone	1.88	−0.30
Styrene	2.95	−0.03
1-Phenyl-3-hydroxypropane	1.95	−0.48
Methyl styryl ketone	2.07	−0.09
1,1,2-Trichloroethylene	2.29	−0.01
2-Methoxyanisole	2.08	−0.08
Ethyl vinyl ether	1.04	−0.06
Pyrazole	0.13	0.11
1,1-Difluoroethylene	1.24	−0.12
1,2,3,4-Tetrahydroquinoline	2.29	0.19
Average absolute Δ		0.14
Maximum Δ		−0.94

a. See note 3 on page 1-12.

Source: Chou and Jurs [3]. *(Reprinted with permission from the American Chemical Society.)*

of log K_{ow} >6 are likely to be overestimates of the measured log K_{ow}, perhaps by one or more log units.[4]

It is recommended that the user consider the method error to be in log K_{ow} units (e.g., the average uncertainty of ± 0.12 log K_{ow} unit from the two test sets described) and apply this to any log K_{ow} value calculated from scratch. For example, if an estimate of 2.86 is obtained for log K_{ow} (for a structurally simple compound), report the estimate as log K_{ow} = 2.86 $\pm$ 0.12. In terms of K_{ow}, this would be written as $K_{ow} = 720^{-170}_{+280}$.

If an estimated value of log K_{ow} is derived by modification of a reliable measured value for a structurally related compound (as in Eq.1-4), the method error is likely to be less than if log K_{ow} were calculated "from scratch." To a large extent, the method error will reflect the uncertainty in the specific fragment constants (f values) and factors (F values) that are employed. Data provided in Refs. 11, 28, and 34 indicate that fragment constants (f or π values) and factors for the more common group interactions generally have uncertainties in the range of 0.02 to 0.05 log K_{ow} unit. More complex fragments or factors may have larger uncertainties. For example, an uncertainty of ± 0.08 log K_{ow} unit was assigned to the f-value for the fragment NH_2COO- on the basis of log K_{ow} values for twelve pairs of chemicals (H-R vs NH_2COOR in each pair) [3].

Hansch and Leo [14] do not give the uncertainties for each of their f and F values, but a typical value of 0.03 log K_{ow} unit can be assumed for common fragments and factors and 0.05 for less common ones. The total uncertainty in any estimate derived via the method outlined in Eq. 1-4 can then be calculated by the method outlined in Appendix C of this handbook. Since simple addition and subtraction of terms is involved here, the total method error is the square root of the sum of the squares of the individual uncertainties. Using the example given previously in Eq. 1-5, and assuming that the uncertainties for f_{Cl} and f_{Br} are both 0.03 log K_{ow} unit (Cl and Br are common fragments), the total uncertainty is $(0.03^2 + 0.03^2)^{1/2}$ = 0.04 log K_{ow} unit. Note that this does not consider any uncertainty in the measured value of log K_{ow} for the base chemical. Accordingly, method errors of 0.04 to 0.1 log K_{ow} unit should be expected when this method is used.

---

4. Personal communication from C. Hansch and A. Leo, Pomona College, Claremont, CA, 1980.

Fragments and Factors[5]. The basic fragment constants (f values) and factors (F values) for this method are given in Tables 1-5 and -6, respectively. Table 1-7 provides the F values for one special factor which considers the interaction between halogens and certain polar groups. This chapter does not contain the fragment constants and factors that Hansch and Leo [14] derived for calculating log K_{ow} of ions; the reader is referred to their work for this information. The following paragraphs explain the terms used and refer to specific examples that show how each rule is applied.

• *Fragments.* A fragment is an atom, or string of atoms, whose exterior bonds are to isolating carbon atoms. (An isolating carbon is one that either has four single bonds, at least two of which are to non-hetero atoms, or is multiply bonded to other carbon atoms.) The fragments for which f values are available are listed in the left-hand column of Table 1-5. Superscripts on f denote the type of attachment; superscript symbols are defined on p. 1-24.

A **single-atom fragment** can only be (1) an isolating carbon atom, or (2) a hydrogen or hetero atom, all of whose bonds are isolating carbons. For example:

$$-\overset{\displaystyle |}{\underset{\displaystyle |}{C}}-\ \text{in CH}_4\text{, and }\ \rangle\text{C= in H}_2\text{C=CH}_2$$
$$\text{but not in (CH}_3)_2\,\text{C=NH}$$

$$-\text{H}\quad\text{in H–C}\equiv\text{C–H but not in R}\overset{\displaystyle \text{O}}{\overset{\displaystyle \|}{-\text{C}}}\text{–H or }\ \rangle\text{N–H}$$

$$-\text{F}\quad\text{in CH}_3\text{–F but not in R–SO}_2\text{–F}$$

$$-\text{O}-\ \text{in CH}_3\text{–O–CH}_3\text{ but not in CH}_3\overset{\displaystyle \text{O}}{\overset{\displaystyle \|}{-\text{C}}}\text{–O–CH}_3$$

[See Examples 1-1 and -2]

A **multiple-atom fundamental fragment** can be formed by any combination of (a) a non-isolating carbon, (b) hydrogen, and/or (c) hetero atoms. A fundamental fragment is complete when all its

(continued on p. 1-21)

---

5. The information provided here is a condensation of a much more detailed text by Hansch and Leo [14].

TABLE 1-5

Fragment Constants[a]

Fragment[b]	f	f^ϕ	$f^{\phi\phi}$	Special Types
Without C or H				
—F	−0.38	0.37		$f^{\phi/2} = 0.00$
—Cl	0.06	0.94		$f^{\phi/2} = 0.50$
—Br	0.20	1.09		$f^{1R} = 0.48$, $f^{\phi/2} = 0.64$
—I	0.59	1.35		$f^{\phi/2} = 0.97$
—N<	−2.18	−0.93	−0.50[c]	$f^{1R} = -1.76$
—O—	−1.82[d]	−0.61	0.53	$f^{X1} = -0.22$, $f^{X2} = +0.17$, $f^{\phi/2} = -1.21$
—S—	−0.79	−0.03	0.77	
—NO		0.11		
—NO$_2$	−1.16	−0.03		$f^{X2} = 0.09$
—ONO$_2$	−0.36			
—IO$_2$		−3.23		
—OP(O)O$_2$<	−2.29	−1.71		$f^{X1} = -1.50$
—P(O)<				Triple aromatic = −2.45
—P(O)O$_2$<		−2.33		
—OP(S)O$_2$<		−0.30[c]		
>NP(S)(N<)$_2$	−3.37			
—SP(S)O$_2$<	−2.89			
—SO$_2$F		0.30		
—SO$_2$N<		−2.09		
—S(O)—	−3.01	−2.12	−1.62	
—SO$_2$—	−2.67	−2.17	−1.28	
—SO$_2$O—	−2.11	−2.06	−0.62	$f^{1/\phi} = -1.42$
—SF$_5$		1.45		
—SO$_2$O[−e]	−5.87	−4.53		
—OSO$_3$[−e]	−5.23			
—N=N—			0.14	
—NNN—		0.69		
—N=NN<		−0.85		$f^{X1} = -0.67$
>NNO	−2.40	−0.84		
—O[−e]		−3.64		
—Si<	−0.09[c]	0.65[c]		$f^{1R} = -0.38$[c]

(continued)

TABLE 1-5 (Continued)

Fragment[b]	f	f^ϕ	$f^{\phi\phi}$	Special Types
Without C, with H				
—H	0.23	0.23		
—NH—	−2.15	−1.03	−0.09	f^{X1} = −0.37
—NH$_2$	−1.54	−1.00		f^{X1} = −0.23, f^{1R} = −1.35
—OH	−1.64	−0.44		f^{X1} = 0.32, f^{1R} = −1.34
—SH	−0.23	0.62		
—SO$_2$NH—		−1.75[c]	−1.10	$f^{1/\phi}$ = −1.72
—SO$_2$(NH$_2$)		−1.59		f^{X1} = −1.04
—SO$_2$NH(NH$_2$)	−2.04			
—NHSO$_2$(NH$_2$)		−1.50		
—NH(OH)		−1.11		
—NHNH—			−0.74	f^{1R} = −2.84
—NH(NH$_2$)		−0.65		
—SP(O)(O—)NH—	−2.18[c]			
—SP(O)(NH$_2$)O—	−2.50			
—As(OH)$_2$O—		−1.84		
—As(O)(OH)$_2$		−1.90		
—B(OH)$_2$		−0.32		
With C, without H				
$-\overset{\vert}{\underset{\vert}{C}}-$	0.20	0.20		
—CF$_3$		1.11		
—CN	−1.27	−0.34		f^{1R} = −0.88
—C(O)N<	−3.04	−2.80	−1.93	$f^{1/\phi}$ = −2.20
—SCN	−0.48	0.64		f^{1R} = −0.45
—C(O)—	−1.90	−1.09	−0.50	f^{X1} = −0.83, f^{1R} = −1.77
—C(O)O—	−1.49	−0.56	−0.09	f^{X1} = −0.36, f^{1R} = −1.38, $f^{1/\phi}$= −1.18
—C(O)O^{-e}	−5.19	−4.13		
—N=CCl$_2$		0.64		
—OC(O)N<	−2.54?	−1.84		
—C(O)N—N=N—				$f^{1/\phi}$ = −0.87
—C(=S)O—	−1.11			

(continued)

TABLE 1-5 (Continued)

Fragment[b]	f	f^ϕ	$f^{\phi\phi}$	Special Types
With C and H				
—CH$_3$	0.89	0.89		
—C$_6$H$_5$ (benzene)	1.90			
—C(O)H	—1.10	—0.42		
—C(O)OH	—1.11	—0.03		f^{1R} = —1.03
—C(O)NH—	—2.71	—1.81	—1.06	$f^{1/\phi}$ = —1.51
—C(O)NH$_2$	—2.18	—1.26		f^{X1} = —0.82, f^{1R} = —1.99
—OC(O)NH—	—1.79	—1.46		$f^{1/\phi}$ = —0.91
—OC(O)NH$_2$	—1.58	—0.82		f^{1R} = —1.24
—CH=N—		—1.03	+0.08[c]	
—CH=NOH	—1.02	—0.15		
—CH=NNH—	—2.75			
—NHC(O)NH—	—2.18	—1.57	—0.82	
—NHC(O)NH$_2$	—2.18	—1.07		
>NC(O)NH$_2$		—2.25	—2.15	
>C=NH			—1.29	
>NC(O)H	—2.67	—1.59		
—OC(O)NH—	—1.79	—1.45		$f^{1/\phi}$ = —0.91
—C(=S)NH—	—2.00			$f^{1/\phi}$ = —0.96
—NHCN		—0.03		
—CH=NN<		—1.71		
—NHC(O)N<		—2.29		$f^{1/\phi}$ = —2.42
—NNO(C(O)NH—)	—1.50			$f^{1/\phi}$ = —0.76
—OC(O)H	—1.14	—0.64		
—NHC(O)H		—0.64		
—C=NOH(OH)		—1.64		
—C(=S)NH$_2$		—0.41		
—N(C(O)NH$_2$)—		—2.25	—2.07	
—SO$_2$NHN=CH—				$f^{1/\phi}$ = —1.47
—NHC(=S)NH—		—1.79		
—NNO(C(O)NH$_2$)	—0.95			
—C(O)NHNH$_2$		—1.69		
—NHC(=S)NH$_2$	—1.29	—1.17		
—CNH$_2$(=NH·HCl)		—3.49		
—NHC=NH(NH$_2$)	—5.65			
—C(O)C(O)—	—3.00		—0.30	

(continued)

TABLE 1-5 (Continued)

Fragment[b]	f	f^ϕ	$f^{\phi\phi}$	Special Types
—C(O)NHC(O)—	−3.31		−3.00^c	
—C(O)NHC(O)H	−2.84			
—C(O)NHN=CH—			−1.12	
—C(O)NHC(O)NH$_2$	−1.91			$f^{1R} = -1.57$
—CH(NH$_2$)C(O)OH	−3.97			
—CH=NNHC(O)NH$_2$	−0.63	−0.66		
—CH=NNHC(=S)NH$_2$		−0.05		
—CH=NNHC(O)NHNH$_2$		−1.09		
—C(O)NHC(O)NHC(O)—	−2.38			

Fused in Aromatic Ring^f

Fragments[b] Without C	f^ϕ	Fragments[b] With C	f^ϕ
—N=	−1.12	C	0.13
—N<	−1.10^c	Ċ (ring fusion carbon)	0.22$_5$
—N<$^\phi$	−0.56	C̊* (ring fusion hetero)^g	0.44
—N=N—	−2.14	CH	0.35$_5$
—N≤^{O}	−3.46	—C(O)—	−0.59
—O—	−0.08	—OC(O)—	−1.40
—S—	0.36	—CH=NNH—	−0.47
—S≤^{O}	−2.08	—N=CHNH—	−0.79
—Se—	0.45	—NHC(O)—	−2.00
—NH—	−0.65	—N=CH—O—	−0.71
—NHN=N—	−0.86	—N=CH—S—	−0.29
		—CH=N—O—	−0.63
		—N=CHN=	−1.46
		—NHC(O)NH—	−1.18
		—C(O)NHC(O)—	−1.08^c
		—C(O)NHC(O)NH—	−1.78
		—C(O)NHC(O)NHN=	−1.36

a. The superscript symbols used with f are defined on p. 1-24.

b. The fragment notation is simplified and does not show all bonds. Intrafragment single bonds are not shown. A carbonyl oxygen is always shown in parentheses — e.g., —OC(O)NH$_2$ is

$$-O-\overset{O}{\overset{\|}{C}}-NH_2.$$ A $-\overset{S}{\overset{\|}{C}}-$ group is represented by —C(=S)—. Portions of a fragment shown in

(continued)

TABLE 1-5 (Continued)

parentheses are bonded to a C, N, P or S atom to the left of the parentheses. For example:

$$-N(C(O)NH_2)- \quad \text{is} \quad -N-$$

with

$$
\begin{array}{c}
O{\nwarrow}C-NH_2 \\
|
\end{array}
$$

$$-SP(O)(O-)NH- \quad \text{is} \quad -S-\overset{\displaystyle O}{\underset{\displaystyle O}{\overset{||}{P}}}-NH-$$

$$-NHC=NH(NH_2) \quad \text{is} \quad -NH\overset{\displaystyle NH_2}{\overset{|}{C}}=NH$$

c. Hansch, C., Pomona College, Claremont, CA, personal communication, October 16, 1980.
d. For methyl ethers and ethylene oxide, use −1.54.
e. These fragments are negatively charged ions.
f. A ring system is considered aromatic unless interrupted by a saturated carbon.
g. This factor is also used for fusion to a non-isolating carbon.

Source: Hansch and Leo [14], except as noted in note c. *(Reprinted with permission from John Wiley & Sons, Inc.)*

remaining bonds are to isolating carbons. Some common multiple-atom fundamental fragments are:

$$-\overset{\displaystyle O}{\overset{||}{C}}-O- \qquad -NH_2 \qquad -OH \qquad -\overset{\displaystyle O}{\overset{||}{C}}-NH-$$

[See Examples 1-3 and -4]

Multiple-atom derived fragments can be any combination of single-atom or multiple-atom fundamental fragments that is common or convenient to use. For example, one can *derive* the fragment value for $-CH_3$ (a common fragment): $3(0.23) + 0.20 = 0.89$. Similarly, for the fragment C_6H_5 the derived fragment value is $5(\underline{CH}) + \underline{C} = 5(0.355) + 0.13 = 1.91$. A slightly better value of 1.90 is derived by subtracting 0.23 (for one H) from the measured value of 2.13 for benzene.

[See Examples 1-5 and -6]

An **H-polar fragment** is one that can be expected to participate in hydrogen bonding, either as a donor or an acceptor, such as $-NH_2$, $-OH$, $-O-$, and $-CO_2H$. For such fragments a factor may have to be added that takes into account hydrogen bonding or interactions with nearby halogens.

[See Examples 1-32, -33 and -34]

TABLE 1-6

Summary of Rules for Calculating Factors

● Involving BONDS
Unsaturation

Double	Triple
Normal: $F_{(=)} = -0.55$	$F_{(\equiv)} = -1.42$
Conjugate to ϕ: $F^{\phi}_{(=)} = -0.42$	$F^{\phi\phi}_{(\equiv)} = 0.00$
Conjugate to 2ϕ: $F^{\phi\phi}_{(=)} = -0.00$	
Conjugate to second $=$ in chain: $F_{(=)} = -0.38$	

Geometric

Proportional to length: $x(n-1)$	Branching in short chains: one-time
Chain: $F_b = -0.12$	Alkane chain: $F_{cBr} = -0.13$
Ring:[a] $F_b = -0.09$	H-polar fragment: $F_{gBr} = -0.22$
Branching: $F_{bYN} = -0.20$ (-amine)	Ring cluster: $F_{rCl} = -0.45$
$\quad\quad\quad\quad F_{bYP} = -0.31$ (phosphorus esters)	

● Involving MULTIPLE HALOGENATION

On same carbon (geminal) F_{mhGn}: $\begin{cases} (n=2) = 0.30^b \\ (n=3) = 0.53^b \\ (n=4) = 0.72^b \end{cases}$

On adjacent carbon (vicinal) $F_{mhVn} = 0.28\,(n-1)$

● Involving H-POLAR PROXIMITY

Chain:	
$F_{P1} = -0.42\,(f_1 + f_2)$	Aliphatic ring: $F_{P1} = -0.32\,(f_1 + f_2)$
$F_{P2} = -0.26\,(f_1 + f_2)$	$F_{P2}{}^c = -0.20\,(f_1 + f_2)$
$F_{P3} = -0.10\,(f_1 + f_2)$	

Aromatic ring: $F^{\phi}_{P1} = -0.16\,(f_1 + f_2)$

$\quad\quad\quad\quad\quad F^{\phi}_{P2} = -0.08\,(f_1 + f_2)$

● Involving INTRAMOLECULAR H-BOND

$F_{HBN} = 0.60$ for nitrogen $\quad\quad\quad\quad\quad F_{HBO} = 1.0$ for oxygen

---

a. Aromatic rings excluded.
b. Values per halogen atom, if α or β to H-polar fragment, additional factor required: see Table 1-7.
c. For morpholine and piperazine derivatives, use coef. $= -0.10$

Source: Hansch and Leo [14]. *(Reprinted with permission from John Wiley & Sons, Inc.)*

TABLE 1-7

Aliphatic H/S-Polar Interactions: α-Halogen Factors ($F_{H/SP}$)

Fragment[a]	No. α-F Atoms[b]			No. α-Cl Atoms[b]			No. α-Br Atoms[b]		
	1	2	3	1	2	3	1	2	3
$-SO_2NH-AR$	1.13	1.86	2.70						
$-SO_2-AR$			2.78						
$\overset{\displaystyle O}{\overset{\displaystyle \|}{-C}}-NH_2$	0.97	*1.49*	2.01	1.05	*1.33*	1.61	0.92		
$\overset{\displaystyle O}{\overset{\displaystyle \|}{-C}}-NH-Alk$	0.91	1.31	1.70	0.91	1.15	1.28	0.91	1.09	1.06
$\overset{\displaystyle O}{\overset{\displaystyle \|}{-C}}-NH-AR$			1.65	0.76					
$-O-AR$			1.71						
$-S-AR$			1.47			(0.36)			
$-CH_2OH$[c]	(0.02)		1.22	0.53	0.56	0.88	0.59		
$-CH_2CO_2H$[c]				*0.68*					
$\overset{\displaystyle O}{\overset{\displaystyle \|}{-C}}-AR$			1.17				*0.86*		
$\overset{\displaystyle O}{\overset{\displaystyle \|}{-C}}-Alk$				*0.87*					
$\overset{\displaystyle O}{\overset{\displaystyle \|}{-C}}-O-Alk$			1.07				*0.85*		
$-CO_2H$	*0.89*			*0.90*	*0.98*	*1.20*	*0.85*		

a. Fragment attached to α carbon. AR = aromatic group; Alk = alkane.
b. Italicized values are less reliable; parenthetical values are doubtful.
c. Note β attachment to fragment.

Source: Hansch and Leo [14]. *(Reprinted with permission from John Wiley & Sons, Inc.)*

A **S-polar** (or **σ-polar**) **fragment** is one with strong electron-with-drawing power but little or no tendency to hydrogen bond — i.e., any of the halogens. For such fragments a factor may have to be added that takes into account interactions with nearby H-polar fragments.

[See Examples 1-32 and -33]

Underlining any symbol associated with a fragment constant means the fragment is present in a ring. [See Examples 1-31, -35, -36 and -37.] Factor symbols may also be underlined to show association with a ring, as in $\underline{F_b} = -0.09$. A ring system is considered aromatic unless interrupted by a saturated carbon.

- *Superscripts Denoting Attachment.* The type of isolating carbon atom to which a fragment is attached affects the f value. Superscripts on f (and sometimes F) denote the type of attachment associated with each f value as follows:

Superscript	Attachment	See Examples
None	Aliphatic structural attachment	1-1 to -5
ϕ	Attached to aromatic ring; if bivalent (e.g., $-CO_2-$, $-SO_2N\langle$, $-CH=N-$) the attachment is from the left as written (Ar$-CO_2-$, Ar$-SO_2N\langle$, Ar$-CH=N-$).	1-6 to -8
$1/\phi$	Attached to aromatic ring from right (as written) for bivalent fragments (e.g., $-CO_2-Ar$, $-SO_2N\langle_{Ar}$, $-CH=N-Ar$)	1-9
$\phi\phi$	Bivalent fragment with two aromatic attachments (e.g., Ar$-NH-Ar$)	1-10
X	Aromatic attachment; value enhanced by second, electron-withdrawing substituent ($\sigma_I > +0.50$).[6] f^{X1} used when σ_I for second substituent is between 0.50 and 0.75 (e.g., F, NO_3, CN, SCN, NH_3^+, $OCCl_3$). f^{X2} is used when σ_I for second substituent is > 0.75 (e.g., NO_2, OCN). The other halogens (Cl, Br, I) all have $\sigma_I < 0.5$; the f^{X1} factor may be used if two of these are present to enhance the f value of a third substituent (e.g., use $f^{X1}_{NH_2}$ for 3,4-dichloroaniline).	1-11

---

6. σ_I is a measure of the static inductive effect of a substituent on an aromatic ring. Values are tabulated in Appendix I of Ref. 14.

Superscript	Attachment	See Examples
1R	Benzyl attachment (i.e., attachment to $C_6H_5CH_2-$)	1-12, -13
$\phi/2$	Attachment to vinyl carbon (i.e., to $>C=C<$). Values of $f^{\phi/2}$ are halfway between f and f^ϕ.	

- *Factors.* Many molecular structures require the consideration of *factors* in addition to the fragment values:

$$F_b \quad = \quad \text{bond factor (special cases: } F_{bYN}, F_{bYP})$$
$$F_{cBr} \quad = \quad \text{chain branch factor}$$
$$F_{gBr} \quad = \quad \text{group branch factor}$$
$$F_{rCl} \quad = \quad \text{ring cluster branch factor}$$
$$F_= \quad = \quad \text{double bond factor}$$
$$F_\equiv \quad = \quad \text{triple bond factor}$$
$$F_{mhG} \quad = \quad \text{multiple halogenation, geminal}$$
$$F_{mhV} \quad = \quad \text{multiple halogenation, vicinal}$$
$$F_P \quad = \quad \text{proximity factor for two H-polar fragments; } f_{P1}$$

one-carbon separation; F_{P2}, two-carbon separation; F_{P3}, three-carbon separation

$$F_{H/SP} \quad = \quad \text{proximity factor for H-polar fragment and}$$

S-polar (halogen) fragment. (Values are listed in Table 1-7.)

$$F_{BHN}, F_{BHO} \quad = \quad \text{intramolecular H-bond factors}$$

Each of these factors is briefly explained below with reference to specific examples where these factors are required. The rules for each factor are summarized in Table 1-6.

$\boxed{F_b}$ [See especially Examples 1-1 through -4 and 1-14 through -20.] A bond factor of -0.12 for chains and -0.09 for non-aromatic rings is taken $(n-1)$ times, where n is the number of bonds in the molecule, with the following provisions:

- Do not count bonds between hydrogen and any other atom.
- Do not count any bonds *within* any multi-atom fundamental fragment. [Examples 1-3 and -4]
- Double and triple bonds are considered equivalent to single bonds (only for the calculation of F_b). [Examples 1-14, -15, -17 and -25]
- In ring-chain combinations, consider that the ring stops the count; e.g., in $CH_3-CH_2-Ar-CONH_2$, n=2 on the left side

of the aromatic ring (Ar) and n=1 on the right side. Thus, F_b = 1 (−0.12) + 0(−0.12). [See also Example 1-18.]

- There are special bond factors (Table 1-6) for amines (F_{bYN}) and phosphorus esters (F_{bYP}). These are used for all "counted" bonds in the molecule if the radiating chains are purely hydrophobic (i.e., contain C and H atoms only). [Examples 1-5 and -19] However, if one (and only one) chain contains an H-polar group, use the F_{bY} factor for all bonds up to the one that connects the H-polar fragment; use F_b for any beyond it. [Example 1-20] If two chains contain H-polar fragments, then all bonds are treated as F_b.

No bond factor applies to the bonds in an aromatic ring.

$\boxed{F_{cBr}, \ F_{gBr} \ \text{and} \ F_{rCl}}$ "One-time" chain branching factors are applied whenever there is branching on the molecule. The rules are:

- The length of each branch must be just one or two carbon atoms; or, two or more of the branches must contain hydrophilic groups.
- The factor is required for each branch in the molecule.
- No branching factor is used if either F_{mhG} or F_{mhv} factor is required at that site.
- If the branching is an alkane chain or single S-polar fragment, use $F_{cBr} = -0.13$. [Examples 1-21 and -29] Note that $CH_3CH_2C(CH_3)_3$ has two branches and would require the factor $2F_{cBr}$.
- If the branching is an H-polar fragment, use $F_{gBr} = -0.22$. [Examples 1-22 and -29]
- If the branching is on a ring cluster (peri-fused rings), use $F_{rCl} = -0.45$. This factor is used only once per ring cluster. [Example 1-23]
- If the branching is more than two carbons long, use the factor F_{bYN} (even if the compound is not an amine), which is taken $(n-1)$ times as described above for F_b factors.

$\boxed{F_=}$ For every double bond, *excluding* those contained within fundamental fragments (e.g., $>C=O$, $-C(=O)NH_2$), add the $F_=$ factor according to the following rules:

- Do not count any double bonds in aromatic rings if you are using the fragment constants (f values) for fragments fused in an aromatic ring.

- When a double bond is present, the site of unsaturation should be considered *saturated* for the purposes of fragment constants (i.e., assume $>C=C<$ is $>CH-CH<$) and bond factors (i.e., F_b).

- For normal (isolated) double bonds, use $F_= = -0.55$. [Examples 1-14, -15 and -17]

- For double bonds conjugate to an aromatic ring, use $F_=^\phi = -0.42$. [Example 1-24]

- For all double bonds conjugate to another double bond in a chain (e.g., in 1,3-butadiene), use $F_= = -0.38$.

- For double bonds conjugate to two aromatic rings, use $F_=^{\phi\phi} = 0.00$.

$\boxed{F_\equiv}$ For every triple bond, *excluding* those contained within fundamental fragments (e.g., $-C\equiv N$), add the $F_\equiv$ factor according to the following rules:

- When a triple bond is present, the site of unsaturation should be considered *saturated* for the purposes of fragment constants (i.e., assume $-C\equiv C$ is $-CH_2CH_2-$) and bond factors (F_b).

- For normal (isolated) triple bonds, use $F_\equiv = -1.42$. [Example 1-25]

- For triple bonds conjugate to two aromatic rings, use $F_\equiv^{\phi\phi} = 0.00$.

$\boxed{F_{mhG}}$ When two or more halogens ($-F$, $-Cl$, $-Br$, $-I$) are *bonded to the same carbon atom,* the F_{mhG} factor is applied as follows [Examples 1-14, -15, -26 and -27]:

- Two halogens: $F_{mhG} = 0.30$ per halogen atom.

- Three halogens: $F_{mhG} = 0.53$ per halogen atom.

- Four halogens: $F_{mhG} = 0.72$ per halogen atom.

- No branching factor (F_{cBr} or F_{gBr}) is needed for the carbon-halogen groups requiring the F_{mhG} factor.

$\boxed{F_{mhV}}$ When two or more halogens are *bonded to adjacent carbon atoms*, the F_{mhV} factor (0.28) is taken $n-1$ times, where n is the number of halogens involved. [Examples 1-18 and -27] The following are qualifications:

- This factor applies only when the two carbon atoms are separated by *single* bonds.
- No branching factor (F_{cBr} or F_{gBr}) is needed for carbon-halogen groups requiring the F_{mhv} factor.

$\boxed{F_{P1}, \ F_{P2}, \ F_{P3}}$ When two H-polar fragments (e.g., $-NH_2$, $-OH$, $-O-$, $-CO_2H$) in the molecule are separated by one, two, or three carbon atoms, a correction factor for their interaction is calculated from the sum of the individual H-polar fragment constants ($f_1 + f_2$) as follows:

- If the H-polar fragments are on (or in) a *chain:*

Separation	Factor	
1 carbon	$F_{P1} = -0.42 \, (f_1 + f_2)$	[Example 1-28]
2 carbons	$F_{P2} = -0.26 \, (f_1 + f_2)$	[Example 1-29]
3 carbons	$F_{P3} = -0.10 \, (f_1 + f_2)$	

- If the H-polar fragments are in an *aliphatic ring:*

Separation	Factor	
1 carbon	$\underline{F_{P1}} = -0.32 \, (f_1 + f_2)$	[Example 1-29]
2 carbons	$\underline{F_{P2}} = -0.20 \, (f_1 + f_2)$	[Example 1-29]

[For morpholine, dioxane, and piperazine derivatives use $\underline{F_{P2}} = -0.10 \, (f_1 + f_2)$.]

- If the H-polar fragments are on (or in) an *aromatic ring:*

Separation	Factor	
1 carbon	$F_{P1}^{\phi} = -0.16 \, (f_1 + f_2)$	
2 carbons	$F_{P2}^{\phi} = -0.08 \, (f_1 + f_2)$	[Example 1-31]

Important points for all F_P factors:

- The factor must be applied to each hydrophobic chain connecting two H-polar fragments. [Examples 1-30 and -31]

- The factor must be applied to every pair of H-polar fragments as long as at least one of the hydrocarbon links between them is not otherwise involved. [Example 1-29]

- Fragments on an aliphatic ring are considered to be on a chain. [Example 1-29]

- Where the placement of the two fragments would imply two different coefficients, an average may be used. If, for example, one fragment is on a chain and the other in an aliphatic ring, use $F_{P1} = (-0.42 - 0.32)/2 = -0.37$ and $F_{P2} = (-0.26 - 0.20)/2 = -0.23$. [Example 1-29]

$\boxed{F_{H/SP}}$ If a halogen (S-polar fragment) is located on the same *aliphatic carbon (i.e., an α-halogen)* as an H-polar fragment, add the appropriate factors from the listing provided in Table 1-7. [Examples 1-32 and -33] The $F_{H/SP}$ values in this table also allow a factor to be added for β-halogens (1 carbon separation) for the $-OH$ and $-CO_2H$ fragments.

$\boxed{F_{HBN}, F_{HBO}}$ If intramolecular hydrogen bonding is possible with a nitrogen atom, add the factor $F_{HBN} = 0.60$; if the bonding is with an oxygen atom, add the factor $F_{HBO} = 1.0$. Both factors are per H-bond. [Example 1-34].

Basic Steps

Note: For those unfamiliar with this method, all of the information provided above (i.e., starting at the beginning of §1-3) should be considered prerequisite reading.

(1) Draw the structure of the chemical.

(2) Two estimation pathways are possible:

- Check to see if a measured value of K_{ow} is available for one or more structurally similar compounds.[7] If so, proceed to Step 3. (This approach is preferred for structurally complex chemicals, as it generally provides a more accurate estimate.)

- If no measured value of K_{ow} is available for a structurally similar compound, or if the approach of Step 3 is unworkable or otherwise undesirable, proceed to Step 4.

---

7. Refs. 14 and 28 provide compilations of measured values for thousands of chemicals.

(3) Select the structurally similar compound(s) closest in structure to the problem chemical; if possible, select only those that differ in the number or type of functional groups (fragments) attached to a base molecule. Then, for each structurally similar compound of interest:

* Identify the fragments (f) and factors (F) that have to be added or subtracted to change the similar chemical to the problem chemical. Fragments are listed in Table 1-5 and factors in Tables 1-6 and 1-7 and the text of §1-3.

* Calculate the log K_{ow} value for the new chemical as follows:

$$\log K_{OW} \text{ (new chemical)} = \log K_{OW} \text{ (similar chemical)}$$
$$\pm \text{ fragments (f)} \pm \text{ factors (F)} \qquad (1\text{-}4)$$

A general example of the method is given in Eq. 1-5. Specific examples are provided in the following subsection; see Examples 1-38 to -43.

(4) Identify the fragments (f) and Factors (F) associated with the molecule, considering the rules described previously. Fragments are listed in Table 1-5, and factors are summarized in Tables 1-6 and -7 and the text of §1-3. Obtain log K_{ow} for the chemical by summing the fragment and factor values. Examples 1-1 to -37 illustrate the use of this method.

Several of the examples that follow are taken from Hansch and Leo [14], which is also the source of the observed ("obsd.") values cited. References 14 and 26 contain additional examples that readers may wish to examine.

Example 1-1 $CH_3 CH_2 - O - CH_2 CH_3$

$4f_C$	=	$4(0.20)$	=	0.80
$+10f_H$	=	$10(0.23)$	=	2.3
$+f_{-O-}$	=		-1.82
$+(4\text{-}1)F_b$	=	$3(-0.12)$	=	-0.36
		$\log K_{ow}$	=	0.92

(Obsd. = 0.77, 0.83, 0.89)

Example 1-2 $ClCH_2 CH_2 CH_2 Cl$

$3f_C$	=	$3(0.20)$	=	0.60
$+6f_H$	=	$6(0.23)$	=	1.38
$+2f_{Cl}$	=	$2(0.06)$	=	0.12
$+(4\text{-}1)F_b$	=	$3(-0.12)$	=	-0.36
		$\log K_{ow}$	=	1.74

(Obsd. = 2.00)

Example 1-3 $CH_3\overset{\displaystyle O}{\overset{\|}{C}}-O-CH_3$

$$2f_C = 2(0.20) = 0.40$$
$$+6f_H = 6(0.23) = 1.38$$
$$+f_{CO_2} = -1.49$$
$$+(2-1)F_b = \underline{-0.12}$$
$$\log K_{ow} = 0.17$$
$$(Obsd. = 0.18)$$

Example 1-4 $CH_3CH_2CH_2-\overset{\displaystyle O}{\overset{\|}{C}}NH_2$

$$3f_C = 3(0.20) = 0.60$$
$$+7f_H = 7(0.23) = 1.61$$
$$+f_{CONH_2}^{=} -2.18$$
$$+(3-1)F_b = 2(-0.12) = \underline{-0.24}$$
$$\log K_{ow} = -0.21$$
$$(Obsd. = -0.21)$$

Example 1-5 $(CH_3)_3N$

$$3f_{CH_3} = 3(0.89) = 2.67$$
$$+f_{-N\langle} = -2.18$$
$$+(3-1)F_{bYN} = 2(-0.20) = \underline{-0.40}$$
$$\log K_{ow} = 0.09$$
$$(Obsd. = 0.16, 0.27)$$

Example 1-6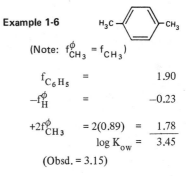

(Note: $f_{CH_3}^{\phi} = f_{CH_3}$)

$$f_{C_6H_5} = 1.90$$
$$-f_H^{\phi} = -0.23$$
$$+2f_{CH_3}^{\phi} = 2(0.89) = \underline{1.78}$$
$$\log K_{ow} = 3.45$$
$$(Obsd. = 3.15)$$

Example 1-7 $O_2N-\langle\rangle-Cl$

$$f_{C_6H_5} = 1.90$$
$$-f_H^{\phi} = -0.23$$
$$+f_{NO_2}^{\phi} = -0.03$$
$$+f_{Cl}^{\phi} = \underline{0.94}$$
$$\log K_{ow} = 2.58$$
$$(Obsd. = 2.39)$$

Example 1-8 $\langle\rangle-\overset{\displaystyle O}{\overset{\|}{C}}-O-CH_3$

$$f_{C_6H_5} = 1.90$$
$$+f_{CO_2}^{\phi} = -0.56$$
$$+f_{CH_3} = 0.89$$
$$+(2-1)F_b = \underline{-0.12}$$
$$\log K_{ow} = 2.11$$
$$(Obsd. = 2.12, 2.23)$$

Example 1-9 $CH_3-\overset{\displaystyle O}{\overset{\|}{C}}-O-\langle\rangle$

$$f_{C_6H_5} = 1.90$$
$$+f_{CO_2}^{1/\phi} = -1.18$$
$$+f_{CH_3} = 0.89$$
$$+(2-1)F_b = \underline{-0.12}$$
$$\log K_{ow} = 1.49$$
$$(Obsd. = 1.49)$$

Example 1-10 $\langle\rangle-NH-\langle\rangle$

$$2f_{C_6H_5} = 2(1.90) = 3.80$$
$$+f_{NH}^{\phi\phi} = -0.09$$
$$+(2-1)F_{bYN} = \underline{-0.20}$$
$$\log K_{ow} \doteq 3.51$$
$$(Obsd. = 3.22, 3.34, 3.50, 3.72)$$

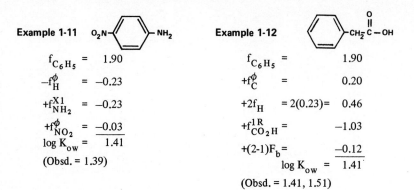

Example 1-11

$$f_{C_6H_5} = 1.90$$
$$-f_H^{\phi} = -0.23$$
$$+f_{NH_2}^{X1} = -0.23$$
$$+f_{NO_2}^{\phi} = \underline{-0.03}$$
$$\log K_{ow} = 1.41$$

(Obsd. = 1.39)

Example 1-12

$$f_{C_6H_5} = 1.90$$
$$+f_C^{\phi} = 0.20$$
$$+2f_H = 2(0.23) = 0.46$$
$$+f_{CO_2H}^{1R} = -1.03$$
$$+(2-1)F_b = \underline{-0.12}$$
$$\log K_{ow} = 1.41$$

(Obsd. = 1.41, 1.51)

Example 1-13

$$f_{C_6H_5} = 1.90$$
$$+f_C^{\phi} = 0.20$$
$$+2f_H = 2(0.23) = 0.46$$
$$+f_{CONH_2}^{1R} = -1.99$$
$$+(2-1)F_b = \underline{-0.12}$$
$$\log K_{ow} = 0.45$$

(Obsd. = 0.45)

Example 1-14

(Note the second rule given for the $F_=$ factor)

$$2f_C = 2(0.20) = 0.40$$
$$+4f_H = 4(0.23) = 0.92$$
$$+2f_F^{\phi/2} = 2(0.00) = 0$$
$$+F_= = -0.55$$
$$+(3-1)F_b = 2(-0.12) = -0.24$$
$$+2F_{mhG_2} = 2(0.30) = \underline{0.60}$$
$$\log K_{ow} = 1.13$$

(Obsd. = 1.24)

Example 1-15

$$2f_C = 2(0.20) = 0.40$$
$$+3f_H = 3(0.23) = 0.69$$
$$+3f_{Cl}^{\phi/2} = 3(0.50) = 1.50$$
$$+F_= = -0.55$$
$$+(4-1)F_b = 3(-0.12) = -0.36$$
$$+2F_{mhG_2} = 2(0.30) = \underline{0.60}$$
$$\log K_{ow} = 2.28$$

(Obsd. = 2.29)

Example 1-16

$$5f_C = 5(0.20) = 1.00$$
$$+10f_H = 10(0.23) = 2.30$$
$$+(5-1)F_b = 4(-0.09) = \underline{-0.36}$$
$$\log K_{ow} = 2.94$$

(Obsd. = 3.00)

Example 1-17

$$6f_C = 6(0.20) = 1.20$$
$$+12f_H = 12(0.23) = 2.76$$
$$+(6-1)\underline{F_b} = 5(-0.09) = -0.45$$
$$+F_= = \underline{-0.55}$$
$$\log K_{ow} = 2.96$$
$$(Obsd. = 2.86)$$

Example 1-18

$$6f_C = 6(0.20) = 1.20$$
$$+6f_H = 6(0.23) = 1.38$$
$$+6f_{Cl} = 6(0.06) = 0.36$$
$$+(6-1)\underline{F_b} = 5(-0.09) = -0.45$$
$$+(6-1)F_{mhV} = 5(0.28) = \underline{1.40}$$
$$\log K_{ow} = 3.89$$
$$(Obsd. = 3.61, 3.72, 3.80 \text{ for } \alpha \text{ isomer}$$
$$= 3.78 \text{ for } \beta \text{ isomer}$$
$$= 4.14 \text{ for } \delta \text{ isomer}$$
$$= 3.72 \text{ for } \gamma \text{ isomer})$$

Example 1-19 $(CH_3CH_2CH_2)_3N$

$$9f_C = 9(0.20) = 1.80$$
$$+21f_H = 21(0.23) = 4.83$$
$$+f_{-N<} = -2.18$$
$$+(9-1)F_{bYN} = 8(-0.20) = \underline{-1.60}$$
$$\log K_{ow} = 2.85$$
$$(Obsd. = 2.79)$$

Example 1-20

$(CH_3CH_2)_2NCH_2CH_2-O-CH_2CH_3$
(Note the final rule given for the F_b factor).

$$8f_C = 8(0.20) = 1.60$$
$$+19f_H = 19(0.23) = 4.37$$
$$+f_{-N<} = -2.18$$
$$+f_{-O-} = -1.82$$
$$+(7-1)F_{bYN} = 6(-0.20) = -1.20$$
$$+(2-1)F_b = \underline{-0.12}$$
$$\log K_{ow} = 0.65$$
$$(No \text{ obsd. value})$$

Example 1-21 $CH_3 \underset{\underset{CH_3}{|}}{\overset{\overset{CH_3}{|}}{CH}CH}CH_3$

$$6f_C = 6(0.20) = 1.20$$
$$+14f_H = 14(0.23) = 3.22$$
$$+(5-1)F_b = 4(-0.12) = -0.48$$
$$+2F_{cBr} = 2(-0.13) = \underline{-0.26}$$
$$\log K_{ow} = 3.68$$
$$(Obsd. = 3.85)$$

Example 1-22 $CH_3 \underset{\underset{OH}{|}}{\overset{\overset{OH}{|}}{CH}CH}CH_3$

$$4f_C = 4(0.20) = 0.80$$
$$+8f_H = 8(0.23) = 1.84$$
$$+2f_{OH} = 2(-1.64) = -3.28$$
$$+(5-1)F_b = 4(-0.12) = -0.48$$
$$+2F_{gBr} = 2(-0.22) = -0.44$$
$$+F_{P2} = (-0.26)(2)(-1.64) = \underline{0.85}$$
$$\log K_{ow} = -0.71$$
$$(Obsd. = -0.92)$$

Example 1-23

$$10f_C \quad = 10(0.20) \quad = \quad 2.00$$

$$+16f_H \quad = 16(0.23) \quad = \quad 3.68$$

$$+(12\text{-}1)F_b \quad = 11(-0.09) = \underline{-0.99}$$

$$+F_{rCl} \qquad\qquad\qquad = \underline{-0.45}$$

$$\log K_{ow} \qquad\quad = \quad 4.24$$

(No Obsd. value)

Example 1-24

$$f_{C_6H_5} \qquad = \qquad\qquad 1.90$$

$$3f_C \qquad = 3(0.20) \quad = \quad 0.60$$

$$+7f_H \qquad = 7(0.23) \quad = \quad 1.61$$

$$+(3\text{-}1)F_b \quad = 2(-0.12) = \quad -0.24$$

$$+F^\phi \qquad = \qquad\qquad \underline{-0.42}$$

$$\log K_{ow} \quad = \quad 3.45$$

(Obsd. = 3.35)

Example 1-25 $CH_3C\equiv CCH_3$

$$4f_C \qquad = 4(0.20) \quad = \quad 0.80$$

$$+10f_H \qquad = 10(0.23) \quad = \quad 2.30$$

$$+(3\text{-}1)F_b \quad = 2(-0.12) = \quad -0.24$$

$$+F_\equiv \qquad = \qquad\qquad \underline{-1.42}$$

$$\log K_{ow} \quad = \quad 1.44$$

(Obsd. = 1.46)

Example 1-26 $F\!-\!\overset{\text{Cl}}{\underset{\text{Cl}}{\overset{|}{\underset{|}{C}}}}\!-\!H$

$$f_C \qquad = \qquad\qquad 0.20$$

$$+f_H \qquad = \qquad\qquad 0.23$$

$$+f_F \qquad = \qquad\qquad -0.38$$

$$+2f_{Cl} \qquad = 2(0.06) \quad = \quad 0.12$$

$$+(3\text{-}1)F_b \quad = 2(-0.12) = \quad -0.24$$

$$+3F_{mhG_3} \quad = 3(0.53) \quad = \quad \underline{1.59}$$

$$\log K_{ow} \quad = \quad 1.52$$

(Obsd. = 1.55)

Example 1-27 $Cl\!-\!\overset{\text{Br}}{\underset{}{\overset{|}{C}}}H\!-\!\overset{\text{F}}{\underset{\text{F}}{\overset{|}{\underset{|}{C}}}}\!-\!F$

$$2f_C \qquad = 2(0.20) \quad = \quad 0.40$$

$$+f_H \qquad = \qquad\qquad 0.23$$

$$+f_{Cl} \qquad = \qquad\qquad 0.06$$

$$+f_{Br} \qquad = \qquad\qquad 0.20$$

$$+3f_F \qquad = 3(-0.38) = \quad -1.14$$

$$+(6\text{-}1)F_b \quad = 5(-0.12) = \quad -0.60$$

$$+2F_{mhG_2} \quad = 2(0.30) \quad = \quad 0.60$$

$$+3F_{mhG_3} \quad = 3(0.53) \quad = \quad 1.59$$

$$+(5\text{-}1)F_{mhV_5} \quad 4(0.28) \quad = \quad \underline{1.12}$$

$$\log K_{ow} \qquad\qquad = \quad 2.46$$

(Obsd. = 2.30)

Example 1-28 $-O\!-\!CH_2\!-\!CO_2H$

$$f_{C_6H_5} \qquad = \qquad\qquad 1.90$$

$$+f^\phi_{-O-} \qquad = \qquad\qquad -0.61$$

$$+f_C \qquad = \qquad\qquad 0.20$$

$$+2f_H \qquad = 2(0.23) \quad = \quad 0.46$$

$$+f_{CO_2H} \qquad = \qquad\qquad -1.11$$

$$+(3\text{-}1)F_b \quad = 2(-0.12) = \quad -0.24$$

$$+F_{P1} \qquad = -0.42(-0.61$$
$$-1.11) \quad = \quad \underline{0.72}$$

$$\log K_{ow} \quad = \quad 1.32$$

(Obsd. = 1.26)

Example 1-29

$$6f_C = 6(0.20) = 1.20$$
$$+7f_H = 7(0.23) = 1.61$$
$$+f_{-O-} = = -1.82$$
$$+5f_{OH} = 5(-1.64) = -8.20$$
$$+(6-1)F_{\underline{b}} = 5(-0.09) = -0.45$$
$$+F_b = = -0.12$$
$$+F_{cBr} = = -0.13$$
$$+4F_{gBr} = 4(-0.22) = -0.88$$
$$+F_{P1} = -0.37(-1.82 \\ -1.64) = 1.28$$
$$+2F_{P2} = 2(-0.23)(-1.82 \\ -1.64) = 1.59$$
$$+3F_{P2} = 3(-0.26)(-1.64 \\ -1.64) = \underline{2.56} \\ \log K_{ow} = -3.36$$
$$(Obsd. = -3.24)$$

Example 1-30

$$4f_C = 4(0.20) = 0.80$$
$$+8f_H = 8(0.23) = 1.84$$
$$+f_{NNO} = = -2.40$$
$$+f_{CO} = = -1.90$$
$$+(6-1)F_{\underline{b}} = 5(-0.09) = -0.45$$
$$+2F_{\underline{P2}} = 2(-0.20)(-2.40 \\ -1.90) = \underline{1.72} \\ \log K_{ow} = -0.39$$
$$(Obsd. = -0.47)$$

Example 1-31

$$4f_{\underline{CH}} = 4(0.355) = 1.42$$
$$+2f_{\underline{C}}^{\phi_*} = 2(0.44) = 0.88$$
$$+2f_H = 2(0.23) = 0.46$$
$$+2f_{-O-}^{\phi} = 2(-0.61) = -1.22$$
$$+f_C = = 0.20$$
$$+(4-1)F_{\underline{b}} = 3(-0.09) = -0.27$$
$$+F_{P1} = (-0.32)(2) \\ (-0.61) = 0.39$$
$$+F_{P2}^{\phi} = (-0.08)(2) \\ (-0.61) = \underline{0.10} \\ \log K_{ow} = 1.96$$
$$(Obsd. = 2.08)$$

Example 1-32

$$3f_F = 3(-0.38) = -1.14$$
$$+f_C = = 0.20$$
$$+f_{SO_2NH}^{1/\phi} = = -1.72$$
$$+f_{C_6H_5} = = 1.90$$
$$+(5-1)F_b = 4(-0.12) = -0.48$$
$$+3F_{mhG_3} = 3(0.53) = 1.59$$
$$+F_{H/SP_3} = = \underline{2.70} \\ \log K_{ow} = 3.05$$
$$(Obsd. = 3.05)$$

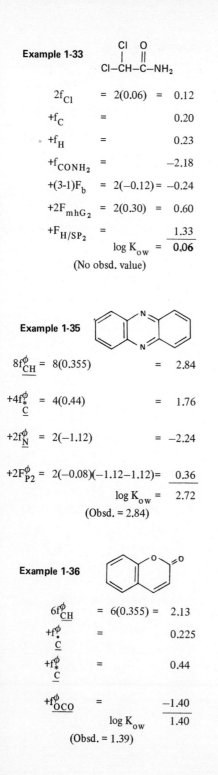

Example 1-33

$$\underset{\substack{|\\ Cl}}{Cl-CH}-\underset{\substack{\|\\ O}}{C}-NH_2$$

$2f_{Cl}$	= 2(0.06)	=	0.12
$+f_C$	=		0.20
$+f_H$	=		0.23
$+f_{CONH_2}$	=		−2.18
$+(3-1)F_b$	= 2(−0.12)	=	−0.24
$+2F_{mhG_2}$	= 2(0.30)	=	0.60
$+F_{H/SP_2}$	=		1.33
		$\log K_{ow}$ =	**0.06**

(No obsd. value)

Example 1-35

$8f_{CH}^{\phi}$	= 8(0.355)	=	2.84
$+4f_{\overset{*}{C}}^{\phi}$	= 4(0.44)	=	1.76
$+2f_{\underline{N}}^{\phi}$	= 2(−1.12)	=	−2.24
$+2F_{P2}^{\phi}$	= 2(−0.08)(−1.12−1.12)=		0.36
		$\log K_{ow}$ =	2.72

(Obsd. = 2.84)

Example 1-36

$6f_{CH}^{\phi}$	= 6(0.355)	=	2.13
$+f_{\overset{\cdot}{C}}^{\phi}$	=		0.225
$+f_{\overset{*}{C}}^{\phi}$	=		0.44
$+f_{OCO}^{\phi}$	=		−1.40
		$\log K_{ow}$	1.40

(Obsd. = 1.39)

Example 1-34   Maleic Acid

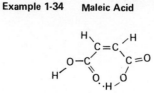

(Dotted line indicates hydrogen bond)

$2f_C$	= 2(0.20)	=	0.40
$+4f_H$	= 4(0.23)	=	0.92
$+2f_{C(O)OH}^{\phi/2}$	= 2(−0.57)	=	−1.14
$+F_=$	=		−0.55
$+(3-1)F_b$	=		−0.24
$+F_{HBO}$	=		1.00
		$\log K_{ow}$	0.39

A similar calculation of $\log K_{ow}$ for the trans isomer of this compound (fumaric acid) yields an estimate of −0.61 since the F_{HBO} factor is not used. No observed values of $\log K_{ow}$ are available for these compounds, but estimates of 0.26 (fumaric) and −0.55 (maleic) may be obtained by the methods in §1-4 using the diethyl ether/water partition coefficients reported in Ref. 14.

Example 1-37

$8f_{CH}^{\phi}$	= 8(0.355)=		2.84
$+2f_{\overset{\cdot}{C}}^{\phi}$	= 2(0.225)=		0.45
$+2f_{\overset{*}{C}}^{\phi}$	= 2(0.44)	=	0.88
$+f_{CO}^{\phi\phi}$	=		−0.50
		$\log K_{ow}$ =	3.67

(Obsd. = 3.58)

Note that the carbonyl carbon is considered to be in an aromatic ring according to footnote f of Table 1-5.

Example 1-38

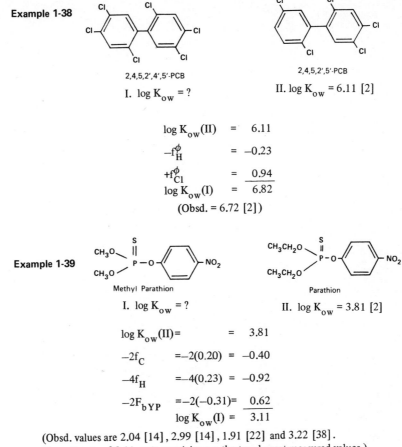

2,4,5,2',4',5'-PCB

I. $\log K_{ow}$ = ?

2,4,5,2',5'-PCB

II. $\log K_{ow}$ = 6.11 [2]

$$
\begin{aligned}
\log K_{ow}(II) &= 6.11 \\
-f_H^{\phi} &= -0.23 \\
+f_{Cl}^{\phi} &= \underline{0.94} \\
\log K_{ow}(I) &= 6.82
\end{aligned}
$$

(Obsd. = 6.72 [2])

Example 1-39

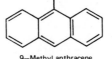

Methyl Parathion

I. $\log K_{ow}$ = ?

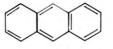

Parathion

II. $\log K_{ow}$ = 3.81 [2]

$$
\begin{aligned}
\log K_{ow}(II) &= & & 3.81 \\
-2f_C &= -2(0.20) &=& -0.40 \\
-4f_H &= -4(0.23) &=& -0.92 \\
-2F_{bYP} &= -2(-0.31) &=& \underline{0.62} \\
\log K_{ow}(I) &= & & 3.11
\end{aligned}
$$

(Obsd. values are 2.04 [14], 2.99 [14], 1.91 [22] and 3.22 [38].
The estimate of 3.11 casts suspicion on the two lowest measured values.)

Example 1-40

9—Methyl anthracene

I. $\log K_{ow}$ = ?

Anthracene

II. $\log K_{ow}$ = 4.54 [21]

$$
\begin{aligned}
\log K_{ow}(II) &= 4.54 \\
-f_H^{\phi} &= -0.23 \\
+f_{CH_3}^{\phi} &= \underline{0.89} \\
\log K_{ow}(I) &= 5.20
\end{aligned}
$$

(Obsd. = 5.07 [21])

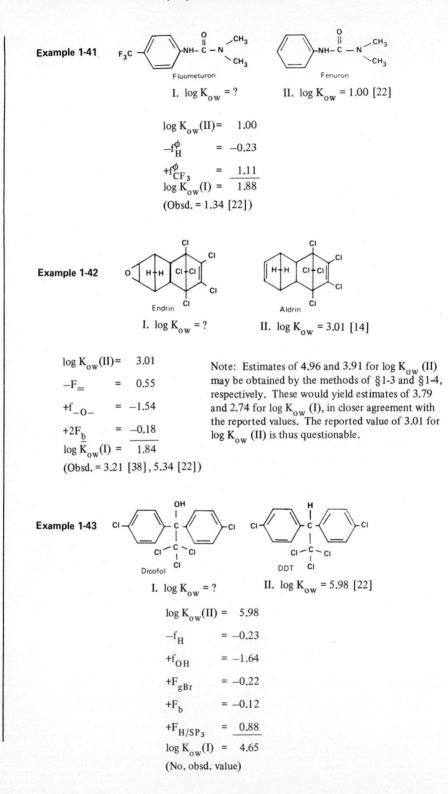

Example 1-41

Fluometuron

Fenuron

I. $\log K_{ow}$ = ?   II. $\log K_{ow}$ = 1.00 [22]

$$\log K_{ow}(II) = 1.00$$

$$-f_H^{\phi} = -0.23$$

$$+f_{CF_3}^{\phi} = \underline{1.11}$$

$$\log K_{ow}(I) = 1.88$$

(Obsd. = 1.34 [22])

Example 1-42

Endrin

Aldrin

I. $\log K_{ow}$ = ?   II. $\log K_{ow}$ = 3.01 [14]

$$\log K_{ow}(II) = 3.01$$

$$-F_{=} = 0.55$$

$$+f_{-O-} = -1.54$$

$$+2F_b = -0.18$$

$$\log K_{ow}(I) = 1.84$$

(Obsd. = 3.21 [38], 5.34 [22])

Note: Estimates of 4.96 and 3.91 for $\log K_{ow}$ (II) may be obtained by the methods of §1-3 and §1-4, respectively. These would yield estimates of 3.79 and 2.74 for $\log K_{ow}$ (I), in closer agreement with the reported values. The reported value of 3.01 for $\log K_{ow}$ (II) is thus questionable.

Example 1-43

Dicofol

DDT

I. $\log K_{ow}$ = ?   II. $\log K_{ow}$ = 5.98 [22]

$$\log K_{ow}(II) = 5.98$$

$$-f_H = -0.23$$

$$+f_{OH} = -1.64$$

$$+F_{gBr} = -0.22$$

$$+F_b = -0.12$$

$$+F_{H/SP_3} = \underline{0.88}$$

$$\log K_{ow}(I) = 4.65$$

(No. obsd. value)

1-4   ESTIMATION WITH SOLVENT REGRESSION EQUATIONS

Principles of Use. Solvent/water partition coefficients (K_{sw}) have been measured for numerous solutes in a large variety of solvent systems. These K_{sw} values, like K_{ow}, have found wide use in structure-activity correlations, especially for pharmaceuticals. Some of the more frequently used organic solvents are ethyl ether, n-butanol, chloroform, cyclohexane, benzene, and vegetable oils. In analogy with Eq. 1-1, K_{sw} is defined as the ratio of the solute's concentration in the organic phase to that in the aqueous phase of the two-phase system at equilibrium. Low solute concentrations are employed in the measurement.

The most comprehensive set of solvent regression equations is that given by Leo and Hansch [27] and repeated in the subsequent publication by Leo, Hansch, and Elkins [28]. Their original equations were written with log K_{sw} as the dependent variable but are restated here in the following form:

$$\log K_{ow} = a \log K_{sw} + b \qquad\qquad (1\text{-}6)$$

Thirty-one such equations are provided, allowing K_{ow} to be calculated if a value of K_{sw} for the solute is available with one or more of approximately twenty different solvents. A modified solvent regression equation, developed by Seiler [41], is provided if the K_{sw} value is for the cyclohexane/water system. Several solvent regression equations are also given by Rekker [39], but most of them involve the use of special fragment constants or correction factors and are thus slightly more difficult to use. Rekker's equations are not included in this chapter.

Solvent Regression Equations. The selection of the appropriate solvent regression equation sometimes depends upon the nature of the solute. Table 1-8 lists a number of solute classes in two basic groups: A (hydrogen donors) and B (hydrogen acceptors). Table 1-9 provides values of a and b for the basic set of solvent regression equations (Eq. 1-7 to -37), all of which are of the form shown in Eq. 1-6. If the solute (the chemical for which K_{ow} is to be calculated) is listed under Group A or B in Table 1-8 *and* if the solvent (associated with the available K_{sw} value) is one of those listed in the first two sections of Table 1-9, then a choice between two equations must be made. For example, if a value of K_{sw} is available from the xylene/water system, one must choose between Eqs. 1-10 and 1-21. The choice depends on where the solute is listed in Table 1-8 — e.g., Eq. 1-10 would be used if the solute were an alcohol, and Eq. 1-21 would be used if it were an ether.

TABLE 1-8

General Solute Classes

Group A — H donors		
		1. Acids
		2. Phenols
		3. Barbiturates
		4. Alcohols
	a	5. Amides (negatively substituted, but not di-N-substituted)
		6. Sulfonamides
		7. Nitriles
		8. Imides
		9. Amides[b]
Group B — H acceptors		
		10. Aromatic amines[b] (not di-N-substituted)
		11. Miscellaneous acceptors
		12. Aliphatic[c] and aromatic hydrocarbons
		13. Intramolecular H bonds[d]
		14. Ethers
		15. Esters
		16. Ketones
		17. Aliphatic amines and imines
		18. Tertiary amines (including ring N compounds)

a. "Neutral" in chloroform and carbon tetrachloride; use "N" equations (Eq. 1-36, -37) for these two solvents.

b. Classes 9 and 10 must be reversed when considering the ether and oils solvent systems.

c. Classification assigned by the author of this chapter.

d. E.g., *o*-nitrophenol.

Source: Leo, Hansch and Elkins [28]. *(Reprinted with permission from the American Chemical Society.)*

TABLE 1-9

Solvent Regression Equations[a]

$$(\log K_{ow} = a \log K_{sw} + b)$$

Eq.	Solvent	a	b	No.	r
	Equations for Group A Solutes				
1-7	Cyclohexane	1.481	2.729	26	0.761
1-8	Heptane	0.947	2.700	10	0.764
1-9	Carbon tetrachloride	0.856	1.852	24	0.974
1-10	Xylene	1.062	1.798	19	0.963
1-11	Toluene	0.881	1.566	22	0.980
1-12	Benzene	0.985	1.381	33	0.962
1-13	Chloroform	0.888	1.193	28	0.967
1-14	Oils[b]	0.910	1.192	65	0.981
1-15	Nitrobenzene	0.850	0.912	9	0.977
1-16	Isopentyl acetate	0.974	−0.070	22	0.986
1-17	Ethyl ether	0.885	0.150	71	0.988
	Equations for Group B Solutes				
1-18	Cyclohexane	0.941	0.690	30	0.957
1-19	Heptane	0.541	1.203	11	0.954
1-20	Carbon tetrachloride	0.829	0.181	11	0.959
1-21	Xylene	0.974	0.579	21	0.986
1-22	Toluene	0.715	0.660	14	0.971
1-23	Benzene	0.818	0.469	19	0.958
1-24	Chloroform	0.784	0.134	21	0.976
1-25	Oils[b]	0.894	0.290	14	0.988
1-26	Ethyl ether	0.876	0.937	32	0.957
	Equation Set C				
1-27	Oleyl alcohol	1.001	0.576	37	0.985
1-28	Methyl isobutyl ketone	0.914	−0.046	17	0.993
1-29	Ethyl acetate	1.073	−0.056	9	0.969
1-30	Cyclohexanone	0.966	−0.866	10	0.972
1-31	Primary pentanols	1.238	−0.335	19	0.987
1-32	sec- and tert-Pentanols	1.121	−0.323	11	0.996
1-33	2-Butanone	2.028	−0.639	9	0.987
1-34	Cyclohexanol	1.342	−1.162	12	0.985
1-35	Primary butanols	1.435	−0.547	57	0.993
	Equations for Group N Solutes				
1-36	Carbon tetrachloride	1.160	0.726	6	0.809
1-37	Chloroform	0.909	0.561	32	0.974

a. The values of a and b are the slope and intercept, respectively, for the solvent regression equation; No. = number of chemicals in data set; r = correlation coefficient for equation as originally written by Leo *et al.* [28] . Their equations were in the form $\log K_{sw} = a' \log K_{ow} + b'$.

b. Most liquid glyceryl triesters fit these equations; olive, cottonseed, and peanut oils were the most frequently used.

Source: Adapted from Leo, Hansch and Elkins [28], Table VIII. *(Reprinted with permission from the American Chemical Society.)*

If the solvent is one of those listed in the third section (set "C") of Table 1-9, the appropriate equation (from Eqs. 1-27 to -35) is selected irrespective of the solute class. Three equations are available for the instances when K_{sw} is from the carbon tetrachloride/water or chloroform/water systems: 1-9, -20, and -36 for the former and 1-13, -24, and -37 for the latter. Footnote "a" of Table 1-8 explains how the correct equation is selected based upon considerations of solute class.

It has been noted in several places [18,27,28,41] that solvent regression equations linking K_{ow} with K_{sw} for such solvents as cyclohexane and heptane have a poorer quality of fit (lower *r* values — see Table 1-9). This is attributed to the effects of hydrogen bonding (solute-solute interactions). Cyclohexane and heptane dissolve very small amounts of water ($\sim 3 \times 10^{-3}$ mol/L); other organic solvents dissolve greater amounts, which tends to inhibit solute-solute hydrogen bonding effects when K_{sw} is measured.

Seiler [41] has proposed a modified solvent regression equation to cover the cyclohexane/water → octanol/water calculation:

$$\log K_{ow} = \log K_{cw} + \Sigma I_H + 0.16$$

$$\text{No.} = 195, \text{r} = 0.967$$

(1-38)

In this equation K_{cw} is the cyclohexane/water partition coefficient and I_H values are hydrogen bonding corrections for specific functional groups in the solute. The I_H values are given in Table 1-10. Note that the correlation coefficient (r) for Eq. 1-38 (0.967) is significantly better than for Eq. 1-7 (0.761) for H donors but is about the same as for Eq. 1-18 (0.957) for H acceptors.

Method Errors. Table 1-11 compares observed and estimated values of K_{ow} for 39 examples, excluding those where K_{sw} was from the cyclohexane/water or heptane/water systems. For the examples listed, the average absolute error is 0.38 log K_{ow} unit and the maximum error is 1.4 log K_{ow} units. The method error for Eq. 1-38 is assumed to be of similar magnitude.

Basic Steps

(1) Obtain a measured value of K_{sw} for any of the solvent systems listed in Table 1-9. (Measured values of K_{sw} for numerous chemicals are tabulated in Refs. 14 and 28. Reference 17 gives some values for the cyclohexane/water system.)

TABLE 1-10

Hydrogen Bonding Corrections, I_H, for Eq. 1-38

Molecular Segment[a]	I_H	No.[b]
—N=N—NH— (triazolo)	4.24	2
$NH_2OC...CONH_2$ (in malonamides)	3.41	4
—C·CO·NH·CO·NH·CO	3.06	2
—COOH, aliphatic	2.88	1
—COOH, aromatic	2.87	4
—OH, aromatic	2.60	33
—CONH—	2.56	7
$—SO_2NH—$	1.93	3
—OH, aliphatic	1.82	11
$—NH_2$, aliphatic	1.33	2
$—NH_2$, aromatic	1.18	21
=N—	1.01	9
$—P(OR)_2=O$, $R \neq H$	0.84	2
—NHR, $R \neq H$	0.61	9
$—CO—CH_2—CO—$	0.59	26
$—NR_1R_2$, $R \neq H$	0.55	10
$—NO_2$	0.45	31
>C=O	0.31	26
—CN	0.23	13
—O—	0.11	34
Ortho substitution to —OH, —COOH, $—NR_1R_2$	−0.62	16
Per ΔpK unit (phenols)[c]	0.30	33
Per ΔpK unit (anilines)[c]	0.16	21

a. Inclusion of the following molecular segments as additional independent variables in the least-squares analysis did not improve the correlation significantly: chlorine (No.=40); bromine (No.=4); iodine (No.=2); fluorine (No.=12); —COOR, $R \neq H$ (No.=55); $—SO_2—$ (No.=1); number of carbon atoms in each molecule; and ΔpK value for benzoic acids (No.=4).

b. Total number of each of the molecular segments used in the least-squares calculation.

c. $\Delta pK = pK_o$ (R=H) − pK_a (R≠H).

Source: Seiler [41]. *(Reprinted with permission from the Societe d'Etudes de Chimie Therapeutique.)*

TABLE 1-11

Comparison of Observed log K_{ow} with Values Estimated from Solvent/Water Partition Coefficients

Chemical[a]	Solvent	log K_{sw}[b]	Eq. Type	log K_{ow} Observed	log K_{ow} Calculated[b]	Error (log K_{ow} units)
Chloroform	Oils	1.86	B	1.97	1.98	0.01
Thiourea	Diethyl ether	-2.15(3)	A	-1.14	-1.48(3)	-0.34
	Chloroform	-3.10	N		-2.38	-1.24
	Oils	-2.92	A		-1.43	-0.29
Chloroacetamide	Diethyl ether	-1.02(2)	A	-0.53	-0.78	-0.25
	Chloroform	-0.96	N		-0.29	0.24
Acetic acid, methyl ester	Diethyl ether	0.43	A	0.18	0.49	0.31
	Oils	-0.37	B		0.20	0.02
	Benzene	0.47	B		0.87	0.69
	Carbon tetrachloride	0.41	B		0.32	0.14
α-Bromobutyric acid	Chloroform	0.08	A	1.42	1.29	-0.13
	Oils	0.14	A		1.12	-0.30
	Benzene	-0.08	A		1.33	-0.09
	Isobutanol	1.46	C		1.55	0.13
	Toluene	-0.27	A		1.32	-0.10
β,β,β-Trichloro-t-butanol	Oils	1.36	A	2.03	2.44	0.41
Piperazine	Diethyl ether	-3.28	B	-1.17	-2.03	-0.86
Diethanolamine	Diethyl ether	-3.27	B	-1.43	-2.02	-0.59
	Isobutanol	-0.70	C		-1.49	-0.06
Pentanol	Oils	0.36	A	1.40	1.52	0.12
	Benzene	0.19	A		1.56	0.16

(continued)

TABLE 1-11 (Continued)

Chemical[a]	Solvent	log K_{sw}[b]	Eq. Type	log K_{ow} Observed	log K_{ow} Calculated[b]	Error (log K_{ow} units)
2,4,6-Trinitrophenol	Chloroform	1.20	A	2.03	2.31	0.28
	Benzene	1.69	A		3.02	0.99
	n-Butanol	0.96	C		0.82	-1.21
	Toluene	1.30(2)	A		2.72(2)	0.69
	Primary pentanols	1.85	C		2.01	-0.02
	sec-Pentanols	0.82	C		0.63	-1.40
m-Bromophenol	Oleyl alcohol	2.02	C	2.63	2.57	-0.06
p-Bromobenzene-sulfonamide	Chloroform	0.39	N	1.36	0.92	-0.44
o-Dihydroxybenzene	Diethyl ether	0.93(3)	A	0.94(2)	0.93(3)	-0.01
	Benzene	-1.19	A		0.21	-0.73
Triethylamine	Benzene	1.13	B	1.44	1.30	-0.14
	Isobutanol	1.32	C		1.47	0.03
	Xylene	1.11	B		1.77	0.33
	Toluene	0.89(3)	B		1.30(3)	-0.14
	Primary pentanols	1.42	C		1.42	-0.02
Phenylthiourea	Diethyl ether	0.23	A	0.73	0.32	-0.41
	Chloroform	0.54	N		0.54	-0.19
2-Naphthol	Diethyl ether	1.77	A	2.84	1.67	-1.17

Average absolute error = 0.38[c]

Maximum error = 1.4

a. Entries were selected from a much larger listing in Table XVII of Ref. 28.

b. Where more than one value of log K_{sw} or log K_{ow} was given, an average was taken for use in this table. The number in parentheses indicates the number of data points in the original source.

c. Higher errors are likely to be associated with K_{ow} values derived from K_{sw} values from the cyclohexane or heptane systems.

Source: Leo, Hansch and Elkins [28]. (Reprinted with permission from the American Chemical Society.)

(2) If the solvent is one of those in "Set C" in Table 1-9, select the appropriate solvent regression equation from this group. Substitute the given values of a and b, along with log K_{sw}, in the generalized equation shown at the head of Table 1-9 (also shown as Eq. 1-6) and solve for log K_{OW}.

(3) If the solvent is listed in the first two sections of Table 1-9, see Table 1-8 to determine whether the solute is in Group A or B and select the regression equation accordingly. (See Step 4 for an alternate method if the solvent is cyclohexane.) Substitute the selected values of a, b and log K_{sw} into the generalized equation shown at the head of Table 1-9 (also shown as Eq. 1-6) and solve for log K_{OW}.

(4) If K_{sw} is from the cyclohexane/water system *and* the chemical is a H-donor, the preferred equation is Eq. 1-38 (given in the text). Values of I_H needed for this equation are obtained from Table 1-10. This method may also be used if the chemical is a H-acceptor.

(5) If K_{sw} values are available for two or more solvents, it is suggested that a value of log K_{OW} be estimated from each K_{sw} value and averaged.

Example 1-44   Estimate K_{ow} for *m*-bromoaniline, given log K_{sw} (benzene) = 2.20 [28].

(1)   There are two benzene equations (Eqs. 1-12 and -23) in Table 1-9. Table 1-8 indicates that the equation in group "B" (Eq. 1-23) should be used for an aromatic amine.

(2)   From Eq. 1-23,

$$\log K_{ow} = 0.818 \,(2.20) + 0.469$$

$$= 2.27 \quad (\therefore K_{ow} = 186)$$

The reported value of log K_{ow} is 2.10 [28]; thus, the error is +0.17 log unit.

Example 1-45   Estimate K_{ow} for *t*-butanol, given log K_{sw} (chloroform) = −0.04 [28].

(1)   There are three chloroform equations in Table 1-9. Table 1-8 indicates that the "N" equation (Eq. 1-37) should be used for alcohols.

(2)   From Eq. 1-37,

$$\log K_{ow} = 0.909 \, (-0.04) + 0.561$$

$$= 0.52 \quad (\therefore K_{ow} = 3.3)$$

The reported value of $\log K_{ow}$ is 0.37 [28]; thus, the error is +0.15 log unit.

Example 1-46   Estimate K_{ow} for *m*-bromophenol, given $\log K_{sw}$ (cyclohexane) = -0.52 and $\log K_{sw}$ (oleyl alcohol) = 2.02 [28].

(1)   For the cyclohexane number, Step 4 (of Basic Steps) states that Eq. 1-38 is preferred. The $-$OH substituent requires a hydrogen bonding correction (I_H) of 2.60, as indicated in Table 1-10. Substituting in Eq. 1-38:

$$\log K_{ow} = -0.52 + 2.60 + 0.16$$

$$= 2.24 \quad (\therefore K_{ow} = 174)$$

(2)   For the oleyl alcohol number, only one equation (Eq. 1-27) is available in Table 1-9; there is no choice to be made based on solute class. Substituting in Eq. 1-27:

$$\log K_{ow} = 1.001 \, (2.02) + 0.576$$

$$= 2.60 \quad (\therefore K_{ow} = 397)$$

The measured value of $\log K_{ow}$ is 2.63 [28]; thus, the errors involved in the two estimates are -0.39 and -0.03 log units, respectively.

1-5   ESTIMATION FROM (ESTIMATED) ACTIVITY COEFFICIENTS

Introduction. This section briefly outlines how K_{OW} values can be estimated with activity coefficients that have been estimated via the methods described in Chapter 11. It is not considered a recommended method, since the calculations (primarily those associated with estimating the activity coefficient) are too complex for those without access to a programmable calculator or small computer. Accordingly, not all details of the procedure are given, there are no step-by-step instructions or examples, and the method error is not stated.

Relating K_{OW} to γ. The activity of a chemical that has been allowed to equilibrate between the phases of the octanol/water system must be the same in each phase. It follows that

$$x_c^o \, \gamma_c^o = x_c^w \, \gamma_c^w \tag{1-39}$$

where

x_c^o = mole fraction of chemical (c) in octanol (o) phase

x_c^w = mole fraction of chemical (c) in water (w) phase

γ_c^o = activity coefficient of chemical in octanol phase

γ_c^w = activity coefficient of chemical in water phase

Thus

$$x_c^o / x_c^w = \gamma_c^w / \gamma_c^o \qquad (1\text{-}40)$$

From the definition of mole fraction, it follows that

$$x_c^w = \frac{n_c^w}{n_w^w + n_c^w + n_o^w} = \frac{C_c^w}{C_w^w + C_c^w + C_o^w} \qquad (1\text{-}41)$$

where   n = number of moles

superscript w = water phase

subscripts w, c, o = water, chemical, and octanol, respectively

C = concentration (mol/L)

In the measurement of K_{ow}, C_c^w is typically $\lesssim 0.01$. C_o^w is the solubility of octanol in water (4.5×10^{-3} M) and C_w^w is 55.5 M. Thus, Eq. 1-41 may be reduced to:

$$C_c^w = 55.5 \ x_c^w \ (\text{mol/L}) \qquad (1\text{-}42)$$

Similarly,

$$x_c^o = \frac{n_c^o}{n_o^o + n_c^o + n_w^o} = \frac{C_c^o}{C_o^o + C_c^o + C_w^o} \qquad (1\text{-}43)$$

where the symbols have the same meaning as described above. Again, C_c^o is typically small ($\lesssim 0.01$ M) and may be neglected. C_w^o is the solubility of water in octanol (2.30 M). C_o^o is then found to be 6.07 M, using a density of 0.825 g/mL for octanol and assuming no volume change upon mixing of the octanol in water. Equation 1-43 may now be reduced to:

$$C_c^o = 8.37 \ x_c^o \ (\text{mol/L}) \qquad (1\text{-}44)$$

Now K_{ow}, by definition (cf. Eq. 1-1), is

$$K_{ow} = C_c^o / C_c^w \qquad (1\text{-}45)$$

Substituting Eqs. 1-42 and -44 into 1-45 we obtain

$$K_{ow} = 0.151 \; x_c^o / x_c^w \qquad\qquad (1\text{-}46)$$

Finally, substituting Eq. 1-40 in the above:

$$K_{ow} = 0.151 \; \gamma_c^w / \gamma_c^o \qquad\qquad (1\text{-}47)$$

Estimating γ^w and γ^o. Chapter 11 provides detailed instructions for calculating activity coefficients in binary systems. These instructions are adequate for estimating γ_c^w for most chemicals, since the presence of 4.5 $\times\ 10^{-3}$M octanol may be ignored.[8] In addition, since $C_c^w < 0.01$M, one can assume that $\gamma_c^w = (\gamma_c^w)^\infty$, where $(\gamma_c^w)^\infty$ is the activity coefficient at infinite dilution. This assumption simplifies the calculations.

The assumption of a binary system cannot, however, be used for the calculation of γ_c^o. The octanol phase is a ternary one, with 0.725 mole fraction octanol, 0.275 mole fraction water, and $\sim 10^{-4}$ mole fraction of the test chemical. Thus, for the calculation of γ_c^o, the instructions of Chapter 11 must be slightly modified (see references cited in Chapter 11) to extend the method to ternary systems.[9] It can again be assumed that $\gamma_c^o = (\gamma_c^o)^\infty$.

1-6 AVAILABLE DATA

A large collection of measured K_{ow} and K_{sw} values (nearly 15,000 data points) is given by Hansch and Leo [14]. This supersedes the list (nearly 6,000 data points) published earlier by Leo, Hansch and Elkins [28]. More up-to-date lists — the result of a continuing project by these researchers — may be purchased from the Pomona College Medicinal Chemistry Project, Pomona College, Claremont, CA 91711. All of these lists are indexed by molecular formula.

Other publications that list substantial amounts of data include the following:

---

8. This was demonstrated by Dec *et al.* [5], who measured the solubility of three chemicals (1,3,5-triaza-1,3,5-trinitrocyclohexane, 1,2,3,5-tetrachlorobenzene, and o-dichlorobenzene) in both pure water and octanol-saturated water and found no significant difference in the results.

9. The general method described in Chapter 11 *is* applicable to ternary systems, and other investigators have carried out numerous calculations to demonstrate this. See Ref. 9 or § 3-4 of Chapter 3 for additional information.

Rekker [39] — Numerous K_{ow} and K_{sw} values

Nys and Rekker [33] — K_{ow} values

Kenaga and Goring [22] — K_{ow} for many pesticides

Rao and Davidson [38] — K_{ow} for many pesticides

Karickhoff, Brown and Scott [21] — K_{ow} for several polynuclear aromatics

Holmes [17] — Some emphasis on K_{sw} for the cyclohexane/water system

Holmes and Lough [18] — K_{ow} for substituted phenols with intramolecular hydrogen bonding.

Additional sources include the references cited in Table 1-2 of this chapter. Appropriate references in Chapters 2, 4, and 5 may also be helpful; these chapters describe regression equations between K_{ow} and (1) solubility, (2) soil adsorption coefficients, and (3) bioconcentration factors.

1-7   SYMBOLS USED

a   = parameter in solvent regression equation (Eq. 1-6)

b   = parameter in solvent regression equation (Eq. 1-6)

BCF   = bioconcentration factor for aquatic life

C   = concentration, mol/L (superscripts o, w, and c for octanol, water, and chemical)

f   = fragment constant (See §1-3 for meaning of superscripts. Subscripts are molecular fragments as identified in Table 1-5. Underline indicates fragment is present in a ring.)

F   = structural factor (See §1-3 for meaning of subscripts and superscripts.)

G   = free energy of solution (subscripts o and w for octanol and water)

HPLC/RT = high-pressure liquid chromatography/retention time

I_H = correction factor for hydrogen bonding in Eq. 1-38

K_{cw} = cyclohexane/water partition coefficient in Eq. 1-38

K_{oc} = soil or sediment adsorption coefficient based on organic carbon

K_{ow} = octanol/water partition coefficient

K_{sw} = solvent/water partition coefficient

n = number of bonds when used with F_b or number of halogens when used with F_{mhG} or F_{mhV} (See Table 1-6)

$n\{\}$ = number of moles in Eqs. 1-41 and -43. (Superscripts o and w for octanol and water phases. Subscripts o, w and c for octanol, water and chemical.)

pK_a = negative log of acid dissociation constant for acid (Table 1-10)

pK_o = negative log of acid dissociation constant for a chemical having a hydrogen atom substituted for the acid function (Table 1-10)

ΔpK = pK_o—pK_a (Table 1-10)

r = correlation coefficient for regression equation

S = solubility in water

X = concentration, mole fraction (Superscripts o and w for octanol and water phases. Subscripts o, w and c for chemical.)

Greek

γ = activity coefficient (Superscripts o and w for octanol and water. Subscript c for chemical.)

$\gamma\infty$ = activity coefficient at infinite dilution

π = "pi" substituent constant

σ_I = static (or sigma) inductive effect parameter

χ = molecular connectivity parameter

1-8 REFERENCES

1. Carlson, R., R. Carlson and H. Kopperman, "Determination of Partition Coefficients by Liquid Chromatography," *J. Chromatogr.*, **107**, 210-23 (1975).

2. Chiou, C.T., V.H. Freed, D.W. Schmedding and R.L. Kohnert, "Partition Coefficient and Bioaccumulation of Selected Organic Chemicals," *Environ. Sci. Technol.*, **11**, 475-78 (1977).

3. Chou, J.T. and P.C. Jurs, "Computer Assisted Computation of Partition Coefficients from Molecular Structures Using Fragment Constants," *J. Chem. Inf. Comput. Sci.*, **19**, 172-78 (1979).

4. Currie, D.J., C.E. Lough, R.F. Silver and H.L. Holmes, "Partition Coefficients of Some Conjugated Heteroenoid Compounds and 1,4-Naphthoquinones," *Can. J. Chem.*, **44**, 1035-43 (1966).

5. Dec, G., S. Banerjee, H.C. Sikka and E.J. Pack, Jr., "Water Solubility and Octanol/Water Partition Coefficients of Organics: Limitations of the Solubility-Partition Coefficient Correlation," preprint (submitted to *Environ. Sci. Technol.*). [Revised version with same title published in *Environ. Sci. Technol.*, **14**, 1227-29 (1980). Authors listed as S. Banerjee, S. H. Yalkowsky, and S.C. Valvani.]

6. Dyott, T.M., A.J. Stuper and G.S. Zander, "MOLY — An Interactive System for Molecular Analysis," *J. Chem. Inf. Comput. Sci.,* **20,** 28-35 (1980).

7. Environmental Protection Agency, "Toxic Substances Control — Discussion of Premanufacture Testing Policy and Technical Issues; Request for Comment," *Fed. Regist.,* **44** (53), 16253-54 (16 March 1979).

8. Fredenslund, A., J.G. Gmehling, M.L. Michelsen, P. Rasmussen and J.M. Prausnitz, "Computerized Design of Multicomponent Distillation Columns Using the UNIFAC Group Contribution Method for Calculating Activity Coefficients," *Ind. Eng. Chem. Process Des. Dev.,* **16**, 450-62 (1977).

9. Fredenslund, A., J.G. Gmehling and P. Rasmussen, *Vapor-Liquid Equilibria Using UNIFAC,* Elsevier Scientific Publishing Co., New York (1977).

10. Fredenslund, A., R.L. Jones and J.M. Prausnitz, "Group-Contribution Estimation of Activity Coefficients in Nonideal Liquid Mixtures," *AIChE J.,* **21,** 1086-99 (1975).

11. Fujita, T., J. Iwasa and C. Hansch, "A New Substituent Constant, π, Derived from Partition Coefficients," *J. Am. Chem. Soc.,* **86,** 5175-80 (1964).

12. Hall L.H., Eastern Nazarene College, Quincy, MA, personal communication (October, 1979).

13. Hall, L.H. and L.B. Kier, "Structure-Activity Studies Using Valence Molecular Connectivity," *J. Pharm. Sci.,* **66,** 642-44 (1977).

14. Hansch, C. and A.J. Leo, *Substituent Constants for Correlation Analysis in Chemistry and Biology,* John Wiley, New York (1979).

15. Hansch, C., A.J. Leo, S.H. Unger, K.H. Kim, D. Nikaitani and E.J. Lien, "'Aromatic' Substituent Constants for Structure-Activity Correlations," *J. Med. Chem.,* **16,** 1207-16 (1973).

16. Hansch, C., J.E. Quinlan and G.L. Lawrence, "The Linear Free-Energy Relationships between Partition Coefficients and the Aquueous Solubility of Organic Liquids," *J. Org. Chem.,* **33,** 347-50 (1968).

17. Holmes, H.L., *Structure-Activity Relationships for Some Conjugated Heteroenoid Compounds, Catechol Monoethers and Morphine Alkaloids,* (2 vols.), Defence Research Establishment, Suffield, Ralston, Alberta, Canada (1975).

18. Holmes, H.L. and C.E. Lough, "Effect of Intramolecular Hydrogen Bonding on Partition Coefficients," Suffield Technical Note No. 365, Defence Research Establishment, Suffield, Ralston, Alberta (July 1976). (NTIS AD A030683).

19. Hopfinger, A.J. and R.D. Battershell, "Application of SCAP to Drug Design: 1. Prediction of Octanol-Water Partition Coefficients Using Solvent-Dependent Conformational Analyses," *J. Med. Chem.,* **19,** 569-73 (1976).

20. Karickhoff, S.W. and D.S. Brown, "Determination of Octanol/Water Distribution Coefficients, Water Solubilities, and Sediment/Water Partition Coefficients for Hydrophobic Organic Pollutants," Report No. EPA-600/4-79-032, U.S. Environmental Protection Agency, Athens, GA (1979).

21. Karickhoff, S.W., D.S. Brown and T.A. Scott, "Sorption of Hydrophobic Pollutants on Natural Sediments," *Water Res.,* **13,** 241-48 (1979).

22. Kenaga, E.E. and C.A.I. Goring, "Relationship Between Water Solubility, Soil Sorption, Octanol-Water Partitioning, and Bioconcentration of Chemicals in Biota," pre-publication copy of paper dated October 13, 1978, given at American Society for Testing and Materials, Third Aquatic Toxicology Symposium, October 17-18, 1978, New Orleans, LA. (Symposium papers to be published by ASTM, Philadelphia, PA, as Special Technical Publication (STP) 707 in 1980).

23. Kier, L.B. and L.H. Hall, *Molecular Connectivity in Chemistry and Drug Research,* Academic Press, New York (1976).

24. Kier, L.B. and L.H. Hall, "Molecular Connectivity VII: Specific Treatment of Heteroatoms," *J. Pharm. Sci.,* **65,** 1806-09 (1976).

25. Kier, L.B. and L.H. Hall, "The Nature of Structure-Activity Relationships and Their Relation to Molecular Connectivity," *Eur. J. Med. Chem. — Chim. Ther.,* **12,** 307-12 (1977).

26. Leo, A.J. "Calculation of Partition Coefficients Useful in the Evaluation of the Relative Hazards of Various Chemicals in the Environment," Chap. 9 in *Structure-Activity Correlations in Studies of Toxicity and Bioconcentration with Aquatic Organisms,* ed. by G.D. Veith and D.F. Konasewich, International Joint Commission, Windsor, Ontario (1975). (NTIS PB-275 670)

27. Leo, A. and C. Hansch, "Linear Free-Energy Relationships Between Partitioning Solvent Systems," *J. Org. Chem.,* **36,** 1539-44 (1971).

28. Leo, A., C. Hansch and D. Elkins, "Partition Coefficients and Their Uses," *Chem. Rev.,* **71,** 525-621 (1971).

29. Mackay, D. and W.Y. Shiu, "Aqueous Solubility of Polynuclear Aromatic Hydrocarbons," *J. Chem. Eng. Data,* **22,** 399-402 (1977).

30. McCall, J., "Liquid-Liquid Partition Coefficients by High Pressure Liquid Chromatography," *J. Med. Chem.,* **18,** 549-52 (1975).

31. Mirrlees, M., S. Moulton, C. Murphy and P. Taylor, "Direct Measurement of Octanol-Water Partition Coefficients by High Pressure Liquid Chromatography," *J. Med. Chem.,* **19,** 615-19 (1976).

32. Murray, W.J., L.H. Hall and L.B. Kier, "Molecular Connectivity III: Relationship to Partition Coefficients," *J. Pharm. Sci.,* **64,** 1978-81 (1975).

33. Nys, G.G. and R.F. Rekker, "Statistical Analysis of a Series of Partition Coefficients with Special Reference to the Predictability of Folding Drug Molecules. The Introduction of Hydrophobic Fragment Constants (f Values)," *Chim. Ther.,* **8,** 521-35 (1973).

34. Nys, G.G. and R.F. Rekker, "The Concept of Hydrophobic Fragmental Constants (f-Values) II. Extension of Its Applicability to the Calculation of Lipophilicities of Aromatic and Heteroaromatic Structures," *Eur. J. Med. Chem. — Chim. Ther.,* **9,** 361-75 (1974).

35. Palmer, D.A., "Predicting Equilibrium Relationships for Maverick Mixtures," *Chem. Eng. (NY).* **12,** 80-85 (June 1975).

36. Pierotti, G.J., C.H. Deal and E.L. Derr, "Activity Coefficients and Molecular Structure," *Ind. Eng. Chem.,* **51,** 95-102 (1959).

37. Potenzone, R., Jr., E. Cavicchi, H.J.R. Weintraub and A.J. Hopfinger, "Molecular Mechanics and the CAMSEQ Processor," *Comput. Chem.,* **1,** 187-94 (1977).

38. Rao, P.S.C. and J.M. Davidson, "Estimation of Pesticide Retention and Transformation Parameters Required in Nonpoint Source Pollution Models," in *Environmental Impact of Nonpoint Source Pollution,* ed. by M.R. Overcash and J.M. Davidson, Ann Arbor Science Publishers, Inc., Ann Arbor, MI (1980).

39. Rekker, R.F., *The Hydrophobic Fragment Constant,* Elsevier Scientific Publishing Co., New York (1977).

40. Saeger, V.W., O. Hicks, R.G. Kaley, P.R. Michael, J.P. Mieure and E.S. Tucker, "Environmental Fate of Selected Phosphate Esters," *Environ. Sci. Technol.,* **13,** 840-44 (1979).

41. Seiler, P., "Interconversion of Lipophilicities from Hydrocarbon/Water Systems into the Octanol/Water System," *Eur. J. Med. Chem. — Chim. Ther.,* **9,** 473-79 (1974).

42. Tsonopoulos, C. and J.M. Prausnitz, "Activity Coefficients of Aromatic Solutes in Dilute Aqueous Solution," *Ind. Eng. Chem. Fundam.,***10,** 593-600 (1971).

43. Unger, S.H., J. Cook and J. Hollenberg, "Simple Procedure for Determining Octanol-Aqueous Partition, Distribution, and Ionization Coefficients by Reversed-Phase High-Pressure Liquid Chromatography," *J. Pharm. Sci.,* **67,** 1364-67 (1978).

44. Veith, G.D., N.M. Austin and R.T. Morris, "A Rapid Method for Estimating Log P for Organic Chemicals," *Water Res.,* **13,** 43-47 (1979).

45. Veith, G.D. and R.T. Morris, "A Rapid Method for Estimating Log P for Organic Chemicals," Report EPA-600/3-78-049, U.S. Environmental Protection Agency, Duluth, MN (1978).

46. Yalkowsky, S.H., R.J. Orr and S.C. Valvani, "Solubility and Partitioning: 3. The Solubility of Halobenzenes in Water," *Ind. Eng. Chem. Fundam.,* **18,** 351-53 (1979).

47. Yalkowsky, S.H. and S.C. Valvani, "Solubilities and Partitioning: 2. Relationships between Aqueous Solubilities, Partition Coefficients, and Molecular Surface Areas of Rigid Aromatic Hydrocarbons," *J. Chem. Eng. Data,* **24,** 127-29 (1979).

48. Yamana, T., "Novel Method for Determination of Partition Coefficients of Penicillins and Cephalosporins by High Pressure Liquid Chromatography," *J. Pharm. Sci.,* **66,** 747-49 (1977).

2

SOLUBILITY IN WATER

Warren J. Lyman

2-1 INTRODUCTION

Of the various parameters that affect the fate and transport of organic chemicals in the environment, water solubility is one of the most important. Highly soluble chemicals are easily and quickly distributed by the hydrologic cycle. These chemicals tend to have relatively low adsorption coefficients for soils and sediments and relatively low bioconcentration factors in aquatic life; they also tend to be more readily biodegradable by microorganisms in soil, surface water, and sewage treatment plants. Other degradation pathways (e.g., photolysis, hydrolysis, and oxidation) and specialized transport pathways (e.g., volatilization from solution and washout from the atmosphere by rain) are also affected by the extent of water solubility.

Water solubility, as an environmental parameter, is much less important for gases than it is for liquids or solids. The solubility of gases is usually measured when the partial pressure of the gas above the solution is one atmosphere, an unlikely situation under most environmental conditions. A more important parameter for gases is Henry's law constant, which describes the ratio of atmospheric to solution concentrations at low partial pressures. (This constant is discussed in Chapters 3, 11, and 15.)

Definition. The solubility of a chemical in water may be defined as the maximum amount of the chemical that will dissolve in pure water at a specified temperature. Above this concentration, two phases will exist if the organic chemical is a solid or a liquid at the system temperature: a saturated aqueous solution and a solid or liquid organic phase.

Units and Range of Values. Aqueous concentrations are usually stated in terms of weight per weight (ppm, ppb, g/kg, etc.) or weight per volume (mg/L, μg/L, moles/L, etc.). Less common units are mole fraction and molal concentration (moles per kg of solvent). At low concentrations all units are proportional to one another. At high concentrations this is not the case, and it becomes important to distinguish if the solubility is per volume of pure water or per volume of solution.

No organic chemical is completely insoluble in water; all are soluble to some extent. At the low end, solubilities below 1 ppb have been measured (e.g., 0.26 ppb for benzo[g,h,i]perylene). The solubilities of most common organic chemicals are in the range of 1 to 100,000 ppm at ambient temperatures, but several are higher, and some compounds (e.g., ethyl alcohol) are infinitely soluble — i.e., miscible with water in all proportions. An overall range covering at least nine orders of magnitude is thus involved.

As will be noted later, the available estimation methods usually yield values that are, on average, uncertain by less than one order of magnitude, but errors of over two orders of magnitude occur in about 10% of the cases with some equations.

Estimation Methods Provided. Five basically different approaches to the estimation of water solubility (S) are given in this handbook (see Table 2-1), but only two are described in this chapter. A more detailed review of available estimation methods, including several not included in this handbook, is presented in §2-2.

It is difficult to recommend any one method as the "best." The method(s) of choice may be determined by, for example, the information available on the chemical, the desired accuracy, and the time available for the calculations. The following are some of the considerations involved:

- Methods 1, 2, 4, and 5 give an estimate of S only for ∼25°C. Only method 3 allows the calculation of S at any temperature.

- Method 1 is probably the most generally applicable (i.e., to various chemical classes and structures) and should provide

TABLE 2-1

Overview of Solubility Estimation Methods Provided in this Handbook

Method No.	Where Described	Basis for Method	Information Required[a]	Comments
1	§ 2-4	Regression equations (several available)	K_{ow}, T_m [b]	K_{ow} easy to estimate from structure (see Chapter 1) Simple calculations
2	§ 2-5	Addition of atomic fragments (only for hydrocarbons and halocarbons)	Structure, T_m [b]	Limited applicability (chemicals with C, H, Cℓ, Br, I, F only)
3	Chapters 3 and 11	Theoretical equations using estimated activity coefficients	Structure, ΔH_f, T_m [b]	Allows calculation of S at any temperature Calculations may be difficult Somewhat limited applicability May be more accurate
4	Chapter 4	Regression equations	K_{oc}	Simple calculations Somewhat less accurate
5	Chapter 5	Regression equations	BCF	Simple calculations Less accurate

a. K_{ow} = octanol/water partition coefficient; T_m = melting point; ΔH_f = heat of fusion; K_{oc} = soil or sediment adsorption coefficient based on organic carbon; BCF = bioconcentration factors for aquatic life.

b. T_m is required only if the chemical is a solid at the system temperature. It is not required at all for some of the regression equations in the first method listed, but some of the equations are applicable to liquids only. Thus, it is only necessary to know if the substance is a liquid or a solid.

a reasonably accurate estimate of S if a value of K_{ow} (measured or estimated) is available.

- Even if measured values of K_{oc} and BCF are available (and no measured value of K_{ow} is available), methods 4 and 5 should not necessarily be considered better than method 1. Measured values of K_{oc} and BCF can have large uncertainties, while K_{ow} values can usually be estimated fairly accurately.

- Method 3, which proceeds via the calculation of activity coefficients, can be tedious. The general method is applicable to solubility in organic solvents as well as in water. Besides allowing the calculation of S at any temperature, it may, in some cases, provide a more accurate estimate.

- Only methods 2 and 3 make a clear distinction between liquids and solids and provide modified procedures for solids to account for their generally lower solubility. Some of the regression equations in method 1 cover both liquids and solids, while others are either limited to liquids or include a correction factor for solids.

If it is important to obtain the most accurate estimate of S, all applicable methods should be investigated in detail. S should be estimated by each method, and one value (or a range of values) should then be reported.

2-2 OVERVIEW OF AVAILABLE ESTIMATION METHODS

The aqueous solubility of organic chemicals can be estimated via numerous pathways. These are shown schematically in Figure 2-1 and described briefly in Table 2-2. To our knowledge, the relative merits, applicability, and accuracy of these pathways have not been reviewed elsewhere. Furthermore, many of the reported correlations and equations have been used primarily to test some theory or to show that two or more parameters were correlated in a certain way; few have actually been presented (and tested) as predictive tools.

The following characteristics of these pathways should be noted:

(1) Most of the pathways can start with only structural information for the chemical, but few can handle complex structures or uncommon functional groups.

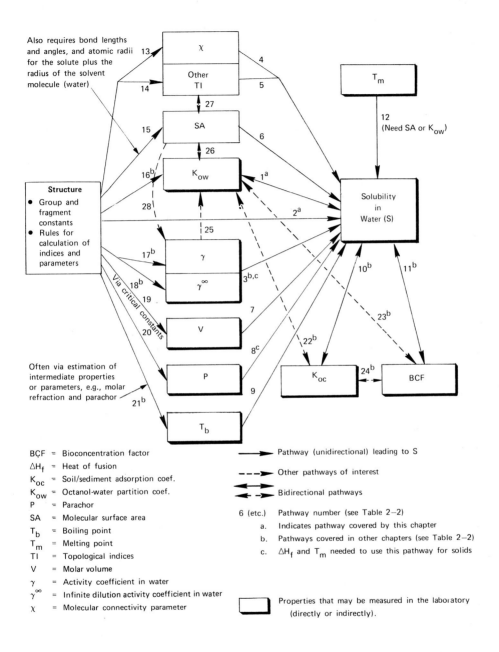

Also requires bond lengths and angles, and atomic radii for the solute plus the radius of the solvent molecule (water)

13

14

15

16[b]

28

17[b]

18[b]

19

20

Often via estimation of intermediate properties or parameters, e.g., molar refraction and parachor

21[b]

Via critical constants

χ

Other TI

27

SA

26

K_{ow}

25

γ

$γ^∞$

V

P

T_b

Structure
• Group and fragment constants
• Rules for calculation of indices and parameters

4

5

6

T_m

12
(Need SA or K_{ow})

1[a]

2[a]

Solubility in Water (S)

3[b,c]

7

8[c]

9

10[b]

11[b]

23[b]

22[b]

K_{oc}

24[b]

BCF

BCF = Bioconcentration factor
$ΔH_f$ = Heat of fusion
K_{oc} = Soil/sediment adsorption coef.
K_{ow} = Octanol-water partition coef.
P = Parachor
SA = Molecular surface area
T_b = Boiling point
T_m = Melting point
TI = Topological indices
V = Molar volume
γ = Activity coefficient in water
$γ^∞$ = Infinite dilution activity coefficient in water
χ = Molecular connectivity parameter

⟶ Pathway (unidirectional) leading to S

- - -▶ Other pathways of interest

◀⟶ Bidirectional pathways

6 (etc.) Pathway number (see Table 2–2)
a. Indicates pathway covered by this chapter
b. Pathways covered in other chapters (see Table 2–2)
c. $ΔH_f$ and T_m needed to use this pathway for solids

☐ Properties that may be measured in the laboratory (directly or indirectly).

FIGURE 2–1   Pathways for Estimating the Aqueous Solubility of Organic Chemicals

TABLE 2-2

Pathways Leading to the Estimation of Aqueous Solubility for Organic Chemicals

Pathway	Reference	Gives[a]	t(°C)[b]	Chemical Classes Covered[c]	Comments
1	[7b,14]	R	25	Mixed	Liquids and solids
	[30]	R	V	Mixed	
	[13]	R	V	Mixed	
	[24]	R	V	(1) Alcohols, (2) ketones, (3) esters, (4) ethers, (5) alkyl halides, (6) alkynes, (7) alkenes, (8) aromatics, (9) alkanes, (10) all exc. alkanes, (11) all	Liquids only
	[56]	R	V	Alkyl and aryl phosphate esters	
	[71]	R	25	Rigid aromatic hydrocarbons	T_m also required
	[70]	R	25	Halobenzenes	T_m also required
2	[27]	G	25	Aliphatic (saturated and unsaturated) and aromatic hydrocarbons plus halogenated hydrocarbons	
	[35]	R	25	n-Alcohols	Correlation with N^d
	[45]	P	25	n-Alkanes	Plot vs N
3	—	I	Any T	(Few restrictions)	Methods described in Chapter 3
4	[22,32]	R	V	(1) Alkanes, (2) alkyl substituted benzenes, (3) alcohols, (4) ethers, (5) esters, (6) combined classes	
5	[4]	R	V	(1) Acyclic saturated hydrocarbons, (2) acyclic and cyclic saturated hydrocarbons	Equations given for three topological indices
				(1) Hydrocarbons, (2) alcohols, (3) ethers, (4) ketones and aldehydes, (5) esters, (6) acids, (7) olefins. (All acyclic monofunctional compounds.)	Uses various "branching indices"
6	[6]	R	V	(1) Aliphatic alcohols, (2) aliphatic alcohols and hydrocarbons	
	[4]	R	V	(1) Hydrocarbons, (2) alcohols, (3) ethers, (4) ketones and aldehydes, (5) esters, (6) acids, (7) olefins	Functional group index also required
	[5]	R	V	Same classes as item above. Two or more equations given for each class using different SA parameters	Functional group index also required

(continued)

TABLE 2-2 (Continued)

Pathway	Reference	Gives[a]	t(°C)[b]	Chemical Classes Covered[c]	Comments
6 (cont.)	[26]	P	25	Hydrocarbons (acyclic, cyclic, aromatic)	T_m also required
	[71]	R	25	Rigid aromatic hydrocarbons	T_m also required
	[70]	R	25	Halobenzenes	
7	[46]	P	25	(1) Alkanes, (2) olefin hydrocarbons, (3) diolefin hydrocarbons, (4) acetylenes, (5) cycloparaffins, (6) aromatic hydrocarbons	
	[36]	R+I	25	C_4 to C_{10} hydrocarbons (all types)	Structure-related factors also used in equation
8	[47]	R	V	(1) Hydrocarbons and halogenated hydrocarbons, (2) alcohols, esters, ethers, ketones, phenols, (3) solids	ΔH_f and T_m also required for (3)
9	[1]	R	25	Aromatic hydrocarbons (including arenes, PAH, and some halogenated benzenes)	
10	[30]	R	V	Mixed classes	Equations given in Chapter 4 for estimation of K_{oc}
	[29]	R	V	Aromatics, mostly PAHs; two chlorinated	
11	[30]	R	V	Mixed classes	Equations given in Chapter 5 for estimation of BCF; other equations may also be listed there
	[13]	R	V	Mixed classes	
	[40]	R	27	Benzene derivatives	
	[49]	R	27	Organochlorine pesticides	
12	[70]	R	25	Halobenzenes	K_{ow} or SA also required
13	[21,32-34]	I	--	Organics which may have one or more of the following functional groups: $-NH_2$, $-NH-$, $-N<$, $=N-$ (pyridine), $C\equiv N$, $-OH$, $-O-$, $=O$, $-C\ell$, $-Br$, $-I$, $-F$, pyridine N, furan O, nitro N, + few other special cases	

(continued)

TABLE 2-2 (Continued)

Pathway	Reference	Gives[a]	t(°C)[b]	Chemical Classes Covered[c]	Comments
14	—	I	—	Hydrocarbons	See Ref. 4 for overview of various methods
15	[4-6,26,70,71]	I	—	Hydrocarbons, alcohols, ethers, ketones, aldehydes, esters, acids, olefins, and few others	
16	[23,37,55]	G	—	Organics (very few restrictions)	Methods covered in Chapter 1
17	[17-19,50]	G+I	Any T	Organics with common functional groups (~25 possibilities); structure must not be too complex	Methods covered in Chapter 11
18	[17-19,50]	G+I	Any T	Same as for pathway 17	Uses N. Methods covered in Chapter 11
	[52]	R	25, 60, or 100	(1) n-Acids, (2) n-primary alcohols, (3) n-sec-alcohols, (4) n-tert-alcohols, (5) alcohols, general, (6) n-allyl alcohols, (7) n-aldehydes, (8) n-alkene aldehydes, (9) n-ketones, (10) n-acetals, (11) n-ethers, (12) n-nitriles, (13) n-alkene nitriles, (14) n-esters, (15) n-formates, (16) n-monoalkylchlorides, (17) n-paraffins, (18) n-alkylbenzenes	
	[62]	R	25	(1) Arenes with straight-chain aliphatic part, (2) polymethylbenzenes, (3) branched arenes, (4) polycyclic aromatic hydrocarbons (PAH). Side chains in (1) and (3) may contain C=C or C≡C bonds. Chemicals in (4) may contain F, Cl, Br, I, OH, COOH, NH_2 or NO_2	Methods covered in Chapter 11; uses N
19	[42]	R	25	Polynuclear aromatic hydrocarbons (PAH)	In Chapter 2
	—	G	—	Organics with fairly common functional groups	See Ref. 54. Estimate via critical constants

(continued)

TABLE 2-2 (Continued)

Pathway	Reference	Gives[a]	t(°C)[b]	Chemical Classes Covered[c]	Comments
20	[53]	G	—	Organics with fairly common functional groups	
21	—	—	—	—	Methods covered in Chapter 12
22	—	R	V	—	Methods to estimate K_{OC} covered in Chapter 4
23	—	R	V	—	Methods to estimate BCF covered in Chapter 5
24	—	R	V	—	
25	—	I	Any T	Same as for pathway 17	Basis for methods given in Refs. 7a, 17, 18, 19
26	[71]	R	25	Rigid aromatic hydrocarbons	
27	[4]	—	V	Acyclic and cyclic saturated hydrocarbons	Correlations between SA and certain topological indices suggested
28	[69]	I	25	n-Alkyl-p-aminobenzoates	Interfacial tension also required

a. R = regression equation(s); P = plot; G = group, fragment and/or structure-related constants; I = other types of instructions or equations.
b. Temperature at which solubility data in data set were measured. V = various temperatures, usually between 15°C and 30°C.
c. Where more than one *numbered* class is given, a separate regression equation is given for each numbered group.
d. N = number of carbon atoms in molecule.

(2) Only one pathway (#2) allows S to be estimated *directly* from structural information; all others go through some intermediate property or parameter. Within pathway #2 only one method [27] has a degree of general applicability.

(3) Most of the pathways that proceed through an intermediate parameter do so with a parameter that is measurable, either directly or indirectly. This is considered to be an advantage, since a measured value of this intermediate property will simplify the estimation of S and usually improve its accuracy.

(4) Most of the pathways leading from structural information to the intermediate parameters do so via the use of group and fragment constants; the remainder (principally 13, 14 and 15) use rules for the calculation of topological indices (TI) and/or surface area (SA).

(5) Most of the pathways leading from some other parameter to S do so via a regression equation which describes the correlation between the two properties.

(6) Also, most of the pathways leading to S, whether through some intermediate parameter or not, either are limited to liquids explicitly or make no distinction between liquids and solids. Other factors (e.g., main structural characteristics) being equal, solids tend to have lower solubilities than do liquids; between solids with similar chemical formulas and melting points, the one with the larger heat of fusion has the lower solubility. Only pathways 3 and 8 (and some of the regression equations) allow the solubility of solids in water to be explicitly considered; both of these pathways require ΔH_f and T_m.

(7) Almost all of the fragment-constant pathways and all of the regression equations (except those in Ref. 52) apply to only one temperature. In some cases the data sets from which the constants or equations were derived were obtained at a single temperature (e.g., 25°C); others used data from several sources and thus represent a mix of temperatures (indicated by V in the table). Since S is a function of temperature, the latter approach reduces the accuracy of the method. Only one general pathway (#17 followed by #3) allows S to be estimated at essentially any temperature.

(8) Some of the pathways are bidirectional; i.e., they are simple regression equations relating two properties or pa-

rameters. In cases where the second parameter cannot be measured as accurately as S or is not commonly available (e.g., pathways 10 and 11), the regression equations are generally better suited for estimating the second parameter from S.

(9) Most of the regression equations between an intermediate property and S are for specific, simple chemical classes (alcohols, esters, alkanes, etc.). Only for pathway #1 are equations given for "mixed chemical classes," where the mix is sufficiently broad to cover almost any chemical. Such equations, understandably, involve larger method errors than those for specific chemical classes.

(10) Of all the pathways leading to an intermediate property, only #16 has been developed sufficiently to be applicable to almost any new chemical. A particular strength of this pathway is that the group/fragment constants can be applied to a measured value of K_{ow} for a structurally related compound,[1] thus significantly reducing the effort involved (and improving the accuracy) in obtaining an estimate of K_{ow} for the new chemical.

2-3 FACTORS INFLUENCING SOLUBILITY

Method of Measurement. Several methods are available for measuring the solubility of organic chemicals, but no single method is usable over the entire range of solubilities in water [16]. Because of special problems encountered in measuring S for very hydrophobic (i.e., low solubility) compounds [28], various methods may yield different values. In general, an excess of the chemical is added to *very pure* water and allowed to equilibrate at constant temperature. Equilibration may take several days for very hydrophobic compounds. The solution is then filtered and/or centrifuged to remove undissolved material before the solution concentration is measured.

Temperature. Solubility in water is a function of temperature, but the strength and direction (i.e., sign) of this function varies. Many, if not most, organic compounds become more soluble as the temperature increases, but some behave in the opposite way. The solubility of benzene, for example, increases with increasing temperature (at temperatures near ambient) [44], but the solubility of p-dichlorobenzene decreases [65]. For

---

1. Compilations of measured values [23,37] contain data for thousands of chemicals.

some chemicals S may either increase or decrease at higher temperatures, depending on the nature of the chemical and the temperature range involved (see Figure 3-1c in Chapter 3); an example is 2-butanone, whose solubility increases with increasing temperature above ~80°C but decreases with increasing temperature between ~ − 6°C and 80°C [59]. If it is necessary to estimate a solubility at some temperature other than ~25°C and the literature does not report the effect of temperature on structurally similar compounds, the combined methods of Chapters 3 and 11 must be used.

Salinity. The presence of dissolved salts or minerals in water leads to moderate decreases in S. For example, the solubilities of several polynuclear aromatics (e.g., naphthalene, biphenyl, anthracene, fluorene) in sea water, which contains about 35 g/L NaCl, are from 30% to 60% below their fresh water solubilities [10]. The general relationship between salinity and solubility can be expressed in the following form [15]:

$$\log (S°/S') = K_s\, C_s \qquad (2\text{-}1)$$

where

$S°$ = molar solubility in pure water

S' = molar solubility in salt solution

K_s = empirical salting parameter

C_s = molar salt concentration

Values of K_s for polynuclear aromatics range from ~ 0.04 to 0.4.

Dissolved Organic Matter. A number of studies have shown that the presence of dissolved organic material, such as the naturally occurring humic and fulvic acids in rivers and other surface waters, leads to an increase in the solubility of many organic chemicals. For example, in one study using the waters of Narragansett Bay and the Providence River, removal of the dissolved organic matter resulted in a 50-99% decrease in the amounts of n-alkanes and isoprenoid hydrocarbons that could be solubilized; the decrease was directly related to the amount of dissolved organic matter removed. The solubilities of aromatic hydrocarbons, however, were unaffected by the process [9]. Another study showed that 500 mg/L of humic material (extracted from soil) increased the solubility of DDT 20 to 40 times [67]. Other studies have shown increases in solubility for 2,2',5,5'-tetrachlorobiphenyl and cholesterol [25], and phthalate esters [43]. Surfactants can also increase the apparent solubility by forming micelles into which the solute partitions.

pH. Hydrogen ion concentration also affects the solubility of organic compounds. Organic acids may be expected to increase in solubility with increasing pH, while organic bases may act in the opposite way. Even the solubility of "neutral" organic chemicals (e.g., alkanes and chlorinated hydrocarbons) may be affected by pH. Significant increases in solubility above pH 8 have been reported for some chemicals [9,48].

2-4 ESTIMATION OF S FROM K_{ow}

Equations Available. Eighteen different regression equations were found that correlate water solubility (S) with the octanol/water partition coefficient (K_{ow}) for different groups of chemicals. Table 2-3 lists these equations (and Eq. 2-20) along with such information as the kind and number of chemicals represented in each data set and the quality of fit (r^2). Table 2-4 gives additional data for each equation, including the range of S and K_{ow} values involved and the temperatures at which the solubility data were obtained. Table 2-4 also refers to subsequent tables (Tables 2-5 to 2-13) which list the actual chemicals used to obtain each regression equation.

One of the listed equations (2-20), which is used for polynuclear aromatic hydrocarbons, does not require a value of K_{ow}; one need only know the number of carbon atoms (N) in the molecule and the melting point (t_m). In its original form [42], this equation was an expression for the infinite dilution activity coefficient ($\gamma\infty$), but since S is directly proportional to $1/\gamma\infty$ at low solubilities (see Chapter 3), the equation has been rewritten here in terms of S in units of mole fraction.[2]

All of these equations provide an estimate of S at ~25°C. Some of the correlations are based on solubility data from a range of temperatures (e.g., 15-30°C) while others use only data measured at 25°C. Clearly, the latter would be expected to provide more accurate estimates.

Note also that Eqs. 2-5 to 2-15 were obtained from data on liquid organics only. Use of these equations for solid solutes will (on average) result in overestimation of the solubility.[3] Most of the other equations include data for both liquid and solid solutes. Equations 2-17, -18 and -20 use a correction factor for solids that requires a knowledge of the melting point; the other equations do not attempt to include any correction factor and may be used for either liquid or solid solutes.

---

2. Mole fraction = number of moles of solute divided by total number of moles (solute plus solvent) present.

3. Correction factors for solids, which require a knowledge of the melting point, are given later in this chapter for use with Eqs. 2-14 and 2-15.

TABLE 2-3

Regression Equations for the Estimation of S

Eq. No.	Equation[a]	Units of S	No.[b]	r^{2c}	Chemical Classes Represented	Ref.
2-2[i]	$\log S = -1.37 \log K_{ow} + 7.26$	μ mol/L	41	0.903	Mixed classes; aromatics and chlorinated hydrocarbons well represented	[14]
2-3	$\log S = -0.922 \log K_{ow} + 4.184$	mg/L	90	0.740	Mixed classes; pesticides well represented	[30]
2-4	$\log S = -1.49 \log K_{ow} + 7.46$	μ mol/L	34	0.970	Mixed classes; several pesticides	[13]
2-5	$\log 1/S = 1.113 \log K_{ow} - 0.926$	mol/L[d]	41	0.935	Alcohols[e]	[24]
2-6	$\log 1/S = 1.229 \log K_{ow} - 0.720$	mol/L[d]	13	0.960	Ketones[e]	[24]
2-7	$\log 1/S = 1.013 \log K_{ow} - 0.520$	mol/L[d]	18	0.980	Esters[e]	[24]
2-8	$\log 1/S = 1.182 \log K_{ow} - 0.935$	mol/L[d]	12	0.880	Ethers[e]	[24]
2-9	$\log 1/S = 1.221 \log K_{ow} - 0.832$	mol/L[d]	20	0.861	Alkyl halides[e]	[24]
2-10	$\log 1/S = 1.294 \log K_{ow} - 1.043$	mol/L[d]	7	0.908	Alkynes[e]	[24]
2-11	$\log 1/S = 1.294 \log K_{ow} - 0.248$	mol/L[d]	12	0.970	Alkenes[e]	[24]
2-12	$\log 1/S = 0.996 \log K_{ow} - 0.339$	mol/L[d]	16	0.951	Aromatics[e] (benzene and benzene derivatives)	[24]
2-13	$\log 1/S = 1.237 \log K_{ow} + 0.248$	mol/L[d]	16	0.908	Alkanes[e]	[24]
2-14	$\log 1/S = 1.214 \log K_{ow} - 0.850$	mol/L[d]	140	0.912	All chemicals represented by Eqs. 2-5 to -12 plus propionitrile[e]	[24]

(continued)

TABLE 2-3 (Continued)

Eq. No.	Equation[a]	Units of S	No.[b]	r^2[c]	Chemical Classes Represented	Ref.
2-15	$\log 1/S = 1.339 \log K_{ow} - 0.978$	mol/L[d]	156	0.874	All chemicals represented by Eqs. 2-5 to -13 plus propionitrile[e]	[24]
2-16	$\log S = -2.38 \log K_{ow} + 12.90$	μ mol/L	11	0.656	Phosphate esters	[56]
2-17[f]	$\log S = -0.9874 \log K_{ow} - 0.0095\, t_m + 0.7178$	mol/L	35	0.990	Halobenzenes	[70]
2-18[f]	$\log S = -0.88 \log K_{ow} - 0.01\, t_m - 0.012$	mol/L	32	0.979	Rigid aromatic hydrocarbons (polynuclear aromatics)	[71]
2-19	$\log S = -0.962 \log K_{ow} + 6.50$	μ mol/L	9	0.878	Halogenated 1- and 2-carbon hydrocarbons (8 with Cl, 1 with Br)	[12]
2-20[g]	$\log S = -0.00987\,(t_m - 25) - 3.5055 - 0.3417\,(N-6) + 0.002640\,(N-6)^2$	mole fraction	21	h	Polynuclear aromatic hydrocarbons (See note g for alkyl-substituted naphthalenes and anthracenes.)	[42]

a. S = aqueous solubility; K_{ow} = octanol/water partition coefficient; t_m = melting point ($^\circ$C), $t_m \geqslant 25^\circ$C; N = number of carbon atoms in molecule.

b. No. = number of compounds in data set used to obtain equation.

c. r^2 = square of correlation coefficient.

d. Actually, moles/1000 g of water (i.e., molar solubility). For most chemicals this is very close to the molar solubility (moles/liter of solution), and no correction need be applied.

e. All chemicals used were liquids. Values of K_{ow} for many of these chemicals were estimated.

f. If t_m is less than 25°C, a value of 25°C should be used for t_m in Eqs. 2-17 and 2-18.

g. If t_m is less than 25°C, the first term on the right side of Eq. 2-20 should be set equal to zero. For alkyl-substituted naphthalenes and anthracenes, multiply the last three terms on the right side of Eq. 2-20 by a factor of 2 before solving for S. This equation is a combination of three equations given in Ref. 42.

h. Not available.

i. [Note added in final proof.] The published version of this equation [7b] differs from the version given here, which was taken from the draft [14]. The revised equation, based on a 27-compound subset of the 41 compounds used for Eq. 2-2, is [7b]: $\log S = -1.12 \log K_{ow} + 7.30 - 0.015\, t_m$. For this equation $r^2 = 0.922$; the melting point of the compound, t_m ($^\circ$C), is set equal to 25°C if the compound is a liquid at 25°C.

TABLE 2-4

Additional Information on Equations for Estimating S

Eq. No.	Range of S Values[a]	Range of K_{ow} Values	t(°C)[b]	Chemicals Used for Regression Listed in Table
2-2	$5 \times 10^{-2} - 2 \times 10^6$	$8 - 2 \times 10^6$	25	2-5
2-3	$5 \times 10^{-4} - 2 \times 10^6$	$1 \times 10^{-3} - 4 \times 10^6$	Var.	2-6
2-4	$1 \times 10^{-3} - 1.7 \times 10^4$	$18 - 5 \times 10^6$	Var.	2-7
2-5	$5 \times 10^{-3} - 1$	$4 - 7 \times 10^2$	Var.	2-8
2-6	$3 \times 10^{-3} - 5$	$2 - 6 \times 10^2$	Var.	2-8
2-7	$8 \times 10^{-5} - 1$	$2 - 5 \times 10^4$	Var.	2-8
2-8	$5 \times 10^{-2} - 1$	$7 - 1 \times 10^2$	Var.	2-8
2-9	$1 \times 10^{-3} - 0.1$	$25 - 1 \times 10^3$	Var.	2-8
2-10	$6 \times 10^{-5} - 3 \times 10^{-2}$	$1 \times 10^2 - 1 \times 10^4$	Var.	2-8
2-11	$2 \times 10^{-5} - 1 \times 10^{-2}$	$56 - 5 \times 10^3$	Var.	2-8
2-12	$4 \times 10^{-4} - 0.4$	$8 - 5 \times 10^3$	Var.	2-8
2-13	$6 \times 10^{-6} - 7 \times 10^{-4}$	$1 \times 10^2 - 1 \times 10^4$	Var.	2-8
2-14	$2 \times 10^{-5} - 5$	$2 - 5 \times 10^4$	Var.	2-8
2-15	$6 \times 10^{-6} - 5$	$2 - 5 \times 10^4$	Var.	2-8
2-16	$(0.36 - 1 \times 10^3)$[c]	$1 \times 10^4 - 5 \times 10^5$	d	2-9
2-17	$2 \times 10^{-8} - 2 \times 10^{-2}$	$1 \times 10^2 - 3 \times 10^6$	25	2-10
2-18	$9 \times 10^{-10} - 9 \times 10^{-4}$	$2 \times 10^3 - 1.3 \times 10^7$	25	2-11
2-19	$1 \times 10^3 - 1 \times 10^5$	$25 - 2400$	20	2-12
2-20	$9 \times 10^{-12} - 1.7 \times 10^{-5}$	(Not pertinent)	25	2-13

a. Units of S are as shown in Table 2-3, except for Eq. 2-16.
b. Temperature at which solubility of chemicals in data set was measured. "Var." indicates that various temperatures are represented by the solubility data, usually in the 15-30°C range.
c. Units are mg/L.
d. Exact value not specified, but is presumably ~ 20°C.

TABLE 2-5

Compounds Used for Regression Equation 2-2

Acenaphthene	*n*-Decane	Nitrobenzene
Acrolein	Dibenzofuran	N-Nitroso-diphenylamine
Acrylonitrile	*o*-Dichlorobenzene	Pentachlorobenzene
Benzene	*m*-Dichlorobenzene	Pentachloroethane
Biphenyl	*p*-Dichlorobenzene	α-Pinene
Butylbenzylphthalate	3,3'-Dichlorobenzidine	Styrene
n-Butylether	1,2-Dichloroethane	1,2,3,5-Tetrachlorobenzene
Camphene	Diethylphthalate	1,1,2,2-Tetrachloroethane
Carbon tetrachloride	2,4-Dimethylphenol	Tetrachloroethylene
Chloroethane	Dimethylphthalate	Toluene
2-Chloroethylether	Diphenylether	1,1,1-Trichloroethane
Chloroform	Docosane	Trichloroethylene
o-Chlorophenol	Hexachloroethane	Vinylidene chloride
p-Cymene	Isophorone	

Source: Dec *et al.* [14]

TABLE 2-6

Compounds Used for Regression Equation 2-3

Halogenated Hydrocarbon **Insecticides**	Naphthalene Phenanthrene Pyrene
DDD DDE	Tetracene
DDT	**Fumigants**
Endrin Methoxychlor	Carbon tetrachloride Tetrachloroethylene
Substituted Benzenes **and Halobenzenes**	**Phosphorus-containing Insecticides**
	Malathion
Bromobenzene	Trichlorfon
Chlorobenzene	Dimethoate
p-Dichlorobenzene	Dichlorvos
Hexachlorobenzene	Crufomate
Pentachlorobenzene	Chlorpyrifos
1,2,3,5-Tetrachlorobenzene	Chlorpyrifos-methyl
1,2,4-Trichlorobenzene	Leptophos
Aniline	Methyl parathion
Diethylaniline	Parathion
Nitrobenzene	Ronnel
Phthalic anhydride	Fenitrothion
Captan	Phosmet
	Phosalone
Halogenated Biphenyls and	Dichlofenthion
Diphenyl Oxides	Dialifor
4-Chlorobiphenyl	**Carbamates, Thiocarbamates and**
4,4'-Dichlorobiphenyl	**Carbamoyl Oximes**
2,4,4'- and 2,2',5-Trichlorobiphenyl	
2,2',4,4'- and 2,2',5,5'-Tetrachlorobiphenyl	Carbaryl
2,2',4,5,5'-Pentachlorobiphenyl	Carbofuran
2,2',4,4',5,5'-Hexachlorobiphenyl	Propoxur
Diphenyloxide	Mexacarbate
4-Chlorodiphenyloxide	Methomyl
x-sec-Butyl-4-chlorodiphenyloxide	
x-hexyl-x'-Chlorodiphenyloxide	**Carboxylic Acids and Esters**
x-dodeca-x'-Chlorodiphenyloxide	
	6-Chloropicolinic acid
	2,4-D acid
Aromatic Hydrocarbons	Dalapon
	Picloram
Anthracene	2,4,5-T
Benzene	Triclopyr (triethylamine salt)
Biphenyl	Triclopyr (butoxyethyl ester)
9-Methylanthracene	Triclopyr
2-Methylnaphthalene	Di-2-ethylhexylphthalate

(continued)

TABLE 2-6 (Continued)

Dinitroanilines	Ipazine
	Propazine
Trifluralin	Simazine
	Trietazine
Ureas and Uracils	
	Miscellaneous Nitrogen Heterocyclics
Diuron	
Fenuron	2-Methoxy-3,5,6-trichloropyridine
Fluometuron	Nitrapyrin
Linuron	3,5,6-Trichloro-2-pyridinol
Monolinuron	
Monuron	**Miscellaneous**
Urea	
	Dinoseb
Symmetrical Triazines	Alachlor
	Propachlor
Atrazine	Bentazon
Cyanazine	

Source: Kenaga and Goring [30]

TABLE 2-7

Compounds Used for Regression Equation 2-4

Benzene	*p,p'*-DDT	Phosmet
Toluene	*p,p'*-DDE	Malathion
Fluorobenzene	Benzoic acid	Fenitrothion
Chlorobenzene	Salicylic acid	Dicapthon
Bromobenzene	Phenylacetic acid	Parathion
Iodobenzene	Phenoxyacetic acid	Phosalone
p-Dichlorobenzene	2,4-D	Methyl chlorpyrifos
Naphthalene	2,4,5,2',5'-PCB	Dialifor
Diphenylether	2,4,5,2',4',5'-PCB	Ronnel
Tetrachloroethylene	4,4'-PCB	Chlorpyrifos
Chloroform		Dichlofenthion
Carbon tetrachloride		Leptophos

Source: Chiou *et al.* [13]

TABLE 2-8

Compounds Used for Regression Equations 2-5 to 2-15[a]

Alcohols (Eq. 2-5)	3-Pentanone	**Alkyl Halides (Eq. 2-9)**
	3-Methyl-2-butanone	
Butanol	2-Hexanone	Chloroethane
2-Methyl-1-propanol	3-Hexanone	Chloropropane
2-Butanol	3-Methyl-2-pentanone	2-Chloropropane
Pentanol	4-Methyl-2-pentanone	Chlorobutane
3-Methyl-1-butanol	4-Methyl-3-pentanone	Isobutyl chloride
Methylbutanol	2-Heptanone	1,3-Dichloropropane
2-Pentanol	4-Heptanone	Chloroform
3-Pentanol	2,4-Dimethyl-3-pentanone	Bromoethane
3-Methyl-2-butanol	5-Nonanone	Bromopropane
2-Methyl-2-butanol		2-Bromopropane
2,2-Dimethylpropanol		Bromobutane
Hexanol	**Esters (Eq. 2-7)**	Isobutyl bromide
2-Hexanol		Isoamyl bromide
3-Hexanol	Ethyl formate	1,3-Dibromopropane
3-Methyl-3-pentanol	Propyl formate	Iodomethane
2-Methyl-2-pentanol	Methyl acetate	Iodoethane
2-Methyl-3-pentanol	Ethyl acetate	Iodopropane
3-Methyl-2-pentanol	Propyl acetate	Iodobutane
4-Methyl-2-pentanol	Isopropyl acetate	Diiodomethane
2,3-Dimethyl-2-butanol	Butyl acetate	$(ClCH_2CH_2)_2S$
3,3-Dimethyl-1-butanol	Isobutyl acetate	
3,3-Dimethyl-2-butanol	Methyl propionate	
Heptanol	Methyl butyrate	**Alkynes (Eq. 2-10)**
2-Methyl-2-hexanol	Ethyl butyrate	
3-Methyl-3-hexanol	Propyl butyrate	1-Pentyne
3-Ethyl-3-pentanol	Ethyl valerate	1-Hexyne
2,3-Dimethyl-2-pentanol	Ethyl hexanoate	1-Heptyne
2,3-Dimethyl-3-pentanol	Ethyl heptanoate	1-Octyne
2,4-Dimethyl-2-pentanol	Ethyl octanoate	1-Nonyne
2,4-Dimethyl-3-pentanol	Ethyl nonanoate	1,8-Nonadiyne
2,2-Dimethyl-3-pentanol	Ethyl decanoate	1,6-Heptadiyne
Octanol		
2,2,3-Trimethyl-3-pentanol	**Ethers (Eq. 2-8)**	**Alkenes (Eq. 2-11)**
Cyclohexanol		
4-Penten-1-ol	Diethyl ether	1-Pentene
3-Penten-2-ol	Methyl butyl ether	2-Pentene
1-Penten-3-ol	Methyl isobutyl ether	1-Hexene
1-Hexen-3-ol	Methyl *sec*-butyl ether	2-Heptene
2-Hexen-4-ol	Methyl *t*-butyl ether	1-Octene
2-Methyl-4-penten-3-ol	Ethyl propyl ether	4-Methyl-1-pentene
Benzyl alcohol	Ethyl isopropyl ether	1,6-Heptadiene
	Dipropyl ether	1,5-Hexadiene
	Propyl isopropyl ether	1,4-Pentadiene
Ketones (Eq. 2-6)	Methyl propyl ether	Cyclopentene
	Methyl isopropyl ether	Cyclohexene
2-Butanone	Cyclopropyl ethyl ether	Cycloheptene
2-Pentanone		

(continued)

TABLE 2-8 (Continued)

Aromatics (Eq. 2-12)	m-Nitrotoluene	Hexane
Benzene	o-Dichlorobenzene	Heptane
Toluene	m-Dichlorobenzene	2,4-Dimethylpentane
Ethylbenzene	Ethyl benzoate	2,2-Dimethylpentane
Propylbenzene	Aniline	Octane
Fluorobenzene		Cyclopentane
Chlorobenzene		Cyclohexane
Bromobenzene	**Alkanes (Eq. 2-13)**	Methylcyclopentane
Nitrobenzene	Pentane	Cycloheptane
1,2,4-Trimethylbenzene	Isopentane	Methylcyclohexane
o-Xylene	2-Methylpentane	Cyclooctane
Isopropylbenzene	3-Methylpentane	1,2-Dimethylcyclohexane

a. Data set for Eq. 2-15 includes all compounds listed plus propionitrile. Data set for Eq. 2-14 does not
include alkane group but is otherwise the same.

Source: Hansch *et al.* [24]

TABLE 2-9

Compounds Used for Regression Equation 2-16

tert-Butylphenyl diphenyl phosphate
Cresyl diphenyl phosphate
Dibutyl phenyl phosphate
2-Ethylhexyl diphenyl phosphate
Isodecyl diphenyl phosphate
Isopropylphenyl diphenyl phosphate
Tributyl phosphate
Tricresyl phosphate
Triphenyl phosphate
Tris(2-ethylhexyl) phosphate
Trixylenyl phosphate

Source: Saeger *et al.* [56]

TABLE 2-10

Compounds Used for Regression Equation 2-17

Hexachlorobenzene	1,2-Difluorobenzene
Pentachlorobenzene	1,3-Difluorobenzene
1,2,3,4-Tetrachlorobenzene	1,4-Difluorobenzene
1,2,3,5-Tetrachlorobenzene	1,2-Diiodobenzene
1,2,4,5-Tetrachlorobenzene	1,3-Diiodobenzene
1,2,4,5-Tetrabromobenzene	1,4-Diiodobenzene
1,2,4-Tribromobenzene	Bromobenzene
1,3,5-Tribromobenzene	Chlorobenzene
1,2,3-Trichlorobenzene	Fluorobenzene
1,2,4-Trichlorobenzene	Iodobenzene
1,3,5-Trichlorobenzene	Benzene
1,2-Dibromobenzene	2-Bromochlorobenzene
1,3-Dibromobenzene	3-Bromochlorobenzene
1,4-Dibromobenzene	4-Bromochlorobenzene
1,2-Dichlorobenzene	4-Bromoiodobenzene
1,3-Dichlorobenzene	2-Chloroiodobenzene
1,4-Dichlorobenzene	3-Chloroiodobenzene
	4-Chloroiodobenzene

Source: Yalkowsky *et al.* [70]

TABLE 2-11

Compounds Used for Regression Equation 2-18

Indan	2-Methylanthracene
Naphthalene	9-Methylanthracene
1-Methylnaphthalene	9,10-Dimethylanthracene
2-Methylnaphthalene	Pyrene
1,3-Dimethylnaphthalene	Fluoranthene
1,4-Dimethylnaphthalene	1,2-Benzofluorene
1,5-Dimethylnaphthalene	2,3-Benzofluorene
2,3-Dimethylnaphthalene	Chrysene
2,6-Dimethylnaphthalene	Triphenylene
1-Ethylnaphthalene	Naphthacene
1,4,5-Trimethylnaphthalene	1,2-Benzanthracene
Biphenyl	7,12-Dimethyl-1,2-benzanthracene
Acenaphthene	Perylene
Fluorene	3,4-Benzopyrene
Phenanthrene	3-Methylcholanthrene
Anthracene	Benzo [g,h,i] perylene

Source: Yalkowsky and Valvani [71]

TABLE 2-12

Compounds Used for Regression Equation 2-19

1,2-Dibromoethane	Trichloroethylene
1,2-Dichloroethane	1,1,1-Trichloroethane
1,2-Dichloropropane	Chloroform
Tetrachloroethylene	Carbon tetrachloride
1,1,2,2-Tetrachloroethane	

Source: Chiou and Freed [12]

TABLE 2-13

Compounds Used for Regression Equation 2-20

Indan	Chrysene
Naphthalene	Triphenylene
Biphenyl	Naphthacene
Acenaphthene	1,2-Benzanthracene
Fluorene	7,12-Dimethyl-1,2-benzanthracene
Phenanthrene	Perylene
Anthracene	3,4-Benzopyrene
Pyrene	3-Methylcholanthrene
Fluoranthene	Benzo[g,h,i] perylene
1,2-Benzofluorene	Coronene
2,3-Benzofluorene	

Source: Mackay and Shiu [42]

The principal input information required for these equations is K_{ow}, the octanol/water partition coefficient, measured at or near room temperature. Compilations of measured (and some estimated) values of K_{ow} are available for thousands of chemicals [23,37]. If the K_{ow} for a particular compound cannot be found, however, it can usually be estimated fairly accurately by the methods described in Chapter 1.

General instructions for selecting the most appropriate equation(s) and for calculating S are given below after an explanation of the basis for the method and a discussion of method errors.

Basis for Estimation Method. The basis for the correlation between S and K_{ow} has been briefly discussed by Mackay [41] and Chiou and Freed [11]. The correlation between log K_{ow} and log S for hydrophobic pollutants is shown actually to be a correlation between

$$(\log \gamma_{w/oct} - \log \gamma_{oct/w} - 0.94) \qquad \text{[log K term]}$$

and

$$(-\log \gamma_w + \log (f_S/f_R) + 7.74) \qquad \text{[log S term]}$$

where

γ_w = activity coefficient of solute in pure water

$\gamma_{w/oct}$ = activity coefficient of solute in octanol-saturated aqueous phase

$\gamma_{oct/w}$ = activity coefficient of solute in water-saturated octanol phase

f_S/f_R = ratio of solid fugacity to reference fugacity (ratio = 1 for liquids)

If one assumes that $\gamma_{w/oct} \approx \gamma_w$ and that γ_w dominates the two terms, then the correlation between log K and log S is a correlation of one quantity (log γ_w) against its reciprocal ($-$log γ_w). With these assumptions, a slope of -1 is predicted for the regression equations of the form log S = a log K_{ow}+b. (The predicted slope is $+1$ for equations of the form log(1/S) = a log K_{ow}+b.) Note that except for Eq. 2-16, most of the equations in Table 2-3 do have slopes (i.e., coefficients of log K_{ow}) close to the predicted value. Chiou and Freed [11] suggest that a slope close to 1 is more likely for a highly soluble liquid solute, since it is expected that $\gamma_{w/oct}/\gamma_w \approx 1$, log $\gamma_{oct/w} \approx$ constant, and $f_S/f_R = 1$.

Dec *et al.* [14] have also pointed out that a plot of log S vs. log K_{ow} will have a slope of -1 only if log ($K_{ow}S$) is a constant. Using the data set from which Eq. 2-2 was obtained, they divided it into five subgroups having similar values of log ($K_{ow}S$). While the equation for the complete data set had a slope of -1.37, the slopes of the equations for the five subgroups were -0.96, -1.02, -1.02, -0.90, and -1.05. The chemicals in these subgroups were frequently structurally dissimilar from one another.

The basis for the correction term for solids (the term involving t_m) in some equations has been explained by Irmann [27] and Yalkowsky and Valvani [71].[4] To account for crystal lattice interactions in solids, the term $-\Delta H_f(T_m - T)/2.30\,RT_mT$ may be added to the right-hand side of equations for log S. (ΔH_f is the heat of fusion, T_m the melting point in K, T the system temperature (K), and R the gas constant.) Since $\Delta H_f = T_m\Delta S_f$ at the melting point (ΔS_f being the entropy change associated with fusion), the previous term becomes $-\Delta S_f(T_m-T)/2.30RT$. At room temperature (25°C), this becomes $-\Delta S_f(t_m - 25)/1360$, where t_m is the melting point in °C and ΔS_f is in cal/mol-°C. If an average value of 13 cal/mol-°C is assumed ΔS_f, the correction factor is $-0.0095\,(t_m-25)$.

Method Errors. The true method error for a linear correlation between log S and log K_{ow} is difficult to determine, since most of the data sets that have been used to date incorporate some erroneous values of S and K_{ow}, and many are also based on values of S measured over a range of temperatures. Estimated values of K_{ow} have been used, in part, in some

---

4. See also §3-5 of Chapter 3 and the references cited therein for additional discussion.

of the data sets. Even under such conditions, values of r^2 (square of the correlation coefficient) are usually above 0.9; one data set covering mixed classes of chemicals (Eq. 2-4) reaches 0.97. On the other hand, a relatively low value of 0.656 for r^2 has been reported for one data set (Eq. 2-16) that is limited to a single class of chemicals (phosphate esters) whose solubilities were all measured at the same temperature. The values of r^2 associated with the equations of Hansch *et al.* [24] (Eqs. 2-5 to -15) also indicate that one should not necessarily expect lower method errors with regression equations derived for a single class of chemicals.

The likely method errors can be visualized from Figure 2-2, which is a plot of the data set and regression equation given by Dec *et al.* [14]. Most data points are well represented by the equation. The data for two chemicals, 1,3,5-triazo-1,3,5-trinitrocyclohexane (RDX) and hexachloro-1,3-butadiene, were not included in the regression analysis; if the K_{ow} values for these chemicals were used to estimate their solubility, the results would be about three orders of magnitude too high. The authors concluded, "While the correlation obviously applies to the majority of the compounds studied, it is *not* universal, and caution is required for the interpretation of results obtained from it." This statement should also be considered applicable to other equations of this kind.

A more quantitative analysis of method errors is provided by Tables 2-14 and -15. The former compares measured values of S for 78 chemicals with estimated values obtained from five selected regression equations. The method errors are summarized by chemical class for each of the five equations in Table 2-15. All of the selected equations were derived from data on mixed chemical classes. The equations were used to estimate a value of S for every chemical, even if it would not appear appropriate to do so normally; for example, some were used outside the range of K_{ow} and S values in their original data sets, and Eqs. 2-14 and 2-15 were used for solids even though the original data sets were limited to liquids. In addition, several of the K_{ow} values used were estimates, although this was limited primarily to simple, monosubstituted compounds for which fairly accurate K_{ow} estimates could be obtained. Although the indicated errors therefore include propagated error in some cases, method errors are presumed to predominate. Note that an estimate is within a factor of 10 of the measured value if the error is between -90% and $+900\%$.

The following general conclusions may be drawn from Tables 2-14 and -15:

- Most equations estimated two thirds of the chemicals within a factor of 10. Equation 2-3 was less accurate.

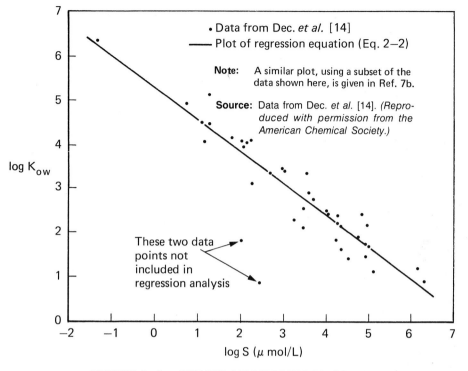

FIGURE 2–2   CORRELATION BETWEEN SOLUBILITY AND OCTANOL/WATER PARTITION COEFFICIENT

- Between 5% and 14% of the estimates were in error by more than a factor of 100. Equation 2-3 was better in this regard. Many of these large errors occurred with the nitrogen-containing compounds, and almost all were overestimates.

- Equations 2-14 and -15 were relatively quite accurate when limited to liquids; approximately 77% of the estimates were within a factor of 10 and 93% within a factor of 100.

- All of the equations showed a significant bias (i.e., tendency to continually overestimate or underestimate S) for selected classes of chemicals. For example, four of the five equations showed strong positive bias in their estimates for nitrogen-containing compounds. The same equations had smaller positive bias when estimating S for miscellaneous pesticides and aliphatic hydrocarbons. Equation 2-3 showed a significant negative bias for halogenated hydrocarbons, oxygen-containing hydrocarbons, and phosphate esters.

TABLE 2-14

Comparison of Measured and Estimated Values of S for Selected Chemicals

Chemical	t_m (°C)[a]	log K_{ow} [b]	Measured S (mg/L)[c]	Percentage Error in Estimated S Using:				
				Eq. 2-2	Eq. 2-3	Eq. 2-4	Eq. 2-14	Eq. 2-15
Aliphatic Hydrocarbons								
2-Butene	—	1.85	78.5	−96	+280	+3500	+2800	+2200
1-Pentene	—	2.20	202	+510	−29	+450	+430	+270
1-Heptene	—	3.20	560	−87	−97	−91	−84	−91
Neopentane	—	3.11	53.5	+35	−60	−10	+60	−12
Cyclohexane	—	2.46	55.6	+1100	+48	+370	+1000	+630
2,2,4-Trimethylpentane	—	5.02	8.46	−97	−96	−98	−92	−98
n-Octane	—	4.00	0.657	+950	+380	+450	+1600	+630
Cyclooctane	—	3.28	7.94	+720	+82	−96	+940	+440
Methyl cyclopentane	—	2.35	42.7	+2100	+140	+1700	+1900	+1300
Isopentane	—	2.30	47.7	+1800	+140	+1500	+1600	+1100
Aromatic Hydrocarbons								
1,2,4-Trimethylbenzene	—	3.65	57.6	−62	−89	−78	−45	−74
Isopropylbenzene	—	3.43	50.1	−13	−79	−46	+16	−42
Indan	—	3.57	109	−75	−93	−85	−64	−83
1-Ethylnaphthalene	—	4.39	10.7	−74	−87	−88	−52	−82
Fluorene	115.8	4.47	1.98	+15	−42	−47	+120	−17
2,3-Benzofluorene	208.8	5.75	0.0020	+2500	+3700	+740	+7800	+1900
3,4-Benzopyrene	175	6.50	0.0038	+50	+310	−60	+500	+25
Benzo[g,h,i]perylene	276.8	7.10	0.00026	+260	+1600	−19	+1700	+210
Halogenated Hydrocarbons								
Vinylidene chloride	—	2.12	273	+700	−38	+610	+570	+390
Trichloroethylene	—	2.42	1,470	−21	−94	−36	−27	−51
Hexachloroethane	—	3.93	27.2	−35	−87	−65	+4	−55
1,3-Dibromopropane	—	2.70	1,680	−56	−97	−67	−55	−72

(continued)

TABLE 2-14 (Continued)

Chemical	t_m (°C)[a]	log K_{ow} [b]	Measured S (mg/L)[c]	Percentage Error in Estimated S Using:				
				Eq. 2-2	Eq. 2-3	Eq. 2-4	Eq. 2-14	Eq. 2-15
Halogenated Hydrocarbons (cont.)								
Diiodomethane	—	2.50	1,220	+50	−94	+19	+43	−7
Fluorobenzene	—	2.27	1,540	−12	−92	−25	−22	−46
1,4-Diiodobenzene	129.4	4.64	1.85	+42	−56	−37	+190	+4
3-Chloroiodobenzene	—	4.12	67.2	−85	−96	−93	−75	−89
1,2,4-Tribromobenzene	44	4.98	9.95	−91	−96	−97	−80	−94
Pentachlorobenzene	86	5.79	0.561	−90	−88	−97	−70	−92
4,4'-Dichlorobiphenyl	148	5.58	0.062	+48	+76	−50	+330	+15
Oxygen-containing Hydrocarbons								
2-Methyl-1-propanol	—	0.61	75,800	+160	−94	+250	+26	+42
3-Hexanol	—	1.61	16,400	−29	−97	−28	−51	−58
2,4-Dimethyl-3-pentanol	<70	1.71	7,050	+36	−94	+35	−2	−20
4-Nonanol	—	3.57	317	−89	−98	−94	−85	−93
2-Ethyl-1-hexanol	—	3.03	880	−81	−97	−87	−78	−88
3-Pentanone	—	0.79	50,500	+160	−94	+230	+33	+42
3-Methyl-2-pentanone	—	1.09	21,300	+170	−93	+220	+58	+55
Cyclopropylethyl ether	—	1.24	20,000	+58	−94	+78	−4	−10
Propylisopropyl ether	—	1.83	4,720	+23	−93	+17	−8	−27
n-Butyl ether	—	2.27	219	+740	−44	+610	+640	+420
Diphenyl ether	28	4.08	18.0	−56	−85	−77	−26	−69
Ethyl formate	—	0.23	88,200	+640	−89	+100	+210	+640
Butyl acetate	—	1.73	23,600	−62	−98	−62	−72	−77
Benzoic acid	122	1.87	2,700	+130	−89	+110	+72	+35
Phenylacetic acid	76.7	1.41	16,600	+75	−95	+87	+13	+1
Diethyl phthalate	—	1.40	7,040	+590	−89	+650	+350	+300
Butylbenzyl phthalate	(?)	4.05	42.2	−62	−93	−80	−36	−73

(continued)

TABLE 2-14 (Continued)

Chemical	t_m (°C)[a]	log K_{ow}[b]	Measured S (mg/L)[c]	Percentage Error in Estimated S Using:				
				Eq. 2-2	Eq. 2-3	Eq. 2-4	Eq. 2-14	Eq. 2-15
Phosphate Esters								
Tributyl phosphate	—	4.00	280	-94	-99	-97	-91	-96
Cresyl diphenyl phosphate	(?)	4.51	2.6	+58	-59	-28	+210	+14
Trixylenyl phosphate	(?)	5.63	0.89	-84	-89	-95	-52	-87
Tricresyl phosphate	77	5.11	0.36	+86	-18	-28	+350	+40
Nitrogen-containing Compounds								
Acrylonitrile	—	-0.92	80,000	$+2.2 \times 10^4$	+35	$+4.5 \times 10^4$	+6000	$+1.1 \times 10^4$
Nitromethane	—	-0.33	95	$+3.3 \times 10^6$	$+3.2 \times 10^4$	$+5.7 \times 10^6$	$+1.1 \times 10^6$	$+1.7 \times 10^6$
1-Nitrobutane	—	1.47	5	$+3.6 \times 10^5$	-43	$+3.8 \times 10^5$	$+2.4 \times 10^5$	$+2.1 \times 10^5$
m-Nitrotoluene	—	2.42	498	+140	-82	+97	+125	+51
m-Nitroaniline	111.8	1.37	890	+3600	-6	+4000	+2300	+2100
Nitrobenzene	—	1.79	1,780	+340	-81	+330	+230	+160
Aniline	—	0.84	36,600	+230	-93	+311	+72	+81
Diethylaniline	—	0.95	670	$+2.0 \times 10^4$	+200	$+2.5 \times 10^4$	$+1.1 \times 10^4$	$+1.1 \times 10^4$
Diphenylamine	53	3.39	36	+94	-68	+20	+150	+29
Triethylamine	—	1.35	15,000	+74	-94	+90	+10	0
Diethanolamine	28	-1.43	954,000	$+1.8 \times 10^4$	-67	$+4.3 \times 10^4$	+4200	+8500
dl-Alanine	295	-2.83	166,000	$+7.4 \times 10^6$	+3600	$+2.6 \times 10^7$	$+1.0 \times 10^6$	$+3.1 \times 10^6$
N-Acetylglycine	206	-1.56	21,700	$+1.3 \times 10^6$	+1800	$+3.3 \times 10^6$	$+3.0 \times 10^5$	$+6.3 \times 10^5$
Acrylamide	84	-1.76	2,050,000	$+1.6 \times 10^4$	-69	$+4.2 \times 10^4$	+3300	+7400
Sulfur-containing Compounds								
Ethyl mercaptan	—	1.20	15,000	+71	-92	+23	+3	-2
Diethyl sulfide	—	1.95	3,130	+12	-62	+3	-12	-33
Thiourea	182	-1.06	91,800 (13°C)	$+4.2 \times 10^4$	+58	$+9.0 \times 10^4$	$+1.1 \times 10^4$	$+2.1 \times 10^5$

(continued)

TABLE 2-14 (Continued)

Chemical (classification)	t_m(°C)[a]	log K_{ow}[b]	Measured S (mg/L)[c]	Percentage Error in Estimated S Using:				
				Eq. 2-2	Eq. 2-3	Eq. 2-4	Eq. 2-14	Eq. 2-15
Miscellaneous Pesticides (classification)								
Malathion (organophosphate)	—	2.89	145	+360	−77	+220	+400	+190
Dichlorvos (organophosphate)	—	1.40	10,000	+380	−92	+420	+210	+180
Parathion (organophosphate)	—	3.81	24	+33	−80	−26	+100	−9
Dichlofenthion (organophosphate)	—	5.14	0.245	+110	+14	−19	+420	+60
Carbaryl (carbamate)	142	2.36	40	+5200	+150	+4300	+4800	+3200
Carbofuran (carbamate)	151	1.60	415	+6100	+23	+6200	+4200	+3500
Dalapon (chlorinated carboxylic acid)	—	0.78	502,000	−56	−99	−43	−77	−76
Fluometuron (substituted urea)	164	1.34	90	$+6.8 \times 10^4$	+890	$+7.5 \times 10^4$	$+4.3 \times 10^4$	$+3.9 \times 10^4$
Atrazine (triazine)	174	2.68	33	+2400	+56	+1800	+2500	+1500
Methoxychlor (organochlorine)	89	4.68	0.003	$+8.1 \times 10^4$	$+2.4 \times 10^4$	$+3.5 \times 10^4$	$+1.7 \times 10^5$	$+5.9 \times 10^4$
DDT (organochlorine)	108	5.98	0.0017	+2300	$+2.7 \times 10^4$	+640	+8000	+1800

a. Melting point of compound if greater than 25°C. A dash (——) indicates the melting point is below 25°C.
b. Many of these values were calculated (i.e., estimated) by the methods described in Chapter 1.
c. From Refs. 5, 6, 13, 14, 24, 30, 42, 56, 64, 69, 70 and 71. All values are for a temperature at or near 25°C.

TABLE 2-15

Analysis of Errors Associated with Methods Using Correlations with K_{ow}

Chemical Class	No.[a]	No. of Calculated Values within a Factor of 10 ($< \times 10$)[b] and Bias in Calculated Values[c]									
		Eq. 2-2		Eq. 2-3		Eq. 2-4		Eq. 2-14[f]		Eq. 2-15[f]	
		$< \times 10$	Bias[c]	$< \times 10$	Bias	$< \times 10$	Bias	$< \times 10$	Bias	$< \times 10$	Bias
Aliphatic hydrocarbons	10 (10)	4	7/3	8	6/4	4	6/4	3	8/2	5	7/3
Aromatic hydrocarbons	8 (4)	7	4/4	5	3/5	8	1/7	6 (4)	5/3 (1/3)	7 (4)	3/5 (0/4)
Halogenated hydrocarbons	11 (7)	10	4/7	5	1/10	8	2/9	11 (7)	5/6 (3/4)	9 (7)	3/8 (1/6)
Oxygen-containing hydrocarbons	17 (12)	17	11/6	4	0/17	16	11/6	17 (12)	8/9 (6/6)	16 (11)	8/9 (6/6)
Phosphate esters	4 (1)	3	2/2	3	0/4	2	0/4	3 (0)	2/2 (0/1)	3 (0)	2/2 (0/1)
Nitrogen-containing compounds[d]	15 (8)	5	15/0	10	6/9	5	15/0	5 (4)	15/0 (8/0)	5 (4)	14/0 (7/0)
Sulfur-containing compounds[e]	2 (2)	2	2/0	1	0/2	2	2/0	2	1/1	2	0/2
Miscellaneous pesticides	11 (5)	5	10/1	7	7/4	6	8/3	5 (5)	10/1 (4/1)	5 (5)	9/2 (3/2)
Total	78 (49)	53	55/23	43	23/55	51	45/33	52 (37)	54/24 (31/18)	52 (38)	46/31 (24/24)
% of total no. of chemicals estimated within a factor of 10		68		55		65		67 (76)		67 (78)	
% of total no. of chemicals estimated within a factor of 100		86		95		86		90 (92)		90 (94)	

a. Number of chemicals in chemical class for which calculations were performed. Number in parentheses is number of chemicals which are known to be liquids at 25°C. (See t_m values in Table 2-14.)

b. A calculated value is within a factor of ten of the measured value if the error is between −90% and +900%.

c. Bias is the tendency of the equation to over- or underestimate the solubility. The number of chemicals with positive and negative errors is given as no. +/no. −. (See Table 2-14 for magnitude of bias.)

d. Includes thiourea.

e. Excludes thiourea.

f. Numbers in parentheses consider the results only with liquid solutes.

A separate analysis of the applicability of Eq. 2-14 to solids was made with $+0.0095(t_m-25)$ as a correction factor on the right side of the equation. The corrected equation was used to estimate the solubilities of the 25 solids in Table 2-14 for which melting points (t_m) were available. The results were as follows:

	Corrected Eq. 2-14	Uncorrected Eq. 2-14
No. overestimated/No. underestimated	+ 14/−11	+ 22/−3
Average bias	+ 2,300%	+ 63,000%
No. (%) within a factor of ± 10	17 (68%)	11 (44%)
No. (%) within a factor of ± 100	24 (96%)	20 (80%)

Similar results could be expected for Eq. 2-15 and for the revised form of Eq. 2-2 given in footnote i of Table 2-3. When the results above are combined with those for the 49 liquids in Table 2-14, the corrected Eq. 2-14 yields estimates within a factor of 10 for 73% of the chemicals.

Selection of Appropriate Equation(s). One or more equations should be selected on the basis of the following considerations:

(1) *Chemical class represented.* If the chemical belongs to a chemical class that is well represented in (or, better, is the sole data base for) a particular data set, select the appropriate equation. Table 2-3 lists the types of chemicals or chemical classes represented in each regression equation. Tables 2-5 to -13 list the specific chemicals in each set. For example:

(a) For benznaphthene, use Eq. 2-18 or 2-20.

(b) For *p*-bromoiodobenzene, use Eq. 2-17.

(c) For isoamyl alcohol, use Eq. 2-5.

Check also items 3 to 5 below.

(2) *Chemical class not well represented.* If the chemical is in a class that is not strongly represented in any of the data sets, use one or more of the equations based on mixed chemical classes (Eqs. 2-2, -3, -4, -14, or -15). Select the equation(s) on the basis of the remaining considerations below.

(3) *Range of values.* Exclude any equation that is incompatible with the input value of K_{ow} or the estimated value of S. (The ranges of K_{ow} and S associated with each data set are given

in Table 2-4.) A small amount of extrapolation may be acceptable, but considerable extrapolation outside the original range can lead to significantly larger errors. For example:

(a) If $K_{ow} = 0.1$ and a mixed-class equation is acceptable, only Eq. 2-3 is appropriate.

(b) If $K_{ow} = 1 \times 10^4$ and the chemical is an alkene, Eq. 2-11 should probably not be used.

(4) *Method errors and bias.* Give appropriate consideration to the quality of fit of the regression equation (r^2 values in Table 2-3) and the likely method error and bias (discussed in the previous subsection). If Tables 2-14 and -15 do not provide sufficient information, errors may be calculated for chemicals of similar structure that have known K_{ow} and S values.

Specifically, if the chemical is a liquid, use any of the equations given by Hansch *et al.* [24], Eqs. 2-5 to -15, other considerations permitting. Conversely, these equations should not be used for solids ($t_m > 25°C$), since much larger errors are involved. For example:

(a) Tripropylamine (liquid): Although Table 2-15 indicates that Eq. 2-3 is best for nitrogen-containing compounds, inspection of Table 2-14 indicates that Eq. 2-15 is probably better for amines, especially those of the form R_3N, and for liquids in general.

(b) 1,3-Dichloro-2-propanol (liquid): According to Table 2-15, Eq. 2-14 estimates S for halogenated and oxygen-containing hydrocarbons slightly better than do the other equations. In addition, Eq. 2-14 is generally better for liquids.

(5) *Solids.* If the chemical is a solid at 25°C and one of the five mixed-class equations must be chosen, first consideration should be given to a corrected version of Eq. 2-14 or -15:

$$\log(1/S) = 1.214 \log K_{ow} - 0.85 + 0.0095 \, (t_m - 25) \qquad \text{(2-14 corr.)}$$

$$\log(1/S) = 1.339 \log K_{ow} - 0.978 + 0.0095 \, (t_m - 25) \qquad \text{(2-15 corr.)}$$

where t_m is the melting point in °C and S is in moles/L. The revised version of Eq. 2-2 (footnote i, Table 2-3) may also be used. Other factors being equal, these equations should provide, on average, more accurate estimates than Eqs. 2-2, -3, or -4. However, the applicability of the latter three equations should be assessed before they are rejected, since all of them included solids in their data sets. If no value of t_m is available, Eqs. 2-2, -3, and -4 are the only mixed-class equations that can be used. For example:

Baygon® (t_m = 91°C) is a carbamate. Table 2-14 indicates that Eq. 2-3 is best, with errors of +150% and +23% for the two carbamates listed. Application of the corrected versions of Eqs. 2-14 and -15 show errors of +270% and +170% (Eq. 2-14) and +160% and +130% (Eq. 2-15) for the two carbamates. Thus, Eq. 2-3 still appears to be best, although all three might be used and the results averaged.

Basic Steps

(1) Obtain the octanol/water partition coefficient, K_{ow}, for the chemical. Large compilations of measured (and some estimated) values are available in Refs. 23 and 37. If no measured value is available, the methods given in Chapter 1 may be used to obtain a reasonable estimate for most chemicals. No value of K_{ow} is required for Eq. 2-20, but this equation is only for polynuclear aromatic hydrocarbons.

(2) Determine if the chemical is a liquid or a solid at 25°C. If it is a solid, it is desirable (but not absolutely necessary) to obtain the melting point, t_m(°C).

(3) Select the most appropriate regression equation(s) on the basis of the considerations discussed in the previous subsection.

(4) Use the value of K_{ow} (and t_m, if required) to calculate a value of the solubility, S, at approximately 25°C. The units of S associated with each equation are given in Table 2-3. Only two significant figures should be reported.

(5) If two or more regression equations were used, and each equation can be presumed to be equally valid (i.e., the likely error in log S is about the same), then it is probably better to calculate a geometric mean than a simple average of the individual answers. To obtain the geometric mean, take the log of each individual estimate (after they have all been converted to the same units), average the logs, and then find the antilog.

Example 2-1 Estimate S for 2-isopropoxyphenyl-N-methylcarbamate (also called Baygon® and Propoxur). It is a solid with $t_m = 91°C$. Measured values of 1.52 and 1.58 have been reported for log K_{ow} [23]. The molecular weight is 209.2 g/mole.

(1) As there is no separate regression equation for carbamates, one of the mixed-class equations must be used. Table 2-14 indicates that, of the three equations that cover both liquids and solids (Eqs. 2-2, -3 and -4), Eq. 2-3 is probably the best with errors of +150% and +23% for the two carbamates listed. However, application of the corrected versions of Eqs. 2-14 and 2-15 to the two carbamates in Table 2-14 shows errors of +270% and +170% (for Eq. 2-14 corr.) and +160% and +130% (for Eq. 2-15 corr.). The differences in these three sets of error values are not significant; thus, it appears appropriate to use all three equations and average the results.

(2a) Using an average value of 1.55 for log K_{ow} in Eq. 2-3,

$$\log S = -0.922 \, (1.55) + 4.184 = 2.755$$

$$S = 570 \text{ mg/L}$$

(2b) Similarly, with Eq. 2-14 (corr.),

$$\log (1/S) = 1.214 \, (1.55) - 0.850 + 0.0095 \, (91\text{-}25) = 1.659$$

$$S = 0.022 \text{ mol/L} = 4600 \text{ mg/L}$$

(2c) And with Eq. 2-15 (corr.),

$$\log (1/S) = 1.339 \, (1.55) - 0.978 + 0.0095 \, (91\text{-}25) = 1.724$$

$$S = 0.019 \text{ mol/L} = 3900 \text{ mg/L}$$

(3) The measured value of S is 2,000 mg/L [30]; the errors associated with each estimate and the geometric mean are:

Eq.	S (mg/L)	% Error
2-3	570	−72%
2-14 (corr.)	4,600	+130%
2-15 (corr.)	3,900	+95%
Geometric Mean	2,200	+10%

Example 2-2 Estimate S for 2-chloroethylether, a liquid. The molecular weight is 108.6 g/mole. The measured value of log K_{ow} is 1.12 [14].

(1) As there is no separate regression equation for chloroethers, it appears that one of the mixed-class equations must be used. Table 2-15 indicates that Eqs. 2-2, -4, -14 and -15 all do well for oxygen- and chlorine-containing compounds. Of these, Eqs. 2-14 and -15 are favored, since they are for

liquids only. However, Table 2-3 shows that the slopes and intercepts of Eqs. 2-8 (for ethers) and -9 (for alkyl halides) are fairly similar to each other; thus, an average of the results from these two equations should also provide a reasonable estimate.

(2a) With $\log K_{ow} = 1.12$ in Eq. 2-14,

$$\log (1/S) = 1.214 (1.12) - 0.850 = 0.510$$

$$S = 0.309 \text{ mol/L}$$

(2b) Similarly with Eq. 2-15,

$$\log (1/S) = 1.339 (1.12) - 0.978 = 0.522$$

$$S = 0.301 \text{ mol/L}$$

(2c) With Eq. 2-8,

$$\log (1/S) = 1.182 (1.12) - 0.935 = 0.389$$

$$S = 0.408 \text{ mol/L}$$

(2d) With Eq. 2-9,

$$\log (1/S) = 1.221 (1.12) - 0.832 = 0.536$$

$$S = 0.291 \text{ mol/L}$$

(3) The measured value of S is 0.120 mol/L [14]; the errors associated with each estimate and the average are:

Eq.	S (mol/L)	% Error
2-14	0.309	+160%
2-15	0.301	+150%
2-8	0.408	+240%
2-9	0.291	+140%
Geometric Mean	0.32	+170%

Example 2-3 Estimate S for 2- chloroiodobenzene, a liquid. An estimated value for $\log K_{ow}$ is 4.12 [70].

(1) Equation 2-17 should be the most appropriate, since it was derived for halobenzenes. Note (Table 2-14) that Eq. 2-2 also did fairly well for halobenzenes. Both equations will be used and the results compared.

(2a) With log K_{ow} = 4.12 and t_m = 25° (Table 2-3, note f), Eq. 2-17 is:

$$\log S = -0.9874\,(4.12) - 0.0095\,(25) + 0.7178 = -3.588$$

$$S = 2.58 \times 10^{-4} \text{ mol/L}$$

(2b) With Eq. 2-2,

$$\log S = -1.37\,(4.12) + 7.26 = 1.616$$

$$S = 41.3\,\mu\text{ mol/L} = 4.13 \times 10^{-5} \text{ mol/L}$$

(3) The measured value of S is 2.88×10^{-4} mol/L [70]. Thus, the errors associated with the use of Eqs. 2-17 and 2-2 are −10% and −86%, respectively.

Example 2-4 Estimate S for naphthacene ($C_{18}H_{12}$), given log K_{ow} = 5.91 and t_m = 357°C [71].

(1) Table 2-3 indicates two equations, 2-18 and -20, that are specifically for polynuclear aromatic hydrocarbons. Each will be used and the errors compared.

(2a) With Eq. 2-18,

$$\log S = -0.88\,(5.91) - 0.01\,(357) - 0.012 = -8.783$$

$$S = 1.65 \times 10^{-9} \text{ mol/L}$$

(2b) With Eq. 2-20,

$$\log S = -0.00987\,(357{-}25) - 3.5055 - 0.3417\,(18{-}6)$$

$$+ 0.002640\,(18{-}6)^2 = -10.50$$

$$S = 3.14 \times 10^{-11} \text{ mole fraction, which is equivalent[5] to}$$
$$1.75 \times 10^{-9} \text{ moles/L.}$$

(3) The measured value of S is 2.05×10^{-9} mol/L [42]. Thus, the errors associated with the use of Eqs. 2-18 and -20 are −20% and −15%, respectively.

---

5. Mole fraction is the ratio of the moles of solute to the total moles present (solute plus water). At very low concentrations, this can be simplified to moles of solute per mole of water. One liter of water is equivalent to 55.49 moles. Thus,

$$\text{Mole fraction} = \frac{\text{solute (moles)}}{\text{water (moles)}} = \frac{\text{solute (moles/L)}}{\text{water (moles/L)}} = \frac{\text{solute (moles/L)}}{55.49}$$

and solute (moles/L) = 55.49 × mole fraction.

2-5 ESTIMATION OF S FROM STRUCTURE (METHOD OF IRMANN)

Irmann [27] developed a means for estimating the aqueous solubilities of hydrocarbons and halo hydrocarbons from structural information alone. The method is intended primarily for organic liquids at 25°C. For solids, the melting point is required.

The basic method involves the substitution of atomic and structural constants, derived from the measured solubilities of nearly 200 compounds, into the following equation:

$$- \log S = x + \Sigma y_i n_i + \Sigma z_j n_j \qquad (2\text{-}21)$$

The negative logarithm of the solubility, S (g/gH$_2$O), is calculated from (1) a basic value, x, which is dependent on the compound type, (2) contributions, y_i, of the various atom types multiplied by their frequency, n_i, in the molecule, and (3) the contributions, z_j, of various structural elements that are present with frequencies, n_j, in the molecule. The x,y, and z values are given in Table 2-16.

For a material that is gaseous under normal pressure (1 atm), the correlation gives the solubility of the liquefied gas at the vapor pressure of both coexisting phases. This can be converted to the approximate value of S at 1 atm by dividing it by the vapor pressure (in atm) of the pure compound.

For a material that is solid at 25°C, Eq. 2-21 gives the solubility of the supercooled liquid. The true solubility of the solid (S_{sol}) can be obtained by the following approximation suggested by Irmann:

$$- \log S_{sol} = - \log S + 0.0095 \, (t_m - 25) \qquad (2\text{-}22)$$

In this equation, $- \log S$ is the value from the right side of Eq. 2-21 and t_m is the melting point of the solid in °C. The 0.0095 factor is based on an assumed melting entropy of 13 cal/mol-°C.

Method Errors. Table 2-17 summarizes the errors involved in the use of Irmann's method for the data set from which the atomic and structural constants were obtained. Over 60% of the estimates were within 25% of the measured values. The solubilities of only three compounds could not be estimated within a factor of 10; all were high-molecular-weight hydrocarbons (octadecane, picene, dibenzanthracene).

TABLE 2-16

Parameters for the Calculation of Water Solubility

a. Values of x

	Type of Compound	No.[a]	x^b
C_6H_6	Aromatic compound	53	
X,H,=C	Halogen[c] derivative, unsaturated aliphatic, with halogen on the unsaturated C, as well as with H in the molecule (no F).	6	0.50
F,H,(Cl),–C	Halogen derivative, saturated aliphatic, containing H besides F	8	
X,H,–C	Halogen derivative, saturated aliphatic (without F)	47	0.90
X,–C or F,(X),–C	Perhalogenated derivative (also with F), saturated aliphatic (without H in molecule)	12	1.25
X,=C	Perhalogenated derivative (no F), unsaturated aliphatic	––	0.90^d
H,C	Hydrocarbon, aliphatic	21	1.50
––	Cycloaliphatic	––	-0.35^d

b. Values of y

Atom	Location	No.[a]	y
C			0.25
H			0.12_5
F	On aromatic C	1	0.19
	On saturated C	19	0.28
Cl	On aromatic and unsaturated C	22	0.67_5
	On saturated C	41	0.37_5
Br	On aromatic and unsaturated C	31	0.79_5
	On saturated C		0.49_5
I	On aromatic and unsaturated C	13	1.12_5
	On saturated C		0.82_5

(continued)

TABLE 2-16 (Continued)

c. Values of z

	Structural Element	No.[a]	z
−C=C−	Double bond (not conjugated) in pure aliphatic compound	16	−0.35
−C=C−C=C−	Two conjugated double bonds in aliphatic compound	−−	−0.55[d]
−C≡C−	Triple bond (individual) in pure aliphatic compound	9	−1.05
>CH, −CH$_2$ (with X substituents)	Group with H besides halogen(s) (also F) on the same saturated C	54	−0.30
−CHX−	Group occurring repeatedly non-terminal	−−	−0.10[d]
−C−C, −C−R (with C branches)	Aliphatic chain branching or non-terminal monosubstitution	17	−0.10

a. Number of compounds available for the determination of the parameter.
b. If more than one compound type is represented in the molecule, use the smallest x value.
c. Unless otherwise specified, X indicates any halogen (Cl,Br,F,I).
d. Approximate value, considered "provisional" by author.

Source: Irmann [27] *(Reproduced with permission from Verlag Chemie International, Inc.)*

A number of the measured solubilities in the data set used by Irmann had uncertainties greater than 20%. A better evaluation of method errors is given in Table 2-18, which lists deviations of calculated solubilities for only those chemicals whose measured solubility was known within 10 — 20%. These data show that the estimates for nearly 90% of the compounds were within ± 15% of the measured values; none deviated by a factor greater than 1.6.

Basic Steps

(1) Draw the molecular structure.
(2) Using Table 2-16, determine the compound type and the appropriate x value.

TABLE 2-17

**Deviations Between Measured and Calculated Solubilities
Using Irmann's Method**

	Deviation in:		No. of	Percentage
	log S	S	Chemicals	of Total
Up to:	± 0.05	± 10%	75	45
	± 0.1	± 25%	103	61
	± 0.2	Factor of 1.6	144	86
	± 0.5	Factor of 3	162	96
	± 1.0	Factor of 10	165	98
Greater than:	± 1.0	Factor of 10	3	2

Source: Irmann [27]. *(Reproduced with permission from Verlag Chemie International, Inc.)*

TABLE 2-18

**Deviations Between Measured and Calculated Solubilities
for Compounds with More Accurately Measured Solubilities**

	Deviation in:		No. of	Percentage
	log S	S	Chemicals	of Total
Up to:	± 0.02	± 5%	24	68
	± 0.06	±15%	31	89
	± 0.2	Factor of 1.6	35[a]	100

a. Included 10 aromatic hydrocarbons, 9 halogenated aromatics, and 15 chlorinated aliphatics. Measured solubilities of all were known within 10-20%.

Source: Irmann [27]. *(Reproduced with permission from Verlag Chemie International, Inc.)*

(3) Using Table 2-16 (and, if necessary, the text following Eq. 2-21) find the appropriate values of y and z and total them in proportion to their frequency (n_i and n_j, respectively) in the molecule.

(4) Substitute the values from steps 2 and 3 in Eq. 2-21 to find S in g/gH_2O at 25°C.

(5) If the compound is a solid at 25°C, use Eq. 2-22 to find the corrected solubility, S_{sol}.

(6) If the compound is a gas at 25°C, note that the solubility obtained from Eq. 2-21 is that of the liquefied gas at the vapor pressure of the two coexisting phases. (See text.)

Example 2-5   Estimate S for o-bromoisopropylbenzene, $C_9H_{11}Br$.

(1)   The structure is

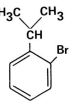

(2)   The basic compound type is aromatic; thus, from Table 2-16, x = 0.50

(3)   The atomic and structural contributions from Table 2-16 are:

$$9C \qquad = 9\,(0.25) \qquad = 2.25$$

$$11H \qquad = 11\,(0.125) \quad = 1.375$$

$$Aromatic\ Br \qquad\qquad\quad = \underline{0.795}$$

$$\Sigma y_i n_i = 4.42$$

$$Aliphatic\ chain\ branching = -0.10 = \Sigma z_j n_j$$

(4)   Substituting in Eq. 2-21,

$$-\log S = 0.50 + 4.42 + (-0.10) = 4.82$$

$$S = 1.51 \times 10^{-5}\ g/g = 15.1\ mg/L$$

The measured value of S = 13 mg/L [27], indicating a deviation of + 16%.

Example 2-6   Estimate S for pyrene, $C_{16}H_{10}$ ($t_m = 150°C$).

(1)   The structure is

(2)   The basic compound type is aromatic; thus, from Table 2-16, x = 0.50.

(3)   The atomic and structural contributions from Table 2-16 are:

$$16C = 16 \, (0.25) \qquad = 4.00$$

$$10H = 10 \, (0.125) \qquad = \underline{1.25}$$

$$\Sigma y_i n_i = 5.25$$

As there are no special structural elements, $\Sigma z_j n_j = 0$.

(4)   Substituting in Eq. 2-21,

$$-\log S = 0.50 + 5.25 + 0 = 5.75$$

(5)   Since the compound is a solid at 25°C, we use Eq. 2-22 to find the corrected solid solubility:

$$-\log S_{sol} = 5.75 + 0.0095 \, (150{-}25) = 6.94$$

$$S_{sol} = 1.15 \times 10^{-7} \text{ g/g} = 0.115 \text{ mg/L}$$

The measured value of $S_{sol} = 0.160$ mg/L [27] or 0.135 [42], indicating deviations of −28% and −15%, respectively.

Example 2-7   Estimate S for DDT, $CCl_3 CH(C_6 H_4 Cl)_2$ ($t_m = 110°C$).

(1)   The structure is

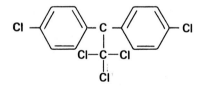

(2)   The basic compound type is aromatic; thus, from Table 2-16, x = 0.50.

(3)   The atomic and structural contributions from Table 2-16 are:

$$14C \qquad\qquad = 14 \, (0.25) \qquad = 3.50$$

$$9H \qquad\qquad = 9 \, (0.125) \qquad = 1.125$$

$$3Cl \text{ on saturated C} = 3 \, (0.375) \qquad = 1.125$$

$$2 \text{ aromatic Cl} \qquad = 2 \, (0.675) \qquad = \underline{1.35}$$

$$\Sigma y_i n_i = 7.10$$

As there are no special structural elements, $\Sigma z_j n_j = 0$.

(4)   Substituting in Eq. 2-21,

$$-\log S = 0.50 + 7.10 + 0 = 7.60$$

(5)   Since the compound is a solid at 25 °C, we use Eq. 2-22 to find the corrected solubility:

$$-\log S_{sol} = 7.60 + 0.0095\ (110-25) = 8.41$$

$$S_{sol} = 3.89 \times 10^{-9}\ g/g = 3.89\ \mu g/L$$

The measured value of $S_{sol} = 1.2\ \mu g/L$ [27] or $1.7\ \mu g/L$ [30], indicating deviations of +220% and +130%, respectively.

Example 2-8   Estimate S for chlorodifluoromethane, $CHClF_2$ (boiling point = −40.8°C, vapor pressure = 10.4 atm at 25°C).

(1)   The structure is

(2)   The basic compound type is "halogen derivative, saturated aliphatic, containing H besides F"; thus, from Table 2-16, x = 0.50.

(3)   The atomic and structural contributions from Table 2-16 are:

1C	= 0.25
1H	= 0.125
1Cl on saturated C	= 0.375
2F on saturated C = 2 (0.28)	= 0.56
	$\Sigma y_i n_i$ = 1.31

Group with H and halogen on same saturated C = $-0.30 = \Sigma z_j n_j$

(4)   Substituting in Eq. 2-21,

$$-\log S = 0.50 + 1.31 + (-0.30) = 1.51$$

$$S = 0.031\ g/g$$

The measured value of S is 0.028 g/g [27], indicating a deviation of +11%.

2-6 AVAILABLE DATA

A number of sources of aqueous solubility data are listed below.

Weast and Astle (1979) [66], *Handbook of Chemistry and Physics*

Perry and Chilton (1973) [51], *Chemical Engineers' Handbook*

Verschueren (1977) [64], *Handbook of Environmental Data on Organic Chemicals*

U.S. Coast Guard (1974) [63], *CHRIS Hazardous Chemical Data*

Wilhelm, *et al.* (1977) [68] — for gases in water

Battino and Clever (1966) [8] — for gases in liquids

American Petroleum Institute (1976) [2] — primarily hydrocarbons

American Petroleum Institute (1969) [3] — focus on hydrocarbons

Freed (1976) [20] — data on pesticides

Linke (1958) [38] and (1965) [39] — inorganic and metal-organic compounds

Seidell (1941) [57] — organic compounds

Seidell and Linke (1952) [58] — organic and inorganic compounds

Stephen and Stephen (1963) [61] — organic and inorganic compounds

A few publications are expected in the near future; these include a new edition (Vol. 3) for inorganic and organic compounds by Stephen [60] and a new *Solubility Data Series* to be published by Pergamon [31].

In addition to the above, the references cited in Tables 2-2 and 2-3 will frequently be helpful, e.g., the work of Kenaga and Goring [30] for pesticides.

2-7 SYMBOLS USED

a	= parameter in Eq. 2-21
b	= parameter in Eq. 2-21
BCF	= bioconcentration factor for aquatic life
c	= parameter in Eq. 2-21
C_s	= molar salt concentration in Eq. 2-1
f_S/f_R	= ratio of solid fugacity to reference fugacity
ΔH_f	= heat of fusion (cal/mol)
K_{oc}	= soil adsorption coefficient based on organic carbon

K_{ow} = octanol/water partition coefficient

K_s = empirical salting parameter in Eq. 2-1

n_i = frequency parameter in Eq. 2-21

n_j = frequency parameter in Eq. 2-21

N = number of carbon atoms in molecule, Eq. 2-20

P = parachor

r = correlation coefficient for regression equation

R = gas constant (1.987 cal/mol-deg)

S = solubility in water

$S°$ = molar solubility in pure water, Eq. 2-1

S' = molar solubility in salt solution, Eq. 2-1

S_{sol} = solubility of a solid, Eq. 2-22

ΔS_f = entropy of fusion, cal/mol-deg

SA = molecular surface area

t = system temperature, °C

T = system temperature (K)

T_b = boiling point (K)

t_m = melting point (°C)

T_m = melting point (K)

TI = topological index

V = molar volume (cm³/mol)

x,y,z = parameters in Eq. 2-21

Greek

γ = activity coefficient

γ^∞ = infinite dilution activity coefficient

γ_w = activity coefficient of solute in water

γ_{oct} = activity coefficient of solute in octanol

$\gamma_w oct$ = activity coefficient of solute in octanol-saturated water

$\gamma_{oct/w}$ = activity coefficient in water-saturated octanol

χ = connectivity parameter

2-8 REFERENCES

1. Almgren, M., F. Greiser, J.R. Powell and J.K. Thomas, "A Correlation Between the Solubility of Aromatic Hydrocarbons in Water and Micellar Solutions, with Their Normal Boiling Points," *J. Chem. Eng. Data,* **24,** 285-87 (1979).

2. American Petroleum Institute, "Phase Equilibria in Water-Hydrocarbon Systems," Chap. 9 of *Technical Data Book — Petroleum Refining,* 3rd ed., Washington, D.C. (1976).

3. American Petroleum Institute, "Solubility and Toxicity Data," Chap. 20 in Volume on Liquid Wastes, *Manual on Disposal of Refinery Wastes,* Washington, D.C. (1969).

4. Amidon, G.L. and S.T. Anik, "Comparison of Several Molecular Topological Indexes with Molecular Surface Area in Aqueous Solubility Estimation," *J. Pharm. Sci.,* **65,** 801-5 (1976).

5. Amidon, G.L., S.H. Yalkowsky, S.T. Anik and S.C. Valvani, "Solubility of Non-electrolytes in Polar Solvents: V. Estimation of the Solubility of Aliphatic Monofunctional Compounds in Water Using a Molecular Surface Area Approach," *J. Phys. Chem.,* **79,** 2239-46 (1975).

6. Amidon, G.L., S.H. Yalkowsky and S. Leung, "Solubility of Non-electrolytes in Polar Solvents: II. Solubility of Aliphatic Alcohols in Water," *J. Pharm. Sci.,* **63,** 1858-66 (1974).

7a. Anderson, T.F. and J.M. Prausnitz, "Application of the UNIQUAC Equation to Calculation of Multicomponent Phase Equilibria: 1. Vapor-Liquid Equilibria," *Ind. Eng. Chem. Process Des. Dev.,* **17,** 552-61 (1978).

7b. Banerjee, S., S.H. Yalkowsky and S.C. Valvani, "Water Solubility and Octanol/Water Partition Coefficients of Organics. Limitations of the Solubility-Partition Coefficient Correlation," *Environ. Sci. Technol.,* **14,** 1227-29 (1980).

8. Battino, R. and H.L. Clever, "The Solubility of Gases in Liquids," *Chem. Rev.,* **66,** 395-463 (1966).

9. Boehm, P.D. and J.G. Quinn, "Solubilization of Hydrocarbons by the Dissolved Organic Matter in Sea Water," *Geochim. Cosmochim. Acta,* **37,** 2459-77 (1973).

10. Brown, R.A., "Fate and Effects of Polynuclear Aromatic Hydrocarbons in the Aquatic Environment," Publication No. 4297, American Petroleum Institute, Washington, D.C. (1978).

11. Chiou, C.T. and V.H. Freed (technical note), *Environ. Sci. Technol.,* **11,** 1219-20 (1977).

12. Chiou, C.T. and V.H. Freed, "Chemodynamic Studies on Bench Mark Industrial Chemicals," Report No. NSF/RA-770286, National Science Foundation, Washington, D.C. (1977). (NTIS PB 274 263)

13. Chiou, C.T., V.H. Freed, D.W. Schmedding and R.L. Kohnert, "Partition Coefficient and Bioaccumulation of Selected Organic Chemicals," *Environ. Sci. Technol.,* **11,** 475-78 (1977).

14. Dec, G., S. Banerjee, H.C. Sikka and E.J. Pack, Jr., "Water Solubility and Octanol/Water Partition Coefficients of Organics: Limitations of the Solubility-Partition Coefficient Correlation," preprint (submitted to *Environ. Sci. Technol.)* [See Ref. 7b for citation of published version.]

15. Eganhouse, R.P. and J.A. Calder. "The Solubility of Medium Molecular Weight Aromatic Hydrocarbons and the Effects of Hydrocarbon Co-solutes and Salinity," *Geochim. Cosmochim. Acta,* **40,** 555-61 (1976).

16. Environmental Protection Agency, "Toxic Substances Control — Discussion of Premanufacture Testing Policy and Technical Issues; Request for Comment," *Fed. Regist.*, **44** (53), 16253-54 (16 March 1979).

17. Fredenslund, A., J.G. Gmehling, M.L. Michelsen, P. Rasmussen and J.M. Prausnitz, "Computerized Design of Multicomponent Distillation Columns Using the UNIFAC Group Contribution Method for Calculating Activity Coefficients," *Ind. Eng. Chem. Process Des. Dev.* , **16**, 450-62 (1977).

18. Fredenslund, A., J.G. Gmehling and P. Rasmussen, *Vapor-Liquid Equilibria Using UNIFAC*, Elsevier Scientific Publishing Co., New York (1977).

19. Fredenslund, A., R.L. Jones and J.M. Prausnitz, "Group-Contribution Estimation of Activity Coefficients in Nonideal Liquid Mixtures," *AIChE J.*, **21**, 1086-99 (1975).

20. Freed, V.H., "Solubility, Hydrolysis, Dissociation Constants, and Other Constants," in *A Literature Survey of Benchmark Pesticides*, unpublished report prepared by George Washington University Medical Center, Dept. of Medical and Public Affairs, Science Communication Division, Washington, D.C., for the U.S. Environmental Protection Agency, Office of Pesticide Programs, Washington,D.C. (1976).

21. Hall, L.H. and L.B. Kier, "Structure-Activity Studies Using Valence Molecular Connectivity," *J. Pharm. Sci.*, **66**, 642-44 (1977).

22. Hall, L.H., L.B. Kier and W.J. Murray, "Molecular Connectivity II: Relationship to Water Solubility and Boiling Point," *J.Pharm. Sci.*, **64**, 1974-77 (1975).

23. Hansch, C. and A.J. Leo, *Substituent Constants for Correlation Analysis in Chemistry and Biology*, John Wiley, New York (1979).

24. Hansch, C., J.E. Quinlan and G.L. Lawrence, "The Linear Free-Energy Relationships between Partition Coefficients and the Aqueous Solubility of Organic Liquids," *J. Org. Chem.*, **33**, 347-50 (1968).

25. Hassett, J.P. and M.A. Anderson, "Association of Hydrophobic Organic Compounds with Dissolved Organic Matter in Aquatic Systems," *Environ. Sci. Technol.*, **13**, 1526-29 (1979).

26. Hermann, R.B., "The Theory of Hydrophobic Bonding: II. The Correlation of Hydrocarbon Solubility in Water with Solvent Cavity Surface Area," *J. Phys. Chem.*, **76**, 2754-59 (1972).

27. Irmann, F., "A Simple Correlation Between Water Solubility and Structure of Hydrocarbons and Halohydrocarbons," *Chem. Ing. Tech.*, **37**, 789-98 (1965). (Translation available from the National Translation Center, The John Crerar Library, 35 West 33rd St., Chicago, Ill. 60616.)

28. Karickhoff, S.W. and D.S. Brown, "Determination of Octanol/Water Distribution Coefficients, Water Solubilities, and Sediment/Water Partition Coefficients for Hydrophobic Organic Pollutants," Report No. EPA-600/4-79-032, U.S. Environmental Protection Agency, Athens, Ga. (1979).

29. Karickhoff, S.W., D.S. Brown and T.A. Scott, "Sorption of Hydrophobic Pollutants on Natural Sediments," *Water Res.*, **13**, 241-48 (1979).

30. Kenaga, E.E. and C.A.I. Goring, "Relationship Between Water Solubility, Soil Sorption, Octanol-Water Partitioning, and Bioconcentration of Chemicals in Biota," pre-publication copy of paper dated October 13, 1978, given at American Society for Testing and Materials, Third Aquatic Toxicology Symposium, October

17-18, New Orleans, La. (Symposium papers were published by ASTM, Philadelphia, Pa., as Special Technical Publication (STP) 707 in 1980.)

31. Kertes, A.S. (ed. in chief), *Solubility Data Series* (approximately 80-100 volumes to be published between 1979 and 1989), Pergamon Press, Inc., Elmsford, N.Y.

32. Kier, L.B. and L.H. Hall, *Molecular Connectivity in Chemistry and Drug Research*, Academic Press, New York (1976).

33. Kier, L.B. and L.H. Hall, "Molecular Connectivity VII: Specific Treatment of Heteroatoms," *J. Pharm. Sci.*, **65**, 1806-9 (1976).

34. Kier, L.B. and L.H. Hall, "The Nature of Structure-Activity Relationships and Their Relation to Molecular Connectivity," *Eur. J. Med. Chem., Chim. Ther.*, **12**, 307-12 (1977).

35. Kinoshita, K., H. Ishikawa and K. Shinoda, "Solubility of Alcohols in Water Determined by the Surface Tension Measurements," *Bull. Chem. Soc. Japan*, **31**, 1081-82 (1958).

36. Leinonen, P.J., D. Mackay and C.R. Phillips, "A Correlation for the Solubility of Hydrocarbons in Water," *Can. J. Chem. Eng.*, **49**, 288-90 (1971).

37. Leo, A., C. Hansch and D. Elkins, "Partition Coefficients and Their Uses," *Chem. Rev.*, **71**, 525-621 (1971).

38. Linke, W.F., *Solubilities of Inorganic and Metal-Organic Compounds*, Vol. 1, D. Van Nostrand Co., New York (1958).

39. Linke, W.F., *Solubilities of Inorganic and Metal-Organic Compounds*, Vol. 2, American Chemical Society, Washington, D.C. (1965).

40. Lu, P.-Y. and R.L. Metcalf, "Environmental Fate and Biodegradability of Benzene Derivatives as Studied in a Model Aquatic Ecosystem," *Environ. Health Perspect.*, **10**, 269-84 (1975).

41. Mackay, D. (technical note), *Environ. Sci. Technol.*, **11**, 1219 (1977).

42. Mackay, D. and W.Y. Shiu, "Aqueous Solubility of Polynuclear Aromatic Hydrocarbons," *J. Chem. Eng. Data*, **22**, 399-402 (1977).

43. Matsuda, K. and M. Schnitzer, "Reactions Between Fulvic Acid, a Soil Humic Material, and Dialkyl Phthalates," *Bull. Environ. Contam. Toxicol.*, **6**, 200-204 (1971).

44. May, W.E. and S.P. Wasik, "Determination of the Solubility Behavior of Some Polycyclic Aromatic Hydrocarbons in Water," *Anal. Chem.*, **50**, 997-1000 (1978).

45. McAuliffe, C., "Solubility in Water of Normal C_9 and C_{10} Alkane Hydrocarbons," *Science*, **163**, 478-79 (1969).

46. McAuliffe, C., "Solubility in Water of Paraffin, Cycloparaffin, Olefin, Acetylene, Cycloolefin, and Aromatic Hydrocarbons," *J. Phys. Chem.*, **70**, 1267-75 (1966).

47. McGowan, J.C., "The Physical Toxicity of Chemicals: IV. Solubilities, Partition Coefficients and Physical Toxicities," *J. Appl. Chem.*, **4**, 41-47 (1954).

48. McNeese, J.A., G.W. Dawson and D.C. Christensen, "Laboratory Studies of Fixation of Kepone ®-Contaminated Sediments," in Vol. 2 of *Toxic and Hazardous Waste Disposal*, ed. by R.B. Pojasek, Ann Arbor Science, Ann Arbor, Mich. (1979).

49. Metcalf, R.L., I.P. Kapoor, P.-Y. Lu, C.K. Schuth and P. Sherman, "Model Ecosystem Studies of the Environmental Fate of Six Organochlorine Pesticides," *Environ. Health Perspect.*, **4**, 35-44 (1973).

50. Palmer, D.A., "Predicting Equilibrium Relationships for Maverick Mixtures," *Chem. Eng. (NY)*, **12**, 80-85 (June 1975).

51. Perry, R.H. and C.H. Chilton (eds.), *Chemical Engineers' Handbook*, 5th ed., McGraw-Hill Book Co., New York (1973).

52. Pierotti, G.J., C.H. Deal and E.L. Derr, "Activity Coefficients and Molecular Structure," *Ind. Eng. Chem.*, **51**, 95-102 (1959).

53. Quayle, O.R., "The Parachors of Organic Compounds: An Interpretation and Catalogue," *Chem. Rev.*, **53**, 439-589 (1953).

54. Reid, R.C., J.M. Prausnitz and T.K. Sherwood, *The Properties of Gases and Liquids*, 3rd ed., McGraw-Hill Book Co., New York (1977).

55. Rekker, R.F., *The Hydrophobic Fragment Constant*, Elsevier Scientific Publishing Co., New York (1977).

56. Saeger, V.W., O. Hicks, R.G. Kaley, P.R. Michael, J.P. Mieure and E.S. Tucker, "Environmental Fate of Selected Phosphate Esters," *Environ. Sci. Technol.*, **13**, 840-44 (1979).

57. Seidell, A., *Solubilities of Organic Compounds*, 3rd ed., Vol. 2, D. Van Nostrand Co., New York (1941).

58. Seidell, A. and W.F. Linke, *Solubilities of Inorganic and Organic Compounds*, supplement to 3rd ed., D. Van Nostrand Co., New York (1952).

59. Siegelman, I. and C.H. Sorum, "Phase Equilibrium Relationships in the Binary System Methyl Ethyl Ketone-Water," *Can. J. Chem.*, **38**, 2015-23 (1960).

60. Stephen, *Solubility of Inorganic and Organic Compounds*, Vol. 3, Pergamon Press, Inc., Elmsford, N.Y. (date not set).

61. Stephen, H. and T. Stephen (eds.), *Solubilities of Inorganic and Organic Compounds*, Vol. 1, Macmillan, New York (1963).

62. Tsonopoulos, C. and J.M. Prausnitz, "Activity Coefficients of Aromatic Solutes in Dilute Aqueous Solution," *Ind. Eng. Chem. Fundam.*, **10**, 593-600 (1971).

63. U.S. Coast Guard, *CHRIS Hazardous Chemical Data*, Report No. CG-446-2, Washington, D.C. (1974).

64. Verschueren, K., *Handbook of Environmental Data on Organic Chemicals*, Van Nostrand Reinhold Co., New York (1977).

65. Wauchope, R.D. and F.W. Getzen, "Temperature Dependence of Solubilities in Water and Heats of Fusion of Solid Aromatic Hydrocarbons," *J. Chem. Eng. Data*, **17**, 38-41 (1972).

66. Weast, R.C. and M.J. Astle (eds.), *Handbook of Chemistry and Physics*, 60th ed., CRC Press, Boca Raton, Fla. (1979).

67. Wershaw, R.L., P.J. Burcar and M.C. Goldberg, "Interaction of Pesticides with Natural Organic Material," *Environ. Sci. Technol.*, **3**, 271-3 (1969).

68. Wilhelm, E., R. Battino and R.J. Wilcock, "Low-Pressure Solubility of Gases in Liquid Water," *Chem. Rev.*, **77**, 219-62 (1977).

69. Yalkowsky, S.H., G.L. Amidon, G. Zografi and G.L. Flynn, "Solubility of Nonelectrolytes in Polar Solvents: III. Alkyl p-Aminobenzoates in Polar and Mixed Solvents," *J. Pharm. Sci.*, **64**, 48-52 (1975).

70. Yalkowsky, S.H., R.J. Orr and S.C. Valvani, "Solubility and Partitioning: 3. The Solubility of Halobenzenes in Water," *Ind. Eng. Chem. Fundam.*, **18**, 351-53 (1979).

71. Yalkowsky, S.H. and S.C. Valvani, "Solubilities and Partitioning: 2. Relationships between Aqueous Solubilities, Partition Coefficients, and Molecular Surface Areas of Rigid Aromatic Hydrocarbons," *J. Chem. Eng. Data*, **24**, 127-29 (1979).

3

SOLUBILITY IN VARIOUS SOLVENTS

Warren J. Lyman

3-1 INTRODUCTION

This chapter provides methods for estimating the solubility limits in liquid-liquid and solid (solute)-liquid binary (i.e., two-chemical) systems. Methods for estimating the solubility of gases in liquids are not provided. All of the methods require knowledge of the activity coefficient for the solute and/or solvent at one or more points on the composition diagram. Measured activity coefficients are available for a relatively small number of chemical systems, but a fairly generalized process for estimating these coefficients from structural information alone is given in Chapter 11. The most powerful estimation method described in Chapter 11 (UNIFAC) relies on the availability of group volume and surface area parameters plus group interaction parameters; these are available for only a limited number of functional groups and can be applied only to molecules with relatively simple structures. The reader should verify that an activity coefficient can be estimated from Chapter 11 before attempting to use the methods described below.

Liquid-Liquid Systems. The methods outlined in § 3-4 allow the estimation of the solubility of one liquid in another at any temperature. One of the liquids may be water. One method is presented for estimating solubility from activity coefficients alone; this is subdivided into three modifications of increasing accuracy but also of increasing calculational difficulty. The methods involving the simpler calculations are limited to chemicals with a low solubility in the solvent. While similar approaches are available for calculating the phase diagram of a system of three liquids, estimation methods are not provided here. The approach provided for the estimation of solubilities may be easily extended to the calculation of vapor-liquid equilibria for binary systems and partition coefficients (at infinite dilution) for ternary systems.

Solid-Liquid Systems. The method recommended in §3-5 allows estimation of the solubility of a solid solute in a liquid solvent at any temperature. The liquid may be water. The activity coefficient, heat of fusion, and melting point of the solute must be known. As with the liquid-liquid systems, three versions of the basic method are provided that offer increasing accuracy with increasing calculational difficulty. The methods involving the simpler calculations are limited to chemicals with a low solubility in the solvent.

Gas-Liquid Systems. The estimation of the solubility of gases in liquids using (estimated) activity coefficients has not been sufficiently investigated for a method to be included here. References 12 and 25 provide some guidance on this subject.

3-2 BASIC APPROACH

All of the methods described in this chapter require the use of measured or estimated values of the activity coefficient of the solute and, in some cases, of the solvent as well. The activity coefficient, γ, is a measure of the nonideal behavior of a chemical in solution. If measured values of the activity coefficient are not available, they may be estimated by the methods described in Chapter 11. In some cases of limited solubility, only the activity coefficient at infinite dilution, γ^∞, is required; this parameter can be obtained from relatively simple regression equations for a very few pairs of solvents and solute chemical classes [24, 26, 34].

The estimation of solubilities from estimated activity coefficients is a fairly recent development. (Solubility data are more frequently used to obtain activity coefficients.) The approach, based upon theoretical considerations, has been made possible by recent advances in the estimation of activity coefficients from structural information alone. Because these methods are still being developed, and the use of activity coefficients to estimate solubility has been little studied, the range of applicability and accuracy of the overall method have not been established. Future investigations of this approach will probably show that it can be used for a wide range of chemicals or chemical classes over the full range of possible solubility values and that its accuracy is more than adequate for questions relating to environmental concerns.

The theoretical basis for the approaches described derives from a consideration of the free energy of mixing (ΔG^M) in binary systems. This subject is well treated in Refs. 12, 13, 25, 26, and 30.

In keeping with most of the publications cited in this chapter, the concentration of one chemical in another is represented by the symbol x in units of mole fraction. The solute is represented by subscript 1 (x_1) and the solvent by subscript 2 (x_2). The mole fraction ranges from 0 to 1. In a binary system, $x_1 + x_2 = 1$.

3-3 OTHER ESTIMATION METHODS CONSIDERED

Three other approaches to the estimation of solubility have been suggested:

(1) Correlations with solvent/water partition coefficients, and the use of this coefficient plus the solubility in water;

(2) Various formulas based upon the use of solubility parameters; and

(3) Method of Cysewski and Prausnitz for gases in liquids.

The first approach has never been demonstrated and would require a significant amount of work to develop. The second approach, while well developed, is (in theory) limited to nonpolar systems and is often limited in other respects as well. The third approach, while valid for polar and non-polar systems, applies only to gas-liquid systems, and has other severe limitations. Each of these three approaches is further discussed below.

Use of Solvent/Water Partition Coefficients. The partition coefficient for an organic solute between some solvent(s) and water(w), K_{sw}, is frequently measured for studies of the relation between structure and activity. In such cases, K_{sw} is measured at very low solute concentrations; thus, one would expect that this parameter would not be equal to the ratio of the solute's solubility in the two phases, i.e.,

$$K_{sw} \neq \frac{x_s}{x_w} \qquad (3\text{-}1)$$

unless x_s and x_w are very small. If both x_s and x_w are believed to be small, Eq. 3-1 could provide a reasonable estimate of x_s given known or estimated values of x_w and K_{sw}.

It has been shown, however, that the solubility in water, x_w, is inversely proportional to the octanol-water partition coefficient, K_{ow}:

$$\log x_w = -a \log K_{ow} + b \tag{3-2}$$

As numerous equations of this form have been reported, one might expect a similar relationship between x_s (the solubility in solvents) and K_{sw}:

$$\log x_s = c \log K_{sw} + d \tag{3-3}$$

It is known [17, 18] that values of K_{ow} and K_{sw} can be related for a number of solvent systems by equations of the form:

$$\log K_{ow} = e \log K_{sw} + f \tag{3-4}$$

Subtracting Eq. 3-2 from Eq. 3-3 and then substituting Eq. 3-4 in the resulting equality yields

$$\log(x_s/x_w) = (c + ae)\log K_{sw} + (d + af - b) \tag{3-5}$$

or

$$x_s/x_w = k(K_{sw})^{k'} \tag{3-6}$$

where k = antilog (d+af-b) and k' = c+ae. Thus, if adequate data were available to obtain (via regression equations) values of k and k' for each solvent of interest, x_s could be estimated. This would be an extremely simple estimation method. Even better would be regression equations giving the constants c and d in Eq. 3-3 for a variety of solvents.

As previously mentioned, the author is not aware that any expressions similar to Eqs. 3-3, 3-5, and 3-6 have been published. Attempts have been made to test the validity of such equations by the use of solubility data for solutes in ether along with the ether-water partition coefficient, but the attempts were unsuccessful because of the limited quality and quantity of the data.

This approach would clearly not be applicable for solvents (e.g., ethyl alcohol) that are miscible with water. In addition, a different set of constants would have to be provided for each temperature of interest.

Use of Solubility Parameters. The use of solubility parameters (represented by the symbol δ) in estimating solubility has been treated in several publications [12, 13, 25, 30]. Equations are typically of the form shown below for the solubility of gases in liquids (subscript 1 refers to the solute, 2 to the solvent):

$$-\ln x_1 = \ln(f_1^L/f_1^V) + \frac{V_1 \phi_2{}^2 (\delta_1 - \delta_2)^2}{RT} \qquad (3\text{-}7)$$

where x_1 = mole fraction of solute

f_1^L = fugacity of pure solute as a liquid (may be estimated from critical temperature and pressure)

f_1^V = fugacity of solute in vapor phase

V_1 = solute molar volume

ϕ_2 = volume fraction of solvent = $x_2 V_2/(x_1 V_1 + x_2 V_2)$ (≈ 1 if solubility is very low)

δ_1 = solubility parameter for solute

δ_2 = solubility parameter for solvent

R = gas constant

T = temperature

The second term on the right-hand side of Eq. 3-7 appears in similar equations for liquid-liquid and solid-liquid systems.

Data for such parameters as f_1^L, V_1, and δ_1 normally come from solubility data in some solute-solvent system and are available for a relatively small number of chemicals. Most of these parameters can be estimated if necessary, but the use of estimated values for all three (as would frequently be necessary) could lead to significant errors. Estimation methods for δ are given in Refs. 5, 12 and 30. Values of δ_2 have been compiled for numerous solvents [5].

In addition to the significant data requirements of this method (e.g., f_1^L, f_1^V, V_1, and δ_1 for the solute, and V_2 and δ_2 for the solvent), another drawback is its limitation to nonpolar systems, which derives from the theoretical basis for the equations. Various attempts have been made to add correction factors for polar systems, but this has necessarily resulted in more complex equations.

Gas Solubilities in Polar and Non-Polar Solvents. Cysewski and Prausnitz [7] have derived a semiempirical correlation which may be useful for estimating gas solubilities in limited cases. The accuracy of the method is not high, but prediction is usually within a factor of 2. The equation allows one to calculate Henry's Law constant (the reciprocal of the solubility when the partial pressure of the solute is 1 atm) of a solute in a solvent provided one knows the molar volume of the solvent and two characteristic, temperature-independent parameters T_{12}^* and v_{12}^*. The latter two parameters must be determined empirically. For v_{12}^*, correlations are given to allow this parameter to be estimated if the critical volumes of the solvent and (for a relatively small, second-order term) the solute are known or can be estimated. Estimation of T_{12}^* is much more difficult and, at present, is possible for only a very few solutes. The derived equation also involves some lengthy calculations. Because of these limitations, the method is not included in this handbook.

3-4 LIQUID-LIQUID BINARY SOLUTIONS

Basis for Estimation Method. The solubility of one liquid in another is a function of temperature. Most binary solutions have a phase-temperature diagram like that of Figure 3-1a, but some are characterized by the curves in Figures 3-1b and -1c. For binaries of the first kind, there is a temperature (called the upper consolute temperature) above which only one phase can exist. Below this temperature two phases can exist; in this region, component 1 has a limited solubility in component 2 and vice versa.

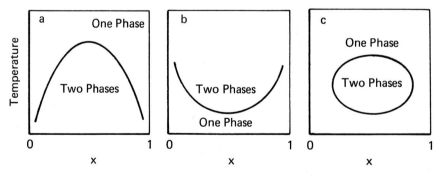

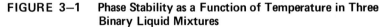

FIGURE 3–1   **Phase Stability as a Function of Temperature in Three Binary Liquid Mixtures**

One measure of phase stability in solutions is ΔG^M, the Gibbs free energy of mixing. This parameter may be expressed as a function of the mole fraction (x) and activity (a) of each component. For a binary solution of liquids

$$\Delta G^M = RT\,(x_1\,\ln a_1 + x_2\,\ln a_2) \tag{3-8}$$

where R is the gas constant and T the temperature in K. Since the activity is related to the activity coefficient, γ, by $a = x\gamma$, Eq. 3-8 may be written as

$$\Delta G^M/RT = x_1 \ln \gamma_1 + x_2 \ln \gamma_2 + x_1 \ln x_1 + x_2 \ln x_2 \qquad (3\text{-}9)$$

The plot of Eq. 3-9 as a function of x has a single minimum (Fig. 3-2a) if only one phase is present at the temperature in question.

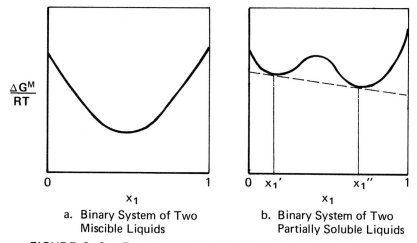

a. Binary System of Two
   Miscible Liquids

b. Binary System of Two
   Partially Soluble Liquids

FIGURE 3–2   Free Energy of Mixing Curves for Binary Solutions

If two phases are present, the curve has two minima (Fig. 3-2b), and it is possible to draw a straight line (dashed line in Fig. 3-2b) that is simultaneously tangent to the curve at two points. The values of x_1 at these two points of tangency, x_1' and x_1'', are the limits of phase stability; i.e., between these two values of x_1 two phases are present, x_1' and x_1'' being the concentrations of component 1 in the two phases. (Note that the points of tangency do not exactly coincide with the points of minima in the $\Delta G^M/RT$ curve unless the two minima are at the same value of $\Delta G^M/RT$.) Following the first minimum, the curve reverses curvature and becomes concave downward — i.e., the second derivative of the equation is negative in this region. Accordingly, if $(\partial^2 \Delta G^M/\partial x^2)_{T,P} < 0$ for any portion of the $\Delta G^M/RT$ curve, some region of phase instability will exist; this region is between x_1' and x_1''. The two minima occur when $(\partial \Delta G^M/\partial x)_{T,P} = 0$ and $(\partial^2 \Delta G^M/\partial x^2)_{T,P} > 0$. If we use the two-suffix Margules equation[1] as the

---

1. The two-suffix Margules equation is $RT \ln \gamma_1 = Ax_2^2$ (or $RT \ln \gamma_2 = Ax_1^2$ where A is an empirically derived constant. Evaluating this equation at x_1 (or x_2) $= 0$, one obtains $A = RT \ln \gamma_1^\infty$ (or $A = RT \ln \gamma_2^\infty$). See Ref. 25 or 26, or §11-2 of Chapter 11, for additional information on this equation and its limitations. It is reasonably valid for simple liquid mixtures, i.e., where the molecules are of similar size, shape, and chemical nature.

expression for the excess Gibbs energy of the binary solution, we can predict phase instability from a knowledge of the infinite dilution activity coefficient, $\gamma\infty$. In particular, two phases are likely to be present (at the appropriate mole fractions) at the temperature in question if

$$\ln \gamma\infty > 2 \qquad (\text{or } \gamma\infty > 7.4) \qquad (3\text{-}10)$$

If $\gamma\infty$ for either component in the binary is greater than 7.4, phase instability is likely at some point; as the value of $\gamma\infty$ increases, instability will exist over a wider range of x_1 (or x_2).

If $\gamma\infty$ is very large (>1000) for either binary component, and if the chemical does not dissociate (or associate with itself) to any significant extent in very dilute solutions, a reasonable estimate of the solubility limits may be obtained from

$$x_1 = 1/\gamma_1{}^\infty \qquad (\text{for } \gamma_1\infty > 1000) \qquad (3\text{-}11)$$

and

$$x_2 = 1/\gamma_2\infty \qquad (\text{for } \gamma_2\infty > 1000) \qquad (3\text{-}12)$$

These two equations may be derived from Eqs. 3-14 and -15 (given below) by assuming that $x \ll 1$.

For values of $\gamma\infty$ between about 50 and 1000 an acceptable estimate of x_1' may be obtained from the equation derived by taking the partial derivative of Eq. 3-9 with respect to x and setting the result equal to 0. The result is:

$$(1-4x_1 + 3x_1^2) \ln\gamma_1\infty + (2x_1 -3x_1^2) \ln\gamma_2\infty + \ln x_1 -\ln(1-x_1) = 0 \quad (3\text{-}13)$$

This equation has three solutions, two of which (the ones with the lowest and highest values of x_1) correspond to the two minima in the $\Delta G^M/RT$ diagram (Fig. 3-2b). If it is not possible to obtain $\gamma\infty$ for both components *and* it is likely that $\gamma_1\infty \approx \gamma_2\infty$, then Eq. 3-13 may be reduced to two simpler equations for the calculation of one solubility limit or the other. If only $\gamma_1\infty$ is known, the expression for x_1' is:

$$(1-2x_1) \ln\gamma_1\infty + \ln x_1 - \ln(1-x_1) = 0 \qquad (3\text{-}14)$$

Similarly, if only $\gamma_2\infty$ is known, the expression for x_1'' is:

$$(1-2x_1) \ln\gamma_2\infty + \ln x_1 - \ln (1-x_1) = 0 \qquad (3\text{-}15)$$

Equation 3-13 was derived using the three-suffix Margules equation which is described in § 11-2 of Chapter 11; Eqs. 3-14 and -15 may be derived using the two-suffix Margules equation. These three equations allow the solubility limits to be estimated from only two input parameters, the infinite dilution activity coefficients for the two components in the binary. Whenever possible, it is clearly better to use the general equation (Eq. 3-13) than Eqs. 3-14 and -15, which require simplifying assumptions. Given a value of γ_1^∞ and/or γ_2^∞, these equations may be solved by trial and error using $1/\gamma^\infty$ as the first trial point. If Eq. 3-14 or 3-15 is used, an approximate solution may be obtained from the plot of γ^∞ vs x_1 given in Figure 3-3. The figure also shows, for comparison, a plot of $1/\gamma^\infty$ versus x_1.

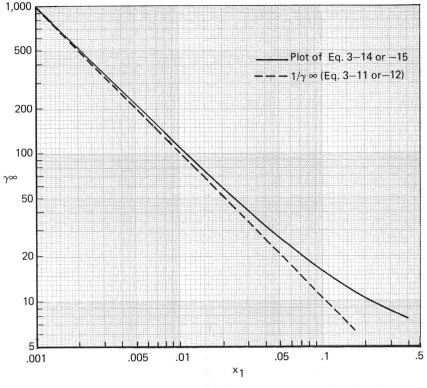

FIGURE 3–3   Plot of x_1 vs γ^∞ from Equations 3–14 and 3–15

When using the trial-and-error method, check that the left-hand side of Eqs. 3-13, -14 and -15 goes from negative to positive as x increases and passes through the value where the equality holds; if the sign changes in the opposite direction, it indicates the point at the maximum of the hump in Fig. 3-2b. Note that Eqs. 3-14 and -15 cannot predict values of x_1' above 0.5 (or x_1'' below 0.5).

For decreasing values of γ^∞ below about 50, the use of Eqs. 3-13 to -15 is likely to give increasingly erroneous results. If the activity coefficients of both components are available (e.g., estimated by the methods in Chapter 11) over the whole range of x, they may be used directly in Eq. 3-9 to plot $\Delta G^M/RT$ versus x. If two phases are present, a curve like that shown in Figure 3-2b will result, and the values of x_1' and x_1'' may be obtained from the two points tangent to the dashed straight line. This approach will always be the most accurate one, irrespective of the value of γ^∞, but the significant increase in the calculational effort is probably justified only when the values of γ^∞ are below 50.

The general method outlined above may be extended to ternary and higher systems. Examples of such calculations are given in Refs. 3 and 9.

Calculation of Vapor-Liquid Equilibria. If activity coefficients are available over the whole range of x (as required above for plotting $\Delta G^M/RT$ versus x), the composition of the vapor phase above any binary liquid solution may also be estimated if the vapor pressures of the two pure components, P_{vp_1} and P_{vp_2} are known. At any (total) system pressure, P, up to a few atmospheres [9]:

$$y_1 P = \gamma_1 x_1 P_{vp_1} \tag{3-16}$$

and

$$y_2 P = \gamma_2 x_2 P_{vp_2} \tag{3-17}$$

where y_1 and y_2 are the vapor-phase mole fractions of components 1 and 2, respectively. Note that the total system pressure (P) is a function of composition. Then, since $y_1 + y_2 = 1$ and $x_1 + x_2 = 1$, for any value of x_1 Eqs. 13-16 and -17 are a set of simultaneous equations with two unknowns and may be easily solved. Some examples of calculated vapor-liquid equilibria using estimated activity coefficients are given in Refs. 1, 2, 8, 9, 10, 23, and 31. The method is quite accurate.

Method Errors. Table 3-1 compares some observed infinite-dilution activity coefficients (γ^∞) and solubilities with the estimated values. Wherever possible, the estimated values of γ^∞ were used to estimate the solubility so that the combined error of the method could be examined. Because of the lack of data for organic-organic systems, several binaries including water are listed. Solubilities estimated from plots of $\Delta G^M/RT$ versus x are not included because of the laborious calculation required.

While most of the tabulated estimates of γ^∞ are within about 10% of the observed values, some errors are as large as 100%. Other comparisons of estimated and observed values of γ^∞ show average errors of about 10-20% [8, 9, 10, 36].

TABLE 3-1

Comparison of Observed and Estimated Activity Coefficients and Solubilities for Liquid-Liquid Binaries

Solute (1)	Solvent (2)	Activity Coefficient, γ_1^∞				Observed Solubility[b]			Estimated Solubility, x_1[b,c]		
		T(°C)	Calc.[a]	Obs.	Ref.	T(°C)	x_1	Ref.	$1/\gamma_1^\infty$	Eq. 3-14,-15	Eq. 3-13
Heptane	Acetonitrile	20	41.0	~46	[9]	20	~.04[f]	[9]	.024	.030	.030
Acetonitrile	Heptane	20	31.4	~30	[9]	20	~.05[f]	[9]	.032	.040	.039
Water	Butanol	NS[d]	4.61	~3.7	[9]	NS	~.63[f]	[9]	.22	>.50	.15
Butanol	Water	NS	80.6	~70	[9]	20; NS; NS	.51; ~.02[f]; .019	[33]; [9]; [14]	.012	.015	.016
2-Butanone	Water	NS	69.4	NA[g]	[9]	20; NS; 25	.020; ~.03[f]; .068	[33]; [9]; [33]	.014	.016	.018
Water	Butanone	NS	9.58	NA	[9]	NS; 25	~.35[f]; .30	[9]; [33]	.10	.22	.11
Water	Hexadiene	20	105	226	[10]		NA[g]		.0095	.010	.0093
Hexadiene	Water	25	30,600	26,900	[10]	15	.000088	[4]	.000033	.000033	.000033
n-Pentane	Acetonitrile	25; 50	18; 13.8	20; NA	[10]; [36]		NA		.056; .072	.082; .12	—; .11
Acetonitrile	n-Pentane	50	17	NA	[36]		NA		.059	.089	.094
Aniline	Water	100	115	80	[10]	90	.012	[14]	.0087	.0094	.0094
Benzene	Water	25	455	458	[10]	25	.00042	[4]	.0022	.0023	.0023
Water	Benzene	25	359	430	[10]	25; 25	.00042; .00016	[33]; [33]	.0028	.0029	.0029
Nonane	Methylamine	0; 20	13.2; 7.97	10.7; 7.90	[10]; [8]		NA		.076[h]; .13[h]	.13; .34	.22; .34
Methylamine	Nonane	0; 20	5.5; 3.72	4.8; 3.55	[10]; [8]				.18[h]; .27[h]	>.5; >.5	—; —
n-Hexane	Methylamine	0	8.6	8.3	[10]		NA		.12	.28	—
n-Propanol	1,2-Dichloroethane	84	14	23	[10]		NA		.071	.12	.22
1,2-Dichloroethane	n-Propanol	97	6.7	12.9	[10]		NA		.15	>.5	.21
n-Heptane	Ethyl alcohol	45; 30	~9.1[f]	14.1[f]	[36]; [36]	NS	.15	[14]	.11	.24	.12

(Continued)

TABLE 3-1 (Continued)

Solute (1)	Solvent (2)	Activity Coefficient, γ_1^∞				Observed Solubility[b]			Estimated Solubility, x_1 [b,c]		
		T(°C)	Calc.[a]	Obs.	Ref.	T(°C)	x_1	Ref.	$1/\gamma_1^\infty$	Eq. 3-14,-15	Eq. 3-13
Ethyl alcohol	n-Heptane	45 / 30	~50.1[f]	~42.7[f]	[36] / [36]		NA		.020	.023	.026
1-Ethylnaphthalene	Water	25	NA	806,000	[22]	NS	1.16×10^{-6}	[22]	1.24×10^{-6}	1.24×10^{-6}	--
n-Heptane	N-Methylpyrrolidone	25	NA	16.2	[19]	25	.145	[19]	.062	.095	--
n-Hexane	Aniline	25	NA	26.6	[19]	25	.089	[16]	.038	.049	--

a. Calculated using the UNIFAC method, which requires only a knowledge of the structures of the solute and solvent.

b. Units are in mole fraction.

c. Estimated solubilities are based upon the calculated value of γ_1^∞ whenever possible. All estimated values of x_1 are at the temperature given for the value of γ_1^∞.

d. NS = Not specified.

e. At 350 mm Hg.

f. Observed values taken from data points in published graphs; uncertainty in cited values is due to difficulty in reading the plots accurately.

g. NA = Not available.

h. Two phases are likely at 0°C, since γ_1^∞ for nonane is significantly above 7.4. However, at 20°C the value is only slightly above 7.4 and only a single phase may exist.

Errors in the estimated values of $\gamma\infty$ will add to the method error if they are used in the calculation of x_1' or x_1''. A comparison of observed and estimated solubilities in Table 3-1 shows a wide range of errors, but they should be tolerable for environmental considerations; since only a semiquantitative evaluation of a chemical's solubility in various solvents may be needed for such purposes, the method presented in this chapter is probably more than adequate. In one study of 50 ternary liquid-liquid equilibria where two liquid phases were known to be present in some regions, the use of estimated activity coefficients[2] was shown to give reasonable predictions of phase splitting (i.e., values of mutual solubilities) for most systems [9]. The quality of the predictions was described as follows:

Rank	No. of Systems	Quality of Predictions
0	3	No phase splitting could be predicted.
1	11	Agreement between predicted values and liquid-liquid solubility curves was poor.
2	28	Predictions agreed qualitatively with experimental values.
3	8	Predictions agreed quantitively with experimental values.

Basic Steps

(1) Check that both the solute (component 1) and solvent (component 2) are liquids at the temperature of interest.

(2) Obtain the infinite dilution activity coefficient[3] for the solute ($\gamma_1\infty$) and the solvent ($\gamma_2\infty$) and proceed as follows:

- If $\gamma_1\infty > 1000$, go to Step (3);
- If $\gamma_1\infty$ is between about 50 and 1000, go to Step (4);
- If $\gamma_1\infty$ is between about 7.4 and 50, go to Step (5);
- If both $\gamma_1\infty$ and $\gamma_2\infty$ are less than 7.4, the two liquids can be assumed to be miscible in all proportions at the temperature considered.

---

2. Values of γ (as a function of x). As explained earlier, this is more accurate than the shorthand method using only $\gamma\infty$.

3. Methods for estimating activity coefficients are given in Chapter 11. These coefficients are a function of temperature, and the methods described allow a calculation at any temperature.

(3) Calculate x_1 (the solubility of 1 in 2, in units of mole fraction) from Eq. 3-11. Similarly, if the solubility of 2 in 1 is desired (x_2), and $\gamma_2^\infty > 1000$, use Eq. 3-12.

(4) Calculate x_1 (the solubility of 1 in 2, in units of mole fraction) from Eq. 3-13; the equation may be solved by trial and error using $1/\gamma^\infty$ as the first approximation for x_1.

Note: Of the three solutions to this equation, the solubility of 1 in 2 is given by the lowest value of x_1 that satisfies the equality. If the left-hand side of the equation goes from negative to positive as x increases through the value where the equality holds, a minimum in the Gibbs free energy curve has been found. (See Figure 3-2b and related text for additional discussion.)

Similarly, x_2, the solubility of 2 in 1, may be calculated from Eq. 3-13; this time the solution with the *highest* value of x_1 is found and subtracted from 1 to obtain x_2.

If only γ_1^∞ is known, and if it can be assumed that $\gamma_1^\infty \approx \gamma_2^\infty$,[4] the solubility of 1 in 2 may be obtained from Eq. 3-14. Similarly, if only γ_2^∞ is known, Eq. 3-15 may be used to calculate the solubility of 2 in 1; i.e., solve for x_1 and then obtain x_2 from $x_2 = 1-x_1$. Both equations may be solved by trial and error, using $1/\gamma^\infty$ as the first approximation or with the plot of these equations given in Figure 3-3. The note above is equally applicable to these two equations.

(5) Calculate x_1 (the solubility of 1 in 2, in units of mole fraction) by using Eq. 3-9 to plot $\Delta G^M/RT$ as a function of x_1. (This requires the calculation of activity coefficients at a number of points over the range of x_1, as described in Chapter 11.) If two phases are predicted, a curve with two minima, such as that shown in Figure 3-2b, will result; if only a single minimum is obtained, the two liquids may be assumed to be miscible in all proportions. Draw a straight line (e.g., the dashed line in Figure 3-2b) that is simultaneously tangent to the curve at the two points. The points of tangency corresponds to the solubility limits for 1 in 2 and 2 in 1 (x_1' and $1-x_1''$, respectively).

Example 3-1   Estimate the solubility of 1-ethylnaphthalene(1) in water(2) at 25°C, given $\gamma_1^\infty = 806{,}000$ [22].

---

4. This is seldom a good assumption unless the solute and solvent are chemically similar. However, the assumption should not lead to excessive errors unless the values are orders of magnitude apart.

(1)   1-Ethylnaphthalene melts below $-14°C$ and decomposes at $258°C$. Therefore, it is liquid at the temperature in question.

(2)   Since $\gamma_1{}^\infty$ is greater than 1000, we may use Eq. 3-11:

$$x_1 = 1/\gamma_1{}^\infty = 1.24 \times 10^{-6} \text{ mole fraction}$$

This is equivalent to 10.7 mg/L, which compares well with a literature value of 10.0 mg/L [22].

Example 3-2   Estimate the solubility of 2-butanone(1) in water(2) at $\sim25°C$ and 350 mm Hg, given $\gamma_1{}^\infty = 69.4$ and $\gamma_2{}^\infty = 9.58$ [9].

(1)   Both the solute and the solvent are liquids at this temperature.

(2)   Since $\gamma_1{}^\infty$ is between 50 and 1000, and $\gamma_2{}^\infty$ is less than 50, it is probably best to use Eq. 3-13:

$$(1-4x_1+3x_1^2) \ln(69.4) + (2x_1-3x_1^2) \ln(9.58) + \ln x_1 - \ln(1-x_1) = 0.$$

(3)   Solve the above equation by trial and error, using $1/\gamma_1{}^\infty$ (=0.014) or the value from Figure 3-3 (=0.016) as the first approximation. The solution is found at $x_1 = 0.018$ mole fraction. The data given in Ref. 9 indicate an observed value of roughly 0.03 mole fraction for these conditions.

Example 3-3   Estimate the mutual solubilities for the heptane(1)-acetonitrile(2) system at $20°C$, given activity coefficients (estimated) as a function of composition, and $\gamma_1{}^\infty = 41.0$ and $\gamma_2{}^\infty = 31.4$ [9].

(1)   Since both $\gamma_1{}^\infty$ and $\gamma_2{}^\infty$ are below 50, it would seem preferable to plot $\Delta G^M/RT$ vs x, as described in Step 5 of the "Basic Steps." However, in this case the values are not much below 50 and, in addition, are approximately equal. Thus, one can probably use Eqs. 3-14 and 3-15:

$$(1-2x_1) \ln(41.0) + \ln x_1 - \ln(1-x_1) = 0$$

$$(1-2x_1) \ln(31.4) + \ln x_1 - \ln(1-x_1) = 0$$

(2)   As a first approximation in solving the first equation, use $1/\gamma_1{}^\infty = 0.024$ or Figure 3-3 (0.030). For the second equation, use $1/\gamma_2{}^\infty = 0.032$ or 0.040 from Figure 3-3. Obtain x_2 from $x_2 = 1 - x_1$.

(3)   Alternatively, Eq. 3-13 can be used:

$$(1-4x_1+3x_1^2) \ln(41.0) + (2x_1-3x_1^2) \ln(31.4) + \ln x_1 - \ln(1-x_1) = 0$$

To solve this equation, use $1/\gamma_1{}^\infty = 0.024$ or Figure 3-3 (0.030) as a first approximation for $x_1{}'$ and use $1-(1/\gamma_2{}^\infty) = 0.968$ or $1-x_1$ (from Figure 3-3) = 0.96 as a first approximation for $x_1{}''$.

(4)   Obtain x_2 from $x_2 = 1-x_1{}''$.

Trial-and-error solutions give a value for the solubility of heptane in acetonitrile (x_1) as 0.030 mole fraction from both Eqs. 3-14 and 3-13. The solubility of acetonitrile in heptane (x_2) is found to be 0.040 mole fraction from Eq. 3-15 and 0.039 mole fraction from Eq. 3-13. The observed values for x_1 and x_2 are about 0.04 and 0.05 mole fraction, respectively [9].

A plot of $\Delta G^M/RT$ vs x would probably look like the curve in Figure 3-4.[5] Note that the minima occur at the points predicted by Eq. 3-13.

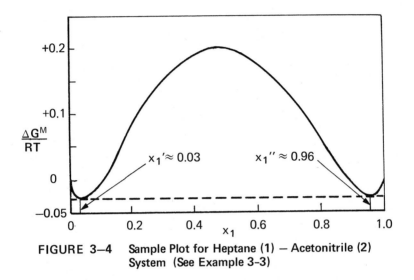

FIGURE 3—4   Sample Plot for Heptane (1) — Acetonitrile (2) System (See Example 3-3)

Example 3-4   Estimate the mutual solubilities for the water(1) butanol(2) system at ~20°C, given activity coefficients (estimated) as a function of composition and $\gamma_1^\infty = 4.61$ and $\gamma_2^\infty = 80.6$ [9].

(1)   In this case, one value of γ^∞ (4.61) would imply a very high solubility or a completely miscible system, but the other value is high enough to indicate that some region of phase instability is likely. Because γ_1^∞ is so low, only Eq. 3-9 (involving a plot of $\Delta G^M/RT$ vs x_1) can be recommended for the calculation of x_1. The relatively high value of γ_2^∞ implies that Eq. 3-13, or even Eq. 3-15, might be appropriate for the calculation of x_2; however, since a plot is necessary for x_1, x_2 will be obtained from the same plot as a matter of convenience.

(2)   Following Step 5 (in "Basic Steps"), a curve like the one shown in Figure 3-5 would be obtained.[5]

5. Activity coefficient values were not estimated directly for each value of x_1 to obtain the curves in Figures 3-4 and 3-5. It was assumed that $\ln\gamma_1 = Ax_2^2/RT$ and $\ln\gamma_2 = A'x_1^2/RT$, where $A = RT\ln\gamma_1^\infty$ and $A' = RT\ln\gamma_2^\infty$. Substituting these equations into Eq. 3-9 gives an expression for $\Delta G^M/RT$ (as a function of x_1) with only γ_2^∞ as the input parameter. The curves shown are a plot of this resulting equation.

(3)   Draw a straight line (shown dashed) tangent to the curve at two points. The values of x_1' and x_1'', the points of tangency, are read as 0.185 and 0.98 mole fraction, respectively. Since $x_2 = 1-x_1''$, we predict $x_2 = 0.02$. The observed values listed in Table 3-1 for x_1 and x_2 are ~0.6 and 0.02, respectively. Note in Figure 3-5 that the point of tangency from which x_1' is found (~0.185) is not the same as the minimum in the $\Delta G^M/RT$ curve (~0.15); the latter value is the one predicted by Eq. 3-13.

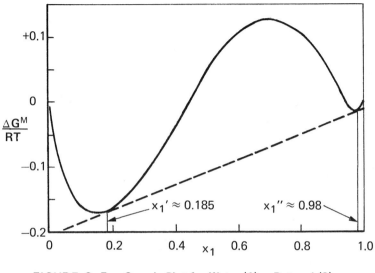

FIGURE 3–5   Sample Plot for Water (1) — Butanol (2) System (See Example 3-4)

3-5 SOLID-LIQUID BINARY SOLUTIONS

Basis for Estimation Method. The solubility of a solid in a liquid solvent is, as in liquid-liquid systems, a function of temperature. In addition, however, the heat of fusion of the solid solute must be considered, since energy is required to overcome the intermolecular forces of the molecules in the solid while it is dissolving. Accordingly, for two chemicals of similar structure (more specifically, with similar melting points), the chemical with the higher heat of fusion will have the lower solubility in any specified solvent.

The nature of the effect of temperature on solubility is shown by the two schematic phase diagrams in Figure 3-6. The point of minimum temperature is called the eutectic point; below this temperature it is not possible to have a single-phase system.

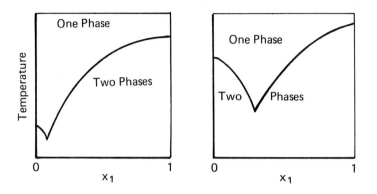

FIGURE 3—6   Phase Stability as a Function of Temperature
in Two Binary Solid (1) — Solvent (2) Systems

The estimation method recommended here was proposed and evaluated by Gmehling *et al.* [11]. If the heat of fusion, ΔH_f, the melting point, $T_m(K)$, and the activity coefficient (as a function of composition), γ_1, are known or can be estimated, then the solubility of the solid solute, x_1, (in mole fraction) may be obtained from

$$\ln \gamma_1 x_1 = \frac{\Delta H_f}{RT} \left(\frac{T}{T_m} -1 \right) \tag{3-18}$$

where T is the system temperature (K) and R is the gas constant. If ΔH_f is in units of calories/mole, R = 1.987 calories/mole·degree. This equation is based upon theoretical considerations which are discussed in Refs. 12, 25, 26, and 34. It neglects certain correction terms proportional to Δc_p (specific heat difference between liquid and solid), because the required c_p data are unlikely to be available; however, the uncertainties associated with neglecting the Δc_p term are expected to be small in comparison with the uncertainties in the estimated activity coefficients which will be required for Eq. 3-18.

The use of Eq. 3-18 requires three input parameters, ΔH_f, T_m and γ_1 (as a function of x_1), in order for x_1 to be calculated at a given temperature (T). Note that γ_1 is also a function of temperature. Estimation methods for activity coefficients (as a function of temperature) are provided in Chapter 11; since measured values of γ_1 are unlikely to be available, it is assumed throughout this section that estimated values are used. Thus, measured values of only ΔH_f and T_m are required. Neither of these two properties can be accurately estimated by methods that have some general applicability. Estimation methods for ΔH_f have been given by Yalkowsky [37] for organic molecules of intermediate size. If ΔH_f is not available for a compound of interest, the term $6.54 (T-T_m)/T$ may be used as a rough approximation of the term $\Delta H_f/RT \cdot (T/T_m - 1)$ in Eq. 3-18 (and also in Eqs. 3-19 and -20 below). The basis for this approximation is explained in §2-4 (see Basis for Estimation Method) of Chapter 2; an average value of 13 cal/mole°C for ΔS_f is assumed for all organic compounds, and $\Delta H_f = T_m \Delta S_f$.

The solution of Eq. 3-18 for x_1 must be by trial and error, since γ_1 is a function of x_1 and, if the UNIFAC method of Chapter 11 is used, this function cannot be expressed in a simple, closed form. The estimation procedure thus involves three steps:

(1) Estimating γ_1 at various values of x_1 (according to the instructions in Chapter 11) so that a plot of γ_1 vs x_1 may be obtained;

(2) Calculating the value of the right-hand side of Eq. 3-18; and

(3) Using calculated and interpolated values of γ_1 (from step 1) to find the value of $\ln\gamma_1 x_1$ that matches the value from step 2. If no value of $\ln\gamma_1 x_1$ matches the value of the right side of the equation, then the two chemicals are completely miscible at the temperature in question. A first estimate of x_1 may be obtained by setting $\gamma_1 = 1$; the value of x_1 obtained with this assumption is called the "ideal solubility." The actual value of x_1 may be above or below this value.

Equation 3-18 may be simplified if one assumes that the change in γ_1 with x_1 may be described by the two-suffix Margules equation. (See footnote 1 for qualifications.) The modified form of Eq. 3-18 is:

$$\ln x_1 + (1-x_1)^2 \ln \gamma_1 \infty = \frac{\Delta H_f}{RT} \left(\frac{T}{T_m} - 1 \right) \tag{3-19}$$

where, as before, $\gamma_1\infty$ is the infinite dilution activity coefficient for the solute. The use of Eq. 3-19 requires only that the activity coefficient of the solute be obtained (estimated) at one point — infinite dilution — which significantly reduces the calculational effort over that associated with Eq. 3-18. The loss of accuracy in using Eq. 3-19 rather than 3-18 should not be significant if x_1 is less than 0.1, although this assumption has not been tested.

The estimation procedure may be even further simplified if x_1 is less than 0.01: the factor $(1-x_1)^2$ in Eq. 3-19 then approaches 1, and we may write

$$\ln x_1 = \frac{\Delta H_f}{RT} \left(\frac{T}{T_m} - 1 \right) - \ln\gamma_1 \infty \tag{3-20}$$

This equation may be solved directly for x_1 (once the input parameters on the right side are obtained) rather than by trial and error as required for Eqs. 3-18 and 3-19.

Although the methods described in this section are limited to binary systems, the basic method may be relatively easily expanded to the prediction of (solid) solute solubilities in solvent mixtures [11]. In all cases, single or mixed solvents, it is necessary to assume that the solvents are insoluble in the solid phase of the solute.

Method Errors. Although the author has not independently tested the procedure for establishing method errors using estimated activity coefficients, several calculations reported by Gmehling *et al.* [11] indicate that estimated and observed solubilities are generally in close agreement. Some of their results using Eq. 3-18 are shown in Tables 3-2 to 3-5. The first three of these tables show average errors (calculated on x_1) in the range of 10-50%, the larger errors being associated with small values of x_1. To predict the eutectic point, one must estimate the mutual solubilities of the two chemicals at several temperatures and plot a curve like those in Figure 3-6; the minimum in the curve corresponds to the eutectic point. As shown in Table 3-5, the results agree reasonably well with the observed values, reflecting the general accuracy of the estimation method.

Table 3-6 compares observed and estimated aqueous solubilities for a number of compounds using the simplified estimation method involving the use of Eq. 3-20. The errors shown, which average 36% based on mole fraction, are not directly comparable with those in the previous tables, since observed rather than estimated values of γ_1^∞ were used. As pointed out in §3-4, however, average errors in estimated values of γ^∞ are typically only 10-20%. Thus, the use of Eq. 3-20 with estimated values of γ^∞ should not involve errors much greater than those shown in Table 3-6.

It is interesting to compare the results given for naphthalene in Table 3-2 with other solubility predictions for naphthalene using the Scatchard-Hildebrand approach, which involves the use of what are now called solubility parameters (see §3-3). Scatchard [27] was able to predict the solubility of naphthalene in five nonpolar or only slightly polar solvents[6] (at 20°C) with an average error of about 7% based on mole

---

6. Benzene, toluene, chlorobenzene, carbon tetrachloride, and hexane.

TABLE 3-2

Observed and Estimated Solubility of Naphthalene in Various Solvents at 40°C

(T_m = 353.4K, ΔH_f = 4494 cal/mol)

Solvent	x_1 (mole fraction) Observed	x_1 (mole fraction) Estimated[a]	Error (%)
Methanol	.044	.048	+9.1
Ethanol	.073	.054	−26.
1-Propanol	.094	.093	−1.1
2-Propanol	.076	.093	+22.
1-Butanol	.116	.111	−4.3
n-Hexane	.222	.259	+17.
Cyclohexanol	.225	.205	−8.9
Acetic acid	.117	.125	+6.8
Acetone	.378	.358	−5.3
Chloroform	.473	.470	−0.6
		Average error =	10.1%

a. Eq. 3-18 used.

Source: Gmehling *et al.* [11]. *(Reprinted with permission from the American Chemical Society.)*

fraction. However, for six polar solvents,[7] the average error in the predicted solubilities was 96%; this relatively high error is not unexpected, since, as mentioned in §3-3, the basic theory is derived from a consideration of nonpolar molecules.

Basic Steps

(1) Check that the solute is a solid at the temperature in question.

(2) Obtain the heat of fusion, ΔH_f (cal/mole), melting point, T_m(°K) and — from Chapter 11 — the infinite dilution activity coefficient, $\gamma_1\infty$, for the solute. Express the system temperature, T, in K. Use 1.987 cal/mol·deg for R. If no value of ΔH_f is available, substitute 6.54 $(T-T_m)/T$ for the term $\Delta H_f/RT·(T/T_m-1)$ in the equation selected in step 3 below.

---

7. Aniline, nitrobenzene, acetone, n-butyl alcohol, methanol, and acetic acid.

TABLE 3-3

**Observed and Estimated Solubility of Anthracene in
Various Solvents at 20°C**

$(T_m = 489.7K, \Delta H_f = 6898 \text{ cal/mol})$

| | x_1 (mole fraction) | | |
Solvent	Observed	Estimated[a]	Error (%)
Acetone	.0031	.0025	−19.
Diethyl ether	.0029	.0045	+55.
Chloroform	.0094	.0182	+94.
Ethanol	.0005	.0004	−20.
Carbon tetrachloride	.0041	.0053	+29.
Phenol[b]	.0099	.0113	+14.
Cyclohexane	.0012	.0031	+160.
Methanol	.0002	.0003	+50.
1-Propanol	.0006	.0006	0
2-Propanol	.0004	.0006	+50.
Aniline	.0035	.0027	−23.
n-Hexane[c]	.0018	.0024	+33.
		Average error =	46%

a. Eq. 3-18 used.
b. Solubility at 60°C.
c. Solubility at 25°C.

Source: Gmehling *et al.* [11]. *(Reprinted with permission from the American Chemical Society.)*

(3) Use $1/\gamma_1^\infty$ as a first approximation for x_1 and proceed as follows:

- If x_1 is less than 0.01, go to Step (4).
- If x_1 is between 0.01 and 0.1, go to Step (5).
- If x_1 is greater than 0.1, go to Step (6).

(4) Calculate x_1 (the solubility of 1 in 2, in units of mole fraction) from Eq. 3-20.

(5) Calculate x_1 (the solubility of 1 in 2, in units of mole fraction) from Eq. 3-19. Trial and error must be used to find a value of x_1 that satisfies this equation.

TABLE 3-4

Observed and Estimated Solubility of Phenanthrene in Various Solvents at 20°C

$$(T_m = 369.5K, \Delta H_f = 4456 \text{ cal/mol})$$

Solvent	x_1 (mole fraction) Observed	x_1 (mole fraction) Estimated[a]	Error (%)
Diethyl ether	.133	.138	+3.8
n-Hexane[b]	.048	.070	+46.
Acetone	.145	.097	−33.
Chloroform	.238	.264	+11.
Ethanol	.0123	.0102	−17.
Carbon tetrachloride	.145	.158	+9.0
Acetic acid	.0192	.0255	+33.
Methanol	.0064	.0091	+42.
Carbon disulfide	.235	.185	−21.
		Average error =	24%

a. Eq. 3-18 used.
b. Solubility at 25°C.

Source: Gmehling *et al.* [11]. *(Reprinted with permission from the American Chemical Society.)*

(6) Obtain (from Chapter 11) values of γ_1 at several values of x_1 over the range of 0 to 1 and plot γ_1 vs x_1 so that interpolation between calculated values is possible. Then, using the information from step (2), obtain a value for the right side of Eq. 3-18. Next, with the calculated values and plot of γ_1 vs x_1, use trial and error to find the value of $\ln \gamma_1 x_1$ that satisfies the equality of Eq. 3-18. (If no value of $\ln \gamma_1 x_1$ satisfies the equality, then the solute and solvent are miscible in all proportions at the system temperature.) The value of x_1, in units of mole fraction, is obtained directly from this procedure.

Example 3-5 Estimate the solubility of naphthalene in 1-butanol at 40°C, given $T_m = 353.4K$ (80.2°C), $\Delta H_f = 4494$ cal/mol [11], and values of γ_1 vs x_1 (estimated from Chapter 11[8]). The value of x_1 is presumed to be greater than 0.1, thus requiring the use of Eq. 3-18.

---

8. These values were assumed to be available for this example; they were not actually calculated.

TABLE 3-5

Observed and Calculated Eutectics in Binary Mixtures

Solute (1)	Solvent(2)	x_1 at Eutectic (mole fraction)			Temp. at Eutectic (°C)		
		Observed	Calculated	% Error	Observed	Calculated	ΔT(°C)
Acetone	Diethyl ether	.240	.320	+33	-126	-123	+3
Acetone	Ethanol	.210	.244	+16	-119	-124	-5
Benzene	1,2-Dichloroethane	.320	.316	-1.3	-55	-55	0
Benzene	Phenol	.625	.654	+4.6	-6	-7	-1
Benzene	Ethanol	.013	.009	-31	-115	-114	+1
Benzene	Chloroform	.260	.268	+3.1	-77	-82	-5
Benzene	1,4-Dioxane	.565	.569	+0.7	-26	-26	0
Benzene	Acetonitrile	.050	.126	+152.	-51	-49	+2
Benzene	Cyclohexane	.265	.266	+0.4	-44	-48	-4
Benzene	Nitrobenzene	.500	.517	+3.4	-26	-27	-1
Phenol	p-Xylene	.425	.409	-3.8	4	0	-4
Phenol	Naphthalene	.838	.837	-0.1	29	30	+1
Ethanol	Ethyl acetate	.850	.897	+5.5	-118	-118	0
Acetic acid	p-Xylene	.616	.595	-3.4	1	5	+4
Acetic acid	Benzene	.409	.453	+10.	-8	-8	0
Acetic acid	Cyclohexane	.074	.074	0	-1	-6	-5
Nitrobenzene	Carbon tetrachloride	.186	.126	-32.	-35	-37	-2

Avg.
Error = 18%

Avg.
ΔT = 2.2°C

Source: Gmehling et al. [11]. (Reprinted with permission from the American Chemical Society.)

TABLE 3-6

Observed and Estimated Aqueous Solubilities for
Solid-Liquid Binaries

Solute[b]	x_1 (mole fraction)[a]		Error (%)
	Observed[c]	Estimated[d]	
2,4,6-Trinitrotoluene	1.19×10^{-5}	1.14×10^{-5}	-4.2
1,4-Dichlorobenzene	1.02×10^{-5}	8.64×10^{-6}	-16.
1,4-Dibromobenzene	1.53×10^{-6}	1.41×10^{-6}	-7.8
1,4-Diiodobenzene	1.01×10^{-7}	5.82×10^{-8}	-43.
Naphthalene	4.38×10^{-6}	1.48×10^{-6}	-66.
Acenaphthene	4.53×10^{-7}	4.69×10^{-7}	+3.5
Biphenyl	8.27×10^{-7}	7.51×10^{-7}	-9.2
Fluorene	2.06×10^{-7}	2.26×10^{-7}	+9.7
Phenanthrene	1.19×10^{-7}	1.25×10^{-7}	+5.0
Anthracene	7.58×10^{-9}	5.56×10^{-9}	-27.
Pyrene	1.29×10^{-8}	3.87×10^{-8}	+200.

Average error =   36%

a. At 25°C.
b. Solvent is water in all cases.
c. From Refs. 33, 35, and 38.
d. Eq. 3-20 used. All input data, including values for $\gamma_1 \infty$, are measured values. Data for $\gamma_1 \infty$ from Refs. 22 and 34, ΔH_f from Refs. 14 and 35, T_m from Ref. 14.

(1)   The right side of Eq. 3-18 is:

$$\frac{4494}{(1.987)(313.2)} \left(\frac{313.2}{353.4} - 1 \right) = -0.8214 \text{ (no units)}$$

(2)   From the data set (and plot) of paired γ_1 and x_1 values, it is found that the value of x_1 that satisfies the equation $\ln \gamma_1 x_1 = -0.8214$ is 0.111 mole fraction. (At this point $\gamma_1 \approx 3.85$.)

Example 3-6   Estimate the solubility of 1,4-diiodobenzene in water at 25°C, given $T_m = 402.6K$ [14], $\Delta H_f = 5340$ cal/mol [14], and $\gamma_1 \infty = 1,660,000$ [34].

(1)   A first approximation of x_1 is ~6 x 10^{-7} ($1/\gamma_1 \infty$); since this is much less than 0.01, we may use Eq. 3-20.

(2) $\ln x_1 = \dfrac{5340}{(1.987)(298.2)} \left(\dfrac{298.2}{402.6} - 1 \right) - \ln (1,660,000)$

$$= -2.337 - 14.322 = -16.659$$

$$\therefore \ x_1 = 5.8 \ x \ 10^{-8} \ \text{mole fraction}$$

The reported value for x_1 at this temperature is $1.01 \ x \ 10^{-7}$ mole fraction [38].

Example 3-7   Estimate the solubility of 4-chloro-1,3-dinitrobenzene in water at 50°C, given $T_m \simeq 328K$ [14] and $\gamma_1^\infty = 27,500$ [34]. Assume that no value for ΔH_f is available.

(1)   A first approximation of x_1 is $3.6 \ x \ 10^{-5}$ $(1/\gamma_1^\infty)$; since this is much less than 0.01, we may use Eq. 3-20 after substituting $6.54 \ (T-T_m) \ T$ for the term containing ΔH_f.

(2) $\ln x_1 = 6.54 \ (323-328)/323 \ - \ln (27,500)$

$$= -0.101 - 10.222 = -10.323$$

$$\therefore x_1 = 3.3 \ x \ 10^{-5} \ \text{mole fraction.}$$

3-6 AVAILABLE DATA

There are, unfortunately, no comprehensive, up-to-date compilations of the solubilities of organic compounds in organic solvents.[9] A few compilations that may be useful are listed below.

Seidell (1941) [28] — organic compounds.

Seidell and Linke (1952) [29] — organic and inorganic compounds.

Stephen and Stephen (1963) [33] — inorganic and organic compounds.

Linke (1958 [20] and 1965 [21]) — inorganic and metal-organic compounds.

Battino and Clever (1966) [6] — gases in liquids.

A few new publications are expected in the near future; these include a new edition (Vol. 3) for inorganic and organic compounds by Stephen [32] and a new *Solubility Data Series* to be published by Pergamon [15].

Sources of data for melting points (T_m) and heats of fusion (ΔH_f) are listed in Appendix A.

---

9. More recent compilations of solubilities in water are listed in Chapter 2.

3-7   SYMBOLS USED

a,b,c,d,e,g = parameters in Eqs. 3-1 to -4

a = activity in Eq. 3-8 (unitless)

A = empirical constant in two-suffix Margules Eq. (Footnotes 1 and 5)

f^L = fugacity of a liquid solute in Eq. 3-7

f^V = fugacity of a solute in the vapor phase in Eq. 3-7

ΔG^M = Gibbs free energy of mixing in Eqs. 3-8, -9,

ΔH_f = heat of fusion in Eq. 3-18, -19, -20 (cal/mol)

k,k' = parameters in Eq. 3-6

K_{ow} = octanol-water partition coefficient in Eqs. 3-3, -4,

K_{sw} = partition coefficient for substance between some solvent(s) and water(w); see §3-1

P = total pressure on system in Eqs. 3-16, -17 (e.g., atm or mm Hg)

P_{vp} = vapor pressure of pure substance in Eqs. 3-16, -17 (e.g., atm or mm Hg)

R = gas constant (1.987 cal/mol·deg)

T = temperature (K)

T_m = melting point in Eqs. 3-18, -19, -20 (K)

$T_{12}{}^*$ = parameter in estimation method for gas solubilities; see §3-1

$v_{12}{}^*$ = parameter in estimation method for gas solubilities; see §3-1

V = molar volume in Eq. 3-7

x = solubility (mole fraction); also used more generally to specify composition in a binary solution. Equal to ratio of moles of solute to total number of moles of solute and solvent

$x_1{}',x_1{}''$ = limiting solubility points in two-phase, liquid-liquid system; see Figure 3-2b

y = mole fraction of component in vapor phase over binary solution in Eqs. 3-16, -17

Greek

δ = solubility parameter in Eq. 3-7

γ = activity coefficient (mole fraction^{-1})

$\gamma\infty$ = infinite dilution activity coefficient (mole fraction^{-1})

ϕ = volume fraction of a component in a binary mixture in Eq. 3-7

Subscripts

1 = solute
2 = solvent
s = solvent
w = water

3-8 REFERENCES

1. Abrams, D.S. and J.M. Prausnitz, "Statistical Thermodynamics of Liquid Mixtures: A New Expression for the Excess Gibbs Energy of Partly or Completely Miscible Systems," *AIChE J.*, **21**, 116-28 (1975).

2. Anderson, T.F. and J.M. Prausnitz, "Application of the UNIQUAC Equation to Calculation of Multicomponent Phase Equilibria: 1. Vapor-Liquid Equilibria," *Ind. Eng. Chem. Process Des. Dev.*, **17**, 552-61 (1978).

3. Anderson, T.F. and J.M. Prausnitz, "Application of the UNIQUAC Equation to Calculation of Multicomponent Phase Equilibria: 2. Liquid-Liquid Equilibria," *Ind. Eng. Chem. Process Des. Dev.*, **17**, 561-67 (1978).

4. American Petroleum Institute, *Manual on Disposal of Refinery Wastes, Volume on Liquid Wastes*, 1st ed. (see esp. Ch. 20 — "Solubility and Toxicity Data"), Washington, D.C. (1969).

5. Barton, A.F., "Solubility Parameters," *Chem. Rev.*, **75**, 731-53 (1975).

6. Battino, R. and H.L. Clever, "The Solubility of Gases in Liquids," *Chem. Rev.*, **66**, 395-463 (1966).

7. Cysewski, G.R. and J.M. Prausnitz, "Estimation of Gas Solubilities in Polar and Nonpolar Solvents," *Ind. Eng. Chem. Fundam.*, **15**, 304-09 (1976).

8. Fredenslund, A., J.G. Gmehling, M.L. Michelsen, P. Rasmussen and J.M. Prausnitz, "Computerized Design of Multicomponent Distillation Columns Using the UNIFAC Group Contribution Method for Calculation of Activity Coefficients," *Ind. Eng. Chem. Process Des. Dev.*, **16**, 450-62 (1977).

9. Fredenslund, A., J.G. Gmehling and P. Rasmussen, *Vapor-Liquid Equilibria Using UNIFAC*, Elsevier Scientific Publishing Co., New York (1977).

10. Fredenslund, A., R.L. Jones and J.M. Prausnitz, "Group-Contribution Estimation of Activity Coefficients in Nonideal Liquid Mixtures," *AIChE J.*, **21**, 1086-99 (1975).

11. Gmehling, J.G., T.F. Anderson and J.M. Prausnitz, "Solid-Liquid Equilibria Using UNIFAC," *Ind. Eng. Chem. Fundam.*, **17**, 269-73 (1978).

12. Hildebrand, J.H., J.M. Prausnitz and R.L. Scott, *Regular and Related Solutions*, Van Nostrand Reinhold Co., New York (1970).

13. Hildebrand, J.H. and R.L. Scott, *The Solubility of Nonelectrolytes*, 3rd ed., Reinhold Publishing Corp., New York (1950).

14. Hodgman, C.D. (ed.), *Handbook of Chemistry and Physics*, 44th ed., The Chemical Rubber Publishing Co., Cleveland (1963).

15. Kertes, A.S. (ed. in chief), *Solubility Data Series* (Approximately 80-100 volumes to be published between 1979 and 1989), Pergamon Press, Inc., Elmsford, N.Y.

16. Keyes, D.B. and J.H. Hildebrand, "A Study of the System Aniline — Hexane," *J. Am. Chem. Soc.*, **39**, 2126-37 (1919).

17. Leo, A. and C. Hansch, "Linear Free-Energy Relationships between Partitioning Solvent Systems," *J. Org. Chem.*, **36**, 1539-44 (1971).

18. Leo, A., C. Hansch and D. Elkins, "Partition Coefficients and Their Uses," *Chem. Rev.*, **71**, 525-621 (1971).

19. Leroi, J.C., J.C. Masson, H. Renon, J.-F. Fabries and H. Sannier, "Accurate Measurement of Activity Coefficients at Infinite Dilution by Inert Gas Stripping and Gas Chromatography," *Ind. Eng. Chem. Process Des. Dev.*, **16**, 139-44 (1977).

20. Linke, W.F., *Solubilities of Inorganic and Metal-Organic Compounds*, Vol. 1, D. Van Nostrand Co., New York (1958).

21. Linke, W.F., *Solubilities of Inorganic and Metal-Organic Compounds*, Vol. 2, American Chemical Society, Washington, D.C. (1965).

22. Mackay, D. and W.Y. Shiu, "Aqueous Solubility of Polynuclear Aromatic Hydrocarbons," *J. Chem. Eng. Data*, **22**, 399-402 (1977).

23. Palmer, D.A., "Predicting Equilibrium Relationships for Maverick Mixtures," *Chem. Eng (NY)*, **12**, 80-85 (June 1975).

24. Pierotti, G.J., C.H. Deal and E.L. Derr, "Activity Coefficients and Molecular Structure," *Ind. Eng. Chem.*, **51**, 95-102 (1959).

25. Prausnitz, J.M., *Molecular Thermodynamics of Fluid-Phase Equilibria*, Prentice-Hall, Inc., Englewood Cliffs, N.J. (1969).

26. Reid, R.C., J.M. Prausnitz and T.K. Sherwood, *The Properties of Gases and Liquids*, 3rd ed., McGraw-Hill Book Co., New York (1977).

27. Scatchard, G., "Equilibria in Non-Electrolyte Solutions in Relation to the Vapor Pressures and Densities of the Components," *Chem. Rev.*, **8**, 321-33 (1931).

28. Seidell, A., *Solubilities of Organic Compounds*, Vol. 2, 3rd ed., D. Van Nostrand Co., New York (1941).

29. Seidell, A. and W.F. Linke, *Solubilities of Inorganic and Organic Compounds*, supplement to 3rd ed., D. Van Nostrand Co., New York (1952).

30. Shinoda, K., *Principles of Solution and Solubility*, Marcel Dekker, Inc., New York (1978).

31. Skjold-Jorgensen, S., B. Kolbe, J. Gmehling and P. Rasmussen, "Vapor-Liquid Equilibria by UNIFAC Group Contribution: Revision and Extension," *Ind. Eng. Chem. Process Des. Dev.*, **18**, 714-22 (1979).

32. Stephen, *Solubility of Inorganic and Organic Compounds*, Vol. 3, Pergamon Press, Inc., Elmsford, N.Y. (date not set).

33. Stephen, H. and T. Stephen (eds.), *Solubilities of Inorganic and Organic Compounds*, Vol. 1, Macmillan, New York (1963).

34. Tsonopoulos, C. and J.M. Prausnitz, "Activity Coefficients of Aromatic Solutes in Dilute Aqueous Solution," *Ind. Eng. Chem. Fundam.*, **10**, 593-600 (1971).

35. Wauchope, R.D. and F.W. Getzen, "Temperature Dependence of Solubilities in Water and Heats of Fusion of Solid Aromatic Hydrocarbons," *J. Chem. Eng. Data*, **17**, 38-41 (1972).

36. Wilson, G.M. and C.H. Deal, "Activity Coefficients and Molecular Structure," *Ind. Eng. Chem. Fundam.*, **1**, 20-23 (1962).

37. Yalkowsky, S.H., "Estimation of Entropies of Fusion of Organic Compounds," *Ind. Eng. Chem. Fundam.*, **18**, 108-11 (1979).

38. Yalkowsky, S.H., R.J. Orr and S.C. Valvani, "Solubility and Partitioning: 3. The Solubility of Halobenzenes in Water," *Ind. Eng. Chem. Fundam.*, **18**, 351-53 (1979).

4

ADSORPTION COEFFICIENT FOR SOILS AND SEDIMENTS

Warren J. Lyman

4-1 INTRODUCTION

The Adsorption Coefficient, K_{oc}. The extent to which an organic chemical partitions itself between the solid and solution phases of a water-saturated or unsaturated soil, or runoff water and sediment, is determined by several physical and chemical properties of both the chemical and the soil (or sediment). In most cases, however, it is possible to express the tendency of a chemical to be adsorbed in terms of a parameter, K_{oc}, which is largely independent of the properties of the soil or sediment. K_{oc} may be thought of as the ratio of the amount of chemical adsorbed per unit weight of organic carbon (oc) in the soil or sediment to the concentration of the chemical in solution at equilibrium:

$$K_{oc} = \frac{\mu\text{g adsorbed/g organic carbon}}{\mu\text{g/mL solution}} \tag{4-1}$$

Values of K_{oc} (in the above units) may range from 1 to 10,000,000.[1]

The existence of this chemical-specific adsorption parameter has an important bearing on assessments of the fate and transport of chemicals in soils and sediments. K_{oc} is commonly used in river models, runoff models, and soil/groundwater models where the transport of a specific chemical is being investigated. The degree of adsorption may not only

---

1. See Table 4-9 for K_{oc} values for selected chemicals.

affect a chemical's mobility but may also be an important parameter in fate processes such as volatilization, photolysis, hydrolysis, and biodegradation. A value of K_{oc} for use in such assessments or models may be easily estimated by the methods described in this chapter.

Since the known methods for estimation are approximate at best, measured values should be used if they are available. The preferred method for measuring adsorption coefficients is to determine an adsorption isotherm with at least one soil or one sediment [12]. Specific soil:solution ratios of the soil and sediment are prepared using six different initial concentrations of the chemical being studied. After the solutions are shaken for about 48 hours to achieve equilibrium, the concentrations in both the solution and solid phases are measured. The amount adsorbed, x/m (μg adsorbed/g of soil or sediment), and the solution concentration, C (μg/mL of solution) are fitted to the Freundlich equation (Eq. 4-2) to determine the adsorption coefficient, K, and the parameter n.[2]

$$x/m = KC^{1/n} \qquad (4\text{-}2)$$

Values of 1/n in this equation are generally found to range from 0.7 to 1.1 although values as low as 0.3 and as high as 1.7 have been reported [16]. Rao and Davidson [36] compiled measured values for 26 chemicals (mostly pesticides) and found the mean value of 1/n to be 0.87 with a coefficient of variation of $\pm$ 15%. No methods are available for estimating n; if a measured value is not available, it is frequently assumed, for convenience, to be equal to 1.

Once a value of K has been determined for a particular soil or sediment, a value of K_{oc} is calculated as follows:

$$K_{oc} = \frac{K}{\%\ oc} \cdot 100 \qquad (4\text{-}3)$$

where % oc is the percentage of organic carbon contained in the soil or sediment. Numerous studies have shown that values of K_{oc} obtained in this manner (for a specific chemical) are relatively constant and reasonably independent of the soil or sediment used [16,26,36]. The spread of values obtained from a number of different soils and sediments generally results in an uncertainty (coefficient of variation) of 10% to 140%.[3]

---

2. The Freundlich equation is frequently written as $x/m = KC^n$. Thus, care should be taken to determine the form of the equation used before any value of n obtained from the literature is used. Note also that K is not the same adsorption coefficient as K_{oc}.

3. See Table 4-9 for examples of uncertainty values.

Some care must be taken with the definitions of K_{oc} implied by Eqs. 4-1 and 4-3. If the adsorption isotherm is nonlinear, the K_{oc} value obtained from Eq. 4-3 would not be the same as the one obtained from a single data point and Eq. 4-1. Both values would differ from the one obtained from an isotherm (several data points) where adsorption was measured in units of $\mu g/g$ of organic carbon.

Some earlier investigations of soil adsorption coefficients reported the results on a soil-organic matter basis (K_{om}) rather than on a soil-organic carbon basis (K_{oc}). Since the organic carbon content of a soil or sediment can be measured more directly, reporting values as K_{oc} is preferred. The ratio of organic matter to organic carbon varies somewhat from soil to soil, but a value of 1.724 is often assumed when conversion is necessary; i.e., $K_{oc} \simeq 1.724\, K_{om}$.

Overview of Estimation Methods. All of the available methods for estimating K_{oc} involve empirical relationships with some other property of the chemical — water solubility (S), octanol-water partition coefficient (K_{ow}), bioconcentration factor for aquatic life (BCF), or parachor (P). The relationships are regression equations obtained from various data sets and are usually expressed in log-log form:

$$\log K_{oc} = a \log (S, K_{ow}, \text{or BCF}) + b \qquad (4\text{-}4)$$

where a and b are constants. Parachor (P) is regressed directly with log K_{oc}.

Although K_{ow} has been used most frequently, about a dozen equations of the above form have been reported. Each was derived from a different data set representing different chemicals (sometimes just one or two chemical classes) and ranges of the parameters involved. Many of the chemicals are insecticides, herbicides, fungicides, or compounds of related structure. Aromatic and polynuclear aromatic hydrocarbons are also well represented.

Many of the available K_{oc} values appear to have been used in one or more of the reported regression equations; thus, there is no independent data set (covering a range of chemical classes and K_{oc} values) with which to test the reported equations for accuracy or general applicability and to determine which are the best. Some guidance can be given, however, on the basis of (1) the chemicals or chemical classes used in the regression equation, (2) the range of K_{oc} values covered by the equation, and (3) the quality of fit (represented by the coefficient, r^2, reported for the equation).

The uncertainty associated with any value of K_{oc} estimated by one of these equations is generally less than one order of magnitude (i.e., less than ± a factor of 10). This assumes that the estimated K_{oc} is to be used for an environmental system that does not differ significantly from one implied by the normal conditions of test (temperature, soil pH, chemical concentration, salinity, etc.).[4] Attempts to extrapolate much beyond these conditions will invite additional errors. Potential errors in estimated values of K_{oc} are discussed in greater detail in §4-2.

Not all procedures for estimating K_{oc} were considered appropriate for inclusion in this handbook. One general method correlates K_{oc} with R_f values obtained from soil thin-layer chromatography tests.[5] Various authors [5,15,17,29] have shown a reasonably good correlation between these properties, but three of them gave no regression equations and the fourth [15] required an additional parameter (the pore fraction of the soil). Furthermore, only a few chemical classes were represented in these studies.

A correlation between K_{oc} and linear free-energy parameters such as Hammet and Taft constants might be expected. Briggs [4] described such a correlation for a series of substituted phenylureas, but his study is the only one reported in the literature, and it gave no regression equations. Therefore, this general approach must be excluded for the present.

Other, more theoretical approaches to the estimation of K_{oc} have been proposed (see, for example, Ref. 32), but none offers any practical solution with readily available data.

Factors Influencing the Values of K and K_{oc}. Numerous publications provide reviews and informative discussions of the various factors influencing the values of K and K_{oc} for organic chemicals.[6] The information given below provides an overview.

---

4. The EPA's test recommendations [12] include the following: use of soils with pH between 4 and 8, organic matter content between 1% and 8%, cation exchange capacity greater than 7 MEQ/100g, and a sand composition less than 70%; use of distilled-deionized water adjusted to pH 7; soil:solution ratio of 1:5; initial chemical concentrations from 0.05 to about 30 μg/mL; temperature at 20°C; equilibrium conditions (48 hours). Most of the reported data derived from the shake-slurry method have used conditions roughly similar to the above.

5. The EPA's proposed protocol for this test is described in Ref. 12.

6. In particular, Refs. 1, 2, 3, 6, 10, 14, 16, 23, 25, 26, 29, 33, 36, and 42.

By basing the adsorption coefficients on soil (or sediment) organic carbon (K_{oc}) rather than on total mass (K), one can eliminate much, but not all, of the variation in sorption coefficients between different soils, sediments, etc. The remaining variation may be due to other characteristics of soils (clay content and surface area, cation exchange capacity, pH, etc.), the nature of the organic matter present, and/or variations in the test methods. Numerous studies of the correlation of K with all of these variables have found that the organic carbon content usually gives the most significant correlation. Furthermore, this correlation often extends over a wide range of organic carbon content — from $\sim$ 0.1% to nearly 20% of the soil in some cases [42].

The emphasis in this chapter on the K_{oc} parameter should not be taken to imply that organic chemicals will not adsorb on minerals free of organic matter. Some adsorption will always take place, and it may be significant under certain conditions such as (1) in clays with very high surface area, (2) where cation exchange (e.g., for dissociated organic bases) occurs, (3) where clay-colloid-induced polymerization occurs, and/or (4) where chemisorption is a factor. Thus, the use of K_{oc} values (measured or estimated) may be completely inappropriate in soils or sediments that are essentially free of organic matter. Methods for estimating adsorption coefficients under these conditions are not currently available.

Other factors that affect the measured value of K_{oc} or are operative under actual environmental conditions are listed below and then discussed individually. Differences in laboratory procedures can also have a significant effect.

Temperature

pH of soil and water

Particle size distribution and surface area of solids

Salinity of water

Concentration of dissolved organic matter in water

Suspended particulate matter in surface water

Non-equilibrium adsorption mechanisms or failure to reach equilibrium conditions

Solids to solution ratio

Loss of chemical (in test) due to volatilization, chemical or
biological degradation, adsorption on flask walls, etc.

Nonlinear isotherm

- *Temperature.* As adsorption is an exothermic process, values of K
(or K_{oc}) usually decrease with increasing temperature. Heats of
adsorption associated with physical adsorption are typically a few hundred calories per degree per mole [16]. With a heat of adsorption of -500
cal/degree · mole), one would expect about a 10% decrease in K (or K_{oc})
with a temperature rise from 20°C to 30°C; an 18% increase would be
expected for a temperature drop from 20°C to 5°C. Care should be taken
in predicting such changes; with some chemicals, temperature has no
effect, or even the opposite effect, on adsorption [3].

- *pH.* Only chemicals that tend to ionize are much affected by pH;
neutral chemicals are little influenced [2,16,33]. Weak acids and weak
bases show the greatest sensitivity to pH changes in the range normally
encountered in soils and surface waters (pH 5-9). The general rule is that
the neutral species of an acid adsorbs much more strongly than the anion.
For organic acids, adsorption begins to be appreciable when the pH of the
bulk solution is approximately 1.0 to 1.5 units above the pK_a value of the
acid [33]. The cations resulting from the dissociation of an organic base
may be strongly sorbed on soils carrying net negative charges.

- *Particle size distribution and surface area.* The fine silt and clay
fractions of soils and sediments have the greatest tendency to adsorb
chemicals. Variation in adsorption between different size fractions is
mostly a reflection of their organic carbon content, but surface area and
other factors may also be involved [24,25,36,41].

- *Salinity.* An increase in salinity can significantly lower the
adsorption coefficient of basic materials that are in the cation form. This
may result from a displacement of the cations from the soil matrix
(cation exchange) or some other action related to the lower activity of the
chemical as the ionic strength of the solution increases [16]. The
adsorption of some acid herbicides increases with greater salinity at pH
values above the pK_a of the acid [33]. Clearly, pH significantly influences the direction and magnitude of salinity effects for organic acids
and bases. Neutral molecules are generally less affected by salinity but
may show increased adsorption with increasing salt concentration. For
example, the adsorption of pyrene on a silt fraction of a stream sediment
was found to increase 15% over the no-salt solution when 20 mg/mL of
NaCl was added [25]. This salt concentration is close to that of seawater.

- *Dissolved organic matter.* The presence of dissolved organic matter commonly reduces the adsorption of a chemical. This may be due to the increased solubility of the chemical in such a solution or to competitive adsorption [19,34,44].

- *Suspended particulate matter.* In surface waters, suspended particles adsorb organic chemicals from the surrounding solution. This can increase the apparent "solution concentration," depending on the degree of filtration used to define the solution phase. In fast-flowing streams and rivers, the suspended particulate matter may not differ much in composition or nature from the bottom sediments, so K_{oc} values from soil or sediment measurements may be used to estimate the amount of chemical adsorbed on this matter. In ponds, lakes, and oceans, however, a large fraction of the suspended particulate matter may be made up of microorganisms with significantly different adsorption characteristics from those of soils and sediments. Some information on the subject of adsorption of organic chemicals on microorganisms in natural waters is given in Refs. 21, 22, and 31. Swisher [43] has compiled numerous coefficients for the adsorption of surfactants onto the solids in sewage sludge, which contains a high fraction of microorganisms.

- *Non-equilibrium adsorption.* Non-equilibrium adsorption commonly occurs when a chemical moves through an environmental compartment so rapidly that equilibrium cannot be achieved. Less commonly, it can be the result of hysteresis, which causes the adsorption and desorption processes for a chemical to follow different isotherms.[7] This usually indicates some degree of irreversible adsorption. Studies reported in the literature are sometimes conflicting; Rao and Davidson [36] have critically reviewed the available information and concluded that, while hysteresis in adsorption isotherms is often an artifact of the laboratory test methods used, it can be real and significant for some compounds.

- *Solids to solution ratio.* Changes in the water content of a soil or sediment will change the fraction of a chemical that is adsorbed: as the water content is lowered, the fraction adsorbed will increase, as does that in solution. Whether or not a change in K (or K_{oc}) is also to be expected with a change in water content is not clear, as conflicting results have been reported [16].

---

7. Values of K and K_{oc} are almost always measured after an adsorption process (i.e., starting with an adsorbent free of the chemical) rather than a desorption process (i.e., starting with excess chemical on the adsorbent).

• *Loss of chemical during test*. Measurements of adsorption coefficient can obviously be distorted by losses of a chemical due to volatilization, chemical or biological degradation, adsorption on walls, etc. Some chemicals may undergo clay-colloid-induced hydrolysis and polymerization. Since similar processes may alter the amount of adsorption measured in the environment, this possibility should be considered when laboratory test data are reviewed and when such data (measured or estimated) are used in environmental assessments.

• *Nonlinear isotherm*. If the adsorption isotherm is nonlinear, the reported value of K_{oc} will depend on the range of chemical concentrations used in the tests.

4-2   AVAILABLE ESTIMATION METHODS

Regression Equations. All available methods for estimating K_{oc} involve correlations with one other property of the chemical: water solubility (S), octanol-water partition coefficient (K_{ow}), bioconcentration factors for aquatic life (BCF), or parachor (P). Twelve regression equations (4-5 through 4-16) are given in Table 4-1 along with some basic information on the data set used to derive each equation. Table 4-2 provides more detailed information on the ranges of the two parameters associated with each data set and indicates the subsequent table or figure in which the chemicals and data are shown.

(Continued on p. 4-19)

TABLE 4-1

Regression Equations for the Estimation of K_{oc}

Eq. No.	Equation[a]	No.[b]	r^2 [c]	Chemical Classes Represented	Ref.
4-5	$\log K_{oc} = -0.55 \log S + 3.64$ (S in mg/L)	106	0.71	Wide variety, mostly pesticides	[26]
4-6	$\log K_{oc} = -0.54 \log S + 0.44$ (S in mole fraction)	10	0.94	Mostly aromatic or polynuclear aromatics; two chlorinated	[25]
4-7[d]	$\log K_{oc} = -0.557 \log S + 4.277$ (S in μ moles/L)	15	0.99	Chlorinated hydrocarbons	[11]
4-8	$\log K_{oc} = 0.544 \log K_{ow} + 1.377$	45	0.74	Wide variety, mostly pesticides	[26]
4-9	$\log K_{oc} = 0.937 \log K_{ow} - 0.006$	19	0.95	Aromatics, polynuclear aromatics, triazines and dinitro-aniline herbicides	[9]
4-10	$\log K_{oc} = 1.00 \log K_{ow} - 0.21$	10	1.00	Mostly aromatic or polynuclear aromatics; two chlorinated	[25]
4-11	$\log K_{oc} = 0.94 \log K_{ow} + 0.02$	9	e	s-Triazines and dinitroaniline herbicides	[7]
4-12	$\log K_{oc} = 1.029 \log K_{ow} - 0.18$	13	0.91	Variety of insecticides, herbicides and fungicides	[36]
4-13[d]	$\log K_{oc} = 0.524 \log K_{ow} + 0.855$	30	0.84	Substituted phenylureas and alkyl-N-phenylcarbamates	[5]
4-14[d,f]	$\log K_{oc} = 0.0067 (P - 45N) + 0.237$	29	0.69	Aromatic compounds: ureas, 1,3,5-triazines, carbamates, and uracils	[18]
4-15	$\log K_{oc} = 0.681 \log BCF(f) + 1.963$	13	0.76	Wide variety, mostly pesticides	[26]
4-16	$\log K_{oc} = 0.681 \log BCF(t) + 1.886$	22	0.83	Wide variety, mostly pesticides	[26]

a. K_{oc} = soil (or sediment) adsorption coefficient; S = water solubility; K_{ow} = octanol-water partition coefficient; BCF(f) = bioconcentration factor from flowing-water tests; BCF(t) = bioconcentration factor from model ecosystems; P = parachor; N = number of sites in molecule which can participate in the formation of a hydrogen bond.

b. No. = number of chemicals used to obtain regression equation.

c. r^2 = correlation coefficient for regression equation.

d. Equation originally given in terms of K_{om}. The relationship $K_{om} = K_{oc}/1.724$ was used to rewrite the equation in terms of K_{oc}.

e. Not available.

f. Specific chemicals used to obtain regression equation not specified.

TABLE 4-2

Information on Equations Given for Estimation of K_{oc}

Eq. No.	Parameter Required[a]	Range of Parameter[b]	Range of K_{oc} Values[b]	Chemicals Used for Regression Listed in Table	Data and/or Regression Eq. Plotted in Figure
4-5	S (mg/L)	0.0005 – 1,000,000	1 – 1,000,000	4-3	4-1
4-6	S (mole fraction)	(0.03 – 410,000) × 10^{-9}	80 – 1,000,000	4-4	4-1
4-7	S (μ moles/L)	0.002 – 100,000	30 – 380,000	4-5	4-1
4-8	K_{ow}	0.001 – 4,000,000	10 – 1,000,000	4-3	4-2
4-9	K_{ow}	100 – 4,000,000	100 – 1,000,000	4-4, 4-6	4-2
4-10	K_{ow}	100 – 4,000,000	100 – 1,000,000	4-4	4-2[e]
4-11	K_{ow}	150 – 200,000	180 – 31,000	4-6	4-2[f]
4-12	K_{ow}	0.3 – 400,000	2 – 250,000	4-7	4-2
4-13	K_{ow}[c]	3 – 2,200	10 – 400	4-8	4-2 (eq. only)
4-14	P[c,d]	g	g	g	g
4-15	BCF(f)	1 – 100,000	30 – 1,000,000	4-3	4-3
4-16	BCF(t)	0.02 – 100,000	10 – 250,000	4-3	4-3

a. S = water solubility; K_{ow} = octanol-water partition coefficient; P = parachor; BCF(f) = bioconcentration factor from flowing-water tests; BCF(t) = bioconcentration factor from model ecosystems.

b. Approximate range of data used for regression equation.

c. Equation was originally given in terms of K_{om} and rewritten in terms of K_{oc} using the relationship $K_{om} = K_{oc}/1.724$.

d. One additional parameter, obtained by visual inspection of chemical's structural formula, is required: N = number of sites in the molecule which can participate in the formation of hydrogen bonds.

e. Data included in the set used for Eq. 4-9 and are generally those with the higher K_{oc} and K_{ow} values. Regression equation not shown.

f. Data included in the set used for Eq. 4-9 and are generally those with the lower K_{oc} and K_{ow} values. Regression equation not shown.

g. The specific chemicals and data used to obtain the regression equation were not given.

TABLE 4-3

Compounds Used by Kenaga and Goring [26] for Regression Equations[a]

Compound	Used in Correlation With:[b]			
	S	K_{ow}	BCF(f)	BCF(t)
Halogenated Hydrocarbon Insecticides				
Aldrin	X			X
DDT	X	X	X	X
Lindane	X		X	X
Methoxychlor	X	X	X	X
Substituted Benzenes and Halobenzenes				
Hexachlorobenzene	X	X	X	X
Chloroneb	X			
Chlorthiamid	X			
Dichlobenil	X			X
Methazole	X			
Norflurazon	X			
Oxadiazon	X			
Halogenated Biphenyls and Diphenyl Oxides				
2,2',4,5,5'-Pentachlorobiphenyl (Aroclor 1254)	X	X	X	X
2,2',4,4',5,5'-Hexachlorobiphenyl	X	X	X	
Aromatic Hydrocarbons				
Anthracene	X	X		
Benzene	X	X		
9-Methylanthracene	X	X		
2-Methylnaphthalene	X	X		
Naphthalene	X	X		
Phenanthrene	X	X		
Pyrene	X	X		
Tetracene	X	X		
Fumigants				
cis-1,3-Dichloropropene	X			
trans-1,3-Dichloropropene	X			
Dibromochloropropane	X			
Ethylene dibromide	X			
Methyl isothiocyanate	X			
Phosphorus-Containing Insecticides				
Crotoxyphos	X			
Disulfoton	X			
Phorate	X			
Diamidaphos	X		X	
Carbophenothion	X			
Chlorpyrifos	X	X	X	X
Chlorpyrifos-methyl	X	X		X
Ethion	X			
Leptophos	X	X	X	X
Methyl parathion	X	X		X
Parathion	X	X		X

a. Eqs. 4-5, -8, -15, -16. b. For symbols, see footnote a of Table 4-2.

(continued)

TABLE 4-3 (Continued)

Compound	S	Used in Correlation With: K_{ow}	BCF(f)	BCF(t)
Carbamates, Thiocarbamates, and Carbamoyl Oximes				
Carbaryl	X	X		
Chlorpropham	X			
Propham	X			
Cycloate	X			
Diallate	X			
EPTC	X			
Pebulate	X			
Triallate	X			
Methomyl	X	X		
Carboxylic Acids and Esters				
Chloramben	X			
Chloramben, methyl ester	X			
6-Chloropicolinic acid	X	X		X
2,4-D acid	X	X		
Dicamba	X			
3,6-Dichloropicolinic acid	X			
Picloram	X	X		X
Silvex	X			
2,4,5-T	X	X		X
Triclopyr	X	X		X
Dinitroanilines				
Benefin	X			
Butralin	X			
Dinitramine	X			
Fluchloralin	X			
Isopropalin	X			
Nitralin	X			
Profluralin	X			
Trifluralin	X	X	X	X
Ureas and Uracils				
Chlorbromuron	X			
Chloroxuron	X			
Diflubenzuron	X			
Diuron	X	X		
Fenuron	X	X		
Fluometuron	X	X		
Linuron	X	X		
Metobromuron	X			
Monolinuron	X	X		
Monuron	X	X		
Neburon	X			
Tebuthiuron	X			
Urea	X	X		
Bromacil	X			
Isocil	X			
Terbacil	X			

(continued)

Compound	Used in Correlation With:			
	S	K_{ow}	BCF(f)	BCF(t)
Symmetrical Triazines				
Ametryn	X			
Atrazine	X	X		X
Cyanazine	X	X		
Dipropetryn	X			
sec-Bumeton	X			
Ipazine	X	X		
Prometon	X			
Prometryn	X			
Propazine	X	X		
Simazine	X	X	X	
Terbutryn	X			
Trietazine	X	X		
Miscellaneous Nitrogen Heterocyclics				
2-Methoxy-3,5,6-trichloropyridine	X	X		
Nitrapyrin	X	X		
Pyroxychlor	X			X
3,5,6-Trichloro-2-pyridinol	X	X	X	X
Metribuzin	X			
Pyrazon	X			
Thiabendazole	X			
Miscellaneous				
Dinoseb	X	X		
Pentachlorophenol	X		X	X
Phenol	X			
Aroclor	X	X		
Napropamide	X			
Pronamide	X			
Propachlor	X	X		X
Asulam	X			

Source: Kenaga and Goring [26]. *(Reproduced with permission from the American Society for Testing and Materials.)*

TABLE 4-4

Compounds Used by Karickhoff *et al.* [25] for Regression Equations[a]

Compounds	
Anthracene	2-Methylnaphthalene
Benzene	Naphthalene
Hexachlorobiphenyl	Phenanthrene
Methoxychlor	Pyrene
9-Methylanthracene	Tetracene

a. Eqs. 4-6, 4-10, and (in part) 4-9.

TABLE 4-5

Compounds Used by Chiou *et al.* [11] for Regression Equation[a]

Compounds	
β-BHC	Parathion
1,2-Bromo-3-chloropropane	2,4'-PCB
DDT	2,5,2',5'-PCB
1,2-Dibromomethane	2,4,5,2',4',5'-PCB
1,2-Dichlorobenzene	1,1,2,2-Tetrachloroethane
1,2-Dichloroethane	Tetrachloroethene
1,2-Dichloropropane	1,1,1-Trichloroethane
Lindane	

a. Eq. 4-7.

TABLE 4-6

Compounds Used by Brown *et al.* [7,9] for Regression Equations[a]

Compounds		
Atrazine	Propazine	Trifluralin
Cyanazine	Simazine	Two photodegradation products of trifluralin
Ipazine	Trietazine	

a. Eqs. 4-9 (in part) and 4-11.

TABLE 4-7

Compounds Used by Rao and Davidson [36] for Regression Equation[a]

Compounds		
Atrazine	Dicamba	Malathion
Bromacil	Dichlobenil	Methylparathion
Carbofuran	Diuron	Simazine
2,4-D	Lindane	Terbacil
DDT		

a. Eq. 4-12.

TABLE 4-8

Compounds Used by Briggs [5] for Regression Equation[a,b]

Substituted Phenylureas

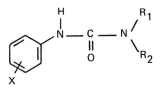

X	R_1	R_2	X	R_1	R_2
4-Cl	CH_3	CH_3	3-Cl	H	H
3,4-Cl	CH_3	CH_3	3,4-Cl	H	H
3-CF_3	CH_3	CH_3	3-Cl, 4-OCH_3	H	H
3-Cl, 4-OCH_3	CH_3	CH_3	3-F	H	H
4-Cl	CH_3	OCH_3	4-F	H	H
3,4-Cl	CH_3	OCH_3	3-CF_3	H	H
4-Br	CH_3	OCH_3	3-Br	H	H
3-Cl, 4-Br	CH_3	OCH_3	4-Br	H	H
3-Cl	CH_3	H	3-OH	H	H
3,4-Cl	CH_3	H	4-SO_3^-	H	H
3-Cl, 4-OCH_3	CH_3	H	H	H	H

Alkyl-N-Phenylcarbamates

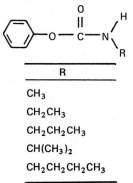

R
CH_3
CH_2CH_3
$CH_2CH_2CH_3$
$CH(CH_3)_2$
$CH_2CH_2CH_2CH_3$

a. Eq. 4-13.

b. Although Briggs [5] states that 30 compounds were used to derive the regression equation, only 28 were listed in the reference cited [4] for the original data.

Source: Briggs [5]. *(Reprinted with permission from Macmillan Journals Ltd.)*

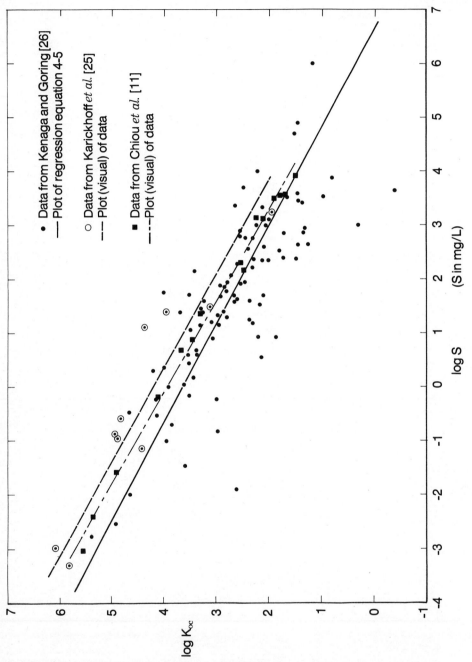

FIGURE 4-1 Correlation Between Adsorption Coefficient and Water Solubility

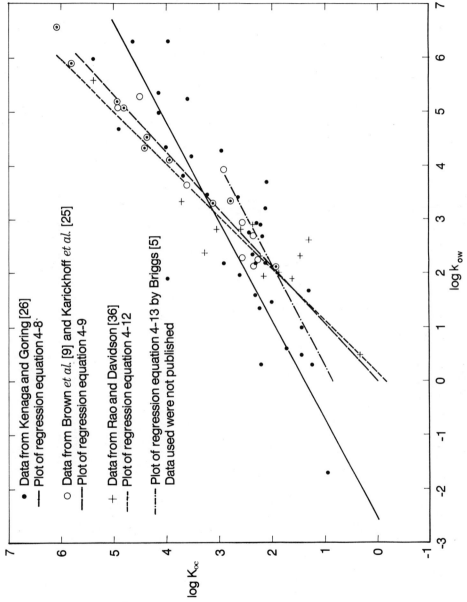

FIGURE 4-2 Correlation Between Adsorption Coefficient and Octanol-Water Partition Coefficient

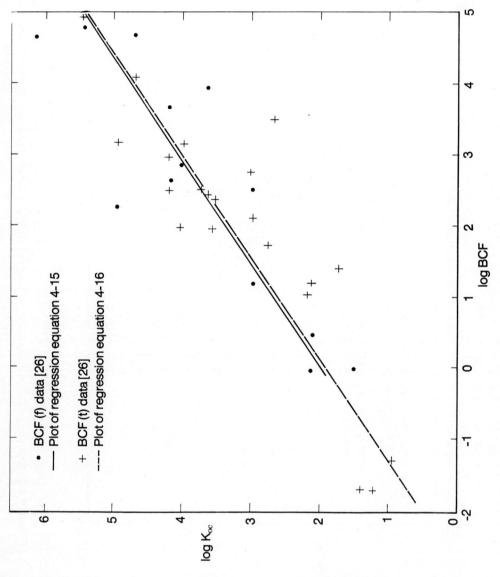

FIGURE 4-3 Correlation Between Adsorption Coefficient and Bioconcentration Factors

Except for parachor, all of the input parameters for these equations are commonly encountered with organic chemicals of environmental concern. Parachor is a constitutive and additive function of molecular structure and, for liquids, is defined as

$$P = \frac{M\sigma^{1/4}}{\rho - \rho^{\circ}} \tag{4-17}$$

where M is the molecular weight, σ the surface tension, ρ the liquid density and ρ° the vapor density [28]. Since ρ° is much smaller than ρ, Eq. 4-17 may be rewritten as

$$P = \frac{M\sigma^{1/4}}{\rho} \quad \text{or} \quad P = V\sigma^{1/4} \tag{4-18}$$

where V is the molar volume. Values of P may be obtained: (1) from Eq. 4-18 (if data are available),[8] (2) from tables of measured values in the literature (such as Ref. 35), or (3) estimated via the addition of fragment constants[8] [35,38]. Parachor values, in units of $g^{1/4} \cdot cm^3/(sec^{1/2} \cdot mol)$, are generally in the range of 100-600.

The parachor regression equation (4-14 in Table 4-1) given by Hance [18] is an extension of the work of Lambert [28], who also found a correlation between the adsorption coefficient and P for two chemical classes. The extension involved reducing the parachor value by 45N, where N is the number of proton- or electron-donating sites on the molecule that could conceivably participate in hydrogen bond formation. For the classes of compounds studied by Hance, such sites were taken to include the following groups: primary, secondary, and tertiary amines (RNH_2, R_2NH, R_3N), carbonyl ($R_2C=O$), heterocyclic nitrogen (nitrogen in a ring), and ether oxygen (R-O-R'). Two examples of the determination of N are shown below. The sites counted are marked with an asterisk.

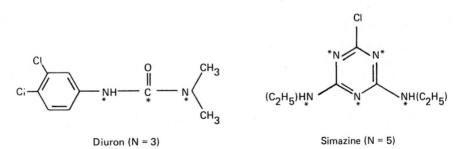

Diuron (N = 3)   Simazine (N = 5)

---

8. Methods for estimating liquid density and surface tension are given in Chapters 19 and 20, respectively. Tables 12-3, 12-4 and 20-2 contain fragment constants for P.

Other structural groups (not in the chemicals studied by Hance) that should probably be counted in N are: hydroxyl (ROH), acid (RCOOH), ester (RCOOR), peroxide (R-O-O-R), and any other group to which hydrogen might bond.

Selection of the Most Appropriate Equation(s). One or more equations should be selected on the basis of (1) the data available on the chemical, (2) the chemical classes covered by each regression equation, and (3) the range of K_{oc} (and input parameter) values covered by each regression equation.

- *Available data.* (See Table 4-2 for input data requirements.) Highest priority should be given to the most accurate data from actual measurements. If, for example, only the solubility is known, then either Eq. 4-5, 4-6 or 4-7 must be chosen. If data are available for all input parameters, and if the other two criteria are not decisive, the following priorities can be applied: $K_{ow} \gtrsim S > P > BCF(t) \gtrsim BCF(f)$. If no measured data are available, estimated values may be used with the following suggested priority: $K_{ow} > S > P$.[9]

- *Chemical classes.* If the chemical for which K_{oc} is to be estimated is in a class covered by one or more of the regression equations (see Tables 4-1 and 4-2), these equations should be given priority. If there is no clear match of chemical classes, it is suggested that Eq. 4-5 or 4-8 be selected, as these were derived from the widest variety of chemicals.

- *Ranges.* In general, a regression equation should not be used for estimation if the value of either the input parameter (K_{ow}, S, P, BCF) or K_{oc} is outside the range covered by the data set from which the equation was obtained. (Ranges shown in Table 4-2.) Otherwise, the estimate may be subject to significant additional uncertainty.

Basic Steps for Estimating K_{oc}, K, and x/m

(1) Select from Table 4-1 the most appropriate equation(s) for estimating K_{oc}, using the criteria discussed in the subsection above.

(2) Calculate the value of K_{oc} using the selected equations. (Examples are given in the following subsection.) If all equations used appear equally applicable and the input

---

9. Estimation methods for K_{ow} are given in Chapter 1, and for S in Chapter 2. Estimation methods for P are given in Refs. 35 and 38. Simple tables of fragment constants for the estimation of P are given in Tables 12-3, 12-4 and 20-2.

data used for each equation have approximately the same uncertainty, an average value may be reported. In this case it is probably better to use the geometric mean than a simple average. To obtain the geometric mean, take the log of each estimate, average the logs, and find the antilog.

(3) If desired, an adsorption coefficient for a particular soil (K) with a known organic carbon content (% oc) may be calculated (cf. Eq. 4-3) from the expression $K = K_{oc}(\% \text{ oc})/100$.

(4) The amount of the chemical that would be adsorbed (x/m, in μg adsorbed per g of adsorbent) in a solution with a known equilibrium concentration (C, in μg/mL) is calculated as follows (cf. Eq. 4-2):

$$x/m = KC^{1/n}$$

The exponent $1/n$ may be taken as: (a) 1.0 if a linear isotherm is likely, (b) 0.87 (the average for 26 pesticides; see §4-1) if a nonlinear isotherm is likely, or (c) any other value that appears reasonable, considering values measured for similar compounds in equivalent concentration ranges.[10]

Example 4-1   Estimate K_{oc}, K, and x/m for hexachlorobenzene, given a water solubility (S) of 0.035 mg/L, a soil organic carbon content of 2%, and a solution concentration (C) of 0.01 mg/L. The molecular weight of hexachlorobenzene is 284.8.

(1a)   From Eq. 4-5, with S in mg/L

$$\log K_{oc} = -0.55 \log (0.035) + 3.64 = 4.44$$

$$K_{oc} = 27{,}600$$

(1b)   From Eq. 4-6, with S in mole fraction

$$\log K_{oc} = -0.54 \log \left(\frac{0.035 \times 10^{-3}}{284.8 \times 55.51} \right) + 0.44 = 5.11$$

$$K_{oc} = 130{,}000$$

(1c)   From Eq. 4-7, with S in μ moles/L

$$\log K_{oc} = -0.557 \log \left(\frac{0.035 \times 10^{3}}{284.8} \right) + 4.277 = 4.78$$

---

10. The reader should be aware that the use of Eq. 4-2 outside the normal concentration range used for measurements of K, as well as the use of an assumed value for $1/n$, will result in an additional uncertainty in the estimated value of x/m. This is discussed in a following subsection.

$$K_{oc} = 60,800$$

Note: The geometric mean of these three estimates, with the assumption of equal method error for each equation, is 60,200. However, based upon considerations of chemical class, Eq. 4-7 is probably the most applicable; Eqs. 4-6 and 4-5 are second and third choice respectively. A (subjective) weighted average estimate would thus be ~80,000 for K_{oc}. Recent measurements of K_{oc} for this chemical gave values in the range of 80,000-100,000, depending on whether the measurement was for a whole sediment or only the fines fraction [8]; thus, the value listed in Table 4-9 (~3,900) may be considered suspect.

(2) Using Eq. 4-3 and 80,000 for K_{oc}

$$K = 80,000 \ (2)/100 = 1,600$$

(3) Using Eq. 4-2 with n=1 and C in $\mu g/mL$

$$x/m = 1,600 \ (0.01)^{1/1} = 16 \ \mu g/g$$

Example 4-2   Estimate K_{oc}, K, and x/m for trichloroethylene, given an octanol-water partition coefficient (K_{ow}) of 195, a soil organic carbon content of 5%, and a solution concentration of 10 mg/L.

(1a) From Eq. 4-8

$$\log K_{oc} = 0.544 \log (195) + 1.377 = 2.62$$

$$K_{oc} = 420$$

(1b) From Eq. 4-9

$$\log K_{oc} = 0.937 \log (195) - 0.006 = 2.14$$

$$K_{oc} = 140$$

(1c) Similarly, Eqs. 4-10 to 13 yield K_{oc} values of 120, 150, 150, and 110, respectively. The agreement is relatively good, considering that chlorinated hydrocarbons are poorly represented, if at all, in some of the equations.

(2) Using Eq. 4-3 and a geometric mean of 160 for K_{oc}

$$K = 160 \ (5)/100 = 8.0$$

(3) Using Eq. 4-2 with n=1 and C in $\mu g/mL$

$$x/m = 8 \ (10)^{1/1} = 80 \ \mu g/g$$

Example 4-3   Estimate K_{oc} for methylphenylaminoacetone, given a parachor (P) of 382.9 [35].

(1)   The number (N) of potential hydrogen bonding sites is 2
(1 amino and 1 ketone)

(2)   Using Eq. 4-14

$$\log K_{oc} = 0.0067\ [382.9 - 45\ (2)] + 0.237 = 2.20$$

$$K_{oc} = 160$$

Example 4-4   Estimate K_{oc} for heptachlor, given BCF (f) = 17,400 and BCF (t) = 2,150 [26].

(1)   Using Eq. 4-15

$$\log K_{oc} = 0.681 \log (17,400) + 1.963 = 4.85$$

$$K_{oc} = 71,000$$

(2)   Using Eq. 16

$$\log K_{oc} = 0.681 \log (2,150) + 1.886 = 4.16$$

$$K_{oc} = 14,000$$

(3)   The geometric mean is 31,500.

Uncertainty in Estimated Values. The uncertainty in values of K_{oc}, K, and x/m estimated from the equations given is related to a number of factors, including: (1) method errors, (2) uncertainty in the input data, (3) variability in environmental factors (e.g., temperature, pH, salinity), and (4) errors resulting from extrapolation based on assumptions of a linear isotherm and reversible adsorption. Method errors in the estimation of K_{oc} are typically less than one order of magnitude; a worst-case combination of the other factors can combine to make the real error in the estimated value of K_{oc}, K, or x/m over two orders of magnitude. However, an uncertainty of this extent should be relatively rare; even when it does occur, it is not completely unreasonable, as the values of K_{oc} can range over seven orders of magnitude. These error factors are discussed below in more detail.

• *Method error.* A good indication of method errors (i.e., those relating to the quality of fit of the various data sets to the associated regression equations) may be obtained from Figures 4-1, -2, and -3. Table 4-9 lists the method error (expressed as the ratio of the estimated to the measured value of K_{oc}) for a number of chemicals, using, where input data were available, four of the regression equations listed in Table 4-1. The chemicals in Table 4-9 are listed in order of increasing

TABLE 4-9

Comparison of Measured and Estimated Values of K_{oc} for Selected Chemicals

Chemical	Measured Values			Ratio of Estimated/Measured Value of K_{oc}[c]		
	K_{oc}^a	% CV[b]	Reference	From S (Eq. 4-5)	From K_{ow} (Eq. 4-8)	From BCF (f) or (t) (Eq. 4-15 or -16)
Dicamba	2.2	74	[36]	19	18	
2,4-D	20	72	[36]	5.2	4.6	
Picloram	26	140	[36]	6.0	1.3	0.21 (t)
Carbofuran	29	30	[36]	6.1		
Acetophenone	43		[27]	0.90	4.1	
Ethylene dibromide	44		[26]	1.1		
Benzene	83		[25]	0.86	4.1	
Chlorthiamid	98	28	[36]	1.0		
Simazine	140	13	[36]	16	2.6	0.66 (f)
Atrazine	160	49	[36]	4.0	4.3	2.5 (t)
Fluometron	175		[26]	2.1	0.73	
Carbaryl	230		[26]	2.5	2.0	
p-Cresol	<500		[40]	>0.14	>0.09	
Quinoline	570		[40]	0.06	0.53	
Linuron	860	72	[36]	0.47	0.43	
Nitralin	960		[26]	6.0		
Lindane	1,080	13	[36]	11	0.74	4.4 (f), 5.3 (t)
Disulfoton	1,600	140	[36]	0.46		

(continued)

Chemical	Measured Values			Ratio of Estimated/Measured Values of K_{oc} [c]		
	K_{oc} [a]	% CV [b]	Reference	From S (Eq. 4-5)	From K_{ow} (Eq. 4-8)	From BCF (f) or (t) (Eq. 4-15 or -16)
Malathion	1,800	66	[36]	0.16	0.50	
Diallate	1,900		[26]	0.54		
Neburon	3,100	24	[36]	0.59		
Hexachlorobenzene	3,900		[26]	7.1	4.2	11 (f), 0.94 (t)
Parathion	10,600	75	[36]	0.07	0.26	0.38 (t)
Dibenzothiophene	11,200		[20]	0.32	0.51	
Trifluralin	13,700		[26]	0.42	1.4	2.1 (f), 0.59 (t)
2,2',4,5,5'-Pentachlorobiphenyl	42,500		[26]	1.3	1.5	3.2 (f), 1.1 (t)
Methoxychlor	80,000		[25]	1.3	0.10	0.04 (f), 0.14 (t)
DDT	243,000	65	[36]	0.60	0.18	0.69 (f), 0.72 (t)
7,12-Dimethylbenz[a] anthracene	476,000		[30]	0.07	0.09	
Benz[a] anthracene	1,380,000		[40]	0.04	0.03	
Mirex	24,000,000		[40]	0.04		0.0001 (t)

a. **Values** given are often the mean of measurements on a variety of soils and/or sediments. All values may be assumed to be uncertain by at least 10%. Units of K_{oc} are as shown in Eq. 4-1. In several cases the values of K_{oc} have been rounded off from those given in the original references.

b. % CV = % Coefficient of variation = (standard deviation/mean) x 100. Unlisted values are not available.

c. Input data (values for S, K_{ow}, BCF) obtained from Refs. 20, 25, 26, 30, 36, and 40. No estimates were made if measured values of these input parameters were not available.

K_{oc}; a listing by chemical class would not show any clear differences in uncertainty with chemical class.

It should be noted that data for many of the chemicals listed in Table 4-9 were used in one or more of the reported regression equations. This is because the number of independent values of K_{oc} available was too small and too limited in chemical class coverage to use as a test set. In addition, a regression equation was sometimes used outside the range of values covered by its data set.

As shown by the ratios of estimated to measured values in Table 4-9, 56 of the 71 estimated values (about 80% of the estimates) had errors of less than one order of magnitude; only one estimate (for mirex) was off by more than a factor of $\pm$ 35. It is also apparent that Eqs. 4-5 and -8 have a tendency to overestimate low values of K_{oc} and underestimate high values.

• *Uncertainty in input data.* Values of K_{ow}, S, and P can generally be measured with more accuracy than can BCF values. Also, the uncertainty in these parameters generally increases with decreasing S and P, and with increasing K_{ow} and BCF. The uncertainty in the input data used should, if known, be carried through and incorporated with the uncertainty in the reported estimate. The propagation of errors is discussed in Appendix C.

• *Environmental factors.* See "Factors Influencing the Values of K and K_{oc}" in §4-1.

• *Assumption of linear isotherm and reversible adsorption.* No method is available for estimating the exponential fraction in the Freundlich equation (Eq. 4-2). Thus, when values of x/m are calculated for specific soil-solution situations, some value of 1/n must be assumed. The magnitude of the associated error will depend not only on the difference between the real and assumed value of 1/n but also on how far one must extrapolate from the concentration range in which K was measured. Table 4-10 indicates the errors to be expected with the use of a linear isotherm (i.e., 1/n = 1), assuming that the value of K is known at a concentration of 1 mg/L.

Errors associated with the assumption of reversible adsorption when irreversible adsorption is taking place have been discussed by Rao and Davidson [36]. Assuming that the value of 1/n found for the adsorption isotherm was 2.3 times the value of 1/n found for the

TABLE 4-10

Deviation from Linearity for the Freundlich Adsorption Isotherm (Eq. 4-2)

$$\frac{(x/m) \text{ Freundlich}}{(x/m) \text{ Linear Distribution}}$$

1/n	Equilibrium Concentration (µg/mL)						
	0.001	0.01	0.1	1	10	100	1000
0.95	1.41	1.26	1.12	1	0.891	0.794	0.708
0.90	2.00	1.59	1.26	1	0.794	0.631	0.501
0.85	2.82	2.00	1.41	1	0.708	0.501	0.355
0.80	3.98	2.51	1.59	1	0.631	0.398	0.251
0.75	5.62	3.16	1.78	1	0.562	0.316	0.178
0.70	7.94	3.98	2.00	1	0.501	0.251	0.126
0.65	11.2	5.01	2.24	1	0.447	0.200	0.089
0.60	15.9	6.31	2.51	1	0.398	0.159	0.063
0.55	22.4	7.94	2.82	1	0.355	0.126	0.045
0.50	31.6	10.0	3.16	1	0.316	0.100	0.032

Source: Hamaker and Thompson [16]. *(Reprinted with permission from Marcel Dekker, Inc.)*

desorption isotherm, and that the desorption coefficient (K_d) was related to the maximum solution concentration prior to desorption (C_m), they found that the ratio of x/m for adsorption, $(x/m)_a$, and desorption, $(x/m)_d$, could be expressed as follows:

$$\frac{(x/m)_d}{(x/m)_a} \simeq C_m^{0.565(1/n)} \cdot C^{-0.565(1/n)} \tag{4-19}$$

where 1/n is the value obtained from the *adsorption* isotherm and C is the equilibrium solution concentration.

Table 4-11 shows the values of $(x/m)_d / (x/m)_a$ obtained with different values of 1/n and C and an assumed concentration of 10 µg/mL for C_m. As the solution concentration decreases and as 1/n increases, an increasing degree of error is seen to result from assuming the isotherm to be reversible.

TABLE 4-11

Errors Associated with Assumption of
Reversible Adsorption[a]

1/n	$(x/m)_d/(x/m)_a$ Solution Concentration, C (μg/mL)		
	1.0	0.1	0.01
1.1	4.19	17.5	73.3
1.0	3.67	13.5	49.6
0.9	3.23	10.4	33.6
0.8	2.83	8.01	22.7
0.7	2.49	6.20	15.4
0.6	2.18	4.77	10.4
0.5	1.92	3.68	7.07

a. Calculated from Eq. 4-19 using $C_m = 10\,\mu$g/mL.

4-3 AVAILABLE DATA

Numerous compilations of K_{oc} values have been published. Most of them focus on pesticides and, to a lesser degree, on aromatic and polycyclic aromatic ("energy-related") compounds. The following references are recommended:

Chiou *et al.* [11] — Reported measurements for 15 chlorinated hydrocarbons.

Farmer, W.J. [13] — Data from a literature search on 49 pesticides.

Freed and Haque [14] — Data from the literature for 16 chemicals. Temperature effects shown.

Hamaker, J.W. and J.M. Thompson [16] — Data from the literature for about 36 pesticides.

Karickhoff, S.W. *et al.* [25] — Reported measurements for 10 chemicals, mostly polycyclic aromatic compounds.

Kenaga, E.E. and C.A.I. Goring [26] — Data from the literature for 109 chemicals, mostly pesticides and some polycyclic aromatics.

Rao, P.S.C. and J.M. Davidson [36] — Data from the literature for 44 pesticides.

Rao, P.S.C. and J.M. Davidson [37] — Data from a literature search on pesticides.

Reinbold *et al.* [39] — Data from a literature review of energy-related organic pollutants.

Smith *et al.* [40] — Reported measurements for nine aromatic or polycyclic aromatic compounds and two pesticides.

4-4 SYMBOLS USED

a	=	parameter in Eq. 4-4
b	=	parameter in Eq. 4-4
BCF	=	bioconcentration factor for aquatic life, obtained from tests in flowing water (f) or tests in model ecosystems (t)
C	=	concentration of chemical in solution at equilibrium (μg/mL)
C_m	=	maximum concentration of chemical in solution prior to desorption, Eq. 4-19 (μg/mL)
%CV	=	% coefficient of variation
K	=	Freundlich adsorption coefficient in Eq. 4-2 ((μg/g)/(μg/mL)); K_d = desorption coefficient
K_{oc}	=	adsorption coefficient based on organic carbon (oc) content of solid phase, Eq. 4-3 ((μg/g oc)/(μg/mL))
K_{om}	=	adsorption coefficient based on organic matter (om) content of a soil ((μg/g)/(μg/mL))
K_{ow}	=	octanol-water partition coefficient
M	=	molecular weight in Eqs. 4-17, -18 (g/mol)
m	=	mass of adsorbent in Eq. 4-2 (g)
N	=	number of potential hydrogen bonding sites on molecule, Eq. 4-14
n	=	parameter in Eq. 4-2
%oc	=	percentage of organic carbon in soil or sediment
P	=	parachor in Eqs. 4-17, -18 (($g^{1/4} \cdot cm^3$)/($sec^{1/2} \cdot mol$))
pK	=	negative log of acid dissociation constant
r	=	correlation coefficient of regression equation (usually reported as r^2)
R_f	=	degree of retention of chemical in soil thin-layer chromatography tests
S	=	water solubility of chemical (mg/L for Eq. 4-5, mole fraction for Eq. 4-6, μ mol/L for Eq. 4-7)
V	=	molar volume in Eq. 4-18 (cm^3/mol)
x	=	amount of chemical adsorbed on soil or sediment, Eq. 4-2 (μg)

Greek

ρ =   liquid density in Eqs. 4-17, -18 (g/cm³)

ρ° =   vapor density in Eq. 4-17 (g/cm³)

σ =   surface tension in Eqs. 4-17, -18 (g·sec⁻²)

Subscripts

a   =   adsorption; used with x/m

d   =   desorption; used with x/m and K

m   =   maximum; used with C as indicated above

oc   =   organic carbon; used with K

om   =   organic matter; used with K

ow   =   octanol-water; used with K

4-5   REFERENCES

1. Adams, R.S., Jr., "Effect of Soil Organic Matter on the Movement and Activity of Pesticides in the Environment," in *Trace Substances in Environmental Health-V*, Proc. Conf. Univ. of Missouri, D.D. Hemphill (ed.), University of Missouri (1972).

2. Ahlrichs, J.L., "The Soil Environment," in *Organic Chemicals in the Soil Environment*, Vol. 1, C.A.I. Goring and J.W. Hamaker (eds.), Marcel Dekker, Inc., New York (1972).

3. Bailey, G.W. and J.L. White, "Soil-Pesticide Relationships. Review of Adsorption and Desorption of Organic Pesticides by Soil Colloids, with Implications Concerning Pesticide Bioactivity," *J. Agric. Food Chem.*, **12**, 324-32 (1964).

4. Briggs, G.G., "Molecular Structure of Herbicides and Their Sorption by Soils," *Nature (London)*, **223**, 1288 (1969).

5. Briggs, G.G., "A Simple Relationship Between Soil Adsorption of Organic Chemicals and Their Octanol/Water Partition Coefficients," *Proc. 7th British Insecticide and Fungicide Conf.*, Vol. 1, The Boots Company Ltd., Nottingham, G.B. (1973).

6. Browman, M.G. and G. Chesters, "The Solid-Water Interface: Transfer of Organic Pollutants Across the Solid-Water Interface," in *Fate of Pollutants in the Air and Water Environments*, Part 1, I.H. Suffet (ed.), John Wiley & Sons, New York (1977).

7. Brown, D.S., U.S. Environmental Protection Agency, Athens, GA, personal communication, September 20, 1979.

8. Brown, D.S., U.S. Environmental Protection Agency, Athens, GA, personal communication, November 29, 1979.

9. Brown, D.S., S.W. Karickhoff and E.W. Flagg, "Empirical Prediction of Organic Pollutant Sorption in Natural Sediments," to be submitted to *J. Environ. Qual.*

10. Carter, F.L. and C.A. Stringer, "Soil Moisture and Soil Type Influence Initial Penetration by Organochlorine Insecticides," *Bull Environ. Contam. Toxicol.*, **5**, 422-28 (1970).

11. Chiou, C.T., L.J. Peters and V.H. Freed, "A Physical Concept of Soil-Water Equilibria for Nonionic Organic Compounds," *Science,* **206**, 831-32 (1979).

12. Environment Protection Agency, "Toxic Substances Control Act Premanufacture Testing of New Chemical Substances," *Fed. Regist.,* **44**, 16257-59 (16 March 1979).

13. Farmer, W.J., "Leaching, Diffusion, and Sorption of Benchmark Pesticides," in *A Literature Survey of Benchmark Pesticides,* report by the Department of Medical and Public Affairs of the George Washington University Medical Center, Washington, DC, under Contract 68-01-2889 for the Office of Pesticide Programs, U.S. Environmental Protection Agency (March 1976).

14. Freed, V.H. and R. Haque, "Adsorption, Movement, and Distribution of Pesticides in Soils," in *Pesticide Formulations,* W. Van Valkenburg (ed.), Marcel Dekker, Inc., New York (1973).

15. Hamaker, J.W., "Interpretation of Soil Leaching Experiments," in *Environmental Dynamics of Pesticides,* R. Haque and V.H. Freed (eds.), Plenum Press, New York (1975).

16. Hamaker, J.W. and J.M. Thompson, "Adsorption," in *Organic Chemicals in the Soil Environment,* Vol. 1, C.A.I. Goring and J.W. Hamaker (eds.), Marcel Dekker, Inc., New York (1972).

17. Hance, R.J., "Relationship Between Partition Data and the Adsorption of Some Herbicides by Soils," *Nature (London),* **214**, 630-31 (1967).

18. Hance, R.J., "An Empirical Relationship Between Chemical Structure and the Sorption of Some Herbicides by Soils," *J. Agric. Food Chem.,* **17**, 667-68 (1969).

19. Hassett, J.J. and M.A. Anderson, "Association of Sterols and PCBs with Dissolved Organic Matter and Effects on Their Solvent Extraction and Adsorption," in *Preprints of Papers Presented at the 176th National Meeting,* Miami Beach, FL, September 10-15, 1978; Vol. 18, American Chemical Society, Division of Environmental Chemistry, pp. 490-92 (1978).

20. Hassett, J.J., J.C. Means, W.L. Banwart, S.G. Wood, S. Ali and A. Khan, "Sorption of Dibenzothiophene by Soils and Sediments," *J. Environ. Qual.* (in press).

21. Herbes, S.E., "Partitioning of Polycyclic Aromatic Hydrocarbons Between Dissolved and Particulate Phases in Natural Waters," *Water Res.,* **11**, 493-96 (1977).

22. Hiraizumi, Y., M. Takahashi and H. Nishimura, "Adsorption of Polychlorinated Biphenyl onto Sea Bed Sediment, Marine Plankton, and Other Adsorbing Agents," *Environ. Sci. Technol.,* **13**, 580-84 (1979).

23. Huang, J.-C., "Water-Sediment Distribution of Chlorinated Hydrocarbon Pesticides in Various Environmental Conditions," in *Proc. Intern. Conf. Transport of Persistent Chemicals,* National Research Council, Ottawa, Canada, pp. II-23 to II-30 (1974).

24. Karickhoff, S.W. and D.S. Brown, "Paraquat Sorption as a Function of Particle Size in Natural Sediments," *J. Environ. Qual.,* **7**, 246-52 (1978).

25. Karickhoff, S.W., D.S. Brown and T.A. Scott, "Sorption of Hydrophobic Pollutants on Natural Sediments," *Water Res.,* **13**, 241-48 (1979).

26. Kenaga, E.E. and C.A.I. Goring, "Relationship Between Water Solubility, Soil-Sorption, Octanol-Water Partitioning, and Bioconcentration of Chemicals in Biota," pre-publication copy of paper dated October 13, 1978, given at the American Society for Testing and Materials, Third Aquatic Toxicology Sym-

posium, October 17-18, 1978, New Orleans, LA. (Symposium papers to be published by ASTM, Philadelphia, PA, as Special Technical Publication (STP) 707 in March 1980.)

27. Khan, A., J.J. Hassett, W.L. Banwart, J.C. Means and S.G. Wood, "Sorption of Acetophenone by Sediments and Soils," *Soil Sci.,* **128**, 297-302 (1979).

28. Lambert, S.M., "Functional Relationship Between Sorption in Soil and Chemical Structure," *J. Agric. Food Chem.,* **15**, 572-76 (1967).

29. Lambert, S.M., "Omega (Ω), a Useful Index of Soil Sorption Equilibria," *J. Agric. Food Chem.,* **16**, 340-43 (1968).

30. Means, J.C., J.J. Hassett, S.G. Wood and W.L. Banwart, "Sorption Properties of Energy-Related Pollutants and Sediments," *Carcinogensis — A Comprehensive Survey 5* (in review).

31. Pavlou, S.P. and R.N. Dexter, "Distribution of Polychlorinated Biphenyls (PCB) in Estuarine Ecosystems. Testing the Concept of Equilibrium Partitioning in the Marine Environment," *Environ. Sci. Technol.,* **13**, 65-71 (1979).

32. Pavlou, S.P. and R.N. Dexter, "Thermodynamic Aspects of Equilibrium Sorption of Persistent Organic Molecules at the Sediment-Seawater Interface. A Framework for Predicting Distributions in the Aquatic Environment," in *Preprints of Papers Presented at the 177th National Meeting,* Vol. 19, No. 1, Honolulu, April 1-6, 1979, American Chemical Society, Division of Environmental Chemistry (1979).

33. Poinke, H.B. and G. Chesters, "Pesticide-Sediment-Water Interactions," *J. Environ. Qual.,* **2**, 29-45 (1973).

34. Poirrier, M.A., B.R. Bordelon and J.L. Laseter, "Adsorption and Concentration of Dissolved Carbon-14 DDT by Coloring Colloids in Surface Waters," *Environ. Sci. Technol.,* **6**, 1033-35 (1972).

35. Quayle, O.R., "The Parachors of Organic Compounds. An Interpretation and Catalogue," *Chem. Rev.,* **53**, 439-589 (1953).

36. Rao, P.S.C. and J.M. Davidson, "Estimation of Pesticide Retention and Transformation Parameters Required in Nonpoint Source Pollution Models," in *Environmental Impact of Nonpoint Source Pollution,* M.R. Overcash and J.M. Davidson (eds.), Ann Arbor Science Publishers, Inc., Ann Arbor, MI (1980).

37. Rao, P.S.C. and J.M. Davidson, "Estimation of Partition Coefficients for Adsorption-Desorption of Pesticides in Soil-Water Systems," in *Retention and Transformation of Pesticides and Phosphorus in Soil-Water Systems: A Review of Available Data Base,* U.S. Environmental Protection Agency (to be published, 1980).

38. Reid, R.C., J.M. Prausnitz and T.K. Sherwood, *The Properties of Gases and Liquids,* 3rd ed., McGraw-Hill Book Co., New York, p. 604 (1977).

39. Reinbold, K.A., J.J. Hassett, J.C. Means and W.L. Banwart, "Adsorption of Energy-Related Organic Pollutants: A Literature Review," Report No. EPA-600/3-79-086, U.S. Environmental Protection Agency, Athens, GA (August 1979).

40. Smith, J.-H., W.R. Mabey, N. Bohonos, B.R. Holt, S.S. Lee, T.W. Chou D.C. Bomberger and T. Mill, "Environmental Pathways of Selected Chemicals in Freshwater Systems, Part II. Laboratory Studies," Report No. EPA-600/7-78-074, U.S. Environmental Protection Agency, Athens, GA (May 1978).

41. Steen, W.C., D.F. Paris and G.L. Baughman, "Partitioning of Selected Polychlorinated Biphenyls to Natural Sediments," *Water Res.,* **12**, 655-57 (1978).

42. Stevenson, F.M., "Organic Matter Reactions Involving Pesticides in Soil," in *Bound and Conjugated Pesticide Residues,* D.D. Kaufman, G.G. Still, G.D. Paulson and S.K. Blandal (eds.), ACS Symposium Series **29**, American Chemical Society, Washington, DC (1976).

43. Swisher, R.D., *Surfactant Biodegradation,* Marcel Dekker, Inc., New York, pp. 115-16 (1970).

44. Wershaw, R.L., P.J. Bucar and M.C. Goldberg, "Interaction of Pesticides with Natural Organic Material," *Environ. Sci. Technol.,* **3**, 271-73 (1969).

5

BIOCONCENTRATION FACTOR IN AQUATIC ORGANISMS

Sara E. Bysshe

5-1 INTRODUCTION

The accumulation of certain chemicals in aquatic organisms has become of increasing concern as an environmental hazard. Concentrations of some compounds that appear safe for organisms (according to bioassay criteria for acute or even chronic exposure) can accumulate to levels that are harmful to the consumers of such organisms or, ultimately, to the organisms themselves. A classic example is the accumulation of pesticide residues in fish, which has led to eventual decreases in the reproductive success of certain fish-eating birds. Further, when acute toxicity thresholds are high, chronic effects from residue-forming chemicals may not be noticed until after significant amounts have been released into the environment; this is especially true of organic compounds that are expensive to monitor. Reliable correlations between concentrations of chemicals in ambient media (e.g., water) and organisms can help to reduce monitoring requirements and provide early warning of potential contamination problems. Thus, it has become evident that methods are needed for screening chemical substances for potential hazard due to accumulation.

This chapter focuses on the use and limitations of methods for estimating the degree to which organic compounds may accumulate in aquatic species. The methods presented are similar, in that they are derived from observed correlations between the physical properties of such organic compounds and their accumulation under laboratory test conditions. Estimates based on currently available regression equations must be assumed to have an uncertainty of about one order of magnitude.

The *bioconcentration factor* (BCF) indicates the degree to which a chemical residue may accumulate in aquatic organisms (usually fish), coincident with ambient concentrations of the chemical in water. Specifically, it is defined here as:

$$BCF = \frac{\text{Concentration of chemical at equilibrium in organism (wet weight)}}{\text{Mean concentration of chemical in water}} \tag{5-1}$$

The units of both numerator and denominator must be the same (e.g., μg/g). Values of BCF range from about 1 to over 1,000,000.

To measure equilibrium residue concentrations in organisms, it is necessary to determine their uptake and depuration rates. Alternatively, measurements of the chemical residue concentration must be made over a sufficient period to ensure that equilibrium conditions exist [37].[1] Flow-through bioassay systems should be used, so that chemical concentrations remain relatively constant during the test.

The above "measurement-specific" definition of bioconcentration factor must be distinguished from other terms commonly used to describe increases in the concentration of chemicals in an organism, such as biomagnification, bioaccumulation, and ecological magnification. These other terms are associated with increasing concentrations of a chemical along a food chain, which could result in higher concentration factors in top-order consumers. They also imply that dietary uptake of contaminated food is additive to, or more significant than, direct exposure to the same contaminants in the water. The term bioconcentration is used in this chapter with the assumption that uptake across external membranous surfaces from water is the chief source of the material that is concentrated in the organism.

The primary significance of this water-to-organism pathway in bioconcentration is supported by numerous investigators [1,7,11,14,15,22]; however, there is also evidence that biomagnification via aquatic food chains can be important under certain environmental circumstances

---

1. In its technical guidelines, the Environmental Protection Agency states that a BCF should either be measured at equilibrium or in a test extending for more than 27 days [41].

[5,7]. Because many aspects of these phenomena are not yet fully understood, this chapter deals primarily with the nature of the estimation methods themselves and the limitations to their use.

The accumulation of organic chemicals in aquatic organisms can be predicted by several methods. This chapter focuses on techniques for estimation that are based on known relationships between bioconcentration factors and other, readily available properties of organic chemicals. Correlations between bioconcentration and octanol-water partition coefficients, water solubility, and soil adsorption coefficients in flow-through bioassays are highlighted in §5-3. In §5-4 the uses and limitations of the estimates are discussed. Correlations based on data from model ecosystems and from static bioassays are given in §5-5, which also describes some additional approaches to projecting the accumulation of organic material in the environment by aquatic organisms.

5-2 BASIC APPROACH

To obtain an estimated bioconcentration factor for a selected chemical, use the following procedure:

(1) Check the tables in §5-3 and any recent review articles to see if a BCF has already been measured by flow-through tests.

(2) If the BCF has not been measured or is not readily available, assess the existing physical and chemical data to see if the water solubility (S), octanol-water partition coefficient (K_{ow}), or soil adsorption coefficient (K_{oc}) is known. Any of these parameters can be used to estimate the BCF. Of the three, correlations for K_{ow} are currently based on the largest body of bioassay data; this parameter is relevant to the estimation of BCF because lipophilic organic chemicals are generally found to be more readily accumulated in organisms. Values of S, on the other hand, may be easier to obtain. Soil adsorption coefficients are likely to be the least available of the three.

(3) If measured values of K_{ow}, S, or K_{oc} are not available, they can be estimated by methods described in Chapters 1, 2, and 4 respectively. First preference should be given to K_{ow}, followed by S and K_{oc}, in that order.

(4) Use the appropriate regression equation in §5-3 to calculate the BCF for the chemical.

(5) See §5-4 for information on how estimated bioconcentration factors can be used, as well as the degree of uncertainty and sources of error associated with these estimates.

5-3 METHODS OF ESTIMATION

All methods described in this section for estimating bioconcentration factors in aquatic organisms are based on data from laboratory experiments that were designed to maintain relatively constant amounts of the chemical in the water environment, and where equilibrium concentrations of the chemical could be ascertained.

Tables 5-1 and 5-2 summarize some of the regression equations that have been developed. Lists of the chemical compounds from which these equations were derived accompany the descriptions of the individual methods.

Estimation from Octanol-Water Partition Coefficient. If the octanol-water partition coefficient (K_{ow}) for the organic chemical in question is available, the following equation is recommended for estimating the bioconcentration factor:

$$\log BCF = 0.76 \log K_{ow} - 0.23 \tag{5-2}$$

This regression equation was derived by Veith *et al.* [39] from the results of laboratory experiments by several investigators with a variety of fish species and 84 different organic chemicals (see Table 5-3). To estimate the BCF with this equation, proceed as follows:

(1) Obtain a measured or estimated value for K_{ow} (see Chapter 1 for estimation techniques).

(2) Substitute this value in Eq. 5-2 and solve for log BCF.

(3) The antilog is the approximate degree to which a chemical will concentrate in aquatic organisms, relative to its ambient concentration in the water. No more than two significant figures should be reported for BCF.

TABLE 5-1

Recommended Regression Equations for Estimating log BCF, Based on Flow-through Laboratory Studies

Eq. No.	Equation[a]	N[b]	r^2 [c]	Chemical Classes Represented	Range of Independent Variable	Species Used	Ref.
5-2[d]	$\log BCF = 0.76 \log K_{ow} - 0.23$	84	0.823	Wide (Table 5-3)	7.9 to 8.1×10^6	Fathead minnow Bluegill sunfish Rainbow trout Mosquitofish	[39]
5-3[e]	$\log BCF = 2.791 - 0.564 \log S$	36	0.49	Wide (Table 5-4)	0.001 to 50,000 ppm	Brook trout Rainbow trout Bluegill sunfish Fathead minnow Carp	[18]
5-4	$\log BCF = 1.119 \log K_{oc} - 1.579$	13	0.757	Wide (Table 5-5)	$<$ 1 to 1.2×10^6	Various	[18]

a. BCF = bioconcentration factor; K_{ow} = octanol-water partition coefficient; S = water solubility (ppm); K_{oc} = soil (or sediment) adsorption coefficient.

b. N = number of chemicals used to obtain regression equation.

c. r = correlation coefficient for regression equation.

d. In the original equation, the octanol-water partition coefficient was represented by "P" instead of "K_{ow}."

e. The original equation used "WS" (water solubility) instead of "S."

TABLE 5-2

Additional Regression Equations for Estimating log BCF, Based on Flow-through Laboratory Studies

Eq. No.	Equation[a]	N[b]	r^2[c]	Chemical Classes Represented	Range of Independent Variable	Species Used to Approximate	Ref.
5-5[d]	log BCF = 0.542 log K_{ow} + 0.124	8	0.899	Table 5-5	437 to 41.6x10⁶ (meas. or calc.)	Rainbow trout (muscle)	[29]
5-6[d]	log BCF = 0.85 log K_{ow} − 0.70	55	0.897	Table 5-6	7.9 to 87,000+	Bluegill Fathead minnow Rainbow trout Mosquitofish	[37]
5-7	log BCF = 0.935 log K_{ow} − 1.495	26	0.757	Table 5-4	436 to 3.7x10⁶	Various	[18]
5-8[d,e]	log BCF = 0.819 log K_{ow} − 1.146	3	0.995	Table 5-7	66 to 28,100 (calc.)	Daphnia pulex (invertebrate)	[33]
5-9[d,e]	log BCF = 0.7520 log K_{ow} − 0.4362	7	0.85	Table 5-8	2000 to 1.5x10⁶ (calc.)	Daphnia pulex	[34]
5-10[e]	log BCF = 3.41 − 0.508 Log S	7	0.930	Table 5-9	437 to 41.6x10⁶	Rainbow trout (muscle) Used data from Ref. 29	[6]

a. BCF = bioconcentration factor; K_{ow} = octanol-water partition coefficient; S = water solubility (μmoles/L).

b. N = number of chemicals used to obtain regression equation.

c. r = correlation coefficient for regression equations.

d. Original equation used "P" rather than "K_{ow}" for the octanol-water partition coefficient.

e. Original equation used "CF" or "BF" instead of "BCF."

TABLE 5-3

Compounds Used to Derive Regression Equation 5-2

Compound	Species[a]	Exposure (days)	Bioconcentration Factor (BCF)[b]
Acenaphthene	BS	28	387
Acrolein	BS	28	344
Acrylonitrile	BS	28	48 (day 28)[d]
Aroclor 1016	FM	32	42,500
Aroclor 1248	FM	32	70,500
Aroclor 1254	FM	32	100,000
Aroclor 1260	FM	32	194,000
Atrazine	FM	276	<7.9
Benzene	––	––	12.6 (calc.)
Biphenyl	RT	4	437
p-Biphenyl phenyl ether	RT	4	550
Bis(2-chloroethyl)ether	BS	14	11
5-Bromoindole	FM	32	14
BSB[c]	BS	50	<2.1
Butylbenzylphthalate	BS	21	772
Carbon tetrachloride	{ BS	21	30
	{ RT	4	17.4
Chlordane	FM	32	37,800
Chlorinated ecosane	FM	32	49
Chlorobenzene	FM	28	450
Chloroform	BS	14	6
2-Chlorophenanthrene	FM	28	4270
2-Chlorophenol	BS	28	214
Chlorpyrifos	M	35	470
DASC-3[c]	BS	30	<2.1
DASC-4[c]	BS	30	<2.1
p,p'-DDT	FM	32	29,400
o,p'-DDT	FM	32	37,000
p,p'-DDE	FM	32	51,000
Dibenzoturon	FM	28	1,350
1,2-Dichlorobenzene	BS	14	89
1,3-Dichlorobenzene	BS	14	66
1,4-Dichlorobenzene	{ BS	14	60
	{ RT	4	215
1,2-Dichloroethane	BS	14	2
Diethylphthalate	BS	21	117
2,4-Dimethylphenol	BS	28	150

(Continued)

TABLE 5-3 (Continued)

Compound	Species[a]	Exposure (days)	Bioconcentration Factor (BCF)[b]
Dimethylphthalate	BS	21	57
Diphenylamine	FM	32	30
Diphenylether	RT	4	195
Endrin	FM	300	4,600
	M	35	1,480
2-Ethylhexylphthalate	FM	56	850
Fluorene	FM	28	1,300
FWA-2-A[c]	B	105	< 2.1
FWA-3-A[c]	B	105	< 2.1
FWA-4-A[c]	B	105	< 2.1
Heptachlor	FM	276	20,000
		32	9,500
Heptachlor epoxide	FM	32	14,400
Heptachloronorbornene	FM	32	11,100
Hexabromobiphenyl	FM	32	18,100
Hexabromocyclododecane	FM	32	18,200
Hexachlorobenzene	RT	4	7,760
	FM	32	18,500
Hexachlorocyclopentadiene	FM	32	29
Hexachloroethane	BS	28	139
Hexachloronorbornadiene	FM	32	6,400
Isophorone	BS	14	7
Lindane	FM	304	470
		32	180
Methoxychlor	FM	32	8,300
2-Methylphenanthrene	FM	4	3,000
Mirex	FM	32	18,100
Naphthalene	FM	28	430
Nitrobenzene	FM	28	15
p-Nitrophenol	FM	28	126
N-Nitrosodiphenylamine	BS	14	217[d]
NTS-1[c]	BS	35	2.1–10
Octachlorostyrene	FM	32	33,000
Pentachlorobenzene	BS	28	3,400
Pentachloroethane	BS	14	67
Pentachlorophenol	FM	32	770
Phenanthrene	FM	4	2,630
N-Phenyl-2-naphthylamine	FM	32	147
1,2,3,5-Tetrachlorobenzene	BS	28	1,800 (days 21-28)[d]

(Continued)

Compound	Species[a]	Exposure (days)	Bioconcentration Factor (BCF)[b]
1,1,2,2-Tetrachloroethane	BS	14	8
Tetrachloroethylene	BS	21	49
Toluene	——	——	15-70 (calc.)
Toluenediamine	FM	32	91
2,4,6-Tribromoanisole	FM	32	865
1,2,4-Trichlorobenzene	FM	32	2,800
1,1,1-Trichloroethane	BS	28	9
1,1,2-Trichloroethylene	{ RT	4	39
	{ BS	14	17
2,4,5-Trichlorophenol	FM	28	1,900
2,5,6-Trichloropyridinol	M	35	3.1
Tricresyl phosphate	FM	32	165
Tris(2,3-dibromopropyl)phosphate	FM	32	2.7

a. BS = bluegill sunfish, FM = fathead minnow, M = mosquitofish, RT = rainbow trout.

b. These values represent either those measured directly by the authors [39] or reported from sources where similar test conditions were used. In some cases, only log BCF was reported; these have been converted to BCF here for convenience.

c. Designations for sulfonated stilbene fluorescent whitening agents.

d. Maximum BCF value.

Source: Veith *et al.* [39]. *(Reprinted with permission from the Canadian Department of Fisheries and Oceans, Ottawa, Ontario.)*

Example 5-1 Estimate the bioconcentration factor for 4,4′-dichlorobiphenyl in fish, given an octanol-water partition coefficient (K_{ow}) of 380,000.

(1) From Eq. 5-2 and the given value for K_{ow},

$$\log BCF = 0.76 \log (380,000) - 0.23$$

(2) $\log BCF = 4.01$

(3) $BCF = 10,000$

A measured value reported for this compound is more than an order of magnitude lower than the above estimate (see Table 5-4).

Example 5-2 Estimate the bioconcentration factor for methoxychlor in fish, given a K_{ow} of 19,950.

(1) From Eq. 5-2 and the given K_{ow} value,

$$\log BCF = 0.76 \log (19,950) - 0.23$$

(2)   log BCF = 3.04

(3)   BCF = 1,100

A comparison of this estimate with laboratory measurements (see § 5-4) shows that Eq. 5-2 underestimated the measured BCF for methoxychlor by a factor of about 7.

Estimation from Water Solubility. If water solubility (S) in parts per million is available for the organic chemical in question, the following equation is recommended for estimating the BCF:

$$\log BCF = 2.791 - 0.564 \log S \qquad (5\text{-}3)$$

This regression equation was derived by Kenaga and Goring [18] from laboratory experiments by a number of investigators with a variety of fish species and 36 organic chemicals (see Table 5-4). Note the reciprocal nature of the relationship between water solubility and bioconcentration.

To estimate a bioconcentration factor from Eq. 5-3, the procedure is as described above for Eq. 5-2, except that the required physical/chemical parameter is S, which must be expressed in parts per million (ppm).

Example 5-3   Estimate the bioconcentration factor for diphenyl oxide in fish, given a water solubility of 21 ppm.

(1)   From Eq. 5-3 and the given value of S,

   log BCF = 2.791 − 0.564 log (21)

(2)   log BCF = 2.04

(3)   BCF = 110

This is reasonably close to the measured value (∼ 196) reported in Ref. 18.

Example 5-4   Estimate the bioconcentration factor for heptachlor in fish, given a water solubility of 0.030 ppm.

(1)   From Eq. 5-3 and the given value of S,

   log BCF = 2.791 − 0.564 log (0.030)

(2)   log BCF = 3.65

(3)   BCF = 4,500

The measured value of BCF is 9,500 (see § 5-4).

TABLE 5-4

Compounds Used to Derive Regression Equations 5-3, 5-4, and 5-7

Compound	Water Solubility (S) for Eq. 5-3	Soil Adsorption Coefficient (K_{oc}) for Eq. 5-4	Octanol-water Partition Coefficient (K_{ow}) for Eq. 5-7[a]	Biocon- centration Factor (BCF)[b]
Halogenated Hydrocarbon Insecticides				
Chlordane	X			11,400
DDT	X	X	X	61,600
Dieldrin	X			5,800
Endrin	X		X	4,050
Heptachlor	X			17,400
Lindane	X	X		325
Methoxychlor	X	X	X	185
Toxaphene	X			26,400
Kepone®	X			8,400
Substituted Benzenes and Halobenzenes				
Chlorobenzene	X		X	12
p-Dichlorobenzene	X		X	215
Hexachlorobenzene	X	X	X	8,600
Pentachlorobenzene	X		X	~ 5,000
1,2,4,5-Tetrachlorobenzene	X		X	4,500
1,2,4-Trichlorobenzene	X		X	491
Halogenated Biphenyls and Diphenyl Oxides				
4-Chlorobiphenyl	X		X	490
4,4'-Dichlorobiphenyl	X		X	215
2,4,4' and 2,2',5-Trichloro- biphenyl (Aroclor 1016, 1242)	X		X	48,980
2,2',4,4' and 2,2',5,5'-Tetra- chlorobiphenyl (Aroclor 1248)	X		X	72,950
2,2',4,5,5'-Pentachloro- biphenyl (Aroclor 1254)	X	X	X	45,600
2,2',4,4',5,5'-Hexachloro- biphenyl	X	X	X	46,000(est)

(Continued)

TABLE 5-4 (Continued)

Compound	Water Solubility (S) for Eq. 5-3	Soil Adsorption Coefficient (K_{oc}) for Eq. 5-4	Octanol-water Partition Coefficient (K_{ow}) for Eq. 5-7[a]	Biocon-centration Factor (BCF) [b]
Halogenated Biphenyls and Diphenyl Oxides (Cont'd.)				
Diphenyloxide	X		X	196
4-Chlorodiphenyloxide	X		X	736
X-sec-Butyl-4-chlorodiphenyl-oxide	X		X	298
X-Hexyl-X'-chlorodiphenyl-oxide	X		X	18,000
X-Dodeca-X'-chlorodiphenyl-oxide	X		X	12
Aromatic Hydrocarbons				
Biphenyl	X		X	340
Phosphorus-containing Insecticides				
Diamidaphos	X	X		1
Chlorpyrifos	X	X	X	450
Liptophos	X	X	X	750
Diazinon	X			35
Carboxylic Acids and Esters				
Di-2-ethylhexylphthalate	X		X	380
Dinitroanilines				
Trifluralin	X	X	X	4,570
Symmetrical Triazines				
Atrazine	X	X	X	—
Simazine	X	X	X	1
Miscellaneous Nitrogen Heterocyclics				
3,5,6-Trichloro-2-pyridinol	X	X	X	3

a. Although the authors stated that 26 compounds were used in deriving Eq. 5-7, they listed the 28 compounds checked here. It is not clear which two were not used.

b. In many cases, the authors used more than one source, but they did not explain how they combined them to arrive at a single value.

Source: Kenaga and Goring [18]. *(Reproduced with permission from the American Society for Testing and Materials.)*

Estimation from Soil Adsorption Coefficients. The relationship between soil adsorption coefficients (K_{oc}) and bioconcentration appears to be essentially empirical, although soil affinity for certain types of organic chemicals may, in fact, be related to the affinity of the same types of chemicals for certain parts of biological systems. Equation 5-4 was derived by Kenaga and Goring [18] from a relatively small number of measurements of soil adsorption coefficients (see Table 5-4). Nonetheless, the correlation between K_{oc} and measured values of BCF appears to be quite good; the derived regression equation could be utilized to estimate BCF if only soil adsorption information is available, or for comparison with estimates based on K_{ow} or S.

$$\log \text{BCF} = 1.119 \log K_{oc} - 1.579 \tag{5-4}$$

The procedure for using this regression equation to estimate BCF is the same as described above for Eqs. 5-2 and 5-3.

Example 5-5 Estimate the bioconcentration factor for DDT in fish, given a soil adsorption coefficient of 238,000.

(1) From Eq. 5-4 and the given value for K_{oc},

   $\log \text{BCF} = 1.119 \log (238,000) - 1.579$

(2) $\log \text{BCF} = 4.44$

(3) $\text{BCF} = 27,000$

This agrees closely with the measured value of 29,400 (see § 5-4).

Other Regression Equations. In addition to the above, various other correlations have been observed between bioconcentration in fish or certain aquatic invertebrates (i.e., *Daphnia*) and the physical/chemical characteristics of a more limited set of organic chemicals. The corresponding regression equations are listed in Table 5-2 and are used in the same way as described for Eqs. 5-2, -3, and -4.

Equation 5-5 is of particular interest, as it represents one of the earliest and most widely known uses of octanol-water partition coefficients for estimating bioconcentration potential in fish. This correlation, which was developed by Neely *et al.* [29], was based on a very small number of measured bioconcentration values in trout muscle (Table 5-5).

TABLE 5-5

Compounds Used to Derive Regression Equation 5-5[a,b]

2-Biphenyl phenyl ether
Carbon tetrachloride
p-Dichlorobenzene
Diphenyl
Diphenyl oxide
Hexachlorobenzene
2,2',4,4'-Tetrachlorodiphenyl oxide
1,1,2,2-Tetrachloroethylene

---

a. Table 5-3 lists log BCF values for these compounds.
b. Partition coefficients were calculated, not measured.

Source: Neely, Branson and Blau [29].

The correlation described by Eq. 5-2 is considered better, because it is based on a much larger number of measured values, including those used by Neely *et al.*

Similarly, Eq. 5-6 represents an earlier relationship observed by Veith, DeFoe and Bergstedt [37] among some of the organic compounds later used to derive Eq. 5-2. These compounds are listed in Table 5-6.

In addition to their regression equation based on solubility (Eq. 5-3), Kenaga and Goring [18] also developed a correlation with octanol-water coefficients (see Eq. 5-7). The compounds on which it is based are listed in Table 5-4.

Equations 5-8 and 5-9 were developed by Southworth *et al.* [33,34] to describe relationships between K_{ow} and bioconcentration in *Daphnia pulex*. The organic compounds that they used are listed in Tables 5-7 and 5-8 respectively.

The relationship between water solubility and bioconcentration described by Eq. 5-10 was developed by Chiou *et al.* [6], who used the uptake data reported by Neely *et al.* Table 5-9 lists the compounds used to derive this equation.

TABLE 5-6

Compounds Used to Derive Regression Equation 5-6[a]

Aroclor 1016	FWA-4-A
Aroclor 1248	Heptachlor
Aroclor 1254	Heptachlor epoxide
Aroclor 1260	Heptachloronorbornene
Atrazine	Hexabromobiphenyl
Biphenyl	Hexabromocyclododecane
p-Biphenyl phenyl ether	Hexachlorobenzene
5-Bromoindole	Hexachloronorbornadiene
BSB	Lindane
Carbon tetrachloride	Methoxychlor
Chlordane	Methylphenanthrene
Chlorobenzene	Mirex
2-Chlorophenanthrene	Naphthalene
Chloropyrifos	Nitrobenzene
DASC-3	p-Nitrophenol
DASC-4	NTS-1
p,p'-DDE	Octachlorostyrene
p,p'-DDT	Pentachlorophenol
o,p'-DDT	Phenanthrene
Dibenzoturon	N-Phenyl-2-naphthylamine
p-Dichlorobenzene	1,1,2,2-Tetrachloroethylene
Diphenylamine	Toluene diamine
Diphenylether	2,4,6-Tribromoanisol
Endrin	1,2,4-Trichlorobenzene
2-Ethylhexylphthalate	2,4,5-Trichlorophenol
Fluorene	2,5,6-Trichloropyridinol
FWA-2-A	Tricresyl phosphate
FWA-3-A	

a. Table 5-3 lists BCF values for these compounds.

Source: Veith, DeFoe and Bergstedt [37].

TABLE 5-7

Compounds Used to Derive Regression Equation 5-8

Compound	BCF in *Daphnia pulex*
Acridine	29.6
Benz(a)acridine	352
Isoquinoline	2.41

Source: Southworth, Beauchamp and Schmieder [33]. *(Reprinted in part with permission from the American Chemical Society.)*

TABLE 5-8

Compounds Used to Derive Regression Equation 5-9

Compound	BCF in *Daphnia pulex*
Anthracene	917
Benz(a)anthracene	10,100
9-Methyl anthracene	4,580
Naphthalene	131
Perylene	7,190
Phenanthrene	325
Pyrene	2,700

Source: Southworth, Beauchamp and Schmieder [34]. *(Reprinted with permission from Pergamon Press, Ltd.)*

TABLE 5-9

Compounds Used to Derive Regression Equation 5-10[a,b]

Biphenyl
Carbon tetrachloride
p-Dichlorobenzene
Diphenyl ether
Hexachlorobenzene
2,4,2',4'-PCB
Tetrachloroethylene

a. BCF values used were those of Neely *et al.* and are listed in Table 5-3.
b. Partition coefficients were calculated.

Source: Chiou *et al.* [6]. *(Reprinted with permission of the American Chemical Society.)*

5-4   USES AND LIMITATIONS OF ESTIMATED VALUES

All of the estimation techniques described in the previous section are based on correlations between a measured or calculated physical/chemical property of an organic chemical and the observed bioconcentration, usually in fish. The accuracy of these estimates is limited by the accuracy of the measurement techniques used for the various correlation parameters. Although efforts to improve and standardize these techniques continue [39], the number of variables that affect the physical/chemical properties and bioconcentration factors make it unlikely that estimates of this kind will ever provide highly accurate projections, particularly with respect to the ambient environment. Accordingly, estimates of BCF based on relationships described in this chapter should be used *to gain understanding of the potential* for an organic chemical to be taken up and stored in aquatic biota and to indicate whether further research into its environmental fate may be warranted. An estimated bioconcentration factor for a compound can be compared with the BCFs for known problematic accumulators such as DDT or Aroclor, a mixture of polychlorbiphenyls, and with the BCFs for compounds like carbon tetrachloride that have *not* been implicated in residue formation in biological organisms. (See Table 5-10.)

It is also important to emphasize that, overall, bioconcentration factors can presently be estimated *only to within an order of magnitude* for most of the correlations listed (see correlation coefficients expressed as r^2 in Tables 5-1 and 5-2). Within this bound, estimates from Eqs. 5-2, 5-3, and 5-7 appear to have greater relative uncertainty. This is probably related to the broader range of chemical classes from which they were derived, in addition to the problems of measurement variability associated with all such correlations.

The so-called error sources that reduce the levels of confidence in BCF estimates are of several kinds. In the subsections that follow, a distinction is made between (1) those that create discrepancies between BCF estimates and laboratory measurements of bioconcentration and (2) those that affect bioconcentration in the ambient environment.

Sources of Discrepancies Between BCF Estimates and Laboratory Data. Table 5-10 compares values of BCF measured in the laboratory with estimates derived from correlations based on K_{ow}, S, and K_{oc}. It illustrates the degree of discrepancy that can be expected from such estimates. Both the variability inherent in biological responses and factors responsible for measurement inaccuracy contribute to the observed differences between estimated and measured values.

TABLE 5-10

Comparison of Estimated Values with Laboratory Measurements of BCF[a]

Compound	Physical/Chemical Parameter for Estimate			Estimated BCF			Laboratory Measurement	
	K_{ow} [37, 39]	S (ppm) [18]	K_{oc} [18]	From K_{ow} (Eq. 5-2)	From S (Eq. 5-3)	From K_{oc} (Eq. 5-4)	BCF	Ref.
Nitrobenzene	851	1,780	—	99	9.1	—	15.1	[37]
Carbon tetrachloride	437	800	—	17	14	—	30	[38]
p-Dichlorobenzene	2,400	79	—	220	53	—	215[b]	[29]
Atrazine	427	33	149	59	86	—	<7.94	[39]
1,2,4-Trichlorobenzene	17,000	30	—	970	91	—	2,800	[37]
Methoxychlor	20,000	0.003	80,000	1,100	16,000	8,100	8,300	[37]
Naphthalene	50,100	31.7	1,300	2,200	88	80	427	[39]
Pentachlorophenol	126,000	14	900	4,400	140	53	770	[37]
Hexachlorobenzene	170,000	0.035	3,910	5,600	4,100	280	18,500	[37]
Heptachlor	275,000	0.030	—	8,000	4,500	—	9,500	[37]
Biphenyl	5,750	7.5	—	420	71	—	437[b]	[29]
DDT	562,000	0.0017	23,800	14,000	23,000	27,000	29,400	[37]
Aroclor 1254	2,950,000	0.01	42,500	49,000	8,300	4,000	100,000	[37]
Chlordane	1,000,000	0.056	—	21,000	120	—	37,800	[37]

a. It is interesting to note that all the examples of estimates based on K_{ow} fall within an order of magnitude of the cited measured values. This is not true for all the examples based on K_{oc} or S. Caution should be exercised at this time in drawing any conclusions concerning the relative usefulness/accuracy of these three physical-chemical parameters for BCF estimates: much of the laboratory data cited, and that used to derive Eq. 5-2, were the result of somewhat standardized procedures. The same situation likely did not exist for the data generated to derive equations 5-3 and 5-4.

b. Muscle only.

Errors in measuring the physical/chemical properties correlated with BCF are also responsible for some of the discrepancies observed. As discussed more thoroughly in Chapters 1, 2, and 4, several different methods can be used to measure K_{ow}, S, and K_{oc}. Although certain methods are becoming accepted as more reliable, the measurements of properties are limited by the accuracy and precision of the techniques and the way in which the results are interpreted. *Estimates* of these parameters, to the extent that they have been used (mostly for K_{ow}), represent an additional source of error for the relationships presented here. *Test conditions,* such as pH and temperature, have a marked effect on measurements of solubility and partition coefficient for some chemical species. Solubility is difficult to measure (and thus is subject to greater unreliability) for very insoluble compounds. K_{ow} is similarly less reliable for very soluble compounds. Because highly water-soluble compounds tend not to form problematic residues in fish, error associated with very soluble compounds is less important when used in BCF estimates [38]. (Such compounds may present other environmental hazards, such as acute toxicity.)

Errors in measurement of BCF itself under laboratory conditions are also a source of discrepancies. Although the regression relationships presented in §5-3 are reported to be based on BCFs measured by flow-through methods, various bioconcentration phenomena can affect the results:

- As stated at the beginning of this chapter, it is important that BCF be measured under equilibrium conditions between uptake and depuration. However, many compounds with high partition coefficients (log BCF > 6) move across membranes very slowly and may not reach equilibrium in 20 or even 30 days; as a result, their BCF measurements and predictions based on them may be too low [10].

- For some organic compounds, particularly those of high solubility, equilibrium can be reached in a few days or less. Highly soluble compounds are also more susceptible to degradation or excretion, both of which tend to cause artificially high BCF measurements in a study of short duration.

- Test temperatures, dissolved oxygen, and the size of test organisms also affect the time required for equilibrium to be established and, thus, the test duration that should be used for measurement of bioconcentration factors [39].

- The relative lipid content of fish species used in tests may well alter residue formation potential [7,17,19]. Even within a given species, lipid content can be affected by growth stage and position in the reproductive cycle [17,39]. Furthermore, measurements of residue concentrations in some specialized groups of tissues can easily produce different results from whole-organism analyses [13,26].

Thus, a number of significant variables can affect the accuracy of measurements from which the correlations are derived. The regression equations are based on straight-line correlations that have been developed from data with a fair degree of scatter. Figure 5-1, the plot of data points from which Eq. 5-2 was derived, illustrates this. It becomes obvious that the order-of-magnitude level of confidence associated with BCF estimates derived in this fashion represents a reasonable level of accuracy. While standardization of methods is likely to improve the reliability of BCF estimates under laboratory conditions (as in Eq. 5-6), there will always be some uncertainty because of the large number of variables. If estimates based on such data are used with an awareness of their limitations and only as a rapid means for identifying a potential for bioconcentration, the issue of absolute accuracy becomes less important.

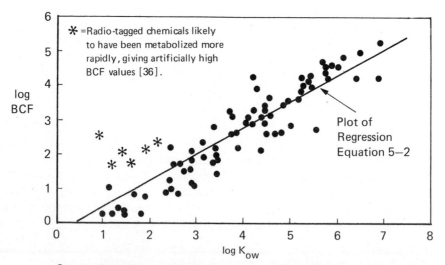

Source: Adapted from Ref. 39. *(Reprinted with permission from the Canadian Department of Fisheries and Oceans, Ottawa, Ontario.)*

FIGURE 5—1   Correlation Between Bioconcentration and Octanol-Water Partition Coefficient

Application of BCF Estimates to Field Situations. The BCF estimation techniques illustrated here could be used to identify environmental situations where further investigation of organic residue formation is warranted. However, attempts to define the fate of an organic compound *without field verification* are subject to many potentially significant errors. Some measures of reliability of correlations based on existing laboratory data have been calculated, but this is not yet possible for applications to field situations. Thus, while we can estimate the bioconcentration potential of an organic compound, it tells us little about the fate of that compound in the variety of potential environments. Table 5-11 compares some estimated BCFs with those measured in the ambient environment. Note that most of the estimates agree within an order of magnitude.

The factors that lead to discrepancies in measurements of bioconcentration under laboratory conditions also apply to ambient conditions, and there is likely to be a wider range of variability in the ambient exposure history of the organism. That exposure history is frequently unknown; for example, fish may move in and out of contact with varying concentrations of a contaminant, so that the bioconcentration depends not only on straightforward uptake and depuration rates, but also on the amount of time that the fish are in contact with the contamination source(s) before they are caught. The exposure of fish in the same body of water may vary with the strength, number, and location(s) of the source(s) and whether the species is pelagic or bottom dwelling, migratory, or remains in a relatively confined habitat. In addition, seasonal variations in major physical factors (e.g., temperature, salinity, dissolved oxygen) are usually uncontrolled in the ambient environment.

Laboratory studies have shown that the water is far more significant than food as a source of organic compounds for fish bioconcentration [1,7,11,14,15]. In the actual environment, however, this may not necessarily be the case. One explanation for the very high BCF observed in the Great Lakes for particularly persistent organics (see Table 5-11) may be that in situations where ambient concentrations are very low, as in the ng/L range illustrated, food may represent a more significant source than water; this would cause the values of BCF to be higher than water concentrations would indicate [37]. The range and significance of contamination from ingested sediment and the differences in feeding behavior between species are additional unquantified uncertainties in the field situation.

TABLE 5-11

Comparison of Estimated BCF with Observed BCF from Field Data

Compound and Location	Equation Used	Log K_{ow}	Log S (ppm)	Estimated BCF	Ambient Water Concentration	Concentration in Fish (Species, Duration)	Observed BCF	Ref.
Aroclor 1016 (Hudson River)	5-2	5.88 [37]	—	17,000	Mean of 0.17μg/L (may be closer to 0.10μg/L)	2.6μg/g (mean of 18 fish, 3 species of various sizes, 14 days' exposure)	15,000 (up to 26,000)	[32]
DDT (Hamilton, Lake Ontario)	5-2	5.75 [37]	—	14,000	4.5 ng/L	0.14μg/g (Alewife) 0.23μg/g (Smelt)	31,000 51,000	[40]
DDE (Hamilton, Lake Ontario)	5-2	5.69 [37]	—	12,000	37.4 ng/L	0.46μg/g (Alewife) 1.36μg/g (Smelt) 0.94μg/g (Sculpin)	12,000 36,000 25,000	[40]
Dieldrin (Hamilton, Lake Ontario)	5-3	—	-1.66 [18]	5,300	3.1 ng/L	0.04μg/g (Alewife)	13,000	[40]
PCB (Aroclor 1254) (Two lakes in South Dakota)	5-2	6.47 [37]	—	49,000	<0.5μg/L	0.11μg/g (one measurement for 10 carp)	>220	[9]
Lindane (Limestone quarry)	5-2	3.89 [37]	—	530	25-13 ng/L (over 235 days)	~27.3-13.3 ng/g (Trout, 3-7 fish per sample)	~1,090 to ~1,020	[12]
Trifluralin (Wabash River)	5-2	5.33 [35]	—	6,600	~1.8μg/L	10.46μg/g (237 Sauger) residue in fat	5,800	[35]

As a final caveat to extrapolation, it appears that estimates for BCF based on fish uptake may not be applicable to some other families of aquatic organisms. The work of Lu, Metcalf, and others [1,23,24,25] has revealed differences in the ability of aquatic organisms (e.g., fish vs. mollusk) to metabolize concentrated organic materials. These differences may be related to phylogenetic differences in enzyme systems. The question remains whether or not species differences can result in potential cumulative discrepancies of more than an order of magnitude between estimated and measured BCF values.

5-5 OTHER APPROACHES TO ESTIMATING THE ACCUMULATION OF ORGANIC COMPOUNDS

There is a significant body of data on bioconcentration measured in organisms in laboratory situations other than flow-through systems. These include data from model ecosystems (both aquatic and terrestrial/aquatic) and from other types of static tests. Table 5-12 lists some regression equations and correlation coefficients developed from such studies.[2] These estimates should be distinguished from those based on single-species flow-through tests, because model ecosystem test conditions do not always represent maximum bioconcentration potential. The reasons are as follows:

(1) The degree of variability in test compound concentrations is likely to be greater in model ecosystem water than in flow-through systems. Equilibrium may not be established at all, or it may exist at an inadequate exposure level. This could be the result of decomposition or other reactions that affect concentrations following a single dose [1,38] and/or the method by which such compounds are introduced.

(2) In a number of cases (e.g., Ref. 19), the duration of the ecosystem tests has been so brief that equilibrium conditions may not have been established for even a short time in the system. This situation would be more problematic with compounds of high BCF than with those at the other end of the spectrum. The relatively low coefficients of correlation given in Table 5-12 for Eqs. 5-15 and 5-19 may additionally be due to the methods used to measure K_{ow} [10,19,20].

---

2. To indicate that model ecosystem rather than flow-through tests were used, the abbreviation "E.M." (for "ecological magnification") is often used instead of "BCF." Some authors prefer BCF(t).

TABLE 5-12

Regression Equations Derived from Model Ecosystem Tests and Static Bioassays[a]

Equation Number	Equation[b]	N[c]	r²[d]	Chemical Classes	Ref.
5-11	$\log$ E.M. = 2.6204 − 0.3339 $\log$ S (ppm)	7	0.6744	Benzene derivatives	[19]
5-12	$\log$ E.M. = 0.2060 + 0.6669 $\log K_{ow}$	7	0.6673		
5-13	$\log$ E.M. = 1.4578 + 0.4112 x	7	0.7302		
5-14	$\log$ E.M. = 2.5580 − 0.4275 $\log$ S (ppm)	19	0.6584	Benzene derivatives & others (a number of pesticides)	[19]
5-15[e]	$\log$ E.M. = 0.0300 + 0.7481 $\log K_{ow}$	19	0.3659		
5-16	$\log$ E.M. = 0.1394 + 0.6308 $\log K_{ow}$	4	0.8477	Veterinary drugs: Sulfamethazine, Clopidol, Diethylstilbestrol, Phenothiazine	[8]
5-17	$\log$ E.M. = 5.99 − 1.176 $\log$ S (ppb)	11	0.7569	Aldrin, Dieldrin, Endrin, Mirex, Lindane, DDT, DDE, DDD, Methoxychlor, Methylchlor, Ethoxychlor	[24]
5-18	$\log$ E.M. = 3.9950 − 0.3891 $\log$ S (ppb)	11	0.8516	Aldrin, DDT, Hexachlorobenzene, Chlorobenzene, Nitrobenzene, Benzoic acid, Aniline, Anisole, 2,6-Diethyl aniline, 3,5,6-Trichloro-2-pyridinol, Pentachlorophenol	[20]
5-19	$\log$ E.M. = 0.7285 + 0.6335 $\log K_{ow}$	11	0.6208		

(Continued)

Equation Number	Equation[b]	N[c]	r²[d]	Chemical Classes	Ref.
5-20[f]	$\log$ E.M. $= 5.328 - 1.059 \log S$ (ppb)	13	0.4761	Three types of insecticides	[25]
5-21[f]	$\log$ E.M. $= 4.529 - 0.627 \log S$ (ppb)	16	0.6889	Range of chemicals (PCBs, benzene derivatives, pesticides, others)	[25]
5-22	$\log$ BCF(t) $= 2.183 - 0.629 \log S$ (ppm)	50	0.4356		[18]
5-23	$\log$ BCF(t) $= 1.225 \log K_{oc} - 2.024$	22	0.8281		[18]
5-24	$\log$ BCF(t) $= 0.767 \log K_{ow} - 0.973$	36	0.5776		[18]

a. All correlations listed in this table were based on tests with mosquitofish.

b. E.M. = ecological magnification, as measured in a terrestrial/aquatic or aquatic model ecosystem; S = water solubility; K_{ow} = octanol–water partition coefficient; x= pi-constant from Fujita et al. 1964 (in Ref. 19); BCF(t) = bioconcentration factor, as measured in a model ecosystem or static bioassay; K_{oc} = soil adsorption coefficient. In several cases, the symbols "x" and "y" used in the original publication have been replaced by the "EM," "K_{ow}," or "S" symbols used throughout this text.

c. N = number of chemicals used to obtain regression equation.

d. r = correlation coefficient for regression equation.

e. Original source gave a minus sign for the log K_{ow} term. The data indicate, however, that the term should have a positive sign.

f. Original equations were expressed in terms of log S and are rearranged here.

(3) Although model ecosystems have been designed to represent more closely a total environmental system, they are still subject to the same potential sources of error as other laboratory tests — size and age of fish, temperature and specificity of dissolved oxygen, etc.

Nevertheless, model ecosystems do provide useful reference data on the fate of doses of specific chemicals and their degradation products. They could be used, for example, to simulate uptake in the aftermath of a chemical spill in a small lake. In view of the present state of the art of estimating BCF for organic chemicals, however, estimates based on data from flow-through systems are generally more reliable, because these systems are more likely to achieve and maximize equilibrium accumulation. Accordingly, Eqs. 5-1 through 5-9 probably provide more safely conservative estimates than those based on static studies (Eqs. 5-10 through 5-24).

Some relatively simple models have been developed to describe bioconcentration in aquatic organisms under different environmental and/or exposure scenarios. (See, for example, Refs. 3-6, 29, and 30.) These models are not described here, primarily because they require more and different information than do the other estimation techniques, including measured uptake and depuration rates and measured or estimated values for organism metabolism, temperature, dissolved oxygen, sediment binding and release, and other environmental factors. Such models attempt to describe more accurately the various factors that affect the accumulation of organic residues in aquatic organisms. Because they are relatively new and untested, the underlying assumptions and their usefulness must be calibrated and/or validated by further laboratory and field experimentation.

5-6   AVAILABLE DATA

This chapter utilizes three major compilations of BCF values based on flow-through bioassay tests, namely, those by Kenaga and Goring [18], Veith, DeFoe, and Bergstedt [37], and Veith, Macek, Petrocelli, and Carroll [39].

5-7   SYMBOLS USED

BCF   = bioconcentration factor for aquatic organisms obtained from flow-through bioassay tests

$BCF(t)$ = bioconcentration factor for aquatic organisms obtained from static bioassay or model ecosystem tests

E.M. = ecological magnification for aquatic organisms based on model ecosystem tests

K_{oc} = soil (sediment) adsorption coefficient

K_{ow} = octanol-water partition coefficient

N = number of chemicals used to derive regression equation

r = correlation coefficient

S = solubility in water (ppm or ppb)

x = pi-constant, used in Eq. 5-13.

5-8 REFERENCES[3]

1. Bahner, L. H., A. J. Wilson, J. M. Sheppard, J. M. Patrick, L. R. Goodman and G. E. Walsh, "Kepone® Bioconcentration, Accumulation, Loss, and Transfer Through Estuarine Food Chains," *Chesapeake Sci.*, **18**, 299-308 (1977), NTIS PB-277 183.

2. Blanchard, F.A., I.T. Takahashi, H. C. Alexander and E. A. Bartlett, "Uptake, Clearance and Bioconcentration of ^{14}C-sec-Butyl-4-Chlorodiphenyl Oxide in Rainbow Trout," in *Aquatic Toxicology and Hazard Evaluation,* ASTM STP 634, ed. by F. L. Mayer and J. L. Hamelink, American Society for Testing and Materials, Philadelphia, Pa., 162-77 (1977).

3. Branson, D. R., "A New Capacitor Fluid — A Case Study in Product Stewardship," in *Aquatic Toxicology and Hazard Evaluation*, ASTM STP 634, ed. by F. L. Mayer and J. L. Hamelink, American Society for Testing and Materials, Philadelphia, Pa., 44-61 (1977).

4. Branson, D. R., "Predicting the Fate of Chemicals in the Aquatic Environment from Laboratory Data," in *Estimating the Hazard of Chemical Substances to Aquatic Life,* ASTM STP 657, ed. by J. Cairns, Jr., K. L. Dixon, and A. W. Maki, American Society for Testing and Materials, Philadelphia, Pa., 55-70 (1978).

5. Branson, D. R., G. E. Blau, H. C. Alexander and W. B. Neely, "Bioconcentration of 2,2',4,4'-Tetrachlorobiphenyl in Rainbow Trout as Measured by an Accelerated Test," *Trans. Am. Fish. Soc.*, **104**, 785-92 (1975).

6. Chiou, C. T., V. H. Freed, D. W. Schmedding and R. L. Kohnert, "Partition Coefficient and Bioaccumulation of Selected Organic Chemicals," *Environ. Sci. Technol.*, **11**, 475-78 (1977).

7. Clayton, J. R., Jr., S. P. Pavlou and N. F. Breitner, "Polychlorinated Biphenyls in Coastal Marine Zooplankton: Bioaccumulation by Equilibrium Partitioning," *Environ. Sci. Technol.*, **11**, 676-82 (1977).

---

3. Several references in this list are not specifically cited in the text but are included as they may be of general interest.

8. Coats, J. R., R. L. Metcalf, P. Lu, D. D. Brown, J. F. Williams and L. G. Hansen, "Model Ecosystem Evaluation of the Environmental Impacts of the Veterinary Drugs Phenothiazine, Sulfamethazine, Clopidol and Diethylstilbestrol," *Environ. Health Perspect.*, **18**, 167-79 (1976).

9. Greichus, Y. A., A. Greichus and R. J. Emerick, "Insecticides, Polychlorinated Biphenyls and Mercury in Wild Cormorants, Pelicans, Their Eggs, Food and Environment," *Bull. Environ. Contam. Toxicol.*, **9**, 321-28 (1973).

10. Hamelink, J. L. and A. Spacie, "Fish and Chemicals: The Process of Accumulation,"*Ann. Rev. Pharmacol. Toxicol.*, **17**, 167-77 (1977).

11. Hamelink, J. L., R. C. Waybrant and R. C. Ball, "A Proposal: Exchange Equilibria Control the Degree Chlorinated Hydrocarbons are Biologically Magnified in Lentic Environments," *Trans. Am. Fish Soc.*, **100**, 207-14 (1971).

12. Hamelink, J. L., R. C. Waybrant and P.R. Yant, "Mechanisms of Bioaccumulation of Mercury and Chlorinated Hydrocarbon Pesticides by Fish in Lentic Ecosystems," in *Fate of Pollutants in the Air and Water Environments*, Part 2, ed. by R. H. Suffet, John Wiley & Sons, New York (1977).

13. Hiltibran, R. C., D. L. Underwood and J. S. Fickle, WRC Research Report No. 52, Water Resource Center, University of Illinois, Urbana-Champaign, Illinois (1972). NTIS PB-208 598.

14. Isensee, A. R. and G. E. Jones, "Distribution of 2,3,7,8-Tetrachlorodibenzo-p-dioxin (TCDD) in Aquatic Model Ecosystem," *Environ. Sci. Technol.*, **9**, 668-72 (1975).

15. Kanazawa, J., A. R. Isensee and T. C. Kearney, "Distribution of Carbaryl and 3,5-Xylyl Methylcarbamate in an Aquatic Model Ecosystem," *J. Agric. Food Chem.*, **23**, 760-63 (1975).

16. Kapoor, I. P., R. L. Metcalf, A. S. Hirwe, J. R. Coats and M. S. Khalsa, "Structure Activity Correlations of Biodegradability of DDT Analogs,"*J. Agric. Food Chem.*, **21**, 310-15 (1973).

17. Kenaga, E. E., "Chlorinated Hydrocarbon Insecticides in the Environment: Factors Related to Bioconcentration of Pesticides," in *Environmental Toxicology of Pesticides*, Part III, ed. by F. Matsumara and G. M. Boush, Academic Press, New York (1972).

18. Kenaga, E. E. and C. A. I. Goring, "Relationship Between Water Solubility, Soil-Sorption, Octanol-Water Partitioning, and Bioconcentration of Chemicals in Biota," prepublication copy of paper dated Oct. 13, 1978, given at the Third Aquatic Toxicology Symposium, American Society for Testing and Materials, New Orleans, La, October 17-18 1978. [Symposium papers to be published by ASTM, Philadelphia, Pa., as Special Technical Publication (STP) 707 in 1980.]

19. Lu, P.-Y., "Model Aquatic Ecosystem Studies of the Environmental Fate and Biodegradability of Industrial Compounds" (unpublished Ph.D. dissertation, University of Illinois, Urbana-Champaign, Ill., 1974). Order No. 74-14,576, Univ. Microfilms, Ann Arbor, Mich.

20. Lu, P.-Y. and R. L. Metcalf, "Environmental Fate and Biodegradability of Benzene Derivatives as Studied in a Model Aquatic Ecosystem," *Environ. Health Perspect.*, **10**, 269-84 (1975).

21. Lu, P.-Y., R. L. Metcalf, N. Plummer and D. Mandel, "The Environmental Fate of Three Carcinogens: Benzo-(α)-pyrene, Benzidine, and Vinyl Chloride Evaluated in Laboratory Model Ecosystems," *Arch. Environ. Contam. Toxicol.*, **6**, 129-42 (1977).

22. Macek, K. J., S. R. Petrocelli and B. H. Sleight, III, "Considerations in Assessing the Potential for and Significance of Biomagnification of Chemical Residues in Aquatic Food Chains," in *Aquatic Toxicology,* ASTM STP 667, L. L. Marking and R. A. Kimerle, Eds., American Society for Testing and Materials, 251-268 (1979).

23. Metcalf, R. L., "Biological Fate and Transformation of Pollutants in Water," in *Fate of Pollutants in the Air and Water Environments*, Part 2, ed. by I. H. Suffet, John Wiley and Sons, New York (1977).

24. Metcalf, R. L., I. P. Kapoor, P.-Y. Lu, C. K. Schuth and P. Sherman, "Model Ecosystem Studies of the Environmental Fate of Six Organo-chlorine Pesticides," *Environ. Health Perspect.*, **4**, 35-44 (1973).

25. Metcalf, R. L., P.-Y. Lu and I. P. Kapoor, "Environmental Distribution and Metabolic Fate of Key Industrial Pollutants and Pesticides in a Model Ecosystem," WRC Research Report No. 69, Water Resource Center, University of Illinois, Urbana-Champaign, Ill. (June 1973). NTIS PB-225 479.

26. Narbonne, J. F., "Accumulation of Polychlorinated Biphenyl (Phenochlor DP6) by Estuarine Fish," *Bull. Environ. Contam. Toxicol.*, **22**, 60-64 (1979).

27. Narbonne, J. F., "Polychlorinated Biphenyl Accumulation in Gray Mullets *(Chelon labrosus)*: Effect of Age," *Bull. Environ. Contam. Toxicol.*, **22**, 65-68 (1979).

28. Neely, W. B., "Estimating Rate Constants for Uptake and Clearance of Chemicals by Fish," *Environ. Sci. Technol.*, **13**, 1506 (1979).

29. Neely, W. B., D. R. Branson and G. E. Blau, "Partition Coefficient to Measure Bioconcentration Potential of Organic Chemicals in Fish," *Environ. Sci. Technol.*, **8**, 1113-15 (1974).

30. Norstrom, R. J., A. E. McKinnon and A. S. W. deFreitas, "A Bioenergics-Based Model for Pollutant Accumulation by Fish. Simulation of PCB and Methyl Mercury Residue Levels in Ottawa River Yellow Perch *(Perca flavescens),"J. Fish. Res. Board Can.*, **33**, 248-67 (1976).

31. Pierce, R. H., Jr., "The Fate and Impact of Pentachlorophenol in a Fresh Water Ecosystem," EPA-600/3-78-063 U.S. Environmental Protection Agency, Athens, Georgia (July 1978).

32. Skea, J. C., H. A. Simonin, H. J. Dean, J. R. Colquhoun, J. J. Spagnoli and G. D. Veith, "Bioaccumulation of Aroclor 1016 in Hudson River Fish," *Bull. Environ. Contam. Toxicol.*, **22**, 332-36 (1979).

33. Southworth, G. R., J. J. Beauchamp and P. K. Schmieder, "Bioaccumulation Potential and Acute Toxicity of Synthetic Fuels Effluents in Fresh Water Biota: Azaarenes," *Environ. Sci. Technol.*, **12**, 1062-66 (1978).

34. Southworth, G. R., J. J. Beauchamp and P. K. Schmieder, "Bioaccumulation Potential of Polycyclic Aromatic Hydrocarbons in *Daphnia pulex*," *Water Res.*, **12**, 973-77 (1978).

35. Spacie, A. and J. L. Hamelink, "Dynamics of Trifluralin Accumulation in River Fishes." *Environ. Sci. Technol.*, **13**, 817-22 (1979).

36. Veith, G. D., U.S. Environmental Protection Agency, Environmental Research Laboratory, Duluth, MN, personal communication, June 5, 1980.

37. Veith, G. D., D. L. DeFoe and B. J. Bergstedt, "Measuring and Estimating the Bioconcentration Factor of Chemicals in Fish," *J. Fish. Res. Board Can.*, **36**, 1040-48 (1979).

38. Veith, G. D. and D. E. Konasewich (eds.), *Structure-Activity Correlations in Studies of Toxicity and Bioconcentration with Aquatic Organisms*, Proceedings of a Symposium held in Burlington, Ontario, at the Canada Center for Inland Waters, March 1975.

39. Veith, G. D., K. J. Macek, S. R. Petrocelli and J. Carroll, "An Evaluation of Using Partition Coefficients and Water Solubility to Estimate Bioconcentration Factors for Organic Chemicals in Fish," *J. Fish. Res. Board Can.* (1980) (Prepublication copy).

40. Waller, W. T. and G. F. Lee, "Evaluation of Observations of Hazardous Chemicals in Lake Ontario During the International Field Year for the Great Lakes," *Environ. Sci. Technol.*, **13**, 79-85 (1979).

41. "Water Quality Criteria: Guidelines for Deriving Water Quality Criteria for the Protection of Aquatic Life," *Fed. Regist.*, **43**, No. 97, 21506-518 (May 18, 1978).

6

ACID DISSOCIATION CONSTANT

Judith C. Harris and Michael J. Hayes

6-1 INTRODUCTION

The extent to which an organic chemical is partitioned among the gaseous, solid, and solution compartments of a given environment is determined by several physical and chemical properties of both the chemical and the environment. Among the chemical properties, those that determine acid-base interactions between the chemical and the aqueous or soil/sediment components of the environment exert a major influence on partitioning. An organic acid or base that is extensively ionized may be markedly different from the corresponding neutral molecule in solubility, adsorption, bioconcentration, and toxicity characteristics. For example, the ionized species of an organic acid is generally adsorbed by sediments to a much lesser degree than is the neutral form.

The significance of acid-base chemistry in this regard is most clearly reflected in the acid dissociation constant, K_a, of the chemical. For an organic chemical, HA, that is weakly acidic, K_a is defined as the equilibrium constant for the reaction:

$$HA + H_2O \rightarrow H_3O^+ + A^- \qquad (6\text{-}1)$$

A chemical, H_xA, that has more than one acidic proton undergoes successive dissociations as follows:

$$H_x A + H_2 O \rightleftharpoons H_3 O^+ + H_{x-1} A^- \qquad (6\text{-}1a)$$

$$H_{x-1} A^- + H_2 O \rightleftharpoons H_3 O^+ + H_{x-2} A^{-2} \qquad (6\text{-}1b)$$

$$\vdots$$

$$HA^{-(x-1)} + H_2 O \rightleftharpoons H_3 O^+ + A^{-x} \qquad (6\text{-}1c)$$

The equilibrium constants for the successive dissociation reactions are referred to as K_1 (Eq. 6-1a), K_2 (Eq. 6-1b) . . . K_x (Eq. 6-1c), respectively, of $H_x A$.

Thus,

$$K_a = \frac{a_{H_3O^+} \, a_{A^-}}{a_{HA} \, a_{H_2O}} \qquad (6\text{-}2)$$

where a_i is the activity of species i in an aqueous solution. It is conventional to assign unit activity to solvent water; this is equivalent to choosing pure water as the standard state and assuming that the solution is sufficiently dilute that the activity of the water is unaffected by the presence of solute(s). This assumption is generally true for solute concentrations in the millimolar range and below (i.e., $< 0.1M$). Equation 6-2 then reduces to

$$K_a = \frac{a_{H_3O^+} \, a_{A^-}}{a_{HA}} = \frac{\left(\gamma_{H_3O^+} \, M_{H_3O^+}\right)\left(\gamma_{A^-} \, M_{A^-}\right)}{\gamma_{HA} \, M_{HA}} \qquad (6\text{-}3)$$

in which γ_i is the molar activity coefficient and M_i is the molar concentration of species i. The value of K_a defined in terms of activities is referred to as the "true" or "thermodynamic" dissociation constant. Equation 6-3 is often further simplified by applying the approximation that all activity coefficients are unity, yielding the following equation:

$$K_a \approx \frac{M_{H^+} \, M_{A^-}}{M_{HA}} \qquad (6\text{-}4)$$

Note that K_a as defined in Eq. 6-4 has dimensions of concentration and units of mol/L. The concentration dissociation constant of Eq. 6-4 is generally a good approximation of the thermodynamic constant for solute concentrations below 0.01 M.

An alternative simplification is to retain the activity expression for H_3O^+ but to assume unit activity coefficients for A^- and HA. This yields the expression for the "mixed" or "Bjerrum" acid dissociation constant:

$$K_a \approx \frac{a_{H^+} M_{A^-}}{M_{HA}} \qquad (6\text{-}5)$$

Taking the negative logarithm of both sides of Eq. 6-5 gives:

$$-\log K_a = -\log a_{H^+} - \log \frac{M_{A^-}}{M_{HA}}$$

or

$$pK_a = pH - \log \frac{M_{A^-}}{M_{HA}} \qquad \qquad (6\text{-}6)$$

Rearranging Eq. 6-6 yields a form of the dissociation constant expression that is particularly useful in describing the aqueous solution behavior of the weak organic acid.

$$\log \frac{M_{A^-}}{M_{HA}} = pH - pK_a \qquad (6\text{-}7)$$

According to Eq. 6-7, the concentrations of organic acid in the dissociated (A^-) and free (HA) forms are equal when $pH = pK_a$, and the ratio of A^- to HA increases by an order of magnitude for each unit of pH above pK_a. Thus, for example, acetic acid with $K_a = 1.8 \times 10^{-5}$ and $pK_a = 4.75$ is

1% dissociated at pH 2.75
10% dissociated at pH 3.75
50% dissociated at pH 4.75
90% dissociated at pH 5.75
99% dissociated at pH 6.75

A comparison of the pK_a of an organic acid with the pH of the aqueous system of concern quickly reveals the potential importance of acid dissociation of the organic compound in determining environmental distribution.

The acid-base behavior of weakly basic organic compounds could be treated in an exactly analogous fashion; the base dissociation constant (K_b) for a base, B, can be defined as follows:

$$B + H_2O \rightarrow BH^+ + HO^- \tag{6-8}$$

$$K_b = \frac{a_{HB^+} \, a_{OH^-}}{a_B \, a_{H_2O}} \tag{6-9}$$

For convenience and consistency, however, it seems preferable to address the behavior of weak bases in terms of the K_a or pK_a values of their respective conjugate acids, as in Eqs. 6-10 and 6-11.

$$BH^+ + H_2O \rightarrow H_3O^+ + B \tag{6-10}$$

$$K_a = \frac{a_{H_3O^+} \, a_B}{a_{BH^+} \, a_{H_2O}} \tag{6-11}$$

K_a for the conjugate acid, BH^+, and K_b for the base, B, are related by a constant — the autodissociation constant of water, K_w:

$$K_a = \frac{K_w}{K_b} \tag{6-12}$$

Thus, a decrease in K_a for BH^+ is automatically reflected in an increase in K_b for B; a stronger base corresponds to a weaker conjugate acid. If the pK_a of the conjugate acid is used as a measure of base strength, a uniform scale can be applied over the entire range of organic acid-base behavior in aqueous media.

Table 6-1, which lists pK_a values for various organic compounds, illustrates the wide range of acid and base strengths that can be encountered. Note that acid strength can vary over about 50 orders of magnitude. In aqueous media, however, the range of interest is restricted to acids with pK_a's of 0-14. Acids with $pK_a < 0$ will be completely dissociated to the corresponding conjugate base, while those with $pK_a > 14$ will be completely associated.

If the focus is further narrowed to aqueous media within the normal environmental pH range of 5-8, the range of acidities that are of concern is even more restricted. Equation 6-7 implies that the acidity range of principal interest corresponds to pK_a of 3 to 10. If an organic species has a pK_a outside these limits, it is expected to be either completely (>99%)

TABLE 6-1

Range of pK$_a$ Values for Organic Acids

Conjugate Acid	pK$_a$	Conjugate Base
Methane	40	$CH_3:^-$
Toluene	35	$C_6H_5CH_2:^-$
Aniline	27	$C_6H_5NH:^-$
t-Butanol	19	$C_4H_9O:^-$
Water	15.7	$HO:^-$
Phenol	10	$C_6H_5O:^-$
RNH_3^+	~ 10	RNH_2
p-Nitrophenol	7.2	$p\text{-}NO_2\text{-}C_6H_4O:^-$
Pyridinium ion	5.2	Pyridine
Carboxylic acids	4.5 ± 0.5	Carboxylate anions
p-Nitroanilinium	1.0	*p*-Nitroaniline
$CH_3OH_2^+$	−2	Methanol
$C_6H_5OH_2^+$	−6.7	Phenol

Source: Hendrickson, Cram, and Hammond [3]. *(Reprinted with permission from McGraw-Hill Book Co.)*

dissociated (pK$_a$ of organic acid <3) or completely undissociated (pK$_a$ of conjugate acid >10) in an aqueous environment.

6-2 EXPERIMENTAL MEASUREMENT OF K$_a$

Because acidity has long been recognized as an important property of organic compounds, methods for experimental measurement of pK$_a$ values are well established. The principal procedures, which have been summarized by Kortüm *et al.* [6], are based on the determination of the pH and the concentrations (or ratio of concentrations) of the acid and conjugate base (Eq. 6-6). Dissociation constants are determined by conductance methods, electrometric methods, spectrophotometric methods, magnetic resonance, and measurements of catalytic effects on well-characterized reactions.

If the true thermodynamic K$_a$ (Eq. 6-3) is sought, one must apply activity coefficient corrections to the measured concentrations of acid and conjugate base. This can be done by making measurements at a series of ionic strengths and extrapolating to zero ionic strength. Alternatively, activity coefficients may be calculated from relationships such as the Davies equation:

$$\log \gamma_i = -(Z_i)^2 \left(\frac{0.15\sqrt{I}}{1 + \sqrt{I}} - 0.30\,I \right) \qquad (6\text{-}13)$$

in which Z_i is the charge on the *i*th species and I is the ionic strength [1].

Acid dissociation constants in the range of pK_a 3 to pK_a 11 can generally be measured with a high degree of accuracy and precision. The state of the art of these measurements is illustrated by the subjective description of uncertainties in tabulated values provided by the compendia of Kortüm *et al.* [6] and of Sergeant and Dempsey [10]:

"Very reliable"	± 0.0005 in pK_a
"Reliable"	± 0.005
"Approximate"	± 0.04
"Uncertain"	> 0.04

The equilibrium constant for acid dissociation is affected by several parameters in addition to the structure of the organic acid. Increasing ionic strength of the aqueous medium influences pK_a by favoring the ionic form of the conjugate acid/base pair. The temperature of the medium also influences the magnitude of K_a. However, both of these effects are generally small compared with those related to molecular structure. For instance, K_a values for typical organic acids change by much less than an order of magnitude, typically less than 10%, between 5°C and 60°C [6,9,10]. In general, K_a decreases with increasing temperatures; some anomalous results in the 0-5°C range may reflect variations in properties of the solvent water within this range.

6-3 OVERVIEW OF ESTIMATION METHOD

The dissociation constant of an organic acid (or conjugate acid of an organic base) can be estimated by applying a linear free energy relationship (LFER). An LFER is an empirical correlation between the standard free energies of reaction (ΔF°) or activation ($\Delta F^{\ddagger}$) for two series of reactions, both subjected to the same variations in reactant structures or reaction conditions. Since $\Delta F^{\circ} = -RT \ln K$ for an equilibrium process and $\Delta F^{\ddagger} \propto -RT \ln k$ for a kinetic process, a linear free energy relationship is also a linear relationship between logarithms of equilibrium/rate constants. Wells [11] expressed the basic LFER for two reaction series, A and B, as

$$\log k_i^B = m \log k_i^A + C \qquad (6\text{-}14)$$

where k may stand for either a rate or an equilibrium constant. Several excellent treatments of both the theoretical aspects and the broad applicability of LFERs have been written [2,4,5,7,11], and no attempt will be made here to present a detailed explanation or derivation.

As applied to the estimation of acid dissociation constants, the LFER method is basically a substituent-effect approach. One member of the "A" series, typically an unsubstituted prototype with dissociation constant K_a^o (A), is taken as a reference point. The similarly unsubstituted member of the "B" series may be regarded as the parent compound of the acid whose dissociation constant, $K_a^x(B)$, is to be determined. Equation 6-14 may then be written as:

$$\log \frac{K_a^X \ (B)}{K_a^o \ (B)} = m \log \frac{K_a^X \ (A)}{K_a^o \ (A)} \tag{6-15}$$

where

$K_a^o(A)$ = dissociation constant of reference acid in A series
(e.g., benzoic acid)
$K_a^x(A)$ = dissociation of substituted acid in A series
(e.g., *p*-chlorobenzoic acid)
$K_a^o(B)$ = dissociation constant of parent acid in B series
(e.g., phenol)
$K_a^x(B)$ = dissociation constant of substituted acid in B series
(e.g., *p*-chlorophenol).

The proportionality constant, m, in Eq. 6-15 is a measure of the relative sensitivity of the B-series reactions to substituent changes, compared with the A series. The term $\log[K_a^x(A)/K_a^o \ (A)]$ may be considered an indication of the intrinsic effect of the substituent change. This concept of the separability of a "reaction parameter" and a "substituent effect" constitutes the major practical strength (and perhaps a theoretical weakness) of the LFER approach.

The choice of the "A" series used to define the substituent parameters is the principal difference between one LFER system and another. Table 6-2 summarizes four of the more familiar LFERs. Hansch and Leo [2] describe several additional systems that have been evolved for special purposes. The Hammett relationship for aromatic systems and the Taft relationship for aliphatics are the most generally applicable LFERs for estimating acid dissociation constants. Although one of the

TABLE 6-2

Commonly Encountered LFERs and Substituent Parameters

Name	Substituent Symbol	Parameter Defining Reaction Series	K or k Basis[a]	Special Feature(s)
Hammett	σ	Dissociation of benzoic acids	K	
Hammett	σ^-	Dissociation of anilinium ions	K	Accounts for "through resonance" between reaction center and electron-withdrawing substituents
Brown	σ^+	Hydrolysis of cumyl chlorides	k	Accounts for "through resonance" between reaction center and electron-donating substituents
Taft	σ^*	Hydrolysis of carboxylate esters — difference between base- and acid-catalyzed rates	k	Corrects for steric effects; inductive effects of substituent dominate σ^*

a. K indicates an equilibrium constant and k indicates a reaction rate constant as the basis for the LFER.

special-purpose equations might give somewhat better results in a particular instance, it is generally not possible to predict which LFER should be used. Moreover, the special-purpose LFERs do not expand the range of substituent types or acid types that can be considered. The Hammett and Taft correlations should provide estimates of dissociation constants that are adequately accurate for evaluation of probable environmental partitioning behavior.

6-4 ESTIMATION OF K_a FOR AROMATIC ACIDS — HAMMETT CORRELATION

The Hammett correlation is most commonly written as follows:

$$\log \frac{K_a^x}{K_a^o} = \sigma\rho \tag{6-16}$$

where:

K_a^x = acid dissociation constant of substituted compound
K_a^o = acid dissociation constant of parent compound
σ = substituent constant, sigma
ρ = reaction constant, rho

The correlation can be rewritten for convenience in solving explicitly for K_a^x (Eq. 6-17) or pK_a^x (Eq. 6-18).

$$K_a^x = K_a^o \; 10^{\sigma\rho} \tag{6-17}$$

$$pK_a^x = pK_a^o - \sigma\rho \tag{6-18}$$

Three steps are involved in estimating the dissociation of a substituted acid:

(1) Selection of an appropriate parent compound for which K_a^o and ρ values are available,

(2) Selection of the substituent constant value(s), and

(3) Calculation of K_a^x or pK_a^x.

Both the complexity of the selection processes and the accuracy of the estimated dissociation constant vary, depending on the type of compound under consideration. The procedure is described in detail below.

The use of the Hammett correlation is simplest when estimating K_a for derivatives of benzoic acids containing only meta and para substituents, so this procedure is given first. The procedure used for more complex aromatic compounds is described in the following subsection.

Basic Steps for Substituted Benzoic Acids

(1) The parent compound of all species involved here is benzoic acid; its K_a is used as K_a^o ($pK_a^o = 4.203$, $K_a^o = 6.26 \times 10^{-5}$) [5], and the value of ρ is defined as 1.

(2) Find the value of σ as follows:

- If the compound is a monosubstituted benzoic acid, find the appropriate substituent constant in Table 6-3.

- If more than one substituent is present, see Table 6-4 for the appropriate multi-substituent σ value.

- If the correct combination is not found in Table 6-4, locate the individual substituents in Table 6-3 and sum their σ values. Use this sum (σ_T) in place of a single σ value to calculate K_a^x. If Table 6-3 does not list one or more of the substituents, find a default value of σ in Table 6-5.

- A substituent that is neither covered by the generalized categories of Table 6-5 nor found in Table 6-3 cannot be assigned a σ value; therefore, K_a^x for the corresponding acid cannot be calculated from the Hammett equation.

(3) Substitute the appropriate values into Eq. 6-17 and solve for K_a^x.

Example 6-1   Estimate the dissociation constant for *p-tert* butyl benzoic acid.

(1)   As described above, K_a^o for benzoic acid is 6.26×10^{-5} and $\rho \equiv 1$.

(2)   From Table 6-3, σ para for the *tert* butyl group, $C(CH_3)_3$, is -0.197.

(3)   Substitute the above values in Eq. 6-17:

$$K_a^x = (6.26 \times 10^{-5})\ 10^{(-0.197)(1)}$$

$$= 3.98 \times 10^{-5}$$

This estimate is identical to the measured value as tabulated by Kortüm *et al.* [6].

TABLE 6-3

Hammett Substituent Constants

Substituent	σ Meta	σ Para	σ⁻ Para[a]	Ref.
Hydrocarbon Groups				
CH_3	-0.069	-0.170		[8]
CH_2CH_3	-0.07	-0.151		[8]
$CH_2CH_2CH_3$		-0.126		[5]
$CH_2(CH_2)_2CH_3$		-0.161		[5]
$CH_2CH(CH_3)_2$		-0.115		[5]
$CH(CH_3)CH_2CH_3$		-0.123		[5]
$CH_2CH_2CH(CH_3)_2$		-0.225		[5]
$C(CH_3)_2CH_2CH_3$		-0.190		[5]
$CH(CH_3)_2$		-0.151		[8]
$C(CH_3)_3$	-0.10	-0.197		[8]
C_6H_5	0.06	-0.01		[8]
$CH=CHC_6H_5$	0.141		0.619	[5]
$C{\equiv}CC_6H_5$	0.14	0.16		[11]
Carbonyl-Containing Groups				
$COCH_3$	0.376	0.502		[8]
$COOH$	0.355	0.265	0.728	[5]
$COOCH_3$	0.315		0.636	[5]
$COOCH_2CH_3$	0.398	0.522	0.678	[5]
$COO(CH_2)_3CH_3$			0.674	[5]
$COOCH_2C_6H_5$			0.667	[5]
$CONH_2$	0.280		0.267	[5]
CHO	0.35	0.22	1.126	[11]
COC_6H_5		0.459		[5]
COO^-	-0.1	0.0		[8]
$OCOCH_3$	0.39	0.31		[8]
CH_2CH_2COOH	-0.027	-0.066		[5]
Nitrogen-Containing Groups				
$N(CH_3)_2$		-0.83		[8]
$NHCOCH_3$	0.21	0.00		[8]
$NHCOC_6H_5$	0.217	0.078		[5]
$NHNH_2$	-0.1	-0.4		[11]
$NHOH$	-0.044	-0.339		[5]
NH_3^+	0.634			[5]
$NH_2CH_3^+$	0.958			[5]
$NH_2CH_2CH_3^+$	0.958			[5]
$N(CH_3)_3^+$	0.88	0.82		[8]
$N(CF_3)_2$	0.45	0.53		[11]
NO_2	0.710	0.778	1.270	[8]
CH_2CN		0.01		[11]
CN	0.56	0.660	1.00	[8,5]
$C_6H_4N{=}NC_6H_5$			1.088	[5]
$N{=}NC_6H_5$		0.64		[5]
N_2^+	1.76	1.91		[8]
NH_2	-0.16	-0.66		[8]
$NHCH_3$		-0.84		[8]
N_3	0.33	0.08		[11]
$NHCH_2CH_3$	-0.240			[5]
$NH(CH_2)_3CH_3$	-0.344			[5]

(Continued)

TABLE 6-3 (Continued)

Substituent	σ Meta	σ Para	σ⁻ Para[a]	Ref.
Halogens and Alkyl Halide Groups				
F	0.337	0.062		[8]
Cl	0.373	0.227		[8]
Br	0.391	0.232		[8]
I	0.352	0.18		[8]
IO$_2$	0.70	0.76		[8]
CF$_3$	0.43	0.54		[8]
CCl$_3$	0.40	0.46		[11]
CH$_2$Cl		0.184		[5]
Hydroxy and Alkoxy Groups				
OCH$_3$	0.115	-0.236		[8]
OCH$_2$CH$_3$	0.1	-0.24		[8]
O(CH$_2$)$_2$CH$_3$	0.1	-0.25		[8]
OCH(CH$_3$)$_2$	0.1	-0.45		[8]
O(CH$_2$)$_3$CH$_3$	0.1	-0.32		[8]
O(CH$_2$)$_4$CH$_3$	0.1	-0.34		[8]
O(CH$_2$)$_5$CH(CH$_3$)$_2$	—	-0.265		[5]
OC$_6$H$_5$	0.252	-0.320		[8]
OCH$_2$C$_6$H$_5$	—	-0.415		[5]
OCF$_3$	0.4	0.5		[11]
O⁻	-0.71	-0.52		[11]
OH	0.121	-0.37		[8]
Sulfur-Containing Groups				
SCH$_3$	0.15	0.00		[8]
SCH$_2$CH$_3$		0.03		[8]
SCH(CH$_3$)$_2$		0.07		[8]
SH	0.25	0.15		[8]

Substituent	σ Meta	σ Para	σ⁻ Para[a]	Ref.
SCOCH$_3$	0.39	0.31		[8]
SCN		0.52		[8]
SOCH$_3$	0.52	0.49		[8]
SO$_2$CH$_3$	0.60	0.72	1.049	[8,5]
SO$_2$NH$_2$	0.46	0.57		[8]
S(CH$_3$)$_2$⁺	1.00	0.90		[8]
SO$_3$⁻	0.05	0.09		[8]
SO$_3$H		0.5		[11]
SCF$_3$	0.44	0.57		[5]
SCF$_5$	0.6	0.7		[5]
Phosphorus-Containing Groups				
P(CH$_3$)$_2$	0.1	0.05		[11]
P(CF$_3$)$_2$	0.6	0.7		[11]
P(CH$_3$)$_3$⁺	0.8	0.9		[11]
PO$_3$H⁻	0.2	0.26		[11]
Miscellaneous Groups				
CH$_2$Si(CH$_3$)$_3$	-0.16	-0.21		[11]
Si(CH$_3$)$_3$	-0.04	-0.07		[8]
Si(CH$_2$CH$_3$)$_3$		0.0		[11]
Ge(CH$_3$)$_3$		0.0		[8]
Ge(CH$_2$CH$_3$)$_3$		0.0		[8]
Sn(CH$_3$)$_3$		0.0		[11]
Sn(CH$_2$CH$_3$)$_3$		0.0		[8]
AsO$_3$H⁻		-0.02		[8]
SeCH$_3$	0.1	0.0		[8]
SeCN	0.6	0.664		[8]
B(OH)$_2$	0.006	0.454		[5]

a. See Table 6-2 for definition.

TABLE 6-4

Values of σ for Multiple Substituents

R_1	R_2	$\Sigma\sigma_R$ [a]	σ Measured[b]

3,4-Disubstitution

(structure: benzene ring with Y at top, R_1 and R_2 positions)

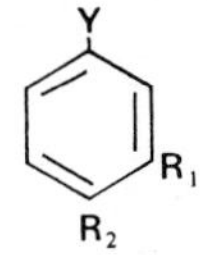

R_1	R_2	$\Sigma\sigma_R$ [a]	σ Measured[b]
Cl	Cl	0.600	0.525
Cl	OH	−0.016	−0.049
Cl	CH_3	0.203	0.235
Cl	OCH_3	0.105	0.268
CH_3	CH_3	−0.239	−0.303
CH_3	NO_2 (c)	0.709	0.694
CH_3	OCH_3	−0.337	−0.265
CH_3	$N(CH_3)_2$	−0.669	−0.302
CH_3	Cl	0.158	0.174
CH_3	NH_2	−0.720	−0.716
OCH_3	OCH_3	−0.153	−0.117
OCH_3	OH	−0.242	−0.329
OCH_3	Cl	0.342	0.338
NO_2	NO_2 (c)	1.488	1.379
NO_2	Cl	0.937	0.901
NO_2	Br	0.942	0.826
NO_2	OCH_3	0.442	0.414
NO_2	CH_3	0.540	0.505
NO_2	NO_2 (c)	1.980	2.036
OH	OH	−0.359	−0.278
NH_2	CH_3	−0.331	−0.209
$N(CH_3)_2$	CH_3	−0.381	−0.176
Br	CH_3	0.221	0.150
Br	OCH_3	0.123	0.088

3,5-Disubstitution

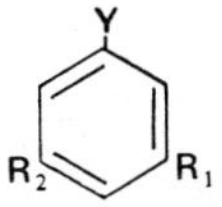

(structure: benzene ring with Y at top, R_2 and R_1 positions)

R_1	R_2	$\Sigma\sigma_R$ [a]	σ Measured[b]
NO_2	NO_2	1.420	1.395
NO_2	Cl	1.083	1.073
OCH_3	OCH_3	0.230	0.050
OCH_3	Cl	0.488	0.439
CH_3	CH_3	−0.138	−0.173
CH_3	Cl	0.304	0.347
Br	Br	0.782	0.720
Cl	Cl	0.746	0.746
OH	OH	0.242	0.162

(Continued)

TABLE 6-4 (Continued)

Trisubstituted Compounds

R_1	R_2	R_3	$\Sigma \sigma_R{}^a$	σ Measured[b]
OCH_3	OCH_3	OCH_3	−0.038	0.075
OCH_3	OH	NO_2	0.468	0.433
OH	OCH_3	NO_2	0.444	0.634

Fused Ring Systems

R	σ Measured[b]
$3,4\text{-}(CH_2)_3$	−0.259
$3,4\text{-}(CH_2)_4$	−0.477
$3,4\text{-}(CH)_4$	0.170
$3,4\text{-}CH_2O_2$	−0.159

a. The sums of the individual σ values from Table 6-3 are listed here to permit comparison with the measured σ for a multiple substitution, in order to demonstrate the magnitude of the uncertainty that may be incurred by assuming additivity of σ values.
b. To be used in calculations instead of $\Sigma \sigma$.
c. In calculating $\Sigma \sigma$, the σ value is used for the 4-nitro group.

Source: Jaffé [5]. *(Reprinted with permission from the Williams & Wilkins Co.)*

TABLE 6-5

Default Values of σ for Substituents in Generalized Categories for Aromatic Acids

Substituent Category	Atom or Group Attached to Parent Acid (Ar)	Default Value[a]	
		σ Meta	σ Para
Alkyl (C,H only)	$Ar - \overset{\mid}{\underset{\mid}{C}} -$	−0.08	−0.16
Alkyl with α Halogen	$Ar - \overset{\mid}{\underset{\mid}{C}} - X$	0.14[b]	0.18
Alkyl with α Carbonyl	$Ar - \overset{O}{\overset{\parallel}{C}} -$	0.35	0.43
Amine[c]	$Ar - N{<}$	−0.30[c]	−0.83[c]
Aryl amide	$Ar - \overset{O}{\underset{\parallel}{N} - C} - R$	0.21	0.05
Ammonium	$Ar - \overset{\mid}{\underset{\mid}{N^+}} - R$	0.85	0.8
Ether	$Ar - O - R$	0.1	0.31
Mercaptan or Sulfide	$Ar - S - R$	0.15	0.03
Sulfoxy	$Ar - \overset{O}{\underset{\parallel}{S}} - R_2$	0.6	0.7
Phosphorous	$Ar - P\,(III) - R$	0.1	0.05
Phosphoric	$Ar - P\,(V) - R$	0.2	0.26
Organo-Silicon	$Ar - Si\,R_3$	−0.04	−0.07
Organo-Germanium	$Ar - Ge\,R_3$	NA	0.0
Organo-Tin	$Ar - Sn\,R_3$	NA	0.0

a. Default values are arithmetic averages of values reported in Table 6-3 for each substituent category, except as noted.
b. Calculated as one third of the average of the Table 6-3 σ meta values for CCl_3 and CF_3.
c. Excluding $NOCOCH_3$, $NHCOC_6H_5$, and $N(CH_3)_2$.

Example 6-2 Estimate the dissociation constant for 4-methyl-3, 5-dinitrobenzoic acid.

(1) As above, K_a^o for benzoic acid is 6.26×10^{-5} and $\rho \equiv 1$.

(2) The exact desired substituent pattern is not included in Table 6-4, but the total σ value can be estimated by summing the value for 3,5-dinitro substitution (Table 6-4) and the value for 4-methyl substitution from Table 6-3.

	σ
m,m-diNO$_2$	$=$ 1.395
p-CH$_3$	$=$ -0.170
σ_T	$=$ 1.225

(3) Substitute the above values in Eq. 6-17:

$$K_a^x = (6.26 \times 10^{-5}) \, 10^{(1.225)\,(1)} = 1.05 \times 10^{-3}$$

This estimate deviates -1.8% from the measured value of 1.07×10^{-3} [4].

Basic Steps for Other Aromatic Acids

(1) Locate in Table 6-6 the parent compound that is the most suitable model for the compound of interest and substitute the reaction constant (ρ) and dissociation constant (K_a^o) for that compound in Eq. 6-17. Use the measured value of K_a^o, not the calculated value, from Table 6-6. Use the following criteria in choosing the parent compound:

   • Choose a parent compound that contains the same acid function (carboxylic acid, phenol) as the compound of interest.

   • Choose a parent compound that contains any ortho substituents present in the compound of interest.

   • Choose the parent so that substituent constants are available for the remaining substituents.

   • If two comparable routes are available, calculate both and average the results.

(2) Search Tables 6-3 and 6-4, if necessary, for the substituents required to complete the structure of interest. For derivatives of phenol and aniline, use values of σ^-, if available. Obtain the value of σ for each substituent needed. If more than one is needed, sum the constants before substituting in the equation. If an exact substituent cannot be found, choose a default value from Table 6-5.

TABLE 6-6

Hammett Reaction Constants (ρ) for Equilibrium Reactions in Water

Parent Compound	pK_a^o (Measured)		T (°C)	ρ	n^a	r^b	pK_a^o (Calc.)^c
Benzoic Acids							
Benzoic	4.205	[10]	25	1.000			
o-Nitrobenzoic	2.21	[10]	25	0.905	6	0.992	2.206
Toluic (o-methyl)	3.93	[10]	25	1.430	4	0.955	3.875
Salicylic (o-hydroxy)	3.08	[10]	25	1.103	6	0.978	3.997
o-Chlorobenzoic	2.900	[10]	25	0.855	4	0.970	3.693
o,o-Dimethylbenzoic	3.354	[10]	26	1.116^d	4	0.999	3.974^d
p-Phenylbenzoic^e	—		25	0.482^e	9	0.981	5.636^e
Other Carboxylic Acids							
Phenylacetic	4.307	[10]	25	0.489	14	0.981	4.297
3-Phenylpropanoic	4.664	[10]	25	0.212	8	0.979	4.551
3-Phenylpropenoic (trans)	4.42	[10]	25	0.466	9	0.977	4.447
4-Phenyl-2-keto-but-3-enoic	—		25	-0.054	5	0.903	1.971
2-Furancarboxylic	3.16	[10]	24	1.396^f	6	0.988	2.819
Other Acids							
Phenol	9.994	[10]	25	2.113^g	17	0.990	9.847
Catechol (o-hydroxyphenol)	9.45	[10]	25	3.512^{g,h,i}	8	0.991	11.012^{h,i}
Anilinium ion	4.603	[9]	25	2.767^g	14	0.995	4.557

(Continued)

TABLE 6-6 (Continued)

Parent Compound	pK$_a^o$ (Measured)		T (°C)	ρ	n[a]	r[b]	pK$_a^o$ (Calc.)[c]
Other Acids (Cont.)							
Dimethylanilinium ion	5.068	[9]	20	3.426[g,i,l]	5	0.986	3.285[i,l]
Benzylammonium ion	9.33	[9]	25	0.732	5	0.942	9.315
Thiophenol	6.615	[10]	20-22	2.236[i,l]	12	0.975	7.666[j,l]
Phenylboronic acid	8.84	[10]	25	2.146[k,l]	14	0.990	9.700[k,l]
Phenylphosphonic acid	1.83	[6] (pK$_1$)	25	0.755	10	0.995	1.836
	7.43	[10] (pK$_2$)	25	0.949	12	0.990	6.965
o-Chloro- or o-bromo- phenylphosphonic acid	1.63	[6] (pK$_1$)	25	0.995	4	0.998	2.942
	6.98	[6] (pK$_2$)	25	0.908[i,l]	6	0.964	6.901
Phenylarsonic acid	3.65	[10] (pK$_1$)	18-25	1.050	9	0.975	3.540
	8.77	[10] (pK$_2$)	22	0.874	11	0.965	8.491
Benzeneseleninic acid	4.70	[10]	25	0.905	16	0.906	4.740

a. Number of acids used in calculation of ρ.
b. Correlation coefficient.
c. Calculated pK$_a$ for parent acid = intercept of regression line with the ordinate (σ=0).
d. Value for 20% dioxane/water solvent.
e. For substituents in the p-phenyl ring (not the aromatic ring bearing the —COOH). Values are for 50% butyl cellosolve/water solvent.
f. For substituents at the 5-position in the furan ring.
g. Use σ⁻ para values (Table 6-3), if available, for para substituents.

h. Value for 40% dioxane/water solvent.
i. Value for 30% ethanol/water solvent.
j. Value for 50% ethanol/water solvent.
k. Value for 25% ethanol/water solvent.
l. In general, ρ values are inversely related to dielectric constant of the medium; thus, ρ would be expected to be smaller for pure water than for mixed solvent. However, there are some known exceptions [5], and available data are insufficient for estimation of the magnitude of solvent effects on ρ values.

Source: Jaffé [5]. *(Reprinted with permission from the Williams & Wilkins Co.)*

(3) Substitute the above data in Eq. 6-17 and solve for K_a. If K_b is needed, find K_a as above and calculate K_b from either of the alternative forms of Eq. 6-2:

$$K_b = 10^{-14}/K_a$$
$$pK_b = 14 - pK_a$$

Example 6-3   Estimate the dissociation constant for 3-chloro-4-methoxyphenyl-phosphonic acid.

(1) The most appropriate parent compound is phenylphosphonic acid. From Table 6-6

ρ $= 0.755$

$pK_a^o = 1.83$

$K_a^o = 1.46 \times 10^{-2}$

(2) From Table 6-4, the substituent constant for 3-chloro-4-methoxy is

σ found $= 0.268$

(3) Substitute the above values in Eq. 6-17 and solve:

$K_a^x = K_a^o \, 10^{\sigma\rho}$

$= (1.46 \times 10^{-2}) \, 10^{(0.263)\,(0.755)}$

$= 2.32 \times 10^{-2}$

This estimate deviates +314% from the experimentally measured value of 0.56×10^{-2} [4].

Example 6-4   Estimate the dissociation constant for 4-*t*-butylphenylacetic acid.

(1) The parent compound from Table 6-6 is phenylacetic acid.

ρ $= 0.489$

$pK_a^o = 4.307$

$K_a^o = 4.93 \times 10^{-5}$

(2) From Table 6-3, the substituent constant for the *para-tert* butyl group is $\sigma = -0.197$.

(3) Substitute the above values in Eq. 6-17.

$$K_a^x = (4.93 \times 10^{-5}) \, 10^{(-0.197)(0.489)}$$

$$= 3.95 \times 10^{-5}$$

This estimate deviates +3.9% from the measured K_a of 3.8×10^{-5} [4].

Example 6-5 Estimate the dissociation constant for 3,4-dimethylaniline.

(1) The parent compound is aniline. From Table 6-6 (for anilinium ion)

$$\rho \quad = \quad 2.767$$

$$pK_a^o \quad = \quad 4.603$$

(2) From Table 6-4, the combined sigma constant is −0.303.

(3) Substitute the above values in Eq. 6-18.

$$pK_a^x = 4.603 - (-0.303)(2.767)$$

$$pK_a^x \quad = \quad 5.44$$

This estimate deviates −37% (in K_a) from the average measured value $pK_a = 5.24$ [5].

To calculate pK_b:

$$pK_b = 14 - 5.40 = 8.60$$

6-5 ESTIMATION OF K_a FOR ALIPHATIC ACIDS — TAFT CORRELATION

Correlation data have been collected, primarily from the work of Taft, for estimating the dissociation constants of aliphatic acids. The procedure used to estimate K_a parallels that for aromatic systems and uses the Taft equation, which, although derived differently, is similar to the Hammett equation. Any of the following three forms can be used:

$$\log \frac{K_a^x}{K_a^o} = \sigma^* \rho^* \tag{6-19}$$

$$K_a^x = K_a^o \, 10^{\sigma^* \rho^*} \tag{6-20}$$

$$pK_a^x = pK_a^o - \sigma^* \rho^* \tag{6-21}$$

Basic Steps

(1) Obtain from Table 6-7 the value of ρ^* and K_a^o for the parent compound corresponding to the species of interest.

TABLE 6-7

Reaction Parameters for Acid Dissociation

Acid	ρ^* [11]	pK_a^o
RCH_2COOH	1.75	4.76 [6]
$RCH_2PO_3H_2$	1.16	2.38 (pK_2 = 7.72) [6]
RCH_2OH	3.47	-2 [3]
RCH_2SH	3.73	
$RCH_2PH_3^+$	2.64	
$(RCH_2)_2PH_2^+$	2.61	
$(RCH_2)_3PH^+$	2.67	8.80 [9]
$RCH_2NH_3^+$	3.80	11.08 [9]
$(RCH_2)_2NH_2^+$	3.90	10.8 [9]
$(RCH_2)_3NH^+$	4.29	9.80 [9]

(2) Obtain from Table 6-8 the substituent constant σ^* for the group that completes the structure.

(3) Substitute the above values into the Taft equation (6-19, -20, or -21) and solve for K_a.

Example 6-6 Estimate the dissociation constant for isovaleric acid, $(CH_3)_2CHCH_2COOH$.

(1) The parent compound, from Table 6-7, is RCH_2COOH.

$$\rho^* = 1.75 \qquad pK_a^o = 4.76$$

(2) The substituent that completes the structure, from Table 6-8, is $i\text{-}C_3H_7$, which is $(CH_3)_2CH-$. The substituent constant is $\sigma^* = -0.13$.

(3) Substitute the above values into Eq. 6-21.

$$pK_a^x = 4.76 - (-0.13)(1.75) = 4.99$$

This estimate deviates -37% in K_a from the measured value of $pK_a = 4.79$.

TABLE 6-8

Substituent Constants for Taft Equation

R[a]	σ^*	R[a]	σ^*	R[a]	σ^*
$(CH_3)_3N^+$	2.00	$OCOCH_3$	0.89	$N(CH_3)_2$	0.22
NO_2	1.40	OCH_3	0.66	C_6H_5	0.22
CH_3SO_2	1.38	CO_2R	0.66	$CH=CH_2$	0.12
CH_3SO	1.33	$COCH_3$	0.62	$C_6H_4CH_2$	0.08
CN	1.25	$NH\overset{\overset{O}{\|}}{C}CH_3$	0.60	H	0.00
F	1.10	OH	0.55	CH_3	−0.10
Cl	1.05	SH	0.47	C_2H_5	−0.12
Br	1.02	SCH_3	0.42	i-C_3H_7	−0.13
CF_3	0.92	NH_2	0.40	i-C_4H_9	−0.17
I	0.88	O^-	0.27	$Si(CH_3)_3$	−0.25

a. Substituent R in RCH_2

Source: Wells [11]

6-6 UNCERTAINTY IN ESTIMATED VALUES

It is difficult to gauge the probable uncertainties in values of K_a estimated by the methods described here. An indication of the inherent uncertainty in the LFER approach can be deduced from the data in Table 6-9. On the basis of observed ranges of standard deviations and correlation coefficients, Jaffé [5] has estimated that an average error of ±15% can be expected for a prediction based on a given LFER. Fundamental sources of error in predicted values include deviations from the basic LFER assumption of separability of substituent constants (σ values) and reaction constants (ρ values). These deviations, and hence the errors in the predicted values, are likely to be largest for strongly interacting substituents (large absolute value of σ). Some additional assumptions behind the LFER concept, such as the presumed constancy of the reaction mechanism, may safely be considered valid in the case of dissociation reactions. However, the uncertainty in the estimated K_a values using the procedures described in this chapter may be somewhat higher than the 15% indicated by Jaffé or that implied by Table 6-9.

The greater uncertainty is associated with difficulties in selecting an appropriate parent compound as well as in extending the correlations

TABLE 6-9

Some Correlation Data for Linear Free Energy Relationships of Hammett and Taft

Reaction Series	n^a	ρ^b	r^c	s^d
Hammett				
Dissociation of phenylacetic acids	5	0.56 ± 0.16	0.982	
Dissociation of phenylphosphonic acids	5	0.75 ± 0.00	1.000	
Dissociation of phenols	7	2.26 ± 0.07	0.997	
Dissociation of anilinium ions	7	2.94 ± 0.06	0.999	
Taft				
Dissociation of acetic acids	16	1.72 ± 0.03		0.06
Dissociation of alcohols	8	1.36 ± 0.09		0.09

a. Number of points in correlation
b. Reaction constant ± standard deviation
c. Correlation coefficient
d. Probable error of fit of single observation

Source: Wells [11]

beyond the range of substituents used in defining ρ for that parent. Additional uncertainty arises when value of pK_a° measured for the parent acid differs substantially from that calculated from the intercept of the correlation equation. We have recommended use of the measured pK_a°. Wolfe [12] has suggested that it seems more appropriate to use the pK_a° value calculated from the regression equation. Our recommendation was made on the basis that the value of pK_a° (calculated) depends on which particular compounds were used to develop the correlation, while pK_a° (measured) is an intrinsic property of the parent acid. The recommended approach will give estimates that are more accurate for low absolute values of σ and less accurate for high values. Following Wolfe's suggestion will give more accurate estimates for large substituent changes and less accurate estimates for substituents with low values of σ.

A further source of uncertainty in estimated values results from the fact that the substituent and reaction constants given in Tables 6-3 through 6-7 are obtained from a variety of sources, rather than from one consistent set of measurements. Although the reliability of the estimated K_a's cannot be determined quantitatively, values estimated according to the procedures given in this chapter can confidently be regarded as reliable to within an order of magnitude. Most estimates are probably good to within a factor of 2 or 3 in K_a or ± 0.3-0.5 in pK_a.

Table 6-10 compares some measured and estimated values of K_a. It seems obvious from this small sample that errors are smallest for aromatic species with a single acid functionality. Errors are larger for aliphatic species and for compounds containing more than one acid functional group.

6-7 AVAILABLE DATA

There are a number of compilations of acid dissociation constants which cover a wide range of organic chemical compounds. The following are especially useful:

Kortüm, G., *et al.* [6] — Critical compilation of literature values of acid dissociation constants through 1955.

Perrin, D.D., [9] — Critical compilation of literature values of base dissociation constants through 1961.

Sergeant, E.P. and B. Dempsey, [10] — Critical compilation of literature values of acid dissociation constants through 1972.

6-8 SYMBOLS USED

a_i	=	activity of species i
A^-	=	conjugate anion of neutral organic acid
B	=	organic base
BH^+	=	conjugate acid of organic base
C	=	parameter in Eq. 6-14
ΔF^o	=	standard free energy of reaction
$\Delta F^\ddagger$	=	standard free energy of activation
HA	=	organic acid
I	=	ionic strength
k_i	=	rate or equilibrium constant for reaction i
K_a	=	acid dissociation constant in Eq. 6-2
K_b	=	base dissociation constant in Eq. 6-9
K_w	=	autodissociation constant of water

TABLE 6-10

Comparison of Measured and Estimated Values of Dissociation Constants

Aromatic Compounds	K_a			Error in Estimated Value (%)
	Measured	Ref.	Estimated	
p-Aminobenzoic acid				
K_1 (NH_3^+ group)	5.13×10^{-3}	[6]	2.58×10^{-3}	-49
K_2 (COOH group)	1.37×10^{-5}	[6]	1.36×10^{-5}	-0.7
m-Aminobenzoic acid				
K_1 (NH_3^+ group)	8.51×10^{-4}	[6]	2.39×10^{-4}	-72
K_2 (COOH group)	1.86×10^{-5}	[6]	4.32×10^{-5}	$+132$
p-Methoxybenzoic acid	3.38×10^{-5}	[6]	3.62×10^{-5}	$+7$
m-Phenoxybenzoic acid	1.12×10^{-4}	[6]	1.11×10^{-4}	-0.9
m-Methylsulfonylbenzoic acid	3.02×10^{-4}	[6]	2.49×10^{-4}	-17
p-Tolylacetic acid	4.27×10^{-5}	[6]	4.07×10^{-5}	-5
p-Nitrophenylarsonic acid, K_1	1.27×10^{-3}	[6]	1.47×10^{-3}	$+16$
p-Cyanophenol	1.12×10^{-8}	[6]	1.31×10^{-8}	$+17$
Tetralol-2	3.31×10^{-11}	[6]	9.96×10^{-12}	-70
1,3,5-Trihydroxybenzene, K_1	3.56×10^{-9}	[6]	2.23×10^{-10}	-94^a
m-Aminophenol, K_1 (NH_3^+ group)	6.76×10^{-5}	[9]	5.39×10^{-5}	-20
m-Aminophenol, K_2 (OH group)	1.35×10^{-10}	[6]	4.66×10^{-11}	-65
3-Bromo-4-methoxy anilinium ion	8.32×10^{-5}	[9]	4.37×10^{-5}	-47
4-Chloro-3-nitroanilinium ion	1.26×10^{-2}	[9]	7.76×10^{-3}	-39

(Continued)

TABLE 6-10 (Continued)

| Aliphatic Compounds | K_a | | | Error in Estimated Value (%) |
	Measured	Ref.	Estimated	
Bromoacetic acid	1.25×10^{-3}	[6]	1.06×10^{-3}	− 15
Dichloroacetic acid	5.53×10^{-2}	[6]	8.22×10^{-2}	+ 49
Trifluoroacetic acid	0.59	[6]	10.35	+1600
Cyanoacetic acid	3.36×10^{-3}	[6]	2.68×10^{-3}	− 20
But-3-enoic acid	4.62×10^{-5}	[6]	2.82×10^{-5}	− 39
Chloromethylphosphonic acid	3.98×10^{-2}	[6]	6.89×10^{-2}	+ 73
Hydroxymethylphosphonic acid	1.23×10^{-2}	[6]	1.81×10^{-2}	+ 47
Glycine				
K_1	4.47×10^{-3}	[6]	8.7×10^{-5} (COOH group)	b
K_2	1.66×10^{-10}	[6]	2.6×10^{-9} (NH_3^+ group)	b
Aminocyanomethane	4.57×10^{-6}	[9]	4.67×10^{-7}	90

a. Estimated value not corrected for symmetry effects on K_a.
b. Estimation method fails seriously for amino acids; this is due, at least in part, to the existence of Zwitterions as the dominant form of the neutral molecule.

LFER = linear free energy relationship
m   = proportionality constant in Eqs. 6-14 and 6-15
M_i = molar concentration of component i
pK_a = negative logarithm of acid dissociation constant
pK_a^o = pK_a of unsubstituted parent acid
pK_a^x = pK_a of substituted acid
r   = correlation coefficient
R   = gas constant
s   = probable error of fit of single observation in Table 6-9
T   = temperature
Z_i = charge on ith species in Eq. 6-13

Greek

γ_i = activity coefficient for ith species
ρ = reaction constant in Hammett correlation
ρ^* = reaction constant in Taft correlation
σ = substituent constant in Hammett correlation
σ^- = substituent constant in Hammett correlation especially for anilinium ions and phenols
σ^+ = substituent constant in Brown correlation
σ^* = substituent constant in Taft correlation

6-9   REFERENCES

1. Davies, C.W., *Ion Association,* Butterworths, London, p. 41 (1962).

2. Hansch, C. and A. Leo, *Substituent Constants for Correlation Analysis in Chemistry and Biology,* John Wiley and Sons, New York (1979).

3. Hendrickson, J.B., D.J. Cram and G.S. Hammond, *Organic Chemistry,* 3rd ed., McGraw-Hill, New York, pp. 303-307 (1970).

4. Hine, J., *Structural Effects on Equilibria in Organic Chemistry,* Wiley-Interscience, New York (1975).

5. Jaffé, H.H., "A Reexamination of the Hammett Equation," *Chem. Rev.,* **53**, 191-261 (1953).

6. Kortüm, G., W. Vogel and K. Andrussow, *Dissociation Constants of Organic Acids in Aqueous Solution,* Butterworths, London (1961).

7. Leffler, J.E. and E. Grunwald, *Rates and Equilibria of Organic Reactions,* John Wiley and Sons, New York (1963).

8. McDaniel, D.H. and H.C. Brown, "An Extended Table of Hammett Substituent Constants Based on Ionization of Substituted Benzoic Acids," *J. Org. Chem.,* **23**, 420 (1958).

9. Perrin, D.D., *Dissociation Constants of Organic Bases in Aqueous Solution,* Butterworths, London (1965).

10. Sergeant, E.P. and B. Dempsey, *Ionization Constants of Organic Acids in Aqueous Solution,* Pergamon Press, New York (1979).

11. Wells, P.R., *Linear Free Energy Relationships,* Academic Press, New York (1968).

12. Wolfe, N.L., U.S. Environmental Protection Agency, Athens, GA, personal communication, September 23, 1980.

7

RATE OF HYDROLYSIS

Judith C. Harris

7-1 INTRODUCTION

Hydrolysis is a chemical transformation process in which an organic molecule, RX, reacts with water, forming a new carbon-oxygen bond and cleaving a carbon-X bond in the original molecule. The net reaction is most commonly a direct displacement of X by OH:

$$R-X \xrightarrow{\text{H}_2\text{O}} R-OH + X^- + H^+ \tag{7-1}$$

This process can be distinguished from several other possible reactions between organic chemicals and water such as acid:base reactions (Eq. 7-2), hydration of carbonyls (Eq. 7-3), addition to carbon-carbon bonds (Eq. 7-4), and elimination (Eq. 7-5):

Acid: Base

$$
\left\{
\begin{array}{c}
R-COOH + H_2O \rightleftharpoons RCOO^- + H_3O^+ \\
\text{organic} \qquad\qquad \text{conjugate} \\
\text{acid} \qquad\qquad\quad \text{base} \\
\\
R-NH_2 + H_2O \rightleftharpoons RNH_3^+ + OH^- \\
\text{organic} \qquad\qquad \text{conjugate} \\
\text{base} \qquad\qquad\quad \text{acid}
\end{array}
\right\}
\tag{7-2}
$$

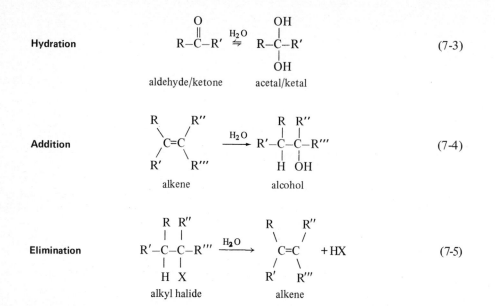

Hydration
$$\underset{\text{aldehyde/ketone}}{R-\overset{\displaystyle O}{\overset{\|}{C}}-R'} \overset{H_2O}{\rightleftharpoons} \underset{\text{acetal/ketal}}{R-\overset{\displaystyle OH}{\underset{\displaystyle OH}{\overset{|}{\underset{|}{C}}}}-R'} \qquad (7\text{-}3)$$

Addition

alkene → alcohol (7-4)

Elimination

alkyl halide → alkene + HX (7-5)

Acid-base equilibria of the type shown in Eq. 7-2 are discussed in Chapter 6. Hydration reactions as shown in Eq. 7-3 are reversible and therefore do not lead to a permanent chemical transformation of the organic species; these reactions are not considered further in this chapter. Addition reactions of the type shown in Eq. 7-4 are also excluded from further consideration, as they generally require reaction conditions that are unlikely to occur in the environment.

Detailed consideration of elimination reactions is also beyond the scope of this chapter. Reactions of this type are generally favored by higher temperatures and more strongly basic conditions than are commonly found in aqueous environments. However, elimination may be competitive with hydrolysis for organic compounds that contain good leaving groups (X of Eq. 7-5) such as a halide or sulfonate. For example, the hydrolysis of the nematocide 1,2-dibromo-3-chloropropane at 85°C and pH 9 has been reported [3b] to proceed via elimination of hydrogen bromide (major pathway) or hydrogen chloride (minor pathway) with subsequent further hydrolysis to 2-bromoallyl alcohol (Eq. 7-6). It is important that the possibility of competitive elimination be taken into account when one attempts to predict the hydrolytic behavior of organic chemicals in aqueous environments.

$$CH_2-CH-CH_2 \xrightarrow[\text{(elimination)}]{OH^-,\,H_2O} CH_2=C-CH_2 + CH_2-C=CH_2$$

with substituents Br, Br, Cl on the reactant; Br, Cl on the first product; Br, Br on the second product.

$$\downarrow \begin{array}{c} H_2O \\ \text{(hydrolysis)} \end{array}$$

$$CH_2=C-CH_2$$

with substituents Br, OH.

$$(7\text{-}6)$$

Hydrolysis (Eq. 7-1) is likely to be the most important reaction of organic compounds with water in aqueous environments and is a significant environmental fate process for many organic chemicals. It is actually not one reaction but a family of reactions involving compound types as diverse as alkyl halides, carboxylic acid esters, organophosphonates, carbamates, epoxides, and nitriles. Equations 7-7 through 7-12 illustrate some of these possible hydrolysis reactions and products.

$$CH_3CH_2CH_2CHCH_3 \xrightarrow{H_2O} CH_3CH_2CH_2CH-CH_3 + Br^- + H^+ \qquad (7\text{-}7)$$

with Br substituent on reactant (alkyl halide); OH substituent on product (alcohol); anion.

carboxylic acid ester → carboxylic acid + CH_3OH (alcohol)   (7-8)

$$CH_3\overset{\displaystyle O}{\overset{\|}{P}}(OCH_3)_2 \xrightarrow{H_2O} CH_3\overset{\displaystyle O}{\overset{\|}{\underset{OH}{P}}}OCH_3 + CH_3OH \qquad (7\text{-}9)$$

phosphonic acid diester → phosphonic acid monoester + alcohol

$$CH_3O\overset{\displaystyle O}{\overset{\|}{C}}NHC_6H_5 \xrightarrow{H_2O} CH_3OH + CO_2 + NH_2C_6H_5 \qquad (7\text{-}10)$$

carbamate → alcohol + + amine

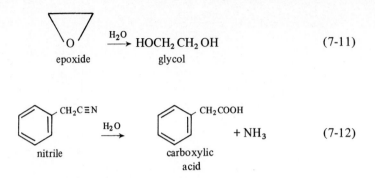

$$\text{epoxide} \xrightarrow{\text{H}_2\text{O}} \text{HOCH}_2\text{CH}_2\text{OH} \quad \text{glycol} \tag{7-11}$$

$$\text{nitrile} \xrightarrow{\text{H}_2\text{O}} \text{carboxylic acid} + \text{NH}_3 \tag{7-12}$$

Many organic functional groups (Table 7-1) are relatively or completely inert with respect to hydrolysis. Other functional groups that may hydrolyze under environmental conditions are listed in Table 7-2. Figure 7-1 gives examples of the range of hydrolysis half-lives that may be encountered for several categories of compounds.

TABLE 7-1

Types of Organic Functional Groups That Are Generally Resistant to Hydrolysis[a]

Alkanes	Aromatic nitro compounds
Alkenes	Aromatic amines
Alkynes	Alcohols
Benzenes/biphenyls	Phenols
Polycyclic aromatic hydrocarbons	Glycols
Heterocyclic polycyclic aromatic hydrocarbons	Ethers
	Aldehydes
Halogenated aromatics/PCBs	Ketones
Dieldrin/aldrin and related halogenated hydrocarbon pesticides	Carboxylic acids
	Sulfonic acids

a. Multifunctional organic compounds in these categories may, of course, be hydrolytically reactive if they contain a hydrolyzable functional group in addition to the alcohol, acid, etc., functionality.

7-2 CHARACTERISTICS OF HYDROLYSIS

Hydrolysis Mechanism. When an organic compound undergoes hydrolysis, a *nucleophile*[1] (water or hydroxide ion) attacks an *electrophile*[2] (carbon atom, phosphorus atom, etc.) and displaces a *leaving*

---

1. Nucleophile = nucleus-seeker
2. Electrophile = electron-seeker

TABLE 7-2

Types of Organic Functional Groups That Are
Potentially Susceptible to Hydrolysis

Alkyl halides	Nitriles
Amides	Phosphonic acid esters
Amines	Phosphoric acid esters
Carbamates	Sulfonic acid esters
Carboxylic acid esters	Sulfuric acid esters
Epoxides	

group (chloride, phenoxide, etc.). As early as 1933, it was recognized that nucleophilic displacement reactions usually fit one of two distinct kinetic patterns, which were named S_N1 (Substitution, Nucleophilic, Unimolecular) and S_N2 (Substitution, Nucleophilic, Bimolecular) [11]. Although a detailed discussion of reaction mechanisms would be beyond the scope of this chapter, it is important to review briefly the important hydrolysis mechanisms. As Exner [5] has noted, "The fundamental condition for all correlations of rate data is a simple and constant reaction mechanism. . . . The condition of a constant reaction mechanism remains one of the most serious problems for correlation equations."

Kinetically, the "unimolecular" S_N1 process is characterized by a rate independent of the concentration and nature of the nucleophile, formation of racemic products from optically active material, and enhancement of rate by electron-donating substituents on the central atom. It is postulated that the rate-determining step is the ionization of RX to give a planar carbonium ion (Eq. 7-13a), which then undergoes a relatively rapid nucleophilic attack (Eq. 7-13b).

$$RX \xrightarrow{\text{slow}} R^+ + X^- \tag{7-13a}$$

$$R^+ + H_2O \xrightarrow{\text{fast}} ROH + H^+ \tag{7-13b}$$

In an S_N2 process, on the other hand, the rate depends on the concentration and identity of the nucleophile, and an optically active starting material gives a product of inverted configuration. This is postulated as a one-step bimolecular process involving nucleophilic attack on the central atom at the side opposite the leaving group:

$$H_2O + R-X \longrightarrow [H_2O \cdots R \cdots X] \longrightarrow H^+ + HO-R + X^- \tag{7-14}$$

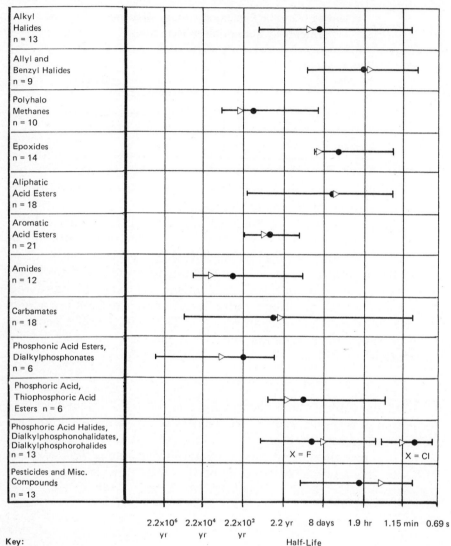

Key:

● Average
▷ Median
n No. of Compounds Represented

Source: Adapted [7] from data of Mabey and Mill [16].

FIGURE 7—1   **Examples of the Range of Hydrolysis Half-Lives for Various Types of Organic Compounds in Water at pH 7 and 25° C**

Hydrolysis of species such as carboxylic acid esters, amides, or organophosphorus compounds generally involves bimolecular nucleophilic attack. As the example in Eq. 7-15 indicates, such reactions are analogous to the S_N2 (rather than S_N1) mechanism of attack on saturated carbon.

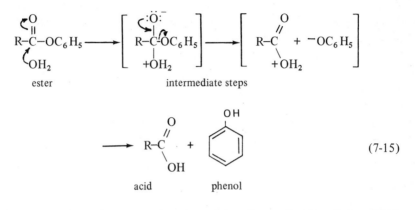

ester   intermediate steps

$$(7\text{-}15)$$

acid   phenol

Many reactions appear to involve either "pure S_N1" or "pure S_N2" mechanisms. The limiting S_N1 case is favored by R- systems that form stable carbonium ions (e.g., tributyl and triphenyl methyl systems), by X- systems that are good leaving groups (e.g., halide ions, *p*-toluenesulfonate ions), and by high-dielectric-constant solvents such as water. Conversely, the limiting S_N2 case is favored by R- systems with low steric hindrance and low carbonium ion stability (e.g., methyl and other primary alkyl systems), by X- systems that are poor leaving groups (e.g., NH_2^- or $CH_3CH_2O^-$), and by organic solvents such as acetone. However, there probably exists in nature a continuum of mechanisms between these two extremes [8]. In estimating rates of hydrolysis, it is important to consider whether the reaction of interest and the available "model reactions" involve similar mechanisms. Convincing evidence as to similarity of mechanism is available only from measured kinetic data — the form of the rate law, thermodynamic activation parameters, isotope effects, etc. — which are normally unavailable for the compound whose hydrolysis rate must be estimated. The only general guidance that can be provided is that one should select model reactions in which both R-and X- groups are as similar as possible to those of the compound whose hydrolysis rate is unknown.

Hydrolysis Rate Law. It is generally observed that hydrolysis of organic chemicals in water is first-order in the concentration of the organic species (Eq. 7-16) [1,6,16,17,30,31,33]; the rate of disappearance of RX, $-d[RX]/dt$, is directly proportional to the concentration of the compound, [RX]:

$$- d[RX]/dt = k_T [RX] \tag{7-16}$$

where k_T = hydrolysis rate constant. The first-order dependence is important, because it implies that the hydrolysis half-life of RX (Eq. 7-17) is independent of the RX concentration and, thus, that results obtained at relatively high RX concentration can be extrapolated to low concentrations of RX, assuming other reaction conditions (e.g., temperature, pH) are constant.

$$t_{1/2} = 0.693/k_T \tag{7-17}$$

The rate expression presented in Eq. 7-16 is an oversimplification for most organic hydrolysis reactions. The rate constant k_T is a pseudo first-order rate constant that may include contributions from acid- and base-catalyzed hydrolysis as well as nucleophilic attack by water. The following equation explicitly recognizes these possibilities:

$$k_T = k_H [H^+] + k_0 + k_{OH}[OH^-] + \sum_i k_{HA_i} [HA] + \sum_j k_{B_j} [B_j] \tag{7-18}$$

where:

k_T	= total hydrolysis rate constant
k_H	= rate constant for specific acid-catalyzed hydrolysis
k_0	= rate constant for neutral hydrolysis
k_{OH}	= rate constant for specific base-catalyzed hydrolysis
k_{HA}	= rate constant for general acid-catalyzed hydrolysis
k_B	= rate constant for general base-catalyzed hydrolysis
$[H^+]$	= hydrogen ion concentration
$[OH^-]$	= hydroxyl ion concentration
$[HA]$	= general acid concentration
$[B]$	= general base concentration
i,j	= indices to identify different acids and bases that may be present

The first term in Eq. 7-18 represents specific acid catalysis by hydronium ion, H^+. Such catalysis is common in both S_N1 and S_N2 reactions of those compounds, RX, which can be protonated at a site that either makes X a better leaving group (as in the case of amine hydrolysis), or makes the central carbon in R more electrophilic (as in the case of ester hydrolysis), or both.

The second term, which corresponds to neutral hydrolysis, k_0, could be written in terms of a second-order rate constant:

$$k_0 = k_{H_2O} [H_2O] \qquad\qquad (7\text{-}19)$$

For S_N2 reactions, however, k_0 is more simply treated as pseudo first-order, since the concentration of water in aqueous systems is essentially constant at 55.5 M.

The third term in Eq. 7-18 represents specific base catalysis by hydroxide ion, OH^-. A contribution of this type to k_T is observed for virtually all substrates that undergo S_N2-type hydrolysis. The hydroxide ion "catalysis" reflects the fact that OH^- is a much stronger nucleophile (typically by a factor of about 10^4) than is water [22]. Specific base catalysis is not a feature of S_N1 reactions, because no nucleophile is involved in the rate-determining step (see Eq. 7-13a).

The two final terms in Eq. 7-18 reflect the possibility of general acid/base catalysis by acids/bases other than H^+ and OH^-. These processes can make significant contributions to values of k_T for some types and compounds when the hydrolysis rate constant is measured in aqueous buffer solutions. However, because it is impossible to predict the types and concentrations of acidic and basic species that may be present in aqueous environments, it is not possible to estimate the importance of general acid/base catalysis. Therefore, the two final terms of Eq. 7-18 are generally dropped [16,21], and the expression for k_T is written as:

$$k_T = k_H [H^+] + k_0 + k_{OH} [OH^-] \qquad\qquad (7\text{-}20)$$

Mabey and Mill [16] have neatly summarized the pH dependence of the hydrolysis rate implicit in Eq. 7-20 (Figure 7-2). They point out that pH-rate profiles for hydrolysis may be U-shaped (solid line) or V-shaped (dashed line), depending on the magnitude of the neutral hydrolysis rate constant compared with those of the specific acid/base-catalyzed process. The three transition points marked I_{AN}, I_{AB}, and I_{NB} in Figure 7-2 correspond to values of pH at which the acid- or base-catalyzed processes begin to make significant contributions to k_T. If such transition points fall within the aquatic environmental pH range of 5-8 for a particular organic compound or class, acid or base catalysis must be considered in predicting rates of aqueous hydrolysis. Table 7-3 summarizes Mabey and Mill's data for a number of categories of hydrolyzable organics.

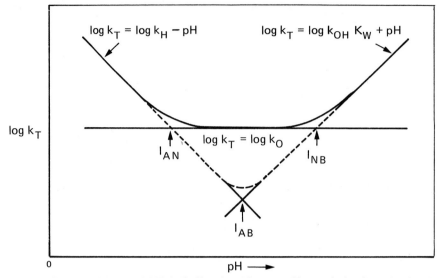

$$\log k_T = \log k_H - pH \qquad \log k_T = \log k_{OH}\ K_W + pH$$

$$\log k_T = \log k_O$$

I_{AN} I_{NB}

I_{AB}

log k_T

0   pH $\longrightarrow$

Source: Mabey and Mill [16]. *(Reprinted in part with permission from the American Chemical Society.)*

FIGURE 7—2   **pH Dependence of k_T for Hydrolysis by Acid-, Water-, and Base-Promoted Processes**

TABLE 7-3

pH Regimes in Which Specific Acid/Base Catalysis Is Significant for Organic Functional Groups

Category	Acid Catalysis	Base Catalysis
Organic halides	none	> 11
Epoxides	< 3-8[a]	> 10
Aliphatic acid esters	< 1.2-3.1	> 5.2-7.1[b]
Aromatic acid esters	< 3.9-5.2[a]	> 3.9-5.0[b]
Amides	< 4.9-7[a]	> 4.9-7[b]
Carbamates	< 2	> 6.2-9[b]
Phosphonic acid esters	< 2.8-3.6	> 2.8-3.6

a. Acid catalysis may be important within the typical aquatic-environment pH range of $5 < pH < 8$.

b. Base catalysis may be important within the typical aquatic-environment pH range of $5 < pH < 8$.

Source: Based on data of Mabey and Mill [16].

More complex pH-rate profiles may be observed for organic species, such as mono esters of phosphoric/phosphonic acids, that undergo acid/base dissociation in the environmental pH range. This possibility should be kept in mind when estimating hydrolysis rate constants.

Measurement of Hydrolysis Rate. Experimental measurement of the rate of hydrolysis should involve the determination of four conceptually distinct components:

* The form of the rate law,
* The magnitude of the rate constant(s),
* The products of reaction, and
* Temperature dependence (energy of activation).

Because of the simplicity of handling first-order kinetic data, experimental reaction conditions are usually selected so that the reaction is expected to be (pseudo) first-order in RX (Eq. 7-16). For a reaction that actually follows a rate law of the form shown by Eq. 7-21, the pseudo first-order condition is achieved by using dilute aqueous buffer to fix H^+ and OH^- at constant concentrations for the duration of the experiment.

$$- d[RX]/dt = k_H [H^+] [RX] + k_0 [RX] + k_{OH} [OH^-] [RX] \qquad (7\text{-}21)$$

The decrease in concentration of RX as a function of time is then monitored by any convenient method, such as withdrawal of aliquots for extraction and chromatographic analysis of RX, measurement of visible or ultraviolet light absorbance at a frequency characteristic of RX, or determination of concentration of X released by the hydrolysis. The presumed first-order dependence of the reaction rate on [RX] is confirmed by plotting ln[RX] versus time; the plot should be linear with an intercept at the initial RX concentration. The slope is equal to $- k_T$, as shown below.

$$- d[RX]/dt = k_T [RT] \qquad (7\text{-}16)$$

$$- d[RX]/[RX] = k_T dt \qquad (7\text{-}22)$$

$$- \ln[RX] = k_T t + constant \qquad (7\text{-}23)$$

$$\ln[RX] = -k_T t - \ln[RX]_0 \qquad (7\text{-}24)$$

It is desirable that kinetic measurements be conducted over one or preferably two half-lives to ensure that any deviations from the presumed first-order linearity are detectable, but this may not be feasible for slow reactions.

Most kinetic studies are carried out at organic compound concentrations of at least 0.001 M, which is substantially higher than those typically encountered in the environment. The studies have therefore been criticized as unrealistic. Mabey and Mill [16] point out that such criticism is misplaced, since "it is axiomatic that rate processes found to be simple at high concentrations remain so at low concentrations [of the organic species]."

To determine whether Eq. 7-21 is the correct rate law and to determine the magnitude of k_H and/or k_{OH}, k_T is determined in separate experiments at various pH values. The k_H is determined from low-pH experiments, while k_{OH} is determined from results of high-pH experiments. It should be noted that uncertainties in the experimental measurement of pH may be a significant source of error in determination of the value of k_H or k_{OH}. For example, an uncertainty of ± 0.02 in the measured pH value corresponds to an uncertainty of about 5% in the concentration of H^+ (or OH^-) and thus in the value of k_H (or k_{OH}) calculated from k_T.

The identity of the product(s) of hydrolysis is important for confirming the mechanism and for selecting appropriate model reactions to estimate the rate of hydrolysis of additional organic chemicals of the type RX. Unfortunately, the products are frequently unknown, because many kinetic studies monitor only the disappearance of the starting material, RX, or appearance of one possible reaction product such as X^- (e.g., chloride ion). In such studies it is not possible to determine whether elimination reactions (Eq. 7-5) may be occurring instead of, or in addition to, the presumed hydrolysis.

An example of the complexities that can be encountered and that can be addressed only by careful examination of reaction products is provided by studies of the hydrolysis of malathion [6,30,31]. Wolfe and co-workers studied the pseudo first-order hydrolysis over a range of pHs and temperatures [30,31]. Under basic conditions at 27°C after one half-life, the relative product distribution was as shown in Eq. 7-25.

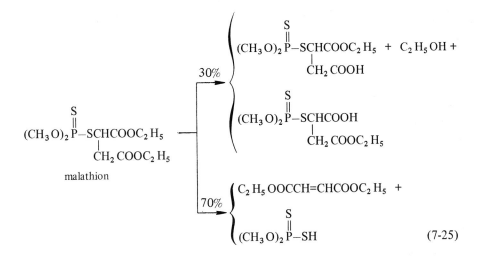

$$30\% \quad
\begin{cases}
\underset{\underset{\underset{CH_2COOH}{|}}{}}{(CH_3O)_2\overset{\overset{S}{\parallel}}{P}{-}SCHCOOC_2H_5} + C_2H_5OH + \\[2em]
\underset{\underset{\underset{CH_2COOC_2H_5}{|}}{}}{(CH_3O)_2\overset{\overset{S}{\parallel}}{P}{-}SCHCOOH}
\end{cases}$$

malathion

$$70\% \quad
\begin{cases}
C_2H_5OOCCH{=}CHCOOC_2H_5 + \\[1em]
(CH_3O)_2\overset{\overset{S}{\parallel}}{P}{-}SH
\end{cases} \qquad (7\text{-}25)$$

The relative importance of the two pathways — carboxylate ester cleavage versus phosphorodithioate ester cleavage — was found to be temperature-dependent. Furthermore, a different set of reaction pathways was found to be operative under acid conditions (Eq. 7-26).

$$\underset{\underset{CH_2COOC_2H_5}{|}}{(CH_3O)_2\overset{\overset{S}{\parallel}}{P}{-}SCHCOOC_2H_5} \xrightarrow{\ H^+/H_2O\ } C_2H_5OH +$$

$$\underset{\underset{CH_2COOC_2H_5}{|}}{(CH_3O)_2\overset{\overset{S}{\parallel}}{P}{-}SCHCOOH} \ + \ \underset{\underset{CH_2COOH}{|}}{(CH_3O)_2\overset{\overset{S}{\parallel}}{P}{-}SCHCOOC_2H_5} \xrightarrow{\ H^+/H_2O\ }$$

$$\underset{\underset{CH_2COOH}{|}}{HS{-}CHCOOH} + (CH_3O)_2\overset{\overset{S}{\parallel}}{P}{-}OH + C_2H_5OH \qquad (7\text{-}26)$$

Estimation of reaction rates for complex hydrolytic pathways such as these is well beyond the state of the art. Conversely, malathion hydrolysis is unlikely to provide a useful model for predicting reactivities of other organophosphorus compounds. *The estimation of hydrolysis rates is feasible only when the hydrolysis pathway is reasonably simple and straightforward and the product(s) can be predicted.*

Temperature Dependence of k. The rate of hydrolysis of organic chemicals increases with temperature. The quantitative relationship between the rate constant and temperature is frequently expressed by the Arrhenius equation,

$$k = Ae^{-E_A/RT} \qquad (7\text{-}27)$$

in which E_A is the Arrhenius activation energy (kcal/mol), R is the gas constant (1.987 cal/deg·mol), and T is the temperature (K) [15]. The pre-exponential factor, A, has the same units as the rate constant. According to Eq. 7-27, a plot of log k versus l/T is linear, with slope equal to $-E_A/2.303R$ and intercept equal to log A:

$$\log k = \log A - \frac{E_A}{2.303RT} \qquad (7\text{-}28)$$

An alternative form of temperature dependence derived from the Eyring reaction rate theory (transition state theory) is

$$k = \frac{kT}{h} e^{-\Delta H^{\ddagger}/RT} \, e^{\Delta S^{\ddagger}/R} \qquad (7\text{-}29)$$

in which k is Boltzmann's constant, h is Planck's constant, and $\Delta H^{\ddagger}$ and $\Delta S^{\ddagger}$ are enthalpy of activation and entropy of activation, respectively [18]. According to Eq. 7-29, $\Delta H^{\ddagger}$ can be calculated from the slope of a plot of log k/T versus l/T, and $\Delta S^{\ddagger}$ can be calculated from the intercept of the following equation:[3]

$$\log \frac{k}{T} = \log \frac{k}{h} - \frac{\Delta H^{\ddagger}}{2.303RT} + \frac{\Delta S^{\ddagger}}{R} \qquad (7\text{-}30)$$

Some investigators [16] choose to fit the k,T data using a relationship of the following form:

$$\log k = -\frac{A}{T} + B \log T + C \qquad (7\text{-}31)$$

In practice, Eqs. 7-28, -30 and -31 usually give equally good fit to experimental data because of the small number of data points (typically no more than 3 to 5) and the uncertainties in the individual measured values. In theory, the temperature dependence of k is more

---

3. The thermodynamic parameters of activation, $\Delta H^{\ddagger}$ and $\Delta S^{\ddagger}$, can be interpreted in terms of reaction mechanism as well as temperature dependence of k. Comparisons with ΔH° and ΔS° for equilibria and with $\Delta H^{\ddagger}$ and $\Delta S^{\ddagger}$ for other reactions are helpful. In particular, a change in sign of $\Delta S^{\ddagger}$ between one reaction and another is generally a reliable indicator that the two reactions involve different mechanisms.

complex than either equation would suggest, because E_A and A (Eq. 7-27), $\Delta H^{\ddagger}$ and $\Delta S^{\ddagger}$ (Eq. 7-29), and constants B and C (Eq. 7-31) are themselves temperature-dependent. These second-order temperature dependencies can be accounted for (in the rare instances where it is warranted by the data) by incorporating heat capacity corrections to $\Delta H^{\ddagger}$ and $\Delta S^{\ddagger}$, for example [15].

Equations 7-28, -30 and -31 are appropriately applied to k_H, k_0 and k_{OH} separately, rather than to the overall hydrolysis rate constant, k_T. Since the catalyzed and uncatalyzed reaction pathways generally show quite different temperature dependencies, plots based on k_T would probably be distinctly nonlinear. Furthermore, the slopes and intercepts of such plots would have no physical significance.

The values of E_A and $\Delta H^{\ddagger}$ for hydrolysis of organics in water usually fall in the range of 12-25 kcal/mol, with values of 17-20 kcal/mol most common. Some useful rules of thumb for temperatures in the vicinity of 300K (0-50°C) are that:

- a 1° change in temperature causes a 10% change in k,
- a 10° change in temperature causes a factor of 2.5 change in k, and
- a 25° change in temperature causes a factor of 10 change in k.

These rules are based on a 17-18 kcal/mol value of E_A or $\Delta H^{\ddagger}$.

The rather high sensitivity of k to changes in temperature has three important consequences:

(1) In the experimental measurement of k, the temperature must be controlled both accurately and precisely. For example, an uncertainty of $\pm 0.2°C$ in T corresponds to $\pm 2\%$ in k; $\pm 1°$ in T corresponds to $\pm 10\%$ in k.

(2) As pointed out by Mabey and Mill [16], $\pm 2\%$ in k leads to an uncertainty of $\pm 5\%$ in E_A ($\Delta H^{\ddagger}$), while $\pm 10\%$ in k leads to $\pm 100\%$ (factor of 2) uncertainty in E_A ($\Delta H^{\ddagger}$). The uncertainty in E_A ($\Delta H^{\ddagger}$) is magnified when laboratory rate data are extrapolated over 25° or larger intervals to estimate values of k under environmental conditions. For example, a 5% error in E_A ($\Delta H^{\ddagger}$) will give rise to a 30% error in the estimated k for an extrapolation from 50°C to 25°C. This inherent propagation of error (e.g., from 2% in

the laboratory rate data to 5% in E_A to 30% in the extrapolated k) is quite independent of the possibility that a typical small k-versus-T data set may contain at least one outlier. It is probably prudent to regard rate data obtained by extrapolation as order-of-magnitude estimates.

(3) A 10° seasonal temperature variation, a 1° diurnal variation, or a 5° spatial temperature gradient in an aquatic ecosystem would be associated with a corresponding 10% to 250% variation in hydrolysis rate for a compound with $\Delta H^{\ddagger} \approx 18$ kcal/mol. To adequately model the effects of these temperature variations on the hydrolysis rate, more accurate and precise kinetic data than are generally available would be required. On the other hand, the rough order-of-magnitude estimates of k that can be drawn from the available data are probably quite compatible with existing black-box, homogeneous-isothermal-compartment models of the environment.

Effect of Reaction Medium. Hydrolysis reactions, which frequently involve ionic species as reactants, intermediates, and/or products, are affected by changes in the solvating power of the reaction medium. Both changes in ionic strength and the presence of organic solvents can affect the solvating power and thus alter the hydrolysis rate. Specific medium effects due to general acid/base and trace-metal catalysis are also possible.

Freshwater environmental media are characterized by low organic content and by low (<0.01 M [16]) ionic strength. The dilute aqueous buffer solutions employed by most current workers for determining rate constants and developing empirical correlations may be adequate approximations of actual freshwater conditions. Wolfe *et al.* [31] have reported good agreement between laboratory data for distilled water solutions and results of hydrolysis in natural river water for malathion. Preliminary data suggest that this is also true for hydrolysis of hexachlorocyclopentadiene in three natural water samples [27].

Salt effects of buffer components will generally not alter k by more than 5-10% as long as the total ionic strength is ≤0.10 M [16]. However, the possibility of general acid or base catalysis (Eq. 7-18) by buffer components should be considered for reactions that show significant effects of specific acid (H^+) or base (OH^-) catalysis. Current practice is to use low (0.01-0.001 M) total buffer concentrations, which are adequate when RX concentrations are in the range of 100-1000 ppm, but the general acid/base species are still substantially higher in concentration than the 10^{-5} to 10^{-8} H^+/OH^- concentrations of the environmental pH range. This effect would lead to overestimation of the rate of hydrolysis unless the laboratory data were corrected by extrapolation to zero buffer concentration. On the other hand, it is also possible that trace-metal species present in natural waters but not in the laboratory systems could catalyze hydrolysis and lead to reaction rates higher than predicted. An effect of this sort was postulated by Meikle and Youngson [17] to explain an observed rate enhancement of about 15 fold for hydrolysis of chlorpyrifos in canal water over that measured in phosphate buffer solutions in distilled water.

Sediment is another medium effect that may be important in comparing hydrolysis rates estimated from laboratory data with those for natural waters. Work is under way at EPA's Athens (Georgia) Environmental Research Laboratory [29] and elsewhere to develop a better understanding of the effect of sediment on chemical transformations of organics in aquatic environments. The literature gives little information on the possibility of significant effects beyond the removal of some fraction of the organic chemical by adsorption on sediment.

The present state of the art does not enable us to predict the influence of potential general acid/base catalysts, trace metal catalysts, or sediments present in natural water systems on the rate of hydrolysis of organic chemicals. The reader should, however, be aware of these potential complications in extrapolating estimated hydrolysis rates to aquatic environments.

The aqueous solvent influences the rate and mechanism of hydrolysis reactions in a number of ways: as a nucleophilic reagent, as a high-dielectric-constant continuum in which reaction takes place, and as a specific solvating agent for organic reactants and products (leaving groups). Therefore, it is highly desirable to base estimates of hydrolysis rates on kinetic data and empirical correlations developed from 100% aqueous solvent systems. Most present-day experimental programs approach this ideal; stock solutions of the organic are commonly prepared (for convenience) in a solvent such as methanol, acetone, or acetonitrile and then diluted with water to <1% organic solvent for kinetic runs.

Unfortunately, many of the rate-constant tabulations, such as those of Ref. 24, and empirical rate-constant correlations

[2,5,13,15,20,22,25] in the older literature refer to kinetic experiments in mixed organic-aqueous solvents. To achieve adequate solubility for determination of [RX] by the methods available at the time, solvent systems containing 50% to 90% of a polar organic solvent (such as methanol, ethanol, acetone, or dioxane) were commonly used. The influence of solvent composition on organic reactivity has been reviewed [13,25]. The subject is complex and only poorly understood in theory. Mabey and Mill [16] observed that "although extrapolation of rate data from mixed solvents to water can be done with moderate success using schemes like the Winstein-Grunwald relation [15], combined extrapolations of temperature and solvent composition together with the questionable meaning of pH in mixed solvents introduce sufficient error in the final estimate to make such effort of questionable value" for purposes of comparing rates of hydrolysis of organic compounds in water under environmental conditions.

We believe that this observation is equally true for purposes of this chapter and have therefore not discussed approaches to correcting for solvent effects. However, in the absence of any other data, it may be appropriate to apply an existing empirical correlation to estimate the rate constant for hydrolysis in a particular mixed organic-water solvent and treat this as an estimated *lower limit* of the hydrolysis rate constant in water. It has been noted that rate constants for hydrolysis in water may be 20 to 2500 times higher than in 50% organic solvent [16].

7-3   OVERVIEW OF ESTIMATION METHODS

Table 7-4 summarizes the estimation methods described in this chapter. These approaches are first discussed in general terms below; instructions are then given for: (1) initiating the estimation process, (2) calculating an overall hydrolysis rate, (3) correcting for temperature, and (4) calculating a hydrolysis half-life.

The fundamental approach for estimating the rate of hydrolysis of organic chemicals in water is the application of linear free energy relationships to the estimation of the hydrolysis rate constant(s). An LFER is an empirical correlation between the standard free energies of reaction (ΔF°) or activation ($\Delta F^\ddagger$) for two series of reactions, both subjected to the same variations in reactant structures or reaction conditions. Several excellent treatments of the theoretical aspects and the broad applicability of LFERs have been written [4,10,15,25], and no attempt will be made here to present a detailed explanation or

TABLE 7-4

Characteristics of Estimation Methods Described

Section	Estimate[a]	Basis	Chemical Classes Covered [b]
7-5	k_H	Hammett Correlation	Ring-substituted benzamides; ethyl benzoates
7-6	k_H	Taft Correlation	Ortho-substituted benzamides
7-7	k_0	Hammett Correlation	Benzyl halides; dimethyl benzyl halides; benzyl tosylates. (All in mixed organic/aqueous solvents.)
7-8	k_{OH}	Hammett Correlation	Benzene ring-substituted compounds based on $ArCOOCH_3$, $ArCOOCH_2CH_3$, $ArCH_2COOCH_2CH_3$, $ArCH=CHCOOCH_2CH_3$, $ArCONH_2$, $ArOCOCH_3$, $ArCH_2OCOCH_3$, $ArCON(CH_3)_2$, $ArCONHCH_3$, $ArCH_2Cl$, and $ArOSi(CH_2CH_3)_3$. (All in mixed organic/aqueous solvents.)
7-9	k_{OH}	Taft Correlation	Dialkyl phthalate esters
7-10	k_{OH}	Correlation with pK_a of leaving group	Aryl esters of methylphosphonic acid $((CH_3)_2CHOP(O)(CH_3)OAr)$; carbamates of the form: (1) $(C_6H_5)NHCOOAr$; (2) $CH_3N(C_6H_5)COOAr$; (3) $CH_3NHCOOAr$; or (4) $(CH_3)_2NCOOAr$.

a. k_H = rate constant for acid-catalyzed hydrolysis.

   k_0 = rate constant for neutral hydrolysis.

   k_{OH} = rate constant for base-catalyzed hydrolysis.

b. Ar = aromatic group.

derivation. The potential utility of LFERs in estimating environmental reaction rates has also been described by others [18,32].

Use of the LFER method to estimate hydrolysis reaction rates is basically a substituent-effect approach. It is essentially the same in concept, though somewhat more complex in practice, as the approach to estimation of acid dissociation constants presented in Chapter 6. The reader who is not familiar with LFER concepts and approaches is urged to read through Chapter 6 and work out some simple acid dissociation constant examples before undertaking the estimation of hydrolysis rate constants.

Substituent changes in the hydrolyzable molecule, RX, may be made either in the central, R, portion of the molecule or in the leaving group, X. When substituent changes are made in R, and R is aromatic, the hydrolysis rate constant(s) may be correlated with the Hammett σ substituent constants as shown below (see also §6-4, Eq. 6-16 and Table 6-3).

$$\log k = \rho\sigma + \log k_0 \qquad (7\text{-}32)$$

As in the case of acid dissociation constants, ρ is a reaction constant that reflects the sensitivity of the particular reaction series to substituent effects.

If RX is aliphatic and substituent changes are made in the R group, the Taft $\rho^*\sigma^*$ is used in place of the Hammett equation (see §6-5). For estimation of hydrolysis reaction rate constants, it is recognized [20,25,30] that a single Taft substituent parameter, σ^*, sometimes does not give good correlations. Improved correlations are achieved by using a two-parameter Taft equation (Eq. 7-33) in which σ^* is a measure of the *polar* effects and E_s is a measure of the *steric* effects of the substituent.

$$\log k = \rho^*\sigma^* + \delta E_s + \log k_0 \qquad (7\text{-}33)$$

ρ^* and δ are reaction constants. Although there is a danger that the apparent improvement in correlation is simply an artifact due to inclusion of a second term in the equation, it is plausible that hydrolysis rates should be susceptible to both steric and polar effects of substituents and that these *might* be separable. In some cases, the steric effect is apparently dominant, so that $\rho = 0$ and Eq. 7-34 holds [20].

$$\log k = \delta E_s + \log k_0 \qquad (7\text{-}34)$$

If substituent changes are made in the leaving group, X, the Hammett and Taft correlations are potentially applicable to aromatic and aliphatic moieties, respectively. An alternative, conceptually equivalent to the Hammett relationship, is to attempt to apply a correlation between the rate constant(s) and the pK_a of the leaving group [9,32].

The data base of established LFER correlations for prediction of hydrolysis rate constants is very limited. In fact, if one were to apply the criterion suggested by Exner [5] that "simple regressions with less than 10 points and multiple regressions with less than 20 may not be worthwhile," the data base would disappear altogether, save for one carbamate data set [32]. The correlations that have been published are described later in this chapter. The number of sample calculations provided is quite limited, because there are few independently measured rate constants with which estimated values can be compared. Essentially, all available data were used in developing the correlation equations.

The general procedure for estimating the rate of hydrolysis of an organic chemical is described below. Sections 7-5 through 7-10 describe the specific steps for estimating k_H, k_0, or k_{OH} from a particular correlation equation.

(1) Categorize the organic chemical in terms of functional groups present. Consult Tables 7-1 and 7-2 to identify hydrolyzable groups.

(2) Check Table 7-3 to determine whether k_H and/or k_{OH} in the hydrolyzable groups are potentially significant in the environmental pH range of 5-8.

(3) If k_H is required, estimate it from correlations in §7-5 or 7-6.

(4) Estimate k_0 from correlations in §7-7.

(5) If k_{OH} is required, estimate it from correlations in §7-8, 7-9 or 7-10.

(6) If k_H, k_0, and/or k_{OH} refers to a temperature, T_2, other than 298K (25°C), convert to 25°C as follows:

$$\log k_{25°C} = \log k_{T_2} - 3830\left(\frac{T_2 - 298}{298T_2}\right) \qquad (7\text{-}35)$$

in which T_2 is in K and an average $\Delta H^{\ddagger}$ or E_A value of 17.5 kcal/mol has been assumed.

(7) Calculate k_T for pH(s) of interest according to Eq. 7-36 (cf. Eq. 7-20):

$$k_T = (k_H \times 10^{-pH}) + k_0 + (k_{OH} \times 10^{pH-14}) \qquad (7\text{-}36)$$

(8) Calculate the hydrolysis half-life ($t_{1/2}$) according to Eq. 7-37:

$$t_{1/2} = \frac{0.693}{k_T} \qquad (7\text{-}37)$$

7-4   UNCERTAINTY IN ESTIMATING VALUES

Hydrolysis rate constants that are estimated by the methods described here are subject to the following major sources of uncertainty:

(1) The correlation equations are typically based on three to six data points. This reduces confidence in the validity of extrapolating to compounds outside the original data set.

(2) Substituent and reaction constants are obtained from a variety of sources and may refer to temperatures and reaction media that differ from those of the ambient aquatic environment.

(3) Changes in reaction mechanism across a series of related organic compounds is a real possibility.

(4) Correlation equations apply to k_H, k_0, and k_{OH} individually; it may be impossible to estimate all of the rate constants required for calculation of k_T and hence the hydrolysis half-life.

While it is not possible to quantify the probable uncertainties, a qualitative review would suggest that estimated k's be considered order-of-magnitude estimates. If an estimated k is within one or two orders of magnitude of the value considered critical in a given context, a sufficiently reliable value would probably be obtainable only by experimental measurement.

7-5   ESTIMATION OF k_H FROM THE HAMMETT CORRELATION

Data presently available limit the strict applicability of this method to two reaction series for which Hammett reaction constants for water solvent have been determined. These data (Table 7-5) are for hydrolysis of ring-substituted benzamides and ethyl benzoates. The tabulated ρ values (but *not* the tabulated rate constants) could be

TABLE 7-5

Data for Estimation of k_H from the Hammett Correlation

Compound	T (°C)	k_H^o (M^{-1}s^{-1})[a]	Ref.	ρ	Ref.
(benzamide structure, Z-substituted, C=O, NH$_2$)	100	3.1×10^{-4}	[30]	0.12[b]	[25]
(benzoate ester structure, Z-substituted, C=O, OC$_2$H$_5$)	25	1×10^{-7}	[12]	0.11[b]	[25]
(benzene ring, Z-substituted, OSO$_3$H)	49	1.2×10^{-4}	[30]	0.60[c]	[12]

a. k_H for parent compound with Z=H.
b. For displacement of —NH$_2$ or —OC$_2$H$_5$.
c. For displacement of substituted phenoxide.

assumed to apply as well to closely related reaction series, such as acid-catalyzed hydrolysis of ring-substituted N-methylbenzamides or other alkyl benzoates. The range of applicability could thus be extended somewhat if values of the rate constants for parent compounds of interest (designated as k° in this chapter) were measured or found in the literature.

Basic Steps

(1) From Table 7-5 or other (literature) sources, obtain the value of k_H^o for an unsubstituted parent compound.

(2) From Table 7-5 or other literature source, obtain the value of ρ for the applicable reaction series.

(3) Find the value of σ as follows:

- If the compound of interest is a monosubstituted benzamide or benzoate ester, find the appropriate substituent constant in Table 6-3 of Chapter 6.

- If more than one substituent is present, see Table 6-4 (Chapter 6) for the appropriate multi-substituent σ value.

- If the correct combination is not found in Table 6-4, locate the individual substituents in Table 6-3 and sum their σ values. Use this sum (σ_T) in place of a single σ value to calculate k. If Table 6-3 does not list one or more of the substituents, find a default value of σ in Table 6-5.

- A substituent that is neither covered by the generalized categories of Table 6-5 nor found in Table 6-3 cannot be assigned a σ value; therefore, k for the corresponding organic compound cannot be calculated from the Hammett equation.

(4) Calculate k_H from Eq. 7-32.

(5) If k_H is for $T \neq 25°C$ and temperature coefficient data for k_H are available from the literature, calculate k_H (25°C) according to Eq. 7-27, -29, or -31. In lieu of these, use Eq. 7-35.

Example 7-1   Estimate k_H for ethyl p-nitrobenzoate. Also, estimate the hydrolysis half-life (considering only the acid-catalyzed reaction) at pH = 6.

(1) $k_H^o = 1 \times 10^{-7}$ $M^{-1}s^{-1}$ for ethyl benzoate in water at 25° (Table 7-5).

(2)   From Table 7-5, $\rho = 0.11$ for ethyl benzoate hydrolysis.

(3)   From Table 6-3, $\sigma = 0.778$ for a p-nitro substituent.

(4)   From Eq. 7-32,

$$\log k_H = (0.11)(0.778) - 7.00$$

$$\log k_H = -6.92$$

$$k_H = 1.2 \times 10^{-7} \text{ } M^{-1}s^{-1}$$

The literature value is 1.4×10^{-7} $M^{-1}s^{-1}$ [16], and the error is -14%.

(5)   From Eq. 7-36 (assuming k_0 and $k_{OH} = 0$),

$$k_T = (1.2 \times 10^{-7})(10^{-6}) = 1.2 \times 10^{-13} \text{ } s^{-1}$$

(6)   From Eq. 7-37,

$$t_{1/2} = 0.693/(1.2 \times 10^{-13} \text{ } s^{-1}) = 5.8 \times 10^{12} \text{ } s$$

$$= 1.8 \times 10^5 \text{ yr}$$

7-6 ESTIMATION OF k_H FROM THE TAFT CORRELATION

Data presently available limit the strict applicability of this method to ortho-substituted benzamides. For this series of compounds, the rate constant was reported [20] to be correlated with the Taft steric substituent constant, E_s, according to Eq. 7-34. The published δ value (but not the rate constant) could be assumed to apply as well to closely related reactions, such as acid-catalyzed hydrolysis of N-methylbenzamides.

Basic Steps

(1) Set $k_H^\circ = 3.3 \times 10^{-5}$ (water, 100°C) [23] for o-methyl-benzamide or use an appropriate literature value for some other (N-substituted) parent benzamide compound.

(2) Set $\delta = 0.81$ [20] for benzamide series, or use an appropriate literature value for some other (N-substituted) benzamide series.

(3) From Table 7-6, select the appropriate E_s value for the ortho-substituent.

TABLE 7-6

Taft, σ^*, and Steric, E_s, Substituent Constants for Taft Correlation[a]

Z	σ^*	E_s	Z	σ^*	E_s
F	1.10	−0.24	H	0.00	0.00
Cl	1.05	−0.24	CH_3	−0.10	−0.07
Br	1.00	−0.27	CH_2CH_3	−0.115	−0.36
I	0.85	−0.37	$i\text{-}CH(CH_3)_2$	−0.125	−0.93
			$t\text{-}C_4H_9$	−0.165	−1.74

a. Values are for Z substituent in ZCH_2-; therefore Z=H corresponds to a $-CH_3$ group.

Source: Calculated from data of Shorter [20].

(4) Calculate k_H at 100°C from Eq. 7-38 (cf. Eq. 7-34).

$$\log k_H = \log k_H^\circ + \delta E_s \qquad (7\text{-}38)$$

(5) If k_H is for $T \neq 25°C$ and temperature coefficient data for k_H are available from the literature, calculate k_H (25°C) by use of Eq. 7-27, -29, or -31. In lieu of such data, use Eq. 7-35.

7-7   ESTIMATION OF k_0 FROM THE HAMMETT EQUATION

Data presently available (Table 7-7) are based on rates of hydrolysis of benzyl and dimethyl benzyl halides and benzyl tosylates in mixed-organic aqueous solvents. This method is applicable only to these compound types. Furthermore, the estimated k_0 should be regarded as the lower limit of the value that might be observed in an aquatic environment.

Basic Steps

(1) Choose an appropriate value of k_0° from Table 7-7 (or from the literature).

(2) Find the appropriate value of the reaction parameter, ρ, from Table 7-7 (or from the literature).

(3) Find the value of σ from Table 6-3, -4, or -5 (in Chapter 6) as directed in §7-5.

(4) Calculate k_0 by the following equation (cf. Eq. 7-32):

$$\log k_0 = \rho\sigma + \log k_0^\circ \qquad (7\text{-}39)$$

(5) If k_0 is for $T \neq 25°C$ and temperature coefficient data for k_0 are available from the literature, calculate k_0 (25°C) according to Eq. 7-27, -29, or -31. In lieu of such data, use Eq. 7-35.

Example 7-2   Estimate k_0 for *p*-methylbenzyl chloride.

(1) $k_0^\circ = 6.2 \times 10^{-6}$ s^{-1} at 25°C (Table 7-7)

(2) $\rho = -1.31$ (Table 7-7)

(3) $\sigma = -0.170$ (Table 6-3)

(4)   Substituting in Eq. 7-39, $\log k_0 = (-1.31)(-0.170) - 5.21$

$$\log k_0 = -4.99$$

$$k_0 = 1.0 \times 10^{-5} \text{ s}^{-1}$$

The literature value is 2.97×10^{-3} s^{-1} at 30°C for water [16] indicating an error of about a factor of 300.

TABLE 7-7

Correlation of Neutral Hydrolysis Rate Constant With Hammett Substituent Constant

Compound/Conditions	n^a	$k_o^0 (s^{-1})^b$	Correlation Equation[c] $(\log k_o = \rho\sigma + \log k_o^0)$
⌬–CH$_2$Cl (Z) 25°C 50% Acetone	7	6.25×10^{-6}	$\log k_o = -1.31\,\sigma -5.21$ $r = 0.964$
⌬–C(CH$_3$)$_2$Cl (Z) 25°C 90% Acetone	8	1.1×10^{-4}	$\log k_o = -4.48\,\sigma -3.95^d$ $r = 0.998$
⌬–CH$_2$OC$_6$H$_5$OSO$_2$H (Z) 25°C 50% Acetone	7	2.6×10^{-4}	$\log k_o = -2.32\,\sigma -3.58$ $r = 0.972$

a. Number of compounds used to establish correlation.
b. Calculated from intercept of correlation equation.
c. Equation and correlation coefficient cited by Wells [25].
d. σ^+ values [25] used for electron-donating ρ substituents.

Source: Wells [25]. *(Reprinted with permission from Academic Press, Inc.)*

7-8   ESTIMATION OF k_{OH} FROM THE HAMMETT EQUATION

A number of studies have established correlations between alkaline hydrolysis rates and Hammet σ constants. However, almost all of the data were obtained from systems containing mixtures of organic and aqueous solvents. A value of k_{OH} estimated from one of these correlations should be regarded as a lower limit of the value that might be observed in an aquatic environment.

Basic Steps

(1) Choose an appropriate value of k_{OH}° from Table 7-8 (or from the literature).

(2) Choose the appropriate value of ρ (coefficient of σ) from Table 7-8 (or from the literature).

(3) Find the value of σ from Table 6-3, 6-4 or 6-5, as directed in §7-5.

(4) Calculate k_{OH} as follows (cf. Eq. 7-32):

$$\log k_{OH} = \rho\sigma + \log k_{OH}^{\circ} \qquad (7\text{-}40)$$

(5) If k_{OH} is for $T \neq 25°C$ and temperature coefficient data for k_{OH} are available from the literature, calculate k_{OH} (25°C) by use of Eq. 7-27, -29, or -31. In lieu of such data, use Eq. 7-35.

Example 7-3   Estimate the rate constant for hydrolysis of methyl *p*-nitrobenzoate.

(1) k_{OH}° (60% acetone) = 7.2×10^{-3} $M^{-1}s^{-1}$ (Table 7-8)

(2) $\rho = 2.38$ (Table 7-8)

(3) $\sigma = 0.778$ (Table 6-3)

(4)   Substituting in Eq. 7-40,

$$\log k_{OH} = (2.38)\,(0.778) + \log\,(7.2 \times 10^{-3})$$

$$= 1.85 - 2.14 = -0.29$$

$$k_{OH} = 5.1 \times 10^{-1} \ M^{-1}s^{-1} \ (60\% \ \text{acetone})$$

Literature vaues are 7.4×10^{-2} $M^{-1}s^{-1}$ in 55% methanol and 6.4×10^{-1} $M^{-1}s^{-1}$ in 56% acetone [16].

TABLE 7-8

Correlation of Alkaline Hydrolysis Rate Constant With Hammett Substituent Constant

Compound/Conditions	n^a	k_{OH}° $(M^{-1} s^{-1})$	Correlation Equation[b]
$ArCOOCH_3$ 60% Acetone, 25°	5	7.2×10^{-3} [c]	$\log k_{OH} = 2.38\ \sigma - 2.14$ $r = 0.991$
3% Ethanol, 25° [19a]	18	2.75	$\log k_{OH} = 1.17\ \sigma + 2.26$ $r = 0.996$
$ArCOOCH_2CH_3$ 60% Acetone, 25°	7	2.4×10^{-3} [c]	$\log k_{OH} = 2.47\ \sigma - 2.62$ $r = 0.996$
85% Ethanol, 25°	5	5.5×10^{-4} [c]	$\log k_{OH} = 2.61\ \sigma - 3.26$ $r = 0.999$
$ArCH_2COOCH_2CH_3$ 60% Acetone, 25°	8	4.4×10^{-2} [c]	$\log k_{OH} = 1.00\ \sigma - 1.36$
$ArCH=CHCOOCH_2CH_3$ 85% Ethanol, 25°C	5	1.4×10^{-3} [c]	$\log k_{OH} = 1.24\ \sigma - 2.86$ $r = 1.000$
$ArCONH_2$ 60% Ethanol, 53°C	4	7.6×10^{-6} [c]	$\log k_{OH} = 1.40\ \sigma - 5.12$ $r = 0.998$
$ArOCOCH_3$ 60% Acetone, 15°C	3	2.4×10^{-1} [c]	$\log k_{OH} = 1.51\ \sigma - 0.62$ $r = 1.000$

(continued)

TABLE 7-8 (Continued)

Compound/Conditions	n^a	k_{OH}° $(M^{-1}s^{-1})$	Correlation Equation[b]
$ArCH_2OCOCH_3$ 60% Acetone, 25°C	4	6.6×10^{-2} [c]	$\log k_{OH} = 0.75\,\sigma - 1.18$ $r = 0.996$
$ArCON(CH_3)_2$ [19a] 10% Ethanol, 25°C	14	3.8×10^{-5}	$\log k_{OH} = 1.14\,\sigma - 2.59$ [d] $r = 0.991$
$ArCONHCH_3$ [19a] 10% Ethanol, 38°C	17	4.00	$\log k_{OH} = 2.69\,\sigma^- + 2.440$
$ArCH_2Cl$ [12]	6	3.3×10^{-6}	$\log k_{OH} = -0.333\,\sigma - 5.484$ [12]
$ArOSI(CH_2CH_3)_3$ 50% Ethanol, 25°C	3	1.4×10	$\log k_{OH} = 1.99\,\sigma + 0.15$ $r = 0.999$
$(ArO)_3P{=}O$ [26] Water, 30°C	4	2.7×10^{-1}	$\log k_{OH} = 1.4\,\Sigma\sigma - 0.47$ $r = 0.995$

a. Number of compounds used in establishing correlation.

b. $\log k_{OH} = \rho\sigma + \log k_{OH}^\circ$; units of k_{OH} are $M^{-1}s^{-1}$.

c. Calculated from correlation equation, $\sigma = 0$.

d. Correlation used σ values tabulated by Exner [15].

Source: Wells [25] except as noted. *(Reprinted with permission from Academic Press, Inc.)*

These figures cannot be directly compared with the estimated value because of variations in the solvent. Therefore, percentage deviations have not been calculated. It should be noted that the rate of hydrolysis in the aquatic environment will probably be considerably higher than the estimate based on the 60% acetone solvent. Compare the much higher rate constant reported for 95% water-5% ethanol in Table 7-8.

7-9   ESTIMATION OF k_{OH} FROM THE TAFT EQUATION

The applicability of this method has been described for only one compound class, the alkyl phthalate esters. Wolfe *et al.* [28] have presented a correlation between rates of alkaline hydrolysis of five phthalate esters and the Taft σ^* and E_s constants. The correlation equation, for water at 30°C, is (cf. Eq. 7-33):

$$\log k_{OH} = 4.59\,\sigma^* + 1.52 E_s - 1.02 \tag{7-41}$$

and $r^2 = 0.975$. Wolfe *et al.* make no claims of generality for this relationship, which was developed for a relatively narrow range of compounds (dialkyl phthalate esters). The number of data points available is also small for evaluation of a two-parameter correlation equation.

Basic Steps

(1) If the compound is a dialkyl phthalate ester, whose parent compound is dimethyl phthalate:

- Find σ^* and E_s from Table 7-6.
- Calculate k_{OH} according to Eq. 7-41. This is the rate constant at T = 30°C.
- If k_{OH} at 25°C is desired, substitute the above value of k_{OH} in Eq. 7-35, using $T_2 = 303$.

(2) For other compounds (or to find k_{OH} for dialkyl phthalate esters at a temperature other than 25°C or 30°C), use Eqs. 7-27, -29, or -31.

Example 7-4   Estimate k_{OH} for diisobutyl phthalate at 25°C.

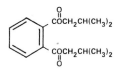

(1) From Table 7-6, $\sigma^* = -0.13$ and $E_s = -0.93$. (Note that values in Table 7-6 are for Z in $CH_2 Z$; therefore, the correct Z to use for an isobutyl ester is the value tabulated for $i\text{-}CH(CH_3)_2$.)

(2) Substituting in Eq. 7-41,

$$\log k_{OH} = 4.59\,(-0.13) + 1.52\,(-0.93) - 1.02$$

$$= -3.03$$

$$k_{OH} = 9.4 \times 10^{-4}\ M^{-1}s^{-1}\ \text{at}\ 30°C$$

(3) Equation 7-35 is used to find k_{OH} at 25°C:

$$\log k_{OH} = -3.03 - 3830 \left(\frac{303 - 298}{298 \times 303} \right)$$

$$= -3.24$$

$$k_{OH} = 5.7 \times 10^{-4}\ M^{-1}s^{-1}$$

The literature value is $1.4 \times 10^{-3}\ M^{-1}s^{-1}$ [28], indicating an error of approximately a factor of two.

7-10   ESTIMATION OF k_{OH} FROM THE pK_a OF THE LEAVING GROUP

This method is applicable, based on presently available data, to alkaline hydrolysis of a series of aryl esters of methylphosphonic acid and to four carbamate series. It is based upon correlations between log k_{OH} and the pK_a (negative log of the acid dissociation constant) of the phenol which is the conjugate acid of the leaving group. Table 7-9 presents the available correlation data. Figure 7-3 illustrates the fit of the correlation equations for two of the organophosphate ester series.

Basic Steps

(1) From the literature or Chapter 6, find pK_a for the phenol which is the conjugate acid of the leaving group, X.

(2) Choose the appropriate correlation equation from Table 7-9 (or from the literature).

(3) Calculate k_{OH} according to the correlation equation.

(4) If k_{OH} is for $T \neq 25°C$ and temperature coefficient data for k_{OH} are available from the literature, calculate k_{OH} (25°C) according to Eq. 7-27, -29, or -31. In lieu of these, use Eq. 7-35.

TABLE 7-9

Correlations of Alkaline Hydrolysis Rate Constant with pK$_a$ of Leaving Group

Compound/Conditions	n[a]	k$^\circ_{OH}$ (M^{-1}s^{-1})[b]	Correlation Equation[c]	Ref.
$(CH_3)_2\,CHOP(=O)(CH_3)-O-Ar$ (pH 10.2, 65°C)	4	3.5×10^{-2}	$\log k_{OH} = -0.68\ pK_a + 8.9$	9
$H-N(C_6H_5)-C(=O)-O-Ar$ (25°C)	20	5.1×10^{1}	$\log k_{OH} = -1.15\ pK_a + 13.6$ r = 0.994	32
$CH_3-N(C_6H_5)-C(=O)-O-Ar$ (25°C)	3	1.6×10^{-4}	$\log k_{OH} = -0.26\ pK_a - 1.3$ r = 1.00	32
$CH_3-N(H)-C(=O)-O-Ar$ (25°C)	6	1.6×10^{0}	$\log k_{OH} = -0.91\ pK_a + 9.3$ r = 0.994	32

(Continued)

TABLE 7-9 (Continued)

Compound/Conditions	n[a]	k°_{OH} $(M^{-1}s^{-1})$[b]	Correlation Equation[c]	Ref.
$CH_3-\overset{\overset{\displaystyle O}{\|\|}}{N}-C-O-Ar$ $\underset{CH_3}{\|}$ (25°C)	7	7.9×10^{-5}	$\log k_{OH} = -0.17\ pK_a - 2.6$ $r = 0.89$	32
$(CH_3O)_2\overset{\overset{\displaystyle O}{\|\|}}{P}-OR$ (27°C)	3	N/A	$\log k_{OH} = -0.28\ pK_a + 0.50$ $r = 0.987$	26
$(CH_3CH_2O)_2\overset{\overset{\displaystyle O}{\|\|}}{P}-OR$ (27°C)	4	N/A	$\log k_{OH} = -0.28\ pK_a - 0.22$ $r = 0.962$	26
$(CH_3O)_2\overset{\overset{\displaystyle S}{\|\|}}{P}-OR$ (27°C)	5	N/A	$\log k_{OH} = -0.25\ pK_a + 0.34$ $r = 0.982$	26
$(CH_3CH_2O)_2\overset{\overset{\displaystyle S}{\|\|}}{P}-OR$ (27°C)	4	N/A	$\log k_{OH} = -0.21\ pK_a - 1.6$ $r = 0.972$	26

a. Number of compounds studied to establish correlation.
b. For aromatic carbamate series, k°_{OH} is tabulated for phenyl compound; for mixed alkyl/aryl phosphate ester series, no k°_{OH} value is tabulated.
c. Equation presented in cited reference.

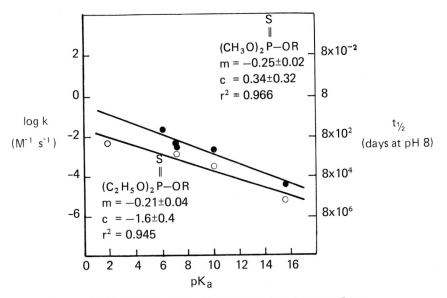

Source: Wolfe [26]. *(Reprinted with permission from Pergamon Press, Ltd.)*

FIGURE 7–3   **Linear Free Energy Relationships for the Alkaline Hydrolysis of O,O–Dimethyl and O,O–Diethyl–O–alkyl and Aryl Phosphorothioate in Water at 27°C**

Example 7-5   Calculate the alkaline hydrolysis rate constant for *p*-cyanophenyl-N-phenylcarbamate.

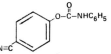

(1)   Conjugate acid of leaving group is *p*-cyanophenol.

$$K_a = 1.12 \times 10^{-8} \text{ [14]}; pK_a = 7.95$$

(2)   From Table 7-9, the correlation equation for N-phenyl carbamates at 25°C is

$$\log k_{OH} = -1.15 \, pK_a + 13.6$$

(3) $\log k_{OH} = (-1.15)(7.95) + 13.6$

$$= 4.46$$

$$k_{OH} = 2.87 \times 10^4 \text{ M}^{-1}\text{s}^{-1}$$

The literature value is 3.7×10^4 M^{-1}s^{-1} [32], indicating an error of -22%.

7-11 AVAILABLE DATA

Rate constant data are compiled in the sources listed below. The primary literature may provide additional, more recent values for particular compounds of interest. Tables 7-10 and -11 are compilations of some literature data for a variety of organic compound types.

National Bureau of Standards (1951) [24], *Tables of Chemical Kinetics: Homogeneous Reactions* — Older data, largely for organic solvent media. Supplements 1 (1956) and 3 (1961) provide additional data. Supplement 2 (1960) is an index.

Mabey and Mill (1977) [16], "Critical Review of Hydrolysis of Organic Compounds in Water Under Environmental Conditions" —Data mostly for reactions in water, some in mixed solvents, which have been critically reviewed. Temperature and pH dependence of hydrolysis rates frequently noted.

Freed, V.H., "Solubility, Hydrolysis, Dissociation Constants and Other Constants of Benchmark Pesticides," Chap. 1 in *A Literature Survey of Benchmark Pesticides,* Department of Medical and Public Affairs, George Washington University Medical Center, Washington, DC, under Contract 68-01-2889 for Office of Pesticide Programs, U.S. Environmental Protection Agency (March 1976) — Primarily data on half-life and/or percent persistence after a specified time for one reaction condition.

Palm, V.A. (1975-1979) [19b], *Tables of Rate and Equilibrium Constants of Heterolytic Organic Reactions,* Vols. 1-5. Volume 5, Part 2, is a table of correlation constants which may be useful.

7-12 SYMBOLS USED

Λ = parameter in Eq. 7-27

B = a base

E_A = Arrhenius activation energy in Eq. 7-27 (kcal/mol)

E_s = steric effects constant in Taft equation, Eq. 7-33

ΔF° = standard free energy of reaction

$\Delta F^\ddagger$ = free energy of activation

ΔH° = enthalpy of reaction for an equilibrium process

$\Delta H^\ddagger$ = enthalpy of activation for a rate process in Eq. 7-29

(Continued on p. 7-45)

TABLE 7-10

Example Rate Data and Estimated Half-Lives (25°, pH 7) for Organic Compounds of Various Types in Aqueous Solution

	k_H[a] $(M^{-1}s^{-1})$	k_0[a] (s^{-1})	k_{OH}[a] $(M^{-1}s^{-1})$	T (°C)	Temperature Dependence[b]	$t_{1/2}$[c]
Organohalides						
CH_3F	—	4.4×10^{-6}	9.0×10^{-4}	100 / 95	$\Delta H^{\ddagger} = 25.6;\ \Delta S^{\ddagger} = -12.2$ $E_A = 21.6;\ \log A = 9.607$	30y
CH_3Cl	—	5.6×10^{-5}	2.4×10^{-2}	90 / 95	$\Delta H^{\ddagger} = 25.3;\ \Delta S^{\ddagger} = -8.9$ $E_A = 24.3;\ \log A = 12.614$	0.9y
CH_3Br	—	1.1×10^{-4}	3.5×10^{-1}	71 / 95	$\Delta H^{\ddagger} = 24.1;\ \Delta S^{\ddagger} = -6.6$ $E_A = 23.0;\ \log A = 13.017$	20d
CH_3I	—	8.2×10^{-5}	1.2×10^{-1}	80 / 95	$\Delta H^{\ddagger} = 26.0;\ \Delta S^{\ddagger} = -5.2$ $E_A = 22.2;\ \log A = 12.093$	110d
$CH_3CH_2CH_2Br$	—	1.6×10^{-4}	—	80	$\Delta H^{\ddagger} = 23.3;\ \Delta S^{\ddagger} = -10.1$	26d
$(CH_3)_2CHCl$	—	1.0×10^{-3}	—	98	$\Delta H^{\ddagger} = 24.9;\ \Delta S^{\ddagger} = -5.3$	38d
$(CH_3)_3CF$	—	5.2×10^{-4}	—	90	$\Delta H^{\ddagger} = 22.6;\ \Delta S^{\ddagger} = -2.7$	50d
$CH_2{=}CHCH_2I$	—	4.1×10^{-6}	—	25		2d
$CHCl_3$	—	1.6×10^{-8} 7.3×10^{-8}	6.0×10^{-5}	25 100	(See Ref. 12)	3500y
$BrCHCl_2$	—	—	1.6×10^{-3}	25		140y
$C_6H_5CH_2Cl$	—	7.0×10^{-6}	—	20		15h
$C_6H_5CCl_3$	—	3.9×10^{-3}	—	5		19s

(continued)

TABLE 7-10 (Continued)

	k_H[a] $(M^{-1}s^{-1})$	k_0[a] (s^{-1})	k_{OH}[a] $(M^{-1}s^{-1})$	T (°C)	Temperature Dependence[b]	$t_{1/2}$[c]
Epoxides						
CH_2-CH_2 (epoxide)	9.3×10^{-3}	6.8×10^{-7}	1.0×10^{-4}	25	For k_H, $\Delta H^{\ddagger} = 19.0$; $\Delta S^{\ddagger} = -6.1$ For k_0, $E_A = 19.0$; log A = 7.726 For k_{OH}, $E_A = 18.0$; log A = 9.312	12d
(cyclohexene oxide structure)	1.6×10^7	2.2×10^3	1.5×10^1	25		6m
(dihydronaphthalene oxide structure)	4.7×10^2 [3a] 1.4×10^2 [3a]	3.0×10^{-3} 2.9×10^{-3f} [3a]	—	30 30		4m
(phenanthrene oxide structure)	1.0×10^3 [3a]	3.1×10^{-2} [3a]	—	30		22s[e]
Esters						
$HCOOCH_3$	—	—	3.7×10^1	25	$E_A = 40.0$; log A = 8.57	2d[e]
$CH_3COOCH_2CH_3$	1.1×10^{-4}	1.5×10^{-10}	1.1×10^{-1}	25		2y
$CH_3COOCH_2C_6H_5$	1.1×10^{-4}	—	2.0×10^{-1}	25		1.1y
$CH_3COOC_6H_5$	—	6.6×10^{-8}	—	25		38d
$Cl_2CHCOOC_6H_5$	—	1.8×10^{-3}	1.3×10^4	25		3.7m
$Cl_2CHCOOCH_3$	2.3×10^{-4}	1.5×10^{-5}	2.8×10^3	25		38m
$C_6H_5OCH_2COOCH_2CH_3$	—	—	3.0×10^1	28	$\Delta H^{\ddagger} = 20.1$; $\Delta S^{\ddagger} = 14.8$	2.4d[e]

(continued)

	k_H^a $(M^{-1}s^{-1})$	k_0^a (s^{-1})	k_{OH}^a $(M^{-1}s^{-1})$	T (°C)	Temperature Dependence[b]	$t_{1/2}^c$
$C_6H_5COOCH_3$	1.7×10^{-4} d	—	1.9×10^{-3} d 9.0×10^{-3} d	100 25 25	$E_A = 17.7$; log A = 6.57 $E_A = 14.2$; log A = 8.38	118y
benzene-1,2-(COOCH₃)₂	—	—	6.9×10^{-2} [28]	30		3.2y[e]
$C_6H_5COOCH_2C_6H_5$	—	—	5.9×10^{-3} to 1.2×10^{-2} d	30	$E_A = 12.5 - 14.5$ log A = 7.07 − 8.25	27y
pyridine-N—$COOCH_2CH_3$	—	—	5.4×10^{-3}	25		41y[e]
Amides						
CH_3CONH_2	1.0×10^{-3}	—	1.4×10^{-3}	75 75	$\Delta H^{\ddagger} = 19.2$; $\Delta S^{\ddagger} = -17.4$ $\Delta H^{\ddagger} = 13.2$; $\Delta S^{\ddagger} = -33.9$	3950y
$CH_3CH_2CONH_2$	1.2×10^{-3}	—	1.3×10^{-3}	75 75	$\Delta H^{\ddagger} = 18.1$; $\Delta S^{\ddagger} = -20.2$ $\Delta H^{\ddagger} = 14.7$; $\Delta S^{\ddagger} = -29.7$	
benzene-CH_2CONH_2	5.2×10^{-4}	—	1.8×10^{-3}	75 75	$\Delta H^{\ddagger} = 18.0$; $\Delta S^{\ddagger} = -21.9$ $\Delta H^{\ddagger} = 11.8$; $\Delta S^{\ddagger} = -39.0$	
$ClCH_2CONH_2$	1.2×10^{-3}	—	1.4×10^{-1}	75 75	$\Delta H^{\ddagger} = 18.7$; $\Delta S^{\ddagger} = -18.8$	1.46y
Cl_3CCONH_2	—	—	1.4×10^{-1}	75		84d
$CH_3CONH(CH_3)$	5.8×10^{-5}	—	3.6×10^{-4}	75 75	$\Delta H^{\ddagger} = 20.7$; $\Delta S^{\ddagger} = -18.6$ $\Delta H^{\ddagger} = 16.6$; $\Delta S^{\ddagger} = -29.1$	38,000y
$CH_3CON(CH_2CH_3)_2$	2.3×10^{-6}	—	1.2×10^{-5}	75 75	$\Delta H^{\ddagger} = 18.0$; $\Delta S^{\ddagger} = -31.1$	

(continued)

TABLE 7-10 (Continued)

	k_H^a $(M^{-1}s^{-1})$	k_0^a (s^{-1})	k_{OH}^a $(M^{-1}s^{-1})$	T (°C)	Temperature Dependence[b]	$t_{1/2}^c$
Carbamates						
$CH_3CH_2O\overset{O}{\overset{\|}{C}}NHC_6H_5$	—	—	3.3×10^{-5}	25	$\Delta H^{\ddagger} = 15.9;\ \Delta S^{\ddagger} = -25.5$	$6{,}700y^e$
$CH_3CH_2O\overset{O}{\overset{\|}{C}}N(CH_3)C_6H_5$	—	—	5.0×10^{-6}	25	$\Delta H^{\ddagger} = 12.9;\ \Delta S^{\ddagger} = -39.3$	$44{,}000y$
$C_6H_5O\overset{O}{\overset{\|}{C}}NHC_6H_5$	—	—	4.7×10^{-1}		$\Delta H^{\ddagger} = 16.6;\ \Delta S^{\ddagger} = -28.0$	$170d^e$
$C_{10}H_9OCNHCH_3$ (α-naphthyl)	1.4×10^{-7}	3.7×10^{-5}	3.4×10^0	25 80 23	$E_A = 12.7;\ \log A = 2.48$ $E_A = 15.0;\ \log A = 4.82$ $E_A = 16.8;\ \log A = 11.95$	$8.5d$
$CCl_3O\overset{O}{\overset{\|}{C}}NHC_6H_5$	—	—	3.2×10^{-1}	25		$250d$
Organophosphorus Compounds						
$CH_3P(O)(OCH_3)_2$	1.1×10^{-5}	—	1.5×10^{-3}	99 50	$E_A = 26.7;\ \log A = 10.7$ $E_A = 13.5;\ \log A = 7.3$	$88y$
$CH_3P(O)(OC_6H_5)_2$	5.0×10^{-6}	—	—	110	$E_A = 26.6;\ \log A = 10.7$	
$C_6H_5P(O)(OCH_2CH_3)_2$	1.0×10^{-5}	—	4.5×10^{-3}	59		$440y$
$(CH_3O)_3PO$	—	1.6×10^{-7}	1.3×10^{-4}	45 25	$E_A = 22.7;\ \log A = 8.9$ $E_A = 16.2;\ \log A = 8.1$	$1.2y$

(continued)

	k_H[a] $(M^{-1}s^{-1})$	k_0[a] (s^{-1})	k_{OH}[a] $(M^{-1}s^{-1})$	T (°C)	Temperature Dependence[b]	$t_{1/2}$[c]
(CH₃CH₂S)₃PO	—	7×10^{-7}[d]	—	82 / 25	$E_A = 24.4$; $\log A = 9.25$ / $E_A = 15.0$; $\log A = 7.94$	8.5y
(C₆H₅O)₃PO	—	3×10^{-9}[d]	1.2×10^{-2}[d]	100 / 25	$E_A = 10.9$; $\log A = 6.11$	1.3y
(p-NO₂C₆H₄O)₃PO	—	1.0×10^{-3}[d]	1.1×10^{-2} / 3.4×10^{1}[d]	25 / 25	$E_A = 4.1$; $\log A = 0.062$ / $E_A = 4.1$; $\log A = 4.56$	11m
(p-NO₂C₆H₄O)₃PS	—	2.1×10^{-1}	—	25		3.3s[e]
(CH₃O)₂P(O)Cl	—	2.9×10^{-5}	—	0	$E_A = 10.6$; $\log A = 5.7$	1.3m
Miscellaneous						
CH₂—CH₂ \ NH	5.2×10^{-7}	—	—	25		154d
CH₂—C=O | CH₂—O	—	3.3×10^{-3}	—	25		3.5m
(CH₃O)₂SO₂	—	1.7×10^{-4}	1.5×10^{-2}	25		1.2m
ClCH₂OCH₂Cl	—	1.8×10^{-2}	—	20		25s
C₆H₅COCl	—	4.2×10^{-2}	—	25		16s
CH₃OSO₂—C₆H₅	—	1.2×10^{-5}	8.8×10^{-4}	25		16h[e]
CH₃CH₂OSO₂C₆H₅	—	1.1×10^{-5} [22]	1.2×10^{-4} [22]	25 [22]		
NaOSO₂(OCH₃)	—	7.7×10^{-6}	4.6×10^{-4}	138		
NaOSO₂(OCH₂CH₃)	—	4.6×10^{-6}	9.2×10^{-5}	138		

(Footnotes are listed on the following page)

a. Rate constant for acid (k_H), neutral (k_0) or basic (k_{OH}) hydrolysis. The overall hydrolysis rate in $M^{-1}s^{-1}$ is equal to k_T [O], where [O] is the molar concentration of the organic chemical and $k_T = k_H$ [H^+] + k_0 + k_{OH} [OH^-].

E_A = Arrhenius activation energy in kcal/mol;
log A = Arrhenius pre-exponential factor; units of A same as for k

b. $\Delta H^{\ddagger}$ = enthalpy of activation in kcal/mol
$\Delta S^{\ddagger}$ = entropy of activation in entropy units (cal/deg K•mol).

c. Estimated half-life of organic chemical in water at pH7 and 25°C according to Mabey and Mill [16]. y = years; d = days; h = hours; m = minutes; s = seconds. Note that $k_T = 0.693/t_{1/2}$.

d. Mixed organic, aqueous solvent; > 50% organic.

e. Calculated (this work) as $0.693/k_T$; k_T at pH 7 calculated from the tabulated rate constants as indicated in footnote a.

Source: Mabey and Mill [16] unless otherwise noted. *(Reprinted in part with permission from the American Chemical Society.)*

TABLE 7-11

Disappearance Rate Constants for Acid, Neutral and Alkaline Hydrolyses of Common Pesticides

Compound	T (°C)	$k_H{}^a$ (M^{-1}s^{-1})	$k_0{}^a$ (s^{-1})	$k_{OH}{}^a$ (M^{-1}s^{-1})	Comments	Ref.
Organophosphates:						
Malathion	27	4.8×10^{-5}	7.7×10^{-9}	5.5×10^0	10^{-4} M in 1% acetonitrile: water; overall $\Delta H^{\ddagger}$ = 22.3 kcal/mol; $\Delta S^{\ddagger}$ = −4.1 eu	31
Parathion	20	—^{b}	4.5×10^{-8}	2.3×10^{-2}	10^{-5} M in aqueous buffer	6

(Continued)

Compound	T (°C)	$k_H{}^a$ ($M^{-1}s^{-1}$)	$k_0{}^a$ (s^{-1})	$k_{OH}{}^a$ ($M^{-1}s^{-1}$)	Comments	Ref.
Paraoxon	20	—[b]	4.1×10^{-8}	1.3×10^{-1}	10^{-5} M in aqueous buffer	6
Diazinon	20	2.1×10^{-2}	4.3×10^{-8}	5.3×10^{-3}	10^{-5} M in aqueous buffer	6
Diazoxon	20	6.4×10^{-1}	2.8×10^{-7}	7.6×10^{-2}	10^{-5} M in aqueous buffer	6
Chlorpyrifos	25	—	1×10^{-7}	1×10^{-1}	10^{-7} to 10^{-9} M in aqueous buffer; $E_A = 21.1$ kcal/mol	17
Carbamates: Sevin	20			7.7×10^{0}	10^{-5} M in aqueous buffer; $E_A = 16.9$ kcal/mol	6
	23			3.4×10^{0}	$2.5 \times 10^{-6} - 1 \times 10^{-4}$ M in water; $E_A = 16.9$ kcal/mol	1
Baygon	20			5.0×10^{-1}	$2.4 \times 10^{-6} - 9.7 \times 10^{-5}$ M in water; $E_A = 15.8$ kcal/mol	1
	20			4.6×10^{-1}	10^{-5} M in aqueous buffer; $E_A = 15.8$ kcal/mol	6
Pyrolam	20			1.1×10^{-2}	$2.0 \times 10^{-6} - 8.2 \times 10^{-5}$ M in water; $E_A = 13.7$ kcal/mol	1

(Continued)

TABLE 7-11 (Continued)

Compound	T (°C)	k_H^a (M⁻¹s⁻¹)	k_0^a (s⁻¹)	k_{OH}^a (M⁻¹s⁻¹)	Comments	Ref.
Dimetilan	20			5.7×10^{-5}	$2.1 \times 10^{-6} - 8.3 \times 10^{-5}$ M in water; E_A = 14.0 kcal/mol	1
p-Nitrophenyl-N-methyl carbamate	25		$< 4 \times 10^{-5}$	3.0×10^{3}	3×10^{-5} M in aqueous buffer	2
2,4-D Esters:						
n-Butoxyethyl	28	2.0×10^{-5}	2.0×10^{-5}	3.02×10^{1}	1×10^{-5} M in water. For k_H, $\Delta H^{\ddagger}$ = 17.6 kcal/mol and $\Delta S^{\ddagger}$ = 14.8 eu. For k_{OH}, $\Delta H^{\ddagger}$ = 20.1 kcal/mol, $\Delta S^{\ddagger}$ = -21.3 eu	34
	67	6.6×10^{-4}	2.7×10^{-7}	5.0×10^{3}		
Methyl	28			1.7×10^{1}		34
Methoxychlor	27	—	2.2×10^{-8}	3.8×10^{-4}	1×10^{-8} M in water	33
DDT	27	—	1.9×10^{-9}	9.9×10^{-3}	1×10^{-8} M in water	33

a. Rate constant for acid (k_H), neutral (k_0), or basic (k_{OH}) hydrolysis. Overall hydrolysis rate in M⁻¹ s⁻¹ = k_T [O] where [O] = molar concentration of the organic chemical and $k_T = k_H$ [H⁺] + k_0 + k_{OH} [OH⁻].

b. A dash in the k_H or k_{OH} column indicates that acid- or base-catalyzed hydrolysis is slow and may be insignificant. A blank simply indicates that no rate constant was reported.

c. Calculated from data presented in the cited reference on pH dependence of k_T.

h	=	Planck's constant in Eq. 7-29
HA	=	an acid
I_{AN}, I_{AB}, I_{NB}	=	transition points in Figure 7-2
k	=	Boltzmann's constant in Eq. 7-29
k	=	rate constant
k°	=	rate constant for parent compound in a class of compounds
k_0	=	rate constant for neutral hydrolysis (s^{-1})
k_B	=	rate constant for general base-catalyzed hydrolysis in Eq. 7-18 ($M^{-1}s^{-1}$)
k_H	=	rate constant for specific acid-catalyzed hydrolysis ($M^{-1}s^{-1}$)
k_{HA}	=	rate constant for general acid-catalyzed hydrolysis in Eq. 7-18 ($M^{-1}s^{-1}$)
k_{H_2O}	=	rate constant for neutral hydrolysis in Eq. 7-18 ($M^{-1}s^{-1}$)
k_{OH}	=	rate constant for specific base-catalyzed hydrolysis ($M^{-1}s^{-1}$)
k_T	=	total, or overall, hydrolysis rate constant (s^{-1})
LFER	=	linear free energy relationship
n	=	number of compounds
pH	=	$-\log[H^+]$
pK_a	=	$-\log K_a$, where K_a = acid dissociation constant
R	=	gas constant (1.987 cal/mol·deg). Also used (in drawings of chemical structures) to represent an unspecified organic group
$\Delta S°$	=	standard entropy of reaction for an equilibrium process
$\Delta S^{\ddagger}$	=	entropy of activation for a rate process in Eq. 7-29
S_N1	=	substitution, nucleophilic, unimolecular (see §7-2)
S_N2	=	substitution, nucleophilic, bimolecular (see §7-2)
T	=	temperature (K)
$t_{1/2}$	=	half-life due to hydrolysis in Eq. 7-37
t	=	time in Eqs. 7-16 and -21 to -24
X	=	leaving group on a molecule
Z	=	general designation for a substituent group on a compound

Greek

δ	=	reaction constant in Taft equation, Eq. 7-33
ρ	=	Hammett reaction constant in Eq. 7-32
ρ^*	=	Taft reaction constant in Eq. 7-33
σ	=	Hammett substituent constant in Eq. 7-32
σ^*	=	Taft substituent constant in Eq. 7-33
σ_T	=	total σ value (simple sum) due to presence of more than one substituent
σ^-	=	substituent constant in Hammett correlation especially for anilinium ions and phenols

7-13 REFERENCES

1. Aly, O.M. and M.A. El-Dib, "Studies on the Persistence of Some Carbamate Insecticides in the Aquatic Environment," *Water Res.,* **5**, 1191-204 (1971)

2. Bender, M.L. and R.B. Homer, "The Mechanism of the Alkaline Hydrolysis of p-Nitrophenyl N-Methylcarbamate," *J. Org. Chem.,* **30**, 3975-78 (1965).

3a. Bruice, P.Y., T.C. Bruice, P.M. Dansette, H.G. Selander, H. Yagi and D.M. Jenina, "Comparison of the Mechanism of Solvolysis and Rearrangements of K-Region vs. Non-K-Region Arene Oxides of Phenanthrene. Comparative Solvolytic Rate Constants of K-Region and Non-K-Region Arene Oxides," *J. Am. Chem. Soc.,* **98**, 2965-73 (1976).

3b. Burlinson, N.E., L.A. Lee and D.H. Rosenblatt, "Kinetics and Products of Hydrolysis of 1,2-Dibromo-3-chloropropane (DBCP)," prepublication draft dated April 18, 1981. To be submitted to *Environ. Sci. Technol.*

4. Chapman, N.B. and J. Shorter (eds.), *Advances in Linear Free Energy Relationships,* Plenum Press, New York (1972).

5. Exner, O., "The Hammett Equation — The Present Position," Chap. 1 in *Advances in Linear Free Energy Relationships,* ed. by N.B. Chapman and J. Shorter, Plenum Press, New York (1972).

6. Faust, S.D. and H.M. Gomaa, "Chemical Hydrolysis of Some Organic Phosphorus and Carbamate Pesticides in Aquatic Environments," *Environ. Lett.,***3**, 171-201 (1972).

7. Fiksel, J. and M. Segal, "An Approach to Prioritization of Environmental Pollutants: The Action Alert System," Final Draft Report on EPA Contract No. 68-01-3857 (June 1980).

8. Harris, J.C., "Characterization of Transition States for S_N2 Reactions," Ph.D. Thesis, Washington University, St. Louis, MO (January 1970).

9. Higuchi, T., "The Effect of Molecular Structures on Catalysis and Molecular Binding," Progress Reports IV and V on DOD Contract No. DA18-108-405 CML-265 (March 1961).

10. Hine, J., *Structural Effects on Equilibria in Organic Chemistry,* Wiley Interscience, New York (1975).

11. Hughes, E.D., C.K. Ingold and J.C. Patel, *J. Chem. Soc.*, **526** (1963) as cited in Ref. 20.

12. Jaffé, H.H., "A Reexamination of the Hammett Equation," *Chem. Rev.*, **53**, 191-261 (1953).

13. Koppel, I.A. and V.A. Palm, "The Influence of the Solvent on Organic Reactivity," Chap. 5 in *Advances in Linear Free Energy Relationships,* ed. by N.B. Chapman and J. Shorter, Plenum Press, New York (1972).

14. Kortüm, G., W. Vogel and K. Andrussow, *Dissociation Constants of Organic Acids in Aqueous Solution,* Butterworths, London (1961).

15. Leffler, J.E. and E. Grunwald, *Rates and Equilibria of Organic Reactions,* John Wiley & Sons, New York (1963).

16. Mabey, W. and T. Mill, "Critical Review of Hydrolysis of Organic Compounds in Water Under Environmental Conditions," *J. Phys. Chem. Ref. Data,* **7**, 383-415 (1978).

17. Meikle, R.W. and C.R. Youngson, "The Hydrolysis Rate of Chlorpyrifos, O-O-Diethyl-O-(3,5,6-Trichloro-2-Pyridyl)Phosphorothioate, and its Dimethyl Analog, Chlorpyrifos-Methyl, in Dilute Aqueous Solution," *Arch. Environ. Contam. Toxicol.,* **7**, 13-22 (1978).

18. Mill, T., "Structure Reactivity Correlations for Environmental Reactivity," EPA-560/11-79-012 (September 1979). Available through NTIS, PB 80-110323.

19a. Nishioka, T., T. Fujita, K. Kitamura and M. Nakajima, "The Ortho Effect in Hydrolysis of Phenyl Esters," *J. Org. Chem.,* **40**, 2520-25 (1975).

19b. Palm, V.A. (ed.), *Tables of Rate and Equilibrium Constants of Heterolytic Organic Reactions,* Vols. 1-5, Laboratory of Chemical Kinetics and Catalysis, Tartu State University, Moscow, U.S.S.R. (1975-1979).

20. Shorter, J., "The Separation of Polar, Steric and Resonance Effects by the Use of Linear Free Energy Relationships," Chap. 2 in *Advances in Linear Free Energy Relationships,* ed. by N.B. Chapman and J. Shorter, Plenum Press, New York (1972).

21. Smith, J.H., W.R. Mabey, N. Bohonos, B.R. Holt, S.S. Lee, T.-W. Chou, D.C. Bomberger and T. Mill, "Environmental Pathways of Selected Chemicals in Freshwater Systems, Part I: Background and Experimental Procedures," EPA-600/7-77-113 (October 1977).

22. Streitwieser, A., Jr., *Solvolytic Displacement Reactions,* McGraw-Hill Book Co., New York (1962).

23. Taft, R.W., Jr., "Separation of Polar, Steric, and Resonance Effects in Reactivity" ed. by M.S. Newman in *Steric Effects in Organic Chemistry,* John Wiley & Sons, New York, pp. 556-675 (1956).

24. U.S. National Bureau of Standards, "Tables of Chemical Kinetics: Homogeneous Reactions," NBS Circular 510 (1951) and supplements.

25. Wells, P.R., *Linear Free Energy Relationships,* Academic Press, New York (1968).

26. Wolfe, N.L., "Organophosphate and Organophosphorothioate Esters: Application of Linear Free Energy Relationships to Estimate Hydrolysis Rate Constants for Use in Environmental Fate Assessment," *Chemosphere,* **9**, 571-79 (1980).

27. Wolfe, N.L., "Sediment Effects on Hydrolysis Pathways," Quarterly Report, EPA Environmental Research Laboratory, Athens, GA (October-December 1978).

28. Wolfe, N.L., W.C. Steen and L.A. Burns, "Phthalate Ester Hydrolysis: Linear Free Energy Relationships," *Chemosphere,* **9**, 403-408 (1980).

29. Wolfe, N.L. and S.W. Karickhoff, "Kinetics of Sorption and Reaction in Suspended Systems," paper presented at Third International Symposium on Aquatic Pollutants, Jekyll Island, GA (October 1979).

30. Wolfe, N.L., R.G. Zepp, G.L. Baughman and J.A. Gordon, "Kinetic Investigation of Malathion Degradation in Water," *Bull Environ. Contam. Toxicol.,* **13**, 707-13 (1975).

31. Wolfe, N.L., R.G. Zepp, J.A. Gordon, G.L. Baughman and D.M. Cline, "Kinetics of Chemical Degradation of Malathion in Water," *Environ. Sci. Technol.,* **11**, 88-93 (1977).

32. Wolfe, N.L., R.G. Zepp and D.F. Paris, "Use of Structure Reactivity Relationships to Estimate Hydrolytic Persistence of Carbamate Pesticides," *Water Res.,* **12**, 561-63 (1978).

33. Wolfe, N.L., R.G. Zepp, D.F. Paris, G.L. Baughman and R.C. Hollis, "Methoxychlor and DDT Degradation in Water: Rates and Products," *Environ. Sci. Technol.,* **11**, 1077-81 (1977).

34. Zepp, R.G., N.L. Wolfe, J.A. Gordon and G.L. Baughman, "Dynamics of 2,4,-D Esters in Surface Waters; Hydrolysis, Photolysis and Vaporization," *Environ. Sci. Technol.,* **9**, 1144-50 (1975).

8

RATE OF AQUEOUS PHOTOLYSIS

Judith C. Harris

8-1 INTRODUCTION

It is increasingly recognized that photochemical processes may be important in determining the fate of organic pollutants in aqueous environments. Both *direct* photolysis, in which the pollutant itself absorbs solar radiation, and *sensitized* photolysis, in which energy is transferred from some other species in the aquatic solution, may occur. The rates of these processes in a natural water system depend both on the properties of the aquatic environment (intensity and spectrum of solar radiation, presence or absence of sensitizers, quenchers, etc.) and on the properties of the organic chemical (extent of absorption of light and inherent tendency to undergo photochemical reaction).

Existing models for predicting photochemical reactivity in the environment are essentially models for calculating the net rate at which an aqueous solution containing the organic chemical absorbs light. The basic approach is to evaluate the degree of overlap between the ultraviolet/visible absorption spectrum of the organic molecule and the solar radiation to which it is exposed. The necessary compound-specific data must usually be determined in the laboratory, since literature spectra do not contain this information in retrievable form.

The estimated rate of absorption of light by a solution of the molecule of interest is necessary but not sufficient for calculating the rate constant for direct photochemical degradation. To estimate the latter quantity, which is a measure of the inherent photochemical reactivity of the organic chemical, one must also know its quantum yield for photolysis. There are, at present, no methods for estimating the quantum yield of an organic molecule from its chemical structure or from other physical/chemical properties.

Although the state of the art does not include any methods for estimation of photolytic reactivity in aquatic environments, there are some qualitative guidelines to indicate compounds that may be reactive and the types of reactions they may undergo. These guidelines apply to rates of direct photolysis only; the possibility for sensitized photolysis depends at least as much on the chemical characteristics of the aquatic system as on those of the organic chemical of concern and is not addressed in this chapter.

8-2 BASIC PRINCIPLES OF EXCITATION/DEACTIVATION

An organic molecule can undergo photochemical transformations if, and only if, both of the following conditions are met:

- Light energy is absorbed by the molecule to produce an electronically excited state of the molecule, and
- Chemical transformations of the excited state are competitive with deactivation processes.

Excitation. The necessity for light absorption is cited as the first law of photochemistry or the Grotthus-Draper Law [2,27]: *Only the light which is absorbed by a molecule can be effective in producing photochemical change in the molecule.* It is well known that molecules absorb light in several regions of the electromagnetic spectrum, corresponding to different kinds of molecular transitions. Table 8-1 lists the types of transitions and summarizes the energy, wavelength, and frequency regimes for the respective absorption regions. If these data are compared with typical bond dissociation energies (Table 8-2), it becomes clear that only the electronic transitions, corresponding to UV/visible light absorption, are inherently energetic enough to lead to chemical reactions. The regime of importance for photochemical transformations is thus confined to UV/visible light with a wavelength of 110-750 nm.

When we focus on environmental photochemistry at or near the earth's surface, the wavelength regime of importance can be further

TABLE 8-1

Types of Molecular Transition and Associated Energy Levels

Type of Transition	Absorption Region	Energy Range		Time Scale (s)
		E (kcal/mol)[a]	λ (nm)	
Translational	(Thermal)			
Rotational	Microwave	0.01-0.1	$\begin{cases} 1 \times 10^6 - \\ 1 \times 10^7 \end{cases}$	10^{-10}
Vibrational	Infrared	1-10	$\begin{cases} 10,000- \\ 1,000 \end{cases}$	10^{-13}
Electronic	Visible-UV	38-250	750-110	10^{-15}

a. A "mol" of photons = 1 Einstein = 6.023×10^{23} photons of a specified wavelength.

Source: Calvert and Pitts [2], pp. 1-26.

TABLE 8-2

Some Approximate Bond Dissociation Energies

Bond Broken	ΔH_{298K} (kcal/mol)
C-H (alkanes)	91-99
C-H (benzene)	103
C-C (alkanes)	78-84
C-F	114-110
C-Cl	78-82
C-Br	67
C-I	53
O-H (alcohols)	100-102
C-O (alcohols)	89-90
C-N (amines)	79
O-O (peroxides)	35-51

Source: Calvert and Pitts [2], pp. 824-826. *(Reprinted with the permission of the authors.)*

narrowed, because the stratospheric ozone layer effectively prevents UV irradiation of less than 290 nm from reaching the ecosphere. Thus, the first law of [environmental] photochemistry may be restated: *Only the light of 290-750 nm wavelength absorbed by a molecule can potentially lead to photochemical transformations of that molecule in the environment.*

The excitation process is expressed as

$$P \xrightarrow{\;h\upsilon\;} P^*$$ (8-1)

where P is the ground-state molecule, $h\upsilon$ is a quantum of light, and P^* the excited-state molecule. In quantitative terms, the light absorbed in this process is given by the Beer-Lambert Law [5]:

$$A = \log \frac{I_0}{I} = \epsilon c \ell$$ (8-2)

in which A is the absorbance, I_0 is the intensity of the incident light of specified wavelength, I is the intensity of transmitted light of the same wavelength, ϵ is the molar absorptivity (extinction coefficient) of the absorbing species, c is the concentration of the absorbing species, and ℓ is the depth of the absorbing medium. Deviations from Eq. 8-2, which can be derived by assuming a first-order absorption of light by species P, are generally not observed at values of $A < {\sim}0.7$. Deviations due to second-order effects can occur at high concentrations of the absorbing species. Values of ϵ for compounds absorbing light above 290 nm wavelength are typically in the range from 10 to 100,000. Since environmental concentrations of pollutants in aqueous media are usually <10 ppm (w/v) or $<10^{-5}$ M, deviations from the Beer-Lambert Law are not likely under environmental conditions. However, materials that are opaque to UV/visible light (other than the pollutant) may be present in the environmental medium and may alter the effective value of I_0, thus effectively reducing the quantity of light absorbed by P.

Deactivation: Internal Conversion and Intersystem Crossing. The formation of a photochemically excited state, P^* is a necessary but not a sufficient condition for producing a photochemical reaction of molecule P. The probability of accomplishing a net photochemical degradation (photolysis of P) depends on the competition among the primary photophysical processes of radiative (Eq. 8-3) and radiationless (Eq. 8-4) decay to the ground state and any primary photochemical processes (Eq. 8-5) that may occur.

$$P* \longrightarrow P + h\upsilon' \tag{8-3}$$

$$P* \longrightarrow P \tag{8-4}$$

$$P* \longrightarrow products \tag{8-5}$$

The efficiency of each primary process, i, is conventionally expressed in terms of its quantum yield, ϕ_1, which may be defined as

$$\phi_i = \frac{\text{No. of P* excited states undergoing process i}}{\text{No. of quanta of light absorbed by P}} \tag{8-6}$$

The idea that the several deactivation and reaction pathways (Eqs. 8-3 through 8-5) are competitive is expressed in the second law of photochemistry, the Stark-Einstein-Bodenstein Law, which is stated by Calvert and Pitts [2] as: *The absorption of light by a molecule is a one-quantum process, so that the sum of the primary-process quantum yields (ϕ) must be unity.*

The quantum yields for disappearance of organic chemicals in water are generally <0.01. This implies that 90%, 99%, or more of the excited-state molecules undergo photophysical deactivation rather than photo-reaction/photolysis.

One deactivation pathway involves a radiative process, *fluorescence,* in which a quantum of light is emitted during the transition to the ground electronic state and some residual vibrational excitation is rapidly lost via collision processes. This process is the inverse of the absorption process; in fact, fluorescence spectra are often mirror images of absorption spectra [11]. The "natural lifetime," τ, of the state can be estimated, assuming that fluorescence is the only important decay pathway, by Eq. 8-7 [11]:

$$\tau \approx 10^{-4}/\epsilon_{max} \tag{8-7}$$

where ϵ_{max} is the molar absorptivity at the wavelength of maximum absorption.[1] This fluorescence lifetime is on the order of 10^{-9} to 10^{-5} second [11]. Fluorescence quantum yields on the order of 0.3 are observed for simple aromatic hydrocarbons in solution at ordinary temperatures [2].

No other photophysical or photochemical primary process can compete with fluorescence unless it takes place within the brief fluorescence

---

1. τ is the time required for decay to 1/e of the original concentration of the excited state; it is about 44% longer than $t_{1/2}$ (the half-life) for a first-order decay process.

lifetime. One photophysical process that is sufficiently rapid is *internal conversion,* a non-radiative deactivation process in which the energy of the quantum of light originally absorbed is eventually dissipated as increased thermal energy of the reaction medium. The net effect of internal conversion and fluorescence photophysical processes is usually to regenerate the ground state of the organic molecule within 10^{-9} to 10^{-5} second.

It is empirically observed that the quantum yield for fluorescence, ϕ_f, is independent of the precise wavelength of light absorbed [11,28]. This generalization is called "Vavilov's Rule" by Turro *et al.* [28], who also discuss the relatively small number of exceptions known. However, this rule does not necessarily apply to the quantum yields for photochemical reaction when very large differences in the wavelength are involved. Quantum yields for photolysis in the vapor phase are normally wavelength dependent; irradiation with short-wavelength light (high energy quanta) may provide sufficient excitation energy to induce photochemical/physical transformations that are impossible when longer wavelength irradiation is applied. It is important to keep this point in mind when attempting to extrapolate results of labora-tory photolysis experiments (usually at 254 nm UV irradiation) to environmental conditions (>290 nm irradiation). For wavelengths >300 nm, there may be little wavelength dependence of quantum yield in solution because relaxation to the lowest excited state is usually more rapid than reaction from the higher states.

An important additional photophysical process is the transition be-tween singlet (all electron spins paired) and triplet (two unpaired elec-tron spins) electronic states, known as *intersystem crossing.* Experimental evidence suggests that the quantum yield for intersystem crossing is on the order of 0.99 for aromatic ketones (acetophenone, benzophenone) and on the order of 0.2-0.6 for other aromatic species (benzene, naphthalene, quinoline, naphthol, naphthoic acid, and others) in organic solvents at ordinary temperatures.[2] The existence of the triplet state is important from the perspective of potential photochemical trans-formations, primarily because its natural lifetime is much longer than that of the corresponding singlet. Since triplet lifetimes are on the order of 10^{-5} to 10^{-3} s (vs. 10^{-9} to 10^{-5} s for singlets), relatively slow photochemical processes can compete with photophysical deactivation from this state. The latter process can occur by radiative *(phosphorescence)* or non-radiative (intersystem crossing) pathways. The quantum yield for

---

2. See Ref. 2, pp. 293-321.

phosphorescence, ϕ_p, from the lowest triplet excited state is highly sensitive to the medium in which irradiation occurs. Phosphorescence is normally observed only when organic molecules are frozen in a glassy matrix (such as a mixture of ether: pentane:alcohol at $-196°C$). In fluid media (solution or gas phase), radiationless deactivation occurs in time periods shorter than the natural phosphorescence lifetime of 10^{-5} to 10^{-3}s. The presence of other molecules such as oxygen can lead to enhanced rates of intersystem crossing [33]. Kan [11] suggests a time scale on the order of 10^{-6} s for intersystem crossing processes, such as radiationless deactivation from the triplet excited state to the ground singlet electronic state. Photochemical processes with first-order or pseudo first-order rate constants on the order of 10^6 s^{-1} or higher can therefore be expected to compete with photophysical deactivation in solution for systems which have high quantum yields for triplet formation.

Energy Transfer: Sensitization and Quenching. The preceding discussion focused on intramolecular photophysical processes. Triplet excited states, however, are sufficiently long-lived in solution to participate in intermolecular electronic energy transfer processes. (Singlet-singlet energy transfer resulting in enhanced fluorescence is quite possible in solid media but rare in solution or in gas phase reactions.) In such a process, energy is transferred from the triplet state of an excited donor molecule, S*, to the ground state of the acceptor molecule, P, yielding the P* triplet state. This process provides a means of populating an electronically excited state of P with no direct absorption of light by P ground-state molecules. This type of energy transfer can occur whenever the "triplet energy" (energy difference between the triplet excited state and singlet ground state) of S is greater than the corresponding energy difference in the P system. Table 8-3 lists triplet energies for some sample compounds. Energy transfer can either enhance (sensitize) or reduce (quench) the photochemical reactivity of an organic molecule in the aqueous environment.

Photochemical *sensitization* is said to occur when some species in solution, other than the target organic molecule, absorbs light and transfers its excitation energy to the target species. The donor species (the "sensitizer") undergoes no net reaction in the process but has an essentially catalytic effect. Photochemical sensitization is thus distinguished in principle from degradation of the target molecule by photoinitiated free radicals in solution. In practice, these two phenomena may be indistinguishable for complex situations, such as natural water systems. The importance of photochemical sensitization in the aquatic environment has not been well established.

TABLE 8-3

Triplet Energies for Selected Compounds

Compound	Triplet Energy (kcal/mol)
Acetophenone	73.6
Benzaldehyde	71.9
Carbazole	70.1
Triphenylamine	70.1
Benzophenone	68.5
Anthraquinone	62.4
Phenanthrene	62.2
Naphthalene	60.9
Biacetyl	54.9
Fluorenone	53.3
Pyrene	48.7

Source: Calvert and Pitts [2]. *(Reprinted with the permission of the authors.)*

Quenching of a photochemical process is said to occur when excitation energy present in the target organic molecule is transferred to some other species in solution. This process is, in a sense, the inverse of sensitization, as it results in net *de*activation of the organic substance of concern via energy transfer. As noted above, energy can be transferred to any species with lower triplet energy. A very important and effective quencher (acceptor) is molecular oxygen, which has a triplet ground state. The second-order rate constant for oxygen quenching is on the order of 10^{10} L/mol-s [2]. At a dissolved oxygen concentration of 10 mg/L (0.31mM), this corresponds to a pseudo first-order quenching rate constant of 3×10^6 s^{-1}, or a half-time for quenching of 2×10^{-7} s. This triplet energy transfer quenching is essentially a diffusion-controlled process in fluid solution [2] and is thus competitive with any potential bimolecular photochemical degradation processes. Non-photolytic deactivation of P* by oxygen quenching of the excited triplet may therefore be a significant and perhaps dominant fate process in aerobic aqueous environments.

Summary. The first law of photochemistry states that only light which is absorbed by a molecule can result in photochemical reaction. The extent of absorption of light, given by Beer's Law for each incident wavelength, can be calculated if the absorption spectrum of the organic compound and the distribution of intensities/wavelengths of incident light are known. Photochemical excitation processes are thus rather straightforward, except for the potential complication of sensitized excitation by [unknown] sensitizer species in aqueous environments.

Deactivation processes include photophysical transitions among electronically excited and ground states, as well as photochemical degradation (photolysis). Within any particular broad absorption/excitation band, the relative importance of various deactivation processes is likely to be independent of excitation wavelength, since an "equilibrium" population of lowest excited singlet and triplet states is generally established. This is the basis for the common statement [24-36] that quantum yields are independent of wavelength. The generalization may frequently be invalid if large differences in wavelength (254 nm vs. >290 nm) are involved or if the excited states have different character (i.e., $n \rightarrow \pi^*$ vs. $\pi \rightarrow \pi^*$).

Photophysical deactivation processes include fluorescence, quenching, radiationless conversion to the ground state, and phosphorescence. The characteristic times for these processes, with which potential photolysis/photochemical transformations must compete, are on the order of: 10^{-9} to 10^{-5} s, 10^{-7} s, 10^{-6} s, and 10^{-5} to 10^{-3} s, respectively. Photolysis or other photochemical transformation processes must be rapid (pseudo first-order rate constants on the order of 10^5 to 10^9 s^{-1}) in order to compete with photophysical deactivation.

For reactions in fluid solutions, photophysical deactivation to the ground state, with no net chemical degradation, can generally be expected to account for more than 95% of the light energy absorbed.

8-3 ABSORPTION OF LIGHT

Chromophores and Characteristic Absorption Bands. As noted in the previous section, it is a necessary but not sufficient condition for photolysis that the organic species in question absorb light. A comparison of the spectrum of solar radiation with the characteristic absorption spectra of organic molecules will therefore provide a preliminary indication of the potential for photochemical reactivity.

Figure 8-1 represents the spectral distribution of solar energy incident on earth, or insolation [17]. Integration of the area under the curves of Figure 8-1 would show that about 10% of the incident light energy is in the ultraviolet region and 45% each in the visible and infrared

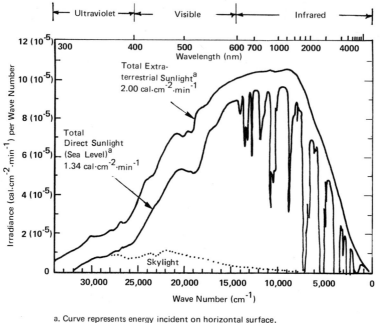

a. Curve represents energy incident on horizontal surface.

Source: Odum [17]. *(Reprinted with permission from Holt, Rinehart and Winston.)*

FIGURE 8–1   **Spectral Distribution of Extraterrestrial Solar Radiation and of Solar Radiation at Sea Level on a Clear Day**

regions [20]. Virtually all of the insolation is of wavelength >300 nm; shorter wavelengths are effectively filtered out by the stratospheric ozone layer.

At the earth's surface, light of < 290 nm wavelength has such a low intensity that direct photochemical activation at these wavelengths is improbable. On the other hand, UV/visible light of wavelength >290 nm (frequency 3.45×10^4 cm^{-1} or 100 kcal/Einstein) is available at moderate intensity. For a temperate zone such as the United States, it has been calculated [20] that the mean incident solar energy on a horizontal surface varies from about 3000 kcal/m²-day (northeast) to about 5000 kcal/m²-day (southwest). This energy input is not constant, of course, but varies diurnally and seasonally. Figure 8-2 indicates the seasonal fluctuations in the incident solar energy.

The solar energy incident on the surface of a natural water body is not uniformly transmitted through the aqueous medium. Figure 8-3 presents some examples showing the attenuation of solar irradiance with

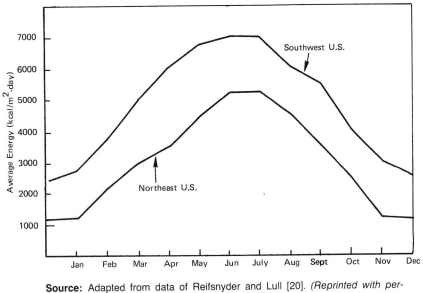

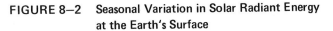

Source: Adapted from data of Reifsnyder and Lull [20]. *(Reprinted with permission from Holt, Rinehart and Winston.)*

FIGURE 8—2 **Seasonal Variation in Solar Radiant Energy at the Earth's Surface**

depth as a function of wavelength. Tyler [29] notes that the long-wavelength absorption of light is due to water itself, while the 400-500 nm absorption in the eutrophic lake can be attributed to phytoplankton and organic degradation products in the water column.

The absorption spectra of organic molecules can be compared with the solar spectra of Figures 8-1 and 8-3 to determine whether absorption of light energy in the environment is likely to be significant. A quantitative approach to such comparisons is presented in the following section of this chapter. Some qualitative rules of thumb that are useful in making preliminary assessments of potential photochemical reactivity are based on the characteristic absorption frequencies of various compounds.

As a first approximation, the electronic spectrum of an organic molecule can be attributed to the presence of one or more *chromophores*, which are functional groups that absorb UV/visible light. Table 8-4 lists typical frequencies of maximum absorption, λ_{max}, and molar absorptivities, ϵ, for particular chromophores in organic molecules that have λ_{max} above the 290-nm solar insolation cutoff. (Some molecules with

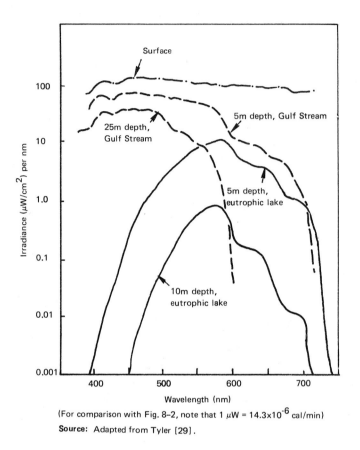

(For comparison with Fig. 8–2, note that 1 μW = 14.3x10^{-6} cal/min)

Source: Adapted from Tyler [29].

FIGURE 8–3   Attenuation of Solar Spectrum in Natural Waters

absorption maxima below 290 nm have been included, because the "tails" of such absorption peaks may lead to light absorption and subsequent photoreactivity.)

Table 8-5 presents comparable information for some specific organic compounds. Note that speciation of ionizable organics such as phenol can affect both the wavelength and the intensity of absorption maxima. In general, organic molecules that have moderate-to-strong absorption in the > 290 nm wavelength range contain either: (1) an extended conjugated hydrocarbon system (as in anthracene or larger fused-ring systems) or (2) a functional group with an unsaturated heteroatom (e.g., carbonyl, azo group, nitro group).

The exact position of the absorption maximum for a chromophore in a particular organic molecule depends on the details of the molecular

TABLE 8-4

Chromophoric Groups and Their Characteristic
Absorption Maxima at $\lambda > 290$ nm

Group	λ_{max} (nm)	Molar Absorptivity, ϵ (L/mol-cm)
$\rangle$C=O (aldehyde, ketone)	295	10
$\rangle$C=S	460	weak
–N=N–	347	15
–NO$_2$	278	10
(naphthalene structure)	311	250
	270	5000
(anthracene structure)	360	6000
O=(ring)=O	440	20
	300	1000
$\rangle$C=C–C=O	330	20

Source: Calvert and Pitts [2], pp. 265-67. *(Reprinted with the permission of the authors.)*

structure. Several compilations of UV/visible spectral data for organic compounds are available [7,9,10,21]. There are also several empirical correlations, such as the Woodward Rules [32] for calculating λ_{max} values for simple conjugated systems (dienes and α,β-unsaturated carbonyls); these involve addition of wavelength increments for each substituent on the simple system to the λ_{max} for the intense, short-wavelength absorption band of the unsubstituted system. In each instance, however, the base λ_{max} for the unsubstituted system is so low (217, 215, 209, 197 nm) [2] that wavelengths near the 290-nm solar cutoff are not achievable, even with extensive substitution. The Woodward type of empirical correlation is, therefore, not relevant to prediction of the environmental photochemical behavior of organic molecules.

TABLE 8-5

UV/Visible Absorption Maxima and Molar Absorptivities for
Selected Organic Compounds

Compound	λ Below 290 nm		λ Above 290 nm	
	λ_{max}[a]	ϵ	λ_{max}[a]	ϵ
Hydrocarbons				
Ethylene	193	10,000		
1,3-Butadiene	217	20,900		
Benzene	255	215		
Styrene	244	12,000		
	282	450		
Biphenyl	246	20,000		
Naphthalene	221	100,000	311	250
	270	5,000		
Anthracene	250	150,000	360	6,000
Pyrene $(C_{16}H_{10})$[b]	231	45,000	295	45,000
	241	83,000	305	11,000
	262	25,000	319	29,000
	272	50,000	335	47,000
			357	420
			362	360
			372	250
Benzo[a] pyrene $(C_{20}H_{10})$[b]			347	13,000
			364	24,000
			384	29,000
β-Carotene[c]			452	139,000
			478	122,000
Substituted Aromatics				
Aniline	230	8,600		
	280	1,430		
Anilinium ion	203	7,500		
	254	160		
Acetophenone	240	13,000	319	50
	278	1,100		
Azobenzene			319	19,500
			445	300
Benzaldehyde	244	15,000	328	20
	280	1,500		
Benzoic acid	230	10,000		
	270	800		

Compound	λ Below 290 nm		λ Above 290 nm	
	λ_{max}[a]	ϵ	λ_{max}[a]	ϵ
Substituted Aromatics				
Benzoquinone	250	15,000	300	1,000
			440	20
Chlorobenzene	210	7,500		
	257	170		
Nitrobenzene	252	10,000	330	125
	280	1,000		
Phenol	210	6,200		
	270	1,450		
Phenolate ion	235	9,400		
	287	2,600		
Benzonitrile	224	13,000		
	271	1,000		
Others				
Acetaldehyde			293	12
Acetone	272	19		
Acrolein	210	11,000	315	26
Furan	252	1		
Acridine	250	107,000	355	10,500
Amyl nitrite	218	1,120	356	56
Azomethane			347	4
Ethyl nitrate	270	12		
Nitrosobutane			300	100
			665	20
Pyridine	250	2,000		
Pyrrole	240	302		
Quinoline	275	3,700	313	2,700
Butanethiol	231	126		
Cyclohexyl methyl sulfoxide	210	1,500		
Di-n-butyl sulfide	210	1,200		
	229	145		
Thiophene	231	5,620		

(Continued)

TABLE 8-5 (Continued)

Footnotes

a. Values generally refer to spectra in organic solvents and should be regarded as approximate absorption maxima for species in aqueous solution.

b. Data from Kamlet [10]

c. Trans-β-carotene, a plant pigment [5] :

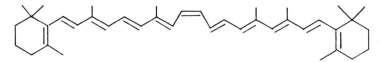

Source: Refs. 2 and 23, except as noted in footnotes b and c. *(Reprinted with permission of John Wiley & Sons, Inc., and the authors of reference 2.)*

Some insight into the inherent photochemical activity of organic target molecules can be obtained by considering the nature of the potential photochemical transitions. The majority of the photochemical activation processes that occur within the 290-750 nm wavelength range involve electronic transitions from non-bonding (n) orbitals — i.e., unshared electron pairs on covalently bonded N or O — or π-bonding (π) orbitals — i.e., electron pairs in double bonds or rings — to antibonding π^* orbitals. Table 8-6, compares some of the features of n $\rightarrow$ π^* and $\pi \rightarrow \pi^*$ transitions. In an α,β-unsaturated carbonyl compound, for example, the $\pi \rightarrow \pi^*$ transition is the intense, shorter wavelength band ($\lambda_{max} \approx 210$ nm, $\epsilon \approx 12,000$), while the n $\rightarrow \pi^*$ is the weak, longer wavelength band ($\lambda_{max} \approx 330$ nm, $\epsilon \approx 20$) [27]. The n $\rightarrow \pi^*$ transitions of carbonyls are particularly likely to generate excited states that undergo photochemical reaction (dissociation, hydrogen abstraction) as opposed to photophysical deactivation.

Quantitative Calculation of Absorption of Solar Energy. The qualitative consideration of chromophores and characteristic absorption bands described in the previous section is very useful for screening out compounds that have no potential for absorption of solar radiation in the environment. Such compounds cannot be active in direct photolysis. For compounds which do have absorption maxima near or above 290 nm, it is necessary to go beyond consideration of λ_{max} and ϵ_{max} to predict the extent and rate of absorption of solar energy.

A quantitative model for calculating the rate of absorption of light in aquatic environments has been developed and described in a series of

TABLE 8-6

Comparison of the Features of $n{\rightarrow}\pi$ and $\pi{\rightarrow}\pi^*$ Transitions

Property	$n{\rightarrow}\pi^*$	$\pi{\rightarrow}\pi^*$
Maximum ϵ	Less than 100	Greater than 1,000
Vibration band structure	Sharp in nonpolar solvents, broad in polar solvents. Possesses localized vibrational progression (e.g., C=O)	Moderately sharp in most solvents. Possesses C=C vibrational progression
τ_f and ϕ_f	$\tau_f > 10^{-6}$ s, $\phi_f < 0.01$	$\tau_f \sim 10^{-9}-10^{-7}$ s $\phi_f \sim 0.5-0.05$
τ_p and ϕ_p	$\tau_p \sim 10^{-3}$ s, $\phi_p \sim 0.5-0.05$	$\tau_p \sim 0.1-10$ s $\phi_p \sim 0.5-0.05$
Direction of transition moment	Perpendicular to molecular plane for singlet-singlet transitions	Parallel to molecular plane for singlet-singlet transitions
Effect of increasing solvent polarity or electron-donating substituents	Transition shifts to shorter wavelengths	Transition shifts to longer wavelengths

Source: Turro [27], p. 46. *(Reprinted with permission from Benjamin/Cummings, Inc.)*

papers [8,14,15,31,34-39] by Zepp and his colleagues at EPA's Environmental Research Laboratory in Athens, Georgia. Some similar procedures have also been proposed by others [12,13]. The essence of the Zepp approach is to evaluate the degree of overlap between the UV/visible absorption spectrum of an organic compound and the incident solar energy in an aquatic environment. In its simplest terms, this approach can be represented by

$$k_{a\lambda} = \frac{2.303}{j}\,\epsilon_\lambda I_\lambda \qquad (8\text{-}8)$$

in which $k_{a\lambda}$ is the rate constant for absorption of light of wavelength λ by the organic chemical, ϵ_λ is the molar absorptivity of the chemical at wavelength λ, I_λ represents the flux of solar energy of wavelength λ

incident on the chemical, and j is a conversion constant numerically equal to 6.023×10^{20}. Clearly, ϵ_λ is characteristic of the chemical, and I_λ is characteristic of the aquatic environment under consideration.

The rate constant for photochemical degradation, $k_{p\lambda}$, is related to the rate constant for absorption of light, $k_{a\lambda}$, by the quantum yield, ϕ_λ:

$$k_{p\lambda} = \phi_\lambda \, k_{a\lambda} \tag{8-9}$$

The overall first-order rate of photochemical degradation is then

$$-\frac{d[P]}{dt} = \int_\lambda k_{p\lambda} \, [P] \, d\lambda \tag{8-10}$$

$$= \int_\lambda \phi_\lambda \, k_{a\lambda} \, [P] \, d\lambda \tag{8-11}$$

where [P] is the molar concentration of the organic chemical. If it can be assumed that the quantum yield for photochemical reaction is independent of wavelength, Eq. 8-11 becomes

$$-\frac{d[P]}{dt} = \phi[P] \int_\lambda k_{a\lambda} \, d\lambda \tag{8-12}$$

$$= \phi[P] \int_\lambda \frac{2.303}{j} \, I_\lambda \, \epsilon_\lambda \, d\lambda \tag{8-13}$$

The assumption that ϕ_λ is invariant with λ has general validity for small changes in wavelength regime (e.g., for different irradiation wavelengths within the same absorption band). On the other hand, it is commonly observed that $\phi_{254 \text{ nm}} \neq \phi_{300 \text{ nm}}$ and, in fact, that entirely different photochemical processes and products are involved at the two wavelengths. Therefore, one should not attempt to estimate the rate of photolysis of P in the environment by using a literature value of ϕ that was measured using 254-nm irradiation (most common in photochemical literature) or any other irradiation wavelength that is not representative of insolation.

It may be valid to argue that the solar spectrum is sufficiently constant in its distribution of wavelengths that the quantum yield for disappearance of P is constant when one solar irradiation situation is compared with another or with a simulated solar source. To emphasize this point, the ϕ's of Eqs. 8-12 and 8-13 are given a subscript:

$$-\frac{d[P]}{dt} = \phi_{solar} [P] \int_\lambda k_{a\lambda} \, d\lambda \qquad (8\text{-}14)$$

$$= \phi_{solar} [P] \frac{2.303}{j} \int_\lambda I_\lambda \, \epsilon_\lambda \, d\lambda \qquad (8\text{-}15)$$

To simplify data input requirements and computation of the rate of absorption of solar energy, the integral of Eq. 8-15 can be approximated as a sum:

$$-\frac{d[P]}{dt} = \phi_{solar} [P] \frac{2.303}{j} \sum_\lambda I_\lambda \, \epsilon_\lambda \qquad (8\text{-}16)$$

Thus, a finite number of pairs of I_a and ϵ values corresponding to a set of discrete wavelengths will suffice for calculation. In the Zepp model, a set of 39 individual wavelengths over the range 297.5 nm to 800 nm is specified [37]. These correspond to narrow wavelength intervals (2.5 nm) in the 295-320 nm region where organic molecules are more likely to absorb strongly and to wider intervals (10-50 nm) in the wavelength region above 330 nm. The use of Eq. 8-16 thus requires 39 compound-specific data inputs, the ϵ_λ values, and 39 ecosystem-specific data inputs, the $I_{a\lambda}$ values, as well as a value for the quantum yield, ϕ_{solar}.

• *Compound-specific Inputs* (ϵ_λ *values*). The molar absorptivities at the specified wavelengths are obtained experimentally by determination of the UV/visible spectrum of the compound. Smith et al. [24,25] have described a procedure for making the required measurements. They note that the spectra should be obtained with a high-quality UV/visible spectrophotometer (such as a Cary Model 14 or 15), using solutions of the compound at 10^{-2} to 10^{-6} M concentration in water. A water-acetonitrile solvent mixture can be used if necessary to achieve sufficiently high concentrations of water-insoluble species for accurate measurement of absorbances. The value of ϵ at a specified wavelength is calculated from the Beer-Lambert Law (Eq. 8-2) as

$$\epsilon_\lambda = \frac{A_\lambda}{\ell c} \qquad (8\text{-}17)$$

It is suggested [24] that the ϵ_λ value for each wavelength specified by Zepp be calculated as the mean of values for the upper and lower bounds of the appropriate interval, e.g.,

$$\epsilon_{297.5} = \frac{\epsilon_{295.0} + \epsilon_{300}}{2} \tag{8-18}$$

Table 8-7 presents ϵ_λ data for eleven compounds investigated by Smith [25] and three pesticides studied by Hautala [8].

TABLE 8-7

Molar Absorptivities, ϵ_λ, of Selected Compounds
as a Function of Wavelength

λ (nm)	ϵ_λ (L/mol-cm) at pH $\approx$ 7[a] (Data of Smith *et al.* [25])						
	p-Cresol	Benz[a] anthra- cene	Benzo[a] pyrene	Quino- line	Benzo[f] quino- line	9H- Carba- zole	7H-Dibenzo [c,g] carba- zole
297.5	18	7,930	46,600	2,910	3,960	5,540	16,500
300	7.2	7,070	27,700	3,050	3,910	3,100	15,900
302.5	3.8	5,880	13,900	2,740	2,140	2,440	12,300
305	2	3,790	6,670	2,480	1,500	2,270	8,760
307.5	2	3,200	4,840	2,050	1,240	2,390	6,480
310	2	3,480	3,970	2,440	1,180	2,530	4,990
312.5	1	3,900	3,890	2,920	1,430	2,600	4,340
315	0	4,200	3,650	1,680	1,670	2,700	4,070
317.5		4,170	3,730	622	1,480	2,920	3,930
320		4,120	3,570	269	1,380	3,190	3,960
323.1		4,800	3,650	119	1,490	3,170	4,260
330		5,450	5,400	26	3,020	2,900	5,830
340		5,390	8,330	9	1,680	1,520	9,220
350		4,850	12,300	0	1,530	166	11,000
360		3,350	18,100		250	23	12,700
370		1,560	19,680		185	13	7,890
380		662	21,910		96	12	770
390		417	15,160		37	2	10
400		17	2,100		0	0	0

	ϵ_λ (L/mol-cm) at pH ≈ 7[a]						
	Data of Smith *et al.* [25]				Data of Hautala [8]		
λ (nm)	Benzo[b] thio- phene	Dibenzo- thio- phene	Methyl Para- thion	Mirex	2,4D Methyl Ester	Sevin	Para- thion
297.5	1,793	1,154	6,040	0[b]	236	1,410	4,800
300	395	1,224	5,460		78.7	930	4,500
302.5	130	1,327	4,930		52.5	737	4,250
305	30	1,499	4,310		39.4	529	3,750
307.5	13	1,782	3,700		39.4	409	3,250
310	7	2,025	3,210		26.2	351	2,750
312.5	3	2,080	2,760		26.2	378	2,350
315		1,939	2,290		13.1	259	2,000
317.5		1,892	1,920		13.1	236	1,600
320		2,119	1,630			112	1,550
323.1		2,394	1,310			29[c]	1,400[d]
330		526	933			13.2	950
340		13.1	568			3.2	550
350		7.5	374				400
360		0	244				
370			145				
380			82				
390			45				
400			9				

a. pH = 4.5 for 7H-dibenzo[c,g] carbazole.

b. $\epsilon_\lambda < 0.01$ for all $\lambda > 290$ nm [22].

c. λ = 325 nm.

d. λ = 322.5 nm.

The necessary ϵ_λ values for use in Eq. 8-16 generally cannot be obtained from the older compilations of UV/visible spectral information [7,9,10,21]. Most of these give only the values of λ and ϵ_λ corresponding to absorption maxima for a given compound rather than the complete spectrum. The Sadtler compilation [21] presents actual spectra and generally includes the concentration and pathlength information but does not always cover the >290 nm range at a sufficient sensitivity to allow calculation of ϵ_λ values for the solar spectral region. Also, most published UV/visible spectra of organics have been obtained with organic solvents (e.g., hexane, ethanol) rather than aqueous solutions. Solvent effects on spectra, both λ_{max} and ϵ_{max}, can significantly degrade the resolution implied in the Zepp approach. Thus, data suitable for use in Eq. 8-16 are not generally available for a large number of compounds beyond those in Table 8-7.

The stepwise procedure for application of Eq. 8-16 therefore begins in the laboratory, as follows:

Step 1: Prepare dilute solutions of the chemical at known concentration in water or water/acetonitrile.

Step 2: Determine the UV/visible spectrum at several concentrations over a tenfold concentration range using cells of 1-cm and 10-cm pathlength.

Step 3: Calculate ϵ_λ values from the measured spectra.

• *Ecosystem-specific Inputs* (I_λ Values). The principles and procedures for computation of I_λ have been described by Zepp and Cline [37], who developed a computer program to accomplish the task. Equation 8-19 describes the contributions to I_λ from direct radiation, I_d, and sky radiation, I_s

$$I_\lambda = \frac{I_{d\lambda}\,(1 - 10^{-\alpha_\lambda \ell_d}) + I_{s\lambda}\,(1 - 10^{-\alpha_\lambda \ell_s})}{D} \tag{8-19}$$

where α_λ = attenuation coefficient for adsorption of light in the aquatic medium itself
  ℓ_d = average pathlength for direct light in the water (cm)
  ℓ_s = average pathlength for skylight in the water (cm)
  D = depth in the water body (cm).

An explicit solution of Eq. 8-19 for a particular location, time of year/day, and body of water requires input information on [34]:

• Attenuation coefficients and refractive index of the aquatic medium,
• Solar declination, solar right ascension, and sidereal time,
• Latitude and longitude,
• Average ozone layer thickness, and
• Solar spectrum.

Zepp and Cline [37] describe two limiting cases in which Eq. 8-19 can be simplified. When both of the $\alpha\ell$ exponents exceed 2, essentially all sunlight is absorbed within the water column. Equation 8-19 then becomes

$$I_\lambda = \frac{I_{d\lambda} + I_{s\lambda}}{D} = \frac{W_\lambda}{D} \tag{8-20}$$

Table 8-8 lists values of W_λ, the solar radiation intensity, calculated by Zepp and Cline for a body of water at a hypothetical 40°N latitude location at midday and midseason.

The other limiting case described by Zepp and Cline is that in which the water column under consideration absorbs very little of the incident light, which is true for a sufficiently shallow surface layer of any natural water body. Equation 8-19 then becomes

$$I_\lambda = 2.303\, Z_\lambda \tag{8-21}$$

where $Z_\lambda = I_{d\lambda} \sec \theta + 1.2\, I_{s\lambda}$

   θ = angle of refraction

Values of Z_λ from Zepp and Cline's work are presented in Table 8-9, again for a hypothetical 40°N latitude location at midday.

Once a phytolysis rate, -d[P]/dt, has been calculated (e.g., from Eq. 8-16), a photolysis half-life, $t_{1/2}$, may be calculated as follows:

$$t_{1/2} = 0.693[P]\,(-d[P]/dt)^{-1}\ s \tag{8-22}$$

The examples of ecosystem-dependent parameters presented in Tables 8-8 and 8-9 are not intended to represent values for a "typical" ecosystem, although they might be used as such. To a considerable extent, the elegance of the Zepp approach to aqueous photolysis lies in the fact that it is almost as easy to model a real situation of interest as it is to compute the behavior of a hypothetical case. The tables have been included here partly because they provide some quantitative information on the distribution of solar energy as a function of wavelength and complement the more qualitative picture provided by Figures 8-1 and 8-3.

Another reason for presenting sample values of W_λ and Z_λ is to show the seasonal variation in solar intensity. This is also illustrated in Figure 8-4, where seasonal variations in log W_λ and log Z_λ are plotted as a function of wavelength. Note that the high (summer) and low (winter) intensities differ by no more than a factor of four for the longer wavelengths ($\geq$320 nm). Within the 295-320 nm range, however, the summer intensity is up to 36 times the winter intensity. Light with a wavelength

TABLE 8-8

Solar Irradiation Intensity, W_λ, from Eq. 8-20[a]

Wavelength (nm)	Spring	Summer	Fall	Winter
Photons (cm^{-2} s^{-1} per 2.5 nm interval)				
297.5	0.240E + 12	0.648E + 12	0.786E + 11	b
300.0	0.105E + 13	0.219E + 13	0.434E + 12	0.601E + 11
302.5	0.369E + 13	0.657E + 13	0.185E + 13	0.300E + 12
305.0	0.106E + 14	0.163E + 14	0.555E + 13	0.139E + 13
307.5	0.195E + 14	0.274E + 14	0.112E + 14	0.369E + 13
310.0	0.325E + 14	0.444E + 14	0.173E + 14	0.698E + 13
312.5	0.510E + 14	0.643E + 14	0.308E + 14	0.145E + 14
315.0	0.683E + 14	0.836E + 14	0.410E + 14	0.222E + 14
317.5	0.867E + 14	0.103E + 15	0.532E + 14	0.296E + 14
320.0	0.103E + 15	0.121E + 15	0.663E + 14	0.408E + 14
Photons (cm^{-2} s^{-1} per 3.75 nm interval)				
323.1	0.193E + 15	0.226E + 15	0.119E + 15	0.740E + 14
Photons (cm^{-2} s^{-1} per 10 nm interval)				
330.0	0.669E + 15	0.762E + 15	0.421E + 15	0.279E + 15
340.0	0.778E + 15	0.875E + 15	0.500E + 15	0.341E + 15
350.0	0.835E + 15	0.938E + 15	0.533E + 15	0.363E + 15
360.0	0.895E + 15	0.100E + 16	0.568E + 15	0.383E + 15
370.0	0.997E + 15	0.112E + 16	0.623E + 15	0.418E + 15
380.0	0.110E + 16	0.124E + 16	0.679E + 15	0.450E + 15
390.0	0.133E + 16	0.148E + 16	0.895E + 15	0.646E + 15
400.0	0.191E + 16	0.212E + 16	0.129E + 16	0.931E + 15
410.0	0.251E + 16	0.279E + 16	0.170E + 16	0.123E + 16
420.0	0.258E + 16	0.287E + 16	0.175E + 16	0.127E + 16
430.0	0.249E + 16	0.277E + 16	0.170E + 16	0.123E + 16
440.0	0.295E + 16	0.327E + 16	0.201E + 16	0.146E + 16
450.0	0.332E + 16	0.368E + 16	0.227E + 16	0.164E + 16
460.0	0.335E + 16	0.372E + 16	0.230E + 16	0.167E + 16
470.0	0.347E + 16	0.384E + 16	0.238E + 16	0.172E + 16
480.0	0.355E + 16	0.394E + 16	0.244E + 16	0.177E + 16
490.0	0.336E + 16	0.372E + 16	0.231E + 16	0.168E + 16

Wavelength (nm)	Spring	Summer	Fall	Winter
		Photons (cm^{-2} s^{-1} per 10 nm interval)		
500.0	0.343E + 16	0.380E + 16	0.236E + 16	0.171E + 16
525.0	0.362E + 16	0.401E + 16	0.251E + 16	0.181E + 16
550.0	0.377E + 16	0.418E + 16	0.262E + 16	0.188E + 16
575.0	0.380E + 16	0.423E + 16	0.265E + 16	0.190E + 16
600.0	0.385E + 16	0.427E + 16	0.268E + 16	0.192E + 16
625.0	0.387E + 16	0.428E + 16	0.271E + 16	0.196E + 16
650.0	0.389E + 16	0.429E + 16	0.273E + 16	0.199E + 16
675.0	0.388E + 16	0.427E + 16	0.273E + 16	0.200E + 16
700.0	0.384E + 16	0.422E + 16	0.272E + 16	0.200E + 16
750.0	0.369E + 16	0.404E + 16	0.261E + 16	0.193E + 16
800.0	0.354E + 16	0.387E + 16	0.252E + 16	0.187E + 16

a. 1.0E + 12 = 1.0 X 10^{12} etc.
b. Irradiation intensity below detection limit.

Source: Zepp and Cline [37]. *(Reprinted with permission from the American Chemical Society.)*

of <320 nm is most likely to overlap the absorption spectra of organic molecules. A tenfold or greater seasonal variation in photolysis half-lives can thus be expected due to variations in insolation intensity. This effect is likely to be comparable to or larger than the effects of seasonal temperature variation.

Finally, the fact that the relative intensities within the solar spectrum, as well as the total intensity of insolation, vary with the season has relevance to a key assumption made in deriving Eq. 8-16. The assumption that the quantum yield, ϕ_λ, of Eq. 8-11 can be factored out of the integral (sum) and treated as a constant, ϕ_{solar}, is less likely to be valid when substantial shifts in the distribution of solar energy are considered.

Because the emphasis of this handbook is on the properties of organic chemicals, rather than on the properties of environments, detailed procedures for calculating I_λ values are not discussed here. A computer program is the most convenient way to compute I_λ values and the $I_\lambda\ e_\lambda$ products of Eq. 8-16. Such a program, accepting as input the 39 compound-specific ϵ_λ values, is incorporated into the EPA's Exposure Analysis Modelling System (EXAMS), which is widely available [6].

TABLE 8-9

Solar Irradiation Intensity, Z_λ, from Eq. 8-21[a]

Wavelength (nm)	Spring	Summer	Fall	Winter
Photons (cm^{-2} s^{-1} per 2.5 nm interval)				
297.5	0.274E + 12	0.716E + 12	0.949E + 11	b
300.0	0.120E + 13	0.240E + 13	0.524E + 12	0.733E + 11
302.5	0.419E + 13	0.723E + 13	0.223E + 13	0.368E + 12
305.0	0.121E + 14	0.181E + 14	0.670E + 13	0.170E + 13
307.5	0.223E + 14	0.305E + 14	0.135E + 14	0.450E + 13
310.0	0.372E + 14	0.495E + 14	0.208E + 14	0.854E + 13
312.5	0.584E + 14	0.717E + 14	0.371E + 14	0.177E + 14
315.0	0.780E + 14	0.933E + 14	0.494E + 14	0.271E + 14
317.5	0.992E + 14	0.115E + 15	0.641E + 14	0.362E + 14
320.0	0.117E + 15	0.135E + 15	0.800E + 14	0.498E + 14
Photons (cm^{-2} s^{-1} per 3.75 nm interval)				
323.1	0.221E + 15	0.252E + 15	0.144E + 15	0.906E + 14
Photons (cm^{-2} s^{-1} per 10 nm interval)				
330.0	0.761E + 15	0.846E + 15	0.508E + 15	0.342E + 15
340.0	0.880E + 15	0.963E + 15	0.604E + 15	0.420E + 15
350.0	0.942E + 15	0.103E + 16	0.645E + 15	0.449E + 15
360.0	0.101E + 16	0.110E + 16	0.687E + 15	0.479E + 15
370.0	0.112E + 16	0.122E + 16	0.754E + 15	0.520E + 15
380.0	0.124E + 16	0.135E + 16	0.822E + 15	0.562E + 15
390.0	0.149E + 16	0.161E + 16	0.108E + 16	0.805E + 15
400.0	0.213E + 16	0.231E + 16	0.156E + 16	0.116E + 16
410.0	0.280E + 16	0.302E + 16	0.206E + 16	0.154E + 16
420.0	0.288E + 16	0.310E + 16	0.212E + 16	0.159E + 16
430.0	0.277E + 16	0.298E + 16	0.205E + 16	0.154E + 16
440.0	0.327E + 16	0.351E + 16	0.244E + 16	0.184E + 16
450.0	0.368E + 16	0.394E + 16	0.275E + 16	0.208E + 16
460.0	0.371E + 16	0.398E + 16	0.279E + 16	0.211E + 16
470.0	0.384E + 16	0.411E + 16	0.289E + 16	0.219E + 16
480.0	0.392E + 16	0.420E + 16	0.296E + 16	0.225E + 16
490.0	0.371E + 16	0.396E + 16	0.281E + 16	0.213E + 16

TABLE 8-9 (Continued)

Wavelength (nm)	Spring	Summer	Fall	Winter
	Photons (cm^{-2} s^{-1} per 10 nm interval)			
500.0	0.378E + 16	0.404E + 16	0.287E + 16	0.218E + 16
525.0	0.398E + 16	0.426E + 16	0.305E + 16	0.232E + 16
550.0	0.413E + 16	0.442E + 16	0.318E + 16	0.241E + 16
575.0	0.417E + 16	0.446E + 16	0.322E + 16	0.243E + 16
600.0	0.421E + 16	0.450E + 16	0.326E + 16	0.247E + 16
625.0	0.422E + 16	0.450E + 16	0.329E + 16	0.252E + 16
650.0	0.424E + 16	0.451E + 16	0.332E + 16	0.256E + 16
675.0	0.423E + 16	0.448E + 16	0.333E + 16	0.259E + 16
700.0	0.419E + 16	0.443E + 16	0.330E + 16	0.258E + 16
750.0	0.401E + 16	0.423E + 16	0.318E + 16	0.250E + 16
800.0	0.385E + 16	0.405E + 16	0.306E + 16	0.242E + 16

a. $1.0 \, E + 12 = 1.0 \times 10^{12}$ etc.
b. Irradiation intensity below detection limit.

Source: Zepp and Cline [37]. *(Reprinted with permission from the American Chemical Society.)*

Example 8-1 illustrates the calculation of the rate of photolysis according to the Zepp procedure.

Example 8-1   Estimate the rate of photolysis for Sevin® (Carbaryl®) in a clear, deep lake. Assume that:

- All sunlight is absorbed in the water column. (Thus, Eq. 8-20 may be used.)

- The lake is 5 m deep. (Thus, $D = 5 \times 10^2$ cm.)

- The pollutant concentration, [P], is 5×10^{-5} M ($\approx$10 mg/L)

- The lake is at 40°N latitude and the time is midday and midsummer. (Thus, values of W_λ in Table 8-8 may be used.)

(1)   The structure of Sevin® is

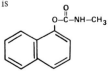

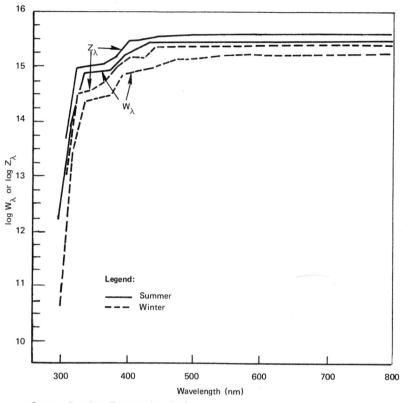

Source: Data from Zepp and Cline [37].

FIGURE 8–4   Seasonal Variations in $\log Z_\lambda$ and $\log W_\lambda$

Note that this structure contains two components (a $>$C=O group and the naphthalene ring system) that are identified in Table 8-4 as chromophoric groups with $\lambda_{max} > 290$ nm. Thus, direct photolysis can take place.

(2) Equation 8-16 provides the basic expression for the rate of photolysis. When Eq. 8-20 ($I_\lambda = W_\lambda/D$) is substituted in Eq. 8-16, the result is:

$$-\frac{d[P]}{dt} = \phi_{solar}\ [P]\ \frac{2.303}{jD}\ \sum_\lambda W_\lambda \epsilon_\lambda \tag{8-23}$$

(3) Using values of W_λ from Table 8-8 (summer) and ϵ_λ from Table 8-7:

$$\sum_\lambda W_\lambda \epsilon_\lambda = (1410 \times 0.648 \times 10^{12}) + (930 \times 0.219 \times 10^{13}) +$$

$$(737 \times 0.657 \times 10^{13}) + \ldots\ldots \text{(next 10 terms of sum not shown)}$$

$$= 1.39 \times 10^{17}\ \frac{\text{L} \cdot \text{photons}}{\text{mol} \cdot \text{cm}^3 \cdot \text{s}}$$

(4)   Then with $\phi_{solar} = 0.01$ (Table 8-11), $D = 5 \times 10^2$ cm, $[P] = 5 \times 10^{-5}$ M, and $j = 6.023 \times 10^{20}$, Eq. 8-23 is evaluated as:

$$-\frac{d[P]}{dt} = (0.01)(5 \times 10^{-5}) \frac{2.303}{(6.023 \times 10^{20})(5 \times 10^2)} (1.39 \times 10^{17})$$

$$= 5.3 \times 10^{-13} \quad mol \cdot L^{-1} \cdot s^{-1}$$

$$= 1.7 \times 10^{-5} \quad mol \cdot L^{-1} \cdot yr^{-1}$$

(5)   From Eq. 8-22, the half-life for photolysis in this lake is

$$t_{\frac{1}{2}} = 0.693 \, (5 \times 10^{-5})/(1.7 \times 10^{-5})$$

$$= 2 \, yr$$

An assessment of the relative importance of photolysis as a removal mechanism in this lake would also require consideration of other degradation processes (e.g., hydrolysis, biodegradation), other removal processes (e.g., volatilization, sedimentation), and the residence time of the water in the lake. Note that if we had considered only the top 5 cm of the lake, the photolysis half-life in this layer would be about 7 days, which is close to the 11-day half-life given by Hautala [8] (see Table 8-12).

(6)   Zepp and Cline [37] have carried out additional calculations for Sevin® to show the effects of time (of day and of year) and latitude on the photolysis rate. Their results are shown in Figures 8-5 and -6. The depth dependence of direct photolysis, for pure water and average river water, is shown in Figure 8-7.

8-4   PHOTOCHEMICAL REACTIONS

General Considerations. Our present understanding of photochemical reaction mechanisms does not allow prediction of either the qualitative product distribution or the quantitative reaction efficiency of chemical transformations that may occur as a result of light absorption by an organic molecule. This is not surprising; photochemical degradation generally competes poorly with the photophysical deactivation processes described in §8-2, so overall quantum yields for photolytic degradation of organics in solution are typically much less than 0.1. Furthermore, those photochemical processes that do take place involve rapid transformations of the short-lived excited states and are, therefore, more difficult to study systematically than the slower thermal processes.

Despite the fact that it is not feasible to develop reliable predictions of the nature and extent of photochemical transformations that may occur under specified conditions, it is useful to review briefly the broad

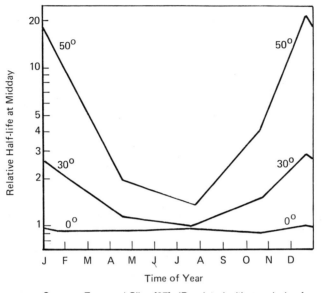

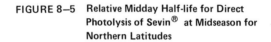

Source: Zepp and Cline [37]. *(Reprinted with permission from the American Chemical Society.)*

FIGURE 8–5 **Relative Midday Half-life for Direct Photolysis of Sevin® at Midseason for Northern Latitudes**

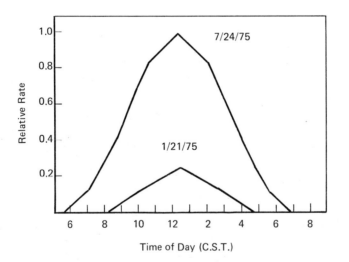

Source: Zepp and Cline [37]. *(Reprinted with permission from the American Chemical Society.)*

FIGURE 8–6 **Diurnal Variation of Direct Photolysis Rate of Sevin® at Latitude 40°N**

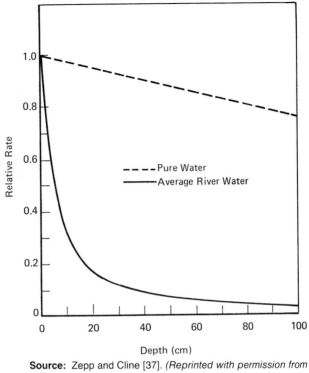

Source: Zepp and Cline [37]. *(Reprinted with permission from the American Chemical Society.)*

FIGURE 8–7   **Computed Depth Dependence of Direct Photolysis of Sevin® at Midday and Midsummer, Latitude 40°N**

categories of reaction types and product types that have been observed. More detailed treatments of these subjects can be found in the excellent texts of Calvert and Pitts [2], Kan [11] and Turro [27].

Primary photochemical processes in organic molecules include fragmentation into free radicals or neutral molecules, rearrangement and isomerization reactions, photoreduction by hydrogen atom abstraction from other molecules, dimerization and related addition reactions, photoionization, and electron transfer reactions. Table 8-10 summarizes primary photochemical reaction processes that are typical of various organic compound categories. This summary has been abstracted from Calvert and Pitts' review [2] of the photochemistry of polyatomic molecules. The entries have been selected to emphasize reactions that are considered important at irradiation wavelengths >290 nm and/or ones that are known to occur in fluid solutions. The

(continued on p. 8-36)

TABLE 8-10

Primary Photochemical Reaction Modes Typical of Various Organic Compound Categories

Reaction Process	Comments
1. Aldehydes, Ketones	
a. $RCHO \longrightarrow R\cdot + HCO\cdot$ $RCOR' \longrightarrow R\cdot + R'CO\cdot$	"Norrish Type I" fragmentation; typical with 313-nm irradiation
b. $RCHO^* \longrightarrow RH + CO$	Fragmentation with 254-nm irradiation
c. $R_2CCR_2CR_2CHO \longrightarrow R_2C=CR_2 + HCR_2CH$ (with $\overset{\mid}{\underset{H}{}}$ and $\overset{O}{\underset{\|}{}}$)	"Norrish Type II" split of aldehydes and ketones; requires γ-hydrogen
d. $RCOR' + SH \longrightarrow R\dot{C}R'OH + S\cdot$ (solvent)	Photoreduction; hydrogen atom abstraction by triplet state of aldehyde or ketone
e. $R_2CCR_2CR_2COR' \longrightarrow \begin{array}{c} OH \\ CR_2-\overset{\mid}{C}-R' \\ \mid \\ CR_2-CR_2 \end{array}$ (with $\overset{\mid}{\underset{H}{}}$)	Intramolecular photoreduction of ketones
f. (cyclohexanone) $\longrightarrow$ (cyclopentane) $+ CO$	Decarbonylation and ring contraction of cyclic ketones
2. Hydrocarbons	
a. $\begin{array}{c} R \\ R' \end{array}C=C\begin{array}{c} R' \\ R \end{array} \longrightarrow \begin{array}{c} R \\ R' \end{array}C=C\begin{array}{c} R \\ R' \end{array}$	Cis-trans isomerization. Occurs at $\lambda > 290$ nm only if R and/or R' aromatic

TABLE 8-10 (Continued)

Reaction Process	Comments
2. Hydrocarbons (cont'd.)	
b. (alkene polymerization structure)	Polymerization. Occurs at $\lambda > 290$ nm only if R and/or R' aromatic
c. $CHR=CR-CR=CHR \longrightarrow R-C-CHR$ / $R-C-CHR$ (cyclobutene ring structure)	Cyclization
d. (o-xylene → m-xylene structure)	Isomerization at 254 nm. Fluorescence at longer wavelength.
e. (anthracene + O_2/Sens. → endoperoxide structure)	Photosensitized oxidation. Not a direct photolysis of the aromatic.
3. Organo Halides	
a. $RX \longrightarrow R\cdot + X\cdot$	Photodissociation. For X=I, can occur at $\lambda > 290$ nm.
b. $R_2CHCH_2X \longrightarrow R_2C=CH_2 + HX$	Dehydrohalogenation. $\lambda < 200$ nm
4. Carboxylic Acids	
a. $RCO_2H \longrightarrow R\cdot + \cdot CO_2H$	Primarily for $\lambda < 254$ nm
b. $RCO_2H \longrightarrow RCO_2\cdot + H\cdot$	Primarily for $\lambda < 254$ nm

TABLE 8-10 (Continued)

Reaction Process	Comments
4. Carboxylic Acids (cont'd.)	
c. $RCO_2H \longrightarrow RH + CO_2$	Decarboxylation; dominant photodecomposition mode in water at 366 nm irradiation
d. $\underset{\overset{\|}{O}}{R}CCO_2H \longrightarrow \underset{\overset{\|}{R}-\overset{\|}{C}-OH}{\overset{COOH}{\|}} \quad \underset{COOH}{\overset{R-C-OH}{\|}}$	Photoreduction + dimerization of α-keto acids
5. Carboxylate Esters	
a. $\underset{\overset{\|}{O}}{R}COR' \longrightarrow \underset{\overset{}{R\cdot + \cdot CO_2R'}}{RCO_2\cdot + R'\cdot}$	Analogs of Norrish Type I splits in carbonyls. Type II splits also occur
b. $\underset{\overset{\|}{O}}{R}COR' \longrightarrow R\cdot + \cdot OR' + CO$	Decarbonylation. Radicals may recombine within solvent cage
6. Peroxides	
a. $ROOR' \longrightarrow RO\cdot + R'O\cdot$	Photodissociation
b. $ROOR' \longrightarrow RO_2\cdot + R'\cdot$	Photodissociation mode found only for < 230 nm irradiation
7. Azo Compounds	
a. $RN{=}NR \longrightarrow 2R\cdot + N_2$	Common, efficient procedure for production of $R\cdot$. Occurs at > 366 nm, but only for aliphatic R.

TABLE 8-10 (Continued)

Reaction Process	Comments
7. Azo Compounds (cont'd.)	
b. $R_2C=\overset{+}{N}=\overset{-}{N} \longrightarrow R_2C: + N_2$	Efficient photoprocess at $\lambda > 300$ nm, but diazo compounds not stable in water
8. Nitroso Compounds	
a. $R-N=O \longrightarrow R\cdot + NO\cdot$	Photodissociation important even at 750-500 nm
9. Nitro Compounds	
a. $R_2CHCH_2NO_2 \longrightarrow R_2CHCH_2\cdot + NO_2\cdot$	Photodissociation of nitroalkane can occur at $\lambda > 300$ nm
b. $R_2CHCH_2NO_2 \longrightarrow R_2C=CH_2 + HONO$	Photoelimination of nitroalkane with β-hydrogen. Can occur at $\lambda > 300$ nm
c.	Principal fragmentation route of nitroarenes.
10. Organic Nitrites/Nitrates	
a. $RONO \longrightarrow RO\cdot + NO\cdot$	Dominant process in solution
b. $RCH_2O\cdot NO_2 \longrightarrow RCH_2O\cdot + NO_2\cdot$	Dominant photodissociation mode
11. Organosulfur compounds	
a. $RSSR \longrightarrow 2\ RS\cdot$	Photodissociation. Can occur at 365 nm
b.	Photooxidation by direct photolysis in presence of oxygen. 588 nm irradiation effective.

Source: Calvert and Pitts [2], pp. 366-579.

latter is a rather severe limitation, since much of the classical literature of organic chemistry refers to gas-phase or frozen-solution (77K) matrices. Because of the selection criteria, Table 8-10 excludes such categories as alcohols, ethers, amines, and nitriles, which do not absorb light in the solar spectral region.

The reactions indicated in Table 8-10 are *primary* photochemical processes. Keep in mind that more than one of these processes can occur on direct photolysis of an organic chemical in water and that they compete with each other and with photophysical deactivation processes in determining the fate of the excited state. Furthermore, the intermediate species produced in the primary photochemical processes, such as free radicals, undergo various secondary chemical reactions until thermally stable products are formed. When the considerable variations in composition of aquatic media are also taken into consideration, it becomes clear that predicting the course of a direct photolysis reaction is generally impractical and frequently not feasible.

Because of the complex network of possible pathways for an excited-state molecule, the photochemical fate of organic compounds in the environment is usually treated simply in terms of disappearance of the pollutant, P. The rate constant, k, and half-life, $t_{1/2}$, are derived for assumed first-order or pseudo first-order degradation of P, and the quantum yield is, similarly, a disappearance quantum yield. Thus, k, $t_{1/2}$, and ϕ reflect the combined effects of all processes other than deactivation of the excited state.

The complexity of the network of possible reaction pathways also makes development of structure/reactivity correlations a formidable task. To date, such correlations have not been derived, even for relatively restricted sets of organic chemicals.

Some Specific Examples. Within the past five years, numerous experimental investigations have been made of the photolysis of organic chemicals in aqueous media. (See, for example, Refs. 1, 3, 4, 8, 14, 19, 22, 25, 26, 30, and 34-39.) The compounds studied have been primarily pesticides and polycyclic aromatic hydrocarbons, although a few other compound categories have been included.

Tables 8-11 and 8-12 present the quantum yield data and photolysis half-life, respectively, for a representative subset of the available data. These were selected primarily from the results of systematic investigations conducted and/or sponsored by the EPA's Environmental

TABLE 8-11

Disappearance Quantum Yield, ϕ, for Photolysis in Aqueous Media

Compound	λ[a] (nm)	ϕ	Ref.
Pesticides			
Carbaryl	313	0.005	31
2,4-D, butoxyethyl ester	313	0.05	31
2,4-D, methyl ester	290	0.06	8
DDE	b	0.30	36
Methoxychlor	>288	0.30	31
Methyl parathion	313	0.00017	25
N-nitrosoatrazine	b	0.30	36
Parathion	313	0.0002	8
	>280	< 0.001	31
	b	0.00015	36
Sevin	290	0.01	8
Trifluralin	b	0.0020	36
Polycyclic Aromatic Hydrocarbons (PAH)			
Anthracene	366	0.003	39
Benzo[a] anthracene	313,366	0.0033	25
Benzo[a] pyrene	313	0.00089	25
Chrysene	313	0.003	39
9,10-Dimethylanthracene	366	0.004	39
Fluoranthene	313	0.0002	39
Naphthalene	313	0.015	39
Phenanthrene	313	0.010	39
Pyrene	313,366	0.002	39
Miscellaneous			
Benzo[f] quinoline	313	0.014	25
Benzophenone	>300	0.02	27
p-Cresol	313	0.079	25
3,4-Dichloroaniline	313	0.052	14
9H-Dibenzocarbazole	366	0.0028	25
Dibenzothiophene	313	0.00050	25
Quinoline	313	0.00033	25

a. Wavelength for which ϕ was determined
b. Sunlight

TABLE 8-12

Half-Life for Disappearance via Direct
Photolysis in Aqueous Media

Compound	λ^a (nm)	$t_{1/2}$	Ref.
Pesticides			
Carbaryl	b	50 h	31
2,4-D, butoxyethyl ester	b	12 d	31
2,4-D, methyl ester	b	62 d	8
DDE	b	22 h (calc)	36
Malathion	b	15 h	31
Methoxychlor	b	29 d	31
Methyl parathion	b	30 d	25
Mirex	b	1 y	25
N-Nitrosoatrazine	b	0.22 h (calc)	36
Parathion	b	10 d (calc)	36
	b	9.2 d	8
Sevin	b	11 d	8
Trifluralin	b	0.94 h (calc)	36
Polycyclic Aromatic Hydrocarbons (PAH)			
Anthracene	366	0.75 h	39
Benz[a] anthracene	b	3.3 h	25
Benzo[a] pyrene	b	1 h	25
Chrysene	313	4.4 h	39
9,10-Dimethylanthracene	366	0.35 h	39
Fluoranthene	313	21 h	39
Naphthalene	313	70 h	39
Phenanthrene	313	8.4 h	39
Pyrene	313,366	0.68 h	39
Miscellaneous			
Benzo[f] quinoline	b	1 h	25
9H-Carbazole	b	3 h	25
p-Cresol	b	35 d	25
9H-Dibenzocarbazole	b	0.3 h	25
Dibenzothiophene	b	4-8 h	25
Quinoline	b	5-21 d	25

a. Wavelength(s) at which photolysis rate was measured
b. Sunlight

Research Laboratory. These studies provide a fairly consistent data base and include all of the types of compounds for which quantitative experimental data are available.

Examination of the data presented in Tables 8-11 and 8-12 confirms the difficulty of predicting photochemical reactivities from molecular structure. Even within a restricted series of similar compounds, such as the polycyclic aromatic hydrocarbons, there is no apparent correlation between $t_{1/2}$ and ϕ. Furthermore, neither $t_{1/2}$ nor ϕ shows a monotonic trend across the series of compounds.

Again, this complex pattern of photochemical reactivities is not unexpected: the data inevitably reflect the influence of a number of interacting properties of the compound and the aqueous system under consideration. Given the present state of the art, the photochemical reactivity of an organic chemical can be "predicted from its chemical structure" only to the extent that direct photolysis can be ruled out for those compounds with no or extremely low absorbance of light at wavelengths of less than 290 nm. For compounds that do absorb at the wavelengths of terrestrial surface solar radiation, the photochemical reactivity can be estimated from the measured UV/visible spectral data and the measured quantum yield, using the approach of Zepp and co-workers as described above. At present, however, there are no known procedures for estimating the compound-specific inputs required by the Zepp model.

Real-World Complications. In the preceding sections of this chapter, it has been assumed implicitly that the photoreactivity of organic molecules is independent of the nature of the aquatic medium. This assumption is not inappropriate in qualitative considerations of potential photochemical reactivity, but any attempts at quantitative prediction of photolysis rates will require more detailed consideration of the medium. There is evidence to suggest that both the rate and the products of photochemical degradation may be influenced by such factors in the environment as suspended sediment [15,16,18], surfactants [8], and sensitizers [40]. Quenchers, such as molecular oxygen, may also influence the rate of photolysis, although one study [39] reported no apparent effect of oxygen on the rate of aqueous photolysis of polycyclic aromatic hydrocarbons. Detailed discussion of ecosystem-specific effects is beyond the scope of this handbook. However, the user should be aware that such effects may complicate attempts to extrapolate data for photolysis rates from one aquatic medium (e.g., distilled water) to a very different medium (e.g., seawater, a eutrophic lake, or a turbid river).

8-5 SYMBOLS USED

A	=	absorbance
c	=	concentration (mol/L)
D	=	depth of water body in Eq. 8-19 (cm)
h	=	Planck constant = 6.6256×10^{-27} erg-s
ΔH_{298K}	=	bond dissociation energy in Table 8-2 (cal/mol)
I	=	intensity of light
$I_{d\lambda}$	=	intensity of direct solar radiation in Eq. 8-19
I_0	=	intensity of incident light
I_λ	=	intensity of monochromatic light of wavelength λ
$I_{s\lambda}$	=	intensity of sky radiation in Eq. 8-19
j	=	conversion constant in Eq. 8-8
$k_{a\lambda}$	=	rate constant for absorption of light of wavelength λ (s^{-1})
$k_{p\lambda}$	=	rate constant for degradation of organic species P by light of wavelength λ (s^{-1})
ℓ_d	=	pathlength of direct light in water
ℓ_s	=	pathlength of skylight in water
P	=	ground-state organic (pollutant) molecule
P*	=	electronically-excited-state organic (pollutant) molecule
[P]	=	pollutant concentration (mol/L)
S*	=	electronically-excited-state sensitizer molecule
$t_{1/2}$	=	half-life for reaction in Table 8-12 and Eq. 8-22
W_λ	=	measure of solar irradiation intensity in Eq. 8-20
Z_λ	=	measure of solar irradiation intensity in Eq. 8-21

Greek

α_λ	=	attenuation coefficient for light in aqueous medium in Eq. 8-19 (cm^{-1})
ϵ	=	molar absorptivity (formerly called extinction coefficient) in Eq. 8-2 (L/mol-cm) (wavelength-dependent)
ϵ_{max}	=	molar absorptivity at a wavelength of maximum absorption
θ	=	angle of refraction in Eq. 8-21
$h\upsilon$	=	quantum of light of frequency υ
λ	=	wavelength of light (nm)
λ_{max}	=	wavelength corresponding to a maximum in the absorption spectrum (nm)
$n \rightarrow \pi^*$	=	electronic transition from non-bonding to π-antibonding orbital
υ	=	frequency of light (s^{-1})

$\pi \rightarrow \pi^* =$ electronic transition from π-bonding to π-antibonding orbital

$\phi \qquad =$ quantum yield in Eq. 8-6

$\phi_\lambda \qquad =$ quantum yield for irradiation with light of wavelength λ

$\phi_f \qquad =$ quantum yield for fluorescence in Table 8-6

$\phi_p \qquad =$ quantum yield for phosphorescence in Table 8-6

$\phi_{solar} \quad =$ quantum yield for disappearance of P by photolysis under solar irradiation in Eq. 8-14

$\tau \qquad =$ natural radiative lifetime of an excited state in Eq. 8-7 (s)

$\tau_f \qquad =$ radiative lifetime for fluorescence in Table 8-6

$\tau_p \qquad =$ radiative lifetime for phosphorescence in Table 8-6

$\omega \qquad =$ wavenumber of light (cm^{-1})

8-6   REFERENCES

1. Aly, O.M. and M.A. El-Dib, "Photodecomposition of Some Carbamate Insecticides in Aquatic Environments," in *Organic Chemicals in Aquatic Environments,* ed. by S.O. Faust and J.W. Hunter, Marcel Dekker, Inc., New York (1971).

2. Calvert, J.G. and J.N. Pitts, Jr., *Photochemistry,* John Wiley & Sons, New York (1966).

3. Crosby, D.G., "Photochemistry of Benchmark Pesticides," in *A Literature Survey of Benchmark Pesticides,* report by the Department of Medical and Public Affairs of the George Washington University Medical Center, Washington, D.C., under Contract 68-01-2889 for the Office of Pesticide Programs, U.S. Environmental Protection Agency (March 1976).

4. Crosby, D.G., E. Leitis and W.L. Winterlin, "Photodecomposition of Carbamate Insecticides," *J. Agric. Food Chem.,* **13,** 204-07 (1976).

5. Dyer, J.R., *Applications of Absorption Spectroscopy of Organic Compounds,* Prentice-Hall, Inc., Englewood Cliffs, N.J, pp. 4-21 (1965).

6. Environmental Protection Agency, "EXAMS: An Exposure Analysis Modeling System," preliminary draft document (February 1980). Available through R.R. Lassiter, Environmental Research Laboratory, U.S. EPA, College Station Road, Athens, GA 30613.

7. Friedel, R.A. and M. Orchin, *Ultraviolet Spectra of Aromatic Compounds,* John Wiley & Sons, New York (1951).

8. Hautala, R.R., "Surfactant Effects on Pesticide Photochemistry in Soil and Water," EPA-600/3-78-060 (June 1978).

9. Hershenson, H.M., *Ultraviolet and Visible Absorption Spectra,* Academic Press, New York, 1930-1954 (1956), 1955-1959 (1961), 1960-1963 (1966).

10. Kamlet, M.J. (ed.), *Organic Electronic Spectra Data,* Vol. 1, Interscience Publishers, Inc., New York (1960).

11. Kan, R.O., *Organic Photochemistry,* McGraw-Hill Book Co., New York, pp. 1-18 (1966).

12. Kramer, R.F. and G.F. Widhopf, "Evaluation of Daylight or Diurnally-Averaged Photolytic Rate Coefficients in Atmospheric Photochemical Models," *J. Atmos. Sci.*, **35,** 1726-34 (1978).

13. Mancini, J.L., "Analysis Framework for Photodecomposition in Water," *Environ. Sci. Technol.*, **12,** 1274-76 (1978).

14. Miller, G.C., M.J. Mille, D.G. Crosby, S. Sontum and R.G. Zepp, "Photosolvolysis of 3,4-Dichloroaniline in Water," *Tetrahedron,* **35,** 1797-99 (1979).

15. Miller, G.C. and R.G. Zepp, "Effects of Suspended Sediments on Photolysis Rates of Dissolved Pollutants," *Water Res.,* **13,** 453-59 (1979).

16. Miller, G.C. and R.G. Zepp, "Photoreactivity of Aquatic Pollutants Sorbed on Suspended Sediments," *Environ. Sci. Technol.,* **13,** 860-63 (1979).

17. Odum, E.P., *Fundamentals of Ecology,* 3rd ed., W.B. Saunders Co., Philadelphia, pp. 40-43 (1971).

18. Oliver, B.G., E.G. Cosgrove and J.H. Carey, "Effects of Suspended Sediments on the Photolysis of Organics in Water," *Environ. Sci. Technol.,* **13,** 1075-77 (1979).

19. Perry, F.M., E.W. Day, Jr., H.E. Magadanz and D.G. Sanders, "Fate of Nitrosamines in the Environment: Photolysis in Natural Waters," presented at the 176th National Meeting, American Chemical Society, Miami Beach, FL (September 1978).

20. Reifsnyder, W.E. and H.W. Lull, "Radiant Energy in Relation to Forests," Technical Bulletin No. 1344, U.S. Department of Agriculture Forest Service (1965), as cited in Ref. 17.

21. Sadtler Research Laboratories, Inc., *Sadtler Standard Spectra — Ultraviolet Spectra.*

22. Silk, P.J. and I. Unger, "The Photochemistry of Carbamates: I. The Photodecomposition of Zectran: 4-Dimethylamino-3,5-xylyl-N-methyl carbamate," *Int. J. Environ. Anal. Chem.,* **2** (3), 213-20 (1973).

23. Silverstein, R.M. and G.C. Bassler, *Spectrometric Identification of Organic Compounds,* John Wiley & Sons, New York, pp. 90-103 (1963).

24. Smith, J.H., W.R. Mabey, N. Bohonos, B.R. Holt, S.S. Lee, T-W. Chou, D.C. Bomberger and T. Mill, "Environmental Pathways of Selected Chemicals in Freshwater Systems: Part I. Background and Experimental Procedures," EPA-600/7-77-113 (October 1977).

25. Smith, J.H., W.R. Mabey, N. Bohonos, B.R. Holt, S.S. Lee, T-W. Chou, D.C. Bomberger and T. Mill, "Environmental Pathways of Selected Chemicals in Freshwater Systems: Part II. Laboratory Studies," EPA-600/7-78-074 (May 1978).

26. Southworth, G.R. and C.W. Gehrs, "Photolysis of 5-Chlorouracil in Natural Waters," *Water Res.,* **10,** 967-71 (1976).

27. Turro, N.J., *Molecular Photochemistry,* W.A. Benjamin, New York (1965).

28. Turro, N.J., V. Ramamurthy, W. Cherry and W. Farneth, "The Effect of Wavelength on Organo Photoreactions in Solution. Reactions from Upper Excited States," *Chem. Rev.,* **78,** 125-45 (1978).

29. Tyler, J.E., "Transmission of Sunlight in Natural Water Bodies," in the Program and Abstracts, Symposium on *Nonbiological Transport and Transformation of Pollutants on Land and Water: Processes and Critical Data Required for Predictive Description,* National Bureau of Standards (May 11-13, 1976).

30. Ware, G.W., D.G. Crosby and J.W. Giles, "Photodecomposition of DDA," *Arch. Environ. Contam. Toxicol.,* **9,** 135-46 (1980).

31. Wolfe, N.L., R.G. Zepp, G.L. Baughman, R.C. Fincher and J.A. Gordon, "Chemical and Photochemical Transformations of Selected Pesticides in Aquatic Systems," EPA-600/3-76-067 (September 1976). NTIS PB 258 846.

32. Woodward, R. B., "Structure and the Absorption Spectra of α,β-Unsaturated Ketones," *J. Am. Chem. Soc.,* **63,** 1123 (1941).

33. Wu, K.C. and A.M. Trozzolo, "Production of Singlet Molecular Oxygen from the O_2 Quenching of the Lowest Excited State of Rubrene," *J. Phys. Chem.,* **83,** 2823-26 (1979); and "Production of Singlet Molecular Oxygen from the Oxygen Quenching of the Lowest Excited Singlet State of Aromatic Molecules in n-Hexane Solution," *J. Phys. Chem.,* **83,** 3180-83 (1979).

34. Zepp, R.G., "Assessing the Photochemistry of Organic Pollutants in Aquatic Environments," paper presented before the Division of Environmental Chemistry, American Chemical Society, Miami Beach, FL (September 10-15, 1978).

35. Zepp, R.G., "Quantum Yields for Reaction of Pollutants in Dilute Aqueous Solution," *Environ. Sci. Technol.,* **12,** 327-29 (1978).

36. Zepp, R.G. and G.L. Baughman, "Prediction of Photochemical Transformation of Pollutants in the Aquatic Environment," in *Aquatic Pollutants: Transformation and Biological Effects,* ed. by O. Hutzinger, I.H. Van Lelyveld and B.C.J. Zoeteman, Pergamon Press, Oxford, pp. 237-64 (1978).

37. Zepp, R.G. and D.M. Cline, "Rates of Direct Photolysis in the Aqueous Environment," *Environ. Sci. Technol.,* **11,** 359-66 (1977).

38. Zepp, R.G., N.L. Wolfe, J.A. Gordon and R.C. Fincher, "Light-Induced Transformations of Methoxychlor in Aquatic Systems," *J. Agric. Food Chem.,* **24,** 727-33 (1976).

39. Zepp, R.G. and P.F. Schlotzhauer, "Photoreactivity of Selected Aromatic Hydrocarbons in Water," in *Polynuclear Aromatic Hydrocarbons,* ed. by P.W. Jones and P. Leber, Ann Arbor Science Publishers, Inc., Ann Arbor, MI, pp. 141-58 (1979).

40. Zepp, R.G., G.L. Baughman and P.F. Schlotzhauer, "Photosensitization of Pesticide Reactions by Humic Substances," paper presented at the Second Chemical Congress of the North American Continent, San Francisco (August 24-29, 1980).

9

RATE OF BIODEGRADATION

Kate M. Scow

9-1 INTRODUCTION

Biodegradation is one of the most important environmental processes that cause the breakdown of organic compounds. It is a significant loss mechanism in soil and aquatic systems and plays an essential role in wastewater treatment. The eventual *mineralization* of organic compounds — i.e., their conversion to inorganic substances — can be attributed almost entirely to biodegradation [6].

This chapter does not provide a procedure for estimating the rate of biodegradation of organic compounds, because investigations of this complex process are still in the early stages. Most research is descriptive, focusing on identification of the organisms responsible for degradation of specific substances, the metabolic products of such degradation, and classification of metabolic pathways. Quantitative data are scarce and have generally not been compiled in secondary sources to facilitate correlation with other chemical properties. Because experimental methods for measuring biodegradation rates are not standardized, the results are not comparable and apply only to a particular set of experimental conditions. The variables that control rates are not well understood, as they have not been examined across different classes of chemicals. New areas must be explored and existing data must be extensively organized before it will be possible to predict rates of biodegradation.

As an aid in judging the potential for biodegradation of a particular organic compound, this chapter presents background information about the process of biodegradation, standard test procedures, chemical rules of thumb for biodegradability, and attempts by various investigators to estimate rates. Sources of additional information are noted, and suggestions are given for methods of generating necessary data. Despite all this, only qualitative judgments are possible.

9-2   PRINCIPLES OF BIODEGRADATION

Definition: Several definitions of biodegradation have been proposed [53];

- *Primary biodegradation* — any biologically induced structural transformation in the parent compound that changes its molecular integrity;
- *Ultimate biodegradation* — biologically mediated conversion of an organic compound to inorganic compounds and products associated with normal metabolic processes;
- *Acceptable biodegradation* — biological degradation to the extent that toxicity or other undesirable characteristics of a compound are removed.

Other definitions are related to specific test methods or analytical techniques and are therefore not as widely applicable [72].

The rate of reaction varies with the type of biodegradation. For example, a complex compound will undergo a long chain of separate and different reactions to reach ultimate biodegradation, while a simple compound may require only one or two reactions to break it down completely.

Biodegradation is most commonly defined in this chapter as the *primary biodegradation* of organic compounds. Therefore, any structural change in the parent compound falls into this definition if the compound no longer responds to the analytical techniques developed for its identification [155]. Although it is important to identify and follow the breakdown of the products of primary biodegradation, which are sometimes more toxic or biologically accessible than the original compound, many biodegradation studies are concerned only with the first step in degradation. Rules of thumb and correlation of biodegradation rates with other chemical properties are usually derived from primary biodegradation results.

Only microbial degradation is covered in this chapter; higher organisms also metabolize compounds, but they play a less significant role in biodegradation in environmental systems.

Almost all of the reactions involved in biodegradation can be classified as oxidative, reductive, hydrolytic, or conjugative [66]. Examples of the first three kinds of reactions are shown in Table 9-1. At least 26 oxidative, 7 reductive, and 14 hydrolytic transformations of pesticides had been identified as of 1975 [50]. Conjugative reactions such as methylation and acetylation have also been observed in the presence of microorganisms [53]. Reactions take place both in the presence and in the absence of oxygen. Some compounds, such as DDT, are transformed under both aerobic and anaerobic conditions (see Figure 9-1).

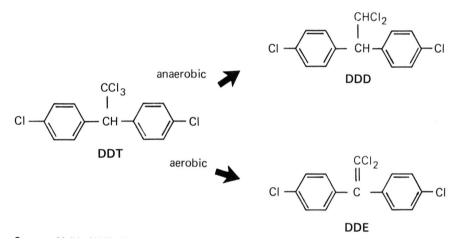

Source: Meikle [105]. *(Reprinted with the permission of Marcel Dekker, Inc.)*

FIGURE 9–1   Anaerobic and Aerobic Biodegradation of DDT

Characterization of the Biological System

• *Organisms Responsible for Biodegradation.* Microorganisms are the most significant group of organisms involved in biodegradation. Although higher organisms, both plant and animal, are capable of metabolizing numerous compounds, microorganisms convert to inorganic substances (H_2O, CO_2, mineral salts) many complex organic molecules that higher organisms are unable to metabolize [3,72]. Furthermore, microorganisms may be the first agents in biodegradation, converting compounds into the simpler forms required by higher organisms [32].

TABLE 9-1

EXAMPLES OF BIODEGRADATION REACTIONS

Type of Reaction		Examples of Chemicals Subject to Reaction

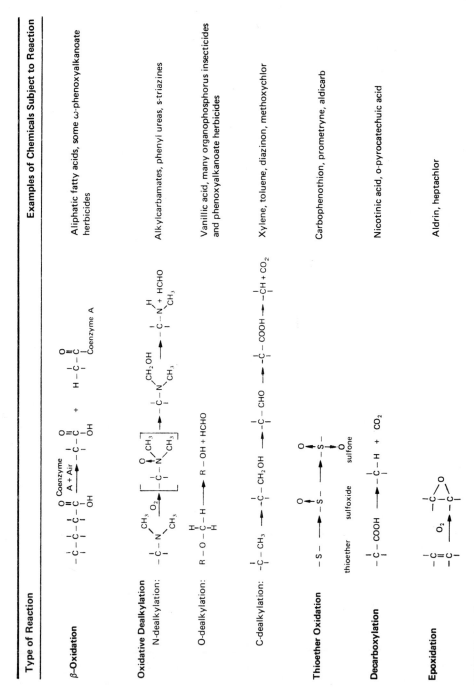

β-Oxidation

Aliphatic fatty acids, some ω-phenoxyalkanoate herbicides

Oxidative Dealkylation

N-dealkylation:

Alkylcarbamates, phenyl ureas, s-triazines

O-dealkylation:

Vanillic acid, many organophosphorus insecticides and phenoxyalkanoate herbicides

C-dealkylation:

Xylene, toluene, diazinon, methoxychlor

Thioether Oxidation

thioether   sulfoxide   sulfone

Carbophenothion, prometryne, aldicarb

Decarboxylation

Nicotinic acid, o-pyrocatechuic acid

Epoxidation

Aldrin, heptachlor

(Continued)

Type of Reaction	Examples of Chemicals Subject to Reaction

Aromatic Hydroxylation

1) Aerobic:

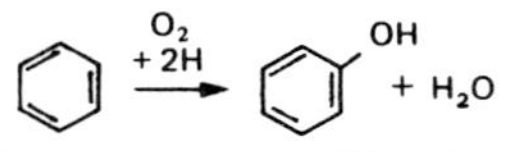

Benzene → Phenol (O_2 + 2H → + H_2O)

Pyridine, nicotinic acid, 2,4-D, some phenylalkanes, benzoic acid

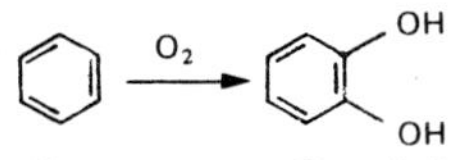

Benzene → Catechol (O_2)

2) Anaerobic:

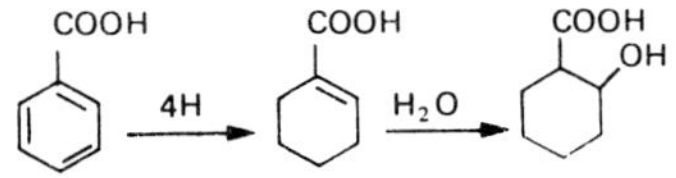

Benzoic acid → Cyclohex-1-ene-1-carboxylate (4H) → 1-Hydroxycyclohexane-carboxylate (H_2O)

Benzoate

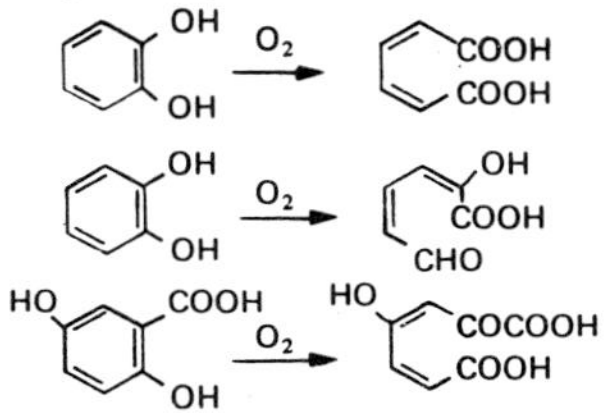

Only first step in degradation pathway is shown.

Aromatic, Non-heterocyclic Ring Cleavage

Ortho fission

(O_2)

Meta fission

(O_2)

"Gentisate" fission

(O_2)

Many catechols and phenols, gentisic acid, hydroxycyclohexanecarboxylate, many phenoxyalkanoate herbicides, carbaryl

(Continued)

TABLE 9-1 (Continued)

Type of Reaction	Examples of Chemicals Subject to Reaction

Aromatic, Heterocyclic Ring Cleavage

1) 5-membered ring

Many heterocyclic pesticides (e.g. paraquat, picloram, amitrole)

2) 6-membered ring

Pyridines, pyrimidines, triazines

Hydrolysis

Ester hydrolysis

Amide hydrolysis

Phosphorus ester hydrolysis

Nitrile hydrolysis

Carbamates, organophosphates, many urea and anilide herbicides

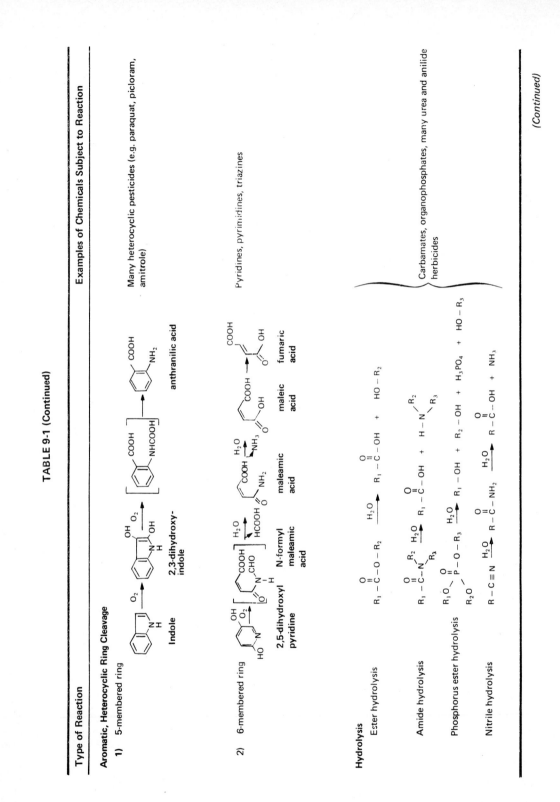

(Continued)

Type of Reaction		Examples of Chemicals Subject to Reaction
Halogen Reactions		
Hydrolytic dehalogenation	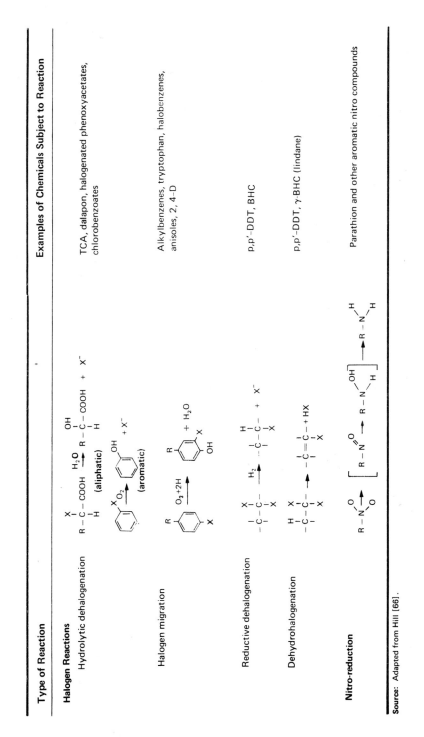	TCA, dalapon, halogenated phenoxyacetates, chlorobenzoates
Halogen migration		Alkylbenzenes, tryptophan, halobenzenes, anisoles, 2, 4–D
Reductive dehalogenation		p,p'–DDT, BHC
Dehydrohalogenation		p,p'–DDT, γ-BHC (lindane)
Nitro-reduction		Parathion and other aromatic nitro compounds

Source: Adapted from Hill [66].

Bacterial metabolism alone can account for 65% of the total metabolism of a soil community because of high bacterial biomass and metabolic rates [98]. Bacteria and fungi utilize energy more efficiently than do higher organisms [100]. The high rates of reproduction and mutation in microorganisms contribute to the considerable diversity of species and adapted strains and, hence, enzyme systems; numerous biochemical pathways for degradation are present in microorganisms as a group [72].

Other lower organisms, such as algae and certain invertebrates, exhibit some of the preceding characteristics. Although they have not been as thoroughly investigated as other microorganisms and higher organisms, their potential as significant degraders of pollutants cannot be discounted. There is some evidence that algal species contribute significantly to biodegradation of substances in the surface layer of water [175].

The microorganisms predominantly responsible for biodegradation in natural systems are heterotrophic bacteria, including the actinomycetes, some autotrophic bacteria, fungi including the basidiomycetes and yeasts, and certain protozoa [2]. A number of detailed reviews describe the biology and ecology of these groups [2,5,53,135,151]. Different conditions favor each group; for example, fungi and *Thiobacillus* are common in acid soils, while most bacteria thrive and apparently have a competitive edge in less acid soils and in alkaline soils (pH >5.5) [2,151]. Fungi are not as important in aquatic systems as in soil [134]. Not only different classes but different genera within classes react to an organic compound with responses ranging from sensitivity to degradation, so it is not possible to categorize the biodegradative ability of microorganisms according to their taxonomic classification.

Anaerobic microorganisms are either obligate anaerobes to which oxygen is toxic (*oxylabile*) or facultative anaerobes that can live with or without oxygen or prefer a reduced oxygen atmosphere (*oxyduric*) [166]. Some species specialize in reducing nitrates or sulfates, and others in reducing various alcohols to methane and other alkanes. As a group, anaerobic organisms are more sensitive and susceptible to inhibition (in sewage treatment, for example) than are aerobic organisms [160].

• *Habitats of Microorganisms.* Soil, water and wastewater treatment systems provide the most important microbial habitats for the biodegradation of pollutants. In all environments, microorganisms are essentially aquatic organisms [151], and certain characteristics are shared by all species. The organisms' habitat has a greater influence on biodegradation than does the similarity of the species [79].

In all three habitats both aerobic and anaerobic conditions exist. Although only one fifth as much free energy is obtained from one electron-mole of a methane-forming reaction as from a complete oxidation reaction, reductive reactions may play a significant role in the environment. The anaerobic habitats of interest in this chapter include some soils, sediments, and certain sewage treatment systems and sludges.

The diversity of microbial populations in soil is attributable to the large variety of food sources and habitats found there [55]. The mobility of microorganisms is decreased in soil, however, because of physical barriers (such as clay aggregates) and patchy distribution of supportive microhabitats. Usually aquatic systems have less diverse microbial populations and support a greater homogeneity in distribution [154], partly because the concentration of nutrients is diluted in the water column. Bottom sediments tend to have high nutrient levels from deposition of decaying organic matter. Growth substrates are potentially more accessible in aquatic than in soil systems, except where removal by adsorption and concentration in bottom sediment occurs [37].

As they are interconnected, soil, freshwater systems and wastewater treatment systems are inhabited by the same major species groups. Populations in the media are related, because the population characteristics of the surrounding soil help define the species make-up of an aquatic system and sewage populations through seeding by soil erosion and runoff [124,173]. Runoff from storm water and sewage overflows also contributes to the mix of species found in natural waters [174].

Air serves primarily as a transport medium for microorganisms rather than as a support system [53]. Organisms are found at low densities in the atmosphere, usually in such non-metabolizing forms as spores. Water availability is low, and extreme fluctuations in temperature and solar radiation discourage growth and activity of populations.

Aquatic systems that support microorganisms vary considerably, encompassing habitats as diverse as streams, small ponds, lakes, estuaries, and open ocean. Although aquatic microbial populations differ by system, they have some common characteristics. An aquatic system can be roughly divided into a sediment fraction, which is suspended or settled in a bottom layer, and a liquid fraction.

Vertical zonation of the liquid fraction (including its suspended sediment) is found in most standing freshwater bodies deeper than two meters with well-distinguished layers differing in temperature, oxygen content, and nutrient distribution. Figure 9-2 shows the microhabitat distribution in a freshwater aquatic system. The stratification of the layers fluctuates seasonally, as the upper layers mix with deeper waters and the sediment layer. The sediment layer is stratified into an upper oxidized zone and a lower reduced zone, each having distinctive bacterial flora (aerobic and anaerobic, respectively).

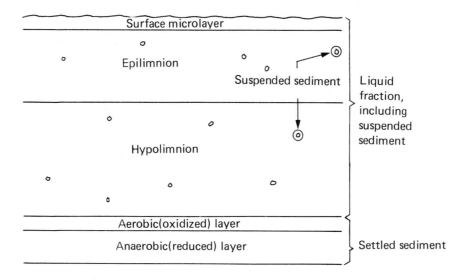

FIGURE 9–2   Microbial Microhabitats in a Generalized
Freshwater System

In marine habitats, increased productivity and biological activity are associated with coastal regions because of upwelling from nutrient-rich deeper waters and the contribution from estuaries [53]. Shore habitats, such as intertidal zones with highly organic muds, support large microbial populations. The continental shelf area (neritic zone) and the open ocean (oceanic or pelagic zone) can be divided into three layers: euphotic, aphotic, and benthic (Figure 9-3). The euphotic layer extends to approximately the point where the light intensity is 1% of that at the surface; the aphotic layer is the deeper water that extends to the benthos, which is the bottom or sediment layer. As in freshwater systems, the sediment contains aerobic and anaerobic zones, which shift according to the availability of oxygen.

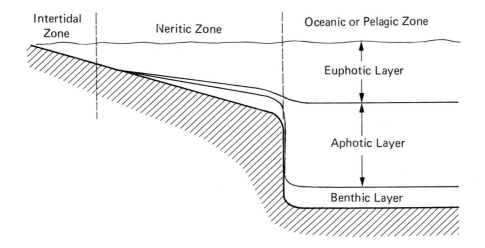

FIGURE 9—3   Microbial Habitats in a Generalized Marine System

Estuaries are among the most productive aquatic systems and are extremely variable because of differences in surrounding topography (which contribute to the degree of silting), ratio of salt to freshwater, tidal activity, and other factors [116]. The sediment layer, stratified into aerobic and anaerobic zones, is well developed biologically with abundant microbial populations [53].

Benthic sediments of both marine and freshwater systems below the surface layer-water interface are usually anaerobic and support microorganisms [42]. Many lakes and some marine areas also have anaerobic bottom waters [41]. Little is known about these environments and their associated species, especially in marine systems [74]. No more than 1% of bacteria observed in these systems will grow under laboratory conditions [166]. Many sediment microorganisms may remain in a dormant stage or at a low rate of activity for long periods because of low temperatures.

Microbial populations are distributed at different densities throughout these various microhabitats. Table 9-2 describes their distribution in the horizontal layers of aquatic systems.

Soil is not as uniform or continuous an environment as most aquatic systems. It consists of discrete compartments, only some of which are suitable as microbial habitats. The majority of the microbial population is located in the top layer of soil (approximately upper 14 cm [19]; see Table 9-3), because nutrient levels and oxygen availability are high

TABLE 9-2

Presence of Microbial Populations in Various Aquatic Systems

	Upper Layer of Water	Lower Layer of Water	Sediment Layer
Fresh Water Lotic (running water)	Microbial population very dependent on stream flow, usually higher in slower streams and rivers.	—	Presence of sediment layer and microbial population dependent on stream flow, surrounding substrate characteristics, and sediment load. Generally higher in slower flowing than in rapid streams.
Lentic (standing water)	*Epilimnion:* Microbial populations primarily associated with this layer, although sediment populations higher under some conditions. Organisms associated with surface area of detritus [41]	*Hypolimnion:* Anaerobic microorganisms may be high in nutrient-rich eutrophic lakes.	Microbial populations vary greatly, depending on depth, bottom substrate, and other factors. In most lakes, populations high near surface of sediment, although sediment investigations are few [53]
Estuary	Biological activity highest in lower bay (from river mouth upstream), in upper basins, and especially in tidal salt marshes and mud flats. Salinity shifts in the headwaters (area of mixing of fresh and salt waters) may be too extreme to support much life.		Microbial populations high, especially in highly organic muds.
Marine Waters Intertidal (Littoral)	—		Microbial populations dependent on substrate: high on organic substrates, low on cobble and shingle beaches.
Neritic	*Euphotic layer:* biological activity high. Organisms associated with surface area of detritus [41]	Most neritic waters fall within the euphotic zone.	Benthic layer: microbial activity high.
Oceanic	*Euphotic layer:* area of greatest microbial activity [70,71,76]	*Aphotic layer:* microbial activity generally lower than in euphotic layer, although specific depths may have higher populations.	Benthic layer: microbial activity low because of cold temperatures and low nutrient levels.

Sources: [53, 116, 134]

TABLE 9-3

Distribution of Microorganisms in
Various Soil Horizons

Depth (cm)	No. Organisms/g soil (x 10^5)	Percentage of Total Organisms Counted
3-8	119.7	79
20-25	24.8	16
35-40	6.3	4
65-75	0.22	<1
135-145	0.04	<1

Source: Adapted from Alexander [5] ; podzol soil.

there [36]. The plant rhizosphere (the area including and surrounding a plant's roots) supports high densities of microorganisms, because root exudates, dead root material, and adhering organic matter provide nutrients for growth (Figure 9-4). Increased microbial activity extends for 1 or 2 millimeters beyond the root surface and is not associated with all locations on the roots [53]. Other nutrient sources, primarily in the form of decomposing organic matter, are scattered throughout the soil in different stages of availability. Some are adsorbed to the mineral fraction or are blocked from access in a clay structure [151]. Although they may represent as little as 15% of the colonizable surface area in soil, organic particles can be populated by 60% (by mass) of the soil bacteria, while mineral particles are only minimally colonized [52]. Microorganisms comprise a large fraction of the living biomass in soil — up to 80% when soil algae are included [116] — although not all organisms are metabolically active at the same time. As would be expected, microbial density is strongly influenced by organic matter content, which can vary from a minimum of 1% in mineral soils to more than 90% in rich, organic soil [151]; the usual range is from 3% to 6% [5]. The density of microorganisms is much lower in the soil water fraction than at soil-water interfaces [2].

In soils submerged in water, oxygen levels and diffusion rates are too low to support aerobic microorganisms. Furthermore, localized anaerobic pockets may be distributed throughout a generally aerobic soil [95,166]. During periods of high microbial activity, such as the early stage of plant residue decomposition, temporary anaerobic conditions may be created when the oxygen demand exceeds the supply [125,166].

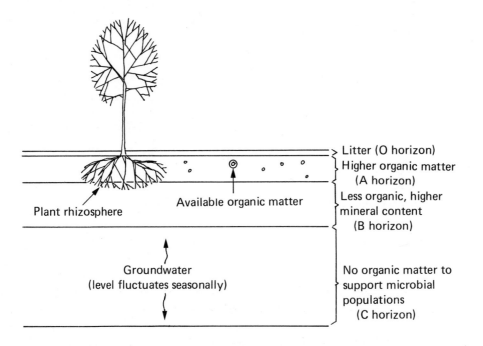

FIGURE 9—4   Microbial Microhabitats in a Generalized Soil System

Wastewater treatment provides a third major system. In the municipal and industrial treatment of organic wastes, two approaches are commonly used, often in combination: aerobic mineralization and anaerobic digestion. The latter takes the form of fermentation to methane and CO_2. Activated sludge treatment and filtration through trickling and/or sand filters are aerobic processes.

Activated sludge is a well-mixed, stirred (for aeration), single-stage process in which organic waste is mixed in a reactor with mixed-species microbial populations that are either growing in flocs or freely suspended in the supernatant liquid. Figure 9-5 depicts the process. Less than 10% of the floc is made up of active organisms; the rest is mostly insoluble organic matter made up of polymeric material [90]. Floc sizes, ranging from 0.02 to 0.2 mm [113], may be rate-limiting because the transport of nutrients to microorganisms in the center of the floc slows the reaction [40]. The activated sludge process has many variations, which differ in degree of aeration, mixing, container size, and procedure.

Slow sand filters (SSF) and trickling filters (TF) are, respectively, two- and three-phase processes in which dissolved organic wastes are passed through a biologically active film colonized by microorganisms. The SSF is slower than the TF and provides no aeration [90].

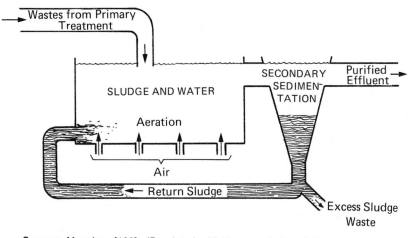

Source: Manahan [100]. *(Reprinted with the permission of Prindle, Weber & Schmidt.)*

FIGURE 9–5   A Conventional Activated Sludge System for Secondary Biological Waste Treatment

Most municipal sewage treatment plants use anaerobic digestion for sludge stabilization [133]. After the settleable matter and supernatant liquid are separated, the process has two stages, starting with an "acid phase" followed by a "methane phase" (Figure 9-6). In the first phase, complex organic solids in the settled material are degraded to acid form, transforming cellulose, starches, proteins, and carbohydrates to simple sugars, amino acids, and volatile acids (formic, acetic, butyric, etc.). During the second phase the acids produced, along with any original long-chain fatty acids, are reduced to methane and carbon dioxide [133].

• *Significant Species.* The microbial species found in natural ecosystems are diverse, but certain groups appear to play prominent roles in biodegradation and are encountered again and again in microbial cultures from natural sources. These genera are able to metabolize a variety of organic substrates. Specificity to certain compounds is more commonly found at the species level, although some species such as *E. coli* are generalists. Table 9-4 lists some genera commonly found in soil, aquatic, sludge, and anaerobic habitats. These are typical only; the table is not intended as a compilation of each system's most prominent genera.

The microorganisms involved in anaerobic digestion are primarily bacteria, both facultative (able to live under aerobic and anaerobic conditions) and obligate (able to live only under anaerobic conditions)

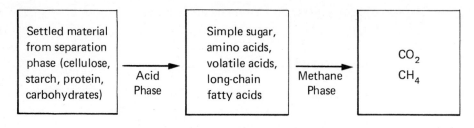

FIGURE 9-6   Anaerobic Wastewater Treatment Process

TABLE 9-4

Representative Microbial and Protozoan Genera Found in Different Environments

Environment	Genera	Source
Freshwater Aquatic and Soil Systems	*Arthrobacter, Aspergillus, Bacillus, Corynebacterium, Flavobacterium, Fusarium, Nocardia, Penicillium, Pseudomonas, Thiobacillus, Torulopsis, Trichoderma, Micromonospora, Streptomyces*	[37,50,51, 53,134]
Marine Aquatic System	*Achromobacter, Flavobacterium, Pseudomonas, Vibrio*	[53]
Sludge and Anaerobic Systems	Acid formers: *Pseudomonas, Flavobacterium, Alcaligenes, Escherichia, Aerobacter, Aeromonas Clostridium, Leptospira, Micrococcus, Sarcina*	[133,161]
	Methane formers: *Methanobacterium, Methanobacillus, Methanococcus, Methanosarcina*	[100,133]
	Activated Sludge: *Achromobacter, Alcaligenes, Arthrobacter, Bacillus, Bacterium, Bdellovibrio, Comomonas, Flavobacterium, Microbacterium, Nitrosomonas; Pseudomonas, Sphaerotilus*	[127,161]
	Others: *Aspergillus, Fusarium, Rhizopus, Penicillium, Cladosporium*	[84,88,133,142, 144,161,167]

anaerobes. The acid-forming bacteria have higher rates of reproduction and tolerate a pH as low as 5.0. Methane-forming bacteria are inhibited at a pH below 6.5 and are generally more sensitive to temperature and substrate concentration [133].

 • *Biodegradation Reactions.* For any one microorganism, organic compounds can be divided into three groups according to their biodegradability: (1) usable immediately as an energy or nutrient source, (2) usable following acclimation by microorganisms, and (3) degraded slowly or not at all [155,17]. Some investigators believe that a fourth group also exists, consisting of compounds subject to cometabolic degradation. Figure 9-7 depicts a generalized disappearance curve for each of the first three groups. A chemical may be classified in more than one category, depending on the response of the microorganisms to which it is exposed; different species may react differently to the same compound.

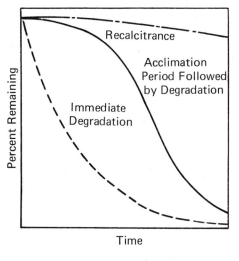

FIGURE 9—7   Degradation of
Organic Compounds

The first group includes certain simple sugars, amino and fatty acids, and compounds in the proper form to enter typical metabolic pathways. The enzymes necessary for taking up or degrading these compounds are constitutive or immediately inducible and thus minimal acclimation is required [26].

The second group requires acclimation, a lag period during which little or no degradation takes place. The delay is usually caused by the following processes which are somewhat interrelated:

(a) Selection of those species in a mixed population that are capable of assimilating the substance, in which case the lag is due to the initial phase of exponential population growth of the favored organism, and

(b) Adaptation of existing microorganisms through induction of enzymes that catalyze degradation.

Lag periods vary from a few hours to days or even weeks, depending on the chemical, the organism, and the medium (see Figure 9-8). A period of more than 50 days has been observed for pyrazon in garden soil [38]. Thus, laboratory experiments conducted over a prescribed period of time, rather than until degradation commences, may not establish whether a substance is biodegradable if the chemical requires a long acclimation period.

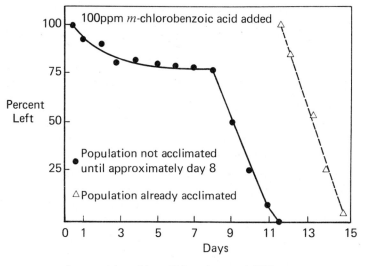

Source: Adapted from DiGeronimo *et al.* [33]

FIGURE 9—8   Lag Period in Biodegradation of
m-Chlorobenzoic Acid

Once acclimation is achieved, the degradation reaction begins. Intensive activity occurs first with primary alteration of the introduced substance; this is usually followed by slower activity as the intermediate products are digested [155]. The microbial population increases at first, levels off, and declines once the substrate has disappeared or has been converted either to non-metabolizable catabolites or to inorganic compounds. The disappearance curve for the parent compound can follow one of several forms, depending upon the kinetics of the reaction. Biodegradation reaction kinetics are discussed in §9-4.

The third group of organic compounds includes such naturally occurring substances as humus and lignin, as well as such anthropogenic substances as some of the organochlorine pesticides [1]. These substances degrade at very slow rates or not at all. Furthermore, they may not be degradable due to factors other than chemical structure — e.g., physical inaccessibility or environmental influences (low O_2, pH, etc.). Alexander [1,3,4] has written extensively on the subject of recalcitrance. Some of the factors responsible for the failure of biodegradation are discussed later in this chapter.

Cometabolism is thought to play a role in the degradation of certain chemicals, although little research has been done on the process. It is defined as the degradation of a compound that does not provide a nutrient or energy source for the degrading organisms but is broken down during the degradation of other substances [7]. Figure 9-9 compares metabolic and cometabolic rate curves. Because cometabolism does not provide a growth substrate, the population increase characteristic of metabolic degradation reactions does not take place [7] and the rate of degradation is often slower. Compounds with chlorine, nitro, or other substituents are sometimes susceptible to cometabolism [7].

• *Microbial Population Densities and Biomasses.* Counting the number of individuals in a population overestimates the significance of microorganisms in a community; measurement of biomass, on the other hand, underestimates their significance [116]. For example, in the benthic community of a small lake, bacteria accounted for 30% of the community respiration but less than 1% of the total biomass [116]. Nevertheless, an actual count is necessary when one is investigating the biodegradation of specific compounds by microorganisms with variable metabolic activity.

Measurement of microbial populations is subject to considerable error because of the characteristics of the organisms and deficiencies in measurement techniques. The problem is greater in soil and activated sludge than in more homogeneous media such as water [151,155]. A population of microorganisms is unlikely to be uniformly active; this is due to species specialization on substrates that are not all equally or consistently available. Furthermore, because of their adaptability and short regeneration time, microbial populations are quite variable and dependent on the conditions at the moment of sampling [116]. It is virtually impossible to distinguish between active and dormant or dead organisms without using respirometry or similar measures of activity such as acridine orange staining and epifluoresence microscopy.

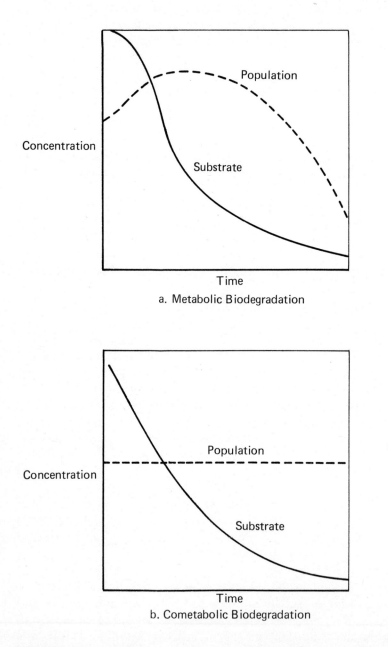

a. Metabolic Biodegradation

b. Cometabolic Biodegradation

FIGURE 9–9   Population and Substrate Concentrations During Biodegradation

Several techniques have been developed to estimate microbial population; each has its own set of disadvantages (see Table 9-5). Numbers determined by direct microscopic count are usually higher than those determined by plate counts, because any one plate culture medium cannot satisfy the requirements of most bacterial species in natural habitats [153]. Differences of the order of 10^3 are often seen between these two most commonly used techniques [53]. A combination of two or more methods is the best approach, although it may be expensive and time-consuming in some cases.

Microbial population densities reported in the literature for aquatic, soil, anaerobic, and sludge systems are compiled in Table 9-6. In Table 9-7 microbial biomasses for soil and water are presented. In all systems, population densities varied more than several orders of magnitude, probably because of environmental factors and differences in sampling techniques.

Variables in Biodegradation. The variables that influence the rate of biodegradation fall into two general categories: (1) those that determine the availability and concentration of the compound to be degraded (e.g., propensity for adsorption) or that affect the microbial population size and activity (e.g., population interactions) and (2) those that directly control the reaction rate itself (e.g., population size, temperature). Both direct and indirect variables can be classified as substrate-related, organism-related, or environment-related. Table 9-8 lists all the variables discussed in this section. Most of the information is presently qualitative, primarily as observations of variable influences on degradation by specific species. Because of considerable variation in species, habitat, and chemical environment, not all variables will influence all situations in the same way. For example, low pH is likely to decrease metabolic activity in most bacteria, but it favors activity in fungi [21].

An important characteristic of environmental variables is their degree of interrelation. For example: (1) temperature and moisture content in soil are interdependent; (2) where there are high levels of organic matter, the pH is usually low; and (3) pH affects adsorption [55]. In this section, each variable is considered separately. Both direct and indirect variables are discussed under each of the three main categories.

- *Substrate-Related.* Certain properties of a compound that serves as a substrate for biodegradation affect microbial reactivity. Correlations between various physicochemical properties and biodegradability of organic compounds have been reported. These observations are discussed in §9-5.

TABLE 9-5

Methods for Estimating Microbial Populations and Biomass

Type	Method	Problems
Viable Counts • Pour plates • Spread plates • Membrane filter plates	Cells to be counted are cultured on a growth-supporting medium; each cell divides, forming a colony that indicates presence of a cell.	No one medium will support all species; organisms tend to clump together around detritus, making separation difficult.
Direct Microscopic Count	Visual counting of cells under a microscope	Does not differentiate between dead/inactive and living cells; dye concentration difficult to maintain; cells clump together; difficult to detect individuals at low concentrations ($<10^6$ cells/mL)
Turbidity	Measurement of light transmitted through bacterial suspension	Interference of shape of cell with passage of light; only works in dilute microbial concentrations; mass of micro-organism can change while number of cells remains constant.
Measurement of Cell Constituent	Content of ATP (adenosine triphosphate) or DNA is measured to indicate total biomass.	Concentrations of many indicator consitituents are too low to measure. Method requires that: (1) concentration of constituent being measured is constant in relation to biomass; and (2) constituent is rapidly degraded when released outside of cell wall. Turnover time and concentration of ATP per cell varies by cell and species (50-fold range of ATP per unit weight for variety of organisms).
Respirometry	Oxygen uptake or CO_2 generation	Some CO_2 may be non-biological in origin. Oxygen removed may be adsorbed or consumed by chemical reaction.
	Electron transport (tetrazolium salts)	Reaction can interfere with normal electron transport process.

Source: Adapted from Refs. 2, 75, 116, and 151.

TABLE 9-6

Microbial Population Density in Various Environments

System	Population Density	Comments	Source
Aquatic	10^3-10^6/mL	Surface water	[92]
	$< 10^{-3}$-10^8/mL	Oceanic waters (from open water to inshore respectively)	[177]
	10^{13}-10^{14}/m^2	Open pond water	[116]
	10^9-10^{10}/mL	Laboratory culture	[2]
	10^6(10^5 active)[a]/mL	Bacteria –pond	[12]
	10^5(10^4 active)/mL	Bacteria →stream	[12]
	10^6(10^5 active)/mL	Bacteria –eutrophic lake	[12]
	10^2(10 active)/mL	Bacteria –oligotrophic lake	[12]
	10^5-10^{10}/mL	Natural waters	[2]
Soil	10^8/g	Bacteria only	[52]
	10^5/g	Actinomycete spores	[52]
	5m/g[b]	Fungal mycelium	[52]
	10^7/g	Upper 3-8 cm of soil	[5]
	10^{14}-10^{15}/m^2	Meadow or old field	[116]
Anaerobic	10^1-10^8/g	Marine sediment	[177]
	10^3-10^6/g	Gram-negative motile bacilli --marine sediment	[167]
	10^9/g	Feedlot waste	[137]
Activated Sludge	10^{10}-10^{12}/g dry wt	In floc	[127]
Sewage Treatment	10^7/mL	Sewage entering	[172]
	10^8/mL	Mixed liquor	[172]
	10^6-10^7/mL	Effluent	[172]
	10^6-10^7/mL	Supernatant	[92]
Anaerobic Treatment	10^8-10^{10}/mL	Nonmethanogenic obligate anaerobes	[162, 30]
	10^5-10^{10}/mL	Methanogenic bacteria	[91]
	10^4-10^9/mL	Sulfur-reducing bacteria	[129, 167]

a. "Active" = bacteria not in dormant stage

b. Length of mycelium (m) is measured

TABLE 9-7

Microbial Biomass in Various Environments

System	Biomass	Comment	Source
Aquatic	1-10g/m^{2}[a]	Open pond water	[116]
Soil	100-1000 kg/ha[a]	Meadow or old field	[116]
	6.0 x 10^4 g/g	0.06% of soil mass	[52]
	300-3000 kg/ha	0.015-0.05% of soil mass	[2]
	37 kg/ha (living)	Bacteria in woodland soil	[53]
	9113 kg/ha (dead)	Bacteria in woodland soil	[53]
	110 kg/ha (living)	Fungi in woodland soil	[53]
	566 kg/ha (dead)	Fungi in woodland soil	[53]

a. Dry weight; other measures are assumed to be wet weight.

Another influential factor is the substrate concentration. If it is too low, biodegradation may be limited, possibly from lack of sufficient stimulus to initiate enzymatic response [3]. There is some evidence that compounds that are usually easily degradable are persistent at very low concentrations [33,74]. On the other hand, high concentrations may be toxic or inhibitory to metabolism. The optimum concentration is chemical- and species-specific. Several discussions of the deleterious effects of introduced chemicals on microbial populations have been published [11,35,124]. Concentrations greater than a compound's solubility in water may result in a lower rate constant than concentrations below the solubility limit, as observed for chlorodiphenyl oxide [20]. Reaction kinetics may shift in order and rate as the substrate is depleted and its concentration decreases during the biodegradation process [92].

• *Organism-Related.* Biological influences include the species composition of the microbial population, their concentration and distribution, their past history, and intra- and interspecies interactions among population members. Another significant factor is the ability of the species to synthesize the enzyme systems required for the breakdown of organic compounds.

Species variability is exhibited in the metabolic response of a microbial population to a newly introduced organic compound. Some of the

TABLE 9-8

Variables Potentially Affecting Rate
of Biodegradation

Substrate-Related
- Physico-chemical properties
- Concentration

Organism-Related
- Species composition of population
- Spatial distribution
- Population density (concentration)
- Previous history
- Interspecies interactions
- Intraspecies interactions
- Enzymatic make-up and activity

Environment-Related
- Temperature
- pH
- Moisture
- Oxygen availability
- Salinity
- Other substances

---

simpler molecules, such as glucose, are immediately degradable and support growth of numerous species [155]. Complex organics requiring more extensive metabolic pathways are likely to support fewer species — specifically, only those that have evolved mechanisms for induction of adaptive enzymes matched to the chemical. Therefore, glucose is rapidly metabolized in most biologically active environments, but many hydrocarbons support few microbial species (e.g., *Nocardia, Pseudomonas, Mycobacterium*) and often require acclimation periods before degradation proceeds [21].

The distribution of microorganisms in the medium in which a potential substrate is contained is an important factor in biodegradation. Either environmental parameters (see below) or the presence of toxic substances may limit microbial colonization of the site. Soil is such a heterogeneous environment that the distribution of microorganisms is patchy. Soil microhabitats immediately adjacent to one another commonly differ in numbers of microorganisms because of a wide temporal and spatial distribution of organic matter available for microbial digestion, variations as high as three units in pH around growth sites, differences in moisture retention ability, and other factors [53].

Over long periods of time, microbial concentration is not as important as the other factors described because of the rapid response of, and numerical increase in, populations of a species capable of metabolizing the substrate. If short time periods are of concern, however, microbial concentration can have a significant effect. For example, the time for complete metabolic oxidation of glucose (including intermediates) may range from a few hours in a concentrated bacterial culture (100 to 1000 ppm in activated sludge, assuming 10% of mass by weight is active) to days in a dilute culture (10^3 to 10^5 cells/mL)[1] [155].

The previous history of microorganisms in relation to the particular compound undergoing degradation may be reflected in the reaction rate. If a compound is continually introduced into a system, as are some agricultural pesticides, often the microorganisms soon acclimate to the substance and begin degradation immediately, without a lag period. The difference is noticeable even in regard to simple, readily degradable substances. A glucose-adapted laboratory culture was found to degrade sugar at a rate three times higher than a culture of fresh-water isolates [154]. For more complex compounds the significance of prior acclimation on biodegradation rates is well known (e.g., see Figure 9-8).

Inter- and intraspecific interactions among species may indirectly affect the rate of biodegradation in the initial period through their effects on microbial activity in general. These effects can be positive or negative and are quite specific to each population mix. Processes common to all mixed-species groups, such as competition and predation, determine which species will succeed in growing on a substrate compound. The presence of other species, such as protozoa and rotifers, can increase the degradation rate of a population through selective predation on weak or inactive members [24]. The metabolic activity of a microbial population is not necessarily equal to the additive effects of each species; metabolism may be cooperative, with successive species degrading the initial substrate in sequential steps [153]. Extracellular enzymes of one organism may break down a compound such as polysaccharide sufficiently for uptake and metabolism by another organism [3]. Dissimilar species may have to attack different sites on a branched compound, such as melanin, before degradation can take place [3]. *Arthrobacter* and *Streptomyces* can degrade the pesticide diazinon together but are unable to do so independently [54].

---

1. Equivalent to 2-200 mg/L by mass, assuming one cell = 2×10^{-13} g [44].

Enzymes are so substrate-specific that a compound subjected to structural alteration may require a different enzyme catalyst. Specialization is so precise that enzymes can distinguish between amino acid stereoisomers and between such geometrical isomers as fumaric and maleic acid [3]. Microorganisms without the enzymatic make-up required by a compound will be unable to degrade it. In some cases the necessary enzymes can be induced during a period of acclimation following contact with a substance.

Once enzymes are activated, other factors may prevent their catalyzing a degradation reaction. Inhibition of the enzyme or repression of its synthesis by a substrate or its catabolites can complicate initiation or continuation of a degradation reaction [153]. Extracellular enzymes, such as hydrolytic enzymes, can be inhibited or inactivated by clay or other colloids, humic acids, and other substances [3]. Because of cross linkages, coiling, folding, etc., enzymes may be unable to complement a compound's particular steric configuration and reach the activation site [3]. The absence of appropriate enzymes and physical interference are responsible for the recalcitrance of various compounds, such as some of the synthetic high-molecular-weight polymers and certain proteins [3].

- *Environment-Related.* Environmental variables control microbial metabolic activity in general rather than biodegradation specifically. The significance of particular parameters varies with each ecosystem. Also, as expected from the considerable genetic variability in microorganisms as a group, certain species have evolved to function in extreme environmental conditions.

Microbial growth has been observed in environmental temperatures ranging from -12 to $100°C$ [21]. Individual species are usually adapted to a 30-40 degree range somewhere between these extremes. Depending on the temperature in which microorganisms have a competitive advantage over other species, they are commonly classified in one of three groups: psychrophiles ($< 25°C$), mesophiles (between $25°$ and $40°C$), and thermophiles (above $40°C$) [21]. Organisms that degrade chitin in tropical soils (ranging from $28°$ to $30°C$) are primarily actinomycetes, protozoa, and higher organisms, while in temperate soils, fungi and *eubacteria* are responsible [117]. Temperatures outside a microorganism's range are not necessarily lethal; many species (e.g., spore-formers) have a dormant state that permits survival until conditions supportive of growth return.

Rates of biological reactions increase with increasing temperature within the range tolerated by the organism. This is illustrated in Figure 9-10, which is a plot of the degradation of a chemical at two temperatures. The relationship can be described by the Arrhenius equation:

$$Y = Ae^{-E_a/RT} \qquad (9\text{-}1)$$

where

Y	=	temperature-corrected rate of reaction
A	=	initial rate of reaction
E_a	=	activation energy
R	=	gas constant
T	=	absolute temperature

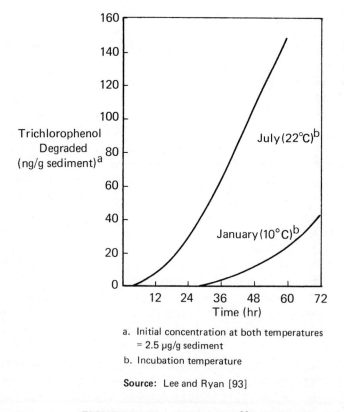

a. Initial concentration at both temperatures
   = 2.5 µg/g sediment

b. Incubation temperature

Source: Lee and Ryan [93]

FIGURE 9—10 Amount of ^{14}C—trichlorophenol Degraded with Time at Two Temperatures

Reduced bacterial activity was observed in several river-water tests when the incubation temperature was decreased: a 75% reduction in maximum breakdown rate of 2,4-D occurred in a river die-away test when the temperature was reduced from 25°C to 15°C [163]. The Arrhenius plots in Figure 9-11 show the relationship between temperature and biodegradation.

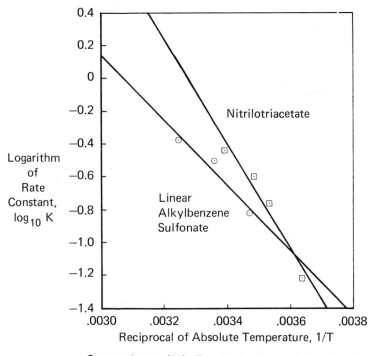

Source: Larson [92]. *(Reprinted with permission from the American Society for Microbiology.)*

FIGURE 9—11   Arrhenius Temperature Plots for Biodegradation in Water

Wastewater biological treatment processes are dependent on temperature, functioning optimally between approximately 20°C and 35°C [27]. Figure 9-12 shows the relationship between temperature and efficiency of carbon removal in an activated sludge system. Degradation took place at both the low and high temperatures tested.

Populations that are adapted to temperature extremes may deviate from rates predicted by Eq. 9-1 — e.g., psychrophilic populations may show increased efficiency during winter [92]. The relationship between reaction rate and temperature may also be complicated by

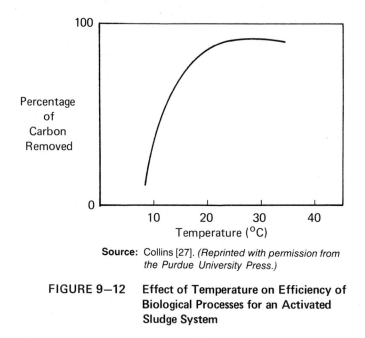

Source: Collins [27]. *(Reprinted with permission from the Purdue University Press.)*

FIGURE 9—12   **Effect of Temperature on Efficiency of Biological Processes for an Activated Sludge System**

other factors. For example, although temperatures are lower in winter, upwelling or lower water flow rates may increase microbial population density and, therefore, the rate of degradation [174].

Temperature is interrelated with other environmental parameters, especially in soil. Moist soils conduct heat more efficiently and thus have a smaller temperature gradient over a given depth than do dry soils [130]. Adsorption of some pesticides on clay particles increases with decreasing temperature, while the reverse is expected for organic matter [147].

Microorganisms as a group have adapted to the entire pH range normally encountered in natural systems. Optimal growth for fungi usually occurs under slightly acidic conditions, between pH 5 and 6, but activity continues at a pH less than 3. Bacterial growth is favored by slightly alkaline conditions and, except in acidophilic species, is inhibited when the pH drops to approximately 5 [21,53]. Microbial oxidation is most rapid between pH 6 and 8 [72].

Moisture is an important variable in the soil habitat. First, moisture controls soil oxygen levels by competing with oxygen for soil pore spaces. Second, most microorganisms require water because of their relatively permeable cell membranes and large surface-to-volume ratio. Many do not survive drying, although some wait for more favorable conditions in such dormant forms as endospores [52,106]. Some filamentous fungi can tolerate dry soil conditions by extracting moisture from air, but their activity increases if water is present [21]. The measure of soil moisture that is most relevant to microorganisms is not moisture content but "water potential." This is the difference between the energy state of soil water and of free water and represents the total contributions of gravity, soil matrix, and osmotic pressure. Water potential levels tolerated by various microbial groups are listed in Table 9-9.

TABLE 9-9

Effect of Soil Water Potential on Microorganisms

Aspergillus penicillium (fungus)	Predominates at less than –145 bar[a]
Most other fungi	Lower limit approximately –40 bar
Most bacteria	Lower limit –80 bar, upper limit approximately –5 bar.

a. 1 bar = 10^6 dynes/cm^2 = 0.987 atm.

Source: Gray [53]

Biodegradation can occur in both aerobic and anaerobic environments; the type and rate of the reaction is affected by the amount of oxygen present. In aerobic environments, oxygen is used as a terminal electron acceptor for many degradation reactions (e.g., for many aliphatic hydrocarbons). Some organisms also need oxygen for the dissimilative process [1]. Lack of air commonly limits the growth of bacteria in laboratory cultures in closed systems [21], but it does not become rate-limiting until the concentration of dissolved oxygen drops below about 1 mg/L [92]. Oxygen levels are reduced by microbial depletion of non-replaceable oxygen during metabolism or, in soil, by encroachment of water into pore spaces containing oxygen, which can reduce oxygen diffusion rates by as much as two thirds [55]. When the gas-filled pore spaces represent less than approximately 10-20% of the total pore space, conditions shift from aerobic to anaerobic [136]; in this environment, even small amounts of oxygen inhibit microbial activity [21].

The effects of salt vary with the species of microorganism. Salt in a concentrated solution causes dehydration of living cells, but some species, such as those adapted to marine or other saline environments, require salt at typical seawater concentrations (approximately 3.5%) for membrane stability and enzyme activation [21]. Organisms in freshwater or non-saline soil environments, on the other hand, have not required evolution of salt tolerance, so their activity may be limited or temporarily repressed under saline conditions.

The presence or absence of substances other than the substrate can influence the rate of biodegradation. Some metabolic reactions require compounds in addition to the substrate for the induction of enzymes necessary for degradation or as nutrient sources. The lack of an essential nutrient can retard or limit biodegradation [3,114b,115a]; for example, insufficient nitrogen and phosphorus in certain estuarine systems retards the degradation of glucose at concentrations above 1 mg/L [92]. Marine, oligotrophic (low productivity) lake, and some soil systems may show similar limitations. The importance of oxygen for aerobic biodegradation has already been mentioned.

Another controlling parameter is the presence and concentration of substances onto which the substrate can be adsorbed or with which it can form complexes, making the substrate inaccessible to biological activity. Adsorption may be the primary factor preventing, or significantly delaying, the degradation in soil of some usually metabolizable compounds [1]. Some compounds are physically trapped within the lattice structure of clay in pores too small for penetration by microorganisms; alternatively, by combining with clay or other material, a compound may become unable to penetrate cell membranes [3]. New, recalcitrant compounds may appear when the substrate forms complexes with resistant organic compounds such as lignins, melanins, and tannins [3]. In aquatic systems, similar interactions between a compound and particulate matter can occur. Adsorption onto suspended solids and biological matter in sludge may significantly reduce the concentration of substrate available for biodegradation [155]. Rate equations have been adapted to account for adsorption in natural waters. An increase in the sediment-water ratio by 100 was associated with an equivalent decrease in the second-order rate contant for biodegradation of chlorpropham and di-n-butyl phthalate [149].

The importance of sediment in aquatic systems as a source of nutrients and microhabitats for microbial populations has also been indicated [114b,115a]. In a biodegradation study of 2,4-D in river water, 50% was degraded at 40 days in samples with sediment added as compared with only 10% degraded at the same time in unadulterated samples [115a]. The mechanisms responsible for this phenomenon were not isolated in this study.

9-3   STANDARD TEST METHODS

Principles of Use. To estimate biodegradability and to generate a usable measure of the loss rate of an organic compound in the environment, it is necessary to conduct experiments under controlled conditions. Such experiments are especially important in biodegradation studies because so many variables may influence the process, as described in the preceding section. The three biological habitats of interest in this chapter are aquatic, soil, and wastewater treatment systems. In addition, anaerobic conditions exist within these habitats (in soil, sediment, and sludge). A particular set of biological, physical, and chemical properties is associated with each system, and the testing procedures must replicate these controlling parameters. The text that follows describes, first, biodegradation tests in general and then specific tests that best simulate each of the environments.

Tests for measuring biodegradation generally follow a standard procedure:

(1) A microbial population is collected from an environmental source (e.g., river water, agricultural soil) or is isolated through use of an enrichment culture. The substrate may be introduced early as a carbon source, to ensure the presence of a population capable of biodegradation before the experiment begins.

(2) The population is incubated with the substrate in some medium (e.g., water, soil), with or without additional nutrient or energy supply.

(3) The rate of disappearance of the substrate is monitored through indirect or direct analytical techniques.

A standard reference compound should be tested under the same conditions concurrently [163], but this is not always possible.

The time required for a given test depends on the nature of the chemical being tested, the source of the microbial inoculum, and the procedures used. Tests can take from a few days (river die-away) to 14 weeks (trickling filter) [72,163]. The assimilation of organic compounds by microorganisms may depend on acclimation of the cells to the test compound, which requires varying amounts of time for different organisms (3-30 days) [72,163] until synthesis of the necessary enzymes for species selection takes place. Acclimation time also depends on the temperature and the source of seed.

The quantity of chemical compound required per test varies from 5 to 200 mg total organic carbon/L for many tests. Lower concentrations may simulate the natural environment more accurately, but they can also make it more difficult to obtain conclusive results. Biodegradation should be measured in several separate runs, each with a different initial concentration.

Many biodegradation experiments employ sterilized or poisoned controls to compare with biologically active samples; this is essential for differentiating between chemical and biological reactions. The drawback to sterilization by chemicals, heat or radiation is that this may alter the system chemically or trigger other reactions [50]. Filtration is an alternative.

Mixed-species microbial cultures are preferable to single-species cultures, because they better reflect the microbial diversity found in nature. The main disadvantage is the difficulty in replicating results; different species may play dominant roles in different runs, or dominance may shift during a single run.

Several analytical techniques are available to monitor changes in chemical concentration over time. Both direct and nonspecific (for the parent compound) methods are used in biodegradation experiments. Table 9-10 lists analytical techniques commonly used in biodegradation tests and their main disadvantages.

Nonspecific methods include bioassays, O_2 uptake, measure of a constituent of the compound that becomes available during degradation (such as chlorine), CO_2 evolution, and increases in bacterial populations. There are arguments against all of these methods; each assumes a consistent relationship between the effect measured and the compound concentration. CO_2 evolution assumes that all carbon liberated originates from the test compound and not from the death of the original organisms. Bacterial counts assume that the substrate is the only growth-supporting medium; furthermore, it is difficult to measure microbial population accurately.

Direct methods are analytical procedures sensitive to the parent compound, including chromatography, spectrophotometry, and radio-labeling with carbon-14. The latter is the most accurate method; by permitting a total mass balance of the parent compound and its metabolites, it accounts for all losses due to biodegradation [72]. Also, lower initial concentrations can be used. Radio-labelled material should be checked by GLC or TLC (see Table 9-10) for impurities. The primary drawback to ^{14}C-labelling is expense.

TABLE 9-10

Analytical Techniques Commonly Used in Biodegradation Tests

Technique	Direct (D) or Indirect (I)	Potential for Metabolite Identification	Problems
Chromatography ● Paper ● Thin-layer (TLC) ● Column ● Gas (GC or GLC)	D	Yes (with co-chromatography)	Analytical techniques must be developed specifically for a chemical or chemical group; only volatile substances can be measured with GLC.
Radiotracers ● Assay for loss of ^{14}C in parent compound	D	Yes	Expensive; label must be attached to site of rate-determining step unless $^{14}CO_2$ evolution is measured; complex equipment required; lack of $^{14}CO_2$ evolution may only mean incomplete mineralization; must be combined with TLC or GLC as analytical tools.
Colorimetry	D	Poor	Interference from other compounds in medium; not very sensitive.
Spectrometry ● UV absorption ● Infra-red (IR)	D	Yes	Not as sensitive as GLC & TLC; potential for interference from other substances; fails to reveal minor modifications in parent compound; UV requires large amount of compound to be measured.
CO_2 Evolution	I	No	Not all released carbon goes to CO_2, so results not precise; used to measure ultimate biodegradation (i.e. mineralization).

(Continued)

TABLE 9-10 (Continued)

Technique	Direct (D) or Indirect (I)	Potential for Metabolite Identification	Problems
O$_2$ Consumption ● BOD ● Respirometer	I	No	Reaction must be oxidation; O$_2$ may be utilized for other reasons than oxidation of substrate.
Total Carbon determination ● Chemical Oxygen Demand (COD) determination ● Combustion ● Dissolved Organic Carbon (DOC) removal	I	No	Substrate must be sole carbon source; differences in susceptibility of different chemicals to analytical technique (combustion); interference by other impurities.

Source: Howard [72] and Swisher [155].

Many tests use relatively inexpensive nonspecific analytical techniques that do not measure changes in concentration of the parent compound or identify degradation products. Nonspecific tests do not yield any quantifiable data on the biodegradation reaction rate of the substance's disappearance *per se*. Some quantification of biodegradation can be obtained, however, by measuring the rate of CO$_2$ evolution or by other processes. Such data cannot substitute directly for a measured biodegradation rate. In many cases more than one analytical technique can be chosen for a given test procedure: any of five different techniques might be used in a soil perfusion test, for example. The following section summarizes the test methods commonly used to screen for biodegradability of organic compounds and discusses how the choice of test and analytical technique can affect the results.

The type of test selected can greatly influence the biodegradation measurements, as shown in Table 9-11. Some methods, such as semicontinuous sludge and trickling filters, may provide better conditions for biodegradation than others.

Table 9-12 lists methods that have been recommended by various groups for screening organic compounds for biodegradability.

TABLE 9-11

Comparison of Biodegradation Test Methods
(Percent removal of MBAS[a] after 15 days)

	Surfactant[b]			
	A	B	C	D
Continuous sludge	61 ± 5.2	66 ± 2.9	75 ± 5.0	34 ± 5.5
Slope culture[c]	74 ± 8.8	89 ± 1.6	0–66	20 ± 7.3
River water	88 ± 0.9	93 ± 0.6	96 ± 0.3	29 ± 1.9
Shake culture	88	96	91	34
Semicontinuous sludge	89 ± 0.4	96 ± 0.3	98 ± 0.3	70 ± 4.0
Recycle trickling filter	92 ± 1.6	96 ± 0.7	97 ± 0.4	83 ± 1.5

a. MBAS = methylene blue active substances, which include anionic surfactants and/or certain natural materials detected by this method.
b. A=Dobane JNX, B=Dobane JNQ, C=Dobane 055, D=ABS.
c. Die-away test using activated sludge inoculum in aerated BOD dilution water.

Source: Swisher [155]. *(Reprinted with permission from Marcel Dekker, Inc.)*

Characteristics of Typical Tests. Table 9-13 lists some biodegradation tests that are commonly used for each of the four environments described above. Table 9-14 describes in greater detail each of these test methods for each environment. The tests were selected for the table on the basis of EPA recommendations under the Toxic Substances Control Act (TSCA) [162]. Additional information was obtained from reviews of biodegradation testing procedures [72,155].

  • *Surface Water.* Several tests are commonly used to estimate biodegradation in surface water. The TSCA guidelines [162] recommend the shake flask, CO_2 evolution, and BOD dilution tests. Howard *et al.* [72] described the river die-away and BOD respirometer tests. Many different seed sources can be used for the shake flask test, and more complete information about biodegradation can be obtained if both acclimated and unacclimated seed is used [72]. The shake flask test has better reproducibility than the river die-away. The BOD (with dilution technique) is used most frequently in testing surface water, but there are a number of problems with this method [72].

  • *Soils.* Three test methods are used to simulate the aerobic soil environment. The $^{14}CO_2$ evolution test recommended by the EPA un-

TABLE 9-12

Biodegradation Tests Recommended for Screening Organic Compounds

Recommended by	Test Methods	Ref.
Task Group on Methodological Criteria for Biodegradation	• Activated sludge method (batch and continuous) • River die-away	[132]
EPA under TSCA[a]	• Shake flask method • Activated sludge method • Methane and CO_2 production in anaerobic digestion • CO_2 evolution[b] • BOD method No soil tests recommended	[162]
Monsanto[a]	• River die-away • Semi-continuous activated sludge • CO_2 evolution[b] • Anaerobic	[47]

a. Recommended tests are meant for screening purposes; quantification of rates of disappearance applicable to environmental conditions requires radiolabeling or other direct techniques.
b. Measures ultimate biodegradation.

TABLE 9-13

Summary List of Standard Tests for Measuring Biodegradation

Aquatic	Soil	Anaerobic	Activated Sludge
• Shake Flask	• Soil perfusion	• Anaerobic digestion	• Semi-continuous activated sludge
• River die-away	• Soil incubation	• Closed river die-away	• Trickling filter
• BOD respirometer	• Soil suspended in aqueous solution		• Recirculating filter

der TSCA [162] is considered better than the aqueous solution or perfusion test [72]. The recommended test can be used for sediments, and the transformation products of the test compound can be quantified by thin layer chromatography of acetone extracts [141].

Most of the analytical techniques listed in Table 9-14 for the aqueous solution test method can be used for the other two test methods as well. The O_2 consumption and CO_2 production tests are usually not used for natural soils because of the high endogenous rates of soil respiration [72].

Tests for biodegradability in soil are affected by soil type and amount. Using a soil with a high proportion of organic matter should give higher degradation rates. On the other hand, it might produce a lower rate because cells and enzymes could be adsorbed onto the organic matter. Also, the organic matter present may be degraded by the microbial population in preference to an added substrate, delaying the rate of biodegradation for the compound. Therefore, the type (or degradability) of organic matter present, as well as the concentration, may be an important factor in biodegradation tests. In one study [84], the $^{14}CO_2$ evolved from five soil types receiving ^{14}C-carbaryl varied from 5% to 35% [72]. The high microbial content of most soils allows them to be used as microbial inoculum in degradation studies without adding nutrient amendment or other microorganisms, thus achieving a closer simulation of natural conditions.

*   *Anaerobic Soils or Sludge.* The EPA under TSCA [162] recommends the anaerobic digestion test to assess biodegradation potential in anaerobic sludge. Anaerobic soil and water conditions can be simulated by flooding natural soils or by preventing air from contacting a river die-away system.

*   *Activated Sludge Waste Treatment Plant.* The activated sludge test is most frequently used to simulate an activated sludge waste treatment plant. Acclimation of seed can be an unpredictable parameter in this test. Many attempts have been made to standardize it by using freeze-dried or air-dried sludge [72]. Temperature has a strong effect on the degree of biodegradation [72].

Although a BOD test could be used, it provides less insight into biodegradability under treatment plant conditions than does the continuous activated sludge or trickling filter test. The river die-away test might also be used, seeded with a much lower bacterial concentration than the activated sludge [72].

TABLE 9-14

Environment and Ref.	Test Method	Analytical Technique	Time for Test	Quantity or Conc. of Test Compound	Procedure (pH, nutrient, source, temp., culture)
Surface Water [72, 132, 155, 162]	River Die-Away	1. TLC (chromatography) 2. Spectrophotometry 3. Radiolabeling (^{14}C) 4. GC-MS 5. Colorimetry	Few days to 8 weeks	Varies according to test compound & analytical technique. Concentrations reported from 1-200 mg/L. [72]	Monitor disappearance of compound after it is placed in natural water sample and incubated until degradation ceases.
	BOD (Biochemical Oxygen Demand)	O_2 Dilution	1. 5 days 2. 10 days, or 3. long term (~42 days)	0.2 mg/mL to 4.8 mg/mL	Same as CO_2 Evolution Test with acclimated culture (13 days). Measure D.O. (dissolved oxygen).
	BOD Respirometer	1. Warburg differential manometer 2. O_2 electrolytic respirometer.	Varies	Varies according to test compound & analytical technique. Concentrations reported from 1-320 mg/L [72]	Measure O_2 consumption. High microorganism concentration required. Allows continual introduction of substrate and oxygen.
	Shake flask	Loss of DOC (Dissolved organic carbon)	13 day adaptation & 21 days of testing = 34 days	Relatively low (Supply 10 mg organic carbon per L of basal medium.)	Microorganisms inoculated in flasks with basal compound & test compound & aerated after 4 adaptive transfers; biodegradation is measured by reduction in DOC

Standard Laboratory Test Methods for Measuring Biodegradation

Results Indicate	Calculations and Information Recorded	Problems
Disappearance of parent compound over time.	Plot rate of parent compound disappearance.	Variation in bacterial count & composition of different rivers. Populations from industrial rivers may be acclimated. Small size of inoculum. [132]
O_2 uptake $>$60% of theoretical maximum suggests substantial degradation.	Subtract daily D.O from D.O. on day zero (= depletion value). Subtract depletion value of blanks. Multiply by inverse of dilution factor to get BOD. Use molecular structure of test compound to calculate O_2 needed to oxidize it to CO_2, H_2O, & inorganic molecules. $$\% \text{ theoretical} = 100\left(\frac{BOD_T - BOD_M}{BOD_T}\right)$$ where T = theoretical M = measured	1. No information on nature of degradation products. 2. O_2 also used to make new cell material etc., so unless extremes are noted (0 or 100%) cannot assume biodegradation. 3. Ignores possibility that normal O_2 is ↑ or ↓ by chemical means. 4. If outside carbon source is used, could confuse results. 5. O_2 could be consumed by nitrification & misrepresent results; so determine NO_3 formed. 6. All oxygen required for degradation must be dissolved in water at start of experiment, thus limiting concentration of substrate. 7. Not very accurate
Same as BOD	Same as BOD	See above (except #6 and #7). Also, CO_2 must be continually removed to prevent interference.
Degradation of compound, but not complete conversion to CO_2.	% removal of organic C at time t. Express DOC as mgC/L.	Loss of DOC could be due to cellular uptake, sorption, or loss by evaporation. Use of high concentration of microorganisms permits relatively short test period but makes conditions more favorable for degradation than those encountered in nature.

(Continued)

Environment and Ref.	Test Method	Analytical Technique	Time for Test	Quantity or Conc. of Test Compound	Procedure (pH, nutrient source, temp., culture)
Surface Water (Continued)	CO_2 Evolution	Evolution of CO_2	13 day acclimation + 1 day aeration + 28 day test = 42 days	5-10 mgC/L	In presence of O_2, microorganisms degrade organic compound to CO_2 & inorganic salts. Calculate theoretical maximum evolution of CO_2 & use of O_2 if all C atoms in compound are oxidized to CO_2. CO_2 evolution values >60% of theoretical max. indicate degradation.
Aerobic Soils [72]	1. Soil perfusion 2. Soil incubation 3. Soils suspended in aqueous solution	1. Chromatographic TLC 2. Spectrophotometric 3. Radiolabeling (^{14}C) 4. GC-MS 5. Colorimetry 6. Oxygen consumption methods usually not used because of high endogenous rates of soil respiration.	Continue until rate of disappearance levels off.	5-100 ppm	In general, obtain natural soil and mix with chemical with (Methods 1 and 3) or without (Method 2) addition of water. Many different apparatus used, usually liquid reservoir with soil column and tube to deliver solution and air. Stationary containers aerated for Method 1, not for Methods 2 and 3.

Results Indicate	Calculations and Information Recorded	Problems
Ultimate biodegradability potential. Use this test or BOD (this test preferred). CO_2 evolution >60% of theoretical maximum indicates substantial degradation.	mg CO_2 produced from substrate = $$\frac{(T_b - T_x)\ 220}{V}$$ where T_b = ml of 0.1 N HCl required to titrate aliquot from blank absorber; T_x = ml of 0.1 N HCl required to titrate aliquot from test compound absorber; V = ml of aliquot used in titration. Calculate % of theoretical CO_2 = $$\frac{100\ \sum_1^n [CO_2]}{110}$$ where $\sum [CO_2]$ = sum of CO_2 production values from the absorber samples taken on day 1 through last day (n).	See Table 9-11.
Disappearance of parent compound over time, either in soil (Method 2) or by a soil inoculum (Methods 1 and 3).	Plot rate of disappearance of parent compound.	1. Method 1 has high biodegradability potential so does not simulate most natural environments. 2. In Method 2 analytical techniques more difficult to use; additional extraction and clean-up steps required because of adsorption of chemical onto soil; less uniformity in distribution of compound in soil. 3. When soil used as medium (in Method 2), difficult to replicate results because of high degree of variability. 4. In Methods 1 and 3, moisture content too high to simulate natural soils. 5. Difficulty in handling multi-units of Method 1.

(Continued)

TABLE 9-14

Environment and Ref.	Test Method	Analytical Technique	Time for Test	Quantity or Conc. of Test Compound	Procedure (pH, nutrient, source, temp., culture)
Anaerobic Soils or Sludge or Aquatic [72, 155]	Anaerobic digestion	Compare production of methane & CO_2 by anaerobic bacteria in samples with & without test material.	At least: 3 days to equilibrate & 28 days to test = 31 days	10-200 mg/L	Obtain anaerobic sludge from municipal plant and allow to equilibrate. Put test compound in some containers. Periodically measure gas production and analyze for methane and CO_2 content. Methane production in units receiving test compound compared with controls will provide information on the biodegradability of the substrate under anaerobic conditions.
	Die-away	Same as aerobic river die-away test.	14-60 days (for surfactant).	10-100 ppm (for surfactant)	Initial dissolved O_2 is consumed by aerobic biooxidation processes and system becomes anaerobic; air prevented from contacting the river die-away type system. Different studies put surfactant with sewage in closed jars for 2 weeks, 40 days, & 60 days. Measure test compound left at end.
Activated Sludge Waste Treatment Plant [60, 72, 128, 155, 162]	1. One-batch die-away 2. Semi-continuous activated sludge 3. Trickling filter 4. Recirculating filter	1. GLC 2. Colorimetry 3. Radiolabeling 4. DOC removal	Maximum of: 1. See river die-away 2. 30 days acclimation + 19 days testing = 49 days 3. 4-8 weeks acclimation + 14 weeks to develop mature film 4. 7 days for recirculating filter	1. Moderate to heavy concentration test: 50-100 mg/L as DOC. 2. Low concentration test: 100 mg of compound as DOC. 3. High concentration test: 200 mg of compound as DOC.	After activated sludge has adapted to synthetic sewage & increasing concentrations of test compound, it is: 1. Like river die-away using sludge inoculum (see river die-away test). 2. Exposed to mineral salts medium plus compound in aerated chamber for up to 20 days. Process involves: aeration, settling of sludge, removal of supernatant liquor, filling with fresh sludge and substrate, repeat. Biodegradation is followed by comparing DOC at start with DOC mixed liquor on last day. 3. Passed once through packing medium, on which bacterial populations develop over time. Once population is established, solutes may be adsorbed onto film for long exposure times. 4. Similar to trickling, except that water recirculates throughout test procedures.

(Continued)

Results Indicate	Calculations and Information Recorded	Problems
Excess gas production in units receiving test compound (compared to control) may be related to anaerobic digestion of test compound. Excess CH_4 & CO_2 (as mg of C) produced is compared with theoretical maximum & % theoretical production can be calculated.	Record total gas as well as CH_4 & CO_2 content. % of theoretical $$= \frac{(G_T - G_M) \, 100}{G_T}$$ where G_T = total mg organic C in sample, G_M = mg of C in excess $(CH_4 + CO_2)$.	
Disappearance of parent compound over time.	Plot rate of disappearance of parent compound.	Same as aerobic die-away test.
Depending on analytical technique, either disappearance of parent compound over time or removal of organic carbon (DOC).	% removal of DOC during acclimation & test period or plot rate of disappearance of parent compound.	Difficult to maintain continuous circulation of sludge; long test time; fly nuisance; lack of easy accommodation in constant temp. room or bath; operational conditions are not readily adjusted; large amounts of substrate required; long acclimation period required for Methods 3 and 4.

The activated sludge test recommended by TSCA [162] is semi-continuous. Each cycle is a batch run on a particular unit of feed solution, but the cycle is repeated over and over with fresh feed, which provides an opportunity for acclimation and attainment of a "steady state" [155]. This semi-continuous process is also called fill-and-draw, because after aeration of sludge and feed solution, the sludge settles and the supernatant liquor is drawn off. Continuous systems generally require a much greater investment of time, space, and money than do semi-continuous systems [155].

Effect of Method and Analytical Technique on Measured Rates. In addition to the general variables affecting biodegradation discussed in §9-2, specific variables that characterize each test methodology influence the measured rate of biodegradation. The variables are related to the choice of:

- Chemical (concentration used, position of radiolabels);
- Microorganisms (source, concentration, acclimation time);
- Medium (amount of adsorbing soil or sediment);
- Procedure (pH, temperature, use of agitation); and
- Analytical technique.

The measured biodegradability of a given compound can vary significantly from one test method to another, because some tests may provide a better environment for biodegradation than others. Table 9-11 presents rates of degradation obtained for a chemical using different measuring techniques. Methods with optimum conditions for biodegradation support high microbial activity. Methods with the lowest potential usually have a low bacterial concentration in a synthetic medium (for example, the shake culture test). The higher bacterial concentration and thus high activity rates of the activated sludge test provide a higher potential for biodegradation. Even though bacterial concentrations in the river water test are relatively low, the use of naturally occurring water and microbial species often results in a high potential for biodegradation. Soil systems with unsaturated flow conditions exhibit the highest biodegradation potential [155].

Both continuous and semi-continuous systems can be used to simulate an activated sludge waste treatment plant. Biodegradation of surfactants has been shown to vary greatly in continuous systems, compared with the inherently more stable semi-continuous system. This variation is due to the wide variation from one sludge microbial culture to another and even within a single sludge culture at different times [155].

9-4 BIODEGRADATION RATE CONSTANTS

Derivation. Before the rate of biodegradation can be quantified and a rate constant can be calculated, a kinetic expression must be derived to describe the pattern of loss over time. Two general rate laws have been proposed to describe biodegradation: the power rate law and the hyperbolic rate law. Both are described below.

Depending on whether a chemical is degraded cometabolically or metabolically, is strongly adsorbed or not, is subject to competing reactions simultaneously, and other factors, different rate equations are applicable in deriving the rate constant [50]. One rate law may not adequately describe a chemical over its total degradation curve because of changes in its concentration-dependency and availability over time; in most cases, however, one rate order is assumed to be in effect over the entire biodegradation curve.

The *power rate law* states that the rate is proportional to some power of the substrate concentration [55]:

$$\frac{-d[C]}{dt} = k\,[C]^n \tag{9-2}$$

where

$$
\begin{aligned}
n &= \text{the order of the reaction} \\
[C] &= \text{concentration of substrate} \\
k &= \text{biodegradation rate constant}
\end{aligned}
$$

If first-order kinetics are assumed (i.e., $n = 1$), the rate is simply the product of the rate constant and the substrate concentration. The assumption of first order is most common in homogeneous media [55] or as a first approximation when the relationship between concentration and the variables affecting it are not understood. It can be used to calculate the half-life ($t_{1/2}$) of a chemical subjected to biodegradation ($t_{1/2} = 0.693/k$). Figure 9-13 depicts a typical first-order decay curve due to biodegradation; when the log of the concentration is plotted against time, the curve becomes a straight line.

At low pollutant concentrations, the assumption of first-order kinetics for biodegradation is reasonable [36,55]. For a system as variable and complex as soil, however, there are likely to be many exceptions to this assumption. The measured rate of disappearance of pollutants from soil under natural conditions is commonly lower than would be expected

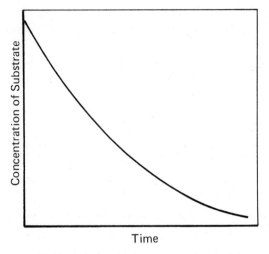

a. Measured Substrate Concentration vs. Time

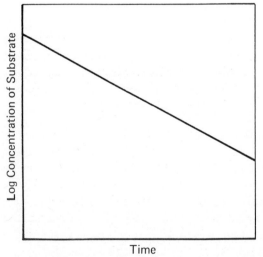

b. Logarithm of Substrate Concentration vs. Time

FIGURE 9—13   First-order Disappearance Curve of a Chemical

based on laboratory results [58,59]. This discrepancy may be partly due to:

- Lower availability of pollutant because of increasing adsorption over time;
- Changes in the microbial population over time; and
- Shut-down of reactive sites because of toxic effects of metabolic products [55].

Further investigation of the influence of these factors on the rate of degradation is needed before they can be expressed as terms in a rate equation.

The *hyperbolic rate law* is commonly used to quantify the growth of microbial populations. Based on Monod kinetics, this law expresses the rate as a hyperbolic saturation function of the substrate concentration [111,112]. Although the measured rate refers to population growth, it can be converted to a term to describe the disappearance of the substrate supporting the growth. The equation is a reasonable first approximation for biodegradation in aquatic systems [12] and in soil [55].

The Monod kinetics rate equation states that the growth rate of a single-species population of microorganisms on a single carbon substrate is dependent on the substrate concentration and, at higher concentrations, on the sum of concentration and other terms (comprising a single constant):

$$U = \frac{U_{max} [C]}{K_c + [C]} \tag{9-3}$$

where:

U = specific growth rate of microorganism
U_{max} = maximum growth rate of microorganism
$[C]$ = concentration of substrate
K_c = concentration of substrate in water supporting a half-maximum growth rate ($U_{max}/2$) (pseudoequilibrium constant).

Equation 9-3 is also applicable to mixed-species populations [101], and it can be modified [13] to a die-away expression through use of a yield coefficient describing the conversion efficiency of substrate to microorganism mass:

$$Y_d = -\frac{d[B]}{d[C]} \qquad (9\text{-}4)$$

where:

Y_d = yield coefficient
$[B]$ = microbial population concentration
$[C]$ = substrate concentration

The expression describing substrate disappearance is written:

$$\frac{-d[C]}{dT} = \frac{U_{max}\,[B]\,[C]}{Y_d\,(K_c + [C])} \qquad (9\text{-}5)$$

Further simplification is possible with the following assumptions:

- K_c values commonly range from 0.1 to 10 mg/L, which is higher than most environmental concentrations of substrates [13]; therefore, [C] in the denominator can be ignored.
- $U_{max}/Y_d K_c$ is equivalent to a second-order constant K [120].

The simplified form becomes a second-order rate expression,

$$\frac{-d[C]}{dt} = K[B]\,[C] \qquad (9\text{-}6)$$

which is a function of both population and substrate concentration. Figure 9-14 depicts the concentration dependence of a second-order decay rate, using the full form of Eq. 9-3.

The use of Monod kinetics to describe biodegradation rates requires that biodegradation be directly measurable in terms of all microbial growth that occurs during the course of the experiment. Although this may be true in cases where the substrate of concern is the sole source of energy or nutrient, the population increase may be partly dependent on other available substrates, which can be controlled in an experiment but not in the field. Larson discusses the following assumptions of Monod kinetics which may not be applicable to environmental conditions [92]:

- *The growth yield is a constant, equal to 50% of the substrate.* This assumption is not applicable to many of the dilute systems found in the environment, where a significant

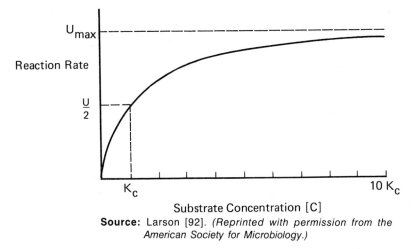

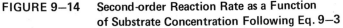

Source: Larson [92]. *(Reprinted with permission from the American Society for Microbiology.)*

FIGURE 9—14   Second-order Reaction Rate as a Function of Substrate Concentration Following Eq. 9—3

amount of the energy derived from the substrate may be used for maintenance rather than for growth.

- *The biomass of microorganisms can be accurately measured by plate counts.* Measurement can be very difficult in natural systems. Even at a substrate concentration of 10 mg/L, the density of microorganisms would reach no higher than 10^6 cells/mL, assuming a growth yield of 50% and dry cell weights of 10^{-12} to 10^{-13} g/cell.

The most accurate analytical technique for monitoring the disappearance of a chemical and collecting data for use in rate equations is one that measures the concentration over time directly and keeps track of the distribution and quantity of the degradation products. A method that employs carbon-14 labeling, while expensive, is the best because it allows subtraction of other losses from the biodegradation rate. Other direct analytical techniques can be used, however, as described in §9-3.

Rate Constants for Various Organic Compounds. The following section tabulates some biodegradation rate constants reported in the literature. These values were measured in laboratory experiments simulating aquatic, soil, activated sludge and anaerobic environments. The constants are both first- and second-order; Table 9-15 lists the units in which each rate order constant is expressed. For some chemicals, half-lives are reported if the authors stated that first-order kinetics were observed.

TABLE 9-15

Units for Biodegradation Rate Constants

Form Reported In	Rate Order	Definition
day^{-1}	1st	Per day
$t_{1/2} = ($) days	1st	$t_{1/2}$ is half-life, i.e., time required for 50% of chemical to be biodegraded
$mL\ (gVS)^{-1}\ day^{-1}$	2nd	Milliliters of substrate per gram of volatile solids (including microorganisms) per day
$mg\ (g\ sludge)^{-1}\ day^{-1}$	2nd	Milligrams of substrate per gram of sludge (dry weight) per day
$mg\ (g\ bacteria)^{-1}\ day^{-1}$	2nd	Milligrams of substrate per gram of bacteria (dry weight) per day
$mL\ (cell)^{-1}\ day^{-1}$	2nd	Milliliters of substrate per bacterial cell (determined by counts or estimation) per day
$mg\ COD\ (g\ biomass)^{-1}\ hr^{-1}$	2nd	Milligrams of COD per gram of initial biomass (dry weight) of inoculum per hour (for Table 9-22)

Although all rate constants describe the disappearance of the chemical over time, only results measured by direct analytical techniques or $^{14}CO_2$ evolution (in the case of soil) are presented. Information on test conditions (pH, temperature) is not given in this compilation, but the test method is stated wherever possible. An assortment of organic compounds is included, to represent the maximum number of chemical groups; for many chemicals, however, there were no data for certain conditions, and very few data were available for anaerobic systems.

Tables 9-16 through 9-19 are not intended to be a compilation of all the rate constants cited in the literature. At best, they provide an assortment of rate constants representative of each environment and illustrate typical ranges of constants.

Extrapolation of Laboratory Results to Field Conditions. The use of laboratory-derived rate constants to predict the persistence of a chemical in the environment must be done cautiously. As discussed in §9-2, many

TABLE 9-16

Biodegradation Rate Constants for Organic Compounds in Aquatic Systems

Compound	Rate Constant[a]	See Note	Ref.
Anthracene	0.007-0.055 day^{-1}	b	[46]
Atrazine (N-phosphorylated)	t$_{1/2}$ 3.21 days	c	[39]
Benzo[a] anthracene	0		[146]
Benzene	0.11 day^{-1}		[93]
Benzo[a] pyrene	0		[146]
Benzo[f] quinoline	8.6 x 10^{-7} mL(cell)$^{-1}$ day^{-1}		[146]
Bis(2-ethylhexyl)phthalate	1.0 x 10^{-9} mL(cell)$^{-1}$ day^{-1}		[168]
Carbaryl	2.4 x 10^{-10} mL(cell)$^{-1}$ day^{-1}		[167]
Carbazole (9H)	1.2 x 10^{1} mL(cell)$^{-1}$ day^{-1}		[146]
Chlorobenzene	0.0045 day^{-1}		[93]
	0.0092 day^{-1}	d	[93]
Chlorodiphenyl oxide	7.2 x 10^{-3} mL(gVS)$^{-1}$ day^{-1}		[20]
p-Chlorophenol	0.23 day^{-1}	d	[93]
Chlorpropham	1.6-1.8 x 10^{-8} mL(cell)$^{-1}$ day^{-1}	e	[149]
	3.6-6.7 x 10^{-10} mL(cell)$^{-1}$ day^{-1}	e,f	[123]
Crotoxyphos	t$_{1/2}$ = 7.5 days (pH 9)		[86]
	= 22.5 days (pH 2)		[86]
2,4-D (Butoxyethyl ester)	6.24-24.0 x 10^{-6} mL(cell)$^{-1}$ day^{-1}	e,f	[123]
	6.2 x 10^{-5} mL(cell)$^{-1}$ day^{-1}	g	[121]
	9.6 x 10^{-7} mL(cell)$^{-1}$ day^{-1}	h	[121]
p,p'-DDE	0.0006 day^{-1}	d	[93]
Diazinon	t$_{1/2}$ = 4.91 days (pH 3.1)		[48]
	= 185 days (pH 7.4)		[48]
Diazoxon	t$_{1/2}$ = 0.016 days (pH 3.1)		[48]
	= 27.9 days (pH 7.4)		[48]
Dibenzo[c,g] carbazole	0		[146]
Dibenzothiophene	1.27 x 10^{-5} mL(cell)$^{-1}$ day^{-1}		[146]
Dimethyl phthalate	1.2 x 10^{-4} mL(cell)$^{-1}$ day^{-1}		[168]
Di-n-butyl phthalate	7.4 x 10^{-7} mL(cell)$^{-1}$ day^{-1}		[149]
Di-n-octyl phthalate	7.4 x 10^{-9} mL(cell)$^{-1}$ day^{-1}		[168]

(Continued)

TABLE 9-16 (Continued)

Compound	Rate Constant[a]	See Note	Ref.
Galactose	$1.2\text{-}10 \times 10^3$ mg(g bacteria)$^{-1}$ day^{-1}	i	[153]
	1.4×10^3 mg(g bacteria)$^{-1}$ day^{-1}	j	[153]
Glucose	0.24 day^{-1}		[92]
	$1.1\text{-}1.6 \times 10^4$ mg(g bacteria)$^{-1}$ day^{-1}	i	[153]
	5.2×10^3 mg(g bacteria)$^{-1}$ day^{-1}	j	[153]
Hexachlorophene	0.0024 day^{-1}	d	[93]
Malathion	$2.6\text{-}16.1 \times 10^{-7}$ mL(cell)$^{-1}$ day^{-1}	e,f	[123]
	6.2×10^{-8} mL(cell)$^{-1}$ day^{-1}		[120]
	5.0×10^{-8} mL(cell)$^{-1}$ day^{-1}		[12]
	1.9×10^{-1} mg(g fungi)$^{-1}$ day^{-1}		[94]
Methyl anisate	1.3×10^{-8} mL(cell)$^{-1}$ day^{-1}		[168]
Methyl benzoate	1.7×10^{-8} mL(cell)$^{-1}$ day^{-1}		[168]
Mirex	0		[146]
Nitrilotriacetate (NTA)	0.05-0.23 day^{-1}	k	[92]
Parathion	$t_{1/2}$ = >4250 days		[166]
Paraoxon	$t_{1/2}$ = >4250 days		[166]
p-Cresol	1.24×10^{-5} mL(cell)$^{-1}$ day^{-1}		[146]
Phenol	0.079 day^{-1}		[93]
Propham (IPC)	0.003-2.1 mg(g bacteria)$^{-1}$ day^{-1}		[167]
Quinoline	7.4×10^{-5} mL(cell)$^{-1}$ day^{-1}		[146]
Triallate	$t_{1/2} \approx$ 680 days 0pH 6,8)		[145]
	$\approx$ 1170 days (pH 7)		[145]
2,4,5-T	0.001 day^{-1}		[93]
	0.01-0.03 day^{-1}	d,l	[93]
1,4,5-Trichlorophenoxy-	0.0005 day^{-1}		[93]
acetic acid	0.0012-0.012 day^{-1}	d,l	[93]

a. All tests assumed to be river die-away.
b. First value is mean for days 0-15; second is for days 20-65.
c. First-order half-life in aqueous solution.
d. In sediment (slurry).
e. Range due to measurement in different samples of river water.
f. Rate constant does not account for lag period.
g. Degradation by yeast culture (*Rhodotorula glutinis*).
h. Degradation by bacterial culture (*Bacillus subtilus*).
i. First value from unacclimated microbial population, second from acclimated population.
j. River water bacterial culture.
k. Dissolved concentrations ranging from 0.2 mg/L to saturation.
l. Temperature range 9-21°C.

TABLE 9-17

Biodegradation Rate Constants
for Organic Compounds in Soil[a]
(day^{-1})

Compound	Test Method	
	Die-Away	$^{14}CO_2$ Evolution
Aldrin, Dieldrin	0.013	
Atrazine	0.019	0.0001
Bromacil	0.0077	0.0024
Carbaryl	0.037	0.0063
Carbofuran	0.047	0.0013
Dalapon	0.047	
DDT	0.00013	
Diazinon	0.023	0.022
Dicamba	0.022	0.0022
Diphenamid	0.123[b]	
Fonofos	0.012	
Glyphosate	0.1	0.0086
Heptachlor	0.011	
Lindane	0.0026	
Linuron	0.0096	
Malathion	1.4	
Methyl parathion	0.16	
Paraquat	0.0016	
Parathion	0.029	
Phorate	0.0084	
Picloram	0.0073	0.0008
Simazine	0.014	
TCA	0.059	
Terbacil	0.015	0.0045
Trifluralin	0.008	0.0013
2,4-D	0.066	0.051
2,4,5-T	0.035	0.029

a. All constants are from soil incubation studies. Except where noted, source is Rao and Davidson [131], a compilation of first-order rate constants derived from data published from other studies.

b. Optimum degradation rate, from Donigan *et al.* [36]. Test method not specified.

TABLE 9-18

Biodegradation Rate Constants
for Organic Compounds in Anaerobic Systems
(day^{-1})

Compound	In Soil[a]		In Sewage Sludge[b]
	Die-Away	$^{14}CO_2$ Evolution	
Carbofuran	0.026		
DDT	0.0035		
Endrin	0.03		
Lindane		0.0046	
PCP		0.07	
Trifluralin	0.025		
Mirex			0.0192
Methoxychlor			9.6
2,3,5,6-Tetrachlorobenzene			12.72
Bifenox			6.27

a. Flooded soil incubation studies as reported in Rao and Davidson [131], a compilation of first-order rate constants derived from data published from other sources.

b. As reported by Geer [45]. Test method not specified.

variables influence biodegradation rates. In a laboratory experiment, most of the variables are controlled, and results derived under the same conditions can be compared; natural habitats, on the other hand, have numerous unpredictable elements, and at least one of these elements is likely to cause the biodegradation rate to differ from the value obtained in the laboratory.

Besides the differences in the control of variables between laboratory and field conditions, certain basic, unavoidable differences caused by the constraints of the laboratory further complicate the extrapolation process:

- The microbial population isolated for an experiment cannot truly reflect the diversity of the environment it represents.

- To save time in the laboratory, experimental nutrient conditions are often better than those found in the environment. Organic matter concentrations are commonly 1-10 g/L in culture media but only 1-10 mg/L in nature [21]

- High substrate and microbial concentrations must be used in most experiments to generate quick results.

TABLE 9-19

Rate Constants for Biodegradation of Organic Compounds by Activated Sludge Cultures

Compound	Rate Constant	Reference
Chlorodiphenyl oxide	8.9×10^2 mL (gVS)$^{-1}$ day^{-1}	[20]
Linear alkyl benzene sulfonate (LAS)	0.10 day^{-1}	[92]
Glucose	0.20 day^{-1}	[92]
	0.36 day^{-1} [a]	[153]
	6.6×10^3 mg (g bacteria)$^{-1}$ day^{-1} [a]	[153]
	1.7-9.1 $\times 10^3$ mg (g sludge)$^{-1}$ day^{-1}	[119]
Galactose	2.6×10^3 mg (g bacteria)$^{-1}$ day^{-1} [a]	[153]
Fructose	1.6-4.4 $\times 10^3$ mg (g sludge)$^{-1}$ day^{-1}	[119]
Sucrose	3.8-16.8 $\times 10^3$ mg (g sludge)$^{-1}$ day^{-1}	[119]
2,4-D	6.9×10^{-2} mL (g bacteria)$^{-1}$ day^{-1}	[64]

a. Specific substrate utilization rate.

Biodegradation rate constants have several applications. One is the comparison of disappearance rates for a series of compounds; another is the comparison with rates measured for other loss processes, such as hydrolysis, for the same chemical. In situations where the conditions of a specific habitat have quantified or are well understood and their effect has been observed in the laboratory, a meaningful extrapolation is possible. Ideally, investigations will continue from this point, analyzing the persistence of a chemical in field conditions under various climatic and habitat regimes, such as was done by Hamaker *et al.* [56] for picloram in soil.

9-5 ESTIMATION OF BIODEGRADATION RATES

Two general methods of estimation are covered in this section:

(1) Rules of thumb for obtaining a qualitative and relative estimate of biodegradation based on structural factors and on chemical class (Table 9-20). These generalizations are applicable only to the specific groups of chemicals in which

TABLE 9-20

RULES OF THUMB FOR BIODEGRADABILITY

Factors	Schematic Example[a]

Branching — Highly branched compounds are more resistant to biodegradation.

Branching

1) Unbranched side chains on phenolic and phenoxy compounds are more easily metabolized than branch alkyl moieties [164].

$$CH_3-CH_2-CH_2-CH_2-CH_3 \; > \; \underset{CH_3}{\underset{|}{CH}}-CH_2-\underset{CH_2CH_3}{\underset{|}{\overset{CH_3}{\overset{|}{C}}}}-CH_3$$

2) 2,4-Dichlorophenoxyalkanates with side chains of 4 or more carbons degraded easily, the propionate more slowly, and the dichlorophenoxyacetate not at all by a *Flavobacterium* sp. [99].

3) Branched alkyl benzene sulfonates degrade more slowly than straight-chain [156].

(phenol with $CH_2CH_2CH_3$) > (phenol with $CH(CH_3)CH_3$)

Chain Length — Short chains are not as quickly degraded as long chains.

Chain Length

1) Rate of oxidation of straight-chain aliphatic hydrocarbons is correlated to length of chain [89].

$$CH_3-CH_2-CH_2-CH_2-CH_2-CH_2-CH_3 \; > \; CH_3-CH_2-CH_2-CH_3$$

2) Soil micros attack long-chain mononuclear aromatics faster than short-chain [152].

3) Micros grow on normal alkanes from n-octane to n-eicosane but not on n-heptane to methane [43].

4) Sulfate-reducing bacteria more rapidly degrade long-length carbon chains (decane to hentriacontane) than short-length carbon chains [22].

5) ABS detergents increase in degradability with increase in chain length from C_6 to C_{12} but not $> C_{12}$ [73, 156].

6) Rate of mineralization of N in urea-formaldehyde complexes declines with increasing ureaform chain [96].

Oxidation — Highly oxidized compounds, like halogenated compounds, may resist further oxidation under aerobic conditions but may be more rapidly degraded under anaerobic conditions [50, 55, 63, 67].

Polarity

$$CH_3-\underset{O}{\overset{\parallel}{C}}-O-NO_2 \; > \; CH_3-\underset{O}{\overset{\parallel}{C}}-O-CH_3$$

Polarity and Ionization — Non-ionic compounds with active halogens are likely to be degraded by nucleophilic displacement reactions like hydrolysis. Same is true of linkages separating highly polar groups [50].

Saturation

$$CH_2=CH_2 \; > \; CH_3-CH_3$$

Saturation — Unsaturated aliphatics are more readily attacked than corresponding saturated hydrocarbons, perhaps because of presence of many ethylene-reducing enzyme systems and few ethane ones.

(Continued)

Factors	Schematic Example[a]
Substituents (Number of) on Simple Organic Molecules	**Substituents (Number of)**
1) Alcohols, aldehydes, acids, esters, amides, and amino acids are more susceptible to biodegradation than the corresponding alkanes, olefins, ketones, dicarboxylic acids, nitriles, amines, and chloroalkanes [118].	(1) $CH_2OH - CH_2 - CH_3$ > $CH_3 - CH_2 - CH_3$ $NH_2 - CH_2 - COOH$ > $NH_2 - CH_3$ $CH_3 - CH$ > $CH_3 - C - CH_3$ $\quad\quad \| O$ $\quad\quad \| O$
2) Increased substitution hinders oxidation responsible for breakdown of alkyl chains [57].	(2) $CH_3 - CH_2 - CH_2 - CH_2 - R$ > $CH_3 - CH - CH - CH - R$ $\quad\quad\quad\quad OH \quad Cl \quad NH_2$
3) No significant oxidation of polycyclic aromatic hydrocarbons containing more than three rings [104].	(3) 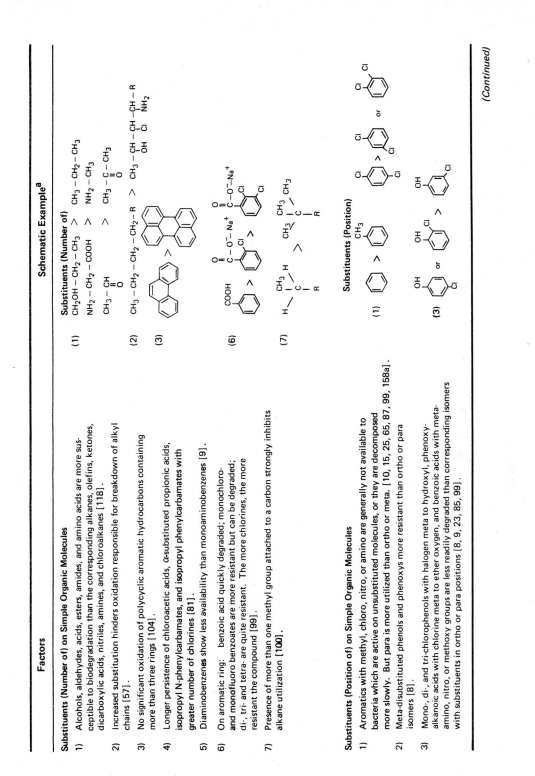
4) Longer persistence of chloroacetic acids, α-substituted propionic acids, isopropyl N-phenylcarbamates, and isopropyl phenylcarbamates with greater number of chlorines [81].	
5) Diaminobenzenes show less availability than monoaminobenzenes [9].	
6) On aromatic ring: benzoic acid quickly degraded; monochloro- and monofluoro benzoates are more resistant but can be degraded; di-, tri- and tetra- are quite resistant. The more chlorines, the more resistant the compound [99].	(6)
7) Presence of more than one methyl group attached to a carbon strongly inhibits alkane utilization [100].	(7)
Substituents (Position of) on Simple Organic Molecules	**Substituents (Position)**
1) Aromatics with methyl, chloro, nitro, or amino are generally not available to bacteria which are active on unsubstituted molecules, or they are decomposed more slowly. But para is more utilized than ortho or meta. [10, 15, 25, 65, 87, 99, 158a].	(1)
2) Meta-disubstituted phenols and phenoxys more resistant than ortho or para isomers [8].	
3) Mono-, di-, and tri-chlorophenols with halogen meta to hydroxyl, phenoxyalkanoic acids with chlorine meta to ether oxygen, and benzoic acids with meta-amino, nitro, or methoxy groups are less readily degraded than corresponding isomers with substituents in ortho or para positions [8, 9, 23, 85, 99].	(3)

(Continued)

TABLE 9-20 (Continued)

Factors	Schematic Example[a]
Substituents (Position of) (Cont.)	
4) On the other hand, ortho isomers of nitrophenols, methylanilines, sulfonates of 1-phenyldodecane, and chlorine-containing isopropyl phenylcarbonates are most persistent [9, 81, 157].	(4) OH-ring with OH / OH, NO_2 (para) or OH, NO_2 (meta) $>$ OH, NO_2 (ortho)
5) In fatty acids, introduction of halogen or phenyl group on alpha carbon reduces rate of degradation as opposed to same group on omega carbon [23, 34].	(5) $\text{CH}_2(\text{CH}_2)_5\text{COOH}$ (phenyl) $>$ $\text{CH}_3(\text{CH}_2)_4\text{CHCOOH}$ (phenyl)
6) Parahydroxybenzoate degrades more rapidly than ortho or meta (fewer micros degrade these) [148].	
7) For ABS, para sulfonates are more readily degraded than ortho sulfonates of phenyldodecane and phenyltetradecane. In diheptylbenzene sulfonates the meta is more susceptible than the para substituent [156, 157].	
8) Neo-pentyl group addition inhibits alkane utilization if carbon atom bonded is next to last on the chain [100].	(8) $\text{CH}_3\text{CH}_2\text{CHCH}_2\text{CH}_2-\text{R}$ with $\text{CH}_2\text{C(CH}_3)_2\text{CH}_3$ $>$ $\text{CH}_3\text{CHCHCH}_2\text{CH}_2\text{CH}_2-\text{R}$ with $\text{CH}_2\text{C(CH}_3)_2\text{CH}_3$
Substituents (Type of) on Simple Organic Molecules	**Substituent (Type)**
1) Mono- and dicarboxylic acids, aliphatic alcohols, and ABS are decreasingly degraded when hydrogen is replaced by methyl groups [34, 57, 155].	(1) $\text{CH}_3-\text{CH}_2-\text{COOH}$ $>$ $\text{CH}_3-\overset{\text{CH}_3}{\underset{\text{CH}_3}{\text{C}}}-\text{COOH}$
2) Aliphatic acids are less easily degraded when chlorine replaces a hydrogen [34].	(2) $\text{Cl}-\text{CH}-\overset{}{\underset{\text{O}}{\text{C}}}-\text{OH}$ (Cl) $>$ $\text{Cl}-\overset{\text{Cl}}{\underset{\text{Cl}}{\text{C}}}-\overset{}{\underset{\text{O}}{\text{C}}}-\text{OH}$
3) Triazines or methoxychlor is less easily degraded when methoxy groups are replaced by chlorines [61].	(3) $\text{CH}_3-\text{O}-$ ring $-\text{CH}-$ ring $-\text{O}-\text{CH}_3$ with $\text{Cl}-\text{C}-\text{Cl, Cl}$ $>$ $\text{Cl}-$ ring $-\text{CH}-$ ring $-\text{O}-\text{CH}_3$ with $\text{Cl}-\text{C}-\text{Cl, Cl}$
4) Degradation of disubstituted benzenes is less when carboxyl or hydroxyl is replaced by nitro, sulfonate or chloro group [9].	(4) OH/OH ring $>$ OH/Cl ring

(Continued)

Factors	Schematic Example[a]
Substituents (Type of) (Cont.)	
5) Successive replacement of hydroxyls of cyanuric acid with amino groups makes compounds less degradable [61].	
6) For naphthalene compounds, nuclei bearing single small alkyl groups (methyl, ethyl or vinyl) oxidize at more rapid rate than those with a phenyl substituent [104].	(6) 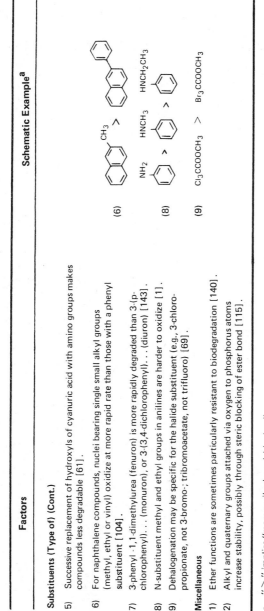
7) 3-phenyl -1,1-dimethylurea (fenuron) is more rapidly degraded than 3-(p-chlorophenyl)... (monuron), or 3-(3,4-dichlorophenyl)... (diuron) [143].	
8) N-substituent methyl and ethyl groups in anilines are harder to oxidize [1].	
9) Dehalogenation may be specific for the halide substituent (e.g., 3-chloro-propionate, not 3-bromo-; tribromoacetate, not trifluoro) [69].	
Miscellaneous	
1) Ether functions are sometimes particularly resistant to biodegradation [140].	
2) Alkyl and quaternary groups attached via oxygen to phosphorus atoms increase stability, possibly through steric blocking of ester bond [115].	

a. ">" implies "more easily degradable than"

they have been observed. They are also dependent on test method, on the species responsible for biodegradation, on the definition of biodegradation, and other variables.

(2) Correlations observed between the biodegradability of certain chemical groups and fundamental properties that have been investigated in a relatively systematic way.

No attempt is made here to correlate the results of these estimation techniques with biodegradation rate constants measured under standard conditions. The reason is the lack of a data base to support quantification of these relationships.

None of the following estimation techniques is recommended for use in predicting biodegradation rates, because (1) it is not consistently valid, being based on gross assumptions (such as that BOD represents biodegradation), or (2) it has not been tested for more than a few chemicals or chemical groups.

Some empirical relationships between biodegradability and molecular characteristics are listed in Table 9-20. Correlations with other chemical properties are described below.

Solubility. Water-insoluble compounds are thought to persist longer than those that are water-soluble [1,3]. Little quantitative work has been done on microbial degradation of the former compounds, but Alexander [3] suggests the following possible reasons for this behavior:

(1) Inability of the compound to reach the reaction site in the microbial cell;

(2) A reduced rate of reaction when biodegradation is regulated by the rate of solubilization; and

(3) The inaccessibility of insoluble compounds because of increased adsorption or trapping in inert material due to insolubility.

A correlation between solubility and the biodegradability index (B.I.) was found [80] for DDT analogs in mosquito fish (*Gambusia affinis*). However, as the focus of this chapter is on microbial biodegradation, this relationship will not be discussed here.

BOD/COD. Several methods have been proposed for estimating the biodegradability of organic compounds, two of which are described below. The limitations of such approaches have been described in the discussion of analytical techniques in §9-3; briefly stated, the concepts of biochemical oxygen demand (BOD) and chemical oxygen demand (COD) assume that all carbon is assimilated into new biomass and that the transformation from substrate to biomass is not inhibited by the compound under investigation or by any other substances in the test medium. COD and BOD may involve different sites of reaction and degrees of reactivity. Moreover, the reactivity of a site in COD depends on the reagent used. The results, with the exception of Pitter's approach [128], provide only an index of degradability, not the rate of degradation.

Several approaches for estimating biodegradability of organic compounds are based on the ratio between BOD and either COD or UOD (ultimate oxygen demand). Lyman [97] classified a number of chemicals into categories of biodegradability based on the BOD/COD ratio. Compounds with a ratio less than 0.01 were classified as relatively undegradable, between 0.01 and 0.1 as moderately degradable, and greater than 0.1 as relatively degradable. Table 9-21 presents BOD_5/COD ratios for various compounds.[2]

In Czechoslovakia, the Department of Water Technology and Environmental Engineering has developed a standard test, based on measurement of COD decrease, to compare the biodegradability of organic compounds [128]. Using activated sludge inoculum with 20 days of adaptation to the substrate, the decrease in the COD of a substance is calculated until no further decrease is observed. The percentage decrease of total COD is calculated as well as the rate, expressed in units of mg COD removed per gram of initial biomass (dry weight) of inoculum per hour. Pitter [128] tabulated these data for 94 aromatic, 15 cycloaliphatic, and 14 aliphatic compounds. Removal rates ranged over two orders of magnitude from 0 (e.g., nitroanilines and dinitrobenzenes) to 180 mg COD/g-hr (for glucose). Table 9-22 lists percentage removal and rate of biodegradation for 123 organic compounds. Pitter considered rates greater than 15 mg COD/g-hr as "readily decomposable" (defined as 90% of initial COD removed in 120 hours of incubation). Rates may be overestimated for some compounds, especially aromatics, due to a lack of control for volatilization.

The refractory index (RI) was developed to indicate the degradability of organic compounds [14,60] and to predict the persistence of a compound and its degradation products after discharge into receiving waters. It is the ratio of ultimate biochemical oxygen demand (BOD_u) to

---

2. BOD_5 refers to the results of a 5-day BOD test.

TABLE 9-21

BOD$_5$/COD Ratios for Various Organic Compounds[a]

Compound	Ratio	Compound	Ratio
Relatively Undegradable		**Moderately Degradable (cont'd.)**	
Butane	~ 0	Mineral spirits	~ 0.02
Butylene	~ 0	Cyclohexanol	0.03
Carbon tetrachloride	~ 0	Acrylonitrile	0.031
Chloroform	~ 0	Nonanol	> 0.033
1,4-Dioxane	~ 0	Undecanol	≤ 0.04
Ethane	~ 0	Methylethylpyridine	0.04-0.75
Heptane	~ 0	1-Hexene	< 0.044
Hexane	~ 0	Methyl isobutyl ketone	≤ 0.044
Isobutane	~ 0	Diethanolamine	≤ 0.049
Isobutylene	~ 0	Formic acid	0.05
Liquefied natural gas	~ 0	Styrene	> 0.06
Liquefied petroleum gas	~ 0	Heptanol	≤ 0.07
Methane	~ 0	sec-Butyl acetate	0.07-0.23
Methyl bromide	~ 0	n-Butyl acetate	0.07-0.24
Methyl chloride	~ 0	Methyl alcohol	0.07-0.73
Monochlorodifluoromethane	~ 0	Acetonitrile	0.079
Nitrobenzene	~ 0	Ethylene glycol	0.081
Propane	~ 0	Ethylene glycol monoethyl ether	< 0.09
Propylene	~ 0	Sodium cyanide	≤ 0.09
Propylene oxide	~ 0	Linear alcohols (12-15 carbons)	> 0.09
Tetrachloroethylene	~ 0	Allyl alcohol	0.091
Tetrahydronaphthalene	~ 0	Dodecanol	0.097
1-Pentene	< 0.002	**Relatively Degradable**	
Ethylene dichloride	0.002	Valeraldehyde	≤ 0.10
1-Octene	> 0.003	n-Decyl alcohol	> 0.10
Morpholine	≤ 0.004	p-Xylene	< 0.11
Ethylenediaminetetracetic acid	0.005	Urea	0.11
Triethanolamine	≤ 0.006	Toluene	< 0.12
o-Xylene	< 0.008	Potassium cyanide	0.12
m-Xylene	< 0.008	Isopropyl acetate	≤ 0.13
Ethylbenzene	< 0.009	Amyl acetate	0.13-0.34
Moderately Degradable		Chlorobenzene	0.15
Ethyl ether	0.012	Jet fuels (various)	~ 0.15
Sodium alkylbenzenesulfonates	~ 0.017	Kerosene	~ 0.15
Monoisopropanolamine	≤ 0.02	Range oil	~ 0.15
Gas oil (cracked)	~ 0.02	Glycerine	≤ 0.16
Gasolines (various)	~ 0.02	Adiponitrile	0.17

Compound	Ratio	Compound	Ratio
Relatively Degradable (cont'd.)		**Relatively Degradable (cont'd.)**	
Furfural	0.17-0.46	Ethyleneimine	0.46
2-Ethyl-3-propylacrolein	< 0.19	Monoethanolamine	0.46
Methylethylpyridine	< 0.20	Pyridine	0.46-0.58
Vinyl acetate	< 0.20	Dimethylformamide	0.48
Diethylene glycol		Dextrose solution	0.50
monomethyl ether	≤ 0.20	Corn syrup	~ 0.50
Naphthalene (molten)	≤ 0.20	Maleic anhydride	≥ 0.51
Dibutyl phthalate	0.20	Propionic acid	0.52
Hexanol	~ 0.20	Acetone	0.55
Soybean oil	~ 0.20	Aniline	0.56
Paraformaldehyde	0.20	Isopropyl alcohol	0.56
n-Propyl alcohol	0.20-0.63	n-Amyl alcohol	0.57
Methyl methacrylate	< 0.24	Isoamyl alcohol	0.57
Acrylic acid	0.26	Cresols	0.57-0.68
Sodium alkyl sulfates	~ 0.30	Crotonaldehyde	< 0.58
Triethylene glycol	0.31	Phthalic anhydride	0.58
Acetic acid	0.31-0.37	Benzaldehyde	0.62
Acetic anhydride	≥ 0.32	Isobutyl alcohol	0.63
Ethylenediamine	≤ 0.35	2,4-Dichlorophenol	0.78
Formaldehyde solution	0.35	Tallow	~ 0.80
Ethyl acetate	≤ 0.36	Phenol	0.81
Octanol	0.37	Benzoic acid	0.84
Sorbitol	≤ 0.38	Carbolic acid	0.84
Benzene	< 0.39	Methyl ethyl ketone	0.88
n-Butyl alcohol	0.42-0.74	Benzoyl chloride	0.94
Propionaldehyde	< 0.43	Hydrazine	1.0
n-Butyraldehyde	≤ 0.43	Oxalic acid	1.1

a. BOD_5 values were not measured under the same conditions for all chemicals.

Source: Lyman [97].

TABLE 9-22

COD Removal and Rate of Removal for Various Compounds

Compound	Percent Removed[a] (based upon COD)	Average Rate of Biodegradation (mg COD g^{-1} hr^{-1})
Aliphatic Compounds		
Ammonium oxalate	92.5	9.3
n-Butanol	98.8	84.0
sec-Butanol	98.5	55.0
tert-Butanol	98.5	30.0
1,4-Butanediol	98.7	40.0
Diethylene glycol	95.0	13.7
Diethanolamine	97.0	19.5
Ethylene diamine	97.5	9.8
Ethylene glycol	96.8	41.7
Glycerol	98.7	85.0
Glucose	98.5	180.0
n-Propanol	98.8	71.0
Isopropanol	99.0	52.0
Triethylene glycol	97.7	27.5
Cycloaliphatic Compounds		
Borneol	90.3	8.9
Caprolactam	94.3	16.0
Cyclohexanol	96.0	28.0
Cyclopentanol	97.0	55.0
Cyclohexanone	96.0	30.0
Cyclopentanone	95.4	57.0
Cyclohexanolone	92.4	51.5
1,2-Cyclohexanediol	95.0	66.0
Dimethylcyclohexanol	92.3	21.6
4-Methylcyclohexanol	94.0	40.0
4-Methylcyclohexanone	96.7	62.5
Menthol	95.1	17.7
Tetrahydrofurfuryl alcohol	96.1	40.0
Tetrahydrophthalimide	0	—
Tetrahydrophthalic acid	0	—
Aromatic Compounds		
Aniline	94.5	19.0
Aminophenolsulfonic acid	64.6	7.1
Acetanilide	94.5	14.7

Compound	Percent Removed[a] (based upon COD)	Average Rate of Biodegradation (mg COD g^{-1} hr^{-1})
Aromatic Compounds (cont'd.)		
p-Aminoacetanilide	93.0	11.3
o-Aminotoluene	97.7	15.1
m-Aminotoluene	97.7	30.0
p-Aminotoluene	97.7	20.0
o-Aminobenzoic acid	97.5	27.1
m-Aminobenzoic acid	97.5	7.0
p-Aminobenzoic acid	96.2	12.5
o-Aminophenol	95.0	21.1
m-Aminophenol	90.5	10.6
p-Aminophenol	87.0	16.7
Benzenesulfonic acid	98.5	10.6
m-Benzenedisulfonic acid	63.5	3.4
Benzaldehyde	99.0	119.0
Benzoic acid	99.0	88.5
o-Cresol	95.0	54.0
m-Cresol	95.5	55.0
p-Cresol	96.0	55.0
d-Chloramphenicol	86.2	3.3
o-Chlorophenol	95.6	25.0
p-Chlorophenol	96.0	11.0
o-Chloroaniline	98.0	16.7
m-Chloroaniline	97.2	6.2
p-Chloroaniline	96.5	5.7
2-Chloro-4-nitrophenol	71.5	5.3
2,4-Dichlorophenol	98.0	10.5
1,3-Dinitrobenzene	0	—
1,4-Dinitrobenzene	0	—
2,3-Dimethylphenol	95.5	35.0
2,4-Dimethylphenol	94.5	28.2
3,4-Dimethylphenol	97.5	13.4
3,5-Dimethylphenol	89.3	11.1
2,5-Dimethylphenol	94.5	10.6
2,6-Dimethylphenol	94.3	9.0
3,4-Dimethylaniline	76.0	30.0
2,3-Dimethylaniline	96.5	12.7
2,5-Dimethylaniline	96.5	3.6
2,4-Diaminophenol	83.0	12.0
2,4-Dinitrophenol	85.0	6.0

(continued)

TABLE 9-22 (Continued)

Compound	Percent Removed[a] (based upon COD)	Average Rate of Biodegradation (mg COD g^{-1} hr^{-1})
Aromatic Compounds (cont'd.)		
3,5-Dinitrobenzoic acid	50.0	—
3,5-Dinitrosalicylic acid	0	—
Furfuryl alcohol	97.3	41.0
Furfurylaldehyde	96.3	37.0
Gallic acid	90.5	20.0
Gentisic acid	97.6	80.0
p-Hydroxybenzoic acid	98.7	100.0
Hydroquinone	90.0	54.2
Isophthalic acid	95.0	76.0
Metol	59.4	0.8
Naphthoic acid	90.2	15.5
1-Naphthol	92.1	38.4
1-Naphthylamine	0	0
1-Naphthalenesulfonic acid	90.5	18.0
1-Naphthol-2-sulfonic acid	91.0	18.0
1-Naphthylamine-6-sulfonic acid	0	0
2-Naphthol	89.0	39.2
p-Nitroacetophenone	98.8	5.2
Nitrobenzene	98.0	14.0
o-Nitrophenol	97.0	14.0
m-Nitrophenol	95.0	17.5
p-Nitrophenol	95.0	17.5
o-Nitrotoluene	98.0	32.5
m-Nitrotoluene	98.5	21.0
p-Nitrotoluene	98.0	32.5
o-Nitrobenzaldehyde	97.0	13.8
m-Nitrobenzaldehyde	94.0	10.0
p-Nitrobenzaldehyde	97.0	13.8
o-Nitrobenzoic acid	93.4	20.0
m-Nitrobenzoic acid	93.4	7.0
p-Nitrobenzoic acid	92.0	19.7
o-Nitroaniline	0	—
m-Nitroaniline	0	—
p-Nitroaniline	0	—
Phthalimide	96.2	20.8
Phthalic acid	96.8	78.4
Phenol	98.5	80.0
Phloroglucinol	92.5	22.1

Compound	Percent Removed[a] (based upon COD)	Average Rate of Biodegradation $(mg\ COD\ g^{-1}\ hr^{-1})$
Aromatic Compounds (cont'd.)		
N-Phenylanthranilic acid	28.0	—
o-Phenylendiamine	33.0	—
m-Phenylendiamine	60.0	—
p-Phenylendiamine	80.0	—
Pyrocatechol	96.0	55.5
Pyrogallol	40.0	—
Resorcinol	90.0	57.5
Salicylic acid	98.8	95.0
Sulfosalicylic acid	98.5	11.3
Sulfanilic acid	95.0	4.0
Thymol	94.6	15.6
p-Toluenesulphonic acid	98.7	8.4
2,4,6-Trinitrophenol	0	—

a. "Percent Removed" represents to what extent the reaction goes before stopping.

Source: Pitter [128]. *(Reprinted with permission from Pergamon Press, Ltd.)*

UOD, indicating the proportion of the theoretical total oxidation of an organic compound that is attributed to bacterial action. An RI approaching 1.0 indicates that a substance is readily degraded to the point of mineralization. An error factor of approximately 13% is associated with the RI because of interactions with microorganisms. Refractory indices for 25 compounds are listed in Table 9-23.

Hydrolysis. Structure-activity relationships between second-order alkaline hydrolysis rate constants (k_{OH}) and microbial degradation rate constants (k_b) have been reported for two groups of esters [13,168]. Figure 9-15 is a plot of the correlation between hydrolysis and biodegradation for the two groups of chemicals. The curve is described by the equation:

$$\log k_b = m \log k_{OH} + c \qquad (9\text{-}7)$$

The specific compounds which were tested are listed in Figure 9-15. The authors discussed similar relationships found by analyzing data for substituted phenols [158a]. Although further study on other chemical groups is needed, these correlations comprise a significant step forward in relating the biodegradation of some substances to other

TABLE 9-23

Refractory Indices for Various Organic Compounds

Compound	RI	Compound	RI
High Degradability		**Low Degradability**	
Biphenyl	1.14	Benzene	0.23
Antifreeze	1.12	Gasoline	0.21
Sevin	1.0	Adenine	0.14, 0.12
d-Glutamic acid	1.00	Vinyl chloride	0
d-Glucose	0.93	Carboxymethyl	
l-Valine	0.93	cellulose	0
Acetone	0.93, 0.71	Humics	0
Phenol	0.87	DDT with carrier	0
Sodium butyrate	0.84	p-Chlorophenol	0
l-Aspartic acid	0.81	Dichlorophenol	<0
Sodium propionate	0.80	DDT	<0
Propylene glycol	0.78, 0.52	Bipyridine	<0
Ethylene glycol	0.76	Chloroform	<0
		Cyanuric Acid	<0
Medium-High Degradability			
Potato Starch	0.72, 0.64		
l-Arginine	0.65		
Acetic acid	0.61		
Aniline	0.58		
Soluble starch	0.54		
l-Histidine	0.52		
l-Lysine	0.52		
Hydroquinone	0.41		

Source: Bedard [14] and Helfgott *et al.* [62] .

chemical properties. The relationship may not be applicable to all classes of chemicals.

9-6   AVAILABLE DATA

Degradability and Degradation Pathways of Organic Compounds

Chapman, P.J. [26] — Classification of common metabolic pathways associated with biodegradation of organics.

Menzie, C.M. [107,108] — Extensive review of the metabolism of pesticides by microorganisms and higher organisms in all systems.

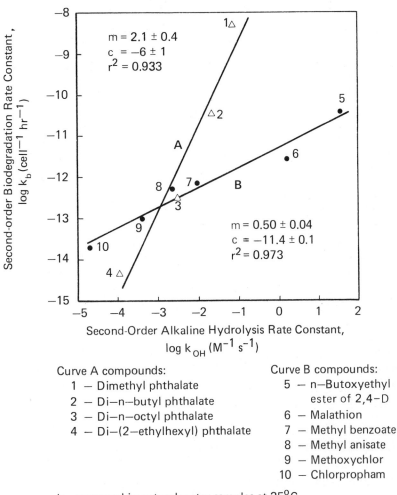

Curve A compounds:
1 — Dimethyl phthalate
2 — Di–n–butyl phthalate
3 — Di–n–octyl phthalate
4 — Di–(2–ethylhexyl) phthalate

Curve B compounds:
5 — n–Butoxyethyl
    ester of 2,4–D
6 — Malathion
7 — Methyl benzoate
8 — Methyl anisate
9 — Methoxychlor
10 — Chlorpropham

k_B measured in natural water samples at 25°C
k_{OH} measured in distilled water at 30°C(A) and 27°C(B)

Source: Wolfe et al. [168]. *(Reprinted with permission from the American Chemical Society.)*

FIGURE 9–15   Correlation of Second-order Alkaline Hydrolysis
Rate Constants with Second-order Biodegradation
Rate Constants for Two Groups of Compounds

Goring, C.A.I., *et al.* [50] — Review of biodegradation pathways.

Matsumara, F. [102] — Review of degradation of pesticides in soil.

Kaufman, D.D. [82] — Review of degradation of pesticides in soil.

Kaufman, D.D. [83] — Review of degradation of pesticides in soil, including tables of half-lives.

Meikle, R.W. [105] — Review of degradation of organic compounds in soil.

NAS [114a] — Collection of papers on the degradation of synthetic organic molecules in the environment.

Miller, M.W. [109] — Handbook compiling all known microbial metabolites, organized by chemical group.

Rao, P.S.C. and J.M. Davidson [131] — Compilation of biodegradation rate constants measured in soil.

Pfister, R.M. [126] — Review of biodegradation of halogenated pesticides.

Sanborn, J.R. *et al.* [139] — Extensive literature review of degradation reactions and metabolites of selected pesticides in soil.

Williams, P.P. [166] — Review of anaerobic metabolic pathways for various pesticides.

Bollag, J.M. [16] — Review of degradation of pesticides by soil fungi.

Swisher, R.D. [155] — Book on biodegradation of surfactants.

Hill, I.R. and S.J.L. Wright [68] — Collection of review articles on microbial degradation of pesticides in all environments.

Matsumara, F. and H.J. Benezet [103] — Review of degradation of insecticides.

Cripps, R.E. and T.R. Roberts [29] — Review of degradation of herbicides.

Woodcock, D. [171] — Microbial degradation of fungicides, fumigants, and nematocides by chemical group and discussion of how application method affects degradation.

Biology and Ecology of Microorganisms

Alexander, M. [2] — Textbook on microbial ecology.

Brock, T.D. [21] — Textbook on microbial biology.

Stotzky, G. [151] — Article on microorganisms in soil environment.

Gaudy, A. and E. Gaudy [44] — Text on environmental microbiology and applications to wastewater treatment.

Gray, T.R.G. *et al.* [53] — Article on microorganisms in soil, aquatic and air environments in biota.

Alexander, M. [5] — Text on soil microorganisms and ecology.

Mitchell, R. [110] — Collection of articles on water pollution microbiology.

Jones, J.G. [77,78] — Ecology of freshwater microorganisms.

Bourquin, A.W. and P.H. Pritchard [18] — Collection of articles on aquatic microbial degradation of pollutants in marine environments.

Colwell, R.R. and R.Y. Morita [28] — Textbook on marine microbiology.

Wood, E.J.F. [169] — Textbook on marine microbiology.

Wood, E.J.F. [170] — Textbook on marine and estuarine microbiology.

Rodina, R.G. [138] — Textbook on aquatic microbiology.

Stevenson, L.H. and R.R. Colwell [150] — Textbook on estuarine microbiology.

Curds, C.R. and H.A. Hawkes [31] — Article on waste water microbiology.

Biodegradation Test Methods

Howard, P.H. *et al.* [72]

Swisher, R.D. [155]

TSCA [162]

Kinetics of Microbial Degradation

Hamaker, J.W. [55] — Discussion of degradation kinetics, degradation in soil, variables affecting degradation, application to field environment.

Larson, R.J. [92] — Discussion of degradation kinetics.

Rules of Thumb

Kearney, P.C. and J.R. Plimmer [85] — Discussion of relation of chemical structure to degradation in pesticides.

Alexander, M. [1,3] — Discussion of factors (environmental, biological, and physico-chemical) responsible for recalcitrance of chemicals in natural systems.

Kaufman, D.D. [81] — Discussion of relation of structure to degradation of pesticides.

Other Sources

Thom, N.S. and A.R. Agg [160] — Classification of over 200 synthetic organic compounds into 3 categories of degradability by biological sewage treatment.

Helfgott, T.B. *et al.* [62] — Review of laboratory techniques to derive refractory index (BOD/UOD ratio) and compilation of indices for 38 organic compounds.

Bedard, R.G. [14] — Compilation of refractory indices for organic compounds.

Tabak, H.H. [158b] — Compilation of results from static flask die-away experiments on 96 organic compounds.

9-7 SYMBOLS AND ABBREVIATIONS

A = initial rate of reaction in Eq. 9-1

ABS = alkyl benzene sulfonate

[B] = microbial population concentration in Eq. 9-4

BI = biodegradability index

BOD = biological (or biochemical) oxygen demand

[C] = substrate concentration

c = parameter in Eq. 9-7

COD = chemical oxygen demand

DO = dissolved oxygen

DOC = dissolved organic carbon

E_a = activation energy in Eq. 9-1

k = biodegradation rate constant in Eq. 9-2

K_c = concentration of substrate supporting a half-maximum growth rate

k_b = second-order biodegradation rate constant in Eq. 9-7

k_{OH} = second-order hydrolysis rate constant in Eq. 9-7

m = parameter in Eq. 9-7

n = order of reaction in Eq. 9-2

PAH = polycyclic aromatic hydrocarbon

R = gas constant

RI = refractory index

T = temperature (absolute)

t   =   time

U   =   microorganism specific growth rate in Eq. 9-3

U_{max} =   microorganism maximum growth rate in Eq. 9-3

UOD =   ultimate oxygen demand

Y   =   rate of reaction (after Arrhenius temperature correction) in Eq. 9-1

Y_d =   yield coefficient in Eq. 9-4

9-8   REFERENCES

1. Alexander, M., "Biodegradation: Problems of Molecular Recalcitrance and Microbial Fallibility," *Adv. Appl. Microbiol.,* **7**, 35 (1965).

2. Alexander, M., *Microbial Ecology,* John Wiley & Sons, New York (1971).

3. Alexander, M., "Nonbiodegradable and Other Recalcitrant Molecules," *Biotechnol. Bioeng.,* **15**, 611 (1973).

4. Alexander, M., "Environmental and Microbiological Problems Arising from Recalcitrant Molecules," *Microb. Ecol.,* **2**, 17 (1975).

5. Alexander, M., *Introduction to Soil Microbiology,* 2nd ed., John Wiley & Sons, New York (1977).

6. Alexander, M., "Biodegradation of Toxic Chemicals in Water and Soil," in *Proc. 176th National Meeting,* Miami Beach, Fla., September 10-15, 1978, Vol. 93, American Chemical Society, Division of Environmental Chemistry (1978).

7. Alexander, M., "Role of Cometabolism," in *Proceedings of the Workshop: Microbial Degradation of Pollutants in Marine Environments,* A.W. Bourquin and P.H. Pritchard (eds.), EPA-600/9-79-012, Washington, D.C. (1979).

8. Alexander, M. and M.I.H. Aleem, "Effect of Chemical Structure on Microbial Decomposition of Aromatic Herbicides," *J. Agric. Food Chem.,* **9**, 44 (1961) as quoted in [1].

9. Alexander, M. and B.K. Lustigman, "Effect of Chemical Structure on Microbial Degradation of Substituted Benzenes," *J. Agric. Food Chem.,* **14**, 410 (1966) as quoted in [3].

10. Ali, D.A., A.G. Callely and M. Hayes, "Ability of a Vibrio Grown in Benzoate to Oxidize p-Fluorobenzoate," *Nature (London),* **196**, 194 (1962) as quoted in [1].

11. Anderson, J.R., "Pesticide Effects on Non-Target Soil Microorganisms," in *Pesticide Microbiology,* ed. by I.R. Hill and S.J.L. Wright, Academic Press, New York (1978).

12. Baughman, G.L. and R.R. Lassiter, "Prediction of Environmental Pollutant Concentration," in *Estimating the Hazard of Chemical Substances to Aquatic Life,* ed. by J. Cairns, Jr., K.L. Dickson, and A.W. Maki, ASTM STP 657, American Society for Testing and Materials, Philadelphia, Pa. (1978).

13. Baughman, G.L., D.F. Paris and W.C. Steen, "Quantitative Expression of Biotransformation Rate," in *Biotransformation and Fate of Chemicals in the Aquatic Environment,* ed. by A. Maki, K. Dickson and J. Cairns, Jr., American Society for Microbiology, Washington, D.C. (1980).

14. Bedard, R.G., "Biodegradability of Organic Compounds," M.S. Thesis, University of Connecticut (Storrs) (PB-264 707) (1976).

15. Bell, C.R., "A Soil Achromobacter Which Degrades 2,4-Dichlorophenoxyacetic Acid," *Can. J. Microbiol.,* **5,** 325 (1960) as quoted in [1].

16. Bollag, J.M., "Biochemical Transformation of Pesticides by Soil Fungi," in *Microbial Ecology,* ed. by A.I. Laskin and H. Lechavalier, CRC Press, Cleveland, Ohio (1974).

17. Bollag, J.M., "Transformation of Xenobiotics by Microbial Activity," in *Proceedings of the Workshop: Microbial Degradation of Pollutants in Marine Environments,* ed. by A.W. Bourquin and P.H. Pritchard, EPA-600/9-79-012, Washington, D.C. (1979).

18. Bourquin, A.W. and P.H. Pritchard (eds.), *Proceedings of the Workshop: Microbial Degradation of Pollutants in Marine Environments,* EPA-600/9-79-012, Washington, D.C. (1979).

19. Brady, N.C., *The Nature and Properties of Soils,* 8th ed., Macmillan, New York (1974).

20. Branson, D.R., "Predicting the Fate of Chemicals in the Aquatic Environment from Laboratory Data," in *Estimating the Hazards of Chemical Substances to Aquatic Life,* ed. by J. Cairns, Jr., K.L. Dickson, and A.W. Maki, ASTM STP 657, American Society for Testing and Materials, Philadelphia, Pa. (1978).

21. Brock, T.D., *Biology of Microorganisms,* Prentice-Hall, Englewood Cliffs, New Jersey (1970).

22. Bokova, E.N., "Oxidation of Ethane and Propane by Mycobacterium," *Mikrobiologiya,* **23,** 15 (1954) as quoted in [1].

23. Burger, K., I.C. MacRae and M. Alexander, "Decomposition of Phenoxyalkyl Carboxylic Acids," *Soil Sci. Soc. Am. Proc.,* **26,** 243 (1962) as quoted in [1].

24. Calaway, W.T., "The Metazoa of Waste Treatment Processes — Rotifers," *J. Water Pollut. Control Fed.,* **40,** R412 (1968) as quoted in [154].

25. Chambers, C.W. and P.W. Kabler, "Biodegradability of Phenols as Related to Chemical Structure," *Dev. Ind. Microbiol.,* **5,** 85 (1964) as quoted in [1].

26. Chapman, P.J., "Degradation Mechanisms," in *Proceedings of the Workshop: Microbial Degradation of Pollutants in Marine Environments,* ed. by A.W. Bourquin and P.H. Pritchard, EPA-600/9-79-012, Washington, D.C. (1979).

27. Collins, C.E., C.P.L. Grady, Jr. and F.P. Incropera, "The Effects of Temperature on Biological Wastewater Treatment Processes," Purdue University Water Resources Center Technical Report No. 98, West Lafayette, Indiana (March 1978).

28. Colwell, R.R. and R.Y. Morita, *Effects of the Ocean Environment on Microbial Activities,* University Park Press, Baltimore (1974).

29. Cripps, R.E. and T.R. Roberts, "Microbial Degradation of Herbicides," in *Pesticide Microbiology,* ed. by I.R. Hill and S.J.L. Wright, Academic Press, New York (1978).

30. Crowther, R.F. and N. Harkness, "Anaerobic Bacteria" in *Ecological Aspects of Used Water Treatment,* Vol. I, ed. by C.R. Curds and H.A. Hawkes, Academic Press, London (1975) as quoted in [91].

31. Curds, C.R. and H.A. Hawkes (eds.), *Ecological Aspects of Used Water Treatment,* Vol. I of *The Organisms and Their Ecology,* Academic Press, London (1975).

32. Dagley, S., "Microbial Degradation of Stable Chemical Structures: General Features of Metabolic Pathways," in *Degradation of Synthetic Organic Molecules in the Biosphere,* National Academy of Sciences, Washington, D.C. (1972).

33. DiGeronimo, M.J., R.S. Boethling and M. Alexander, "Effects of Chemical Structure and Concentration on Microbial Degradation in Model Ecosystems," in *Proceedings of the Workshop: Microbial Degradation of Pollutants in Marine Environments,* ed. by A.W. Bourquin and P.H. Pritchard, EPA-600/9-79-012 (April 1979).

34. Dias, F.F. and M. Alexander, "The Effects of Chemical Structure on the Biodegradability of Aliphatic Acids and Alcohols," *Appl. Microbiol.,* **22,** 1114 (1971) as quoted in [3].

35. Domsch, K.H., "Effects of Fungicides on Microbial Populations in Soil," in *Pesticides in the Soil: Ecology, Degradation and Movement,* presented at International Symposium on Pesticides in the Soil, Michigan State University, East Lansing, Michigan (February 25-27, 1970).

36. Donigan, A.S., Jr., D.C. Beyerlein, H.H. Davis, Jr. and N.H. Crawford, *Agricultural Runoff Management (ARM) Model Version II: Refinement and Testing,* EPA 600/3-77-098, EPA, Athens, Georgia (1977).

37. Edwards, C.A., "Biological Aspects of the Degradation and Behavior of Pesticides in Soil," in *Proc. 7th British Insecticide and Fungicide Conference,* Brighton, England (November 19-22, 1973).

38. Engvild, K.C. and H.L. Jensen, "Microbiological Decomposition of the Herbicide Pyrazon," *Soil Biol. Biochem.,* **1,** 295 (1969) as quoted in [55].

39. Furmidge, C.G. and J.M. Osgerby, "Persistence of Herbicides in Soil," *J. Sci. Food Agric.,* **18,** 269 (1967) as quoted in [55].

40. Faust, S.G. and J.V. Hunter (eds.), *Organic Compounds in Aquatic Environments,* Marcel Dekker, New York (1976).

41. Fenchel, T.M. and B.B. Jorgensen, "Detritus Food Chains of Aquatic Ecosystems: The Role of Bacteria," in *Advances in Microbial Ecology,* ed. by M. Alexander, Plenum Press, New York (1977).

42. Fenchel, T.M. and R.J. Riedl, "The Sulfide System: A New Biotic Community Underneath the Oxidized Layer of Marine Sand Bottoms," *Mar. Biol.,* **7,** 255 (1970) as quoted in [41].

43. Finnerty, W.R., E. Hawtrey and R.E. Kallio, "Alkane-oxidizing Micrococci," *Z. Allg. Mikrobiol.,* **2,** 169 (1962) as quoted in [1].

44. Gaudy, A. and E. Gaudy, *Microbiology for Environmental Scientists and Engineers,* McGraw-Hill Book Co., New York (1980).

45. Geer, R.D., "Predicting the Anaerobic Degradation of Organic Chemical Pollutants in Waste Water Treatment Plants from Their Electrochemical Reduction Behavior," Waste Resources Research Center, Montana State University, MUJWRRC-95, W79-01-OWRT-A-097-MONT(1), PB-289 22478WP, Bozeman, Montana (July 1978).

46. Giddings, J.M., B.T. Walton, G.K. Eddlemon and K.G. Olson, "Transport and Fate of Anthracene in Aquatic Microoganisms," in *Proceedings of the Workshop, Microbial Degradation of Pollutants in Marine Environments,* ed. by A.W. Bourquin and P.H. Pritchard, EPA-600/9-79-012, Washington, D.C. (1979).

47. Gledhill, W.E. and V.W. Saeger, "Microbial Degradation in the Environmental Hazard Evaluation Process," *ibid.*

48. Gomaa, H.M., I.H. Suffett and S.D. Faust, "Kinetics of Hydrolysis of Diazinon and Diazoxon," *Residue Rev.,* **29,** 171 (1969) as quoted in [55].

49. Goring, C.A.I., "Agricultural Chemicals in the Environment: A Quantitative Viewpoint," in *Organic Chemicals in the Soil Environment,* Vol. 2, ed. by C.A.I. Goring and J.W. Hamaker, Marcel Dekker, New York (1972).

50. Goring, C.A.I., D.A. Laskowski, J.W. Hamaker and R.W. Meikle, "Principles of Pesticide Degradation in Soil," in *Environmental Dynamics of Pesticides,* ed. by R. Haque and V.H. Freed, Plenum Press, New York (1975).

51. Graetz, D.A., G. Chesters, T.C. Daniel, L.W. Newland, and C.B. Lee, "Parathion Degradation in Lake Sediments," *J. Water Pollut. Control Fed.,* **12,** 76 (1970).

52. Gray, T.R.G., "Microbial Growth in Soil," in *Pesticides in the Soil: Ecology, Degradation and Movement,* presented at International Symposium on Pesticides in the Soil, Michigan State University, East Lansing, Michigan (February 25-27, 1970).

53. Gray, T.R.G., "Microbial Aspects of the Soil, Plant, Aquatic, Air and Animal Environment," in *Pesticide Microbiology,* ed. by I.N. Hill and S.J.L. Wright, Academic Press, New York (1978).

54. Gunner, H.B. and B.M. Zuckerman, "Nature and Sources of Pollution by Pesticides," Conference on Pollution of Our Environment, Montreal (1978).

55. Hamaker, J.W., "Decomposition: Quantitative Aspects," in *Organic Chemicals in the Soil Environment,* Vol. I, ed. by C.A.I. Goring and J.W. Hamaker, Marcel Dekker, New York (1972).

56. Hamaker, J.W., C.R. Youngson and C.A.I. Goring, "Prediction of the Persistence and Activity of Tordon Herbicide in Soils under Flooded Conditions," *Down Earth,* **23,** 30 (1967) as quoted in [55].

57. Hammond, M.W. and M. Alexander, "Effects of Chemical Structure on Microbial Degradation of Methyl-Substituted Aliphatic Acid," *Environ. Sci. Technol.,* **6,** 732 (1972) as quoted in [3].

58. Hance, R.J., "Decomposition of Herbicides in Soil," *J. Sci. Food Agric.,* **20,** 144 (1969) as quoted in [55].

59. Hance, R.J. and C.E. McKone, "Effect of Concentration on the Decomposition Rates in Soil of Atrazine, Linuron and Picloram," *Pestic. Sci.,* **2,** 31 (1971) as quoted in [55].

60. Hart, F.L. and T. Helfgott, "Bio-Refractory Index for Organics in Water," *Water Res.,* **9,** 1055 (1975).

61. Hauck, R.D. and H.F. Stephenson, "Nitrification of Triazine Nitrogen," *J. Agric. Food Chem.,* **12,** 147 (1964) as quoted in [1].

62. Helfgott, T.B., F.L. Hart and R.G. Bedard, "An Index of Refractory Organics," EPA-600/2-77-174; PB-272 438 (August 1977).

63. Helling, C.S., P.C. Kearney and M. Alexander, "Behavior of Pesticides in Soil," *Adv. Agron.,* **23,** 147 (1971) as quoted in [53].

64. Hemmett, R.B., Jr. and S.D. Faust, "Biodegradation Kinetics of 2,4-Dichlorophenoxyacetic Acid by Aquatic Organisms," *Residue Rev.,* **29,** 191 (1969) as quoted in [55].

65. Henderson, M.E.K., "The Metabolism of Methoxylated Aromatic Compounds by Soil Fungi," *J. Gen. Microbiol.,* **16,** 686 (1975) as quoted in [1].

66. Hill, I.R., "Microbial Transformation of Pesticides," in *Pesticide Microbiology,* ed. by I.R. Hill and S.J.L. Wright, Academic Press, New York (1978).

67. Hill, I.R. and P.L. McCarty, "Anaerobic Degradation of Selected Chlorinated Pesticides," *J. Water Poll. Control Fed.,* **39,** 1259 (1967).

68. Hill, I.R. and S.J.L. Wright, "The Behavior and Fate of Pesticides in Microbial Environments," in *Pesticide Microbiology,* ed. by I.R. Hill and S.J.L. Wright, Academic Press, New York (1978).

69. Hirsch, P. and M. Alexander, "Microbial Decomposition of Halogenated Propionic and Acetic Acids," *Can. J. Microbiol.,* **6,** 241 (1960) as quoted in [1].

70. Hobbie, J.E., O. Holm-Hansen, T.T. Packard, L.R. Pomeroy, R.W. Sheldon, J.P. Thomas and W.J. Wiebe, "A Study of the Distribution and Activity of Microorganisms in Ocean Water," *Limnol. Oceanogr.,* **17,** 544 (1972) as quoted in [41].

71. Holm-Hansen, O., "The Distribution and Chemical Composition of Particulate Material in Marine and Freshwaters," *Mem. Ist. Ital. Idrobiol. Dott. Marco de Marchi,* **29 Suppl.,** 37 (1972) as quoted in [41].

72. Howard, P.H., J. Saxena, P.R. Durkin, and L.-T. Ou, "Review and Evaluation of Available Techniques for Determining Persistence and Routes of Degradation of Chemical Substances in the Environment," EPA 560/5-75-006 (May 1975).

73. Huddleston, R.L. and R.C. Allred, "Microbial Oxidation of Sulphonate Alkylbenzenes," *Dev. Ind. Microbiol.,* **4,** 24 (1963) as quoted in [1].

74. Jannasch, H.W., "Growth of Marine Bacteria at Limiting Concentrations of Organic Carbon in Seawater," *Limnol. Oceanogr.,* **12,** 264 (1967) as quoted in [3].

75. Jannasch, H.W., "New Approaches to Assessment of Microbial Activity in Polluted Waters," in *Water Pollution Microbiology,* ed. by R. Mitchell, Wiley Interscience, New York (1972).

76. Jannasch, H.W., "The Ultimate Sink: Keynote Address," in *Proceedings of the Workshop: Microbial Degradation of Pollutants in Marine Environments,* EPA-600/9-79-012, Washington, D.C. (1979).

77. Jones, J.G., "Studies on Freshwater Bacteria: Factors Which Influence the Population and its Activity," *J. Ecol.,* **59,** 593 (1971) as quoted in [53].

78. Jones, J.G., "Heterotrophic Microorganisms and Their Activity," in *River Ecology,* ed. by B.A. Whitton, Blackwells Scientific Publications, Oxford (1975) as quoted in [53].

79. Kaplan, A.M., "Prediction from Laboratory Studies of Biodegradation of Pollutants in Natural Environments," in *Proceedings of the Workshop: Microbial Degradation of Pollutants in Marine Environments,* ed. by A.W. Bourquin and P.H. Pritchard, EPA-600/9-79-012, Washington, D.C. (1979).

80. Kapoor, I.P., R.L. Metcalf, A.S. Hirwe, J.R. Coats and M.S. Khalsa, "Structure Activity Correlations of Biodegradability of DDT Analogs," *J. Agric. Food Chem.,* **21,** 760 (1973).

81. Kaufman, D.D., "Structure of Pesticides and Decomposition by Microorganisms," *Pesticides and Their Effects in Soil and Water,* ASA Special Publication No. 8, Soil Science Society of America, Madison, Wisconsin (1966) as quoted in [3].

82. Kaufman, D.D., "Degradation of Pesticides by Soil Microorganisms," in *Pesticides in Soil and Water,* ed. by W.D. Guenzi, Soil Science Society, Madison, Wisconsin (1974).

83. Kaufman, D.D., "Soil Degradation and Persistence of Benchmark Pesticides," in *A Literature Survey of Benchmark Pesticides,* George Washington University Medical Center (1976).

84. Kazano, H., P.C. Kearney and D.D. Kaufman, "Metabolism of Methyl Carbamate Insecticides in Soils, *J. Agric. Food Chem.,* **20,** 975 (1972).

85. Kearney, P.C. and J.R. Plimmer, "Relation of Structure to Pesticide Decomposition," in *Pesticides in the Soil,* Michigan State University, East Lansing, Michigan (1970).

86. Konrad, J.G. and G. Chesters, "Degradation in Soils of Ciodrin, an Organophosphate Insecticide," *J. Agric. Food Chem.,* **17,** 226 (1969) as quoted in [55].

87. Kramer, N. and Doetsch, R.N., "Growth of Phenol-Utilizing Bacteria on Aromatic Carbon Sources," *Arch. Biochem. Biophys.,* **26,** 401 (1950) as quoted in [1].

88. Laanio, T.L., P.C. Kearney and D.D. Kaufman, "Microbial Metabolism of Dinitramine," *Pestic. Biochem. Physiol.,* **3,** 271 (1973) as quoted in [66].

89. Ladd, J.N., "The Oxidation of Hydrocarbons by Soil Bacteria: I. Morphological and Biochemical Properties of a Soil Diphtheroid Utilizing Hydrocarbons," *Aust. J. Biol. Sci.,* **9,** 92 (1956) as quoted in [1].

90. LaRiviere, J.W.M., "A Critical View of Waste Treatment," in *Water Pollution Microbiology,* ed. by R. Mitchell, John Wiley and Sons, New York (1972).

91. LaRiviere, J.W.M., "Microbial Ecology of Liquid Waste Treatment," in *Advances in Microbial Ecology,* ed. by M. Alexander, Plenum Press, New York (1977).

92. Larson, R.J., "Role of Biodegradation Kinetics in Predicting Environmental Fate," in *Biotransformation and Fate of Chemicals in the Aquatic Environment,* ed. by A. Maki, K. Dickson and J. Cairns, Jr., American Society for Microbiology, Washington, D.C. (1980).

93. Lee, R.F. and C. Ryan, "Microbial Degradation of Organochlorine Compounds in Estuarine Waters and Sediments," in *Proceedings of the Workshop: Microbial Degradation of Pollutants in Marine Environments,* EPA-600/9-79-012, Washington, D.C. (1979).

94. Lewis, D.L., D.F. Paris and G.L. Baughman, "Transformation of Malathion by Fungus, *Aspergillus Oryzae,* Isolated from a Freshwater Pond," *Bull. Environ. Contam. Toxicol.,* **13** (1975).

95. Lemon, E.R., "Soil Aeration and Plant Root Relations: I. Theory," *Agron. J.,* **54,** 167 (1962).

96. Long, M.I.E. and G.W. Winsor, "Isolation of Some Urea-Formaldehyde Compounds and Their Decomposition in Soil," *J. Sci. Food Agric.,* **11,** 441 (1960) as quoted in [1].

97. Lyman, W., L. Nelson, L. Partridge, A. Kalelkar, J. Everett, D. Allen, L.L. Goodier and G. Pollack, "Survey Study to Select a Limited Number of Hazardous Materials to Define Amelioration Requirements," U.S. Coast Guard, CG-D-46-75 (March 1974).

98. MacFayden, A., *Animal Ecology, Aims and Methods,* 2nd ed., Pitman and Sons, London (1963) as quoted in [72].

99. MacRae, I.C. and M. Alexander, "Metabolism of Phenoxyalkyl Carboxylic Acids by a Flavobacterium Species," *J. Bacteriol.,* **86,** 1231 (1963) as quoted in [1].

100. Manahan, S.E., *Environmental Chemistry,* Willard Grant Press, Inc., Boston (1969).

101. Mateles, R.I. and S.K. Chiam, "Kinetics of Substrate Uptake in Pure and Mixed Culture," *Environ. Sci. Technol.,* **3,** 569 (1969).

102. Matsumara, F., "Degradation of Pesticide Residues in the Environment," in *Environmental Pollution by Pesticides,* Vol. 3, ed. by C.A. Edwards, Plenum Press, New York (1973).

103. Matsumara, F. and H.J. Benezet, "Microbial Degradation of Insecticides," in *Pesticide Microbiology,* ed. by I.R. Hill and S.J.L. Wright, Academic Press, New York (1978).

104. McKenna, E.J. and R.D. Heath, "Biodegradation of Polynuclear Aromatic Hydrocarbon Pollutants by Soil and Water Microorganisms," prepared for U.S. Department of the Interior Water Resources Research Act of 1964 (1976).

105. Meikle, R.W., "Decomposition: Qualitative Relationships," in *Organic Chemicals in the Soil Environment,* Vol. 1, ed. by C.A.I. Goring and J.W. Hamaker, Marcel Dekker, New York (1972).

106. Meiklejohn, J., "Numbers of Bacteria and Actinomycetes in a Kenya Soil," *J. Soil Sci.,* **8,** 240 (1957).

107. Menzie, C.M., "Metabolism of Pesticides," Bureau of Sport Fisheries and Wildlife, U.S. Department of the Interior, Special Scientific Report — Wildlife, No. 127 (1969) as quoted in [50].

108. Menzie, C.M., "Fate of Pesticides in the Environment," *Ann. Rev. Entomol.,* **17,** 199 (1972).

109. Miller, M.W., *The Pfizer Handbook of Microbial Metabolites,* McGraw-Hill Book Co., New York (1961).

110. Mitchell, R. (ed.), *Water Pollution Microbiology,* Wiley Interscience, New York (1972).

111. Monod, J., *Recherches sur la Croissance des Cultures Bactériennes,* Hermann & Cie (Paris) (1942) as quoted in [101].

112. Monod, J., *Ann. Inst. Pasteur (Paris),* **79,** 390 (1950) as quoted in [154].

113. Mueller, J.A., J. Morand and W.C. Boyle, "Floc Sizing Techniques," *Appl. Microbiol.,* **15,** 125 (1967) as quoted in [90].

114a. National Academy of Sciences, *Principles for Evaluating Chemicals in the Environment,* Washington, D.C. (1975).

114b. Nesbitt, H.J. and J.R. Watson, "Degradation of the Herbicide 2,4-D in River Water — I. Description of Study Area and Survey of Rate Determining Factors," *Water Res.,* **14,** 1683 (1980).

115a. Nesbitt, H.J. and J.R. Watson, "Degradation of the Herbicide 2,4-D in River Water — II. The Role of Suspended Sediment, Nutrients and Water Temperature," *Water Res.,* **14,** 1689 (1980).

115b. O'Brien, R.D., *Insecticides, Action and Metabolism,* Academic Press, New York (1967) as quoted in [68].

116. Odum, E.P., *Fundamentals of Ecology,* 3rd ed., W.B. Saunders Co., Philadelphia (1971).

117. Okafor, N., "Ecology of Microorganisms on Chitin Buried in Soil," *J. Gen. Microbiol.,* **44,** 311 (1966) as quoted in [53].

118. Painter, H.A., "Biodegradability," *Proc. R. Soc. London, Ser. B,* **185,** 149 (1974).

119. Painter, H.A., R.S. Denton and C. Quarmby, "Removal of Sugars by Activated Sludge," *Water Res.,* **2,** 427 (1968).

120. Paris, D.F., D.L. Lewis and N.L. Wolfe, "Rates of Degradation of Malathion by Bacteria Isolated from Aquatic System," *Environ. Sci. Technol.,* **9,** 135 (1975).

121. Paris, D.F., D.L. Lewis, J.T. Barnett and G.L. Baughman, "Microbial Degradation and Accumulation of Pesticides in Aquatic Systems," EPA-660/3-75-007 (1975).

122. Paris, D.F., W.C. Steen, and G.L. Baughman, "Role of Physico-Chemical Properties of Aroclors 1016 and 1242 in Determining Their Fate and Transport in Aquatic Environments," *Chemosphere,* **4,** 319 (1978).

123. Paris, D.F., W.C. Steen and G.L. Baughman, "Prediction of Microbial Transformation of Pesticides in Natural Waters" (unpublished), presented before the American Chemical Society, Division of Pesticide Chemistry, Anaheim, California, Environmental Research Laboratory, U.S. EPA, Athens, Ga. (March 1978).

124. Parr, J.F., "Effects of Pesticides on Microorganisms in Soil and Water," in *Pesticides in Soil and Water,"* ed. by W.D. Guenzi, J.L. Ahlrichs, G. Chesters, M.E. Bloodworth and R.G. Nash, Soil Science Society of America, Inc. (1974).

125. Patrick, Z.A. and L.W. Koch, "The Adverse Influence of Phytotoxic Substances from Decomposing Plant Residues on Resistance of Tobacco to Black Root Rot," *Can. J. Bot.,* **41,** 447 (1963).

126. Pfister, R.M. "Interactions of Halogenated Pesticides and Microorganisms: Review," in *Microbial Ecology,* ed. by A.I. Laskin and H. Lechevalier, CRC Press, Cleveland, Ohio (1974).

127. Pike, E.G., "Aerobic Bacteria," in *Ecological Aspects of Used Water Treatment,* Vol. I, ed. by C.R. Curds and H.A. Hawkes, Academic Press, London (1975).

128. Pitter, P., "Determination of Biological Degradability of Organic Substances," *Water Res.,* **10,** 231 (1976).

129. Postgate, J.R., "Recent Advances in the Study of the Sulfate-Reducing Bacteria," *Bacteriol. Proc.,* **29,** 425 (1965) as quoted in [167].

130. Raney, W.A., "Physical Factors of the Soil as They Affect Microorganisms," in *Ecology of Soil-Borne Plant Pathogens,* ed. by K.F. Baker and W.C. Snyder, University of California Press, Berkeley (1965) as quoted in [53].

131. Rao, P.S.C. and J.M. Davidson, "Estimation of Pesticide Retention and Transformation Parameters Required in Non-Point Source Pollution Models," in *Environmental Impact of Non-Point Source Pollution,* Ann Arbor Science Publishers, Inc., Ann Arbor, Mich. (in press, 1979/1980).

132. Raymond, R.L., "Methodological Criteria for Biodegradation," in *Proceedings of the Workshop: Microbial Degradation of Pollutants in Marine Environments,* ed. by A.W. Bourquin and P.H. Pritchard, EPA-600/9-79-012 (1979).

133. Reed, R.C., "The Effects of Alum Sludge on Anaerobic Digestion," M.S. Thesis, Auburn University, NTIS, PB-290 890, Washington, D.C. (March 1975).

134. Reid, G.K. and R.D. Wood, *Ecology of Inland Waters and Estuaries,* 2nd ed., D. Van Nostrand Company, New York (1976).

135. Richards, B.N., *Introduction to the Soil Ecosystem,* Longman Inc., New York (1974).

136. Rixon, A.J. and B.J. Bridge, "Respiration Quotient Arising from Microbial Activity in Relation to Matric Suction and Air-Filled Pore Space of Soil," *Nature (London),* **218,** 961 (1968) as quoted in [53].

137. Rhodes, R.A. and G.R. Hrubant, "Microbial Population of Feedlot Waste and Associated Sites," *Appl. Microbiol.,* **24,** 369 (1972) as quoted in [173].

138. Rodina, A.G., *Methods in Aquatic Microbiology,* Butterworths, London (1972).

139. Sanborn, J.R., B.M. Francis and R.L. Metcalf, "The Degradation of Selected Pesticides in Soil: A Review of the Published Literature," EPA-600/9-77-022, PB-272 353/4 WP (August 1977).

140. Sawyer, C.N. and D.W. Ryckman, "Anionic Synthetic Detergents and Water Supply Problems," *J. Am. Water Works Assoc.,* **49,** 480 (1957) as quoted in [1].

141. Schwall, L.R. and S.E. Herbes, "Methodology for Determination of Rates of Microbial Transformation of Polycyclic Aromatic Hydrocarbons in Sediments," Presentation at ASTM Sediment Microbial Activity and Biomass Symposium (January 1978).

142. Sethunathan, N., "Microbial Degradation of Insecticides in Flooded Soil and in Anaerobic Cultures," *Residue Rev.,* **47,** 153 (1973).

143. Sheets, T.J., "The Comparative Toxicities of Four Phenyl Urea Herbicides in Several Soil Types," *Weeds,* **6,** 413 (1958) as quoted in [1].

144. Sidderamappa, R. and N. Sethunathan, "Persistence of Gamma-BHC and Beta-BHC in Indian Rice Soils Under Flooded Conditions," *Pestic. Sci.,* **6,** 395 (1975) as quoted in [66].

145. Smith, A.E., "Factors Affecting the Loss of Tri-Allate from Soils," *Weed Res.,* **9,** 306 (1969) as quoted in [55].

146. Smith, J.H., W.R. Mabey, N. Bohonos, B.R. Holt, S.S. Lee, T.-W. Chou, D.C. Bomberger and T. Mill, "Environmental Pathways of Selected Chemicals in Freshwater Systems," EPA-600/7-77-113 (October 1977).

147. Spencer, W.F., "Distribution of Pesticides Between Soil, Water and Air," in *Pesticides in the Soil: Ecology, Degradation and Movement,* ed. by G.F. Guyer, Michigan State University, East Lansing, Mich. (1970).

148. Stanier, R.Y., "The Oxidation of Aromatic Compounds by Fluorescent Pseudomonads," *J. Bacteriol.,* **55,** 477 (1948) as quoted in [1].

149. Steen, W.C., D.F. Paris and G.L. Baughman, "Effects of Sediment Sorption on Microbial Degradation of Toxic Substances," in *Proc. 177th National Meeting,* Honolulu, April 1979, American Chemical Society; to be published in the proceedings of the April 1979 symposium on "Processes Involving Contaminants and Sediments."

150. Stevenson, L.H. and R.R. Colwell (eds.), *Estuarine Microbial Ecology,* University of South Carolina Press, Columbia, S.C. (1973).

151. Stotzky, G., "Activity, Ecology, and Population Dynamics of Microorganisms in Soil," in *Microbial Ecology,* ed. by A.I. Laskin and H. Lechevalier, CRC Press, Cleveland, Ohio (1974).

152. Strawinski, R.J., "The Dissimilation of Pure Hydrocarbons by Members of the Genus *Pseudomonas,*" Ph.D. Thesis, Pennsylvania State College, University Park, Pennsylvania (1943) as quoted in [1].

153. Stumm-Zollinger, E., "Substrate Utilization in Heterogeneous Bacterial Communities, *J. Water Pollut. Control Fed.,* **40,** R213 (1968).

154. Stumm-Zollinger, E. and R.H. Harris, "Kinetics of Biologically Mediated Aerobic Oxidation of Organic Compounds," in *Aquatic Environments,* ed. by S.L. Faust and J.V. Hunter, Marcel Dekker, New York (1971).

155. Swisher, R.D., *Surfactant Biodegradation,* Marcel Dekker, New York (1970).

156. Swisher, R.D., "Biodegradation of Alkylbenzene Sulfonates in Relation to Chemical Structure," *J. Water Pollut. Control Fed.,* **35,** 877 (1963) as quoted in [1].

157. Swisher, R.D., "Biodegradation Rates of Isomeric Diheptylbenzene Sulfonates," *Dev. Ind. Microbiol.,* **4,** 39 (1963) as quoted in [1].

158a. Tabak, H.H., C.W. Chambers and P.W. Kabler, "Microbial Metabolism of Aromatic Compounds: I. Decomposition of Phenolic Compounds and Aromatic Hydrocarbons by Phenol-Adapted Bacteria," *J. Bacteriol.,* **87,** 910 (1964).

158b. Tabak, H.H., S.A. Quave, C.I. Mashni and E.F. Barth, "Biodegradability Studies with Priority Organic Compounds," Staff report to U.S. Environmental Protection Agency, Environmental Research Center, Office of Research and Development, Cincinnati, Ohio (March 1980).

159. Thimann, K.V., *The Life of Bacteria,* Macmillan, New York (1966).

160. Thom, N.S. and A.R. Agg, "The Breakdown of Synthetic Organic Compounds in Biological Processes," *Proc. R. Soc. London, Ser. B,* **189,** 347 (1975).

161. Toerien, D.F. and W.H.J. Hattingh, "Anaerobic Digestion: I. The Microbiology of Anaerobic Digestion," *Water Res.,* **3,** 385 (1969).

162. TSCA, "Toxic Substances Control Act Premanufacture Testing of New Chemical Substances, Guidance for Premanufacture Testing," *Fed. Regist.,* **44,** 16240 (1979).

163. Watson, J.R., "Seasonal Variation in the Biodegradation of 2,4-D in River Water," *Water Res.,* **11,** 153 (1977).

164. Webley, D.M., R.B. Duff and V.C. Farmer, "Formation of β-hydroxy Acid as an Intermediate in the Microbial Conversion of Monochlorophenoxybutyric Acids to the Corresponding Acetic Acids," *Nature (London),* **183,** 748 (1959) as quoted in [1].

165. Williams, E.F., "Properties of O,O-diethyl O-p-nitrophenyl Thiophosphate and O,O-diethyl O-p-nitrophenyl Phosphate," *Ind. Eng. Chem.,* **43,** 950 (1951) as quoted in [55].

166. Williams, P.P., "Metabolism of Synthetic Organic Pesticides by Anaerobic Microorganisms," *Residue Rev.,* **66,** 63 (1977).

167. Wolfe, N.L., R.G. Zepp and D.F. Paris, "Carbaryl, Propham, and Chloropropham: A Comparison of the Rates of Hydrolysis and Photolysis with the Rate of Biolysis," *Water Res.,* **12,** 565 (1978).

168. Wolfe, N.L., D.F. Paris, W.C. Steen and G.L. Baughman, "Correlation of Microbial Degradation Rates with Chemical Structure," *Environ. Sci. Technol.,* **14,** 1143 (1980).

169. Wood, E.J.F., *Marine Microbial Ecology,* Chapman and Hall, London (1965).

170. Wood, E.J.F., *Microbiology of Oceans and Estuaries,* Elsevier, Amsterdam (1967).

171. Woodstock, D., "Microbial Degradation of Fungicides, Fumigants and Nematocides," in *Pesticide Microbiology,* ed. by I.R. Hill and S.J.L. Wright, Academic Press, New York (1978).

172. WPCF (Water Pollution Control Federation), Biodegradability Subcommittee, "Required Characteristics and Measurements of Biodegradability," *J. Water Pollut. Control Fed.,* **39,** 1232 (1967).

173. Wuhrmann, K., "River Bacteriology and the Role of Bacteria in Self-Purification of Rivers," in *Principles and Applications in Aquatic Microbiology,* ed. by H. Heukelekian and N.C. Dondero, John Wiley & Sons, New York (1964).

174. Wuhrmann, K., "Stream Purification," in *Water Pollution Microbiology,* ed. by R. Mitchell, Wiley Interscience, New York (1972).

175. Wurtz-Arlet, J., "Disappearance of Detergents in Algal Cultures," *Surfactant Congress No. 4,* **3,** 937 (1967) as quoted in [155].

176. Zobell, C.E., "Assimilation of Hydrocarbons by Microorganisms," *Adv. Enzymol.,* **10,** 443 (1950) as quoted in [1].

177. Zobell, C.E., "Domain of the Marine Microbiologist," in *Symposium on Marine Microbiology,* ed. by C.H. Oppenheimer, Charles C. Thomas, Springfield, Ill. (1963).

10

ATMOSPHERIC RESIDENCE TIME

Warren J. Lyman

10-1 INTRODUCTION

The residence time of a chemical in a specified atmospheric compartment (total atmosphere, troposphere, stratosphere, etc.) is well defined only under steady-state conditions, i.e., when the total mass and the statistical distribution in the compartment do not vary with time. In such cases the residence time, τ, may be simply defined as the ratio of the total mass in the compartment (Q) to the total emission rate (E) or removal rate (R):

$$\tau \equiv Q/E = Q/R \qquad (10\text{-}1)$$

In this equation, E is the sum of all land, fresh water and ocean emissions to the atmosphere plus any other inputs, such as in-situ generation in the atmosphere. Similarly, R is the sum of all losses from the compartment, not only by outflow to land, ocean, and space but also by in-situ degradation. More formal definitions of atmospheric residence time (also called "turn-over time" or "average transit time") are given by Bolin and Rodhe [3] and Slinn [31].

Note that τ is *not* the same as the "average age" of a pollutant molecule in the compartment, nor is it equal to the "half-life" of the pollutant. When the removal rate (R) for a chemical is due solely to some first-order loss process, then the half-life ($t_{1/2}$) and τ are related as follows:

$$t_{1/2} = 0.693 \, \tau \qquad (10\text{-}2)$$

Atmospheric residence time, unlike the other properties discussed in this handbook, cannot be directly measured.[1] It must be calculated or inferred on the basis of a simplified model of the atmosphere. When very simple models are chosen (steady-state, well-mixed atmosphere, uniform distribution of sources and sinks, etc.), the only requisite chemical-specific data may be such numbers as total emission rates, atmospheric concentrations, and/or reactivities.

Since residence time cannot be directly measured, this chapter can give no "measured values" with which estimated values can be compared. In the few cases where adequate data are available, an estimate may be compared with others obtained by different methods, but there are no firm rules for deciding which is the most reliable.

The number of ways by which atmospheric residence times can be estimated is limited only by the imagination of the modeler, the available data, and the computation facilities available. The five methods described in this chapter, which are listed below, can be solved without a computer and do not require voluminous data in the computations.

- Steady-state model [3,31]
- Nonsteady-state model, one compartment [25,30]
- Nonsteady-state model for two compartments [26,30]
- Use of chemical reactivity data [4,31]
- Correlation with mean standard deviation (Junge's correlation) [17]

These methods are intended only for calculating tropospheric residence times,[2] although residence times in other compartments (total atmosphere, stratosphere, portions of the troposphere, etc.) may be calculated if the appropriate data are available. The above methods are applicable to both organic and inorganic chemicals, although reactivity of inorganic chemicals is not addressed here.

---

1. It is possible, if wall interference is negligible, to measure the rate of disappearance of a chemical in a small test chamber designed to simulate atmospheric conditions. This can yield valuable information on losses to be expected via chemical or photochemical reactions, but it may not give a true indication of the residence time of relatively unreactive chemicals.

2. The troposphere, which extends up to about 8-12 km, is a well-mixed compartment (for most pollutants) and contains a large fraction of the atmospheric mass; the overlying stratosphere is not well mixed. The lower part of the troposphere (the earth's boundary layer), which extends from the earth's surface up to approximately one kilometer, generally contains higher levels of some components such as carbon dioxide, water, and suspended particulate matter.

Estimation of the tropospheric residence time for a chemical can yield valuable insights into its atmospheric fate and the effectiveness of the processes by which it is removed. If, for example, a chemical has a residence time of ten years or more, appreciable quantities may enter the stratosphere, where special reactions (e.g., ozone depletion) may be a concern. Such a residence time also indicates that tropospheric degradation (by direct photolysis, reaction with hydroxyl radical or ozone, etc.) and removal via wet and dry fallout occur very slowly relative to the input rate.

However, atmospheric residence time is not an intrinsic property of a chemical, nor is it even well defined for a given chemical in a specified compartment. It is a rough measure, averaged over both space and time, of the input fluxes and removal processes acting on the chemical in a somewhat arbitrarily defined atmospheric compartment.

The residence time of a pollutant is affected by many factors, such as latitude, input flux, and various atmospheric phenomena. Most of the latter are associated with characteristic time scales. For example, precipitation and the temperature and density of the atmosphere are subject to seasonal cycles; the prevalence of OH radicals follows both diurnal and seasonal cycles, because the formation of these radicals is light-induced. Some other time scales are listed in Table 10-1.

10-2 SELECTION OF APPROPRIATE METHOD

Method selection should be based on several considerations, including:

(1) the appropriateness of the method and the assumptions implied,

(2) the nature and quality of the available data,

(3) the value of τ, and, to a lesser degree,

(4) the complexity of the calculations.

For each of the five methods described in this chapter, Table 10-2 lists information on data requirements, most applicable ranges, and advantages and limitations. The data requirements are more explicitly described in Table 10-3. Table 10-2 should be used to make a preliminary selection of the most appropriate methods; a value of τ should be calculated by each of these and the results reviewed before one value is chosen.

TABLE 10-1

Time Scales for Atmospheric Phenomena

Process	Typical Time Scale	Ref.
Precipitation or nucleation scavenging	1 week	[18,31]
Vertical mixing time of troposphere	1 week	[31]
Horizontal mixing time of troposphere	1 year	[31]
Mixing between northern and southern hemispheres	1 year	[26,29]
Movement from troposphere to lower stratosphere	4 years[a]	[18]
Movement from lower stratosphere to troposphere	1 year[a]	[18]

a. Time required for exchange of air between the specified compartments. Movement from the troposphere to the lower stratosphere takes longer than the reverse process because the troposphere contains about four times as much air as the stratosphere.

In addition to the disadvantages listed in Table 10-2, all of the methods except the one based on chemical reactivity data require fairly accurate and extensive measurements of atmospheric concentrations so that a valid average value can be obtained. This average is then used to calculate the atmospheric (or tropospheric) burden, Q, for the chemical.

As previously mentioned, it is not possible to determine the absolute error associated with the estimation of atmospheric residence times, since they cannot be directly measured. Nonetheless, some aspects of the likely errors are suggested by (1) the accuracy of the input data used, (2) the appropriateness of the method selected, and (3) the values of τ obtained from the different methods used. The uncertainty associated with the accuracy of the input data should always be calculated by evaluating the propagation of errors in the equations involved. The uncertainty associated with items (2) and (3) may be assessed with the help of the information in Tables 10-2 and 10-4.

TABLE 10-2

Estimation Methods Considered

Section	Method [Ref.]	Information Required[a]	Most Applicable Range for τ	Advantages	Disadvantages and Limitations
10-3	Steady-state model [3,31]	Q, E (or R)	> 1 yr[b]	— Minimal data requirements — Simple calculation	— Valid only for steady-state conditions — E and R difficult to estimate accurately — Q uncertain, since it derives from a somewhat arbitrarily chosen tropospheric mass
10-4	Nonsteady-state, one-compartment model [25,30]	Q, A, b	> 1 yr[b]	— Allows consideration of non-steady-state conditions — Simple calculation	— Q, A, and b difficult to estimate accurately — Limited to cases where emissions increase exponentially with time
10-5	Nonsteady-state, two-compartment model [26,30]	Q_N, Q_S, A, b, τ_e	> 1 yr[b]	— Allows consideration of non-steady-state conditions — Allows northern and southern hemispheres to be considered as separate compartments	— Data requirements large — Q_N, Q_S, A, and b difficult to estimate accurately — Limited to cases where emissions increase exponentially with time — Calculations relatively difficult; may be no solution for some sets of input data

(Continued)

TABLE 10-2 (Continued)

Section	Method [Ref.]	Information Required[a]	Most Applicable Range for τ	Advantages	Disadvantages and Limitations
10-6	Use of chemical reactivity data [4,31]	k_{OH}, k_{O_3}, etc. [OH], [O_3], etc.	< 1 yr	— Minimal data requirements — Simple calculation — Only method that does not require data on concentrations in atmosphere — Only method for reactive chemicals (i.e., when τ is on the order of hours to a few days)	— Calculated τ must be considered a maximum, since other reactions (for which no rate constant is available) may be important — Rate constants (k) and reactant concentrations may have large uncertainties; k, if measured at high temperature, must be extrapolated to ambient temperature
10-7	Correlation with mean standard deviation (Junge's Correlation) [17]	C, σ	$3 - 10^3$ yr	— Minimum data requirements — Simple calculation	— Probably valid only when steady state conditions are at least approximately fulfilled and when sources and sinks are evenly distributed — Measurements of C must be accurate enough so that σ reflects natural variability and not sampling and analysis errors

a. See Table 10-3 for details.

b. This value assumes the whole troposphere is being considered; the method is appropriate for smaller values of τ if smaller compartments are being considered.

TABLE 10-3

Data Required for Estimation

Section	Method	Required Data
10-3	Steady-state model	(1) Average concentration of chemical in troposphere (C); this is used to estimate total mass of chemical in troposphere (Q).
		(2) Rate of emission of chemical into troposphere (E) or Rate of removal of chemical from troposphere (R).
10-4	Nonsteady-state, one-compartment model	(1) Average concentration of chemical in troposphere (C); this is used to estimate total mass of chemical in troposphere (Q).
		(2) Year-by-year emissions inventory for chemical; this is used to obtain cumulative emissions (A) and the parameter (b) in the exponential expression for the rate of emission in recent years.
10-5	Nonsteady-state, two-compartment model	(1) Average concentrations of chemical in both northern and southern hemispheres; these are used to estimate the total mass of chemical in the northern (Q_N) and southern (Q_S) hemispheres.
		(2) Year-by-year emissions inventory for chemical; this is used to obtain cumulative emissions (A) and the parameter (b) in the exponential expression for the rate of emission in recent years.
		(3) Interhemispheric exchange rate (τ_e); this may be taken as $\sim$ 1.2 years.
10-6	Use of chemical reactivity data	(1) Rate constants for reaction of chemical with hydroxyl radical (k_{OH}), ozone (k_{O_3}), and other reactants, if any. (Tables of measured values of k_{OH} and k_{O_3} are given from which appropriate surrogates may be selected for some chemicals.)
		(2) Concentration of hydroxyl radical, [OH·], ozone [O_3], and any other reactant being considered. (A table of default values is given for [OH·] and [O_3] and may be used if site-specific data are not available.)
10-7	Correlation with mean standard deviation (Junge's correlation)	(1) Average concentration of chemical in troposphere (C).
		(2) Standard deviation (σ) associated with average concentration.

TABLE 10-4

Estimated Atmospheric Residence Times for Selected Chemicals

Compound	Residence Time (τ)	Compartment	Basis for Calculation	Ref.
Methane	3.1 yr 3.8 yr	N. Troposphere S. Troposphere	Junge's correlation with data from [29]	a
	4 yr	Troposphere	Best estimate (used by Junge [17]) after considering data from several sources. Estimate uncertain by factor of 3	[17]
Gaseous non-methane hydrocarbons:				
(a) Considering only anthropogenic sources	2-3 yr	Troposphere	Q/E (Q = 1.02×10^{11} kg, E = 4.5×10^{10} kg/yr)	[11]
(b) Considering anthropogenic plus natural sources	0.5 yr	Troposphere	Q/E (Q = 1.02×10^{11} kg, E = 1.95×10^{11} kg/yr)	[11]
Ethene	3 hr[b]	Troposphere	Reaction with OH radical in polluted atmospheres ([OH·] = 10^7 cm^{-3}); data from [7]	a
Formaldehyde	60 d	Troposphere	Junge's correlation with data from [19]	a
Phenol	50 d	Troposphere	Junge's correlation with data from [19]	a

(continued)

Compound	Residence Time (τ)	Compartment	Basis for Calculation	Ref.
Methyl Chloride	1 yr	N. Troposphere	Junge's correlation with data from [29]	a
	0.8 yr	S. Troposphere	Junge's correlation with data from [29]	a
	0.37 yr[c]	Troposphere	Reaction with OH radical	[6]
	2.3 yr	Troposphere	Reaction with OH radical ($[OH\cdot] = 3 \times 10^5 - 5 \times 10^5$ cm^{-3})	[29]
	~2 yr	Troposphere	Q/E with oceans assumed to be major source of the chemical	[29]
Methyl Iodide	6 d	Troposphere	Junge's correlation with data from [20]	a
	50 hr	Atmosphere	Photolysis assumed to be sole destructive process	[12]
	<~20 – ~150 d	Troposphere	Q/E (with concentration <5 – ~35 ppt)	[9]
	~40 – 230 d	Troposphere	Reaction with OH radical ($[OH\cdot] = 10^6$ cm^{-3})	[9]
	~12 hr – ~1 yr	Troposphere	Rate of disappearance in simulated tropospheric reaction chamber	[9]
Methylene Chloride	110 d[c]	Troposphere	Reaction with OH radical	[6]
	~160 d ~250 d	N. Troposphere S. Troposphere	Junge's correlation with data from [29]	a
	100 d	N. Troposphere	Junge's correlation with data from [29]	a
Chloroform	1.7 yr	Troposphere	One-compartment, nonsteady-state model	[30]

(continued)

TABLE 10-4 (Continued)

Compound	Residence Time (τ)	Compartment	Basis for Calculation	Ref.
	~ 1.4 yr	Troposphere	Junge's correlation	[20]
	3.5 yr	N. Troposphere	Junge's correlation with data from [29]	a
	4.2 yr	S. Troposphere	Junge's correlation with data from [29]	a
Carbon Tetrachloride	~ 20 yr	Troposphere	Q/E with data from [1]	a
	~ 10 – 20 yr	Troposphere	Q/E with data from [27]	a
	> 330 yr[c]	Troposphere	Reaction with OH radical	[6]
	30 – 50 yr[d]	Troposphere and Stratosphere	Q/E, where Q calculated with consideration of photodissociation in stratosphere	[21]
	100 yr	Troposphere	Q/R, where R calculated from consideration of oceans as sink	[29]
	1.2 yr	Troposphere	Junge's correlation with data from [29]	a
Dichlorodifluoromethane	67 – 70 yr	Troposphere	Two-compartment, nonsteady-state model	[29]
	50 yr	N. Troposphere	One-compartment, nonsteady-state model	[30]
	1 yr	Troposphere	Junge's correlation with data from [20]	a
Trichlorofluoromethane	1.4 yr	Troposphere	Junge's correlation with data from [29]	a
	6 yr	Troposphere	Q/E with data from [23]	a
	1000 yr[c]	Troposphere	Reaction with OH radical	[6]
	40 – 45 yr	Troposphere	Two-compartment, nonsteady-state model	[29]

(continued)

Compound	Residence Time (τ)	Compartment	Basis for Calculation	Ref.
Trichlorofluoromethane (cont.)	36 yr	N. Troposphere	One-compartment, nonsteady-state model	[30]
	15 – 20 yr	Troposphere	More complex mathematical model of troposphere and stratosphere	[16]
	8 – 10 yr	Troposphere	Two-compartment, nonsteady-state model	[25,29]
Methyl Chloroform	7.2 yr	N. Troposphere	One-compartment, nonsteady-state model	[30]
	1.1 yr[c]	Troposphere	Reaction with OH radical	[6]
	4 d	Troposphere	Q/E	[22]
	~ 200 d	Troposphere	Junge's correlation with data from [29]	a
	150 d	N. Troposphere	One-compartment, nonsteady-state model	[30]
	150 – 390 d	N. Troposphere	Reaction with OH radical using annual average of $3 - 8 \times 10^{-19}$ moles/cm^3 for [OH·] (Table 10-7). $k_{OH} = 1.0 \times 10^{11}$ cm$^3 \cdot$ mole$^{-1} \cdot$ sec^{-1} (Table 10-5)	a
Tetrachloroethylene	1 – 7 d	Polluted atmosphere in lower troposphere with constant sunlight	Reaction with OH radical using $1.7 - 17 \times 10^{-17}$ moles/cm^3 for [OH·] (Table 10-7). $k_{OH} = 1.0 \times 10^{11}$ cm$^3 \cdot$ mole$^{-1} \cdot$ sec^{-1} (Table 10-5)	a

(continued)

TABLE 10-4 (Continued)

Compound	Residence Time (τ)	Compartment	Basis for Calculation	Ref.
Chloracetyl chlorides:				
$CClH_2COCl$	$13{,}000$ yr[b,e]	Atmosphere	Hydrolysis rate constant	[5]
CCl_3COCl	450 yr[b,e]	($T = 298$ K,	measured at $T = 470\text{-}620$ K	
$COCl_2$	110 yr[b,e]	$P_{H_2O} = 10$ mm Hg)		
	3.3 yr	Troposphere	Q/R, where R based on assumption that rainfall is dominant mechanism for removal from air	[37]
DDT	50 d	Troposphere (over Sargasso Sea)	Q/R, where R based on assumption that major source of DDT input to oceans is via the atmosphere	[2]
Polychlorinated biphenyls	40 d	Troposphere (over Sargasso Sea)	Q/R, where R based on assumption that major source of PCB input to oceans is via the atmosphere	[2]

a. The value listed was calculated by the author of this chapter. In many cases these calculations were carried out even though it was known that the method of calculation might not be applicable. Thus, these numbers should be considered more as examples of the estimation method than as valid estimates of τ.

b. The value listed is the half-life.

c. Other information given in Ref. 6 indicates that the values of τ listed here may be low by as much as a factor of 3 for some chemicals.

d. The estimate of 30-50 years may be too low, according to Ref. 27. The rate of photodissociation assumed in the calculations is considered to be too high.

e. Butler and Snelson [5] concluded from their data that homogeneous gas-phase hydrolysis is not an efficient conversion process. They suggested that heterogeneous hydrolysis by water droplets would be a more efficient scavenging process for these compounds. This could involve typical residence times as short as one week (see Table 10-1).

Table 10-4 lists the calculated residence times for a number of chemicals by one or more methods. For purposes of illustration, some methods of calculation (primarily Junge's correlation) are used in situations where they might not be appropriate. *Thus, the estimates given in this table should not be taken as the best estimates of τ.* Note that Junge's correlation gives values of τ on the order of one year ($\pm$ a factor of 3) for most of the halocarbons; this suggests that the reported values of standard deviation (σ) are dominated more by sampling and analysis errors than by the natural fluctuations in atmospheric concentration.

The possibility that misleading values of τ will be calculated from chemical reactivity data is exemplified in the estimates given for trichlorofluoromethane and carbon tetrachloride in Table 10-4. If one had calculated τ on the assumption that reaction with OH radicals was the only tropospheric loss mechanism, erroneously high values of τ would be predicted. The steady-state model (using Q/E) appears to give lower values for τ than the two nonequilibrium models, which are presumably more accurate; this is seen for trichlorofluoromethane, carbon tetrachloride, and tetrachloroethylene. It is not clear, however, if this result should be generally expected or whether it is unique to these cases.

10-3   STEADY-STATE MODEL

Principles of Use. The use of a steady-state model for estimating tropospheric residence time is limited to those cases where the total "growth" rate of the chemical in the troposphere (caused by emissions from land and oceans plus input, if any, from the stratosphere) may be assumed to equal the total "removal" rate of the chemical in the same compartment via outflow plus chemical degradation. This assumption of steady state will generally be appropriate when the residence time is large compared with the time scale for atmospheric fluctuations. If emissions of the chemical to the atmosphere have been relatively constant for a number of years and τ is expected to be about a year or more, the method is probably appropriate to use for the whole troposphere.

The residence time may be calculated with Eq. 10-3 or 10-4.

$$\tau = Q/E \qquad (10\text{-}3)$$

$$\tau = Q/R \qquad (10\text{-}4)$$

where Q is the total mass of the pollutant in the troposphere, E is the total emissions rate, and R is the total removal rate. Consistent units of mass must be used for Q, E and R (Q has units of mass and E and R units of mass/time), and the data used to obtain these values should come from the same time period.

The value of Q (in grams) may be obtained from

$$Q = C\,(4\times10^{21})/10^9 \qquad\qquad (10\text{-}5)$$

where C is the global average tropospheric concentration (in parts per billion by weight)[3] and 4×10^{21} is the assumed mass [31] of the troposphere in grams. The concentration of man-made pollutants is generally greater in the northern hemisphere than in the southern hemisphere (see, for example, Refs. 28 and 29); therefore, values of Q calculated from northern hemispheric data only may be too large, especially for chemicals with short residence times. Even some naturally occurring chemicals may have different concentrations in the northern and southern hemispheres. Methane, for example, which is presumed to be emitted primarily from land masses, would have a higher concentration in the northern hemisphere, since the ratio of land masses in the northern and southern hemispheres is about 2.4:1.

Equation 10-3 ($\tau = Q/E$) is used more frequently than Eq. 10-4 ($\tau = Q/R$). The value of E must be obtained from as accurate an emissions inventory as possible for the whole earth. There are no easy shortcuts or firm guidelines for estimating E; the user must evaluate all known sources and use the best available emissions data for each. (Some guidance in the basic methodology used to estimate emission rates is given in Appendices A and B of Ref. 4.) Values of R are no easier to estimate. One mechanism that can be considered for calculating a value of R is rainout. (See examples for DDT and PCBs in Table 10-4.) If rainfall is expected to be the major atmospheric loss mechanism, R (in grams per year) may be obtained from

$$R = C_p(4.2\times10^{20})/10^9 = C_p(4.2\times10^{11}) \qquad\qquad (10\text{-}6)$$

where C_p is the global average concentration of the pollutant in the precipitation (in parts per billion by weight) and 4.2×10^{20} is the assumed [34] annual precipitation (grams/year) for the earth.[4]

---

3. Atmospheric concentrations are often reported on a volume per volume basis (v/v), e.g., 50 ppb(v/v). To convert to a weight per weight basis (w/w) for C, multiply this number by the molecular weight of the pollutant and divide by the molecular weight of air (~28.9). If measurements are reported on a weight per volume basis (e.g., 50 ng/m³), convert this to a w/w basis by using 1205 g/m³ (20°C, 76 cm Hg) or 1293 g/m³ (0°C, 76 cm Hg) as the density of air and adjust the scale of the units so that C is expressed in ppb by weight.

4. This is equivalent to an annual rainfall of 81 cm (32 inches).

For unreactive gases, if the average concentration in precipitation is not known, it may be estimated from

$$C_{(precipitation)} \simeq C_{(air)} / H \tag{10-7}$$

where H is Henry's law constant in the appropriate units [32]. Equation 10-7 should be used to estimate concentrations in rain only; snow scavenging of gases can generally be ignored [32]. Estimation methods for H are given in Chapter 15. (See especially §15-6.)

Basic Steps

(1) Estimate the mass (Q) of the chemical in the compartment of interest. If the whole troposphere is being considered, Eq. 10-5 may be used.

(2) If the emission rate (E) can be estimated, use Eq. 10-3 to estimate τ.

(3) If the removal rate (R) can be estimated, use Eq. 10-4 to estimate τ. If rainfall is expected to be the principal removal mechanism, Eqs. 10-6 and 10-7 may be used in situations where the concentration in rain is not known.

Example 10-1   Estimate the tropospheric residence time for methylene chloride, given an estimated global anthropogenic emission rate of 3.5×10^{11} g/yr and a mean tropospheric concentration of 30 ppt (by weight).

(1)   From Eq. 10-5, $Q = .03 (4 \times 10^{21})/10^9 = 1.2 \times 10^{11}$ g.

(2)   Then, with Eq. 10-3,

$$\tau = (1.2 \times 10^{11} \text{ g})/(3.5 \times 10^{11} \text{ g/yr})$$

$$= 0.34 \text{ yr}$$

Example 10-2   Estimate the tropospheric residence time for methane based on the assumption that rainfall is the principal removal mechanism. (This assumption is not valid but is used for purposes of the example.) The concentration of methane in air is about 1.4 ppm (v/v), and the (dimensionless) value of Henry's law constant is 25.7 at 15°C [32].

(1)   Convert air concentration to weight/weight basis.
   C = 1.4 (molecular wt. of CH_4/molecular weight of air)
      = 1.4 (16/28.9)
      = 0.78 ppm (w/w) or 780 ppb (w/w)

(2)   From Eq. 10-7, $C_{(precipitation)}$ = 0.78 ppm/25.7 = 30 ppb (w/w)

(3)   Equation 10-6 gives $R = 30 (4.2 \times 10^{20})/10^9 = 1.3 \times 10^{13}$ g/yr for the removal rate.

(4)   The tropospheric burden is obtained from Eq. 10-5 as
$$Q = 780 \, (4 \times 10^{21})/10^9 = 3.1 \times 10^{15} \text{ g}.$$

(5)   Finally with Eq. 10-4,
$$\tau = (3.1 \times 10^{15} \text{ g})/(1.3 \times 10^{13} \text{ g/yr})$$
$$= 240 \text{ yr}$$

Since this estimate is much larger than Junge's estimate of ~ 4 yrs [17], one can conclude that rainfall is not an important removal mechanism for methane.

10-4   NONSTEADY-STATE, ONE-COMPARTMENT MODEL

Principles of Use. Nonsteady-state methods are appropriate for anthropogenic pollutants for which recent emissions have been increasing exponentially. The one-compartment model considers the whole troposphere as the compartment of interest, making no distinction between the northern and southern hemispheres. The calculation is simpler than with the two-compartment model described in the following section, but the estimates of τ may be less reliable if pollutant concentrations and removal rates differ significantly between the two hemispheres.

With this model the tropospheric residence time is given [25,30] by

$$\tau = \frac{\theta}{b\,(1-\theta)} \tag{10-8}$$

where b is the coefficient in the exponential expression for the emission rate (E) as a function of time (t)

$$E = ae^{bt} \tag{10-9}$$

and θ is the ratio, calculated at a specified time, of the tropospheric burden (Q) to the cumulative emissions (A) of the chemical since it was first produced; i.e.

$$\theta = Q/A \tag{10-10}$$

Basic Steps

(1) Prepare a year-to-year emissions inventory for the chemical, listing the total emissions in each year since substantial amounts were produced.

(2) Determine the cumulative emissions, A, by summing the numbers.

(3) Estimate the tropospheric burden, Q, from the global average tropospheric concentration (C) with Eq. 10-5. Note that Q must be calculated for the same time up to which A was obtained.

(4) Calculate θ with Eq. 10-10.

(5) Plot the emissions vs. time as ln E vs. t, and obtain b from the slope of the line. Note that b has units of time^{-1}.

(6) Calculate τ with Eq. 10-8.

Example 10-3 Estimate the tropospheric residence time for dichlorodifluoromethane (mol. wt. = 120.93), given a measured tropospheric concentration of 0.19 ppb (v/v) and the year-by-year emissions inventory (synthesized for this example) given below.

Year	Emissions (E) g/yr x 10^{-11}	ln E
1961	1.0	25.3
1962	1.4	25.7
1963	1.8	25.9
1964	2.0	26.0
1965	2.2	26.1
1966	2.3	26.2
1967	2.6	26.3
1968	2.7	26.3
1969	3.2	26.5
1970	3.4	26.6
1971	3.6	26.6
1972	4.3	26.8
1973	4.6	26.8
1974	5.0	26.9
1975	5.5	27.0
Total	45.6	

(1) The cumulative emissions (A) of 45.6 x 10^{11} g is the sum of the yearly emissions. Emissions prior to 1961 are assumed to be small compared with this total.

(2) Convert the measured atmospheric concentration to units of ppb by weight (see footnote 3).

$$C = 0.19 \text{ ppb (v/v)} (120.93/28.9) = 0.80 \text{ ppb (w/w)}$$

(3) From Eq. 10-5, the tropospheric burden is

$$Q = 0.80 (4 \times 10^{21})/10^9 = 3.2 \times 10^{12} g$$

(4) From Eq. 10-10, $\theta = 3.2 \times 10^{12} g/4.56 \times 10^{12}$ g = 0.70

(5) Plot ln E (values given above) as shown in Figure 10-1. The slope, b, of the best-fit line through the points (drawn by eye) is found to be 0.10 yr^{-1}.

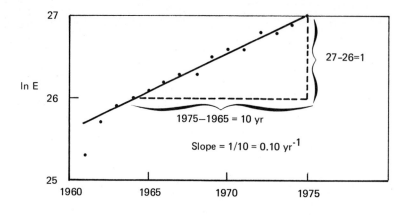

FIGURE 10-1   Plot of Yearly Emissions for Example 10-3

(6)   Then, with Eq. 10-8,

$$\tau = \frac{0.70}{0.10\,(1{-}0.70)} = 23 \text{ yr}$$

10-5 NONSTEADY-STATE, TWO-COMPARTMENT MODEL

Principles of Use. The two-compartment model is appropriate for anthropogenic pollutants for which tropospheric burdens and removal rates in the northern and southern hemispheres are expected to differ. Since tropospheric burdens must be estimated for each hemisphere (Q_N and Q_S), concentration data are required for both. The inter-hemispheric exchange rate (τ_e) is also required; various estimates put τ_e in the range of 1 to 1.4 years, with 1.2 years being an acceptable value for most calculations [23,26,29]. The method allows the estimation of a chemical's residence time in both the northern (τ_N) and southern (τ_S) hemispheres as well as a global average residence time (τ). All of the emissions are assumed to be in the northern hemisphere.

With this model [26,30], the global average residence time (τ) is obtained from

$$\tau = \frac{1 + Q_N/Q_S}{1/\tau_S + (1/\tau_N)\,(Q_N/Q_S)} \qquad (10\text{-}11)$$

τ_S is obtained from

$$\tau_S = \cfrac{1}{\cfrac{1}{\tau_e}\left(\cfrac{Q_N}{Q_S} - 1\right) - b} \qquad (10\text{-}12)$$

where, as in Eq. 10-9, b is the coefficient in the exponential expression for the emissions growth rate. The value of a parameter, D, is next calculated as follows:

$$D = \frac{\theta}{b\,(1 + Q_N/Q_S)} \qquad (10\text{-}13)$$

where $\theta = (Q_N + Q_S)/A$ and A is the cumulative emissions of the chemical since it was first produced. The value of τ_N is then obtained from the following two equations:

$$D = 1/[\tau_e\,(b-\alpha)\,(b-\beta)] \qquad (10\text{-}14)$$

where α and β are the roots[5] of the equation

$$p^2 + p\left(\frac{2}{\tau_e} + \frac{1}{\tau_S} + \frac{1}{\tau_N}\right) + \frac{1}{\tau_e\,\tau_N} + \frac{1}{\tau_e\,\tau_S} + \frac{1}{\tau_N\,\tau_S} = 0 \qquad (10\text{-}15)$$

The calculation of τ from Eq. 10-11, τ_S from Eq. 10-12 and D from Eq. 10-13 is straightforward. However, once the value of D is known, Eqs. 10-14 and -15 must be solved by trial and error; various values of τ_N are used until a matching value of D (calculated with Eq. 10-14) is found. It can be shown that the values of α and β will always be real numbers; i.e., the solution of Eq. 10-15 does not involve complex numbers. The value of τ_N obtained from Eqs. 10-14 and -15 may be quite sensitive to the values of both θ and τ_e; τ_S is somewhat less sensitive, and τ (the global average) is quite insensitive to these parameters. For some values of θ it is possible that no value of D obtained from Eq. 10-14 will match that obtained from Eq. 10-13. In this case the method must be abandoned.

For most chemicals, especially those subject to reaction with OH radicals, τ_N should be greater than τ_S, as the concentration of OH radicals in the southern hemisphere is about twice that in the northern hemisphere. (See Table 10-7 in the following section.) Thus, a reason-

---

5. For a quadratic equation of the form $p^2 + mp + n = 0$, the roots are given by:

$$\alpha = \frac{-m + \sqrt{m^2 - 4n}}{2} \quad \text{and} \quad \beta = \frac{-m - \sqrt{m^2 - 4n}}{2}$$

able first guess for τ_N when solving Eqs. 10-14 and -15 by trial and error is $2\tau_S$. If the first guess for τ_N gives a value of D from Eq. 10-14 larger than that from Eq. 10-13, use a smaller value of τ_N for the second guess. If desired, the values of D from these first two guesses may be plotted against τ_N; the third guess for τ_N can then be taken off the straight line through these two points, since the "correct" value of D (from Eq. 10-13) is known. This iterative procedure should require only a few trial-and-error solutions and should not be taken beyond the point where two significant figures for τ_N have been obtained.

Basic Steps

(1) Prepare a year-by-year emissions inventory for the chemical, listing the total emissions in each year since substantial amounts were produced.

(2) Determine the cumulative emissions, A, by summing the numbers.

(3) Plot emissions vs. time as ln E vs. t and obtain b from the slope of the line.

(4) Calculate the total mass of the chemical in the troposphere $(Q_N + Q_S)$ using the global average tropospheric concentration and Eq. 10-5. Note that Q must be calculated for the same time up to which A was calculated.

(5) Use Eq. 10-12 to calculate τ_S. A value of 1.2 years may be used for τ_e. Q_N/Q_S may simply be obtained from the ratio of the average atmospheric concentrations in the two hemispheres.

(6) Use Eq. 10-13 to calculate D.

(7) Using the values of τ_e and τ_S from above, try different values of τ_N in Eq. 10-15 until the value of D obtained from Eq. 10-14 matches that obtained in step (6). A value of $\tau_N = 2\tau_S$ should provide a reasonable first guess for τ_N. Use a lower value of τ_N for the second (and subsequent) guess(es) if D from Eq. 10-14 is greater than D from Eq. 10-13.

(8) Calculate the global average residence time (τ) from Eq. 10-11.

Example 10-4   Estimate τ_S, τ_N, and τ for methyl chloroform, given b = 0.17 yr^{-1}, τ_e = 1.2 yr, Q_N/Q_S = 1.47, and θ = 0.59

(1)   With Eq. 10-12,

$$\tau_S = \frac{1}{\dfrac{1}{1.2}\,(1.47-1) - 0.17}$$

$$= 4.5 \text{ yr}$$

(2)   From Eq. 10-13,

$$D = \frac{0.59}{0.17\,(1 + 1.47)}$$

$$= 1.41$$

(3)   Using Eq. 10-15 and, subsequently, Eq. 10-14 for various values of τ_N shows $\tau_N \simeq 20$ yr. Values of α, β and D for three values of τ_N are as follows:

τ_N	α	β	D
9 yr	−0.16	−1.84	1.24
15 yr	−0.14	−1.83	1.34
20 yr	−0.13	−1.82	1.40
∞	−0.11	−1.80	1.51

(4)   Then, from Eq. 10-11, the global average residence time is

$$\tau = \frac{1 + 1.47}{1/4.5 + (1/20)\,1.47} = 8 \text{ yr}$$

10-6   USE OF CHEMICAL REACTIVITY DATA

Principles of Use. The residence time of a chemical in the atmosphere may be estimated if the rate constants for one or more destruction or removal processes are known. The reactions most frequently considered are those with hydroxyl radicals and with ozone. Rate constants for reactions with such species are relatively easy to measure in the laboratory, although the experiments must sometimes be carried out at elevated temperatures. The value of τ estimated from such data should be viewed as a maximum, since other reactions for which no data are available (e.g., direct or sensitized photolysis, reaction with other radicals) may proceed more rapidly than the ones considered.

For reactions with such species as ozone and the hydroxyl radical, a reaction that is first-order with respect to the pollutant is generally found (and thus is usually presumed for other chemicals), and the rate constant reported, k, is usually the bimolecular rate constant. Thus, the rate of decrease in the concentration (C) of the chemical is given by:

$$\frac{-dC}{dt} = k_{OH} \, [OH\cdot] \, [C] \qquad\qquad (10\text{-}16)$$

or

$$\frac{-dC}{dt} = k_{O_3} \, [O_3] \, [C] \qquad\qquad (10\text{-}17)$$

Similar equations could be written for other reactions. The rate constant is frequently given in liters·mole^{-1}·sec^{-1}; in such cases [OH·] and [O$_3$] must be expressed in units of moles/liter. Other units encountered include cm^3·mole^{-1}· sec^{-1} and cm^3·sec^{-1}; concentrations of [OH·] and [O$_3$] then must be expressed in moles/cm^3 and number/cm^3, respectively.

To calculate a residence time by this method, a value of τ is first calculated for each reaction of interest. Equations 10-18 and 10-19 show the formulas for reaction with OH· or O$_3$; an identical form may be used for any other reaction that follows similar kinetics.

$$\tau_{OH\cdot} = \frac{1}{k_{OH} \, [OH\cdot]} \qquad\qquad (10\text{-}18)$$

$$\tau_{O_3} = \frac{1}{k_{O_3} \, [O_3]} \qquad\qquad (10\text{-}19)$$

An overall value of τ is then obtained as follows [31]:

$$\frac{1}{\tau} = \frac{1}{\tau_{OH\cdot}} + \frac{1}{\tau_{O_3}} + \frac{1}{\tau_x} + \cdots \qquad\qquad (10\text{-}20)$$

where τ_x refers to the residence time associated with any other reaction for which data are available.

If reaction with OH· or O$_3$ is considered to be likely for a chemical but values of k_{OH} and k_{O_3} are not available, approximate values may be found for a similar chemical or class of chemicals in Table 10-5 (for k_{OH}) or Table 10-6 (for k_{O_3}). Additional assistance in the selection of a surrogate k_{OH} value may be found in Ref. 15 or in the relative (atmospheric) reactivity scales given in Refs. 7 and 33.

Since reaction rate constants are a function of temperature, it is desirable to use values of k that relate to the mean temperature in the compartment and the time scale of concern. Mean temperatures in the

TABLE 10-5

Rate Constants[a] for Reaction of Organic Chemicals with OH Radical at 300K

Compound	k_{OH} $(cm^3 \cdot mol^{-1} \cdot sec^{-1})$	Compound	k_{OH} $(cm^3 \cdot mol^{-1} \cdot sec^{-1})$
Alkanes		**Alcohols**	
Methane	4.8×10^9	Methanol	5.7×10^{11}
Ethane	1.7×10^{11}	Ethanol	1.8×10^{12}
Propane	1.3×10^{12}	Propanol	2.3×10^{12}
Methyl-	1.3×10^{12}	2-Propanol	4.3×10^{12}
Dimethyl-	4.8×10^{11}	Butanol	4.1×10^{12}
n-Butane	1.6×10^{12}	4-Methyl-2-pentanol	4.3×10^{11}
Methyl-	7.8×10^{11}	**O,N,S Substituted Alkanes**	
2,3-Dimethyl-	3.1×10^{12}	Methyl ether	2.1×10^{12}
2,2,3-Trimethyl-	2.3×10^{12}	Ethyl ether	5.4×10^{12}
2,2,3,3-Tetramethyl-	6.6×10^{11}	n-Propyl ether	1.0×10^{13}
n-Pentane	3.9×10^{12}	Tetrahydrofuran	8.8×10^{12}
2-Methyl-	3.2×10^{12}	1-Propylacetate	2.7×10^{12}
3-Methyl-	4.3×10^{12}	2-Butylacetate	3.4×10^{12}
2,2,4-Trimethyl-	2.3×10^{12}	Methylamine	1.3×10^{13}
n-Hexane	3.6×10^{12}	Methyl sulfide	2.0×10^{13}
n-Octane	5.1×10^{12}	Formaldehyde	9.0×10^{12}
Cycloalkanes		Acetaldehyde	9.6×10^{12}
c-Butane	7.2×10^{12}	Propionaldehyde	1.3×10^{13}
c-Pentane	3.7×10^{12}	Benzaldehyde	7.8×10^{12}
Haloalkanes		**Alkenes**	
Methane		Ethene	4.7×10^{12}
Fluoro-	9.6×10^9	Propene	1.5×10^{13}
Difluoro-	4.7×10^9	Methyl-	3.0×10^{13}
Trifluoro-	1.2×10^8	1-Butene	2.1×10^{13}
Tetrafluoro-	$< 2.4 \times 10^8$	2-Methyl-	3.5×10^{13}
Chloro-	2.4×10^{10}	3,3-Dimethyl-	1.7×10^{13}
Dichloro-	8.4×10^{10}	2-Butene	
Trichloro-	6.6×10^{10}	cis-	3.2×10^{13}
Tetrachloro-	$< 2.4 \times 10^8$	trans-	4.2×10^{13}
Bromo-	2.4×10^{10}	2-Methyl-	4.8×10^{13}
Ethane		2,3-Dimethyl-	9.2×10^{13}
Chloro-	2.3×10^{11}	1-Pentene	1.8×10^{13}
1,1-Dichloro-	1.6×10^{11}	cis-2-Pentene	3.9×10^{13}
1,2-Dichloro-	1.3×10^{11}	1-Hexene	1.9×10^{13}
1,1,1-Trichloro-	9.0×10^9	1-Heptene	2.2×10^{13}
1,1,1-Trifluoro-2-chloro-	6.0×10^9	**Cycloalkenes**	
1,1,1-Trifluoro-2,2-dichloro-	1.7×10^{10}	c-Cyclohexene	4.3×10^{13}
1,1,1,2-Tetrafluoro-2-chloro-	7.2×10^9	1-Methyl-	5.8×10^{13}
1,2-Dibromo-	1.5×10^{11}	**Haloalkenes**	
Alkanones		Ethene	
Butanone	2.0×10^{12}	Fluoro-	3.4×10^{12}
2-Methylpentanone	9.0×10^{12}	1,1-Difluoro-	1.2×10^{12}
2,6-Dimethylheptanone	1.5×10^{13}	Chloro-	3.9×10^{12}

(Continued)

TABLE 10-5 (Continued)

Compound	k_{OH} $(cm^3 \cdot mol^{-1} \cdot sec^{-1})$	Compound	k_{OH} $(cm^3 \cdot mol^{-1} \cdot sec^{-1})$
Ethene (Cont.)		**Arenes**	
Trichloro-	1.2×10^{12}	Benzene	8.4×10^{11}
Tetrachloro-	1.0×10^{11}	Methyl-	3.5×10^{12}
Chlorotrifluoro-	4.2×10^{12}	1,2-Dimethyl-	7.8×10^{12}
Bromo-	4.1×10^{12}	1,3-Dimethyl-	1.2×10^{13}
O-Substituted Alkene		1,4-Dimethyl-	6.0×10^{12}
Methoxy	2.0×10^{13}	1,2,3-Trimethyl-	1.5×10^{13}
Alkadienes		1,2,4-Trimethyl-	2.0×10^{13}
Propadiene	2.7×10^{12}	1,3,5-Trimethyl-	3.0×10^{13}
1,3-Butadiene	4.6×10^{13}	Ethyl-	4.5×10^{12}
2-Methyl-	4.7×10^{13}	1,2-Ethylmethyl-	8.2×10^{12}
Terpenes		1,3-Ethylmethyl-	1.2×10^{13}
p-Menthane	4.0×10^{12}	1,4-Ethylmethyl-	7.7×10^{12}
α-Pinene	1.5×10^{13}	Propyl-	3.6×10^{12}
β-Pinene	1.3×10^{13}	2-Propyl-	4.7×10^{12}
3-Carene	1.7×10^{13}	1,4-Methylpropyl-2-	9.1×10^{12}
Carvomenthane	2.5×10^{13}	Hexafluoro-	1.3×10^{11}
β-Phellandrone	2.3×10^{13}	Propylpentafluoro-	1.8×10^{12}
d-Limonene	2.9×10^{13}	**Substituted Arenes**	
Dihydromyrcene	3.6×10^{13}	Methoxybenzene	1.2×10^{13}
Myrcene	4.5×10^{13}	o-Cresol	2.0×10^{13}
cis-Ocimene	6.3×10^{13}		
Alkynes			
Ethyne	9.6×10^{10}		
Methyl	5.7×10^{11}		

a. $k \ (L \ mol^{-1} \ sec^{-1}) = 10^{-3} \ k \ (cm^3 \cdot mol^{-1} \cdot sec^{-1})$

Source: Hendry and Kenley [15].

TABLE 10-6

Rate Constants for Reaction of Organic Chemicals with Ozone at 300K

Compound	k_{O_3} (cm³·mole⁻¹·sec⁻¹)	Compound	k_{O_3} (cm³·mole⁻¹·sec⁻¹)
Alkanes		Trichloro-	3.6×10^3
Methane	0.84	Tetrachloro-	10.0×10^2
Ethane	0.72	Tetrafluoro-	8.1×10^7
Propane	4.1	Propene	
Methyl-	1.2	3-Chloro-	1.1×10^7
n-Butane	5.9	Hexafluoro-	1.3×10^7
Alkenes		**Terpenes**	
Ethene	1.1×10^6	α-Pinene	9.9×10^7
Propene	7.8×10^6	**Alkynes**	
Methyl-	9.1×10^6	Ethyne	4.7×10^4
1-Butene	7.4×10^6	**Aromatic Hydrocarbons**	
2-Butene		Benzene	2.8×10^1
cis-	9.7×10^7	Methyl-	1.6×10^2
trans-	1.6×10^8	1,2-Dimethyl-	9.5×10^2
2-Methyl-	3.0×10^8	1,3-Dimethyl-	7.8×10^2
2,3-Dimethyl-	9.1×10^8	1,4-Dimethyl-	9.5×10^2
1-Pentene	6.4×10^6	1,3,4-Trimethyl-	2.8×10^3
2-Pentene		1,3,5-Trimethyl-	4.2×10^3
cis-	2.7×10^8	1,2,4,5-Tetramethyl-	1.1×10^4
trans-	3.4×10^8	Pentamethyl-	5.0×10^4
1-Hexene	6.7×10^6	Hexamethyl-	2.4×10^5
1-Heptene	4.9×10^6	Ethyl-	3.4×10^2
1-Octene	4.9×10^6	1,3-Diethyl-	1.1×10^3
1-Decene	6.5×10^5	1,3,5-Triethyl-	3.4×10^3
Cyclohexene	1.0×10^8	Pentaethyl-	1.0×10^4
Conjugated Alkenes		Hexaethyl-	3.4×10^3
1,3-Butadiene	5.0×10^6	2-Propyl-	3.5×10^2
Phenylethene	1.0×10^8	t-Butyl-	6.9×10^1
Halogenated Alkenes			
Ethene			
Chloro-	1.2×10^6		
1,1-Dichloro-	2.2×10^4		
1,2-Dichloro-			
cis-	3.7×10^4		
trans-	2.3×10^5		

troposphere are about 17°C near the earth's surface and drop to −40°C to −60°C at an altitude of 10-12 km. Daily and seasonal cycles should obviously be considered; a tropospheric temperature of about −10°C might be appropriate for a mean residence time of months to years, but a value of 20°C would be appropriate for a highly reactive chemical ($\tau \leq 1$ day) being emitted into the atmosphere in a warm climate. The Arrhenius equation[6] may be used to find a temperature-adjusted value of k if laboratory data for two or more temperatures are available or if the energy of activation for the reaction is known. The change in k may be as much as one order of magnitude for a temperature change of about 50°C.

If the concentrations of OH· and/or O_3 are now known for the time period and compartment of interest, appropriate default values may be selected from those listed in Table 10-7.

Basic Steps

(1) Obtain literature values for k_{OH}, k_{O_3}, etc. and, if necessary, correct these rate constants for the temperature of the compartment being considered. If literature values are not available, surrogate values may be selected on the basis of information given in Tables 10-5 and -6 or, preferably, Ref. 15.

(2) Determine the appropriate values to use for [OH·], [O_3], etc. Default values may be selected from Table 10-7.

(3) For each reaction being considered, calculate a residence time as shown by Eqs. 10-18 and -19 for reaction with OH· and O_3.

(4) Use Eq. 10-20 to estimate an overall (maximum) residence time.

Example 10-5   Estimate the atmospheric residence time for 1,3,5-trimethylbenzene, given $k_{OH} = 31 \times 10^9$ liters · mole^{-1} · sec^{-1} [10], $k_{O_3} = 4.2 \times 10^3$ cm^3 · mole^{-1} · sec^{-1} (Table 10-6), [OH·] = 1 × 10^{-15} moles/liter, and [O_3] = 1.6 × 10^{-12} mole/cm^3.

(1)   Using Eq. 10-18, we obtain

$$\tau_{OH} = \frac{1}{(31 \times 10^9)(1 \times 10^{-15})} \text{ sec} = 3.2 \times 10^4 \text{ sec} = 9 \text{ hr}$$

---

6. $k = A \exp(-E_A/RT)$, where k is the rate constant, A a constant, E_A the energy of activation, R the gas constant, and T the temperature. E_A is typically on the order of 10^2 to 10^4 cal/mole for reactions with OH·; with E_A in these units, R is 1.987 cal/mole·K and T must be in K.

TABLE 10-7

Typical Concentrations of OH· and O_3 in the Atmosphere

Situation	Concentration			
	Moles/cm^3	Moles/Liter	Number/cm^3	Ref.
Ozone:				
Annual average	1.6×10^{-12}	1.6×10^{-9}	9.6×10^{11}	[15]
Urban	5.0×10^{-12}	5.0×10^{-9}	3.0×10^{12}	[13]
Rural	1.6×10^{-12}	1.6×10^{-9}	9.6×10^{11}	[13]
Hydroxyl Radical:				
Global annual average	1.8×10^{-18}	1.8×10^{-15}	1.1×10^{6}	[24]
	6.8×10^{-19}	6.8×10^{-16}	4.1×10^{5}	[26]
Northern hemisphere	8.0×10^{-19}	8.0×10^{-16}	4.8×10^{5}	[24]
(annual average)	$3\text{-}5 \times 10^{-19}$	$3\text{-}5 \times 10^{-16}$	$2\text{-}3 \times 10^{5}$	[26]
Southern hemisphere	3.0×10^{-18}	3.0×10^{-15}	1.8×10^{6}	[24]
(annual average)	$8\text{-}10 \times 10^{-19}$	$8\text{-}10 \times 10^{-16}$	$5\text{-}6 \times 10^{5}$	[26]
Atmospheric above boundary layer in N.H. (daytime)	$5.8\text{-}14 \times 10^{-18}$	$5.8\text{-}14 \times 10^{-15}$	$3.5\text{-}8.1 \times 10^{6}$	[8]
Polluted atmospheres (daytime, ground level, with full sunlight)	$1.7\text{-}17 \times 10^{-17}$	$1.7\text{-}17 \times 10^{-14}$	$10^{7}\text{-}10^{8}$	[7,35,36]

(2)   Using Eq. 10-19, we obtain

$$\tau_{O_3} = \frac{1}{(4.2 \times 10^3)(1.6 \times 10^{-12})} \text{ sec} = 1.5 \times 10^8 \text{ sec} = 4.7 \text{ yr}$$

(3)   Finally, Eq. 10-20 is used to calculate an overall value of τ; since in this example $1/\tau_{OH} \gg 1/\tau_{O_3}$, $\tau \simeq \tau_{OH} = 9$ hr

10-7   CORRELATION WITH MEAN STANDARD DEVIATION (JUNGE'S CORRELATION)

Principles of Use. While this method is enticingly simple, more care must be exercised in determining its appropriateness than with any of the other methods. The only data requirements are an adequate number of measurements of the atmospheric concentration (over appropriate space and time scales) so that an accurate average, C, and standard deviation, σ, can be obtained. Using such data, mostly for trace inorganic gases, Junge has shown [17] that a correlation exists between the mean standard deviation (σ/C) and the tropospheric residence time. Using the data shown in Table 10-8, Junge obtained the following correlation:

$$\tau = \frac{0.14}{\sigma/C} \text{ years} \qquad (10\text{-}21)$$

TABLE 10-8

Data Used in Junge's Correlation

Gas	Residence Time		Standard Spatial Variation	
	τ (yr)	Uncertainty Factor	σ/C	Uncertainty Factor
O_2	5×10^3	3.0	$\leqslant 2.4 \times 10^{-5}$	a
CO_2	15.0	1.5	5.0×10^{-3}	1.5
N_2O	8.0	2.0	8.0×10^{-2}	1.5
H_2	6.5	2.0	8.0×10^{-2}	1.4
CH_4	4.0	3.0	1.0×10^{-1}	2.0
CO	6×10^{-1}	2.0	5.0×10^{-1}	1.3
O_3	2.5×10^{-1}	1.5	4.0×10^{-1}	1.5
H_2O	2.2×10^{-2}	1.3	5.0×10^1	1.2
Rn	1.4×10^{-2}	1.1	1.0×10^1	4.0

a. Only upper limits can be given.

Source: Junge [17]. *(Reprinted with permission from the International Union of Crystallography.)*

The standard deviation of a set of measured C_i values is given by

$$\sigma = \left[\frac{1}{n-1} \left(\sum_{i=1}^{n} C_i^2 - nC^2 \right) \right]^{1/2} \qquad (10\text{-}22)$$

where C is the average value of the individual C_i values and n is the number of data points (i.e., number of C_i values). Sample calculations of standard deviations are shown in Appendix B.

If the data used to obtain C and σ are obtained from a large number of geographically separated stations over a suitable time period, the correlation is probably appropriate as long as the precision of the measurements is high enough so that σ represents the real variability in C rather than measurement errors. In addition, the sources and sinks for the pollutant must be evenly distributed. Even when the above-mentioned requirements are met, Table 10-8 shows that the value of τ estimated from Junge's correlation will have an uncertainty factor of about 2.

Although Eq. 10-21 was obtained as a correlation with existing data, the form of the equation has been validated on a theoretical basis. Junge showed [17] that for 5 days $\leq \tau \leq 1.5$ years a correlation of the following form would be expected:

$$\tau^{0.95} = \frac{0.0216}{\sigma/C} \quad \text{years} \qquad (10\text{-}23)$$

Another model evaluated by Gibbs and Slinn [14] indicated that $\tau^{0.5}$ would be inversely proportional to σ/C.

Basic Steps

(1) Determine appropriateness of method (see text) on the basis of the precision of the measurements, the number and geographic location of the measurement sites, and the time interval of the sampling.

(2) Obtain the average (C) and standard deviation (σ) from the available data on atmospheric concentrations. Use Eq. 10-22 to calculate σ if it is not known.

(3) Solve for τ using Eq. 10-21.

Example 10-6   Estimate τ for methyl chloride, given C = 610 ppt and σ = 90 ppt. (We will assume the method is appropriate.)

Substituting these values in Eq. 10-21, we obtain

$$\tau = \frac{0.14}{90/610} = \frac{0.14}{0.15} = 0.95 \text{ yr}$$

10-8   SYMBOLS USED

Note: The dimensions of the important parameters are indicated in parentheses. The actual units used in calculations can vary as long as they are self-consistent.

A = cumulative emissions of a chemical, Eqs. 10-10, -13 (mass)

a = parameter in Eq. 10-9

b = parameter in Eq. 10-9 (time^{-1})

C = average concentration of chemical in compartment (mass/mass, or mass/volume)

D = parameter in Eqs. 10-13, -14

E = total emission rate, as in Eq. 10-1 (mass/time)

H = Henry's law constant, used in Eq. 10-7 (conc/conc)

k, k_{OH}, k_{O_3} = rate constant for bimolecular reaction; see §10-6 (volume · mass^{-1} · time^{-1} or volume · number^{-1} · time^{-1})

n = number of points in data set, used in Eq. 10-22

P_{H_2O} = partial pressure of water in atmosphere (not used in any equation)

p = parameter in Eq. 10-15

Q = total mass of chemical in compartment, as in Eq. 10-1 (mass)

R = total removal rate, as in Eq. 10-1 (mass/time)

T = temperature

t = time

$t_{1/2}$ = half-life; in Eq. 10-2 (time)

Greek

α = parameter in Eq. 10-14

β = parameter in Eq. 10-14

σ = standard deviation of measured atmospheric concentration, Eq. 10-21 (mass/mass or mass/volume)

τ = residence time (time)

θ = ratio of current atmospheric burden to cumulative emissions, Eq. 10-10 (dimensionless)

Subscripts

e = used with τ (τ_e) to denote exchange rate for air between northern and southern hemispheres

i = individual value, used with C in Eq. 10-22

N = northern hemisphere

p = precipitation, used with C in Eq. 10-6

S = southern hemisphere

10-9   REFERENCES

1. Altshuller, A.P., "Average Tropospheric Concentration of Carbon Tetrachloride Based on Industrial Production, Usage, and Emissions," *Environ. Sci. Technol.,* **10**, 596-98 (1976).

2. Bidleman, T.F. and C.E. Olney, "Chlorinated Hydrocarbons in the Sargasso Sea Atmosphere and Surface Water," *Science,* **183**, 516-18 (1974).

3. Bolin, B. and H. Rodhe, "A Note on the Concepts of Age Distribution and Transit Time in Natural Reservoirs," *Tellus,* **25**, 58-62 (1973).

4. Brown, S.L., B.R. Holt and K.E. McCaleb, "Systems for Rapid Ranking of Environmental Pollutants. Selection of Subjects for Scientific and Technical Assessment Reports," EPA-600/5-78-012, pp. 143, 199, 200 (June 1978).

5. Butler, R. and A. Snelson, "Kinetics of the Homogeneous Gas Phase Hydrolysis of CCl_3COCl, CCl_2HOCl, $CH_2ClCOCl$ and $COCl_2$," *J. Air Pollut. Control Assoc.,* **29**, 833-37 (1979).

6. Council on Environmental Quality, "Fluorocarbons and the Environment. Report of Federal Task Force on Inadvertent Modification of the Stratosphere (IMOS)," U.S. Government Printing Office, Washington, DC (June 1975).

7. Darnall, K.R., A.C. Lloyd, A.M. Winer and J.N. Pitts, Jr., "Reactivity Scale for Atmospheric Hydrocarbons Based on Reaction with Hydroxyl Radical," *Environ. Sci. Technol.,* **10**, 692-96 (1976).

8. Davis, D.D., W. Heaps and T. McGee, "Direct Measurements of Natural Tropospheric Levels of OH via an Aircraft-Borne Tunable Dye Laser," *Geophys. Res. Lett.,* **3**, 331-33 (1976).

9. Dilling, W.L. and H.K. Goersch, "Organic Photochemistry — XVI: Tropospheric Photodecomposition of Methylene Chloride," in Preprints of Papers Presented at the 177th National Meeting, American Chemical Society, Division of Environmental Chemistry, **19**, No. 1 (1979).

10. Doyle, G.J., A.C. Lloyd, K.R. Darnall, A.M. Winer and J.N. Pitts, Jr., "Gas Phase Kinetic Study of Relative Rates of Reaction of Selected Aromatic Compounds with Hydroxyl Radicals in an Environmental Chamber," *Environ. Sci. Technol.,* **9**, 237-41 (1975).

11. Duce, A., G. Quinn and L. Wade, "Residence Time of Non-Methane Hydrocarbons in the Atmosphere," *Mar. Pollut. Bull.,* **5**, 59-61 (1974).

12. Eggleton and Clough (unpublished work), cited by Lovelock *et al.* [20].

13. Eschenroeder, A.Q., E. Irvine, A.C. Lloyd, C. Tashima and K. Tran, "Investigation of Profile Models for Toxic Chemicals in the Environment," Report by Environmental Research and Technology, Inc., Santa Barbara, CA, to the National Science Foundation, Washington, DC (February 1978).

14. Gibbs, A.B. and W.G.N. Slinn, "Fluctuations in Trace Gas Concentrations in the Troposphere," *J. Geophys. Res.,* **78**, 574-76 (1973).

15. Hendry, D.G. and R.A. Kenley, "Atmospheric Reaction Products of Organic Compounds," Report No. EPA-560/12-79-001, U.S. EPA, Office of Toxic Substances, Washington, DC (June 1979).

16. Jesson, J.P., P. Meakin and L.C. Glasgow, "The Fluorocarbon Ozone Theory — II: Tropospheric Lifetimes — An Estimate of the Tropospheric Lifetime of CCl_3F," *Atmos. Environ.,* **11**, 499-508 (1977).

17. Junge, C.E., "Residence Time and Variability of Tropospheric Trace Gases," *Tellus,* **26**, 477-88 (1974).

18. Junge, C.E., "Basic Considerations about Trace Constituents in the Atmosphere as Related to the Fate of Global Pollutants," in Part 1 of *Fate of Pollutants in the Air and Water Environments,* I.H. Suffet (ed.), John Wiley & Sons, New York (1977).

19. Kalpasanov, Y. and G. Kurchatova, "A Study of the Statistical Distribution of Chemical Pollutants in Air," *J. Air Pollut. Control Assoc.,* **26**, 981-85 (1976).

20. Lovelock, J.E., R.J. Maggs and R.J. Wade, "Halogenated Hydrocarbons in and over the Atlantic," *Nature (London),* **241** (5836), 194-96 (1973).

21. Molina, M.J. and F.S. Rowland, "Predicted Present Stratospheric Abundances of Chlorine Species from Photodissociation of Carbon Tetrachloride," *Geophys. Res. Lett.,* **1**, 309-12 (1974).

22. National Academy of Sciences, "Assessing Potential Ocean Pollutants," Washington, DC, pp. 21-23, 117 (1975).

23. Neely, W.B., "Material Balance Analysis of Trichlorofluoromethane and Carbon Tetrachloride in the Atmosphere," *Sci. Total Environ.,* **8**, 267-74 (1977).

24. Neely, W.B. and J.H. Plonka, "Estimation of Time-Averaged Hydroxyl Radical Concentration in the Troposphere," *Environ. Sci. Technol.,* **12**, 317-321 (1978).

25. Singh, H.B., "Atmospheric Halocarbons: Evidence in Favor of Reduced Average Hydroxyl Radical Concentration in the Troposphere," *Geophys. Res. Lett.,* **4**, 101-104 (1977).

26. Singh, H.B., "Preliminary Estimation of Average Tropospheric HO Concentrations in the Northern and Southern Hemispheres," *Geophys. Res. Lett.,* **4**, 453-56 (1977).

27. Singh, H.B., D.P. Fowler and T.O. Peyton, "Atmospheric Carbon Tetrachloride: Another Man-Made Pollutant," *Science,* **192**, 1231-34 (1976).

28. Singh, H.B., L.J. Salas, H. Shigeishi and E. Scribner, "Global Distribution of Selected Halocarbons, Hydrocarbons, SF_6 and N_2O," EPA-600/3-78-100 (December 1978).

29. Singh, H.B., L.J. Salas, H. Shigeishi and E. Scribner, "Atmospheric Halocarbons, Hydrocarbons and Sulfur Hexafluoride: Global Distribution, Sources and Sinks," *Science,* **203**, 899-903 (1979).

30. Singh, H.B., L.J. Salas, H. Shigeishi and A.H. Smith, "Fate of Halogenated Compounds in the Atmosphere — Interim Report," EPA-600/3-78-017 (January 1978).

31. Slinn, W.G.N., "Relationship Between Removal Processes and Residence Times for Atmospheric Pollutants," NTIS CONF-780611-3 (March 1978).

32. Slinn, W.G.N., L. Hasse, B.B. Hicks, A.W. Hogan, O. Lal, P.S. Liss, K.O. Munnich, G.A. Schmel and O. Vittori, "Some Aspects of the Transfer of Atmospheric Trace Constituents Past the Air Sea Interface," *Atmos. Environ.,* **12**, 2055-87 (1978).

33. State of California Air Resources Board, "Adoption of a System for the Classification of Organic Compounds According to Photochemical Reactivity," Staff Report 76-3-4 (February 1976).

34. Todd, D.K. (ed.), *The Water Encyclopedia,* Water Information Center, Port Washington, N.Y. (1970).

35. Wang, C.C. and L.I. Davis, Jr., "Measurement of Hydroxyl Concentration in Air Using a Tunable UV Laser Beam," *Phys. Rev. Lett.,* **32**, 349-52 (1974).

36. Wang, C.C., L.I. Davis, Jr., C.H. Wu, S. Japar, H. Niki, and B. Weinstock, "Hydroxyl Radical Concentrations Measured in Ambient Air," *Science,* **189**, 797-800 (1975).

37. Woodwell, G.M., P.P. Craig and H.A. Johnson, "DDT in the Biosphere: Where Does It Go?" *Science,* **174**, 1101-07 (1971).

11

ACTIVITY COEFFICIENT

Clark F. Grain

11-1 INTRODUCTION

The purpose of this chapter is to provide methods of estimating the activity coefficients of components in solution. The discussion will be limited to binary systems although, in principle, the methods are applicable to multicomponent equilibria. The introductory remarks below may be supplemented by material in Refs. 8 and 21.

As applied to solutions, an activity coefficient, γ, is a correction factor compensating for non-ideal behavior. Consider, for example, an ideal binary solution of two organic liquids at a given temperature, where the number of moles of the first compound is n_1 and that of the second, n_2. The mole fractions of the two are, respectively, $x_1 = n_1/(n_1 + n_2)$ and $x_2 = n_2/(n_1 + n_2)$. If the vapor pressures of the pure liquids are P_1° and P_2°, then, for any mixture, the partial pressure of component 1 is expressed as

$$P_1 = x_1 P_1^\circ \qquad (11\text{-}1)$$

Similarly,

$$P_2 = x_2 P_2^\circ \qquad (11\text{-}2)$$

If the two liquids form a non-ideal mixture, however, we should apply a variable correction factor, γ (the activity coefficient), so that

$$P_1 = \gamma_1 x_1 P_1^\circ \qquad (11\text{-}3)$$

and

$$P_2 = \gamma_2 \, x_2 \, P_2^\circ \qquad (11\text{-}4)$$

We use the convention that $\gamma = 1$ for the pure components 1 and 2. In general, $\gamma > 1$ for dilute solutions of a given component in a given solvent. As the solution of this component becomes increasingly dilute, the value of γ asymptotically approaches a limiting value, γ^∞. The product $\gamma_1 x_1$ is referred to as the activity, a_1, and may be applied to properties other than vapor pressure.

Knowledge of the activity coefficients of the various components is useful in estimating quantities important to the environmental scientist. Examples of such quantities include solubility limits,[1] Henry's law constants, octanol-water partition coefficients[2] and solution equilibria in general. As an example, Lyman (see §1-5, Chapter 1) has shown that the octanol-water partition coefficient, K_{ow}, is proportional to the ratio of activity coefficients of a component A in water and octanol; thus,

$$K_{ow} = 0.151 \frac{\left(\gamma_w^A\right)^\infty}{\left(\gamma_o^A\right)^\infty} \qquad (11\text{-}5)$$

where $(\gamma_w^A)^\infty$, $(\gamma_o^A)^\infty$ = infinite dilution activity coefficients of component A in water and octanol, respectively.[3] For the process engineer a knowledge of activity coefficients is invaluable in calculations involving multicomponent phase equilibria.

Activity coefficients are, of course, dimensionless and range in value from about 0.4 to upwards of 10^7.

11-2 AVAILABLE METHODS

The fundamental equation that defines the activity coefficient for a binary system is[4]

---

1. Methods of estimating solubilities, using activity coefficients, are described in Chapter 3.
2. Methods of estimating K_{ow}, using activity coefficients, are described in Chapter 1.
3. A similar relationship is given by Mackay [13], except that the coefficient is given as 0.115 rather than 0.151.
4. Detailed discussion of the defining equations 11-1 through 11-8 are given in standard texts such as *Physical Chemistry* by N.K. Adam, Oxford University Press, London (1958) and in Refs. 8, 19, and 21.

$$G^M = RT(n_1 \ln a_1 + n_2 \ln a_2)$$

$$= \underbrace{RT(n_1 \ln x_1 + n_2 \ln x_2)}_{G^I} + \underbrace{RT(n_1 \ln \gamma_1 + n_2 \ln \gamma_2)}_{G^E} \qquad (11\text{-}6)$$

where G^M is the energy of mixing, R is the gas constant, and T is the temperature. The two terms on the right side of this equation correspond to the Gibbs free energy of mixing for an ideal mixture (G^I) and the "excess" Gibbs free energy (G^E). From this expression for G^E, we obtain

$$\left(\frac{\partial G^E}{\partial n_1}\right) = RT \ln \gamma_1 \qquad (11\text{-}7)$$

and

$$\left(\frac{\partial G^E}{\partial n_2}\right) = RT \ln \gamma_2 \qquad (11\text{-}8)$$

Thus, in principle, we may relate activity coefficients to composition in mole fraction. All methods of accomplishing this are based upon assuming some analytical form for G^E subject to the conditions $G^E \propto n_1 + n_2$ at fixed composition and $G^E = 0$ when x_1 or $x_2 = 0$. The simplest expression satisfying these conditions is

$$G^E = (n_1 + n_2) A_1 x_1 x_2 \qquad (11\text{-}9)$$

Using Eqs. 11-7 and 11-8 in Eq. 11-9, we obtain

$$\ln \gamma_1 = \frac{A_1}{RT} x_2^2 \quad (\approx x_2^2 \ln \gamma_1^{\infty}) \qquad (11\text{-}10)$$

$$\ln \gamma_2 = \frac{A_1}{RT} x_1^2 \quad (\approx x_1^2 \ln \gamma_2^{\infty}) \qquad (11\text{-}11)$$

(The equivalency of A_1/RT to $\ln \gamma_1^{\infty}$ follows from the fact that $x_2 \approx 1$ when the concentration of component 1 approaches zero; the same applies to $\ln \gamma_2^{\infty}$.)

The parameter A_1 can be evaluated from binary vapor-liquid data or it may be approximated as $RT \ln \gamma_1^{\infty}$. Although Eq. 11-9 is presented as an example, it has little practical value. It contains only one adjustable parameter and hence is restricted to those systems in which $\gamma_1^{\infty} \approx \gamma_2^{\infty}$.

Equation 11-9 is called the two-suffix Margules equation and is essentially empirical. One failing of this equation is that S^E, the excess entropy of mixing, is not taken into account. Similar equations with the same deficiency are listed in Table 11-1 under the heading of enthalpic equations.

The equations listed can be used with some success in estimating solubility limits, particularly for systems with low solubilities, as described in Chapter 3. Furthermore, the van Laar equation has been used to represent vapor-liquid equilibria with reasonable accuracy [15,19]. However, when used to predict liquid-liquid equilibria, such enthalpic equations are unsatisfactory [15]. Another failing of these equations is that when the adjustable parameters are evaluated using binary data obtained at intermediate compositions, extrapolation to the dilute regions is unsatisfactory. Finally, the enthalpic relations all too often predict immiscibility when, in fact, it does not exist.

In 1964 Wilson [27] introduced the concept of local composition which, in effect, takes into account S^E. Wilson used local volume fractions in his final equation. Subsequent investigations by Orye [16], Renon [20], and Abrams [1] have shown that the local composition concept is a powerful tool in representing and predicting equilibrium behavior in both vapor-liquid and liquid-liquid systems.

The equations that have been derived from the local-composition concept have commonly been termed entropic equations. These are also listed in Table 11-1. The Wilson equation cannot predict immiscibility; i.e., no values of the parameters in this equation allow for the prediction of a miscibility gap.

Recent advances [5,6,9] have led to group contribution techniques in which the solute and solvent molecules are divided into groups such as OH, CH_2, and NH_2. Group parameters are then evaluated through use of experimental data. Once the group parameters have been evaluated they may be used to predict the behavior of other systems containing those groups, but for which there are no experimental data. These group contribution schemes have been successfully applied in the Wilson equation [5] and the UNIQUAC (Universal Quasi Chemical) equation [9].

TABLE 11-1

Some Models for the Excess Gibbs Energy and Subsequent Activity Coefficients for Binary Systems

Name	$g^E = G^E/(n_1 + n_2)$	Binary Parameters	$\ln \gamma_1$ and $\ln \gamma_2$
Enthalpic Equations			
Two-suffix[a] Margules	$g^E = A x_1 x_2$	A	$RT \ln \gamma_1 = A x_2^2$ $RT \ln \gamma_2 = A x_1^2$
Three-suffix[a] Margules	$g^E = x_1 x_2 [A + B(x_1 - x_2)]$	A, B	$RT \ln \gamma_1 = (A + 3B) x_2^2 - 4B x_2^3$ $RT \ln \gamma_2 = (A - 3B) x_1^2 + 4B x_1^3$
van Laar	$g^E = \dfrac{A x_1 x_2}{x_1 (A/B) + x_2}$	A, B	$RT \ln \gamma_1 = A \left(1 + \dfrac{A}{B}\dfrac{x_1}{x_2}\right)^{-2}$ $RT \ln \gamma_2 = B \left(1 + \dfrac{B}{A}\dfrac{x_2}{x_1}\right)^{-2}$
Four-suffix[a] Margules	$g^E = x_1 x_2 [A + B(x_1 - x_2) + C(x_1 - x_2)^2]$	A, B, C	$RT \ln \gamma_1 = (A + 3B + 5C) x_2^2 - 4(B + 4C) x_2^3 + 12C\, x_2^4$ $RT \ln \gamma_2 = (A - 3B + 5C) x_1^2 + 4(B - 4C) x_1^3 + 12C\, x_1^4$
Entropic Equations			
Wilson	$\dfrac{g^E}{RT} = -x_1 \ln(x_1 + \Lambda_{12} x_2) - x_2 \ln(x_2 + \Lambda_{21} x_1)$	$\Lambda_{12}, \Lambda_{21}$	$\ln \gamma_1 = -\ln(x_1 + \Lambda_{12} x_2) + x_2 \left(\dfrac{\Lambda_{12}}{x_1 + \Lambda_{12} x_2} - \dfrac{\Lambda_{21}}{\Lambda_{21} x_1 + x_2} \right)$ $\ln \gamma_2 = -\ln(x_2 + \Lambda_{21} x_1) - x_1 \left(\dfrac{\Lambda_{12}}{x_1 + \Lambda_{12} x_2} - \dfrac{\Lambda_{21}}{\Lambda_{21} x_1 + x_2} \right)$
NRTL[b]	$\dfrac{g^E}{RT} = x_1 x_2 \left(\dfrac{\tau_{21} G_{21}}{x_1 + x_2 G_{21}} + \dfrac{\tau_{12} G_{12}}{x_2 + x_1 G_{12}} \right)$ where $\tau_{12} = \dfrac{\Delta g_{12}}{RT}$ $\quad \tau_{21} = \dfrac{\Delta g_{21}}{RT}$ $\ln G_{12} = -\alpha_{12} \tau_{12} \quad \ln G_{21} = -\alpha_{12} \tau_{21}$	$\Delta g_{12}, \Delta g_{21}, \alpha_{12}$[c]	$\ln \gamma_1 = x_2^2 \left[\tau_{21} \left(\dfrac{G_{21}}{x_1 + x_2 G_{21}} \right)^2 + \dfrac{\tau_{12} G_{12}}{(x_2 + x_1 G_{12})^2} \right]$ $\ln \gamma_2 = x_1^2 \left[\tau_{12} \left(\dfrac{G_{12}}{x_2 + x_1 G_{12}} \right)^2 + \dfrac{\tau_{21} G_{21}}{(x_1 + x_2 G_{21})^2} \right]$

TABLE 11-1 (Continued)

Name	$g^E = G^E/(n_1 + n_2)$	Binary Parameters	$\ln \gamma_1$ and $\ln \gamma_2$

Entropic Equations (continued)

UNIQUAC[d]

$g^E = g^E \text{ (combinatorial)} + g^E \text{ (residual)}$

$$\frac{g^E \text{ (combinatorial)}}{RT} = x_1 \ln \frac{\Phi_1}{x_1} + x_2 \ln \frac{\Phi_2}{x_2}$$

$$+ \frac{z}{2} \left(q_1 x_1 \ln \frac{\theta_1}{\Phi_1} + q_2 x_2 \ln \frac{\theta_2}{\Phi_2} \right)$$

$$\frac{g^E \text{ (residual)}}{RT} = -q_1 x_1 \ln \left[\theta_1 + \theta_2 \tau_{21} \right] - q_2 x_2 \ln \left[\theta_2 + \theta_1 \tau_{12} \right]$$

$$\Phi_1 = \frac{x_1 r_1}{x_1 r_1 + x_2 r_2} \qquad \theta_1 = \frac{x_1 q_1}{x_1 q_1 + x_2 q_2}$$

$$\ln \tau_{21} = -\frac{\Delta u_{21}}{RT} \qquad \ln \tau_{12} = -\frac{\Delta u_{12}}{RT}$$

r and q are pure-component parameters and coordination number z = 10.

Binary Parameters: Δu_{12} and Δu_{21}[c]

$$\ln \gamma_1 = \ln \frac{\Phi_1}{x_1} + \frac{z}{2} q_1 \ln \frac{\theta_1}{\Phi_1} + \Phi_j \left(\ell_i - \frac{r_i}{r_j} \ell_j \right)$$

$$- q_i \ln (\theta_i + \theta_j \tau_{ji}) + \theta_j q_i \left(\frac{\tau_{ji}}{\theta_i + \theta_j \tau_{ji}} - \frac{\tau_{ij}}{\theta_j + \theta_i \tau_{ij}} \right)$$

where i = 1 j = 2 or i = 2 j = 1

$$\ell_i = \frac{z}{2} (r_i - q_i) - (r_i - 1)$$

$$\ell_j = \frac{z}{2} (r_j - q_j) - (r_j - 1)$$

a. Two-suffix signifies that the expansion for g^E is quadratic in mole fraction. Three-suffix signifies a third-order, and four-suffix signifies a fourth-order equation.

b. NRTL = Non Random Two Liquid.

c. $\Delta g_{12} = g_{12} - g_{22}$; $\Delta g_{21} = g_{21} - g_{11}$.

d. UNIQUAC = Universal Quasi Chemical.

e. $\Delta u_{12} = u_{12} - u_{22}$; $\Delta u_{21} = u_{21} - u_{11}$.

Source: Reid, Prausnitz and Sherwood [21]. *(Reprinted with permission from McGraw-Hill Book Co.)*

In what follows we will concentrate on two methods of estimating activity coefficients. The first of these will involve the estimation of γ^∞, the infinite dilution activity coefficient. The method to be used was introduced by Pierotti *et al.* [18] and will be described in detail in §11-4.

The infinite dilution activity coefficient alone can be used to estimate solubility limits (Chapter 3), octanol-water partition coefficients (Chapter 1), and Henry's law constants.[5] It may also be used to estimate the parameters in any two-parameter equation, e.g., the van Laar equation, which will then allow us to estimate the activity coefficient at any composition.

The second method, which estimates the concentration dependence of γ_1 directly, uses the UNIQUAC equation combined with group contribution techniques. It has been called the UNIFAC (UNIQUAC Functional Group Activity Coefficients) method [15]. Other group contribution schemes such as ASOG (Analytical Solution of Groups) [13] have been used successfully. However, we chose the UNIFAC method because of its stronger theoretical base. Table 11-2 gives an overview of the methods chosen.

11-3 METHOD ERRORS

Large percentage errors can be tolerated in estimates of activity coefficients, particularly with respect to γ^∞. It has been found [21], for example, that a $\pm 10\%$ variation in γ^∞ does not affect predictions of vapor-liquid equilibria. Furthermore, errors as high as $\pm 50\%$ in γ^∞ often do not affect γ in the high concentration range by more than $\pm 10\%$. Generally speaking, the correlations discussed in this chapter are capable of predicting γ^∞ to within $\pm 25\%$ of the true value. For example, for Method 1, Pierotti *et al.* [18] gave an overall average deviation of 8% in γ^∞, although in isolated instances errors as high as 350% were obtained.

For the UNIFAC method (Method 2) Table 11-3 lists results obtained for a number of systems with water. The average error in γ is 23.5% and the maximum error is 90%.

---

5. Henry's Law constant (H) $\approx$ (vapor pressure)/(solubility). Thus the vapor pressure must also be known to calculate H. (See Example 11-2.)

TABLE 11-2

Characteristics of Methods 1 and 2

Method	Gives	Input Required	Ease of Calculation	Method Error	Limitations
Method 1 Pierotti *et al.* [18] (§11-4)	γ^∞ for $25°C \leqslant T \leqslant 100°C$ (See Note a)	Chemical structure	Easy	Avg. 8% Max. 350%[b]	Limited to certain solute and solvent classes. See Table 11-4.
Method 2 UNIFAC [9] (§11-5)	γ at any solute concentration and temperature	Chemical structure	Difficult	Avg. 23.5% Max. 95%[c]	Limited to chemicals with common structures and functional groups. See Tables 11-6 and -7.

a. Although Method 1 yields only γ^∞, equations are provided in §11-4 to allow estimation of γ at any solute concentration once γ^∞ has been found.
b. Based on information in Ref. 18.
c. Based on information in Ref. 9. See Table 11-3 for details.

TABLE 11-3

Comparison of Experimental and Calculated Activity Coefficients
(Method 2)

System	T (K)	X_1 [a]	γ (exp.)	γ (calc.)
Hydrocarbons				
(These activity coefficients are based on liquid-liquid equilibria)				
Water-Hexane	293.2	0.000	1,880	2,040
Hexane-Water	298.2	0.000	489,000	402,000
Water-Benzene	313.2	0.000	226.3	308.2
Benzene-Water	333.2	0.000	1,730	1,670
Water-Toluene	333.2	0.000	3,320	391
Toluene-Water	333.2	0.000	3,390	7,820
Oxygenated Hydrocarbons				
Water-Ethanol	313.2	0.000	4.748	3.184
Ethanol-Water	313.2	0.043	2.398	2.210
Water-1-Propanol	363.2	0.117	2.845	3.079
1-Propanol-Water	363.2	0.111	4.594	5.066
Water-2-Propanol	328.2	0.141	2.965	2.993
2-Propanol-Water	328.2	0.046	7.706	10.572
Water-1-Butanol	333.2	0.050	4.237	4.339
1-Butanol-Water	333.2	0.016	38.61	54.66
Water-Methanol	340.2	0.154	1.505	1.543
Methanol-Water	368.2	0.029	2.097	2.126
Water-Phenol	348.2	0.321	1.842	2.185
Phenol-Water	348.2	0.015	20.43	23.42
Water-Acetic acid	388.6	0.050	1.769	1.697
Acetic acid-Water	373.7	0.100	2.027	2.300
Water-Propanoic acid	394.3	0.115	2.382	2.125
Propanoic acid-Water	373.1	0.108	3.462	3.809
Water-Acetaldehyde	293.2	0.100	2.982	2.688
Acetaldehyde-Water	293.2	0.100	2.653	2.606
Water-Methyl acetate	308.2	0.019	21.25	20.17
Methyl acetate-Water	298.2	0.005	21.84	24.94

(Continued)

TABLE 11-3 (Continued)

System	T (K)	$X_1{}^a$	γ (exp.)	γ (calc.)
Oxygenated Hydrocarbons (cont'd.)				
Water-Ethyl acetate	343.2	0.032	8.025	8.378
Ethyl acetate-Water	343.2	0.000	21.97	14.71
Water-Acetone	333.2	0.263	2.551	2.362
Acetone-Water	333.2	0.072	5.638	6.736
Water-Tetrahydrofuran	323.2	0.078	6.800	5.690
Tetrahydrofuran-Water	323.2	0.075	10.12	14.67
Water-1,4-Dioxane	365.9	0.120	2.430	3.400
1,4-Dioxane-Water	365.1	0.073	4.670	9.840
Nitrogen Compounds				
Water-Aniline	433.2	0.011	4.985	4.277
Aniline-Water	371.8	0.021	34.16	50.65
Water-Butylamine	350.2	0.045	1.720	1.587
Butylamine-Water	361.3	0.020	12.58	12.95
Water-Diethylamine	311.5	0.100	3.074	1.665
Diethylamine-Water	311.5	0.100	2.485	3.398
Water-Acetonitrile	303.2	0.080	6.778	6.041
Acetonitrile-Water	303.2	0.070	9.666	8.201

a. Mole fraction of the solute (i.e., the first chemical listed for each system). Note that when $x_1 = 0$, $\gamma = \gamma^\infty$.

Source: Fredenslund, Gmehling and Rasmussen [8]. *(Reprinted with permission from Elsevier Scientific Publishing Co.)*

11-4   METHOD 1 — INFINITE DILUTION ACTIVITY COEFFICIENTS

This method requires a knowledge of the molecular structure. As introduced by Pierotti *et al.* [18], the technique relates γ^∞ to the molecular structures of the solute and solvent molecules through an equation that contains the number of carbon atoms for both species. Thus, in general,

$$\log \gamma_1^\infty = A_{1,2} + B_2 \frac{N_1}{N_2} + \frac{C_1}{N_1} + D (N_1 - N_2)^2 + \frac{F_2}{N_2} \quad (11\text{-}12)$$

where subscript 1 refers to solute and 2 to solvent and

$A_{1,2}$ = coefficient which depends on nature of solute and solvent functional groups

B_2 = coefficient which depends only on nature of solvent functional group

C_1 = coefficient which depends only on solute functional group

D = coefficient independent of solute and solvent functional groups

F_2 = coefficient which essentially depends only on nature of solvent functional group

N_1, N_2 = number of carbon atoms in solute and solvent, respectively

For secondary and tertiary alcohols the C_1 term is modified such that it becomes $C_1 (1/N_1' + 1/N_1'')$ for secondary alcohols and $C_1 (1/N_1' + 1/N_1'' + 1/N_1''')$ for tertiary alcohols, where the primed N's are carbon numbers of the respective branches counted from the polar groupings. Thus, for *tert*-butyl alcohol, the central carbon is counted in each branch, i.e., $N_1' = N_1'' = N_1''' = 2$ even though the total carbon number is four. If the alcohol is the solvent, a similar argument holds for the F_2 term.

Alternatively, one may consider all alcohols as a single homologous series by using terms such as $C_1(1/N_1 - 1)$, $C_1[(1/N_1' - 1) + (1/N_1'' - 1)]$ and $C_1[(1/N_1' - 1) + (1/N_1'' - 1) + (1/N_1''' - 1)]$ for primary, secondary and tertiary alcohols, respectively. Thus, a single value of C_1 covers all alcohols with little loss in accuracy. The F_2 term may be represented similarly if the alcohol is the solvent. Other modifications are needed for ketones, acetals, and cyclic hydrocarbons. Table 11-4, which is taken from Ref. 18, footnotes the necessary modifications and gives values of the coefficients in Eq. 11-12. Three additional systems (paraffins-*n* alcohols, water-paraffins, and water-alkyl benzenes), determined by this writer, are included in the table.

The Pierotti correlations were later extended by Tsonopoulos and Prausnitz [25] to include the effect of substituents on alkyl benzenes in dilute aqueous solutions. A detailed analysis of log $\gamma\infty$ data resulted in a number of correction factors to be added to calculated values of log $\gamma\infty$. Their table is reproduced here as Table 11-5, supplemented by correction factors for condensed-ring derivatives, biphenyl derivatives, and polymethyl substituents. The authors derived separate equations for each of the above-named groups. However, we have found that the single equation for alkylbenzenes in water, with the correction factors listed, is adequate.

TABLE 11-4

Correlating Constants for Activity Coefficients at Infinite Dilution, Homologous Series of Solutes and Solvents

Solute (1)	Solvent (2)	Temp. (°C)	$A_{1,2}$	B_2	C_1	D	F_2	See Note
n-Acids	Water	25	−1.00	0.622	0.490	0	0	a
		50	−0.80	0.590	0.290	0	0	a
		100	−0.620	0.517	0.140	0	0	a
n-Primary alcohols	Water	25	−0.995	0.622	0.558	0	0	a
		60	−0.755	0.583	0.460	0	0	a
		100	−0.420	0.517	0.230	0	0	a
Secondary alcohols	Water	25	−1.220	0.622	0.170	0	0	b
		60	−1.023	0.583	0.252	0	0	b
		100	−0.870	0.517	0.400	0	0	b
Tertiary alcohols	Water	25	−1.740	0.622	0.170	0	0	c
		60	−1.477	0.583	0.252	0	0	c
		100	−1.291	0.517	0.400	0	0	c
Alcohols, general	Water	25	−0.525	0.622	0.475	0	0	d
		60	−0.33	0.583	0.39	0	0	d
		100	−0.15	0.517	0.34	0	0	d
n-Allyl alcohols	Water	25	−1.180	0.622	0.558	0	0	a
		60	−0.929	0.583	0.460	0	0	a
		100	−0.650	0.517	0.230	0	0	a
n-Aldehydes	Water	25	−0.780	0.622	0.320	0	0	a
		60	−0.400	0.583	0.210	0	0	a
		100	−0.3	0.517	0	0	0	a
n-Alkene aldehydes	Water	25	−0.720	0.622	0.320	0	0	a
		60	−0.540	0.583	0.210	0	0	a
		100	−0.298	0.517	0	0	0	a
n-Ketones	Water	25	−1.475	0.622	0.500	0	0	b
		60	−1.040	0.583	0.330	0	0	b
		100	−0.621	0.517	0.200	0	0	b
n-Acetals	Water	25	−2.556	0.622	0.486	0	0	e
		60	−2.184	0.583	0.451	0	0	e
		100	−1.780	0.517	0.426	0	0	e
n-Ethers	Water	20	−0.770	0.640	0.195	0	0	b
n-Nitriles	Water	25	−0.587	0.622	0.760	0	0	a
		60	−0.368	0.583	0.413	0	0	a
		100	−0.095	0.517	0	0	0	a
n-Alkene nitriles	Water	25	−0.520	0.622	0.760	0	0	a
		60	−0.323	0.583	0.413	0	0	a
		100	−0.074	0.517	0	0	0	a

Solute (1)	Solvent (2)	Temp. (°C)	$A_{1,2}$	B_2	C_1	D	F_2	See Note
n-Esters	Water	20	−0.930	0.640	0.260	0	0	b
n-Formates	Water	20	−0.585	0.640	0.260	0	0	a
n-Monoalkyl chlorides	Water	20	1.265	0.640	0.073	0	0	a
n-Paraffins	Water	16	0.688	0.642	0	0	0	a
n-Alkylbenzenes	Water	25	3.554	0.622	−0.466	0	0	f
Alcohols, general	Paraffins	25	1.960	0	0.475	−0.00049	0	d
		60	1.460	0	0.390	−0.00057	0	d
		100	1.070	0	0.340	−0.00061	0	d
n-Ketones	Paraffins	25	0.0877	0	0.757	−0.00049	0	b
		60	0.016	0	0.680	−0.00057	0	b
		100	−0.067	0	0.605	−0.00061	0	b
Water	*n*-Alcohols	25	0.760	0	0	0	−0.630	a
		60	0.680	0	0	0	−0.440	a
		100	0.617	0	0	0	−0.280	a
Water	*sec*-Alcohols	80	1.208	0	0	0	−0.690	c
Water	*n*-Ketones	25	1.857	0	0	0	−1.019	c
		60	1.493	0	0	0	−0.73	c
		100	1.231	0	0	0	−0.557	c
Ketones	*n*-Alcohols	25	−0.088	0.176	0.50	−0.00049	−0.630	g
		60	−0.035	0.138	0.33	−0.00057	−0.440	g
		100	−0.035	0.112	0.20	−0.00061	−0.280	g
Aldehydes	*n*-Alcohols	25	−0.701	0.176	0.320	−0.00049	−0.630	
		60	−0.239	0.138	0.210	−0.00057	−0.440	
Esters	*n*-Alcohols	25	0.212	0.176	0.260	−0.00049	−0.630	g
		60	0.055	0.138	0.240	−0.00057	−0.440	g
		100	0	0.112	0.220	−0.00061	−0.280	g
Acetals	*n*-Alcohols	60	−1.10	0.138	0.451	−0.00057	−0.440	h
Paraffins	Ketones	25	0	0.1821	0	−0.00049	0.402	i
		60	0	0.1145	0	−0.00057	0.402	i
		90	0	0.0746	0	−0.00061	0.402	i
Paraffins[j]	*n*-Alcohols	25	0.87	0.176	0	−0.00049	−0.630	
		60	0.80	0.138	0	−0.00057	−0.440	
		100	0.72	0.112	0	−0.00061	−0.280	
Water[j]	Paraffins	25	2.55	0	0	0	3.88	
Water[j]	*n*-Alkylbenzenes	25	3.04	0	0	0	−3.14	

(Continued)

TABLE 11-4 (Continued)

Modification of terms in Eq. 11-12:

a. $B_2 N_1$

b. $B_2 N_1, C_1 \left(\dfrac{1}{N'_1} + \dfrac{1}{N''_1} \right)$

c. $B_2 N_1, C_1 \left(\dfrac{1}{N'_1} + \dfrac{1}{N''_1} + \dfrac{1}{N'''_1} \right), F_2 \left(\dfrac{1}{N'_2} + \dfrac{1}{N''_2} \right)$

d. $B_2 N_1, C_1 \left(\dfrac{1}{N'_1} + \dfrac{1}{N''_1} + \dfrac{1}{N'''_1} - 3 \right)$

e. $B_2 N_1, C_1 \left(\dfrac{1}{N'_1} + \dfrac{1}{N''_1} + \dfrac{2}{N'''_1} \right); N'''_1$ relates to R''' in R' (R") C (OR''')$_2$

f. $B_2 (N_1 - 6), C_1 \left(\dfrac{1}{N_1 - 4} \right)$

g. $C_1 \left(\dfrac{1}{N'_1} + \dfrac{1}{N''_1} \right)$

h. $C_1 \left(\dfrac{1}{N'_1} + \dfrac{1}{N''_1} + \dfrac{2}{N'''_1} \right); N'''_1$ relates to R''' in R' (R") C (OR''')$_2$

i. $F_2 \left(\dfrac{1}{N'_2} + \dfrac{1}{N''_2} \right)$

N_1, N_2 = total number of carbon atoms in molecules 1 and 2, respectively.
N', N'', N''' = number of carbon atoms in respective branches of branched compounds, counting the polar grouping; thus, for *t*-butanol, $N' = N'' = N''' = 2$.

j. Entries contributed by the author of this chapter.

Source: Pierotti, Deal and Derr [18] as modified by Reid, Prausnitz and Sherwood [21] and this author (see note j). *(Reprinted with permission from the American Chemical Society and McGraw-Hill Book Co.)*

Some examples illustrating the estimation of γ^∞ for a variety of materials should be helpful at this point. The basic steps are as follows:

Basic Steps

(1) Draw the structures of the chemical involved.

(2) Use Table 11-4 to obtain the appropriate correlation constants and any modification of terms required.

(3) Substitute the constants and modifications into Eq. 11-12 and calculate $\log \gamma_1^\infty$.

(4) If the solute is an aromatic compound and water is the solvent, calculate $\log \gamma_1^\infty$ for the unsubstituted hydrocarbon; then use Table 11-5, as appropriate, to obtain corrections to $\log \gamma_1^\infty$.

(5) The antilog of $\log \gamma_1^\infty$ (corrected, if appropriate) yields γ_1^∞.

TABLE 11-5

Correction Factors for log γ_1^∞, per Group[a]

Group	Δ	Group	Δ
F	0.14	NH_2	−1.35
Cl	0.70	NO_2	
Br	0.92	(hydrocarbons)	0.15
I	1.40	(*m*-, *p*-phenols)	0.30
		(*m*-, *p*-anilines)	1.00
OH		CH_3 ($N_1 \geqslant 8$)	−0.25
(alcohols)	−1.90		
(phenols)	−1.70	C=C (in side chain)	−0.30
		C≡C (in side chain)	−0.46
COOH		Polycyclic hydrocarbons	−1.11 per addi-
(in side chain)	−1.70	(naphthenes and biphenyls)	tional ring
(on ring)	−0.70		

a. Groups are attached to ring unless otherwise specified.

Source: Tsonopoulos and Prausnitz [25]. *(Reprinted with permission from the American Chemical Society and McGraw-Hill Book Co.)*

Example 11-1 Estimate the infinite-dilution activity coefficients in the system ethanol(1)-n-hexane(2) at 70°C [25].

(1) The structure of ethanol is CH_3 CH_2 OH; for hexane the structure is $CH_3(CH_2)_4 CH_3$. Thus $N_1 = 2, N_2 = 6$.

(2) From Table 11-4 the correlation coefficients for alcohols in paraffin solvents at 25°C, 60°C and 100°C are given. The changes in all the coefficients except $A_{1,2}$ are small enough such that the value at 60°C can be used without loss of accuracy. For $A_{1,2}$ interpolation between 60°C and 100°C should be used for highest accuracy. Thus $A_{1,2}$ (interpolated) = 1.40, $B_2 = 0, C_1 = 0.390$ and D = −0.00057. Under modification (d) the equation for log γ_1^∞ is

$$\log \gamma_1^\infty = A_{1,2} + B_2 N_1 + C_1 (1/N_1 - 1) + D (N_1 - N_2)^2$$

(3) $\log \gamma_1^\infty = 1.400 + 0 + 0.39 (1/2 - 1) - 0.00057 (2 - 6)^2$

$$= 1.400 - 0.195 - 0.00912$$

$$= 1.196$$

(4) γ_1^∞ = antilog (1.196) = 15.7. The experimental value is 18.1 [21] for an error of −13.3%.

For γ_2^∞ the indices are interchanged; i.e., we are now interested in the system hexane(1)-ethanol(2) with $N_1 = 6$ and $N_2 = 2$.

(5)   Repeating Step 2 from Table 11-4, for the paraffin-alcohol correlation, we note that the changes in all of the coefficients with respect to temperature are small enough such that the values at 60°C can be used with sufficient accuracy. Thus $A_{1,2} = 0.80$, $B_2 = 0.138$, $D = -0.00057$ and $F_2 = -0.440$. Equation 11-12 is used as is, with $C_1 = 0$; thus,

(6) $\log \gamma_2^\infty = 0.80 + 0.138\ (6/2) - 0.00057\ (6-2)^2 - 0.440/2$

$= 0.80 + 0.414 - 0.00912 - 0.220$

$= 0.985$

(7) $\gamma_2^\infty = $ antilog $(0.985) = 9.66$. The experimental value is 9.05 [22] for an error of +6.7%.

Example 11-2   Estimate the infinite dilution activity coefficients for the system benzene-water at 25°C. From these values estimate the solubility of benzene in water and water in benzene and Henry's law constant for benzene in water at 25°C.

(1)   Structure of benzene

N₁ = 6. For water $N_2 = 0$. However, this does not lead to an infinite B_2 term because footnote f of Table 11-4 dictates the use of a modified B_2 term.

(2)   From Table 11-4 the correlation coefficients for alkylbenzenes in water are $A_{1,2} = 3.554$, $B_2 = 0.622$ and $C_1 = -0.466$. There are no D and F_2 terms. Under modification (f) the equation for $\log \gamma_1^\infty$ is

$\log \gamma_1^\infty = A_{1,2} + B_2\ (N_1 - 6) + C_1\ [1/(N_1 - 4)]$

(3) $\log \gamma_1^\infty = 3.554 + 0.622\ (6 - 6) - 0.466\ [1/(6 - 4)]$

$= 3.321$

(4) $\gamma_1^\infty = $ antilog $(3.321) = 2,094$. The experimental value is 2400 [4] for an error of −12.8%.

(5)   Repeating Step 2 with water as solute and benzene as solvent the correlation coefficients are $A_{1,2} = 3.04$ and $F_2 = -3.14$. Equation 11-12 is used as is, with $B_2 = C_1 = D = 0$.

(6) $\log \gamma_2^\infty = 3.04 - \dfrac{3.14}{6}$

$= 2.517$

(7) γ_2^∞ = antilog (2.517) = 329. The experimental value is 430 [9] for an error of −23.5%.

(8)   The methods of Chapter 3 allow us now to estimate the mutual solubilities in the benzene-water system. For benzene in water the solubility is inversely proportional to γ_1^∞ as $\gamma_1^\infty > 1000$; hence $x_1 = 1/\gamma_1^\infty = 4.8 \times 10^{-4}$ mole fraction. The experimental value is 4.2×10^{-4} [7]. For water in benzene γ_2^∞ is < 1000 but > 50; hence, Figure 3-5 of Chapter 3 is applicable, and $x_2 = 3.5 \times 10^{-3}$ mole fraction. The experimental value is 3.1×10^{-3} [2].

(9)   The Henry's law constant is defined as $H \approx P_{vp_1}/S$ where P_{vp_1} (atm) is the vapor pressure of pure component 1 at the system temperature (25°C) and S is the solubility in moles/m^3. The vapor pressure of benzene at 25°C is estimated (Method 2, Chapter 14) as 95.5 mm or 0.126 atm. The estimated solubility S in moles/m^3 of water is $S \approx 4.8 \times 10^{-4} \times 55.5 \times 10^3 = 26.6$ moles/m^3. Thus $H = 0.126/26.1 = 4.7 \times 10^{-3}$ atm·m^3/mol. The experimental value is 5.5×10^{-3} atm·m^3/mole [26].

The estimated infinite-dilution activity coefficients can be used to determine the parameters in any two-parameter equation. For example, the parameters in the van Laar equation (Table 11-1) are given by $A = RT \ln \gamma_1^\infty$ and $B = RT \ln \gamma_2^\infty$; hence, at any value of x_1 and x_2, the activity coefficients are given by

$$\ln \gamma_1 = \ln \gamma_1^\infty \left(1 + \frac{x_1 \ln \gamma_1^\infty}{x_2 \ln \gamma_2^\infty}\right)^{-2} \tag{11-13}$$

and

$$\ln \gamma_2 = \ln \gamma_2^\infty \left(1 + \frac{x_2 \ln \gamma_2^\infty}{x_1 \ln \gamma_1^\infty}\right)^{-2} \tag{11-14}$$

In the next example the activity coefficients at $x_1 = 0.292$ will be estimated.

Example 11-3   Estimate the infinite-dilution activity coefficients for the system n-pentane(1)-acetone(2) at their respective boiling points, $T_b = 309K$ (36°C) for pentane and 330K (57°C) for acetone. Use the infinite-dilution activity coefficients to estimate the activity coefficients at a composition of 0.292 mole fraction of n-pentane.

(1)   The structure of n-pentane is $CH_3 (CH_2)_3 CH_3$. For acetone the structure is $(CH_3)_2 C = O$.

(2)   From Table 11-4, there are no $A_{1,2}$ or C_1 terms for Eq. 11-12. The remaining correlation coefficients for paraffins in ketone solvents were interpolated (for

values at 36°C) and are: $B_2 = 0.1608$, $D = -0.00051$ and $F_2 = 0.402$. Under modification (i) the F_2 term is $F_2 (1/N'_2 + 1/N''_2)$ where the N_2s refer to the number of carbons on either side of the carbonyl group plus the carbonyl carbon. Thus $N'_2 = N''_2 = 2$.

(3)   Substituting in Eq. 11-12

$$\log \gamma_1^\infty = 0.1608 \ (5/3) - 0.00051 \ (5-3)^2 + 0.402 \ (1/2 + 1/2)$$

$$= 0.668$$

(4) $\gamma_1^\infty = $ antilog $(0.668) = 4.66$

(5)   Repeating Step 2 for acetone in pentane (using the coefficient values for 60°C) the correlation coefficients are $A_{1,2} = 0.016$, $B_2 = 0$, $C_1 = 0.680$, $D = -0.00057$ and there is no F_2 term. (Interpolation is unnecessary as the boiling point of acetone, 57°C, is close enough to the values given at 60°C.) Under modification (b) the C_1 term is changed to $C_1 (1/N'_1 + 1/N''_1)$ where the N_1s have the same meaning given above when acetone was the solvent.

(6)   Substituting in Eq. 11-12

$$\log \gamma_2^\infty = 0.016 + 0.680 \ (1/2 + 1/2) - 0.00057 \ (3-5)^2$$

$$= 0.694$$

(7) $\gamma_2^\infty = $ antilog $(0.694) = 4.94$

(8)   To estimate γ_1 at $x_1 = 0.292$ we use Eq. 11-13 with $x_2 = 1 - x_1 = 0.708$

$$\ln \gamma_1 = \ln (4.66) \left(1 + \frac{0.292 \ \ln (4.66)}{0.708 \ \ln (4.94)} \right)^{-2}$$

$$= 1.539 \left(1 + \frac{0.292 \times 1.539}{0.708 \times 1.597} \right)^{-2} = 1.539 \ (1.397)^{-2}$$

$$= 0.788$$

$$\gamma_1 = 2.20$$

The experimental value is 2.22 [12] for an error of -0.9%.

(9)   To estimate γ_2 at $x_1 = 0.292$ Eq. 11-14 is used.

$$\ln \gamma_2 = \ln (4.94) \left(1 + \frac{0.708 \ \ln (4.94)}{0.292 \ \ln (4.66)} \right)^{-2}$$

$$= 1.597 \ (3.516)^{-2}$$

$$= 0.129$$

$$\gamma_2 = 1.14$$

The experimental value is 1.18 [12] for an error of − 3.4%.

In Figure 11-1 a comparison is shown between calculated and experimental activity coefficients over the entire range of compositions. The left-most experimental value for the activity coefficient of *n*-pentane almost certainly illustrates the difficulty of measuring the activity coefficient at low concentration.

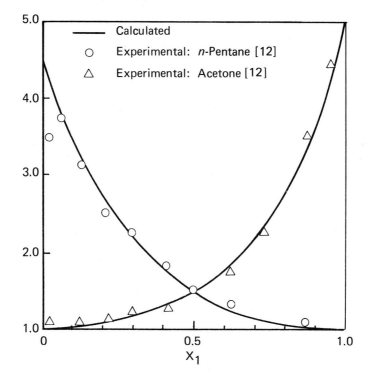

FIGURE 11−1 **Calculated Activity Coefficients in the System *n*-Pentane-Acetone at 760 mm Hg**

The *n*-pentane-acetone system is not too far removed from ideal behavior, and the Van Laar equation may be used to calculate the vapor-liquid equilibrium. This was done following the procedure outlined in Reid *et al.* [21] and using Method 2 of Chapter 14 to estimate pure-component vapor pressures. Figure 11-2 shows the results obtained.

Example 11-4 Estimate the infinite-dilution activity coefficient of aniline in water at 25°C.

(1) The structure of aniline is

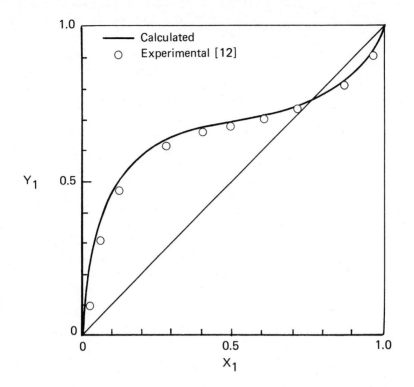

FIGURE 11–2 **Calculated vs. Experimental Vapor-Liquid Equilibrium in the System n-Pentane -Acetone at 760 mm Hg**

(2) In Example 11-2 the log γ_1^∞ for benzene in water was estimated to be 3.321.

(3) From Table 11-5 the correction factor for the NH_2 group is −1.35.

(4) Thus for aniline

$$\log \gamma_1^\infty = (\log \gamma_1^\infty)_{benzene} -1.35 = 1.971$$

$$\gamma_1^\infty = 94$$

The experimental value is 120 [25] for an error of −22%.

11-5 METHOD 2 — UNIFAC

The UNIFAC method is a group contribution concept. The basic idea is that, whereas there are thousands of organic compounds, the number of functional groups that constitute these compounds is much smaller.

Estimation of thermodynamic properties of liquid mixtures from group contributions was first suggested by Langmuir [11]. This suggestion received little attention until Derr *et al.* [5,6] used group contributions to correlate heats of mixing, followed by Wilson and Deal [28], who developed the "solution of groups" method for activity coefficients.

The basic aim of the solution of groups method is to utilize existing phase equilibrium data for predicting phase equilibria of systems for which there are no experimental data. The method entails suitable reduction of experimental data to obtain parameters characteristic of certain groups. These parameters can then be used in other systems containing these groups. A group is defined as any convenient structural fragment — CH_3, CH_2OH, $COOH$, CH_2Cl, etc.

The primary feature of all solution-of-group methods is that the logarithm of the activity coefficient is assumed to be the sum of two contributions — a combinatorial part, essentially due to differences in size and shape of the molecules in the mixture, and a residual part, essentially due to energy interactions. Thus, for molecule i in any solution:

$$\ln \gamma_i = \underset{\text{combinatorial}}{\ln \gamma_i^C} + \underset{\text{residual}}{\ln \gamma_i^R} \tag{11-15}$$

In the ASOG method, developed by Wilson and Deal [28], the combinatorial part is given by the athermal Flory-Huggins [10a] equation, and the residual part is given by the Wilson equation. The practical application of ASOG has been discussed by Palmer [17].

In UNIFAC the combinatorial part of Eq. 11-15 is given [9] for a binary mixture by

$$\ln \gamma_1^C = \ln \frac{\phi_1}{x_1} + \frac{z}{2} q_1 \ln \frac{\theta_1}{\phi_1} + \ell_1 - \frac{\phi_1}{x_1} (x_1 \ell_1 + x_2 \ell_2) \tag{11-16}$$

and

$$\ln \gamma_2^C = \ln \frac{\phi_2}{x_2} + \frac{z}{2} q_2 \ln \frac{\theta_2}{\phi_2} + \ell_2 - \frac{\phi_2}{x_2} (x_1 \ell_1 + x_2 \ell_2) \tag{11-17}$$

where

$$\phi_1 = \frac{r_1 x_1}{r_1 x_1 + r_2 x_2} \qquad\qquad \theta_1 = \frac{q_1 x_1}{q_1 x_1 + q_2 x_2} \tag{11-18a}$$

$$\phi_2 = \frac{r_2\, x_2}{r_1\, x_1 + r_2\, x_2} \qquad\qquad \theta_2 = \frac{q_2\, x_2}{q_1\, x_1 + q_2\, x_2} \qquad (11\text{-}18b)$$

Note that $\phi_1 + \phi_2 = 1$ and $\theta_1 + \theta_2 = 1$.

$$\ell_1 = \frac{z}{2}\,(r_1 - q_1) - (r_1 - 1) \qquad ; \qquad z = 10$$

$$\ell_2 = \frac{z}{2}\,(r_2 - q_2) - (r_2 - 1)$$
$$(11\text{-}19)$$

$$r_1 = \sum_k \nu_k{}^{(1)}\, R_k \qquad ; \qquad r_2 = \sum_k \nu_k{}^{(2)}\, R_k$$

$$q_1 = \sum_k \nu_k{}^{(1)}\, Q_k \qquad ; \qquad q_2 = \sum_k \nu_k{}^{(2)}\, Q_k$$
$$(11\text{-}20)$$

and where

$\begin{aligned}
x_1, x_2 &= \text{mole fraction of components 1 and 2}\\
\nu_k &= 1, 2, \ldots \text{ N (number of groups in molecule 1 or 2)}\\
R_k &= \text{the van der Waals volume for group k}\\
Q_k &= \text{the van der Waals surface area for group k}\\
k &= \text{group number (assigned) from Table 11-6}
\end{aligned}$

In Table 11-6 group volumes and surface areas are given for the 40 main groups that have been studied to date. Note that the main group often contains subgroups with different R_k and Q_k values. These classifications are based on experience gained from fitting the UNIFAC model to vapor-liquid equilibrium data.[6]

---

6. [Note added in final proof.] A new set of UNIFAC parameters (R_k, Q_k and a_{mn} values) has recently been published (Magnussen *et al.*, *Ind. Eng. Chem. Process Des. Dev.*, **20**, 331-339, 1981). The new parameters were determined from binary and ternary liquid-liquid equilibrium data sets rather than vapor-liquid equilibrium data. The new parameters are definitely preferable for solubility calculations (methods of Chapter 3) or other predictions of liquid-liquid equilibria since the method errors will generally be lower. The new parameters should not be used for components with normal boiling points below 300K nor should they be used outside the temperature range of 10 to 40°C.

TABLE 11-6

Group Volume and Surface-Area Parameters

Main Group	Subgroup[a]	k	R_k	Q_k	Sample Group Assignment	
1 "CH$_2$"	CH$_3$ CH$_2$ CH C	1 2 3 4	0.9011 0.6744 0.4469 0.2195	0.848 0.540 0.228 0.000	Butane: 2-Methylpropane: 2,2-Dimethylpropane:	2 CH$_3$, 2 CH$_2$ 3 CH$_3$, 1 CH 4 CH$_3$, 1 C
2 "C=C"	CH$_2$=CH CH=CH CH$_2$=C CH=C C=C	5 6 7 8 9	1.3454 1.1167 1.1173 0.8886 0.6605	1.176 0.867 0.988 0.676 0.485	1-Hexene: 2-Hexene: 2-Methyl-1-butene: 2-Methyl-2-butene: 2,3-Dimethylbutene-2:	1 CH$_3$, 3 CH$_2$, 1 CH=CH 2 CH$_3$, 2 CH$_2$, 1 CH=CH 2 CH$_3$, 1 CH$_2$, 1 CH$_2$=C 3 CH$_3$, 1 CH=C 4 CH$_3$, 1 C=C
3 "ACH"	ACH AC	10 11	0.5313 0.3652	0.400 0.120	Benzene: 6 ACH Styrene: 1 CH$_2$=CH, 5 ACH, 1 AC	
4 "ACCH$_2$"	ACCH$_3$ ACCH$_2$ ACCH	12 13 14	1.2663 1.0396 0.8121	0.968 0.660 0.348	Toluene: Ethylbenzene: Cumene:	5 ACH, 1 ACCH$_3$ 1 CH$_3$, 5 ACH, 1 ACCH$_2$ 2 CH$_3$, 5 ACH, 1 ACCH
5	OH	15	1.000	1.200	1-Propanol: 1 CH$_3$, 2 CH$_2$, 1 OH	
6	CH$_3$OH	16	1.4311	1.432	Methanol: 1 CH$_3$OH	
7	H$_2$O	17	0.92	1.40	Water: 1 H$_2$O	
8	ACOH	18	0.8952	0.680	Phenol: 5 ACH, 1 ACOH	
9 "CH$_2$CO"	CH$_3$CO CH$_2$CO	19 20	1.6724 1.4457	1.488 1.180	Ketone group is 2nd carbon; 2-Butanone: 1 CH$_3$, 1 CH$_2$, 1 CH$_3$CO Ketone group is any other carbon; 3-Pentanone: 2 CH$_3$, 1 CH$_2$, 1 CH$_2$CO	
10	CHO	21	0.9980	0.948	Acetaldehyde: 1 CH$_3$, 1 CHO	
11 "CCOO"	CH$_3$COO CH$_2$COO	22 23	1.9031 1.6764	1.728 1.420	Butyl acetate: 1 CH$_3$, 3 CH$_2$, 1 CH$_3$COO Butyl propanoate: 2 CH$_3$, 3 CH$_2$, 1 CH$_2$COO	
12	HCOO	24	1.2420	1.188	Ethyl formate: 1 CH$_3$, 1 CH$_2$, 1 HCOO	
13 "CH$_2$O"	CH$_3$O CH$_2$O CH—O FCH$_2$O	25 26 27 28	1.1450 0.9183 0.6908 0.9183	1.088 0.780 0.468 1.1	Dimethyl ether: 1 CH$_3$, 1 CH$_3$O Diethyl ether: 2 CH$_3$, 1 CH$_2$, 1 CH$_2$O Diisopropyl ether: 4 CH$_3$, 1 CH, 1 CH—O Tetrahydrofuran: 3 CH$_2$, 1 FCH$_2$O	
14 "CNH$_2$"	CH$_3$NH$_2$ CH$_2$NH$_2$ CHNH$_2$	29 30 31	1.5959 1.3692 1.1417	1.544 1.236 0.924	Methylamine: 1 CH$_3$NH$_2$ Propylamine: 1 CH$_3$, 1 CH$_2$, 1 CH$_2$NH$_2$ Isopropylamine: 2 CH$_3$, 1 CHNH$_2$	
15 "CNH"	CH$_3$NH CH$_2$NH CHNH	32 33 34	1.4337 1.2070 0.9795	1.244 0.936 0.624	Dimethylamine: 1 CH$_3$, 1 CH$_3$NH Diethylamine: 2 CH$_3$, 1 CH$_2$, 1 CH$_2$NH Diisopropylamine: 4 CH$_3$, 1 CH, 1 CHNH	
16 "(C)$_3$N"	CH$_3$N CH$_2$N	35 36	1.1865 0.9597	0.940 0.632	Trimethylamine: 2 CH$_3$, 1 CH$_3$N Triethylamine: 3 CH$_3$, 2 CH$_2$, 1 CH$_2$N	
17	ACNH$_2$	37	1.0600	0.816	Aniline: 5 ACH, 1 ACNH$_2$	
18 "Pyridine"	C$_5$H$_5$N C$_5$H$_4$N C$_5$H$_3$N	38 39 40	2.9993 2.8832 2.667	2.113 1.833 1.558	Pyridine: 1 C$_5$H$_5$N 3-Methyl pyridine: 1 CH$_3$, 1 C$_5$H$_4$N 2,3-Methyl pyridine: 2 CH$_3$, 1 C$_5$H$_3$N	
19 "CCN"	CH$_3$CN CH$_2$CN	41 42	1.8701 1.6434	1.724 1.416	Acetonitrile: 1 CH$_3$CN Propionitrile: 1 CH$_3$, 1 CH$_2$CN	
20 "COOH"	COOH HCOOH	43 44	1.3013 1.5280	1.224 1.532	Acetic acid: 1 CH$_3$, 1 COOH Formic acid: 1 HCOOH	
21 "CCl"	CH$_2$Cl CHCl CCl	45 46 47	1.4654 1.2380 1.0060	1.264 0.952 0.724	1-Chlorobutane: 1 CH$_3$, 2 CH$_2$, 1 CH$_2$Cl 2-Chloropropane: 2 CH$_3$, 1 CHCl 2-Chloro-2-methylpropane: 3 CH$_3$, 1 CCl	
22 "CCl$_2$"	CH$_2$Cl$_2$ CHCl$_2$ CCl$_2$	48 49 50	2.2564 2.0606 1.8016	1.988 1.684 1.448	Dichloromethane: 1 CH$_2$Cl$_2$ 1,1-Dichloroethane: 1 CH$_3$, 1 CHCl$_2$ 2,2-Dichloropropane: 2 CH$_3$, 1 CCl$_2$	

(continued)

TABLE 11-6 (Continued)

Main Group	Subgroup[a]	k	R_k	Q_k	Sample Group Assignment
23 "CCl$_3$"	CHCl$_3$	51	2.8700	2.410	Chloroform: 1 CHCl$_3$
	CCl$_3$	52	2.6401	2.184	1,1,1-Trichloroethane: 1 CH$_3$, 1 CCl$_3$
24	CCl$_4$	53	3.3900	2.910	Tetrachloromethane: 1 CCl$_4$
25	ACCl	54	1.1562	0.844	Chlorobenzene: 5 ACH, 1 ACCl
26 "CNO$_2$"	CH$_3$NO$_2$	55	2.0086	1.868	Nitromethane: 1 CH$_3$NO$_2$
	CH$_2$NO$_2$	56	1.7818	1.560	1-Nitropropane: 1 CH$_3$, 1 CH$_2$, 1 CH$_2$NO$_2$
	CHNO$_2$	57	1.5544	1.248	2-Nitropropane: 2 CH$_3$, 1 CHNO$_2$
27	ACNO$_2$	58	1.4199	1.104	Nitrobenzene: 5 ACH, 1 ACNO$_2$
28	CS$_2$	59	2.057	1.65	Carbon disulfide: 1 CS$_2$
29 "CH$_3$SH"	CH$_3$SH	60	1.8770	1.676	Methanethiol: 1 CH$_3$SH
	CH$_2$SH	61	1.6510	1.368	Ethanethiol: 1 CH$_3$, 1 CH$_2$SH
30	Furfural	62	3.1680	2.484	Furfural: 1 Furfural
31 "DOH"[b]	(CH$_2$OH)$_2$	63	2.4088	2.248	1,2 Ethanediol: 1 (CH$_2$OH)$_2$
32	I	64	1.2640	0.992	Iodoethane: 1 CH$_3$, 1 CH$_2$, 1 I
33	Br	65	0.9492	0.832	Bromo methane: 1 CH$_3$, 1 Br
34 "C≡C"	CH≡C	66	1.2920	1.088	1-Hexyne: 1 CH$_3$, 4 CH$_2$, 1 CH≡C
	C≡C	67	1.0613	0.784	2-Hexyne: 2 CH$_3$, 3 CH$_2$, 1 C≡C
35	DMSO	68	2.8266	2.472	Dimethylsulfoxide: 1 DMSO
36	ACRY	69	2.3144	2.052	Acrylonitrile: 1 ACRY
37 "ClCC"	Cl—(C=C)	70	0.7910	0.724	Trichloroethylene: 1 CH=C, 3 Cl—(C=C)
38	ACF	71	0.6948	0.524	Hexafluorobenzene: 6 ACF
39 "DMF"	DMF-1	72	3.0856	2.736	Dimethylformamid: 1 DMF-1
	DMF-2	73	2.6322	2.120	Diethylformamid: 2 CH$_3$, 1 DMF-2
40 "CF$_2$"	CF$_3$	74	1.4060	1.380	Perfluorohexane: 2 CF$_3$, 4 CF$_2$
	CF$_2$	75	1.0105	0.920	
	CF	76	0.6150	0.460	Perfluorocyclohexane: 1 CH$_3$, 5 CH$_2$, 1 CF

a. "A" refers to an aromatic ring.   b. "DOH" refers to a diol.

Source: Gmehling, Rasmussen and Fredenslund [10b]. *(Reprinted with permission from the American Chemical Society.)*

For a binary mixture, the residual part of Eq. 11-15 is given by:

$$\ln \gamma_1^R = \sum_k \nu_k^{(1)} [\ln\Gamma_k - \ln\Gamma_k^{(1)}]$$

$$\ln \gamma_2^R = \sum_k \nu_k^{(2)} [\ln\Gamma_k - \ln\Gamma_k^{(2)}]$$

(11-21)

where

$$\ln\Gamma_k = Q_k [1 - \ln (\sum_m \Theta_m \psi_{mk}) - \sum_m (\Theta_m \psi_{km} / \sum_n \Theta_n \psi_{nm})] \qquad (11\text{-}22)$$

and m and n = 1,2, . . ., N (all groups)

$$\Theta_m = \frac{Q_m X_m}{\sum_n Q_n X_n} \qquad ; \qquad X_m = \frac{\sum_j \nu_m^{(j)} x_j}{\sum_j \sum_n \nu_n^{(j)} x_j} \qquad (11\text{-}23)$$

j = 1,2, . . . , M; n = 1,2, . . . , N

Similar relationships hold for $\ln\Gamma_k^{(1)}$ and $\ln\Gamma_k^{(2)}$. Note, however, that these refer to the pure components.

Furthermore,

j	=	component number (1 or 2 for a binary mixture)
M	=	total number of components (= 2 for a binary mixture)
n	=	group number
N	=	total groups
Θ_m	=	group surface area fraction
X_m	=	group fraction (For example, for an equimolar mixture of butane and hexane, the group fractions for CH_3 and CH_2 are 4/10 and 6/10, respectively.)
x_j	=	mole fraction of molecule j in the mixture
$\nu_m^{(j)}$	=	number of groups of type m in molecule j
ψ_{nm}	=	$\exp(-a_{nm}/T)$
T	=	temperature (K)

(11-24)

The parameter a_{nm} is a group interaction parameter which is a measure of the difference in interaction energy between groups n and m. In general, a_{nm} is independent of temperature and $a_{nm} \neq a_{mn}$. The interaction parameters have been determined for 40 groups covering a large number of compounds, and are reproduced from Ref. 10b as Table 11-7. Notice that, with respect to a_{nm}, the 40 main groups are not broken down into subgroups as was done for the van der Waals surface areas and volumes.

Basic Steps

(1) Draw the structures of the chemicals involved.

(2) Determine the kind and number of structural groups corresponding to those represented in Table 11-6. *(See footnote 6, p. 11-22.)*

TABLE 11-7

UNIFAC Group Interaction Parameters a_{mn}
(*m* indicates row, *n* indicates column)[a]

		1 CH_2	2 $C=C$	3 ACH	4 $ACCH_2$	5 OH	6 CH_3OH
1	CH_2	0	−200	61.13	76.5	986.5	697.2
2	$C=C$	2520	0	340.7	4102	693.9	1509
3	ACH	−11.12	−94.78	0	167	636.1	637.35
4	$ACCH_2$	−69.7	−269.7	−146.8	0	803.2	603.25
5	OH	156.4	8694	89.6	25.82	0	−137.1
6	CH_3OH	16.51	−52.39	−50	−44.5	249.1	0
7	H_2O	300	692.7	362.3	377.6	−229.1	289.6
8	ACOH	10000	732.2	270.2	10000	−274.5	−111.6
9	CH_2CO	26.76	−82.92	140.1	365.8	164.5	108.7
10	CHO	505.7	n.a.	n.a.	n.a.	−404.8	−340.2
11	CCOO	114.8	362.4	85.84	−170	245.4	249.63
12	HCOO	90.49	91.65	n.a.	n.a.	191.2	155.7
13	CH_2O	83.36	76.44	52.13	65.69	237.7	339.72
14	CNH_2	−30.48	79.4	−44.85	n.a.	−164	−481.7
15	CNH	65.33	−41.32	−22.31	223	−150	−500.4
16	$(C)_3N$	−83.98	−188	−223.9	109.9	28.6	−406.8
17	$ACNH_2$	5339	n.a.	650.4	979.8	529	5.182
18	Pyridine	−101.6	n.a.	31.87	49.8	−132.3	−378.2
19	CCN	24.82	34.78	−22.97	−138.4	185.4	157.8
20	COOH	315.3	349.2	62.32	268.2	−151	1020
21	CCl	91.46	−24.36	4.68	122.9	562.2	529
22	CCl_2	34.01	−52.71	121.3	n.a.	747.7	669.9
23	CCl_3	36.7	−185.1	288.5	33.61	742.1	649.1
24	CCl_4	−78.45	−293.7	−4.7	134.7	856.3	860.1
25	ACCl	−141.3	−203.2	−237.7	375.5	246.9	661.6
26	CNO_2	−32.69	−49.92	10.38	−97.05	341.7	252.6
27	$ACNO_2$	5541	n.a.	1824	−127.8	561.6	n.a.
28	CS_2	−52.65	16.62	21.5	40.68	823.5	914.2
29	CH_3SH	−7.481	n.a.	28.41	n.a.	461.6	382.8
30	Furfural	−25.31	n.a.	157.3	404.3	521.6	n.a.
31	DOH	140	n.a.	221.4	150.6	267.6	n.a.
32	I	128	n.a.	58.68	n.a.	501.3	n.a.
33	Br	−31.52	n.a.	155.6	291.1	721.9	n.a.
34	$C≡C$	−72.88	−184.4	n.a.	n.a.	n.a.	n.a.
35	DMSO	50.49	n.a.	−2.5040	−143.2	−25.87	695
36	ACRY	−165.9	n.a.	n.a.	n.a.	n.a.	n.a.
37	ClCC	41.9	−3.167	−75.67	n.a.	640.9	726.7
38	ACF	−5.132	n.a.	−237.2	−157.3	649.7	645.9
39	DMF	−31.95	37.7	−133.9	−240.2	64.16	172.2
40	CF_2	147.3	n.a.	n.a.	n.a.	n.a.	n.a.

		7 H$_2$O	8 ACOH	9 CH$_2$CO	10 CHO	11 CCOO	12 HCOO
1	CH$_2$	1318	912.2	476.4	677	232.1	741.4
2	C=C	634.2	926.3	524.5	n.a.	4.826	468.7
3	ACH	903.8	1174	25.77	n.a.	5.994	n.a.
4	ACCH$_2$	5965	674.3	−52.1	n.a.	5688	n.a.
5	OH	353.5	−442.1	84	441,8	101.1	193.1
6	CH$_3$OH	−181	−125.1	23.39	306.4	−10.72	193.4
7	H$_2$O	0	125.4	−195.4	−257.3	14.42	n.a.
8	ACOH	531.3	0	−158.8	n.a.	−442.8	n.a.
9	CH$_2$CO	472.5	−246.8	0	−37.36	−213.7	n.a.
10	CHO	232.7	n.a.	128	0	n.a.	n.a.
11	CCOO	10000	−72.58	372.2	n.a.	0	372.9
12	HCOO	n.a.	n.a.	n.a.	n.a.	−261.1	0
13	CH$_2$O	−314.7	n.a.	52.38	−7.838	461.3	n.a.
14	CNH$_2$	−330.4	n.a.	n.a.	n.a.	n.a.	n.a.
15	CNH	−448.2	n.a.	n.a.	n.a.	136	n.a.
16	(C)$_3$N	−598.8	n.a.	n.a.	n.a.	n.a.	n.a.
17	ACNH$_2$	−339.5	n.a.	−399.1	n.a.	n.a.	n.a.
18	Pyridine	−332.9	−362.3	−51.54	n.a.	n.a.	n.a.
19	CCN	242.8	n.a.	−287.5	n.a.	−266.6	n.a.
20	COOH	−66.17	n.a.	−297.8	n.a.	−256.3	312.5
21	CCl	698.2	n.a.	286.3	−47.51	n.a.	n.a.
22	CCl$_2$	708.7	n.a.	423.2	n.a.	−132.9	n.a.
23	CCl$_3$	826.76	n.a.	552.1	n.a.	176.5	488.9
24	CCl$_4$	1201	1953	372	n.a.	129.5	n.a.
25	ACCl	920.4	n.a.	128.1	n.a.	−246.3	n.a.
26	CNO$_2$	417.9	n.a.	−142.6	n.a.	n.a.	n.a.
27	ACNO$_2$	360.7	n.a.	n.a.	n.a.	n.a.	n.a.
28	CS$_2$	1081	n.a.	303.7	n.a.	243.8	n.a.
29	CH$_3$SH	n.a.	n.a.	160.6	n.a.	n.a.	239.8
30	Furfural	23.48	n.a.	317.5	n.a.	−146.3	n.a.
31	DOH	0	772.5	n.a.	n.a.	152	n.a.
32	I	n.a.	n.a.	138	n.a.	21.92	n.a.
33	Br	n.a.	n.a.	−142,6	n.a.	n.a.	n.a.
34	C≡C	n.a.	n.a.	443.6	n.a.	n.a.	n.a.
35	DMSO	−240	n.a.	110.4	n.a.	41.57	n.a.
36	ACRY	386.6	n.a.	n.a.	n.a.	n.a.	n.a.
37	ClCC	n.a.	n.a.	−8.671	n.a.	−34.14	n.a.
38	ACF	n.a.	n.a.	n.a.	n.a.	n.a.	n.a.
39	DMF	−287.1	n.a.	97.04	n.a.	n.a.	n.a.
40	CF$_2$	n.a.	n.a.	n.a.	n.a.	n.a.	n.a.

TABLE 11-7 (Continued)

		13 CH$_2$O	14 CNH$_2$	15 CNH	16 (C)$_3$N	17 ACNH$_2$	18 Pyridine
1	CH$_2$	251.5	391.5	255.7	206.6	1245	287.77
2	C=C	289.3	396	273.6	658.8	n.a.	n.a.
3	ACH	32.14	161.7	122.8	90.49	668.2	−4.449
4	ACCH$_2$	213.1	n.a.	−49.29	23.5	764.7	52.8
5	OH	28.06	83.02	42.70	−323	−348.2	170
6	CH$_3$OH	−180.6	359.3	266	53.9	335.5	580.5
7	H$_2$O	540.5	48.89	168	304	213	459
8	ACOH	n.a.	n.a.	n.a.	n.a.	n.a.	−259.2
9	CH$_2$CO	5.202	n.a.	n.a.	n.a.	937.9	165.1
10	CHO	304.1	n.a.	n.a.	n.a.	n.a.	n.a.
11	CCOO	−235.7	n.a.	−73.5	n.a.	n.a.	n.a.
12	HCOO	n.a.	n.a.	n.a.	n.a.	n.a.	n.a.
13	CH$_2$O	0	n.a.	141.7	n.a.	n.a.	n.a.
14	CNH$_2$	n.a.	0	63.72	−41.11	n.a.	n.a.
15	CNH	−49.3	108.8	0	−189.2	n.a.	n.a.
16	(C)$_3$N	n.a.	38.89	865.9	0	n.a.	n.a.
17	ACNH$_2$	n.a.	n.a.	n.a.	n.a.	0	n.a.
18	Pyridine	n.a.	n.a.	n.a.	n.a.	n.a.	0
19	CCN	n.a.	n.a.	n.a.	n.a.	617.1	134.3
20	COOH	−338.5	n.a.	n.a.	n.a.	n.a.	−313.5
21	CCl	225.4	n.a.	n.a.	n.a.	n.a.	n.a.
22	CCl$_2$	−197.7	n.a.	n.a.	−141.4	n.a.	587.3
23	CCl$_3$	−20.93	n.a.	n.a.	−293.7	n.a.	18.98
24	CCl$_4$	113.9	261.1	91.13	−126	1301	309.2
25	ACCl	n.a.	203.5	−108.4	1088	323.3	n.a.
26	CNO$_2$	−94.49	n.a.	n.a.	n.a.	n.a.	n.a.
27	ACNO$_2$	n.a.	n.a.	n.a.	n.a.	5250	n.a.
28	CS$_2$	112.4	n.a.	n.a.	n.a.	n.a.	n.a.
29	CH$_3$SH	63.71	106.7	n.a.	n.a.	n.a.	n.a.
30	Furfural	n.a.	n.a.	n.a.	n.a.	n.a.	n.a.
31	DOH	9.207	n.a.	n.a.	n.a.	164.4	n.a.
32	I	476.6	n.a.	n.a.	n.a.	n.a.	n.a.
33	Br	736.4	n.a.	n.a.	n.a.	n.a.	n.a.
34	C≡C	n.a.	n.a.	n.a.	n.a.	n.a.	n.a.
35	DMSO	−122.1	n.a.	n.a.	n.a.	n.a.	n.a.
36	ACRY	n.a.	n.a.	n.a.	n.a.	n.a.	n.a.
37	ClCC	−209.3	n.a.	n.a.	n.a.	n.a.	n.a.
38	ACF	n.a.	n.a.	n.a.	n.a.	n.a.	n.a.
39	DMF	−158.2	n.a.	n.a.	n.a.	335.6	n.a.
40	CF$_2$	n.a.	n.a.	n.a.	n.a.	n.a.	n.a.

		19 CCN	20 COOH	21 CCl	22 CCl$_2$	23 CCl$_3$	24 CCl$_4$
1	CH$_2$	597	663.5	35.93	53.76	24.9	104.3
2	C=C	405.9	730.4	99.61	337.1	4584	5831
3	ACH	212.5	537.4	−18.81	−144.4	−231.9	3
4	ACCH$_2$	6096	603.8	−114.1	n.a.	−12.14	−141.3
5	OH	6.712	199	75.62	−112.1	−98.12	143.1
6	CH$_3$OH	36.23	−289.5	−38.32	−102.5	−139.4	−67.8
7	H$_2$O	112.6	−14.09	325.4	370.4	353.7	497.5
8	ACOH	n.a.	n.a.	n.a.	n.a.	n.a.	10000
9	CH$_2$CO	481.7	669.4	−191.7	−284	−354.6	−39.2
10	CHO	n.a.	n.a.	751.9	n.a.	n.a.	n.a.
11	CCOO	494.6	660.2	n.a.	108.9	−209.7	54.47
12	HCOO	n.a.	−356.3	n.a.	n.a.	−287.2	n.a.
13	CH$_2$O	n.a.	664.6	301.1	137.8	−154.3	47.67
14	CNH$_2$	n.a.	n.a.	n.a.	n.a.	n.a.	−99.81
15	CNH	n.a.	n.a.	n.a.	n.a.	n.a.	71.23
16	(C)$_3$N	n.a.	n.a.	n.a.	−73.85	−352.9	−8.283
17	ACNH$_2$	−216.8	n.a.	n.a.	n.a.	n.a.	8455
18	Pyridine	−169.7	−153.7	n.a.	−351.6	−114.7	−165.1
19	CCN	0	n.a.	n.a.	n.a.	−15.62	−54.86
20	COOH	n.a.	0	44.42	−183.4	76.75	212.7
21	CCl	n.a.	326.4	0	108.3	249.2	62.42
22	CCl$_2$	n.a.	1821	−84.53	0	0	⁻56.33
23	CCl$_3$	74.04	1346	−157.1	0	0	−30.1
24	CCl$_4$	492	689	11.8	17.97	51.9	0
25	ACCl	356.9	n.a.	−314.9	n.a.	n.a.	−255.4
26	CNO$_2$	n.a.	n.a.	n.a.	n.a.	n.a.	−34.68
27	ACNO$_2$	n.a.	n.a.	n.a.	n.a.	n.a.	514.6
28	CS$_2$	335.7	n.a.	−73.09	n.a.	−26.06	−60.71
29	CH$_3$SH	125.7	n.a.	−27.94	n.a.	n.a.	n.a.
30	Furfural	n.a.	n.a.	n.a.	n.a.	48.48	−133.16
31	DOH	n.a.	n.a.	n.a.	n.a.	n.a.	n.a.
32	I	n.a.	n.a.	n.a.	−40.82	21.76	48.49
33	Br	n.a.	n.a.	1169	n.a.	n.a.	225.8
34	C≡C	329.1	n.a.	n.a.	n.a.	n.a.	n.a.
35	DMSO	n.a.	n.a.	n.a.	−215	−343.6	−58.43
36	ACRY	−42.31	n.a.	n.a.	n.a.	n.a.	n.a.
37	ClCC	298.4	2344	201.7	n.a.	85.32	143.2
38	ACF	n.a.	n.a.	n.a.	n.a.	n.a.	−124.6
39	DMF	n.a.	n.a.	n.a.	n.a.	n.a.	−186.7
40	CF$_2$	n.a.	n.a.	n.a.	n.a.	n.a.	n.a.

TABLE 11-7 (Continued)

		25 ACCl	26 CNO$_2$	27 ACNO$_2$	28 CS$_2$	29 CH$_3$SH	30 Furfural
1	CH$_2$	321.5	661.5	543	153.6	184.4	354.55
2	C=C	959.7	542.1	n.a.	76.302	n.a.	n.a.
3	ACH	538.2	168	194.9	52.07	−10.43	−64.69
4	ACCH$_2$	−126.9	3629	4448	−9.451	n.a.	−20.36
5	OH	287.8	61.11	157.1	477	147.5	−120.5
6	CH$_3$OH	17.12	75.14	n.a.	−31.09	37.84	n.a.
7	H$_2$O	678.2	220.6	399.5	887.1	n.a.	188
8	ACOH	n.a.	n.a.	n.a.	n.a.	n.a.	n.a.
9	CH$_2$CO	174.5	137.5	n.a.	216.1	−46.28	−163.7
10	CHO	n.a.	n.a.	n.a.	n.a.	n.a.	n.a.
11	CCOO	629	n.a.	n.a.	183	n.a.	202.3
12	HCOO	n.a.	n.a.	n.a.	n.a.	4.339	n.a.
13	CH$_2$O	n.a.	95.18	n.a.	140.9	−8.538	n.a.
14	CNH$_2$	68.81	n.a.	n.a.	n.a.	−70.14	n.a.
15	CNH	4350	n.a.	n.a.	n.a.	n.a.	n.a.
16	(C)$_3$N	−86.36	n.a.	n.a.	n.a.	n.a.	n.a.
17	ACNH$_2$	699.1	n.a.	−62.73	n.a.	n.a.	n.a.
18	Pyridine	n.a.	n.a.	n.a.	n.a.	n.a.	n.a.
19	CCN	52.31	n.a.	n.a.	230.9	21.37	n.a.
20	COOH	n.a.	n.a.	n.a.	n.a.	n.a.	n.a.
21	CCl	464.4	n.a.	n.a.	450.1	59.02	n.a.
22	CCl$_2$	n.a.	n.a.	n.a.	n.a.	n.a.	n.a.
23	CCl$_3$	n.a.	n.a.	n.a.	116.6	n.a.	−64.38
24	CCl$_4$	475.8	490.9	534.7	132.2	n.a.	546.7
25	ACCl	0	−154.5	n.a.	n.a.	n.a.	n.a.
26	CNO$_2$	794.4	0	533.2	n.a.	n.a.	n.a.
27	ACNO$_2$	n.a.	−85.12	0	n.a.	n.a.	n.a.
28	CS$_2$	n.a.	n.a.	n.a.	0	n.a.	n.a.
29	CH$_3$SH	n.a.	n.a.	n.a.	n.a.	0	n.a.
30	Furfural	n.a.	n.a.	n.a.	n.a.	n.a.	0
31	DOH	n.a.	481.3	n.a.	n.a.	n.a.	n.a.
32	I	n.a.	64.28	n.a.	n.a.	n.a.	n.a.
33	Br	224	125.3	n.a.	n.a.	n.a.	n.a.
34	C≡C	n.a.	174.4	n.a.	n.a.	n.a.	n.a.
35	DMSO	n.a.	n.a.	n.a.	n.a.	85.7	n.a.
36	ACRY	n.a.	n.a.	n.a.	n.a.	n.a.	n.a.
37	ClCC	n.a.	313.8	n.a.	167.9	n.a.	n.a.
38	ACF	n.a.	n.a.	n.a.	n.a.	n.a.	n.a.
39	DMF	n.a.	n.a.	n.a.	n.a.	−71	n.a.
40	CF$_2$	n.a.	n.a.	n.a.	n.a.	n.a.	n.a.

		31 DOH	32 I	33 Br	34 C≡C	35 DMSO	36 ACRY
1	CH$_2$	3025	335.8	479.5	298.9	526.5	689
2	C=C	n.a.	n.a.	n.a.	523.6	n.a.	n.a.
3	ACH	210.4	113.3	−13.59	n.a.	169.9	n.a.
4	ACCH$_2$	4975	n.a.	−171.3	n.a.	4284	n.a.
5	OH	−318.9	313.5	133.4	n.a.	−202.1	n.a.
6	CH$_3$OH	n.a.	n.a.	n.a.	n.a.	−399.3	n.a.
7	H$_2$O	0	n.a.	n.a.	n.a.	−139	160.8
8	ACOH	−638.2	n.a.	n.a.	n.a.	n.a.	n.a.
9	CH$_2$CO	n.a.	53.59	245.2	−246.6	−44.58	n.a.
10	CHO	n.a.	n.a.	n.a.	n.a.	n.a.	n.a.
11	CCOO	−101.7	148.3	n.a.	n.a.	−21.13	n.a.
12	HCOO	n.a.	n.a.	n.a.	n.a.	n.a.	n.a.
13	CH$_2$O	−20.11	−149.5	−202.3	n.a	172.1	n.a.
14	CNH$_2$	n.a.	n.a.	n.a.	n.a.	n.a.	n.a.
15	CNH	n.a.	n.a.	n.a.	n.a.	n.a.	n.a.
16	(C)$_3$N	n.a.	n.a.	n.a.	n.a.	n.a.	n.a.
17	ACNH$_2$	125.3	n.a.	n.a.	n.a.	n.a.	n.a.
18	Pyridine	n.a.	n.a.	n.a.	n.a.	n.a.	n.a.
19	CCN	n.a.	n.a.	n.a.	−203	n.a.	81.57
20	COOH	n.a.	n.a.	n.a.	n.a.	n.a.	n.a.
21	CCl	n.a.	n.a.	−125.9	n.a.	n.a.	n.a.
22	CCl$_2$	n.a.	177.6	n.a.	n.a.	215	n.a.
23	CCl$_3$	n.a.	86.4	n.a.	n.a.	363.7	n.a.
24	CCl$_4$	n.a.	247.8	41.94	n.a.	337.7	n.a.
25	ACCl	n.a.	n.a.	−60.7	n.a.	n.a.	n.a.
26	CNO$_2$	139.8	304.3	10.17	−27.7	n.a.	n.a.
27	ACNO$_2$	n.a.	n.a.	n.a.	n.a.	n.a.	n.a.
28	CS$_2$	n.a.	n.a.	n.a.	n.a.	n.a.	n.a.
29	CH$_3$SH	n.a.	n.a.	n.a.	n.a.	31.66	n.a.
30	Furfural	n.a.	n.a.	n.a.	n.a.	n.a.	n.a.
31	DOH	0	n.a.	n.a.	n.a.	−417.2	n.a.
32	I	n.a.	0	n.a.	n.a.	n.a.	n.a.
33	Br	n.a.	n.a.	0	n.a.	n.a.	n.a.
34	C≡C	n.a.	n.a.	n.a.	0	n.a.	n.a.
35	DMSO	535.8	n.a.	n.a.	n.a.	0	n.a.
36	ACRY	n.a.	n.a.	n.a.	n.a.	n.a.	0
37	ClCC	n.a.	n.a.	n.a.	n.a.	n.a.	n.a.
38	ACF	n.a.	n.a.	n.a.	n.a.	n.a.	n.a.
39	DMF	−191.7	n.a.	n.a.	6.699	136.6	n.a.
40	CF$_2$	n.a.	n.a.	n.a.	n.a.	n.a.	n.a.

TABLE 11-7 (Continued)

		37 CICC	38 ACF	39 DMF	40 CF$_2$
1	CH$_2$	−0.505	125.8	485.3	−2.859
2	C=C	237.3	n.a.	320.4	n.a.
3	ACH	69.11	389.3	245.6	n.a.
4	ACCH$_2$	n.a.	101.4	5629	n.a.
5	OH	253.9	44.78	−143.9	n.a.
6	CH$_3$OH	−21.22	−48.25	−172.4	n.a.
7	H$_2$O	n.a.	n.a.	319	n.a.
8	ACOH	n.a.	n.a.	n.a.	n.a.
9	CH$_2$CO	−44.42	n.a.	−61.7	n.a.
10	CHO	n.a.	n.a.	n.a.	n.a.
11	CCOO	−19.28	n.a.	n.a.	n.a.
12	HCOO	n.a.	n.a.	n.a.	n.a.
13	CH$_2$O	145.6	n.a.	254.8	n.a.
14	CNH$_2$	n.a.	n.a.	n.a.	n.a.
15	CNH	n.a.	n.a.	n.a.	n.a.
16	(C)$_3$N	n.a.	n.a.	n.a.	n.a.
17	ACNH$_2$	n.a.	n.a.	−293.1	n.a.
18	Pyridine	n.a.	n.a.	n.a.	n.a.
19	CCN	−19.14	n.a.	n.a.	n.a.
20	COOH	−90.87	n.a.	n.a.	n.a.
21	CCl	−58.77	n.a.	n.a.	n.a.
22	CCl$_2$	n.a.	n.a.	n.a.	n.a.
23	CCl$_3$	−79.54	n.a.	n.a.	n.a.
24	CCl$_4$	−86.85	215.2	498.6	n.a.
25	ACCl	n.a.	n.a.	n.a.	n.a.
26	CNO$_2$	48.4	n.a.	n.a.	n.a.
27	ACNO$_2$	n.a.	n.a.	n.a.	n.a.
28	CS$_2$	−47.37	n.a.	n.a.	n.a.
29	CH$_3$SH	n.a.	n.a.	78.92	n.a.
30	Furfural	n.a.	n.a.	n.a.	n.a.
31	DOH	n.a.	n.a.	302.2	n.a.
32	I	n.a.	n.a.	n.a.	n.a.
33	Br	n.a.	n.a.	n.a.	n.a.
34	C≡C	n.a.	n.a.	−119.8	n.a.
35	DMSO	n.a.	n.a.	−97.71	n.a.
36	ACRY	n.a.	n.a.	n.a.	n.a.
37	CICC	0	n.a.	n.a.	n.a.
38	ACF	n.a.	0	n.a.	n.a.
39	DMF	n.a.	n.a.	0	n.a.
40	CF$_2$	n.a.	n.a.	n.a.	0

a. n.a. = not available. Values of a_{mn} are in units of degrees Kelvin.

Source: Gmehling, Rasmussen and Fredenslund [10b] . *(Reprinted with permission from the American Chemical Society.)*

(3) Calculate r_1, r_2, q_1, and q_2 using Eq. set 11-20 and the R_k and Q_k values from Table 11-6. *(See footnote 6, p. 11-22.)*

(4) Calculate ℓ_1 and ℓ_2 from Eq. set 11-19.

(5) Calculate ϕ_1, ϕ_2, θ_1, and θ_2 from Eq. set 11-18. Note that $\phi_1 + \phi_2 = 1$ and $\theta_1 + \theta_2 = 1$.

(6) Calculate $\ln\gamma_1^c$ and $\ln\gamma_2^c$ from Eqs. 11-16 and 11-17.

(7) Calculate Θ_m and X_m for each group, using Eq. 11-23. Note that $\Sigma\Theta_m = \Sigma X_m = 1$.

(8) Calculate ψ_{mn} for each group using the value of a_{mn} in Table 11-7 and Eq. 11-24. *(See footnote 6, p. 11-22.)* Note that m and n are *subgroup* numbers. In Table 11-7 only *main* group numbers are listed. Hence, care must be exercised in looking up interaction parameters. For example, $a_{1,18}$ is the interaction parameter for *subgroups* 1 and 18, which belong to the *main* groups 1 and 8, respectively. Hence, in Table 11-7 we would search for the values corresponding to main groups 1 and 8.

(9) Calculate $\ln \Gamma_k$ and $\ln \Gamma_k^{(j)}$ for each group, using Eq. set 11-22.

(10) Calculate $\ln \gamma_1^R$ and $\ln \gamma_2^R$ using Eq. set 11-21.

(11) Calculate $\ln \gamma_1$ and $\ln \gamma_2$ using Eq. 11-15.

Example 11-5   Estimate the activity coefficients for the system acetone(1)-n-pentane(2) at $T = 307K$ and $x_1 = 0.047$ (whence $x_2 = 0.953$).

(1)   The structures are:

      acetone $(CH_3)_2 C{=}O$

      n-pentane $CH_3 (CH_2)_3 CH_3$

(2)   Acetone contains one $CH_3 C{=}O$ group and one CH_3 group.

n-Pentane contains two CH_3 groups and three CH_2 groups.

Table 11-8 below lists the appropriate group parameters.

(3)   From Eq. set 11-20

$$r_1 = 1 \times 0.9011 + 1 \times 1.6724 = 2.5735$$

$$q_1 = 1 \times 0.848 + 1 \times 1.488 = 2.336$$

TABLE 11-8

Group Parameters for Example 11-5
(From Table 11-6)

Group	Main Group No.	k	j	ν_k	R_k	Q_k
CH_3	1	1	1	1	0.9011	0.848
$CH_3\ C=O$	9	19	1	1	1.6724	1.488
CH_3	1	1	2	2	0.9011	0.848
CH_2	1	2	2	3	0.6744	0.540

$$r_2 = 2 \times 0.9011 + 3 \times 0.6744 = 3.8254$$

$$q_2 = 2 \times 0.848 + 3 \times 0.540 = 3.316$$

(4)   From Eq. set 11-19

$$\ell_1 = 5\ (2.5735 - 2.336) - 1.5735 = -0.3860$$

$$\ell_2 = 5\ (3.8254 - 3.316) - 2.8254 = -0.2784$$

(5)   From Eq. set 11-18   (Note $x_2 = 1-x_1$)

$$\phi_1 = \frac{2.5735 \times 0.047}{2.5735 \times 0.047 + 3.8254 \times 0.953} = 0.0321$$

$$\phi_2 = 1 - \phi_1 = 0.9679$$

$$\theta_1 = \frac{2.336 \times 0.047}{2.336 \times 0.047 + 3.316 \times 0.953} = 0.0336$$

$$\theta_2 = 1 - \theta_1 = 0.9664$$

(6)   From Eqs. 11-16 and 11-17

$$\ln \gamma_1^c = \ln\left(\frac{0.0321}{0.047}\right) + 5 \times 2.336 \times \ln\left(\frac{0.0336}{0.0321}\right) - 0.3860$$

$$- \left(\frac{0.0321}{0.047}\right) \times [0.047 \times (-0.3860) + 0.953 \times (-0.2784)]$$

$$= -0.040$$

$$\ln \gamma_2^c = \ln\left(\frac{0.9679}{0.953}\right) + 5 \times 3.316 \times \ln\left(\frac{0.9664}{0.9679}\right) - 0.2784$$

$$-\left(\frac{0.9679}{0.953}\right) \times [0.047 \times (-0.3860) + 0.953 \times (-0.2784)]$$

$$= -0.0007$$

(7)   From Eq. set 11-23 for pure acetone:

$$X_1^{(1)} = \frac{\nu_1^{(1)} x_1 + \nu_1^{(2)} x_2}{(\nu_1^{(1)} + \nu_2^{(1)} + \nu_{19}^{(1)}) x_1 + (\nu_1^{(2)} + \nu_2^{(2)} + \nu_{19}^{(2)}) x_2}$$

$$= \frac{1 \times 1 + 2 \times 0}{(1 + 0 + 1) \times 1 + (2 + 3 + 0) \times 0} = 1/2$$

$$X_2^{(1)} = \frac{\nu_2^{(1)} x_1 + \nu_2^{(2)} x_2}{(\nu_1^{(1)} + \nu_2^{(1)} + \nu_{19}^{(1)}) x_1 + (\nu_1^{(2)} + \nu_2^{(2)} + \nu_{19}^{(2)}) x_2}$$

$$= \frac{0 \times 1 + 3 \times 0}{(1 + 0 + 1) \times 1 + (2 + 3 + 0) \times 0} = 0$$

$$X_{19}^{(1)} = \frac{\nu_{19}^{(1)} x_1 + \nu_{19}^{(2)} x_2}{(\nu_1^{(1)} + \nu_2^{(1)} + \nu_{19}^{(1)}) x_1 + (\nu_1^{(2)} + \nu_2^{(2)} + \nu_{19}^{(2)}) x_2}$$

$$= \frac{1 \times 1 + 0 \times 0}{(1 + 0 + 1) \times 1 + (2 + 3 + 0) \times 0} = 1/2$$

Note, however, that

$$X_{19}^{(1)} = 1 - (X_1^{(1)} + X_2^{(1)})$$

$$= 1 - 1/2 - 0 = 1/2 \text{ (See Basic Step 7 above.)}$$

For pure n-pentane

$$X_1^{(2)} = \frac{\nu_1^{(1)} x_1 + \nu_1^{(2)} x_2}{(\nu_1^{(1)} + \nu_2^{(1)} + \nu_{19}^{(1)}) x_1 + (\nu_1^{(2)} + \nu_2^{(2)} + \nu_{19}^{(2)}) x_2}$$

$$= \frac{1 \times 0 + 2 \times 1}{(1 + 0 + 1) \times 0 + (2 + 3 + 0) \times 1} = 2/5$$

$$X_2^{(2)} = \frac{\nu_2^{(1)} x_1 + \nu_2^{(2)} x_2}{\left(\nu_1^{(1)} + \nu_2^{(1)} + \nu_{19}^{(1)}\right) x_1 + \left(\nu_1^{(2)} + \nu_2^{(2)} + \nu_{19}^{(2)}\right) x_2}$$

$$= \frac{0 \times 0 + 3 \times 1}{(1 + 0 + 1) \times 0 + (2 + 3 + 0) \times 1} = 3/5$$

$$X_{19}^{(2)} = 1 - \left(X_1^{(2)} + X_2^{(2)}\right) = 0$$

At $x_1 = 0.047$, $x_2 = 1 - x_1 = 0.953$, and

$$X_1 = \frac{\nu_1^{(1)} x_1 + \nu_1^{(2)} x_2}{\left(\nu_1^{(1)} + \nu_2^{(1)} + \nu_{19}^{(1)}\right) x_1 + \left(\nu_1^{(2)} + \nu_2^{(2)} + \nu_{19}^{(2)}\right) x_2}$$

$$X_1 = \frac{1 \times 0.047 + 2 \times 0.953}{2 \times 0.047 + 5 \times 0.953} = 0.4019$$

$$X_2 = \frac{\nu_2^{(1)} x_1 + \nu_2^{(2)} x_2}{\left(\nu_1^{(1)} + \nu_2^{(1)} + \nu_{19}^{(1)}\right) x_1 + \left(\nu_1^{(2)} + \nu_2^{(2)} + \nu_{19}^{(2)}\right) x_2}$$

$$= \frac{0 \times 0.047 + 3 \times 0.953}{2 \times 0.047 + 5 \times 0.953} = 0.5884$$

$$X_{19} = 1 - (X_1 + X_2) = 0.0097$$

For pure acetone:

$$\Theta_1^{(1)} = \frac{Q_1 X_1^{(1)}}{Q_1 X_1^{(1)} + Q_2 X_2^{(1)} + Q_{19} X_{19}^{(1)}}$$

$$= \frac{0.848 \times 1/2}{0.848 \times 1/2 + 0.540 \times 0 + 1.488 \times 1/2} = 0.3630$$

$$\Theta_2^{(1)} = \frac{Q_2 X_2^{(1)}}{Q_1 X_1^{(1)} + Q_2 X_2^{(1)} + Q_{19} X_{19}^{(1)}} = 0$$

$$\Theta_{19}^{(1)} = 1 - \left(\Theta_1^{(1)} + \Theta_2^{(1)}\right) = 0.6370$$

Similarly for pure pentane:

$$\Theta_1^{(2)} = \frac{0.848 \times 2/5}{0.848 \times 2/5 + 0.540 \times 3/5 + 1.488 \times 0} = 0.5115$$

$$\Theta_2^{(2)} = \frac{0.540 \times 3/5}{0.848 \times 2/5 + 0.540 \times 3/5 + 1.488 \times 0} = 0.4885$$

$$\Theta_{19}^{(2)} = 1 - \left(\Theta_1^{(2)} + \Theta_2^{(2)}\right) = 0$$

At $x_1 = 0.047$

$$\Theta_1 = \frac{Q_1 X_1}{Q_1 X_1 + Q_2 X_2 + Q_{19} X_{19}}$$

$$= \frac{0.848 \times 0.4019}{0.848 \times 0.4019 + 0.540 \times 0.5884 + 1.488 \times 0.0097} = 0.5064$$

$$\Theta_2 = \frac{Q_2 X_2}{Q_1 X_1 + Q_2 X_2 + Q_{19} X_{19}}$$

$$= \frac{0.540 \times 0.5884}{0.848 \times 0.4019 + 0.540 \times 0.5884 + 1.488 \times 0.0097} = 0.4722$$

$$\Theta_{19} = 1 - (\Theta_1 + \Theta_2) = 0.0214$$

(8) From Table 11-7 and Eq. 11-24

$$a_{1,2} = a_{2,1} = a_{1,1} = a_{2,2} = a_{19,19} = 0K$$

$$a_{1,19} = a_{2,19} = 476.4K; a_{19,1} = a_{19,2} = 26.76K$$

$$\psi_{1,19} = \psi_{2,19} = \exp[-476.4/307] = 0.2119$$

$$\psi_{19,1} = \psi_{19,2} = \exp[-26.76/307] = 0.9165$$

$$\psi_{1,2} = \psi_{2,1} = \psi_{1,1} = \psi_{2,2} = \psi_{2,2} = \psi_{19,19} = 1$$

(9) As the coefficients $\nu_2^{(1)}$ and $\nu_{19}^{(2)}$ are both equal to zero, $\ln\Gamma_2$ and $\ln\Gamma_2^{(1)}$, for component 1, and $\ln\Gamma_{19}$, $\ln\Gamma_{19}^{(2)}$ for component 2 need not be calculated. Thus from Eq. 11-22 for pure acetone:

$$\ln\Gamma_1^{(1)} = Q_1 \left[1 - \ln\left(\Theta_1^{(1)} \times \psi_{1,1} + \Theta_2^{(1)} \times \psi_{2,1} + \Theta_{19}^{(1)} \times \psi_{19,1}\right) \right.$$

$$- \frac{\Theta_1^{(1)} \times \psi_{1,1}}{\Theta_1^{(1)} \times \psi_{1,1} + \Theta_2^{(1)} \times \psi_{2,1} + \Theta_{19}^{(1)} \times \psi_{19,1}}$$

$$- \frac{\Theta_2^{(1)} \times \psi_{1,2}}{\Theta_1^{(1)} \times \psi_{1,2} + \Theta_2^{(1)} \times \psi_{2,2} + \Theta_{19}^{(1)} \times \psi_{19,2}}$$

$$\left. - \frac{\Theta_{18}^{(1)} \times \psi_{1,18}}{\Theta_1^{(1)} \times \psi_{1,19} + \Theta_2^{(1)} \times \psi_{2,19} + \Theta_{19}^{(1)} \times \psi_{19,19}} \right]$$

$$= 0.848 \left[1 - \ln\left(0.3630 \times 1 + 0 \times 1 + 0.6370 \times 0.9165\right) \right.$$

$$- \frac{0.3630 \times 1}{0.3630 \times 1 + 0 \times 1 + 0.6370 \times 0.9165}$$

$$- \frac{0 \times 1}{0.3630 \times 1 + 0 \times 1 + 0.6370 \times 0.9165}$$

$$\left. - \frac{0.6370 \times 0.2119}{0.3630 \times 0.2119 + 0 \times 0.2119 + 0.6370 \times 1} \right]$$

$$= 0.4089$$

$$\ln \Gamma_{19}^{(1)} = Q_{19} \left[1 - \ln\left(\Theta_{19}^{(1)} \times \psi_{19,19} + \Theta_1^{(1)} \times \psi_{1,19} + \Theta_2^{(1)} \times \psi_{2,19} \right) \right.$$

$$- \frac{\Theta_{19}^{(1)} \times \psi_{19,19}}{\Theta_{19}^{(1)} \times \psi_{19,19} + \Theta_1^{(1)} \times \psi_{1,19} + \Theta_2^{(1)} \times \psi_{2,19}}$$

$$- \frac{\Theta_1^{(1)} \times \psi_{19,1}}{\Theta_{19}^{(1)} \times \psi_{19,1} + \Theta_1^{(1)} \times \psi_{1,1} + \Theta_2^{(1)} \times \psi_{2,1}}$$

$$\left. - \frac{\Theta_2^{(1)} \times \psi_{19,2}}{\Theta_{19}^{(1)} \times \psi_{19,2} + \Theta_1^{(1)} \times \psi_{1,2} + \Theta_2^{(1)} \times \psi_{2,2}} \right]$$

$$= 1.488 \left[1 - \ln\left(0.6370 \times 1 + 0.3630 \times 0.2119 + 0 \times 0.2119\right) \right.$$

$$- \frac{0.6370 \times 1}{0.6370 \times 1 + 0.3630 \times 0.2119 + 0 \times 0.2119}$$

$$- \frac{0.3630 \times 0.9165}{0.6370 \times 0.9165 + 0.3630 \times 1 + 0 \times 1}$$

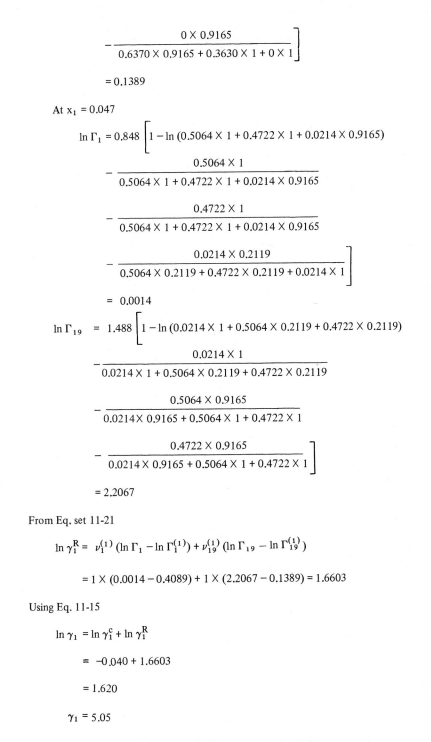

$$-\frac{0 \times 0.9165}{0.6370 \times 0.9165 + 0.3630 \times 1 + 0 \times 1}\Bigg]$$

$$= 0.1389$$

At $x_1 = 0.047$

$$\ln \Gamma_1 = 0.848 \Bigg[1 - \ln (0.5064 \times 1 + 0.4722 \times 1 + 0.0214 \times 0.9165)$$

$$-\frac{0.5064 \times 1}{0.5064 \times 1 + 0.4722 \times 1 + 0.0214 \times 0.9165}$$

$$-\frac{0.4722 \times 1}{0.5064 \times 1 + 0.4722 \times 1 + 0.0214 \times 0.9165}$$

$$-\frac{0.0214 \times 0.2119}{0.5064 \times 0.2119 + 0.4722 \times 0.2119 + 0.0214 \times 1}\Bigg]$$

$$= 0.0014$$

$$\ln \Gamma_{19} = 1.488 \Bigg[1 - \ln (0.0214 \times 1 + 0.5064 \times 0.2119 + 0.4722 \times 0.2119)$$

$$-\frac{0.0214 \times 1}{0.0214 \times 1 + 0.5064 \times 0.2119 + 0.4722 \times 0.2119}$$

$$-\frac{0.5064 \times 0.9165}{0.0214 \times 0.9165 + 0.5064 \times 1 + 0.4722 \times 1}$$

$$-\frac{0.4722 \times 0.9165}{0.0214 \times 0.9165 + 0.5064 \times 1 + 0.4722 \times 1}\Bigg]$$

$$= 2.2067$$

From Eq. set 11-21

$$\ln \gamma_1^R = \nu_1^{(1)} (\ln \Gamma_1 - \ln \Gamma_1^{(1)}) + \nu_{19}^{(1)} (\ln \Gamma_{19} - \ln \Gamma_{19}^{(1)})$$

$$= 1 \times (0.0014 - 0.4089) + 1 \times (2.2067 - 0.1389) = 1.6603$$

Using Eq. 11-15

$$\ln \gamma_1 = \ln \gamma_1^c + \ln \gamma_1^R$$

$$= -0.040 + 1.6603$$

$$= 1.620$$

$$\gamma_1 = 5.05$$

The experimental value is 4.41 [12] for an error of +14.5%

For pure n-pentane:

$$\ln\Gamma_1^{(2)} = Q_1\left[1 - \ln\left(\Theta_1^{(2)} \times \psi_{1,1} + \Theta_2^{(2)} \times \psi_{2,1} + \Theta_{19}^{(2)} \times \psi_{19,1}\right)\right.$$

$$-\frac{\Theta_1^{(2)} \times \psi_{1,1}}{\Theta_1^{(2)} \times \psi_{1,1} + \Theta_2^{(2)} \times \psi_{2,1} + \Theta_{19}^{(2)} \times \psi_{19,1}}$$

$$-\frac{\Theta_2^{(2)} \times \psi_{1,2}}{\Theta_1^{(2)} \times \psi_{1,2} + \Theta_2^{(2)} \times \psi_{2,2} + \Theta_{19}^{(2)} \times \psi_{19,2}}$$

$$\left.-\frac{\Theta_{19}^{(2)} \times \psi_{1,19}}{\Theta_1^{(2)} \times \psi_{1,19} + \Theta_2^{(2)} \times \psi_{2,19} + \Theta_{19}^{(2)} \times \psi_{19,19}}\right]$$

$$\ln\Gamma_1^{(2)} = 0.848\left[1 - \ln\left(0.5115 \times 1 + 0.4885 \times 1 + 0 \times 0.9165\right)\right.$$

$$-\frac{0.5115 \times 1}{0.5115 \times 1 + 0.4885 \times 1 + 0 \times 0.9165}$$

$$-\frac{0.4885 \times 1}{0.5115 \times 1 + 0.4885 \times 1 + 0 \times 0.9165}$$

$$\left.-\frac{0 \times 0.2119}{0.5115 \times 0.2119 + 0.4885 \times 0.2119 + 0 \times 1}\right]$$

$$= 0$$

This must be true as pentane contains only one interaction main group, i.e., "CH_2".

Similarly, $\ln\Gamma_2^{(2)} = 0$

At $x_1 = 0.047$

$$\ln\Gamma_1 = Q_1\left[1 - \ln\left(\Theta_1 \times \psi_{1,1} + \Theta_2 \times \psi_{2,1} + \Theta_{19} \times \psi_{19,1}\right)\right.$$

$$-\frac{\Theta_1 \times \psi_{1,1}}{\Theta_1 \times \psi_{1,1} + \Theta_2 \times \psi_{2,1} + \Theta_{19} \times \psi_{19,1}}$$

$$-\frac{\Theta_2 \times \psi_{1,2}}{\Theta_1 \times \psi_{1,2} + \Theta_2 \times \psi_{2,2} + \Theta_{19} \times \psi_{19,2}}$$

$$- \frac{\Theta_{19} \times \psi_{1,19}}{\Theta_1 \times \psi_{1,19} + \Theta_2 \times \psi_{2,19} + \Theta_{19} \times \psi_{19,19}} \Bigg]$$

$$\ln \Gamma_1 = 0.848 \Bigg[1 - \ln (0.5064 \times 1 + 0.4722 \times 1 + 0.0214 \times 0.9165)$$

$$- \frac{0.5064 \times 1}{0.5064 \times 1 + 0.4722 \times 1 + 0.0214 \times 0.9165}$$

$$- \frac{0.4722 \times 1}{0.5064 \times 1 + 0.4722 \times 1 + 0.0214 \times 0.9165}$$

$$- \frac{0.0214 \times 0.2119}{0.5064 \times 0.2119 + 0.4722 \times 0.2119 + 0.0214 \times 1} \Bigg]$$

$$= 0.0014$$

$$\ln \Gamma_2 = Q_2 \Bigg[1 - \ln \Big(\Theta_1 \times \psi_{1,2} + \Theta_2 \times \psi_{2,2} + \Theta_{19} \times \psi_{19,2} \Big)$$

$$- \frac{\Theta_1 \times \psi_{2,1}}{\Theta_1 \times \psi_{1,1} + \Theta_2 \times \psi_{2,1} + \Theta_{19} \times \psi_{19,1}}$$

$$- \frac{\Theta_2 \times \psi_{2,2}}{\Theta_1 \times \psi_{1,2} + \Theta_2 \times \psi_{2,2} + \Theta_{19} \times \psi_{19,2}}$$

$$- \frac{\Theta_{19} \times \psi_{2,19}}{\Theta_1 \times \psi_{1,19} + \Theta_2 \times \psi_{2,19} + \Theta_{19} \times \psi_{19,19}} \Bigg]$$

$$\ln \Gamma_2 = 0.540 \Bigg[1 - \ln (0.4722 \times 1 + 0.5064 \times 1 + 0.0214 \times 0.9165)$$

$$- \frac{0.4722 \times 1}{0.4722 \times 1 + 0.5064 \times 1 + 0.0214 \times 0.9165}$$

$$- \frac{0.5064 \times 1}{0.5064 \times 1 + 0.4722 \times 1 + 0.0214 \times 0.9165}$$

$$- \frac{0.0214 \times 0.2119}{0.5064 \times 0.2119 + 0.4722 \times 0.2119 + 0.0214 \times 1} \Bigg]$$

$$= 0.00087$$

$$\ln \gamma_2^R = 2\,(0.0014 - 0) + 3\,(0.00087 - 0)$$

$$= 0.0054$$

$$\ln \gamma_2 = \ln \gamma_2^C + \ln \gamma_2^R$$

$$= -0.0007 + 0.0054$$

$$= 0.0047$$

$\gamma_2 = 1.005$; the experimental value is 1.08 [12] for an error of -7.0%.

Example 11-6   Estimate the infinite-dilution activity coefficients for the system n-butanol(1)-water(2) at 25°C. From these values, estimate the solubility limit of n-butanol in water.

(1)   The structure of n-butanol is

$$CH_3\,CH_2\,CH_2\,CH_2\,OH$$

That for water is H_2O.

(2)   n-Butanol contains one CH_3 group, three CH_2 groups and one OH group. Water contains only the H_2O group. Table 11-9 below lists values of the group parameters.

TABLE 11-9

Group Parameters for Example 11-6
(From Table 11-6)

Group	Main Group No.	k	j	ν_k	R_k	Q_k
CH_3	1	1	1	1	0.9011	0.848
CH_2	1	2	1	3	0.6744	0.540
OH	5	15	1	1	1.000	1.200
H_2O	7	17	2	1	0.920	1.400

As $\nu_1^{(2)}$, $\nu_2^{(2)}$, $\nu_{15}^{(2)}$ and $\nu_{17}^{(1)}$ are all equal to zero, no calculations are required for the corresponding $\ln \Gamma_k$s.

(3)   At infinite dilution Eqs. 11-16 and 11-17 are indeterminate; however, note that at $x_1 = 0$:

$$\frac{\phi_1}{x_1} = \frac{r_1}{r_1\,x_1 + r_2\,x_2} = \frac{r_1}{r_2}$$

Similar arguments hold for ϕ_2/x_2, θ_1/ϕ_1 and θ_2/ϕ_2 such that the two equations are altered. For $x_1 = 0$ and $x_2 = 1$, $\ln \gamma_2^c = 0$ and

$$\ln \gamma_1^c = \ln (r_1/r_2) + \frac{z}{2} q_1 \ln \frac{q_1 r_2}{q_2 r_1} + \ell_1 - (r_1/r_2) \ell_2 \qquad (11\text{-}25)$$

For $x_2 = 0$ and $x_1 = 1$, $\ln \gamma_1^c = 0$ and

$$\ln \gamma_2^c = \ln (r_2/r_1) + \frac{z}{2} q_2 \ln \frac{q_2 r_1}{q_1 r_2} + \ell_2 - (r_2/r_1) \ell_1 \qquad (11\text{-}26)$$

From Eq. set 11-20

$$r_1 = 1 \times 0.9011 + 3 \times 0.6744 + 1 \times 1.000 = 3.9243$$

$$q_1 = 1 \times 0.848 + 3 \times 0.540 + 1 \times 1.200 = 3.668$$

$$r_2 = 0.92$$

$$q_2 = 1.40$$

From Eq. set 11-19

$$\ell_1 = \frac{10}{2} \times (3.9243 - 3.668) - (3.9243 - 1) = -1.6428$$

$$\ell_2 = -2.32$$

From Eq. 11-25

$$\ln \gamma_1^c = \ln\left(\frac{3.9243}{0.92}\right) + \frac{10}{2} \times 3.668 \times \ln \left(\frac{3.668 \times 0.92}{1.40 \times 3.9243}\right)$$

$$- 1.6428 - \frac{3.9243}{0.92} \times (-2.32) = 0.766$$

and from Eq. 11-26

$$\ln \gamma_2^c = 0.0262$$

(4)   From Table 11-7

$$a_{1,15} = a_{2,15} = 986.5$$

$$a_{15,1} = a_{15,2} = 156.4$$

$$a_{1,17} = a_{2,17} = 1318$$

$$a_{17,1} = a_{17,2} = 300$$

$$a_{15,17} = 353.5$$

$$a_{17,15} = -229.1$$

$$a_{1,2} = a_{1,1} = a_{2,1} = a_{15,15} = a_{17,17} = 0$$

Using Eq. 11-24

$$\psi_{1,15} = \psi_{2,15} = \exp(-986.5/298) = 0.0365$$

$$\psi_{15,1} = \psi_{15,2} = 0.5917$$

$$\psi_{1,17} = \psi_{2,17} = 0.0120$$

$$\psi_{17,1} = \psi_{17,2} = 0.3654$$

$$\psi_{15,17} = 0.3054$$

$$\psi_{17,15} = 2.157$$

$$\psi_{1,2} = \psi_{1,1} = \psi_{2,1} = \psi_{2,2} = \psi_{15,15} = \psi_{17,17} = 1$$

(5)   For pure n-butanol

$$X_1^{(1)} = \frac{\nu_1^{(1)} x_1 + \nu_1^{(2)} x_2}{(\nu_1^{(1)} + \nu_2^{(1)} + \nu_{15}^{(1)} + \nu_{17}^{(1)}) x_1 + (\nu_1^{(2)} + \nu_2^{(2)} + \nu_{15}^{(2)} + \nu_{17}^{(2)}) x_2}$$

$$= \frac{1 \times 1 + 0 \times 0}{(1 + 3 + 1 + 0) \times 1 + (0 + 0 + 0 + 0) \times 0} = 1/5$$

$$X_2^{(1)} = \frac{\nu_2^{(1)} x_1 + \nu_2^{(2)} x_2}{(\nu_1^{(1)} + \nu_2^{(1)} + \nu_{15}^{(1)} + \nu_{17}^{(1)}) x_1 + (\nu_1^{(2)} + \nu_2^{(2)} + \nu_{15}^{(2)} + \nu_{17}^{(2)}) x_2}$$

$$= \frac{3 \times 1 + 0 \times 0}{(1 + 3 + 1 + 0) \times 1 + (0 + 0 + 0 + 0) \times 0} = 3/5$$

$$X_{15}^{(1)} = \frac{\nu_{15}^{(1)} x_1 + \nu_{15}^{(2)} x_2}{(\nu_1^{(1)} + \nu_2^{(1)} + \nu_{15}^{(1)} + \nu_{17}^{(1)}) x_1 + (\nu_1^{(2)} + \nu_2^{(2)} + \nu_{15}^{(2)} + \nu_{17}^{(2)}) x_2}$$

$$= \frac{1 \times 1 + 0 \times 0}{(1 + 3 + 1 + 0) \times 1 + (0 + 0 + 0 + 0) \times 0} = 1/5$$

$$X_{17}^{(1)} \qquad 1 - (X_1^{(1)} + X_2^{(1)} + X_{15}^{(1)}) = 0$$

From Eq. set 11-23

$$\Theta_1^{(1)} = \frac{0.848 \times 1/5}{0.848 \times 1/5 + 0.540 \times 3/5 + 1.200 \times 1/5}$$

$$= 0.2312$$

Similarly,

$$\Theta_2^{(1)} = 0.4417$$

$$\Theta_{15}^{(1)} = 0.3272$$

$$\Theta_{17}^{(1)} = 0$$

For pure water

$$X_{17}^{(2)} = 1$$

$$\Theta_{17}^{(2)} = 1$$

For pure **n-butanol**, using Eq. 11-22

$$\ln \Gamma_1^{(1)} = Q_1 \left[1 - \ln \left(\Theta_1^{(1)} \psi_{1,1} + \Theta_2^{(1)} \psi_{2,1} + \Theta_{15}^{(1)} \psi_{15,1} \right) \right.$$

$$- \frac{\Theta_1^{(1)} \psi_{1,1}}{\Theta_1^{(1)} \psi_{1,1} + \Theta_2^{(1)} \psi_{2,1} + \Theta_{15}^{(1)} \psi_{15,1}}$$

$$- \frac{\Theta_2^{(1)} \psi_{1,2}}{\Theta_1^{(1)} \psi_{1,2} + \Theta_2^{(1)} \psi_{2,2} + \Theta_{15}^{(1)} \psi_{15,2}}$$

$$\left. - \frac{\Theta_{15}^{(1)} \psi_{1,15}}{\Theta_1^{(1)} \psi_{1,15} + \Theta_2^{(1)} \psi_{2,15} + \Theta_{15}^{(1)} \psi_{15,15}} \right]$$

$$= 0.848 \left[1 - \ln \left(0.2312 + 0.4417 + 0.3272 \times 0.5917 \right) \right.$$

$$- \frac{0.2312}{0.2312 + 0.4417 + 0.3272 \times 0.5917}$$

$$- \frac{0.4417}{0.2312 + 0.4417 + 0.3272 \times 0.5917}$$

$$\left. - \frac{0.3272 \times 0.0315}{(0.2312 + 0.4417) \times 0.0315 + 0.3272} \right]$$

$$= 0.2859$$

$$\ln \Gamma_2^{(1)} = Q_2 \left[1 - \ln \left(\Theta_1^{(1)} \, \psi_{1,2} + \Theta_2^{(1)} \, \psi_{2,2} + \Theta_{15}^{(1)} \, \psi_{15,2} \right) \right.$$

$$- \frac{\Theta_1^{(1)} \, \psi_{2,1}}{\Theta_1^{(1)} \, \psi_{1,1} + \Theta_2^{(1)} \, \psi_{2,1} + \Theta_{15}^{(1)} \, \psi_{15,1}}$$

$$- \frac{\Theta_2^{(1)} \, \psi_{2,2}}{\Theta_1^{(1)} \, \psi_{1,2} + \Theta_2^{(1)} \, \psi_{2,2} + \Theta_{15} \, \psi_{15,2}}$$

$$\left. - \frac{\Theta_{15}^{(1)} \, \psi_{2,15}}{\Theta_1^{(1)} \, \psi_{1,15} + \Theta_2^{(1)} \, \psi_{2,15} + \Theta_{15}^{(1)} \, \psi_{15,15}} \right]$$

$$= 0.540 \left[1 - \ln \left(0.2312 + 0.4417 + 0.3272 \times 0.5917 \right) \right.$$

$$- \frac{0.2312}{0.2312 + 0.4417 + 0.3372 \times 0.5917}$$

$$- \frac{0.4417}{0.2312 + 0.4417 + 0.3372 \times 0.5917}$$

$$\left. - \frac{0.3272 \times 0.0315}{(0.2312 + 0.4417) \times 0.0315 + 0.3272} \right]$$

$$= 0.1821$$

$$\ln \Gamma_{15}^{(1)} = Q_{15} \left[1 - \ln \left(\Theta_1^{(1)} \, \psi_{1,15} + \Theta_2^{(1)} \, \psi_{2,15} + \Theta_{15}^{(1)} \, \psi_{15,15} \right) \right.$$

$$- \frac{\Theta_1^{(1)} \, \psi_{15,1}}{\Theta_1^{(1)} \, \psi_{1,1} + \Theta_2^{(1)} \, \psi_{2,1} + \Theta_{15}^{(1)} \, \psi_{15,1}}$$

$$- \frac{\Theta_2^{(1)} \, \psi_{2,15}}{\Theta_1^{(1)} \, \psi_{1,2} + \Theta_2^{(1)} \, \psi_{2,2} + \Theta_{15}^{(1)} \, \psi_{15,2}}$$

$$\left. - \frac{\Theta_{15}^{(1)} \, \psi_{15,15}}{\Theta_1^{(1)} \, \psi_{1,15} + \Theta_2^{(1)} \, \psi_{2,15} + \Theta_{15}^{(1)} \, \psi_{15,15}} \right]$$

$$= 1.20 \left\{ 1 - \ln \left[(0.2312 + 0.4417) \times 0.0315 + 0.3272 \right] \right.$$

$$- \frac{0.2312 \times 0.5917}{0.2312 + 0.4417 + 0.3272 \times 0.5917}$$

$$- \frac{0.4417 \times 0.0315}{0.2312 + 0.4417 + 0.3272 \times 0.5917}$$

$$-\frac{0.3272}{(0.2312 + 0.4417) \times 0.0315 + 0.3272}\Bigg\}$$

$$= 1.130$$

For pure water at $x_1 = 0$ (infinitely dilute in n-butanol)

$$\ln \Gamma_{17}^{(2)} = 0$$

$$\Theta_1 = \Theta_2 = \Theta_{15} = 0$$

$$\Theta_{17} = 1$$

and Eq. 11-22 reduces to

$$\ln \Gamma_1 = Q_1 (1 - \ln \psi_{17,1} - \psi_{1,17})$$

$$= 0.848 [1 - \ln(0.3654) - 0.0120]$$

$$= 1.692$$

$$\ln \Gamma_2 = Q_2 (1 - \ln \psi_{17,2} - \psi_{2,17})$$

$$= 0.54 [1 - \ln(0.3654) - 0.0120]$$

$$= 1.077$$

$$\ln \Gamma_{15} = Q_{15}(1 - \ln \psi_{17,15} - \psi_{15,17})$$

$$= 1.20 [1 - \ln(2.157) - 0.3054]$$

$$= -0.0889$$

From Eq. set 11-21

$$\ln \gamma_1^R = (1.692 - 0.2859) + 3\,(1.077 - 0.1821) + (-0.0889 - 1.130)$$

$$= 2.871$$

and Eq. 11-15

$$\ln \gamma_1^\infty = 0.766 + 2.871$$

$$= 3.637$$

$$\gamma_1^\infty = 38.0$$

The experimental value is 52.9 [3] for an error of +28.2%. At $x_1 = 1.0$ (infinitely dilute in water)

$$\ln \Gamma_1 \ = \ln \Gamma_1^{(1)}; \ln \Gamma_{17}^{(2)} = 0$$

$$\ln \Gamma_2 \ = \ln \Gamma_2^{(1)}$$

$$\ln \Gamma_{15} = \ln \Gamma_{15}^{(1)}$$

$$\Theta_1 = \Theta_1^{(1)}$$

$$\Theta_2 = \Theta_2^{(2)}$$

$$\Theta_{15} = \Theta_{15}^{(2)}$$

$$\Theta_{17} = 0$$

Using Eq. 11-22

$$
\ln \Gamma_{17} = Q_{17} \left[1 - (\Theta_1 \, \psi_{17,1} + \Theta_2 \, \psi_{17,2} + \Theta_{15} \, \psi_{17,15}) \right.
$$

$$
- \frac{\Theta_1 \, \psi_{1,17}}{\Theta_1 \, \psi_{1,1} + \Theta_2 \, \psi_{2,1} + \Theta_{15} \, \psi_{15,1}}
$$

$$
- \frac{\Theta_2 \, \psi_{2,17}}{\Theta_1 \, \psi_{1,2} + \Theta_2 \, \psi_{2,2} + \Theta_{15} \, \psi_{15,2}}
$$

$$
\left. - \frac{\Theta_{15} \, \psi_{15,17}}{\Theta_1 \, \psi_{1,15} + \Theta_2 \, \psi_{2,15} + \Theta_{15} \, \psi_{15,15}} \right]
$$

$$
= 1.4 \left[1 - \ln (0.2312 + 0.4417) \times 0.3654 + 0.3272 \times 2.157) \right.
$$

$$
- \frac{0.2312 \times 0.0120}{0.2312 + 0.4417 + 0.3272 \times 0.5917}
$$

$$
- \frac{0.4417 \times 0.0120}{0.2312 + 0.4417 + 0.3272 \times 0.5917}
$$

$$
\left. - \frac{0.3272 \times 0.3054}{(0.2312 + 0.4417) \times 0.0315 + 0.3272} \right]
$$

$$
= 1.055
$$

From Eq. set 11-21

$$\ln \gamma_2^R = (1.055 - 0)$$

$$= 1.055$$

$$\ln \gamma_2^\infty = 0.0262 + 1.055$$

$$= 1.081$$

From Eq. 11-15

$$\gamma_2^\infty = 2.95$$

Using Eq. 3-14 of Chapter 3 and the estimated γ_1^∞, a solubility of 0.026 mole fraction is estimated. This can be compared with an experimental value of 0.018 mole fraction [23].

11-6 AVAILABLE DATA

There are no general compilations of activity coefficients. Infinite dilution activity coefficients for several binary systems are given by Nicolaides [14]. In Reference 8 a fairly extensive list of activity coefficients is given for a variety of binary systems.

11-7 SYMBOLS USED

A = parameter in enthalpic equations (Table 11-1)

A_1 = parameter in Margules equation (Eq. 11-9)

$A_{1,2}$ = coefficient in Eq. 11-12

a = activity in Eq. 11-6

a_{mn}, a_{nm} = group interaction parameters

B = parameter in enthalpic equations of Table 11-1

B_2 = coefficient in Eq. 11-12

C = parameter in enthalpic equations of Table 11-1

C_1 = coefficient in Eq. 11-12

D = coefficient in Eq. 11-12

F_2 = coefficient in Eq. 11-12

G^E = excess Gibbs free energy (cal)

G^I = energy of mixing for an ideal mixture (cal)

G^M = energy of mixing in Eq. 11-6 (cal)

G_{21}, G_{12} = parameters in NRTL equation, Table 11-1

g^E = excess Gibbs free energy per mole (cal/mol)

$\Delta g_{12}, \Delta g_{21}$ = parameters in NRTL equation, Table 11-1

H = Henry's law constant (atm·m³/mol)

K_{ow} = octanol-water partition coefficient

ℓ_1 = parameter in UNIQUAC equation (Table 11-1) and Eq. 11-19; i = 1 or 2

N_i = number of carbon atoms in a molecule in Eq. 11-13; i = 1 or 2

N_i', N_i'', N_i''' = number of carbon atoms in respective branches of branched compounds; i = 1 or 2

n_i = number of moles of component i

P_{vp} = vapor pressure

P_i° = pure component vapor pressure in Eqs. 11-1 to -4; i= 1 or 2

P_i =   partial pressure of component i in Eqs. 11-1 to -4; i = 1 or 2

Q_k =   van der Waals surface area for group k

q_i =   molecular van der Waals surface area; i = 1 or 2

R =   gas constant (1.987 cal/K·mol)

R_k =   van der Waals volume for group k

r_i =   molecular van der Waals volume; i = 1 or 2

S =   solubility (mol/m³)

S^E =   excess entropy (e.u.)

T =   temperature (K)

$\Delta u_{12}, \Delta u_{21}$ =   parameters in UNIQUAC equation (Table 11-1)

X_m =   group mole fraction for group m

x =   mole fraction concentration

x_i =   mole fraction of component i; i = 1 or 2

Z =   coordination number (set equal to 10)

Greek

α_{12} =   parameter in NRTL equation (Table 11-1)

Γ_k =   group activity coefficient

$\Gamma_k^{(j)}$ =   group activity coefficient for pure component j; j = 2 or 1

γ =   activity coefficient

γ^∞ =   activity coefficient at infinite dilution

γ_i =   activity coefficient of component i; i = 1 or 2

γ_i^∞ =   activity coefficient of component i at infinite dilution; i = 1 or 2

γ_i^c =   combinatorial activity coefficient for component i; i = 1 or 2

γ_i^R =   residual activity coefficient for component i; i = 1 or 2

$(\gamma_W^A)^\infty$ =   infinite dilution activity coefficient of component A in water

$(\gamma_O^A)^\infty$ =   infinite dilution activity coefficient of component A in octanol

$\Lambda_{12}, \Lambda_{21}$ =   parameters in Wilson equation (Table 11-1)

ν_k =   number of members in group k

$\nu_k^{(j)}$ =   number of members of group k in component j; j = 2 or 1

ϕ_i =   volume fraction of component i; i = 1 or 2

ψ_{mn} =   exp ($-a_{mn}$/T)

ψ_{nm} =   exp ($-a_{nm}$/T)

$\tau_{ij}, \tau_{ji} =$ parameters in NRTL and UNIQAC equation (Table 11-1); i = 1, 2; j = 2, 1

θ_i = molecular surface area fraction (i = 1 or 2)

Θ_m = group surface area fraction

Subscripts

i	=	component i
j	=	component j
k	=	subgroup number
m	=	subgroup number
n	=	subgroup number
mn	=	interaction between groups m and n
nm	=	interaction between groups n and m
o	=	octanol
ow	=	octanol-water
w	=	water

Superscripts

A	=	component A
c	=	combinatorial
E	=	excess
R	=	residual
∞	=	infinite

11-8   REFERENCES

1. Abrams, D.S. and J.M. Prausnitz, "Statistical Thermodynamics of Liquid Mixtures: A New Expression for the Excess Gibbs Energy of Partly or Completely Miscible Systems," *AIChE J.*, **21**, 116 (1975).

2. Black, C., G. Joris and H. Taylor, "The Solubility of Water in Hydrocarbons," *J. Chem. Phys.*, **16**, 537 (1948).

3. Butler, J., C. Ramchandani and D. Thomson, "The Solubility of Non-Electrolytes: Part I. The Free Energy of Hydration of Some Aliphatic Alcohols," *J. Chem. Soc.*, **1935**, 952 (1935).

4. Deal, C. and E.L. Derr, "Selectivity and Solvency in Aromatics Recovery," *Ind. Eng. Chem. Process Des. Dev.*, **3**, 394 (1964).

5. Derr, E.L. and C.H. Deal, "Analytical Solutions of Groups: Correlation of Activity Coefficients Through Structural Group Parameters," *I. Chem. E. Symp. Ser.*, No. 32 (Instn. Chem. Engrs., London) **3**, 40 (1969).

6. Derr, E.L. and C.H. Deal, "Predicted Composition During Mixed Solvent Evaporation from Resin Solutions Using the Analytical Solution of Group Method," *Advan. Chem. Ser.*, **124**, 11 (1973).

7. D
I

8. F
U

9. F
t
(

10a. F
S

10b. G
U
P
o

11. L
S

12. L
A

13. M

14. N
t

15. N
t

16. O
E

17. P
C

18. P
t

19. P
P

20. R
F

21. R
L

22. S
t

23. S
(1

24. S
E
E

25. T
D

26. W
of
R
te

12

BOILING POINT

Carl E. Rechsteiner, Jr.

12-1 INTRODUCTION

The boiling point is defined [15] as the temperature at which the vapor pressure of a liquid is equal to the pressure of the atmosphere on the liquid. For pure compounds, the normal boiling point is defined [15] as the boiling point at one standard atmosphere of pressure on the liquid. Impurities in the liquid, such as a substance in solution, or pressures other than one atmosphere alter the boiling point of the liquid in a predictable fashion from that of the normal boiling point for the pure substance. A discussion of the factors affecting the boiling point of a compound is presented in §12-10.

Besides being an indicator for the physical state (liquid vs. gas) of a chemical, the boiling point also provides an indication of its volatility. This property is required in most hazard assessment forms and is a needed input for some chemical spill models. Other physical properties, such as critical temperature and enthalpy of vaporization, may be predicted for a pure compound through use of its normal boiling point as one of the input values.

In addition to its utility for predicting physical properties and boiling points at various pressures, the normal boiling point is one of the few parameters known for almost every compound. This is partly because normal boiling points are easily measured: a calibrated thermometer suspended in a heated vessel containing the pure compound is sufficient, although slightly more elaborate equipment, such as the Cottrell apparatus [4], permits greater accuracy.

This chapter focuses on methods for estimating normal boiling points for organic compounds. Use of this property to predict other physical properties is discussed in the respective chapters of this handbook. The effect of impurities or different pressures on the boiling point is discussed briefly in §12-10.

Estimates of normal boiling points are based on correlations with physical parameters that are influenced by the molecular structures of the compounds. In most instances, correlations are made initially for compounds within a homologous series; these are later extended to a wider range of compounds, if possible.

- Meissner [14] has related the normal boiling points of many classes of organic compounds to their parachors and molar refractions.

- Since the boiling point and the critical temperature are directly related, additive contribution procedures to estimate the critical temperature [3,7,8] and the ratio between the boiling point and the critical temperature [24] may be combined to estimate normal boiling points.

- Forman and Thodos [6,7] use an additive contribution method to compute the van der Waals equation-of-state constants. These constants may be used to find the critical temperature, which, in combination with the boiling point/critical temperature ratio, yields the approximate boiling point.

- The molal liquid volume of an organic compound is another physical parameter that has been correlated with the normal boiling point [16,32].

- Other physical parameters, such as atomic number [25], carbon number [26], and related numbers [2,11,12] have been closely correlated with normal boiling points for very limited groups of compounds.

Molecular connectivity [9,10] is one of a growing family of competitive techniques for predicting boiling points and other physical properties of compounds on the basis of the topology of their atoms. The general procedure is to define one or more parameters whose values depend on each atom involved and the number of bonds that connect to that atom. Molecular values are computed by summing the individual atomic contributions to the chosen parameters for all of the atoms in the molecule. The molecular values so computed for a series of compounds with known boiling points are then fitted to a regression equation containing a selected number of terms, and that equation is used to estimate the boiling points of similar compounds. A detailed discussion of this and related techniques is not presented here, primarily because most of the regression equations have been reported for aliphatic compounds containing no more than one substituent group, which limits the applicability of this family of techniques. The computation of the individual atomic contributions to the different parameters is also fairly complicated. The other methods presented in this chapter are more amenable to manual calculation.

The methods presented here for the manual estimation of boiling points were developed using sets of fairly simple compounds. To a large extent, the effects of attaching multiple substituent groups to the molecule are assumed to be additive and independent of one another. When estimating boiling points for multifunctional compounds, one should keep in mind that this assumption may not be valid.

12-2   SELECTION OF APPROPRIATE METHOD

Table 12-1 lists recommended methods for estimating normal boiling points, indicates the types of compounds to which each method applies, and lists information concerning input parameters for each method and error specifications from the literature. *None of the estimation methods recommended in this chapter requires any input data beyond the structure of the compound of interest.* This is in contrast to methods in other chapters of this handbook, where knowledge of some related physical properties is essential to the estimation procedures. The methods listed in Table 12-1 are subdivided into two groups according to their range of applicability: the first group comprises the three most broadly useful techniques, and the second group comprises four techniques with limited applications. Detailed descriptions of these methods are presented in the following sections.

For many compounds, more than one method can be used. Table 12-2 compares the estimated boiling points with measured values for some randomly selected compounds. The entries are grouped by compound type and include the absolute and relative errors in terms of

TABLE 12-1

Summary of Methods for Estimating Normal Boiling Points

Sect. No.	Source	Applicability	Basis	Claimed Accuracy
A. Methods with General Application				
12-3	Meissner [14]	Organic compounds containing C, H, N, O, S, and halides. Usable for compounds containing other elements if their molar refraction is measured.	Correlates boiling points with the parachor and the molar refraction of the compound. Adjustment is made for the compound type/class of interest.	Average error of ~2%, maximum error of <8% in K at boiling point.
12-4	Lydersen [24], Forman, Thodos [6,7]	Organic compounds containing C, H, N, O, and halides.	Estimates boiling points from the critical temperature and the ratio between the boiling point and the critical temperature.	Boiling point not expected to be more accurate than 5 to 10 K [24].
12-5	Miller [16]	Most organic compounds.	Estimates boiling points from the boiling point/critical temperature ratio and the critical pressure and critical volume.	
B. Methods with Limited Application				
12-6	Ogata and Tsuchida [20]	Organic compounds containing C, H, N, O, and halides of the form RX, where R is a limited number of hydrocarbon radicals with eight or less carbons.	Estimates boiling points in empirical fashion.	Correlates boiling points for 98% of 600 compounds within 5 K.
12-7	Somayajulu and Palit [25]	Normal alkyl halides, acids, amines, ketones, aldehydes, benzenes, cyclo-hexanes, and cyclo hex-1-enes.	Correlates boiling points with atomic number sum.	Boiling points usually within 5 K
12-8	Kinney [11,12]	Alkanes, alkenes, alkynes, cycloalkanes, and cycloalkenes.	Correlates boiling point with boiling point number.	Boiling point usually within 10 K.
12-9	Stiel and Thodos [26]	Alkanes only.	Correlates boiling point with the number of carbons in saturated aliphatic hydrocarbons.	Average error ~0.5% in K.

	Known T_b	Meissner [14]			Lydersen [24], Forman, Thodos [6,7]			Miller [16]			Ogata and Tsuchida [20]		
		T_b	Error	% Error	T_b	Error	% Error	T_b	Error	% Error	T_b	Error	% Error
E. Oxygen-containing Organics													
Ethanol[a]	351.6	351.1	−0.5	−0.1	345.6	−6	−1.7	353.8	2.2	0.6	346.7	−4.9	−1.4
Benzyl alcohol	478.5	471.5	−7.0	−1.5	497.5	19	4.0	517.6	39.1	8.2			

A. Hydrocarbons

	Known T_b	Meissner [14]			Lydersen [24], Forman, Thodos [6,7]			Miller [16]		
		T_b	Error	% Error	T_b	Error	% Error	T_b	Error	% Error
Methane	109.2	115.9	6.7	6.1	–	–	–	121.5	12.3	11.2
Butane	272.7	273.4	0.7	0.3	273.5	0.8	0.3	281.4	8.7	3.2
Isobutane	261.4	277.8	16.4	6.3	262.5	1.1	0.4	277.3	15.9	6.1
Dodecane	489.4	478	−11.4	−2.3	486.2	−3.2	−0.7	465.2	−24.2	−4.9
Cyclohexane	353.9	338.8	−15.1	−4.3	353.4	−0.5	−0.1	358.2	4.3	1.2
Ethylene	169.4	172.1	2.7	1.6	–	–	–	176.8	7.4	4.4
Benzene	353.2	367.2	14	4.0	353.4	0.2	0.06	355.6	2.4	0.7
m-Xylene	412.2	413.3	1.1	0.3	410.9	−1.3	−0.3	405.2	−7.0	−1.7
Anthracene	613.2	574.4	−38.8	−6.3	613.8	−0.2	0.03	511.7	−101.5	−16.6
Chrysene	721.2	652	−69.2	−9.6	707	−14.2	−2.0	570.0	−151.2	−21.0
Coronene	798.2	764.3	−33.9	−4.2	979.8	181.6	22.8	632.1	−166.1	−20.8
Avg. Absolute Error			19.1	4.1		22.6	3.0		45.5	8.3

	Ogata and Tsuchida [20]			Somayajulu and Palit [25]			Kinney [11,12]			Stiel and Thodos [26]		
	T_b	Error	% Error	T_b	Error	% Error	T_b	Error	% Error	T_b	Error	% Error
Methane	153.4	44.2	40.5	108.7	−0.5	−0.5	118.1	8.9	8.2	126.3	17.1	15.7
Butane	264.9	−7.8	−2.9	276.3	3.6	1.3	273.8	1.1	0.4	271.9	−0.8	−0.3
Isobutane	251.9	−9.5	−3.6	263.3	1.9	0.7	263.3	1.9	0.7	260.4	−1	−0.4
Dodecane	–	–	–	488.0	−1.4	−0.3	486.9	−2.5	−0.5	488.1	−1.3	−0.3
Cyclohexane	–	–	–	–	–	–	349.3	−4.6	−1.3	–	–	–
Ethylene	178.5	9.1	5.4	–	–	–	165.9	−3.5	−2.1	–	–	–
Benzene	382	28.8	8.2	–	–	–	370.8	17.6	5.0	–	–	–
m-Xylene	–	–	–	–	–	–	422.0	9.8	2.4	–	–	–
Avg. Abs. Error		19.9	12.1		1.8	0.7		6.2	2.6		5.0	4.2

(continued)

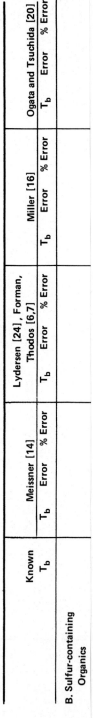

TABLE 12-2 (Continued)

| Known T_b | Meissner [14] | | Lydersen [24], Forman, Thodos [6,7] | | Miller [16] | | Ogata and Tsuchida [20] | |
| | T_b | Error | % Error | T_b | Error | % Error | T_b | Error | % Error | T_b | Error | % Error | T_b | Error | % Error |

B. Sulfur-containing Organics

percentage difference for each entry. From these tables, several generalizations may be made:

- All of the methods with limited applicability (Table 12-1B) yield estimated boiling points that agree closely with measured values; if these methods are applicable to the compound of interest, they should be used in preference to the more general methods listed in Table 12-1A.

- Of the general methods listed, that of Meissner [14] and that of Lydersen [24] and of Forman and Thodos [6,7] give more accurate boiling point estimates than those computed by the method of Miller [16]. Of the former two, Meissner's method is somewhat easier to compute.

- Because of its lower accuracy, Miller's method should only be used for compounds that cannot be estimated by the two other general methods (e.g., nitro-containing compounds).

12-3   MEISSNER'S METHOD

Principles of Use. The method described by Meissner [14] for estimating normal boiling points is based on their correlation with chemical type; molar refraction, $[R_D]$, and parachor, $[P]$, are used as variables in the correlation. Equation 12-1 relates the normal boiling point, T_b, in K to $[R_D]$, $[P]$ and a constant (B) whose value depends upon the chemical type:

$$T_b = \frac{637\,[R_D]^{1.47} + B}{[P]} \qquad (12\text{-}1)$$

The values for substitution into Eq. 12-1 may be obtained by several methods, as described below.

The molar refraction, $[R_D]$, is defined by

$$[R_D] = \left(\frac{M}{\rho_L}\right)\left(\frac{n_D^2 - 1}{n_D^2 + 2}\right) \qquad (12\text{-}2)$$

where M is the molecular weight (g/mole), ρ_L is the liquid density (g/mL), and n_D is the refractive index measured at the wavelength of the sodium D line.[1] All of these values are determined at the same temperature. For

---

1. Estimation methods for liquid density and refractive index are given in Chapters 19 and 26, respectively.

use in estimating boiling point, however, Eq. 12-2 is of limited utility, since the boiling point is normally determined before either the refractive index or the density is determined.

A simpler method for computing $[R_D]$ is the atomic refraction contribution method of Eisenlohr (referenced in [14]), which consists of summing the contributions for each atom and multiple bond in the compound. These contributions are listed in the second column of Table 12-3.

The parachor, [P], is defined by Sugden's equation (referenced in [14]) as

$$[P] = \frac{M \sigma^{1/4}}{\rho_L - \rho_v} \tag{12-3}$$

where M and ρ_L are as defined above, σ is the surface tension (dynes/cm), and ρ_v is the vapor density (g/mL), all measured at the same temperature.[2] One may use the method of McGowan [13] to rapidly estimate the parachor for any compound or use the more accurate additive procedure described by Sugden and updated by Mumford and Phillips [18,19].

- *McGowan's method* involves adding contributions for each atom in the compound (see Table 12-4), followed by subtraction of 19 times the number of bonds present in the compound. This method makes no distinction for single or multiple bonds or for differences in contribution due to bonding between different sets of atoms.
- *Sugden's method* adjusts the parachor values for differences in structural features, such as rings and multiple bonds, and for differences caused by bonding between different atom types. The third column of Table 12-3 lists the contributions to be added.

The computation of molar refraction by Eisenlohr's method and parachor by Sugden's method may be simplified through use of Section B of Table 12-3. Since contributions to either term are additive over all of the atoms present in the compound, substructures may be tabulated and used when necessary. For example, the phenyl contribution in monosubstituted benzenes is listed as 25.207 for $[R_D]$ and 190.0 for [P]; these values represent the total contributions from 6 carbons, 5 hydrogens, the 6-membered ring, and 3 double bonds.

---

2. Estimation methods for surface tension and vapor density are given in Chapters 20 and 19, respectively.

TABLE 12-3

Contributions to Molar Refraction and Parachor

Molecular Feature	Molar Refraction, $[R_D]$ [a]	Parachor, $[P]$ [b]
A. Atomic Contributions		
C (singly bound)	2.418	9.2
H	1.100	15.4[c]
O (hydroxyl)	1.525	20
O (in ethers, esters)	1.643	20
O (carbonyl)	2.211[d]	39[d]
O_2 (esters)	3.736	54.8
F	0.95[e]	25.5
Cl	5.967	55
Br	8.865	69
I	13.900	90
N (primary amine)	2.322	17.5
N (secondary amine)	2.502	17.5
N (tertiary amine or in ring)	2.840	17.5
N (nitrile)	5.516[f]	55.5[f]
S	7.690[g]	50
P	h	40.5
Other elements	h	—
3-member ring	0	12.5
4-member ring	0	6
5-member ring	0	3
6-member ring	0	0.8
7-member ring	0	−4.0
Double bond	1.733	19.0
Semipolar double bond	1.733	0
Triple bond	2.398	38
Singlet linkage	— —	−9.5[i]
B. Structural Contributions		
$-CH_2-$	4.618	40.0
$-CH_3$	5.718	55.4
$-C\equiv C-$	7.234	56.4
$-\overset{\mid}{C}=\overset{\mid}{C}-$	6.569	37.4
$-C\equiv N$	7.934	64.7

TABLE 12-3 (Continued)

Molecular Feature	Molar Refraction, $[R_D]$ [a]	Parachor, $[P]$ [b]
B. Structural Contributions (cont'd.)		
$\overset{\displaystyle O}{\underset{\displaystyle \parallel}{}}$ −C−O−	6.154	64.0
(benzene ring)	25.207	190.0
−OH	2.625	30.0
$\overset{\displaystyle O}{\underset{\displaystyle \parallel}{}}$ −C−	4.629	48.2

a. $[R_D]$ can be estimated within 5% by this method.

b. The strain inherent in some kinds of bonds introduces an additional contribution. In the following list of "strain constants," R is a hydrocarbon radical and X is a negative group. (When the negative group is Br, multiply the strain constant by 1.5.)

Use +3 for carbonyl in the ring.
Use zero for $RCH_2 X$, RCHO, RCOR, $RCH_2 R$, RNH_2, NOR, and NOOR.
Use −3 for $RCHX_2$, RCOOH, RCOOR, RCOCl, $R_2 CHX$, $R_2 CHR$, $RCONH_2$, ROCOOR, ROCOCl, RSOOR, ROSOOR, $R_2 NH$, $NO_2 R$, $NO_2 OR$, and azides.
Use −6 for RCX_3, $R_3 CX$, $RSO_2 Cl$, $R_3 CR$, $RSO_2 R$, $ROSO_2 OR$, $R_3 N$, $R_3 P$, $PO(OR)_3$.
Use −9 for CX_4, $R_4 C$.

c. Hydrogen on oxygen has a parachor of 10; on nitrogen, 12.5; on sulfur and carbon, 15.4.

d. Includes allowance for double bond.

e. Only for one fluorine atom attached to carbon; 1.1 for each F in polyfluorides.

f. Includes allowance for triple bond.

g. As SH, 7.69; as RSR, 7.97; as RCNS, 7.91; as RSSR, 8.11.

h. Value depends on type of compound, apparently differing for various combining forms.

i. Singlet linkage consists of a bond containing an unpaired electron.

Source: R.H. Perry and C.H. Chilton (eds.), *Chemical Engineers' Handbook*, 5th ed., McGraw-Hill Book Co., New York (1973). *(Reprinted with permission from McGraw-Hill Book Co.)*

The parameter B is used to adjust the computation of T_b for compound type. Table 12-5 lists values of B for various chemical compound classes. As most of these values were computed from sets of compounds containing only one functional group, this method may not be applicable to compounds containing more than one functional group. However, the tacit assumption is made throughout the development of this method that parachor and molar refraction are additive by either atomic or substructure groupings, and this should extend to the value of B. This

TABLE 12-4

McGowan Parachor Contributions[a]

Br	76.1	F	30.5	O	36.2
C	47.6	H	24.7	P	73.5
Cl	62.0	I	98.9	S	67.7
		N	41.9		

a. $[P] = \Sigma$ (contributions) $- (19)$ (number of bonds)

Source: McGowan [13]

TABLE 12-5

Constant B for Various Chemical Classes

Compound Class	B
Acids (monocarboxylic)	28,000
Alcohols (monohydroxy), including phenol, cresols, etc.	16,500
Amines:	
Primary	6,500
Secondary	2,000
Tertiary	−3,000
Esters of monocarboxylic acids with monohydroxy alcohols	15,000
Esters of dibasic acids with monohydroxy alcohols	30,000
Ethers and mercaptans	4,000
Hydrocarbons:	
Acetylenic	−500
Aromatic	−2,500
Paraffinic and naphthenic	−2,500
Olefinic	−4,500
Ketones or aldehydes	15,000
Monochlorinated normal paraffins	4,000
Nitriles	20,000

Source: Meissner [14]. *(Reprinted with permission from the American Institute of Chemical Engineers.)*

implication is partially borne out by the entries for esters of mono-carboxylic acids with monohydroxy alcohols, and for esters of dicarboxylic acids with monohydroxy alcohols. The latter compound class, which has two functional groups, has a value of B twice that of the former class, which has a single functional group. For a number of compounds, including those listed in Table 12-2 which contain multiple functional groups and/or hetero elements, use of a summed value for B (exclusive of the hydrocarbon values) gives estimated boiling points closer to the known values than use of a non-summed B value. (The hydrocarbon values of B are excluded because the data used to develop the B values in Table 12-5 include several types of hydrocarbons, and any hydrocarbon contribution has been averaged into the compound class B values.) It appears that use of a summed B value for complicated molecules will yield more accurate boiling point estimates. If, however, a single value is to be used, the value chosen should be that of the functional group with the largest absolute value, since the larger value indicates that the chemical type has more of an effect on the correlation.

Basic Steps

(1) Sketch the structure, showing the location of all atoms, rings, and bonds.

(2) Compute $[R_D]$ by adding the individual atom contributions from Table 12-3A and/or the structural contributions from Table 12-3B. (Additional methods for estimating $[R_D]$ are provided in Chapter 26.)

(3) Use Sugden's method to find [P], adding the individual atom contributions from Table 12-3A and/or the structural contributions from Table 12-3B. Include strain constants (footnote b) where applicable. Alternatively (McGowan method), add the atom contributions of Table 12-4, then subtract 19 times the number of bonds (whether single, double, or triple). An alternative set of values for structural increments for computing parachor values (Quayle [21]) is provided in Chapter 20 (Table 20-2). Those values include more structural information than the values listed in Table 12-3, and parachor values computed with Quayle's increments tend to be slightly more accurate.

(4) Determine B for the appropriate chemical type listed in Table 12-5. For multiple hetero functional groups, add the values of B together, excluding any hydrocarbon contribution.

(5) Compute the normal boiling point from Eq. 12-1 by substituting the derived values of [R$_D$], [P], and B.

Example 12-1   Estimate the normal boiling point of chloroethyl vinyl ether using Meissner's equation.

(1)   Draw the structure:

$$\text{Cl-C-C-O-C=C-H}$$

(2)   From the molar refraction column of Table 12-3,

4 carbons	=	4(2.418)
7 hydrogens bonded to C	=	7(1.100)
1 oxygen (ether)	=	1(1.643)
1 chlorine	=	1(5.967)
1 double bond	=	1(1.733)

$$[R_D] = 26.715$$

(3)   Compute [P]:

- By Sugden's method, using Table 12-3,

4 carbons	=	4(9.2)
7 hydrogens	=	7(15.4)
1 oxygen (ether)	=	1(20)
1 chlorine	=	1(55)
1 double bond	=	1(19)
strain	=	--

$$[P] = 238.6$$

- By McGowan's method, using Table 12-4,

4 carbons	=	4(47.6)
7 hydrogens	=	7(24.7)
1 oxygen	=	36.2
1 chlorine	=	62.0

461.5

12 bonds in compound

$$[P] = 461.5 - 19(12)$$

$$= 233.5$$

(4)   From Table 12-5,

Compound has monochlorinated paraffin	B = 4,000
Compound has ether	B = 4,000

$$\Sigma B = 8,000$$

(5) Compute T_b:

• With Sugden's [P] value

$$T_b = \frac{637(26.715)^{1.47} + 8000}{238.6} = 368\,K$$

$$Error = -14\,K$$

• With McGowan's [P] value

$$T_b = \frac{637(26.715)^{1.47} + 8000}{233.5} = 376\,K$$

$$Error = -6\,K$$

The measured boiling point for chloroethyl vinyl ether is 382.2K; hence this method yields errors of −4% and −2% for boiling points estimated using parachor values computed by the Sugden method and the McGowan method, respectively.

Example 12-2 Estimate the normal boiling point for nicotine.

(1) Draw the structure:

(2) Compute $[R_D]$:

10 carbons	= 10(2.418)
14 hydrogens bonded to carbon	= 14(1.100)
2 nitrogen (tert. amine)	= 2(2.840)
1 6-member ring	= 0
1 5-member ring	= 0
3 double bonds	= 3(1.733)
$[R_D]$ =	50.459

(3) Compute [P]:

• Sugden's method

10 carbons	= 10(9.2)
14 hydrogens	= 14(15.4)
2 nitrogens	= 2(17.5)
1 6-member ring	= 0.8
1 5-member ring	= 3
3 double bonds	= 3(19)
strain[3] $(2 - R_3 N)$	= 2(−6)
[P] =	391.4

---

3. The aromatic ring in this molecule is considered to contain the $R_3 N$ structure in the sense that none of the bonds of the nitrogen atom are connected to a hydrogen atom.

- McGowan's method

  10 carbons $= 10(47.6)$
  14 hydrogens $= 14(24.7)$
  2 nitrogens $= \underline{2(41.9)}$
  905.6

  $[P] = 905.6 - 19(27) = 392.6$

(4)   Compute B:

  2 tertiary amines $B = 2(-3,000) = -6,000$

(5)   Compute T_b:

- Sugden's [P] value

$$T_b = \frac{637(50.459)^{1.47} - 6000}{391.4} = 503K$$

$$\text{Error} = -16K$$

- McGowan [P] value

$$T_b = \frac{637(50.459)^{1.47} - 6000}{392.6} = 502K$$

$$\text{Error} = -17K$$

The measured boiling point for nicotine is 519.2K; hence this method yields errors of −3.1% and −3.3% for boiling points estimated using parachor values computed by the Sugden method and the McGowan method, respectively.

12-4   LYDERSEN-FORMAN-THODOS METHOD

Principles of Use. This method is a combination of two separate methods to allow estimation of boiling point. Forman and Thodos [6,7] and Thodos [28,29] have done extensive work on the estimation of critical temperature (T_c) and other critical properties.

Other workers [24] have defined a parameter $\theta = T_b/T_c$ based upon the observation of Guldberg (ref. in [8]) that the ratio is relatively constant for many organic compounds. Accordingly, boiling points can be estimated as follows:

$$T_b = \theta \, (T_c) \tag{12-4}$$

where θ is the ratio described above.

Calculation of Temperature Ratio, θ. A convenient method of determining θ has been developed by Lydersen [24]. In this method, various ΔT increments relating to the atomic and structural features of the molecule are summed. (See Table 12-6.) $\Sigma\Delta T$ is then substituted into Eq. 12-5 to obtain θ:

$$\theta = 0.567 + \Sigma\Delta T - (\Sigma\Delta T)^2 \qquad (12\text{-}5)$$

Estimation of Critical Temperature. The estimation of T_c is somewhat more involved. The Forman-Thodos method is based on the following correlation of T_c with van der Waal's constants:

$$T_c = \frac{8a}{27bR} = \frac{0.2963a}{bR} \qquad (12\text{-}6)$$

where T_c is the critical temperature in K, a and b are van der Waal's constants for the compound, and R is the universal gas constant, expressed as 82.05 atm-cm^3/mol-K. The constants a and b are related to structure and can be estimated by summing the appropriate increments. One complication is that the additive structural increments contribute to the appropriate van der Waal's constant raised to a power rather than directly to the van der Waal's constant. In the original correlations, the exponents had the values of 0.626 for a and 0.76 for b, which were later rounded to 2/3 and 3/4. The tables included here list the increments in terms of $a^{2/3}$ and $b^{3/4}$; thus, before the critical temperature is computed from Eq. 12-6, the sum of the additive terms must be raised to the appropriate power (3/2 for a, and 4/3 for b).

To use this method, each molecule is considered to be an aggregate of various types of carbon atoms and functional groups; -OH, -NO$_2$, -CO$_2$H are examples of functional groups. Hydrogen atoms are not considered, except for their role in defining carbon atom types or functional group types. The contributions to $a^{2/3}$ and $b^{3/4}$ are summed for each carbon type present in the molecule, adjusted for the presence of rings and multiple bonds, and finally added to the contributions of the other functional groups in the molecule. Figure 12-1 shows the carbon types used with this method. The type number refers to the number of bonds between the central carbon atom and other carbon atoms or functional groups. Letters following some type numbers indicate their use in the compound, i.e., in aliphatics, in olefins, in cyclo- or naphthenic compounds, and in aromatics with or without fused rings. The following paragraphs explain how contributions for different carbon types are computed.

TABLE 12-6

Lydersen's Increments for Calculating $\Sigma \Delta T$

- There are no increments for hydrogen.
- All bonds shown as free are connected to atoms other than H.
- Values in parentheses are based upon too few experimental values to be reliable [24].

Increment	ΔT	Increment	ΔT
Nonring increments:		**Oxygen increments:**	
$-CH_3$	0.020	$-OH$ (alcohols)	0.082
$-CH_2$	0.020	$-OH$ (phenols)	0.031
$-CH$	0.012	$-O-$ (nonring)	0.021
$-C-$	0	$-O-$ (ring)	(0.014)
$=CH_2$	0.018	$-C=O$ (nonring)	0.040
$=CH$	0.018	$-C=O$ (ring)	(0.033)
$=C-$	0	$HC=O$ (aldehyde)	0.048
$=C=$	0	$-COOH$ (acid)	0.085
$\equiv CH$	0.005	$-COO-$ (ester)	0.047
$\equiv C-$	0.005	$=O$ (exc. for above	
Ring increments:		combinations)	(0.02)
$-CH_2-$	0.013	**Nitrogen increments:**	
$-CH$	0.012	$-NH_2$	0.031
C	(−0.007)	$-NH$ (nonring)	0.031
$=CH$	0.011	$-NH$ (ring)	(0.024)
$=C-$	0.011	$-N-$ (nonring)	0.014
$=C=$	0.011	$-N-$ (ring)	(0.007)
$C-H$ (common to 2		$-CN$	(0.060)
saturated con-		$-NO_2$	(0.055)
densed rings)	0.064	**Sulfur increments:**	
Halogen increments:		$-SH$	0.015
$-F$	0.018	$-S-$ (nonring)	0.015
$-Cl$	0.017	$-S-$ (ring)	(0.008)
$-Br$	0.010	$=S$	(0.003)
$-I$	0.012		

Source: Reid and Sherwood [24]. *(Reprinted with permission from McGraw-Hill Book Co.)*

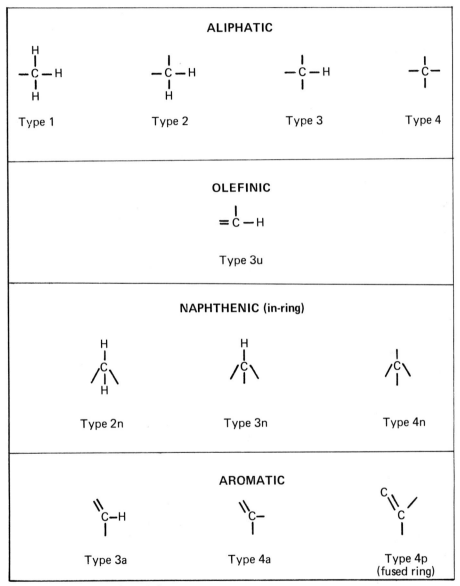

Source: Forman and Thodos [6].

FIGURE 12–1   Carbon Types Used in Forman-Thodos
Estimation of Critical Temperature

Individual contributions to $a^{2/3}$ and $b^{3/4}$ will be referred to as $\Delta a^{2/3}$ and $\Delta b^{3/4}$. These contributions depend on both the carbon type and the number of carbon atoms plus functional atoms.

• *Contributions for Carbon Types.* Table 12-7 lists the specific values for the different carbon types for $2 \leq n \leq 15$, where n is the total number of carbon atoms and functional atoms. Table 12-8 lists the regression equations from which the specific values in Table 12-7 originate. In the following example, which shows only the carbon skeleton, the molecule contains six type 1 carbons (-C), three type 2 carbons (-C-), two type 3 carbons (-C̦-), and one type 4 carbon (-C̦-). (The parenthetical subscripts are the type numbers.)

$$
\begin{array}{c}
 C_{(1)} \\
 | \\
C_{(1)}-C_{(3)}-C_{(2)}-C_{(4)}-C_{(2)}-C_{(3)}-C_{(2)}-C_{(1)} \\
 | | | \\
 C_{(1)} C_{(1)} C_{(1)}
\end{array}
$$

As this compound contains 12 carbon atoms and no functional atoms, n = 12.

• *Corrections for Branching.* An adjustment is made to the summed $\Delta a^{2/3}$ and $\Delta b^{3/4}$ terms if any branching is found in the hydrocarbon (i.e., if any type 3 or type 4 carbon atoms are present). The correction that is made is:

$$\Sigma \Delta a^{2/3} \text{ (correct)} = f_a \, \Sigma \Delta a^{2/3} \text{ (computed)} \tag{12-7}$$

$$\Sigma \Delta b^{3/4} \text{ (correct)} = f_b \, \Sigma \Delta b^{3/4} \text{ (computed)} \tag{12-8}$$

$$f_a = (1/3) \, [(W_i/W_n) + 2 + 0.087 \, m + 0.0045 \sum_{k=1}^{m} k \, (k-1)] \tag{12-9}$$

$$f_b = (1/2) \, [(W_i/W_n + 1 + 0.101 \, m - 0.005 \, m^2] \tag{12-10}$$

m = total number of side chains attached to the main (longest) carbon chain

W = Wiener number

TABLE 12-7

Group Contributions $\Delta a^{2/3}$ and $\Delta b^{3/4}$ for Saturated Aliphatic Hydrocarbons

n^a	$-\overset{\underset{\mid}{H}}{\underset{\underset{\mid}{H}}{C}}H$ Type 1		$-\overset{\mid}{C}H$ Type 2 $\underset{\mid}{H}$		$-\overset{\mid}{\underset{\mid}{C}}-H$ Type 3		$\overset{\diagup}{\underset{\diagdown}{C}}$ Type 4	
	$\Delta a^{2/3}$	$\Delta b^{3/4}$	$\Delta a^{2/3}$	$\Delta b^{3/4}$	$\Delta a^{2/3}$	$\Delta b^{3/4}$	$\Delta a^{2/3}$	$\Delta b^{3/4}$
1								
2	15,577	11.453						
3	15,216	11.453	13,678	6.262				
4	15,035	11.453	13,678	6.262	12,567	2.064		
5	14,927	11.453	13,678	6.262	11,189	0.886	6,181	−4.937
6	14,854	11.453	13,678	6.282	10,270	0.101	4,980	−6.670
7	14,803	11.453	13,678	6.262	9,614	−0.460	4,123	−7.909
8	14,764	11.453	13,678	6.262	9,122	−0.880	3,480	−8.837
9	14,734	11.453	13,678	6.262	8,739	−1.207	2,979	−9.559
10	14,710	11.453	13,678	6.262	8,433	−1.469	2,579	−10.337
11	14,690	11.453	13,678	6.262	8,182	−1.683	2,252	−10.160
12	14,674	11.453	13,678	6.262	7,974	−1.862	1,979	−11.004
13	14,660	11.453	13,678	6.262	7,797	−2.012	1,748	−11.337
14	14,648	11.453	13,678	6.262	7,646	−2.142	1,550	−11.623
15	14,638	11.453	13,678	6.262	7,514	−2.254	1,379	−11.870

a. n = total number of carbon atoms and functional atoms.

Source: Forman and Thodos [6]. *(Reprinted with permission from the American Institute of Chemical Engineers.)*

TABLE 12-8

Equations for $\Delta a^{2/3}$ and $\Delta b^{3/4}$ for Saturated Aliphatic
Hydrocarbons Containing n Carbons ($2 \leqslant n \leqslant 15$)

Type		$\Delta a^{2/3}$	$\Delta b^{3/4}$
1	H | −C−H | H	= (2,168/n) + 14,493	= constant = 11.453
2	| −C−H | H	= constant = 13,678	= constant = 6.262
3	| −C−H |	= (27,560/n) + 5,677	= (23.55/n) − 3.824
4	| −C− |	= (36,013/n) − 1,022	= (52.00/n) − 15.337

Source: Foreman and Thodos [6]

The subscripts i and n refer to the isomer of interest and to the corresponding normal alkane respectively. To compute the Wiener number, the number of carbons on the two sides of any bond are multiplied together and summed for all of the carbon-to-carbon bonds. For convenience, the Wiener number for the normal alkane can be computed from Eq. 12-11.

$$W_n = \frac{(N-1)(N)(N+1)}{6} \qquad (12\text{-}11)$$

where N is the number of carbon atoms in the hydrocarbon.

Example 12-3 illustrates the computation of $a^{2/3}$, $b^{3/4}$, and Wiener numbers for the compound shown previously.

Example 12-3 Compute $a^{2/3}$ and $b^{3/4}$ using the Forman-Thodos method for 2,4,4,6-tetramethyl octane, whose carbon skeleton is shown below. (The bonds are individually numbered for reference.)

(1) Calculate $\Delta a^{2/3}$ and $\Delta b^{3/4}$ values

- Total number of carbons = 12

- From Table 12-7

	$\Delta a^{2/3}$	$\Delta b^{3/4}$
six type 1 carbons	6(14,674)	6(11.453)
three type 2 carbons	3(13,678)	3(6.262)
two type 3 carbons	2(7,974)	2(−1.862)
one type 4 carbon	1(1,979)	1(−11.004)
	$\Sigma\Delta a^{2/3} = 147{,}005$	$\Sigma\Delta b^{3/4} = 72.776$

(2) Compute correction factors

- Total number of side chains (m) = 4

- From Eq. 12-11

$$W_n = \frac{(12-1)(12)(12+1)}{6}$$

$$= 286$$

- Compute W_i as follows:

For bond		
I:	1x11	= 11
II:	1x11	= 11
III:	3x9	= 27
IV:	4x8	= 32
V:	1x11	= 11
VI:	1x11	= 11
VII:	7x5	= 35
VIII:	8x4	= 32
IX:	11x1	= 11
X:	10x2	= 20
XI:	11x1	= 11

$$W_i = 212$$

- From Eq. 12-9

$$f_a = (1/3)[(212/286) + 2 + 0.087 \, (4) + 0.0045 \sum_{k=1}^{4} k \, (k-1)]$$

$$= (1/3)[0.7413 + 2 + 0.348 + 0.0045 \, (20)]$$

$$= 1.060$$

- $$f_b = (1/2)[(212/286) + 1 + 0.101 \, (4) - 0.005 \, (4)^2]$$

$$= (1/2)[0.7413 + 1 + 0.404 - 0.080]$$

$$= 1.033$$

(3)   Correct $\Delta a^{2/3}$ and $\Delta b^{3/4}$ using Eqs. 12-7 and 12-8.

- $\Delta a^{2/3}$ (correct) = 1.060 (147,005) = 155,800

- $\Delta b^{3/4}$ (correct) = 1.033 (72.776) = 75.18

- *Contributions for Multiple Bonds.* For compounds containing multiple bonds, an initial computation of $\Delta a^{2/3}$ and $\Delta b^{3/4}$ is made using the saturated analog of the compound of interest. The multiple bonds are then inserted in the proper location, and appropriate contributions are added to $\Delta a^{2/3}$ and $\Delta b^{3/4}$. Table 12-9 lists the contributions due to unsaturation for various carbon atom types. For a single unsaturated bond, the contribution is found directly from the table using the type numbers shown in Figure 12-1 for the two carbon atoms involved in the saturated bond. Returning to Example 12-3, assume that we wish to make bond III a double bond; the two carbon atoms involved are types 3 and 2, so the contribution for a 3-2 double bond should be used. If we wish to make bond XI a triple bond, the two carbon atoms involved are types 1 and 3, so the contribution for a 3-1 triple bond should be used.

The introduction of more than one double bond alters the contributions that are used for the type 3 carbons and those for a bond that extends the conjugation of the molecule. (Conjugation occurs when there is more than one double bond in a molecule and the double bonds are separated from one another by a single bond.) The carbons joined by double bonds become type 3u, illustrated in Figure 12-1. If an additional double bond joins one of these carbons to another carbon of some type n, a contribution must be incorporated in the $\Delta a^{2/3}$ and $\Delta b^{3/4}$ terms for the 3u-to-n bond. For type 3 carbons, the appropriate contribution from Table 12-9B is used. If the additional double bond is conjugated with a

TABLE 12-9

Double- and Triple-Bond Contributions
in Forman and Thodos Method

	$\Delta a^{2/3}$	$\Delta b^{3/4}$
A. First double bond		
(1-1)	−3,868	−2.021
(2-1)	−3,154	−1.895
(2-2)	−2,551	−2.009
(3-1)	−1,548	−1.706
(3-2)	− 928	−1.820
(3-3)	− 540	−1.930
B. Second double bond		
(3-1)	− 828	−1.259
(3-2)	− 496	−1.343
(3u-1)	−1,332	−1.745
(3u-2)	−1,324	−1.862
(3u-3)	−1,316	−1.979
(3u←2-1)	−1,687	−1.399
(3u←2-2)	− 910	−1.485
C. Triple bond		
(1-1)	−4,269	−3.680
(2-1)	−1,934	−3.008
(2-2)	−1,331	−3.122

Source: Forman and Thodos [6]. *(Reprinted with permission from the American Institute of Chemical Engineers.)*

type 3u carbon (i.e., there is an intervening type 2 carbon), the contributions to the $\Delta a^{2/3}$ and $\Delta b^{3/4}$ terms must reflect the conjugation in the molecule.

An example may help to clarify the multiple-double-bond calculation.

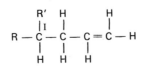

In this partial structure, if R and R' are both hydrogens (i.e., the associated carbon is a type 1), the contributions to $\Delta a^{2/3}$ and $\Delta b^{3/4}$ for a double bond inserted at location I would be for a 3u ← 2-1 double bond, where the arrow indicates conjugation through the type 2 carbon. If R is a hydrocarbon radical and R' is hydrogen (type 2 carbon), the contributions for a double bond at I would be for a 3u ← 2-2 double bond. If both R and R' are hydrocarbon radicals (type 3 carbon), the contributions for a double bond at location I would be for a 3-2 double bond. For other bond locations where an additional double bond is neither conjugated to a previous double bond nor involves a type 3 carbon, the contributions to $\Delta a^{2/3}$ and $\Delta b^{3/4}$ are the same as if the added double bond were the only double bond in the molecule. (Use Table 12-9A.)

• *Naphthenic Compounds.* The contributions for atoms in the rings of naphthenic compounds (cycloalkanes) are different from those for carbons in alkyl sidechains, and an additional contribution is attributable to the ring structure per se. The values for carbon types 2n, 3n, and 4n are listed in Table 12-10 along with the naphthene ring contribution.

For multiple naphthene rings in a compound, a contribution to $\Delta a^{2/3}$ and $\Delta b^{3/4}$ is made for each ring. If the cycloalkane has more than one substituent in the ring, the additional strain on the ring is reflected in positional contributions which vary for *cis* and *trans* configurations; these are also listed in Table 12-10.

In their original work, Forman and Thodos [6] and Thodos [25] obtained data only for 1,2 and 1,3 disubstitutions. For other disubstitutions such as 1,4 or 1,1, they recommended that the contribution for a 1,3 disubstitution be used. While neither author dealt with tri-, tetra-, etc. substitutions, they suggested that the disubstitution rules be used sequentially to account for all of the substitutions. As an example, consider the following tri-substituted cyclohexane:

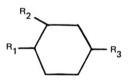

This compound would have a (1,2) disubstitution for the pair R_1R_2, a (1,3) disubstitution for the pair R_2R_3, and a (1,4) disubstitution for the pair R_1R_3. The contribution for the last disubstitution would be taken from the appropriate (1,3) disubstitution entry of Table 12-10. In general, a trisubstituted molecule would have three disubstitution contributions, etc.

TABLE 12-10

**Naphthenic and Aromatic Contributions
in Forman and Thodos Method**

	$\Delta a^{2/3}$	$\Delta b^{3/4}$
Carbon Type		
2n	12,535	5.338
3n	9,910	0.023
4n	2,066	−8.094
3a	11,646	5.991
4a	11,144	1.043
4p	11,561	1.634
Naphthenic ring contribution	2,658	9.073
Aromatic ring contribution	0	0
Position contribution from multiple substitution:		
Naphthene rings		
cis-1,2	− 427	−0.866
1,3	− 2,525	−1.493
trans-1,2	− 2,525	−1.493
1,3	− 4,195	−2.494
Aromatic rings		
1-2	− 830	−1.253
1-3	− 1,597	−0.806
1-4	155	0.212
1-5	279	0.254
1-6	488	0.525

Source: Adapted from Thodos [28,29]

• *Aromatic Compounds.* Aromatic compounds are handled in the same way as cycloalkanes. The three types of aromatic carbons shown in Figure 12-1 are used to account for isolated aromatic rings (types 3a and 4a) and fused rings (type 4p). Their contributions are listed in Table 12-10. There is no additional contribution for the aromatic structure. Multiple substitutions to aromatic rings are accounted for in the same fashion as were multiple substitutions to naphthene rings.

• *Functional Groups.* The final contributions to the $\Delta a^{2/3}$ and $\Delta b^{3/4}$ terms are due to any functional groups present. These contributions depend on the functional group involved and the number of carbon and functional atoms present, according to Eqs. 12-12 and 12-13:

$$\Delta a^{2/3} = K_1/n + K_2 \qquad (12\text{-}12)$$

$$\Delta b^{3/4} = K_3/n + K_4 \qquad (12\text{-}13)$$

Values for K_1, K_2, K_3, and K_4 are listed in Table 12-11, and n is the total number of carbon atoms and functional group atoms. Halogens attached to a single carbon in aliphatic compounds have different contributions to $\Delta a^{2/3}$ and $\Delta b^{3/4}$, depending on the number of halogen atoms attached to that carbon. The procedure for assigning halogen contributions is to use the lower weight halogens first and find their contribution in Table 12-11. For example, in fluorodichloromethane, there would be a "first fluoride" contribution, a "second chloride" contribution, and a "third chloride" contribution to account for the halogens.

Basic Steps

(1) Sketch the molecule, showing the location of the atoms, rings, and double bonds.

(2) Compute ΔT by summing atom contributions from Table 12-6.

(3) Compute θ using Eq. 12-5.

(4) For aliphatic portion(s) of the molecule, sketch completely saturated side chains.

(5) Compute aliphatic contributions to $\Delta a^{2/3}$ and $\Delta b^{3/4}$ using Table 12-7 and Eqs. 12-7 to 12-11.

(6) Compute contribution to $\Delta a^{2/3}$ and $\Delta b^{3/4}$ for any sites of unsaturation using Table 12-9.

(7) Compute contributions to $\Delta a^{2/3}$ and $\Delta b^{3/4}$ for any naphthene and aromatic rings using Table 12-10.

(8) Compute contributions to $\Delta a^{2/3}$ and $\Delta b^{3/4}$ for any functional groups using Table 12-11 and Eqs. 12-12 and 12-13.

(9) Sum contributions to $\Delta a^{2/3}$ and $\Delta b^{3/4}$.

(10) Compute a and b by $(\sum \Delta a^{2/3})^{3/2}$ and $(\sum \Delta b^{3/4})^{4/3}$.

(11) Compute T_c from Eq. 12-6.

(12) Compute T_b from Eq. 12-4.

TABLE 12-11

Constants for Equations that Establish Functional Group Contributions for Organic Compounds in Forman and Thodos Methods

	Functional Group	$\Delta a^{2/3} = K_1/n + K_2$		$\Delta b^{3/4} = K_3/n + K_4$	
		K_1	K_2	K_3	K_4
Alcohols	—OH	30,200	14,000	8.96	7.50
Phenols	—OH	0	8,500	0	4.19
Ethers:					
noncyclic	—O—	14,500	6,500	0	3.26
cyclic	—O—	0	9,440	0	2.74
Ketones	$-\overset{\displaystyle O}{\overset{\|}{C}}-$	62,800	16,700	27.20	4.55
Carboxylic acids	$-\overset{\displaystyle O}{\overset{\|}{C}}OH$	142,670	16,730	66.80	5.10
Acid anhydrides	$-\overset{\displaystyle O}{\overset{\|}{C}}O\overset{\displaystyle O}{\overset{\|}{C}}-$	0	43,880	0	14.78
Esters:					
Formates	$H\overset{\displaystyle O}{\overset{\|}{C}}O-$	35,140	26,800	2.29	15.80
Others	$-\overset{\displaystyle O}{\overset{\|}{C}}O-$	37,430	25,500	− 3.00	12.00
Amines:					
Primary	—NH$_2$	4,800	18,900	0	10.15
Secondary	$-\overset{\displaystyle H}{\underset{}{N}}-$	51,800	0	19.60	− 1.10
Tertiary	$-\overset{\displaystyle \|}{N}-$	60,200	− 4,300	29.20	− 7.90
Nitriles	—CN	86,000	25,900	39.70	12.10
Aliphatic halides:					
Fluorides	—F				
First		2,420	12,240	− 3.70	10.92
Second		−38,500	4,510	−48.50	12.86
Third		0	3,450	0	6.92
Chlorides	—Cl				
First		0	22,580	0	11.54
Second		66,000	− 5,100	19.00	3.90
Third		−60,250	29,100	−40.80	19.40
Fourth		0	16,500	0	11.46
Bromides	—Br				
First		− 2,720	23,550	− 4.35	11.49
Second		0	20,860	0	5.37
Iodides	—I				
First		0	33,590	0	13.91
Aromatic halides:					
Fluoride	—F	0	4,210	0	7.22
Chloride	—Cl	0	17,200	0	10.88
Bromide	—Br	0	24,150	0	12.74
Iodide	—I	0	34,780	0	15.22

Source: Forman and Thodos [7]. *(Reprinted with permission from the American Institute of Chemical Engineers.)*

Example 12-4   Compute the normal boiling point for anthracene by the
Lydersen-Forman-Thodos method.

(1)   Sketch anthracene

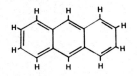

(2)   From Table 12-6:

$$10(=\overset{|}{C}H \text{ ring}) = 10(.011)$$

$$4(=C\!\!\!< \text{ring}) = \underline{4(.011)}$$

$$\Sigma \Delta T = 0.154$$

(3)   From Eq. 12-5:

$$\theta = 0.567 + (0.154) - (0.154)^2$$

$$= 0.697$$

(4)   The only contributions to a and b in this compound are for the carbons in the
aromatic rings. From Table 12-10,

	$\Delta a^{2/3}$	$\Delta b^{3/4}$
10 [=C–H ring, type 3a] =	10(11,646)	10(5.991)
4 [=C– ring, type 4P] =	$\underline{4(11,561)}$	$\underline{4(1.634)}$
Σ =	162,704	66.446

(5) $a = (162,704)^{3/2} = 65,629,000$

$b = (66.446)^{4/3} = 269.138$

(6)   From Eq. 12-6,

$$T_c = \frac{(8)\,(65,629,000)}{(27)\,(269.13)\,(82.05)}$$

$$= 880.6$$

(7)   From Eq. 12-4,

$$T_b = \theta\, T_c$$

$$= (0.697)\,(880.6)$$

$$= 614$$

Error = 1K

The measured boiling point for anthracene is 613.2K; thus, this method shows a relative error of 0.2%.

Example 12-5   Compute the normal boiling point for p-chloro-m-cresol by the Lydersen-Forman-Thodos method.

(1)   Sketch molecule

(2)   From Table 12-6:

$$
\begin{aligned}
-OH &= 0.031 \\
-CH_3 &= 0.020 \\
-Cl &= 0.017 \\
3\ (=\overset{|}{C}H\ \text{ring}) &= 3\ (0.011) \\
3\ (=\overset{|}{C}-\ \text{ring}) &= \underline{3\ (0.011)} \\
\Sigma\Delta T &= 0.134
\end{aligned}
$$

(3)   From Eq. 12-5:

$$
\theta = 0.567 + 0.134 - (0.134)^2
$$

$$
= 0.683
$$

(4)   The "aliphatic side chain" ($-CH_3$) contains a type 1 carbon, and the total number of carbon and functional group atoms, n = 7 (C) + 1 (Cl) + 1 (O) = 9. Therefore, from Table 12-7, the aliphatic contributions to a and b are

$$
\Delta a^{2/3} = 14{,}734
$$

$$
\Delta b^{3/4} \doteqdot 11.453
$$

(5)   The contributions for the aromatic ring (Table 12-10) are:

	$\Delta a^{2/3}$	$\Delta b^{3/4}$
3 (type 3a)	3 (11,646)	3 (5,991)
3 (type 4a)	3 (11,144)	3 (1.043)
Ring	0	0

Multiple substitutions[4]

(1-3)	−1,597	−0.806
(1-2)	−830	−1.253
(1-4)	155	0.212
	$\Delta a^{2/3} = 66{,}098$	$\Delta b^{3/4} = 19.255$

(6)   Obtain the functional group contributions from Table 12-11.

For −OH (phenol) $K_1 = 0$, $K_2 = 8{,}500$, $K_3 = 0$, $K_4 = 4.19$

By Eq. 12-12, $\Delta a^{2/3} = 0/9 + 8{,}500 = 8{,}500$

By Eq. 12-13, $\Delta b^{3/4} = 0/9 + 4.19 = 4.19$

For −Cl (aromatic) $K_1 = 0$, $K_2 = 17{,}200$, $K_3 = 0$, $K_4 = 10.88$

$\Delta a^{2/3} = 0/9 + 17{,}200 = 17{,}200$

$\Delta b^{3/4} = 0/9 + 10.88 = 10.88$

(7)   Totaling the contributions from steps 4, 5, and 6:

$\Delta a^{2/3} = 14{,}734 + 66{,}098 + 8{,}500 + 17{,}200 = 106{,}532$

$\Delta b^{3/4} = 11.453 + 19.255 + 4.190 + 10.880 = 45.778$

(8) $a = (106{,}532)^{3/2} = 34{,}771{,}000$

$b = (45.778)^{4/3} = 163.76$

(9)   From Eq. 12-6,

$$T_c = \frac{(8)\,(34{,}771{,}000)}{(27)\,(163.76)\,(82.05)} = 766.8$$

(10)   From Eq. 12-4,

$$T_b = (0.683)\,(766.8) = 524K$$

$$\text{Error} = 16K$$

The measured boiling point for p-chloro-m-cresol is 508.2K; thus, this method shows an error of 3.1%.

---

4. Recall that the substitution notation describes the relationship between a given substitution and the one that precedes it in a counterclockwise direction. Thus, although CH_3 and Cl are in ring positions 3 and 4 respectively, the Cl is considered a 1-2 substitution in the Forman and Thodos method, because its reference is the CH_3 radical.

12-5 MILLER'S METHOD

Miller [16] combined the empirical work of Rackett [22] and that of Tyn and Calus [30] to produce a method of estimating the normal boiling point of any organic compound from its critical properties — P_c, the critical pressure, V_c, the critical volume, and θ, the ratio of the normal boiling point, T_b, to the critical temperature, T_c. The three required parameters for this method, P_c, V_c, and θ, may be estimated via the techniques of Lydersen [24].

Tyn and Calus proposed Eq. 12-14 to relate the molal liquid volume at the normal boiling point, V_b, to the critical volume, V_c, where both volumes are expressed in terms of cm³ per mole.

$$V_b = 0.285\ V_c^{1.048} \tag{12-14}$$

Rackett developed an empirical equation to relate the reduced volume of saturated liquids to the critical compressibility factor and the reduced temperature. At the normal boiling point, the Rackett equation becomes

$$V_b = \left(\frac{RT_c}{P_c}\right)\ Z_c^{1.0 + (1.0 - \theta)^{2/7}} \tag{12-15}$$

where V_b is the molal liquid volume at the normal boiling point, T_c is the critical temperature in K, P_c is the critical pressure in atm, Z_c is the critical compressibility factor, θ is the ratio of T_b/T_c, and R is the universal gas constant with a value of 82.054 cm³-atm/mol K. On combining Eqs. 12-14 and 12-15 and solving for T_c, one obtains:

$$T_c = \frac{e^\beta}{R} \tag{12-16}$$

where

$$\beta = \frac{[(1 - \theta)^{2/7} - 0.048]\ \ln\,(V_c) + [1 - \theta]^{2/7}\ \ln\,(P_c) + 1.255}{[1 - \theta]^{2/7}} \tag{12-17}$$

The normal boiling point may then be computed from Eq. 12-4.

In §12-4, the Lydersen method for estimating θ was presented. Lydersen has developed analogous methods for estimating P_c and V_c.

Pressure increments (or volume increments) are summed for each atomic and structural feature of the molecule. The increment sums found in Table 12-12 are substituted into Eq. 12-18 or 12-19 to compute the critical pressure or the critical volume, respectively.

$$P_c = \frac{M}{(0.34 + \Sigma \Delta P)^2} \qquad (12\text{-}18)$$

and

$$V_c = 40 + \Sigma \Delta V \qquad (12\text{-}19)$$

where M is the molecular weight in grams per mole.

Basic Steps

(1) Sketch the structure, showing the location of all atoms, rings, and bonds.

(2) Find $\Sigma \Delta T$, $\Sigma \Delta P$, and $\Sigma \Delta V$ by summing the appropriate increments in Tables 12-6 and 12-12.

(3) Compute θ using Eq. 12-5.

(4) Compute P_c using Eq. 12-18.

(5) Compute V_c using Eq. 12-19.

(6) Compute β using 12-17.

(7) Compute T_b from Eqs. 12-16 and 12-4.

Example 12-6   Estimate the normal boiling point of isobutane (molecular weight = 58.13).

(1)   Sketch isobutane:

(2)   Find $\Sigma \Delta T$, $\Sigma \Delta P$, and $\Sigma \Delta V$ from Tables 12-6 and 12-12.

	ΔT	ΔP	ΔV
3(−CH₃)	3(0.020)	3(0.227)	3(55)
1(−CH)	0.012	0.210	51
	$\Sigma \Delta T = 0.072$	$\Sigma \Delta P = 0.891$	$\Sigma \Delta V = 216$

TABLE 12-12

Lydersen's Increments for Calculating $\Sigma\Delta P$ and $\Sigma\Delta V^{a,b,c}$

Increment	ΔP	ΔV
Nonring increments		
$-CH_3$	0.227	55
$-CH_2$	0.227	55
$-CH$	0.210	51
$-C-$	0.210	41
$=CH_2$	0.198	45
$=CH$	0.198	45
$=C-$	0.198	36
$=C=$	0.198	36
$\equiv CH$	0.153	$(36)^c$
$\equiv C-$	0.153	(36)
Ring increments		
$-CH_2-$	0.184	44.5
$-CH$	0.192	46
$\diagup C \diagdown$	(0.154)	(31)
$=CH$	0.154	37
$=C-$	0.154	36
$=C=$	0.154	36

(continued)

TABLE 12-12 (Continued)

Increment	ΔP	ΔV
Halogen increments		
—F	0.224	18
—Cl	0.320	49
—Br	(0.50)	(70)
—I	(0.83)	95
Oxygen increments		
—OH (alcohols)	0.06	(18)
—OH (phenols)	−0.02	(3)
—O— (nonring)	0.16	20
—O— (ring)	(0.12)	(8)
—C=O (nonring)	0.29	60
—C=O (ring)	(0.20)	(50)
HC=O (aldehyde)	0.33	73
—COOH (acid)	0.40	80
—COO— (ester)	0.47	80
=O (exc. for above combinations)	0.12	11
Nitrogen increments		
—NH$_2$	0.095	28
—NH (nonring)	0.135	(37)
—NH (ring)	(0.09)	(27)
—N— (nonring)	0.17	(42)
—N— (ring)	(0.13)	(32)
—CN	(0.36)	(80)
—NO$_2$	(0.42)	(78)

TABLE 12-12 (Continued)

Increment	ΔP	ΔV
Sulfur increments		
—SH	0.27	55
—S— (nonring)	0.27	55
—S— (ring)	(0.24)	(45)
=S	(0.24)	(47)

a. There are no increments for hydrogen.
b. All bonds shown as free are connected to atoms other than H.
c. Values in parentheses are based upon too few experimental values to be reliable [24].

Source: Reid and Sherwood [24]. *(Reprinted with permission from McGraw-Hill Book Co.)*

(3)   From Eq. 12-5

$$\theta = 0.567 + (0.072) - (0.072)^2$$

$$= 0.634$$

(4)   From Eq. 12-18

$$P_c = \frac{58.13}{(0.340 + 0.891)^2}$$

$$= 38.4 \text{ atm}$$

(5)   From Eq. 12-19

$$V_c = 40 + 216$$

$$= 256 \text{ cm}^3/\text{mol}$$

(6)   From Eq. 12-17

$$\beta = \frac{[(1 - 0.634)^{2/7} - 0.048] \ln(256) + [1 - 0.634]^{2/7} \ln(38.4) + 1.255}{[1 - 0.634]^{2/7}}$$

$$= 10.51$$

(7)   From Eqs. 12-16 and 12-4

$$T_b = \frac{(0.634) e^{10.51}}{82.05}$$

$$= 283.4 \text{ K}$$

Error = 20.4 K

The measured boiling point of isobutane is 263.0K; thus, this method shows a relative error of 7.8%.

Example 12-7   Estimate T_b for nicotine (molecular weight = 162.24).

(1)   Sketch nicotine

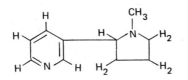

(2)   From Tables 12-6 and 12-12

	ΔT	ΔP	ΔV
1(—CH₃)	0.020	0.227	55
3(—CH₂ ring)	3(0.013)	3(0.184)	3(44.5)
1(—CH ring)	0.012	0.192	46
2(—N ring)	2(0.007)	2(0.13)	2(32)
4(=CH ring)	4(0.011)	4(0.154)	4(37)
1(=C— ring)	0.011	0.154	36
	ΣΔT = 0.140	ΣΔP = 1.871	ΣΔV = 482.5

(3)   From Eq. 12-5

$$\theta = 0.567 + (0.140) - (0.140)^2$$

$$= 0.687$$

(4)   From Eq. 12-18

$$P_c = \frac{162.24}{(0.34 + 1.871)^2}$$

$$= 33.2 \text{ atm}$$

(5)   From Eq. 12-19

$$V_c = 40 + 482.5$$

$$= 522.5$$

(6)   From Eq. 12-17

$$\beta = \frac{[(1-0.687)^{2/7} - 0.048] \ln (522.5) + [1 - 0.687]^{2/7} \ln (33.2) + 1.255}{[1 - 0.687]^{2/7}}$$

$$= 11.09$$

(7)   From Eqs. 12-16 and 12-14

$$T_b = \frac{(0.687) \, e^{11.09}}{82.05}$$

$$= 549K$$

$$Error = +30K$$

The measured boiling point of nicotine is 519.2K; thus, this method shows a relative error of 5.8%.

12-6   METHOD OF OGATA AND TSUCHIDA

Principles of Use. Ogata and Tsuchida [20] found a linear relationship for estimating normal boiling points for compounds of the form RX, where R is a hydrocarbon radical and X is a hydrogen atom or a functional group. The equation takes the form:

$$T_b = py + q \qquad (12\text{-}20)$$

where T_b is the normal boiling point (K), y is a function of the hydrocarbon group(s) contained in the compound, and p and q are constants characteristic of the functional group attached to the hydrocarbon group. For a group of 600 compounds tested by Ogata and Tsuchida, boiling points for 80% deviate by less than 2°C from known values and 98% deviate by less than 5°C.

The method is not without limitations, however: it cannot treat compounds containing more than one functional group, and the range of hydrocarbon fragments for which y values are known is small. For example, it is unsuitable for compounds with condensed or fused rings, such as naphthalene or anthracene, and for dihalogenated compounds. Tables 12-13 and -14 list the values of y, p, and q determined by Ogata and Tsuchida.

TABLE 12-13

Values of y for Equation 12-20

R	y	R	y
Methyl	55.5	t-Amyl	122.0
Ethyl	77.1	Neopentyl	125.0
n-Propyl	102.0	n-Hexyl	171.0
Isopropyl	92.0	Isohexyl	168.0
n-Butyl	124.5	n-Heptyl	191.5
sec-Butyl	118.0	n-Octyl	210.0
Isobutyl	116.5	Vinyl	71.0
t-Butyl	96.0	Allyl	104.0
n-Amyl	149.0	2-Butenyl	127.0
Isoamyl	140.5	Phenyl	197.0

Source: Ogata and Tsuchida [20]. *(Reprinted with permission from the American Chemical Society.)*

Basic Steps

(1) Determine that the compound consists of one or more hydrocarbon radicals and no more than one functional group.

(2) Determine the value of y from Table 12-13, using the largest hydrocarbon radical for R. (If this is not possible, use a different method.)

(3) Determine values of p and q from Table 12-14.

(4) Compute T_b by substituting y, p, and q values in Eq. 12-20.

Example 12-8   Estimate the boiling point of diphenyl sulfide.

(1)   For the phenyl group, from Table 12-13,

$$y = 197.0$$

(2)   For R–S–R, from Table 12-14,

$$p = 1.937, q = 214.4$$

(3)   Using Eq. 12-20,

$$T_b = 1.937 (197.0) + 214.4 = 596.0K$$

TABLE 12-14

Values of Parameters p and q for Equation 12-20

RX	p	q	R Groups with Deviations >5K	RX	p	q	R Groups with Deviations >5K
RH	1.615	63.8	Me, t-Bu	HCOR	1.140	233.8	
RCl	1.348	179.7		MeCOR	1.022	270.6	
RBr	1.260	213.6		EtCOR	0.918	302.2	
RI	1.198	253.4		RCN	0.960	292.2	
				RCOCl	1.040	267.9	
ROH	0.096	277.6	Me, t-Bu	HCOOR	1.073	244.6	
MeOR	1.217	191.2	Me				
				MeCOOR	1.000	273.2	
EtOR	1.137	221.8		EtCOOR	0.963	297.5	
ROR	2.158	143.2	Me, Hep	PhCOOR	0.766	425.9	
				RCOOH	0.903	342.4	
PhOR	0.849	377.4		RCOOMe	1.000	273.2	
RONO$_2$	1.016	280.5		RCOOEt	0.963	297.5	
RSH	1.191	221.0		RCOOPr	0.911	323.4	
				RCOOPh	0.766	425.9	
RSMe	1.146	249.2	Me				
RSEt	1.080	280.0		(RCO)$_2$O	1.286	337.7	Hep
RSR	1.937	214.4	Me, Hep	ClCH$_2$COOR	0.721	359.6	
				Cl$_2$CHCOOR	0.745	372.3	
RNH$_2$	1.194	201.4		BrCH$_2$COOR	0.745	374.4	
RNHMe	1.180	215.2		NCCH$_2$COOR	0.565	433.5	
RNHEt	1.081	247.9		CH$_2$=CHCOOR	0.918	302.2	
RNHPr	0.991	282.8					
RNMe$_2$	1.193	218.7	Me				
RNO$_2$	0.923	308.8	Me, Et				

Source: Ogata and Tsuchida [20]. *(Reprinted with permission from the American Chemical Society.)*

$$\text{Error} = 26.8\text{K}$$

The measured boiling point of diphenyl sulfide is 569.2K; thus, this method shows a relative error of 4.7%.

Example 12-9   Estimate the boiling point of ethylene.

(1)   For CH$_2$=CH–, Table 12-13 lists

$$y = 71.0$$

    (2)   For RH, Table 12-14 lists

$$p = 1.615, q = 63.8$$

    (3) $T_b = 1.615 \, (71.0) + 63.8 = 178.5K$

$$\text{Error} = \quad 9.1K$$

The measured boiling point of ethylene is 169.4K; thus, this method shows a relative error of 5.4%.

12-7 METHOD OF SOMAYAJULU AND PALIT

Principles of Use. Somayajulu and Palit [25] developed a method for estimating the boiling points of normal hydrocarbons that contain no more than one functional group. Their method is a three-parameter fitting of data for homologous series of compounds with respect to the sum of the atomic numbers for individual atoms in the compound. Equation 12-21 describes the correlation:

$$T_b = r \, (\Sigma Z)^t + s \tag{12-21}$$

where T_b is the normal boiling point (K), ΣZ is the sum of the atomic numbers for the individual atoms in the compound, and r, s, and t are parameters dependent upon the homologous series. Table 12-15 lists atomic numbers for some elements commonly found in organic compounds, and Table 12-16 lists values for r, s, and t. Boiling points estimated by this method are about as accurate as those calculated by the method of Ogata and Tsuchida.

Basic Steps

(1) Obtain ΣZ by summing the atomic numbers of all the atoms in the compound.

(2) Obtain values for r, s, and t from Table 12-16.

(3) Compute T_b from Eq. 12-21.

Example 12-10   Estimate the normal boiling point of acetaldehyde and acetic acid.

- For acetaldehyde (CH_3CHO):

    (1) $\Sigma Z = 2 \, (6) + 4 \, (1) + 8 = 24$

    (2)   From Table 12-16, r = 48.87, s = 36.85, t = 1/2

TABLE 12-15

Atomic Numbers of Some Common Elements

Element	At. No.	Element	At. No.
Br	35	I	53
C	6	N	7
Cl	17	O	8
F	9	P	15
H	1	S	16

TABLE 12-16

Values of r, s, and t for Equation 12-21

Homologous Series	r	s	t
n-Alkyl fluorides	51.31	−26.70	½
n-Alkyl chlorides	52.06	−17.05	½
n-Alkyl bromides	57.54	−102.5	½
n-Alkyl iodides	61.10	−164.0	½
n-Alkyl aldehydes	48.87	36.85	½
n-Alkyl ketones	44.11	67.20	½
n-Alkyl acids	36.17	188.0	½
n-Alkyl primary amines	45.43	55.56	½
n-Alkyl secondary amines	46.52	28.10	½
n-Alkyl tertiary amines	47.95	−5.05	½
n-Alkyl benzenes	46.88	52.57	½
n-Alkyl cyclohexanes	49.54	9.16	½
n-Alkyl cyclohex-1-enes	47.24	34.10	½
n-Alkanes	154.40	−223.90	1/3
n-Alk-1-enes	154.45	−222.10	1/3
n-Alk-1-ynes	144.44	−168.78	1/3
Normal alcohols	2.44	288.0	1
n-Alkyl ethers	3.58	157.10	1
n-Alkyl acetates	2.76	220.80	1
n-Alkanethiols	3.87	178.0	1

Source: Somayajulu and Palit [25]. *(Reprinted with permission from the Royal Society of Chemistry, London.)*

(3) From Eq. 12-21, $T_b = 48.87 \ (24)^{1/2} + 36.85 = 276.3K$

$$\text{Error} = -17.7K$$

The measured boiling point of acetaldehyde is 294.0K; thus, this method shows a relative error of −6.0%.

- For acetic acid (CH_3COOH):

    (1) $\Sigma Z = 2 \ (6) + 4 \ (1) + 2 \ (8) = 12$

    (2) From Table 12-16, r = 36.17, s = 188.0 , t = 1/2

    (3) From Eq. 12-21, $T_b = 36.17 \ (32)^{1/2} + 188.0 = 392.6K$

$$\text{Error} = 1.6K \ _\backslash$$

The measured boiling point of acetic acid is 391.0K; thus, this method shows a relative error of 0.4%.

12-8 KINNEY'S METHOD

Principles of Use. Kinney's method [11, 12] can be used to estimate the boiling points of aliphatic hydrocarbons, olefins, and naphthenes. The results deviate from measured values by less than 10K for many compounds tested. In this method, the boiling point is considered to be a function of the "boiling-point number," Y:

$$T_b = 230.1 \ Y^{1/3} - 270 \qquad (12\text{-}22)$$

where T_b is the boiling point (K). The value of Y is determined by adding contributions from the structural groups of the molecule as listed in Table 12-17.

Some care must be exercised in the sequence of contributions added, since the individual contributions to Y are not commutative. The longest chain in the molecule is designated as the main chain or base group for the molecule. The contributions from atoms in the base group are totaled first. Contributions from branching groups are then totaled, followed by olefinic groups and rings. (Rings are always considered to be side groups attached to the base group; for cycloalkanes, such as cyclohexane, the ring is a side group attached to the base group, which is a hydrogen.) An additional consideration is that if two equal chain lengths are possible choices for the base group, the one containing the fewer unsaturated linkages is used.

TABLE 12-17

Atomic and Group Boiling Point Numbers

Structural Group	ΔY	Structural Group	ΔY
Carbon in main chain	0.8	Diolefins:	
Hydrogen attached to		Allenes	4.8
main chain	1.0	Conjugated	a + 0.8
		Not conjugated	a
Radicals, saturated,			
attached to main chain		Triolefins:	
or to cyclic rings:		All bonds conjugated	a + 2.4
Methyl	3.05	Two bonds conjugated	a + 0.8
Ethyl	5.5	No conjugation	a
Propyl	7.0		
Butyl	9.7	Diacetylenes:	
		1,3-Diacetylenes	b
2,2-Dimethyl grouping		All other conjugated	b + 3.0
(in addition to value		No conjugation	b
for each methyl)	−0.4		
		Enynes:	
Multiple alkyls attached		Conjugated	c + 0.8
to adjacent carbons of		No conjugation	c
saturated main chains			
of six carbons or less:		Dienynes:	
2-3 alkyls	0.5	Conjugated	c + 2.4
4 or more alkyls	1.0	No conjugation	c
Olefinic linkages:		Cyclic radicals:	
$CH_2=CH_2$	1.2	Each carbon	0.8
$RCH=CH_2$	1.5	Each hydrogen	1.0
$RCH=CHR$	1.9	Unsaturated linkages	c
$R_2C=CHR$	2.3	Ring:	
$R_2C=CR_2$	2.8	Cyclopropyl	2.1
		Cyclobutyl	2.3
Radicals, unsaturated,		Cyclopentyl	2.5
attached to main chain:		Cyclohexyl	2.7
Methylene	4.4	Cycloheptyl	3.4
Ethylidene	7.0	Cyclooctyl	3.9
Vinyl	5.4	Cyclononyl	4.4
Propylidene	9.0	Cyclodecyl	4.9
Butylidene	10.4	Cyclohendecyl	5.4
		Cyclododecyl	5.9
Acetylenic linkages:		Cyclotridecyl	6.4
HC≡CH	4.0	Cyclotetradecyl	6.9
RC≡CH	4.4	Cyclopentadecyl	7.4
RC≡CCH$_3$	5.4	Cyclohexadecyl	7.9
RC≡CR	4.8	Cycloheptadecyl	8.4

a. Normal value of double bonds.
b. Normal value of triple bonds.
c. Normal value of bonds.

Source: Kinney [11,12]. *(Reprinted with permission from the American Chemical Society.)*

Basic Steps

(1) Draw the compound structure.

(2) Determine the base group or main chain; this is the longest chain in the compound or, if two equal length chains are possible, the one with the fewer unsaturated linkages.

(3) Add the contributions to Y from the main chain, using Table 12-17.

(4) Add the contributions to Y for any branching groups.

(5) Add the contributions to Y for any unsaturated linkages and/or rings.

(6) Compute T_b from Eq. 12-22.

Example 12-11   Estimate the boiling point of isooctane.

(1)   Draw the structure

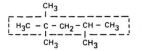

(2)   Determine the main chain (dashed box).

(3)   From Table 12-17, the contributions from the main chain are

$$Y_{main} = 5C + 9H$$

$$= 5\,(0.8) + 9\,(1.0) = 13.0$$

(4)   Contributions from the side chains are

$$Y\,(2,2\text{-dimethyl}) = 2\,(3.05) - 0.4 = 5.7$$

$$Y\,(4\text{-methyl}) = 3.05$$

$$\text{Total } Y_{branch} = 3.05 + 5.7 = 8.75$$

(5) $Y_{total} = Y_{main} + Y_{branch}$

$$= 13.0 + 8.75 = 21.8$$

(6)   From Eq. 12-22,

$$T_b = 230.1\,(21.8)^{1/3} - 270 = 373K$$

$$\text{Error} = 1K$$

The measured boiling point of isooctane is 372.4K; thus, this method shows a relative error of 0.3%.

Example 12-12   Estimate the boiling point of cyclohexene.

(1)   Draw the structure

(2)   The base group is H.

(3) $Y_{base} = 1.0$

(4) $Y_{ring} = 6$ (C) + 9 (H) + cyclohexyl group

$$= 6 (0.8) + 9 (1.0) + 2.7 = 16.5$$

$$Y_{double\ bond} = 1.9$$

(5) $Y_{total} = Y_{base} + Y_{ring} + Y_{double\ bond}$

$$= 1.0 + 16.5 + 1.9 = 19.4$$

(6) $T_b = 230.1\ (19.4)^{1/3} - 270 = 348\,K$

$$Error = -8\,K$$

The measured boiling point of cyclohexene is 356.1K; thus, this method shows a relative error of −2.2%.

12-9   METHOD OF STIEL AND THODOS

Principles of Use. The method of Stiel and Thodos [26] is applicable only to saturated aliphatic hydrocarbons. Their method involves the use of two equations. One of these (Eq. 12-23) is used to estimate the boiling point of normal hydrocarbons:

$$T_{b_n} = 1209 - \frac{1163}{1 + 0.0742\ (N^{0.85})} \qquad (12\text{-}23)$$

where T_{b_n} is the boiling point (K) and N is the number of carbons in the compound. The second equation is used to determine the boiling point of isomeric (branched) saturated hydrocarbons:

$$T_{b_{isomer}} = T_{b_n} - \left[\left(\frac{96.52}{N^2} \right)(W_n - W_i) + 5.45\ (P_n - P_i) \right] \qquad (12\text{-}24)$$

The terms W_n and W_i are the Wiener numbers of the normal hydrocarbon and the isomer respectively, and P_n and P_i are the polarity numbers for the normal hydrocarbon and the isomer respectively.

The Wiener number of a normal hydrocarbon can be computed from the total number of carbons in the molecule by Eq. 12-11 (repeated here for convenience):

$$W_n = \frac{(N-1)\ (N)\ (N+1)}{6} \qquad (12\text{-}11)$$

For an isomer, the Wiener number is computed by multiplying together the number of carbon atoms on each side of a carbon-to-carbon bond, repeating this for all of the carbon-to-carbon bonds in the molecule, and then summing the products.

The polarity number is defined as the number of pairs of carbon atoms in the molecule that are separated by exactly three bonds. For normal hydrocarbons, the calculation of P_n is exceedingly simple:

$$P_n = N-3 \qquad (12\text{-}25)$$

The basic steps and examples below further explain the calculation of W_i and P_i.

Basic Steps

(1) Determine the number of carbons in the compound, N.

(2) Compute T_{b_n} from Eq. 12-23.

(3) If the molecule is a straight chain, T_{b_n} is the boiling point of the compound.

(4) If the molecule is branched, draw the structure.

(5) Compute W_n and P_n from Eqs. 12-11 and 12-25.

(6) Compute P_i by counting the number of pairs of carbons that are separated by three bonds.

(7) Compute W_i by counting carbons on both sides of a single bond (C-C bonds only). Multiply the numbers together and sum the products for all of the single bonds in the molecule.

(8) Compute $T_{b_{isomer}}$ from Eq. 12-24, using the value of T_b estimated in step 2.

Example 12-13   Estimate the boiling point of n-hexane.

(1) $N = 6$

(2)   From Eq. 12-23,

$$T_{b_n} = 1000 \left[1.209 - \frac{1.163}{1 + 0.0742 \, (6)^{0.85}} \right] = 341K$$

$$\text{Error} = -1K$$

The measured boiling point of n-hexane is 342K; thus, this method shows a relative error of −0.3%.

Example 12-14   Estimate the boiling point of isooctane.

(1) $N = 8$

(2) $T_{b_n} = 1000 \left(1.209 - \dfrac{1.163}{1 + 0.0742 \, (8)^{0.85}} \right) = 398K$

(3)   The structure is

$$\begin{array}{c} CH_3 \\ | \\ H_3C - C - CH_2 - CH - CH_3 \\ | \qquad\qquad | \\ CH_3 \qquad CH_3 \end{array}$$

(4) $W_n = \dfrac{7 \times 8 \times 9}{6} = 84$

$P_n = 8 - 3 = 5$

(5) $P_i = 5$

(6) $W_i = (1 \times 7) + (1 \times 7) + (1 \times 7) + (4 \times 4) + (5 \times 3) + (1 \times 7) + (1 \times 7) = 66$

(7) $T_{b_{isomer}} = 398 - \left[\dfrac{(96.52)}{8^2} \, (84 - 66) + 5.45 \, (5 - 5) \right] = 371K$

$$\text{Error} = -1K$$

The measured boiling point of isooctane is 372.4K; thus, this method shows a relative error of −0.4%.

12-10   FACTORS AFFECTING BOILING POINT [1,17]

Each of the preceding seven methods yields a value for the normal boiling point of a pure compound. If the pressure on the liquid differs from one atmosphere, or if the liquid contains impurities, the boiling point observed for the compound differs from that estimated for the pure compound in a predictable fashion. This section briefly discusses the ways in which an estimated normal boiling point can be adjusted to compensate for impurities and/or pressures other than one atmosphere.

Impurities in a compound alter its observed boiling point by changing the vapor pressure of the liquid. If the impurity is non-volatile, such as sugar in water, and the solution is dilute, the vapor pressure of the liquid obeys Raoult's law:

$$P = X_A P_{vp} \tag{12-26}$$

where X_A is the mole fraction of the solvent (i.e., compound of interest), P is the vapor pressure of the solution, and P_{vp} is the vapor pressure of the pure liquid at $X_A = 1$. If X_B is the mole fraction of the solute, Eq. 12-26 may be rewritten as:

$$P = (1 - X_B) P_{vp} \tag{12-27}$$

or

$$\frac{P_{vp} - P}{P_{vp}} = X_B \tag{12-28}$$

The solution cannot boil until its temperature is sufficiently above the boiling point of the pure solvent to compensate for the pressure decrease noted above. An equation that expresses this concept is obtained by combining the Clausius-Clapeyron equation (see Chapters 13 and 14) with Eq. 12-28 in such a way as to balance the vapor pressure decrease due to the presence of the impurity with an increase in the vapor pressure due to temperature:

$$\Delta T_b = \frac{RT_b^{\,2}}{\Delta H_v} \qquad \Delta X_B = \frac{RT_b^{\,2}}{\Delta H_v} X_B \tag{12-29}$$

where T_b is the normal boiling point temperature, ΔT_b is the change in temperature from the normal boiling point due to the solute, R is the universal gas constant (1.9872 cal/mol-K), and ΔH_v is the enthalpy of vaporization (cal/mol). $\Delta X_B = X_B$, since $X_B = 0$ in a pure compound. To estimate the boiling point of a compound containing impurities, first estimate the normal boiling point as outlined previously; then estimate ΔH_{vap} by the methods presented in Chapter 13. Substitute these values and the mole fraction of solute (X_B) in Eq. 12-29 to compute the boiling point elevation. Finally, add ΔT_b to T_b to obtain the boiling point of the impure material.

For compounds that dissociate, such as sodium chloride in water, the situation is somewhat more complicated. The mole fraction term used throughout this discussion is a measure of the number of particles of non-solvent species in the solution, which depends on the extent of dissociation. If the solute dissociates completely into two ions for each molecule, the number of particles contributing to the boiling point elevation would be twice as many as would exist if there were no dissociation. The degree of dissociation depends on the concentration of the impurity and its dissociation constant. In estimating the boiling points of impure compounds, the impurity is assumed to dissociate completely, and its mole fraction in Eq. 12-29 is expressed accordingly. That is, if the impurity compound dissociates into two ions, its mole fraction (X_B) is

$$X_B = \frac{2M_B}{2M_B + M_A} \qquad (12\text{-}30)$$

where M_B = moles of impurity present before dissociation and M_A = moles of compound whose boiling point is being estimated. M_B is multiplied by 2 because each molecule forms two ions when it dissociates. The assumption of complete dissociation causes little error compared with those entailed in estimating the normal boiling point and the enthalpy of vaporization.

Boiling points at pressures other than one atmosphere can be estimated by using the Clausius-Clapeyron equation in the following form:

$$\ln P - \ln P_b = \frac{\Delta H_v}{R} \left(\frac{1}{T_b} - \frac{1}{T} \right) \qquad (12\text{-}31)$$

where P is the pressure on the actual sample (mm Hg) and P_b is the pressure at the normal boiling point (760 mm Hg). Using known values of P and P_b and estimated values of ΔH_v and T_b in Eq. 12-31, one can calculate the observed boiling point for a compound under pressures other than one atmosphere.

12-11 AVAILABLE DATA

Boiling points are listed in many texts. Users should be careful to note the pressures at which the temperatures were measured. Some general references are listed in Appendix A. The following contain complementary information and are recommended:

Dean, J.A. [5] General data including boiling points for many organic and inorganic compounds.

Reid, R.C., *et al.* [23] Appendix with tabulation of normal boiling points and critical constants for many organic compounds.

Stull, D.R. and Prophet, H. [27] Tabulation of thermochemical properties of inorganic compounds.

Vargaftik, N.B. [31] General data on thermochemical properties of liquids and gases.

Weast, R.C. [33] General data including boiling points for many organic and inorganic compounds.

Windholtz, M. [34] General data including boiling points for chemicals and drugs.

12-12 SYMBOLS USED

a = van der Waal's constant in Eq. 12-6

B = chemical class constant in Eq. 12-1

b = van der Waal's constant in Eq. 12-6

f_a, f_b = correction factors for branched hydrocarbons in Forman-Thodos method, Eqs. 12-7 to 12-10

ΔH_v = enthalpy of vaporization in §12-10 (cal/mol)

K_1, K_2, K_3, K_4 = functional group constants in Eqs. 12-12, -13

m = number of side chains in Eqs. 12-9, -10

M = molecular weight in Eqs. 12-2, 12-3 and 12-18 (g/mol)

M_A = moles of solvent or "pure" compound in Eq. 12-30

M_B = moles of impurity in Eq. 12-30

N = number of carbon atoms in saturated aliphatic hydrocarbon in Eqs. 12-23, -24

n = sum of carbon and functional atoms for Forman-Thodos method of estimating critical temperature

n_D = refractive index measured relative to the sodium D line in Eq. 12-2

P = vapor pressure of solution in §12-10 (mm Hg); also (with appropriate subscript) polarity number in Eqs. 12-24, -25

P_c = critical pressure of a compound (atm) in Eqs. 12-15, -17 and -18

P_{vp} = vapor pressure of pure compound in §12-10 (mm Hg)

$[P]$ = parachor in Eqs. 12-1, -3

ΔP = Lydersen pressure increment in Table 12-12 and Eq. 12-18

p = functional group constant in Eq. 12-20

q = functional group constant in Eq. 12-20

R = universal gas constant = 82.05 atm-cm^3/mol-K in Eq. 12-6 = 1.9872 cal/mol-K in Eq. 12-29

$[R_D]$ = molar refraction measured relative to sodium D line in Eqs. 12-1, -2

r = parameter in Eq. 12-21

s = parameter in Eq. 12-21

T_b = normal boiling point (K)

T_c = critical temperature (K)

ΔT = Lydersen increment for computing θ in Eq. 12-5

t = parameter in Eq. 12-21

V_b = saturated liquid molal volume at the boiling point in Eq. 12-14 (cm^3/g-mole)

V_c = critical volume of a compound (cm^3/mol) in Eqs. 12-14, -17 and -19

ΔV = Lydersen volume increment in Table 12-12 and Eq. 12-19

W = Wiener number

X_A = mole fraction of solvent in solution

X_B = mole fraction of impurity in solution

Y = boiling point number in Eq. 12-22

y = parameter in Eq. 12-20

Z = atomic number in Eq. 12-21

Z_c = critical compressibility factor in Eq. 12-15

Greek:

β = parameter in Eqs. 12-16 and 12-17

σ = surface tension in Eq. 12-3 (dynes/cm)

θ = ratio T_b/T_c in Eqs. 12-4, 12-5, 12-15 and 12-17

ρ_L = liquid density in Eqs. 12-2, -3 (g/mL)

ρ_V = vapor density in Eq. 12-3 (g/mL)

Subscripts:

i   = isomeric alkane

n   = normal alkane

12-13   REFERENCES

1. Barrow, G.M., *Physical Chemistry,* McGraw-Hill Book Co., New York (1966).

2. Burnop, V.C.E., "Boiling Point and Chemical Composition. Part I: An Additive Function of Molecular Weight and Boiling Point," *J. Chem. Soc.,* **1938,** 826-29.

3. Costello, J.M. and S.T. Bowden, "The Parachor, Molar Volume, and Critical Constants," *Chem. Ind. (London),* **1956,** 1041-45.

4. Daniels, F.J., W. Williams, P. Bender, R.A. Alberty and C.D. Cornwell, *Experimental Physical Chemistry,* 6th ed., McGraw-Hill Book Co., New York (1962).

5. Dean, J.A. (ed.), *Lange's Handbook of Chemistry,* 12th ed., McGraw-Hill Book Co., New York (1979).

6. Forman, J.C. and G. Thodos, "Critical Temperatures and Pressures of Hydrocarbons," *AIChE J.,* **4,** 356-61 (1958).

7. Forman, J.C. and G. Thodos, "Critical Temperatures and Pressures of Organic Compounds," *AIChE J.,* **6,** 206-9 (1960).

8. Gambill, W.R., "Predict Critical Temperature," *Chem. Eng. (NY),* **66,** 181-4, 1959.

9. Hall, L.H., L.B. Kier and W.J. Murray, "Molecular Connectivity II: Relationship to Water Solubility and Boiling Point," *J. Pharm. Sci.,* **64,** 1974-77 (1975).

10. Kier, L.B. and L.H. Hall, *Molecular Connectivity in Chemistry and Drug Research,* Academic Press, New York (1976).

11. Kinney, C.R., "A System Correlating Molecular Structure of Organic Compounds with Their Boiling Points. I: Aliphatic Boiling Point Numbers," *J. Am. Chem. Soc.,* **60,** 3032-39 (1938).

12. Kinney, C.R., "Calculation of Boiling Points of Aliphatic Hydrocarbons," *Ind. Eng. Chem.,* **32,** 559-62 (1940).

13. McGowan, J.C., "Molecular Volumes and the Periodic Table," *Chem. Ind. (London),* **1952,** 495-6.

14. Meissner, H.P., "Critical Constants from Parachor and Molar Refraction," *Chem. Eng. Prog.,* **45,** 149-53 (1949).

15. Michels, W.C. (ed.), *The International Dictionary of Physics and Electronics,* D. van Nostrand Co., Princeton, N.J. (1956).

16. Miller, C.O.M., Dow Chemical Co., Pittsburg, CA, private communication to Robert C. Reid, Massachusetts Institute of Technology, Cambridge, MA. The essence of this communication was provided by Prof. Reid to this author (C. Rechsteiner) in a personal communication, June 2, 1980.

17. Moore, W.J., *Physical Chemistry,* 3rd ed., Prentice-Hall, Englewood Cliffs, N.J. (1962).

18. Mumford, S.A. and J.W.C. Phillips, "Observations on the Chlorination Products of β,β-Dichlorodiethyl Sulphide," *J. Chem. Soc.,* **1928,** 155-62.

19. Mumford, S.A. and J.W.C. Phillips, "The Evaluation and Interpretation of Parachors," *J. Chem. Soc.,* **1929,** 2112-33.

20. Ogata, Y. and M. Tsuchida, "Linear Boiling Point Relationships," *Ind. Eng. Chem.,* **49,** 415-17 (1957).

21. Quayle, O.R., "The Parachors of Organic Compounds," *Chem. Rev.,* **53,** 439-589 (1953).

22. Rackett, H.G., "Equation of State for Saturated Liquids," *J. Chem. Eng. Data,* **15,** 514-17 (1970).

23. Reid, R.C., J.M. Prausnitz and T.K. Sherwood, *The Properties of Gases and Liquids,* 3rd ed., McGraw-Hill Book Co., New York (1977).

24. Reid, R.C. and T.K. Sherwood, *The Properties of Gases and Liquids — Their Estimation and Correlation,* 2nd ed., McGraw-Hill Book Co., New York (1966).

25. Somayajulu, G.R. and S.R. Palit, "Boiling Points of Homologous Liquids," *J. Chem. Soc.,* **1957,** 2540-44.

26. Stiel, L.I. and G. Thodos, "The Normal Boiling Points and Critical Constants of Saturated Aliphatic Hydrocarbons," *AIChE J.,* **8,** 527-29 (1962).

27. Stull, D.R. and H. Prophet, *JANAF Thermochemical Tables,* 2nd ed., U.S. Government Printing Office, Washington, D.C., Catalog No. C 13.48:37, Stock No. 0303-0872.

28. Thodos, G., "Critical Constants of the Naphthenic Hydrocarbons," *AIChE J.,* **2,** 508-13 (1956).

29. Thodos, G., "Critical Constants of the Aromatic Hydrocarbons," *AIChE J.,* **3,** 428-31 (1957).

30. Tyn, M.T. and W.F. Calus, "Estimating Liquid Molal Volume," *Processing,* 21(4), 16-17 (1975).

31. Vargaftik, N.B., *Tables on the Thermophysical Properties of Liquids and Gases,* Hemisphere Publishing Corp., Washington, D.C. (1975).

32. Watson, K.M., "Prediction of Critical Temperatures and Heats of Vaporization," *Ind. Eng. Chem.,* **23,** 360-64 (1931).

33. Weast, R.C. (ed.), *CRC Handbook of Chemistry and Physics,* 51st ed., Chemical Rubber Co., Cleveland, Ohio (1970).

34. Windholtz, M. (ed.), *The Merck Index,* 9th ed., Merck & Co., Rahway, N.J. (1976).

13

HEAT OF VAPORIZATION

Carl E. Rechsteiner, Jr.

13-1   INTRODUCTION

The heat of vaporization, ΔH_v (also called the latent heat of vaporization, heat of evaporation, or enthalpy of vaporization) is defined as the quantity of heat required to convert a unit mass of liquid into a vapor without a rise in temperature, which implies a constant-pressure process [1]. It is a function of temperature and decreases with increasing temperature; values have been tabulated for many compounds [6,23,34]. The ability to estimate the heat of vaporization is helpful for checking values reported in the literature, but its most important use is in estimating other physico-chemical properties. The parameter is also required in some chemical spill models.

Since the early 1900s, the preferred method for measuring ΔH_v has been to determine the quantity of electrical energy needed to vaporize a unit mass of liquid at a constant temperature. Values obtained with this method are highly reproducible and may be very accurate if all heat losses from the experimental apparatus are accounted for. Prior to 1900, heats of vaporization were determined by measuring the heat of condensation of the vapor; this routinely gave lower values, sometimes exceeding 10%, than those obtained with the present method. Unfortunately, some values obtained by the heat of condensation method are still included in some reference tabulations [8].

The heat of vaporization is a basic thermodynamic quantity and, as such, has theoretical importance. The entropy of vaporization at any temperature may be determined from the ratio of the heat of vaporization at that temperature to the temperature. The vapor pressure of a material at some temperature is frequently estimated by use of the Clapeyron equation or one of its rearrangements, all of which use ΔH_v as an input parameter.

13-2 AVAILABLE ESTIMATION METHODS

Many equations have been proposed for estimating heats of vaporization of organic compounds. Although the methods proposed range from those based on theory to those based on empirical factors, most incorporate a mixture of the two approaches. The most frequent starting point for the proposed relations is some form of the Clapeyron equation, which is modified as necessary and then fit to a set of experimental data to determine the best-fit parameters.

One of the earliest estimation methods for the heat of vaporization of organic compounds was developed by F. Trouton (ca. 1884), who noted that the ratio between the heat of vaporization and the normal boiling point is essentially a constant: 21 cal/deg-mol.

Three authors — Bakker, Batschinsky, and Partington — have proposed relationships for calculating heats of vaporization from critical data using strictly theoretical considerations [20]. Although these methods have shown some success for limited groups of compounds, Fishtine [7] examined the proposed methods with a number of common chemicals and frequently encountered errors of 30%-40%.

Many authors have proposed semiempirical equations for predicting heats of vaporization from critical constants. Haggenmacher [11], Viswanath and Kuloor [28], Meissner [15], Watson [31], Klein [14a], and Giacalone (in Ref. 21) developed correlations based on the Clapeyron equation modified by the compressibility-factor equation of state. The differences among them reflect varying assumptions with respect to the difference between the vapor and liquid compressibilities and the relationship between the temperature and the vapor pressure of the compound. Riedel [25], Lydersen (in Ref. 24), Chen [4], and Pitzer [22] have proposed methods based on Van der Waal's theory of corresponding states; all include a correlating parameter to improve estimation accuracy.

Another group of methods is based on the Thiessen relation, which describes ΔH_v as a function of the reduced temperature (the ratio between the measured temperature and the critical temperature). A number of researchers, such as Watson [32], Nutting [18], Winter [33], and Bowden [2], utilize the ratio of this relation at two temperatures to estimate ΔH_v at one temperature from its value at the other temperature. These methods differ in the value of the exponent in the Thiessen equation. Silverberg and Wenzel [27a] computed the Thiessen exponent for a number of compounds for a range of temperatures below the critical temperature. Although the value ranged from 0.237 to 0.389, the average (0.378) differs by less than 1% from the 0.38 proposed by Watson [32].

Other workers have included Thiessen's equation in developing the "lyoparachor," a temperature-independent additive property of structure analogous to the parachor. Winter [33], Bowden and Jones [3], and Chu [5] have been involved in this effort. Sastri *et al.* [26] have incorporated many of these previous researchers' ideas into a generalized procedure that requires only the compound structure as input for estimating the heat of vaporization at the normal boiling point.

Other attempts have been made to correlate the heat of vaporization with compound structure or structure-related features. Kistiakowsky (in Ref. 10) developed a correlation between ΔH_v and the normal boiling point for nonpolar compounds. Fishtine [10] extended the range of applicability of this equation to other compounds by introducing a constant based on a function of dipole moment. The Kistiakowsky method is described in detail in §14-3 as part of Method 1 for the estimation of vapor pressure. The specific relationship for estimating ΔH_{vb} by the Kistiakowsky method is found in Eq. 14-16 with supporting data in Table 14-4. The accuracy of this method is comparable to those listed in Table 13-2.

Narsimhan [16,17] developed correlations for alcohols, phenols, and fatty acid esters based on their molecular weight, vapor and liquid densities, and molar volume at the critical point. Graph-theory approaches, such as that of Kier and Hall [13] or Ogden and Lielmezs [19], have also been proposed.

The methods recommended in this chapter were chosen for their general accuracy (typically 2% error or less) and ease of calculation. Their input parameters are frequently available in the literature or may be readily estimated.

13-3   SELECTION OF APPROPRIATE METHOD

The methods recommended for estimating heats of vaporization are listed in Table 13-1 along with their theoretical bases, required input parameters, and application ranges. Table 13-2 compares the accuracies of the methods, using various literature references and known values of the required input parameters, such as T_b, T_c, and P_c. All of these methods show average deviations from literature values of less than 5%. If input parameters must be estimated, the uncertainty of ΔH_v may, of course, be higher. Typical errors in estimating T_b and T_c by the methods of Chapter 12 are about 2% to 4%; estimates of P_c usually deviate by about 4% to 8% from measured values.[1]

All methods recommended for predicting heats of vaporization require similar input data. When the boiling point, T_b, the critical temperature, T_c, and the critical pressure, P_c, are known or readily estimated, use the methods described in §13-4. If either P_c or T_c is not available, use the Haggenmacher method (§13-5); in addition, if no vapor pressure data are available, use the Sastri method (§13-6). To estimate heats of vaporization at temperatures other than the boiling point, use the Watson method (§13-7), calculating the heat of vaporization at the boiling point, ΔH_{vb}, by one of the other recommended methods. More reliable estimates will be obtained if ΔH_{vb} is calculated by several of the recommended methods, but this is not always possible.

13-4   ESTIMATION OF ΔH_{vb} FROM CRITICAL CONSTANTS

Principles of Use. A number of methods based on critical temperature and pressure have been proposed for estimating ΔH_{vb}, the heat of vaporization at the normal boiling point. Four of them appear to have the widest applicability and generally estimate ΔH_{vb} within 2% with maximum errors of about 6%. Two of the methods, the modified Klein and the Giacalone, are based on the Clapeyron equation; the other two methods, those of Riedel and Chen, were derived empirically.

The Clapeyron equation (13-1) is the basic expression for describing the shape of the vapor pressure curve for any material as a function of the temperature and certain properties of the material:

$$\frac{dP}{dT} = \frac{\Delta H_v}{T(V_g - V_l)} \qquad (13\text{-}1)$$

---

1. The propagation of errors in the estimation of physico-chemical properties is discussed in Appendix C.

TABLE 13-1

Summary of Recommended Methods for Estimating Heats of Vaporization

Section	Method	Basis	Input Parameters[a]	Application Range	Comments
13-4	Modified Klein [8]	Clapeyron equation	T_b, T_c, P_c	All compounds at T_b	Constant in equation is a function of T_b
	Giacalone [8,21]	Clapeyron equation	T_b, T_c, P_c	Nonhydrocarbon, polar organics at T_b	
	Riedel [25]	Clapeyron equation	T_b, T_c, P_c	Nonpolar and slightly polar organics at T_b	
	Chen [4]	Clapeyron equation, empirical fit	T_b, T_c, P_c	Entire liquid range if vapor pressure at T of interest is known	Primarily used at T_b
13-5	Haggenmacher [9,11]	Clapeyron equation plus Antoine's equation	T_b, T_c, P_c, Antoine's constants B and C	$T \ll T_b$	Antoine's constants B and C may be estimated. ΔH_v can be computed if T_c and P_c unknown.
13-6	Sastri [26]	Lyoparachor	Compound structure	at T_b	
13-7	Watson [31]	Thiessen's equation	T_c and ΔH_v at some temperature	Entire liquid range	

a. T_b, T_c may be estimated by methods of Chapter 12 if unknown. P_c may be estimated, if unknown, by methods provided in §13-4 of this chapter.

TABLE 13-2

Accuracy of Estimation Methods for Heat of Vaporization
(Average percent error)

Literature Reference[a]	Method[b]						
	Modified Klein [14a]	Giacalone [9,17]	Riedel [25]	Chen [4]	Haggenmacher [11,12]	Sastri [26]	Watson [31,32]
Fishtine [7] @ T_b							
Hydrocarbons (30)	0.70	1.43	0.75				
Nonhydrocarbons (51)	1.64	1.68	1.99				
Inorganics (12)	0.45	2.43	1.48				
All compounds (93)	1.29	1.70	1.52				
Chen [4]							
All compounds (165) @ T_b	1.85	2.40	2.02	1.82			
All compounds (12) $T \neq T_b$				2.55			2.18
Haggenmacher [12]							
Hydrocarbons (17) at several temperatures					~1.0		
Viswanath and Kuloor [29] @ T_b							
All compounds (51)		2.20	2.27				
Sastri et al. [26] @ T_b							
Hydrocarbons (42)	1.10	2.30	1.10	0.80		1.00	
Nonhydrocarbons (33)	2.10	2.10	2.10	2.10		2.40	
All compounds (75)	1.60	2.20	1.50	1.40		1.60	

a. Values in parentheses are number of compounds tested.
b. Input parameters for all estimations were known.

where

dP/dT = derivative of the vapor pressure with respect to temperature

ΔH_v = heat of vaporization at temperature T

V_g = saturated molal volume of the vapor phase

V_l = saturated molal volume of the liquid phase

The quantity (V_g-V_l) can be obtained from the compressibility-factor equation of state [1]:

$$V_g - V_l = \frac{RT}{P} (Z_g - Z_l) \qquad (13\text{-}2)$$

where R is the universal gas constant and Z_g and Z_l are the compressibility factors for the vapor and the liquid phases respectively. Substitution of Eq. 13-2 into the Clapeyron equation and rearrangement yields the following:

$$\frac{d\,(\ln P)}{d\,(1/T)} = \frac{-\Delta H_v}{R\,(Z_g - Z_l)} \qquad (13\text{-}3)$$

This equation is frequently the starting point for methods of estimating the heat of vaporization of a material or predicting the shape of its vapor pressure curve.

The Klein method utilizes T_b, T_c, and P_c to estimate heats of vaporization. Starting with Eq. 13-3 and assuming that the right side of this equation is constant, the Klein equation is obtained by integrating Eq. 13-3 between the limits of T_b, 1 and T_c, P_c. When he found that this method often underestimated ΔH_v, Fishtine [8] introduced an empirical correction factor, K_{kl}, whose value is a function of T_b (see Table 13-3). The modified Klein equation that resulted is

$$\Delta H_{vb} = R\,K_{kl}\,T_b\,\frac{(\ln P_c)\,\sqrt{1 - 1/[(P_c)\,(T_b/T_c)^3\,]}}{1 - (T_b/T_c)} \qquad (13\text{-}4)$$

where R is 1.9872 cal/K-mol, ΔH_{vb} is in units of cal/mol, P_c is in atmospheres, and T_b and T_c are in K. The square root term represents an estimate of the quantity $Z_g - Z_l$ in Eq. 13-3.

The Giacalone method was derived from Eq. 13-3 in a similar manner. Rather than estimating a value for the compressibility differences

TABLE 13-3

Values for Klein Constant

T_b (K)	K_{kl}
< 200	1.02
200-300	1.04
> 300	1.045

Source: Fishtine [8]. *(Reprinted with permission from the American Chemical Society.)*

$Z_g - Z_l$, Giacalone proposed use of the value 1. This is a reasonable approximation for values near the boiling point: $Z_g - Z_l$ ranges from 0.95 at T_b to 1.00 at 0.74 T_b. With this assumption, Eq. 13-3 becomes:

$$\Delta H_{vb} = \frac{R\,T_b\,\ln(P_c)}{1 - (T_b/T_c)} \qquad (13\text{-}5)$$

This equation is most accurate for nonhydrocarbons and polar organic compounds, but it can be used for many kinds of compounds.

The Riedel equation is an empirically derived relation for estimating ΔH_{vb} for nonpolar or slightly polar organic compounds. Riedel started with an equation similar to Eq. 13-5 and performed a two-parameter fit to data from nonpolar and slightly polar organic compounds to obtain:

$$\Delta H_{vb} = \frac{T_b\,(5\,\log P_c - 2.17)}{0.930 - (T_b/T_c)} \qquad (13\text{-}6)$$

where 0.930 and 2.17 are the best-fit values from the parameter fitting. As with the Giacalone method, the Riedel method can give reasonably accurate results for compound types other than those for which it was developed.

The final method that uses T_b, T_c, and P_c to estimate ΔH_{vb} is the empirical method of Chen, who graphically and numerically correlated the heat of vaporization with these three parameters using Pitzer's correlation [22]. (Pitzer tabulated sets of values for predicting the entropy of vaporization at reduced temperatures.) By applying thermodynamic considerations to the work of Pitzer, Chen obtained a linear equation relating the logarithm of the reduced pressure to the ratio between the heat of vaporization and temperature. When this equation was subjected to linear regression analysis for heats of vaporization at the normal boiling point, the following result was obtained:[2]

$$\Delta H_{vb} = \frac{T_b \, [7.11 \log P_c - 7.82 + 7.9 \, (T_b/T_c)]}{1.07 - (T_b/T_c)} \tag{13-7}$$

To estimate the heat of vaporization at the normal boiling point by any of the preceding four equations, one must have either measured or estimated values of T_b, T_c, and P_c. Methods for estimating T_b and T_c are described in Chapter 12. The critical pressure, P_c, may be estimated by methods analogous to those given in §12-4 of that chapter.

Lydersen proposed a simple additive atomic group contribution method for estimating P_c with the following equation:

$$P_c = \frac{M}{(0.34 + \Sigma \Delta P)^2} \tag{13-8}$$

where P_c is in atmospheres and M is the molecular weight in grams per mole. $\Sigma \Delta P$ is a dimensionless number found by summing the increments listed in Table 13-4 for each atomic and structural feature of the compound of interest.

Another method for estimating P_c, proposed by Forman and Thodos, is based on correlation of P_c with Van der Waal's constants "a" and "b." These parameters, which can be estimated by the procedures described in §12-4 of Chapter 12, are substituted into Eq. 13-9 to determine the value of P_c in atmospheres.

$$P_c = \frac{a}{27 \, b^2} \tag{13-9}$$

Basic Steps

(1) Obtain T_b and T_c from the literature, or estimate them by methods in Chapter 12.

(2) Obtain P_c from literature, or estimate it using Eq. 13-8 or 13-9.

(3) Find H_{vb} by substituting the above values in one of the following equations:

- Eq. 13-4 (generally applicable)
- Eq. 13-5 (nonhydrocarbon polar organic compounds)

---

2. To use Chen's method for $T \neq T_b$, substitute T for T_b and P_c/P for P_c.

TABLE 13-4

Lydersen's Critical Pressure Increments

- There are no increments for hydrogen.
- All bonds shown as free are connected to atoms other than hydrogen.
- Values in parentheses may be subject to large errors.

Atomic Group	ΔP	Atomic Group	ΔP
Nonring Increments		**Oxygen Increments (cont.)**	
$-CH_3$	0.227	$-O-$ (nonring)	0.16
$-CH_2$	0.227	$-O-$ (ring)	(0.12)
$-CH$	0.210	$-C=O$ (nonring)	0.29
$\diagdown C \diagup$	0.210	$-C=O$ (ring)	(0.2)
$=CH_2$	0.198	$HC=O$ (aldehyde)	0.33
$=C-$	0.198	$-COOH$ (acid)	(0.4)
$=C=$	0.198	$-COO-$ (ester)	0.47
$\equiv CH$	0.153	$=O$ (except for above combinations)	(0.12)
$\equiv C-$	0.153		
Ring Increments		**Nitrogen Increments**	
$-CH_2-$	0.184	$-NH$	0.095
$-CH$	0.192	$-NH$ (nonring)	0.135
$\diagdown C \diagup$	(0.154)	$-NH$ (ring)	(0.09)
$=CH$	0.154	$-N-$ (nonring)	0.17
$=C-$	0.154	$-N-$ (ring)	(0.13)
$=C=$	0.154	$-CN$	(0.36)
Halogen Increments		$-NO_2$	(0.42)
$-F$	0.224	**Sulfur Increments**	
$-Cl$	0.320	$-SH$	0.27
$-Br$	(0.50)	$-S-$ (nonring)	0.27
$-I$	(0.83)	$-S-$ (ring)	(0.24)
Oxygen Increments		$=S$	(0.24)
$-OH$ (alcohols)	0.06	**Miscellaneous**	
$-OH$ (phenols)	(−0.02)	$-Si-$	(0.54)

Source: Lydersen [14b]; also tabulated in Reid and Sherwood [24]. *(Reprinted with permission from McGraw-Hill Book Co.)*

- Eq. 13-6 (nonpolar or slightly polar organic compounds)
- Eq. 13-7 (generally applicable)

Example 13-1   Estimate ΔH_{vb} for anthracene, assuming that no measured values of the input parameters are available.

(1)   Using the methods described in §12-4, the estimated values of T_c and T_b are 880.6K and 614K respectively. (See Example 12-4 for details of this calculation.)

(2)   Estimate P_c as follows:

- Sketch the structure of anthracene

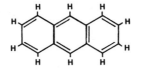

- Refer to Table 13-4 and add the ΔP contributions for the component atomic groups in anthracene.

$$(10)\quad H-\overset{|}{C}= \ = \ 10(0.154) = 1.54$$

$$(4)\quad -\overset{|}{C}= \ = \ 4(0.154) = \underline{0.62}$$

$$\Sigma \Delta P \ = 2.16$$

- Substitute $\Sigma \Delta P$ in Eq. 13-8. (The molecular weight of anthracene is 178.3 g.)

$$P_c = \frac{178.3}{(0.34 + 2.16)^2}$$

$$= \ 28.5 \ \text{atm}$$

(3)   Calculate ΔH_{vb} from the values estimated above. Using the modified Klein method,

- $K_{k1} = 1.045$ from Table 13-3.

- Substituting in Eq. 13-4,

$$\Delta H_{vb} \ = \ 1.987\,(1.045)\,(614) \ \frac{\ln(28.5)\,\sqrt{1 - 1/[(28.5)(614/880.6)^3]}}{1 - (614/880.6)}$$

$$= \ 13,400 \ \text{cal/mol}$$

Since the measured ΔH_{vb} is 13,500 cal/mol, the relative error is -0.74%.

(4)   If the Giacalone method (Eq. 13-5) is used,

$$\Delta H_{vb} = \frac{(1.987)(614) \ln(28.5)}{1 - (614/880.6)}$$

   = 13,500 cal/mol, which agrees with the measured value within the accuracy of the calculation.

(5)   By the Riedel method (Eq. 13-6),

$$\Delta H_{vb} = \frac{(614) [5 \log(28.5) - 2.17]}{0.930 - (614/880.6)}$$

   = 13,500 cal/mol, which again agrees with the measured value.

(6)   Using the Chen method (Eq. 13-7),

$$\Delta H_{vb} = \frac{(614) [7.11 \log(28.5) - 7.82 + 7.9 (614/880.6)]}{1.07 - (614/880.6)}$$

   = 13,230 cal/mol, which deviates -2.0% from the measured value.

13-5   ESTIMATION OF ΔH_{vb} FROM VAPOR PRESSURE DATA

Principles of Use. Equation 13-3 describes the shape of the vapor pressure curve for any material as a function of the heat of vaporization and the difference in the compressibility factors for the liquid and the vapor phases. The shape of the vapor pressure curve may also be described by Antoine's relationship:

$$\log P = A - \frac{B}{t + C} \tag{13-10}$$

where A, B, and C are constants and t is the temperature, all expressed in °C except A, which is dimensionless. Haggenmacher [11,12] combined Eq. 13-10 with Eq. 13-3 and Eq. 13-11, which expresses the compressibility difference in terms of two pressures and two temperatures:

$$Z_g - Z_1 = \sqrt{1 - (P/P_c)/(T/T_c)^3} \tag{13-11}$$

This resulted in the following expression:

$$\Delta H_v = \frac{2.303 BRT^2 \sqrt{1 - (P/P_c)/(T/T_c)^3}}{(t + C)^2} \tag{13-12}$$

where T and T_c are in K, P and P_c are in atmospheres, and B, C, and t are in °C. The constant 2.303 is the natural logarithm of 10, and R is equal to 1.9872 cal/K-mol. At the normal boiling point (where $T = T_b$, $t = t_b$, and $P = 1$ atm), Eq. 13-12 becomes

$$\Delta H_{vb} = \frac{2.303 BRT_b^2 \sqrt{1 - (1/P_c)/(T_b/T_c)^3}}{(t_b + C)^2} \qquad (13\text{-}13)$$

where T_b and T_c are in K, P_c is in atmospheres, and B, C, and t_b are in °C. T_b, T_c, and t_b may be estimated from the methods of Chapter 12, and P_c may be estimated by one of the two procedures outlined in §13-4 of this chapter.

Antoine's constants (A,B,C) have been calculated for many compounds, especially hydrocarbons, and are tabulated in the literature [23,34]. For compounds whose Antoine's constants are not readily available, C may be estimated if the normal boiling point is known; Table 13-5 summarizes the estimation procedure for organic or organometallic compounds. Linear interpolation should be used to estimate C for boiling points that fall between the values listed.

TABLE 13-5

Antoine's Constant C for Organic Compounds

Polyhydric Alcohols (diols, triols, etc.): C = 230°C

Other Organic and Organometallic Compounds:

Boiling Pt. (°C)	C (°C)	Boiling Pt. (°C)	C (°C)
< -150	264 -0.034 t_b	140	212
-150 to -10	240 -0.19 t_b	160	206
-10	238	180	200
0	237	200	195
20	235	220	189
40	232	240	183
60	228	260	177
80	225	280	171
100	221	≥ 300	165
120	217		

Source: Fishtine [9]. (*Reprinted with permission from the American Chemical Society.*)

Antoine's constant B can be estimated from the value of C and two vapor pressure-temperature pairs:

$$B = \frac{(t_2 + C)(t_1 + C)}{t_2 - t_1} \log(P_2/P_1) \qquad (13\text{-}14)$$

Although any two pairs of temperatures and corresponding vapor pressures may be substituted in this equation, Fishtine [9] determined that the boiling points at pressures of 760 mm Hg (1 atm) and 10 mm Hg (0.013 atm) give the most reliable estimates of B, typically within 1%. Since t at 760 mm Hg is the normal boiling point, Eq. 13-14 becomes

$$B = \frac{(t_b + C)(t_{10} + C)}{t_b - t_{10}} (1.881) \qquad (13\text{-}15)$$

where B, C, t_b, and t_{10} are in °C, t_{10} is the temperature corresponding to 10 mm Hg vapor pressure, and 1.881 = log (760/10).

Most tabulations express Antoine's constants in degrees Celsius, but some (such as in Reid *et al.* [31]) use the Kelvin scale. The constant C is converted to Celsius degrees in the usual way (i.e., by subtracting 273.16), but the value of B *should not be changed;* this is because B represents the slope of the curve that relates ΔH_v to temperature, and the slope is not affected by a change from Kelvin to Celsius.

Care should also be taken to establish whether the listed values of B include a factor of 2.303 for converting from base-10 to base-e logarithms. Most tabulations of Antoine's constants list values of B calculated from equations such as 13-14 or 13-15, which use the common logarithm of the pressure ratio; however, some tabulations, such as those of Reid, include in the value of B the conversion factor to natural log. A simple method for determining whether this has been done is to check the value for methane, which is likely to be listed in any tabulation of Antoine's constants. If it is 405.42 (°C or K), the listed values may be used directly in the preceding equations; if it is about 930°, the B values have been multiplied by 2.303.

If T_c and P_c are not known and cannot be estimated, the difference in the compressibility factors $(Z_g - Z_l)$ may be obtained from Table 13-6; the error associated with this approximation is usually 2% or less. The

TABLE 13-6

Estimation of Compressibility-Factor
Difference from Temperature

T/T_b [a]	$Z_g - Z_l$
1.00	0.95
0.98	0.96
0.94	0.97
0.90	0.98
0.84	0.99
0.74	1.00

a. Note that both temperatures must be
   expressed in K.

Source: Fishtine [9]. (*Reprinted with permission from the American Chem-
ical Society.*)

result is then substituted into the following equation, which is equivalent
to Eq. 13-13:

$$\Delta H_{vb} = \frac{2.303 \, BR T_b^2 \, (Z_g - Z_l)}{(t_b + C)^2} \tag{13-16}$$

Basic Steps

(1) Obtain the values of T_b and T_c from the literature, or
estimate them by the methods given in Chapter 12.

(2) Obtain the value of P_c from the literature, or estimate it by
the methods given in §13-4 of this chapter.

(3) Obtain Antoine's constants B and C from the literature, or
estimate C from Table 13-5 and B from Eq. 13-15 (in °C).

(4a) Substitute T_b, T_c, P_c, B, and C into Eq. 13-13 to estimate
ΔH_{vb} (in cal/mol) if T_c and P_c are known.

(4b) If T_c and/or P_c is not available and cannot be estimated,
find the appropriate value of $Z_g - Z_l$ in Table 13-6.

(5) Substitute T_b, B, C, and $Z_g - Z_l$ into Eq. 13-16 to estimate
ΔH_{vb} (in cal/mol).

Example 13-2 Estimate ΔH_{vb} for methyl benzoate using Eqs. 13-13 and 13-16,
given $T_b = 472.2K$, $T_c = 692K$, and $P_c = 36$ atm. (The measured value of ΔH_{vb}
is 10,300 cal/mol.)

(1)   From Table 13-6,

$$Z_g - Z_1 = 0.95$$

From Ref. 34,

$$C = -81.15K$$
$$B = 3751.83K$$

The Celsius equivalents are

$$C = 192.01°C$$
$$B = 3751.83°C$$

For methane, this reference lists $B = 897.84K$, which indicates that the values listed are actually $2.303B$.

(2)   Compute ΔH_{vb} from Eq. 13-13.

$$\Delta H_{vb} = \frac{3751.83\,(1.9872)\,(472.2)^2\,\sqrt{1 - [(1/36)/(472.2/692)^3]}}{[(472.2 - 273.2) + 192.01]^2}$$

$$= 10,400 \text{ cal/mol}$$

The deviation from the measured ΔH_{vb} is 100 cal/mol, indicating a relative error of +1.0%.

(3)   To compute ΔH_{vb} from Eq. 13-16, the value of $Z_g - Z_1$ is needed. Since $T/T_b = 1$ at the boiling point, $Z_g - Z_1 = 0.95$ (Table 13-6). This value will limit the accuracy of the calculation to two significant figures.

(4)   Substituting in Eq. 13-16,

$$\Delta H_{vb} = \frac{(3751.83)\,(1.9872)\,(472.2)^2\,(0.95)}{[(472.2 - 273.2) + 192.01]^2}$$

$$= 10,000 \text{ cal/mol, which deviates from the measured value by } -3\%.$$

13-6 ESTIMATION OF ΔH_{vb} FROM COMPOUND STRUCTURE

Principles of Use. Sastri and coworkers [26] have developed a method for estimating ΔH_{vb} for any organic compound from its molecular structure. This method is based on previous work by Bowden and Jones [3] and by others listed in Ref. 26; these researchers attempted to develop a series of structural constants similar to the parachor but which would be temperature-independent and thereby useful for predicting various thermodynamic properties from the structure of a compound.

The relationship between the heat of vaporization of a compound and the sum of its structural increments is defined by Sastri as

$$\Delta H_{vb} = H_{vo}(1 - T_b/T_c)^n \qquad (13\text{-}17)$$

where H_{vo} is the sum of the structural increments for the heat of vaporization and the constant n is a function of the ratio T_b/T_c (see Table 13-7), which frequently has a value between 0.37 and 0.38.

TABLE 13-7

Values of Exponent n as a Function of T_b/T_c

T_b/T_c	n
< 0.57	0.30
$0.57-0.71$	$0.74(T_b/T_c) - 0.116$
> 0.71	0.41

Source: Fishtine, as cited by Reid *et al.* [24]

The ratio of T_b to T_c may be estimated by the Lydersen method (see §12-4 in the preceding chapter) if either T_b or T_c is unknown. For those cases where the Lydersen method cannot be used, the early work of Guldberg (referenced in [24]) showed that $T_b/T_c \approx 2/3$ for most organic compounds, and this value may be inserted into Eq. 13-17.

The value of H_{vo} is found by summing the increments for each atom or functional group in the molecule. Table 13-8 lists the increments for atoms and functional groups containing C, H, N, O, S, and the halogens. Note that the increments are listed in units of kilocalories, so the resulting value of H_{vb} must be multiplied by 1000 to express it in cal/mol. The values for the halogen atoms vary with the compound structure, and Table 13-9 lists rules to follow for selecting the proper increment. Table 13-10 lists correction increments for three specific compound types, predominantly small or polar organic compounds.

Basic Steps

(1) Draw the structure.

(2) Obtain T_b and T_c from the literature *or* estimate T_b/T_c by Lydersen's method (Eq. 12-5) *or* use $T_b/T_c = 2/3$.

(3) Obtain exponent n from Table 13-7.

TABLE 13-8

Structural Group Increments for H_{vo}

Group	Increment (kcal/mol)	Group	Increment (kcal/mol)	Group	Increment (kcal/mol)
Aliphatic and Alicyclic Hydrocarbons		**Aromatic Hydrocarbons**		**Sulfur Increments**	
—H	0.20	>CH—	1.68	—SH	4.65
—CH₃	2.38	=C—	0.76	—S—	3.50
—CH₂—	1.44	=C< (Bridge carbon in condensed ring system)	1.20	**Nitrogen Increments**	
—CH—	0.08			—C≡N	7.65
—C—	-1.62	**Hydrocarbon Structures, Nonaromatic**		—NH₂	5.60
=CH₂	2.19	3-membered ring	2.20	>NH	3.00 (aliphatics)
=CH—	1.50	4-membered ring	2.55		4.10 (aromatics)
=C—	0.19	5-membered ring	1.85		4.10 (heterocyclics)
=C=	1.85	5-membered ring and monoalkene	1.60	>N—	0.30 (aliphatics)
≡CH	2.35	6-membered ring	1.50		1.75 (aromatics)
≡C—	2.00	6-membered ring and alkene	1.55		3.00 (heterocyclics)
				—NO₂	8.60 (aliphatics)
					5.70 (aromatics)

Oxygen Increments

Group	Increment (kcal/mol)
–OH	9.80 (alcohols, primary bonded)
	9.00 (monohydroxy alcohols, secondary and tertiary bonded)
	8.60 (dihydroxy alcohols, secondary and tertiary bonded)
	3.80 (trihydroxy alcohols, secondary and tertiary bonded)
	6.40 (phenols)
–O–	2.05 (ethers)
	0.80 (aliphatics containing nonhydrocarbon groups other than halogens)
	2.90 (heterocyclics)

Oxygen Increments (cont.)

Group	Increment (kcal/mol)
$\overset{H}{\underset{\,}{-C=O}}$	5.70
>C=O	4.40 (aliphatics)
	2.80 (aromatics)
$\overset{O}{=}\!\!C\text{–OH}$	12.50 (acids with $\leqslant$ 10 C atoms)
	6.70 (acids with > 10 C atoms)
$\overset{O}{=}\!\!C\text{–O–}$	4.20 (esters with $\leqslant$ 8 C atoms)
	3.00 (esters with > 8 C atoms)

Halogen Increments [a]

Group	Increment (kcal/mol)
–F	2.50 (A types)
	1.00 (B types)
	1.75 (aromatics)
–Cl	4.10 (A types)
	2.60 (B types)
	3.10 (aromatics)
–Br	4.60 (A types)
	3.60 (B types)
	3.60 (aromatics)
–I	6.00 (A types)
	5.00 (B types)
	5.00 (aromatics)

a. See Table 13-9 for definitions of types A and B.

Source: Sastri et al. [26]. (Reprinted with permission from IPC Industrial Press, Ltd.)

TABLE 13-9

Rules for Using Type A or Type B Halogen Increments

Compound	Increment Type
Monohalogenated Aliphatics	
Halogen attached to hydrocarbon group (e.g., CH_3Cl)	A
Halogen attached to nonhydrocarbon group (e.g., CH_3COCl or CH_3COOCl)	B
Multihalogenated Aliphatics	
Saturated compounds with x carbons	A for $\leq$ x halogens B for $>$ x halogens
Unsaturated compounds with x carbons	A for $\leq$ x-1 halogens B for $>$ x-1 halogens
Exceptions	
Both halogens in methylene compounds	A
For compounds containing more than one type of halogen, assign type-A increments starting with lowest atomic number halogen until x halogens (if saturated) or x-1 halogens (if unsaturated) have been assigned. Use type-B increments for any additional halogens (e.g., C_2FCl_5 has 1 type-A F, 1 type-A Cl, and 4 type-B Cl)	

Source: Sastri *et al.* [26]. *(Reprinted with permission from IPC Industrial Press, Ltd.)*

(4) Obtain and sum H_{vo} increments from Tables 13-8 to 13-10.

(5) Estimate ΔH_{vb} (in cal/mol) from Eq. 13-17.

Example 13-3 Estimate ΔH_{vb} for anthracene by the Sastri method. (The measured value of ΔH_{vb} is 13,500 cal/mol.)

(1) The structure of anthracene is

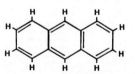

TABLE 13-10

Correction Increments

Compound	Correction Increment (kcal/mol)
Aliphatic compounds (not esters) with single nonhydrocarbon group attached to $-CH_3$, methyl group (e.g., CH_3Cl, $CH_3N(CH_3)_2$)	0.60[a]
Aliphatic compounds formed by the addition of $-H$, single hydrocarbon to the nonhydrocarbon group; includes esters of formic acid (e.g., CH_3COOH)	1.85[a]
Esters with $-CH_3$ attached to $-COO-$	0.90[b]

a. No correction required if compound is associated in the gas phase (e.g., CH_3COOH, $HCOOH$).

b. Increment added for each $-CH_3$ attached to $-COO-$ group; thus, methyl acetate (CH_3COOCH_3) would have correction of 2 x 0.90.
   Formate esters such as $HCOOCH_3$ would have increments of 0.90 + 1.85.

Source: Sastri *et al.* [26]. (*Reprinted with permission from IPC Industrial Press, Ltd.*)

(2) From the literature,

$$T_b = 614.4K$$

$$T_c = 883K$$

$$T_b/T_c = 0.6958$$

(3) From Table 13-7,

$$n = 0.40$$

(4) From Table 13-8,

Structure	Number		Increment		
$=C\overset{'}{\underset{\backslash}{}}$ aromatic	4	X	1.20	=	4.80
$=CH-$ aromatic	10	X	1.68	=	16.80
			ΣH_{vo}	=	21.60

(5)   Compute H_{vb} from Eq. 13-17:

$$\Delta H_{vb} = 21.60(1 - 614.4/883)^{0.40}$$

$$= 13 \text{ kcal/mol}$$

$$= 13,000 \text{ cal/mol}$$

This result is accurate to only two significant figures because of n; compared with the measured value (rounded off to 14,000 cal/mol), it shows a deviation of -7%.

Example 13-4   Compute ΔH_{vb} for 2-methylthiophene using the Sastri method. (The measured value of ΔH_{vb} is 8103 cal/mol.)

(1)   The structure of 2-methylthiophene is

(2)   Using Lydersen's method, Table 12-6 gives the following ΔT increments:

Group	Number		Increment	
$-CH_3$	1	X	0.020	= 0.020
$=CH-$ ring	3	X	0.011	= 0.033
$=\overset{\mid}{C}-$ ring	1	X	0.011	= 0.011
$-S-$ ring	1 .	X	0.008	= 0.008
			$\Sigma\Delta T$	= 0.072

Then, by Eq. 12-5,

$$\theta = T_b/T_c = 0.567 + \Sigma\Delta T - (\Sigma\Delta T)^2$$

$$= 0.567 + 0.072 - (0.072)^2$$

$$= 0.634$$

(3)   From Table 13-7, n = 0.353

(4)   From Table 13-8,

Group	Number		Increment	
$-CH_3$	1	X	2.38	= 2.38
$=CH-$ ring	3	X	1.68	= 5.04
$=\overset{\mid}{C}-$ ring	1	X	0.76	= 0.76
$-S-$	1	X	3.50	= 3.50
			ΣH_{vo}	= 11.68

(5)   From Eq. 13-17:

$$\Delta H_{vb} \;=\; 11.68\,(1-0.634)^{0.353}$$

$$=\; 8.19 \text{ kcal/mol}$$

$$=\; 8,190 \text{ cal/mol, which deviates from the measured value}$$
by +1.1%.

13-7   ESTIMATION OF ΔH_v AT TEMPERATURES OTHER THAN THE BOILING POINT

The previous methods of this chapter permitted estimation of the heat of vaporization at the normal boiling point. Although this is sufficient for most applications, the value of ΔH_v at other temperatures is sometimes needed. The Thiesen correlation [7] is frequently used as a starting point:

$$\Delta H_v \;=\; k(1 - T/T_c)^n \tag{13-18}$$

where k and n are constants, ΔH_v is the heat of vaporization at temperature T, and T_c is the critical temperature. If k, n and T_c are known for a compound, ΔH_v can be computed for any temperature T. Values of k and n have been calculated for some compounds, but most of the methods based on Eq. 13-18 form the ratio of this equation at two temperatures, as shown below.

$$\frac{\Delta H_{v2}}{\Delta H_{v1}} \;=\; \left[\frac{(1 - T_2/T_c)}{(1 - T_1/T_c)} \right]^n \tag{13-19}$$

or

$$\Delta H_{v2} \;=\; \Delta H_{v1} \left[\frac{(1 - T_2/T_c)}{(1 - T_1/T_c)} \right]^n \tag{13-20}$$

where ΔH_{v1} and ΔH_{v2} are the heats of vaporization at temperatures T_1 and T_2, respectively. With Eq. 13-20, the heat of vaporization at any temperature may be estimated as long as one other heat of vaporization is known. It is convenient to choose the normal boiling point as one of the temperatures, since the heat of vaporization at the normal boiling point is either known or may be estimated by the methods presented earlier in this chapter. Therefore, Eq. 13-21 is normally used for estimating the heat of vaporization at any temperature other than the normal boiling point.

$$\Delta H_v = \Delta H_{vb} \left[\frac{(1 - T/T_c)}{(1 - T_b/T_c)} \right]^n \qquad (13\text{-}21)$$

Various values have been used for n. The most common is 0.38, which was proposed by Watson [32] and yields estimates typically within 2%. Fishtine (in Ref. 24) has proposed a correlation of n with the ratio of T_b/T_c, which yields better estimates of ΔH_v; this correlation is summarized in Table 13-7, and corresponding values of n should be used in Eqs. 13-19, -20, and -21.

Basic Steps

(1) Obtain T_b/T_c from the literature, or estimate it by the methods of Chapter 12.

(2) Obtain ΔH_{vb} from the literature, or estimate it as described in §13-3, §13-4, or §13-5.

(3) Obtain n from Table 13-7.

(4) Compute ΔH_v at temperature T from Eq. 13-21.

Example 13-5 Estimate the heat of vaporization for isooctane at 25°C, given $T_b = 372.4\,K$ and $T_c = 543.9\,K$. (The measured value is 8,397 cal/mol.)

(1) $T_b/T_c = 0.6847$

(2) From the literature [34]:

$$\Delta H_{vb} = 7.411 \text{ kcal/mol}$$

(3) From Table 13-7:

$$n = 0.74\,(0.6847) - 0.116$$
$$= 0.39$$

(4) Compute ΔH_v from Eq. 13-21:

$$\Delta H_v = 7.411 \left[\frac{1 - (298/543.9)}{1 - 0.6847} \right]^{0.39}$$

$$= 8.5 \text{ kcal/mol}$$

$$= 8{,}500 \text{ cal/mol, which deviates from the measured value by } +1.2\%.$$

13-8   AVAILABLE DATA

Data on heats of vaporization are available from many sources. The values listed may be measured at a temperature of 25°C or at the normal boiling point. In addition to the general references listed in Appendix A, the following are recommended:

Dean, J.A. [6] — General data, including heats of vaporization for inorganic and organic compounds.

Reid, R.C. *et al.* [23] — Appendix tabulates heats of vaporization and Antoine's constants for many organic compounds.

Perry, R.H. and C.H. Chilton [21] — General chemical engineering information, including heats of vaporization for organic and inorganic compounds.

Zwolinski, B.J. and R.C. Wilhoit [34] — Vapor pressures and heats of vaporization for hydrocarbons and related compounds.

Washburn, E.E. [30] — Critical constants and heats of vaporization for numerous compounds.

Thakore, S.B. *et al.* [27b] — Correlation constants for the Watson relationship (Eq. 13-20) for many common chemicals.

13-9   SYMBOLS USED

a,b	=	Van der Waal's constants in Eq. 13-9
A, B, C	=	Antoine's constants in Eq. 13-10
ΔH_v	=	heat of vaporization at temperature T (cal/mol)
H_{vo}	=	sum of Sastri heat of vaporization structural increments in Eq. 13-17 (kcal/mol)
k	=	constant in Eq. 13-18
K_{kl}	=	Klein constant in Eq. 13-4
M	=	molecular weight in Eq. 13-8 (g/mol)
n	=	exponent in Eqs. 13-17 to 13-21
ΔP	=	Lydersen increment for estimation of P_c in Eq. 13-8
P	=	vapor pressure (atm)
R	=	universal gas constant = 1.9872 cal/mol
t	=	temperature (°C)
t_{10}	=	temperature corresponding to a vapor pressure of 10 mm Hg in Eq. 13-15 (°C)

T	=	temperature (K)
V	=	molal volume in Eqs. 13-1 and -2
Z	=	compressibility factor in Eqs. 13-2, -3, and -18

Subscripts

1	=	temperature T_1
2	=	temperature T_2
b	=	boiling point
c	=	critical property
g	=	vapor phase
l	=	liquid phase

13-10   REFERENCES

1. Barrow, G.M., *Physical Chemistry,* McGraw-Hill Book Co., New York (1966).

2. Bowden, S.T., "Variation of Surface Tension and Heat of Vaporization with Temperature," *J. Chem. Phys.,* **1955**, 2454.

3. Bowden, S.T. and W.J. Jones, "Latent Heats of Vaporization and Composition," *Philos. Mag.,* **37**, 480 (1946).

4. Chen, N.H., "Generalized Correlation for Latent Heat of Vaporization," *J. Chem. Eng. Data,* **10**, 207-10 (1965).

5. Chu, J.C., M. Dmtryszyn, J. Modern and R. Overbeck, "Latent Heat of Vaporization," *Ind. Eng. Chem.,* **41**(1), 131 (1949).

6. Dean, J.A. (ed.), *Lange's Handbook of Chemistry,* 12th ed., McGraw-Hill Book Co., New York (1979).

7. Fishtine, S.H., "How Good Are Latent Heat Calculations?" *Hydrocarbon Process.* **45**, 173-79 (1966).

8. Fishtine, S.H., "Reliable Latent Heats of Vaporization, Pt. 1," *Ind. Eng. Chem.,* **55**(4), 20-28 (1963).

9. Fishtine, S.H., "Reliable Latent Heats of Vaporization, Pt. 2," *Ind. Eng. Chem.,* **55**(5), 49-54 (1963).

10. Fishtine, S.H., "Reliable Latent Heats of Vaporization, Pt. 3," *Ind. Eng. Chem.,* **55**(6), 47-56 (1963).

11. Haggenmacher, J.E., "The Heat of Vaporization as a Function of Pressure and Temperature," *J. Am. Chem. Soc.,* **68**, 1633-34 (1946).

12. Haggenmacher, J.E., "Heat and the External Work of Vaporization of Twenty-two Hydrocarbons," *Ind. Eng. Chem.,* **40**(3), 436-37 (1948).

13. Kier, L.B. and L.H. Hall, *Molecular Connectivity in Chemistry and Drug Research,* Academic Press, New York (1976).

14a. Klein, V.A., "Latent Heats of Vaporization," *Chem. Eng. Prog.,* **45**, 675 (1949).

14b. Lydersen, A.L., "Estimation of Critical Properties of Organic Compounds," College of Engineering, Univ. of Wisconsin, Engineering Experiment Station Report 3, Madison, Wis. (April 1955).

15.   Meissner, H.P., "Latent Heats of Vaporization," *Ind. Eng. Chem.*, **33**(11), 1440-43 (1941).

16.   Narsimhan, G., "A Generalized Expression for Predicting Latent Heats of Vaporization," *Br. Chem. Eng.*, **10**, 253-55 (1965).

17.   Narsimhan, G., "A New Correlation of Latent Heat of Vaporization," *Br. Chem. Eng.*, **12**, 897-99 (1967).

18.   Nutting, P.G., "Vapor Pressure and Heat of Vaporization," *Ind. Eng. Chem.*, **22**(7), 771 (1930).

19.   Ogden, J.M. and J. Lielmezs, "Latent Heat Estimation Using Altenburg's Quadratic Mean Radius," *AIChEJ.*, **15**, 469-70 (1969).

20.   Partington, J.R., "An Advanced Treatise on Physical Chemistry," Vol. II, Longmans, Green and Co., New York (1951).

21.   Perry, R.H. and C.H. Chilton, *Chemical Engineers' Handbook,* 5th ed., McGraw-Hill Book Co., New York (1973).

22.   Pitzer, K.S., D.Z. Lippmann, R.F. Curl, Jr., C.M. Huggins and D.E. Petersen, "The Volumetric and Thermodynamic Properties of Fluids: II. Compressibility Factor, Vapor Pressure and Entropy of Vaporization," *J. Am. Chem. Soc.*, **77**, 3433-40 (1955).

23.   Reid, R.C., J.M. Prausnitz and T.K. Sherwood, *The Properties of Gases and Liquids,* 3rd ed., McGraw-Hill Book Co., New York (1977).

24.   Reid, R.C. and T.K. Sherwood, *The Properties of Gases and Liquids — Their Estimation and Correlation,* 2nd ed., McGraw-Hill Book Co., New York (1966).

25.   Riedel, L., "Critical Constants, Saturated Liquid Density and the Heat of Vaporization," *Chem. Ing. Tech.*, **12**, 679-83 (1954).

26.   Sastri, S.R.S., M.V. Ramana Rao, K.A. Reddy and L.K. Doraiswamy, "A Generalized Method for Estimating the Latent Heat of Vaporization of Organic Compounds," *Br. Chem. Eng.*, **14**, 959-63 (1969).

27a. Silverberg, P.M. and L.A. Wenzel, "The Variation of Latent Heat with Temperature," *J. Chem. Eng. Data*, **10**, 363-66 (1965).

27b. Thakore, S.B., J.W. Miller and C.L. Yaws, "Heats of Vaporization," *Chem. Eng. (NY)*, **83**(12), 85-87 (1976).

28.   Viswanath, D.S. and N.R. Kuloor, "On Latent Heat of Vaporization, Surface Tension and Temperature," *J. Chem. Eng. Data*, **11**, 69-72 (1966).

29.   Viswanath, D.S. and N.R. Kuloor, "Latent Heat of Vaporization," *J. Chem. Eng. Data*, **11**, 544 (1966).

30.   Washburn, E.E. (ed.), *International Critical Tables of Numerical Data, Physics, Chemistry and Technology,* McGraw-Hill Book Co., New York (1926-1933).

31.   Watson, K.M., "Prediction of Critical Temperatures and Heats of Vaporization," *Ind. Eng. Chem.*, **23**(4), 360-64 (1931).

32.   Watson, K.M., "Thermodynamics of the Liquid State: Generalized Prediction of Properties," *Ind. Eng. Chem.*, **35**(4), 398-406 (1943).

33. Winter, R.M., "Latent Heat of Vaporization as a Function of Temperature," *J. Phys. Chem.*, **32**, 576 (1928).

34. Zwolinski, B.J. and R.C. Wilhoit, *Handbook of Vapor Pressures and Heats of Vaporization of Hydrocarbons and Related Compounds*, API44-TRC Publications in Science and Engineering, Pub. No. 101, Thermodynamics Research Center Data Distribution Office, College Station, Texas (1971).

14

VAPOR PRESSURE

Clark F. Grain

14-1 INTRODUCTION

Reliable methods for estimating the vapor pressures of organic materials are of increasing importance as a tool in predicting the behavior and fate of chemicals that are introduced into the environment. When a chemical has been spilled, for example, we must know its approximate vapor pressure in order to estimate its rate of evaporation. The persistence of insecticides, herbicides, and similar substances that have been absorbed in the soil is also highly dependent on this chemical-specific property.

Numerous equations and correlations for estimating vapor pressure are presented in the literature. In general, they require information on at least three of the following properties: (1) the critical temperature, T_c, (2) the critical pressure, P_c, (3) the heat of vaporization, ΔH_v, and/or (4) the vapor pressure (P_{vp}) at some reference temperature. For most liquids at room temperature, the vapor pressure ranges from 10^{-5} to 300 millimeters of mercury.

The equations that relate vapor pressure to temperature are commonly derived by integration of the Clausius-Clapeyron equation

$$\frac{d \ln P_{vp}}{dT} = \frac{\Delta H_v}{\Delta Z \, RT^2} \qquad (14\text{-}1)$$

where P_{vp} is the vapor pressure in *atmospheres,* ΔH_v is the heat of vaporization in cal/mol, R is the gas constant in cal/mol·K, T is the temperature in K, and ΔZ is a compressibility factor, given by

$$\Delta Z = \frac{P_{vp}\, \Delta V}{RT} \qquad (14\text{-}2)$$

where ΔV is the volume difference between vapor and liquid. In Eq. 14-2 R has the units cm³-atm/K, hence, ΔZ is dimensionless and has a value of 1 for an ideal gas.

The simplest equation that can result from integration of Eq. 14-1 is

$$\ln P_{vp} = A_1 - B_1/T \qquad (14\text{-}3)$$

where A_1 and B_1 can be expressed in terms of the parameters in Eq. 14-1. Equation 14-3 is the result obtained when $\Delta H_v/\Delta Z$ is assumed to be constant with changes in temperature; more complex (but more accurate) vapor pressure equations can be derived by assuming an analytical form for the temperature dependence of ΔH_v.

Most of the estimation and correlation methods are designed for greatest accuracy between the normal boiling point, T_b, and the critical temperature, T_c. This range is primarily useful to process engineers, who deal with relatively high temperatures. Environmental studies, however, normally involve temperatures below the boiling point, where such methods are less accurate [8,9]; the need here is for reliable estimates of the vapor pressures of a wide variety of liquids and solids at temperatures in the 10-40°C range. A method for estimating very low vapor pressures (< 1mm Hg) is particularly needed, and the amount of experimental data required must be kept to a minimum.

14-2   SELECTION OF APPROPRIATE METHOD

Evaluation of Parameters. For maximum utility and convenience, any analytical expression that relates vapor pressure to temperature should have as few arbitrary parameters as possible. The general approach that has been used to minimize the number of such parameters is explained below.

Such an expression is first equated to the vapor pressure at the normal boiling point. Since P_{vp} is then equal to one atmosphere, $\ln P_{vp} = 0$. The first derivative of the analytical expression can also be equated to the Clausius-Clapeyron equation at T_b. Thus, in general,

$$\ln P_{vp}\,(\text{atm}) = f(T) \tag{14-4}$$

with boundary conditions

$$\ln P_{vp}\Big|_{T=T_b} = 0 = f(T_b) \tag{14-5}$$

and

$$\frac{d \ln P_{vp}}{dT}\Big|_{T=T_b} = \frac{\Delta H_{vb}}{\Delta Z_b R T_b^2} = f'(T_b) \tag{14-6}$$

By means of Eqs. 14-5 and 14-6, a two-parameter equation can be equated directly to ΔH_{vb} and T_b, while a three-parameter equation can be reduced to an equation with only one adjustable parameter. For example, the parameters in Eq. 14-3 can be evaluated to yield

$$\ln P_{vp}\,(\text{atm}) = \frac{\Delta H_{vb}}{\Delta Z_b R}\left[\frac{1}{T_b} - \frac{1}{T}\right] \tag{14-7}$$

In general, however, this equation is quite inaccurate at temperatures below T_b and is not recommended.

If an experimental vapor pressure datum is available, one could, in principle, use the general procedure just outlined, providing it is noted that

$$\ln P_{vb}\Big|_{T=T_1} \neq 0 = f(T_1) \tag{14-8}$$

where T_1 is the experimental temperature datum.

Equation 14-7 is now generalized such that

$$\ln P_{vp} = \ln P_1 + \frac{\Delta H_{v1}}{\Delta Z_b R}\left[\frac{1}{T_1} - \frac{1}{T}\right] \tag{14-9}$$

Note that if $T_1 = T_b$ then $\ln P_1 = 0$ and Eq. 14-9 becomes identical to Eq. 14-7.

The value of ΔH_{v1} must be known in order to use Eq. 14-9 at temperatures other than T_1. If an experimental value is not available, the methods of Chapter 13 may be used to estimate this quantity. Note, however, that all of the methods presented in Chapter 13 require a knowledge of T_b, the normal boiling point. This means that at least two data points are necessary in order to estimate P_{vp}, whereas Eq. 14-7 requires T_b only. In §14-5, however, a general method of estimating the ratio $\Delta H_{vl}/T_1$ is provided that obviates the necessity for two data points.

Special mention should be made of the problem of estimating the vapor pressures of solids. If a particular vapor pressure equation is extrapolated below the melting point, one obtains the vapor pressure of the supercooled liquid rather than that of the true "crystalline" solid. Generally, the vapor pressure of the solid is considerably lower than that of the supercooled liquid. Mackay [7] has pointed out that the vapor pressure of the solid can be derived from that of the supercooled liquid by applying a correction based upon the assumption that the entropy of fusion is a constant equal to 13.6 cal/mol·K. Thus,[1]

$$\ln P_{vp}\,(s) = \ln P_{vp}\,(\ell) + 6.81\,(1 - T_m/T) \qquad (14\text{-}10)$$

where T_m is the melting point in degrees Kelvin. This correction has been used with limited success with one of the methods recommended in this chapter (Method 1). See Table 14-2.

The other method recommended in this chapter (Method 2) contains one adjustable parameter, which has one value applicable to all liquids and three values for solids, depending upon the ratio of the normal boiling point to the temperature of interest. Thus, no correction is needed and only one experimental input is required.

An interesting correlation has been reported by Chiou and Freed [3], who observed a linear relationship between the logarithm of the vapor pressure at 25°C and the logarithm of the octanol-water partition coefficient, K_{ow}.[2] The relationship was reasonably accurate for five classes of compounds, namely, aromatic hydrocarbons, organo halogens, aliphatic alcohols, aliphatic acids and chlorinated phenols. We do not recommend this method because of its limited applicability with respect to chemical classes and the fact that it is useful at 25°C only. Future development, however, may lead to greater utility for the method.

General Characteristics. Many vapor pressure equations are given in the literature: Partington [10], for example, lists about 56.[3] The two

---

1. Entropy of fusion/R = 6.81.
2. Estimation methods for this parameter are provided in Chapter 1.
3. The literature was reviewed in 1965 by Miller [8] and more recently by Reid, Prausnitz, and Sherwood [12]. Additional information is given in Ref. 11.

methods recommended in this chapter require a minimum of experimental data and are applicable to almost any organic material over a wide pressure range. Their accuracy is quite good over the temperature range of interest to environmental scientists.

Basic information on the two methods is summarized in Table 14-1. Note that a choice may be required, depending upon (1) the physical state of the pure material at the temperature and pressure in question, (2) the value of P_{vp}, and (3) the available input data (i.e., whether T_b or some other temperature datum is available).

TABLE 14-1
Recommended Methods

	Method 1	Method 2
Basis	Antoine Equation [1]	Modified Watson Correlation [17]
Applicable for:		
Physical State	Liquids and Gases	Liquids and Solids
Range of P_{vp} (mm Hg)	$10^{-3}-760$	$10^{-7}-760$
Input Required[a]	T_b	T_b
Method Error (%)[b]	2.7 86.3	2.5 38.7 46.9
for P_{vp} Range (mm Hg) of	10–760 $10^{-3}-10$	10-760 10^{-3}-10 10^{-7}-10^{-3}

a. If a boiling point (T_1) at some reduced pressure (P_1) is available, this information may be used in place of T_b. Instructions are provided in §14-5.

b. From Table 14-3.

In Table 14-2, vapor pressures calculated by both methods are compared with experimental values for a wide variety of materials. The only experimental data required for either method is the normal boiling point, T_b. If T_b is unavailable, it can be estimated by one of the methods described in Chapter 12.

Method 1 is generally applicable over the pressure range from 760 mm to 10^{-3} mm. Method 2 is applicable from 760 mm to at least 10^{-7} mm. Reid et al. [12] state that none of the vapor pressure equations are suitable for estimating pressures below 10 mm within a 10% deviation from experimental data. In many instances, however, particularly for compounds of environmental concern, such a stringent requirement is unnecessary; in estimating the volatilization of a chemical from an open

TABLE 14-2

Calculated vs Experimental Vapor Pressures

Material	T_b (K)	t (°C)	Experimental P_{vp} (mm)[a]	Calculated P_{vp} (mm)	
				Method 1	Method 2
Liquids and Solids using Experimental T_b Data					
Acetone	329	30(ℓ)	270	278	272
Hexane	342	20(ℓ)	120	124	121
Benzene	353	20(ℓ)	76	81	76
Ethanol	351	20(ℓ)	43	45	41
1,4-Dioxane	374	20(ℓ)	30	32	29
2-Ethyl-butyraldehyde	390	20(ℓ)	14	14	13
Acetic acid	391	20(ℓ)	14	14	13
Chloroethanol	402	20(ℓ)	5	3	3
Allyl glycidyl ether	427	20(ℓ)	4	3	3
Furfural	435	20(ℓ)	1	2	1
Aniline	457	20(ℓ)	3×10^{-1}	5×10^{-1}	5×10^{-1}
Phenol	455	20(s)	2×10^{-1}	4×10^{-1}	2×10^{-1}
p-Chloroaniline	505	20(s)	2×10^{-2}	1×10^{-2}	2×10^{-2}
Ethylene glycol	471	20(ℓ)	2×10^{-2}	4×10^{-2}	4×10^{-2}
Glycerol	564	50(ℓ)	3×10^{-3}	1×10^{-3}	2×10^{-3}
Solids and Liquids using Estimated T_b[b]				c	
2,6-Dichlorobenzonitrile	525	20(s)	6×10^{-4}	2×10^{-2}	4×10^{-4}
4,6-Dinitro-o-cresol	582	35(s)	4×10^{-4}	1×10^{-3}	4×10^{-5}
Trifluralin	635	30(ℓ)	2×10^{-4}	6×10^{-5}	3×10^{-4}
Dinoseb	635	25(ℓ)	5×10^{-5}	6×10^{-6}	5×10^{-5}
Carbofuran	570	33(s)	2×10^{-5}	3×10^{-4}	6×10^{-5}
Dicamba	570	25(s)	2×10^{-5}	2×10^{-4}	2×10^{-5}
Lindane	583	20(s)	9×10^{-6}	5×10^{-4}	8×10^{-6}
Aldrin	618	25(s)	6×10^{-6}	4×10^{-5}	8×10^{-7}
DDT	613	20(s)	2×10^{-7}	2×10^{-5}	2×10^{-7}
Endrin	618	20(s)	2×10^{-7}	7×10^{-5}	3×10^{-7}
Dieldrin	623	20(s)	1×10^{-7}	1×10^{-6}	1×10^{-7}

a. From Refs. 2, 5, 13, 14 and 16.

b. T_b estimated using methods of Chapter 12.

c. These values were obtained by using Method 1 to obtain an initial estimate for the supercooled liquid, and then applying the correction factor given in Eq. 14-10.

spill, for example, an order-of-magnitude estimate is usually sufficient. As shown in Tables 14-2 and 14-3, the maximum deviations for both methods are considerably less than this over the entire range of pressures.

TABLE 14-3

Average and Maximum Errors in Estimated Vapor Pressure[a]

Pressure Range (mm)	Average Error (%)	Maximum Error (%)
Method 1		
10–760	2.7	+6.6
10^{-3}–10	86	+100
Method 2		
10–760	2.5	+7.1
10^{-3}–10	39	+50.0
10^{-7}–10^{-3}	47	+200

a. Errors calculated for the chemicals listed in Table 14-2. A larger number of significant figures than shown in Table 14-2 for P_{vp} were used in the calculation of method errors.

14-3 METHOD 1

Derivation. The first method uses the Antoine equation [1], which has the general form

$$\ln P_{vp} = A_2 - \frac{B_2}{T - C_2} \tag{14-11}$$

Applying Eqs. 14-5 and 14-6,

$$A_2 = \frac{B_2}{T_b - C_2} \tag{14-12}$$

and

$$B_2 = \frac{\Delta H_{vb}}{\Delta Z_b RT_b^2} \, [(T_b - C_2)^2] \tag{14-13}$$

Substituting Eqs. 14-12 and 14-13 in Eq. 14-11 yields

$$\ln P_{vp} = \frac{\Delta H_{vb} \, (T_b - C_2)^2}{\Delta Z_b RT_b^{\,2}} \left[\frac{1}{(T_b - C_2)} - \frac{1}{(T - C_2)} \right] \qquad (14\text{-}14)$$

The parameter ΔZ_b is assumed to have the value of 0.97 [8], and T_b is generally known. What remains are techniques for evaluating ΔH_{vb} and C_2. The constant C_2 is estimated via Thomson's rule [15], such that

$$C_2 = -18 + 0.19 \, T_b \qquad (14\text{-}15)$$

The heat of vaporization at the boiling point, ΔH_{vb}, is evaluated using a simple method introduced by Fishtine [4], who modified the Kistiakovskii [6] equation to obtain

$$\frac{\Delta H_{vb}}{T_b} = \Delta S_{vb} = K_F \, (8.75 + R \ln T_b) \qquad (14\text{-}16)$$

where K_F is derived from a consideration of the dipole moments of polar and nonpolar molecules. Table 14-4 lists values of K_F for various compound classes. Thus, the only input data needed is the normal boiling point, T_b. When T_b is not known, the estimation techniques outlined in Chapter 12 may be used.

For polar derivatives of benzene, Fishtine suggested that $K_F = 1 + 2\mu/100$, where μ is the dipole moment of the derivative. For naphthalene derivatives he suggests $K_F = 1 + \mu/100$. Methods of estimating dipole moments are given in Chapter 25. Special consideration must be given to hydrogen bonded systems; Table 14-5 lists values for aromatic phenols, diols, and amine compounds. These values are to be applied no matter what other polar group is present.

Method 1 is applicable only over the normal liquid range. Thus, it should only be used to estimate the vapor pressure of materials that are either in the liquid or vapor state at the temperature of interest. Method 2 (described later) is applicable for estimating the vapor pressures of liquids and solids.

Basic Steps

(1) Obtain the normal boiling point T_b (K). If unavailable, use one of the estimation methods outlined in Chapter 12.

(2) Obtain K_F from Table 14-4 or 14-5 as appropriate.

(3) Calculate $\Delta H_{vb}/T_b$ using Eq. 14-16. The value of R is 1.987 cal/mol·K.

TABLE 14-4

K_F Factors for Aliphatic and Alicyclic[a] Organic Compounds

Compound Type	Number of carbon atoms (N) in compound, including carbon atoms of functional group											
	1	2	3	4	5	6	7	8	9	10	11	12-20
Hydrocarbons												
n-Alkanes	0.97	1.00	1.00	1.00	1.00	1.00	1.00	1.00	1.00	1.00	1.00	1.00
Alkane isomers				0.99	0.99	0.99	0.99	0.99	0.99	0.99	0.99	0.99
Mono- and diolefins and isomers		1.01	1.01	1.01	1.01	1.01	1.01	1.01	1.01	1.01	1.01	1.00
Cyclic saturated hydrocarbons			1.00	1.00	1.00	1.00	1.00	1.00	1.00	1.00	1.00	1.00
Alkyl derivatives of cyclic saturated hydrocarbons				0.99	0.99	0.99	0.99	0.99	0.99	0.99	0.99	0.99
Halides (saturated or unsaturated)												
Monochlorides	1.05	1.04	1.03	1.03	1.03	1.03	1.03	1.03	1.02	1.02	1.02	1.01
Monobromides	1.04	1.03	1.03	1.03	1.03	1.03	1.02	1.02	1.02	1.01	1.01	1.01
Monoiodides	1.03	1.02	1.02	1.02	1.02	1.02	1.01	1.01	1.01	1.01	1.01	1.01
Polyhalides (not entirely halogenated)	1.05	1.05	1.05	1.04	1.04	1.04	1.03	1.03	1.03	1.02	1.02	1.01
Mixed halides (completely halogenated)	1.01	1.01	1.01	1.01	1.01	1.01	1.01	1.01	1.01	1.01	1.01	1.01
Perfluorocarbons	1.00	1.00	1.00	1.00	1.00	1.00	1.00	1.00	1.00	1.00	1.00	1.00
Compounds Containing the Keto Group												
Esters		1.14	1.09	1.08	1.07	1.06	1.05	1.04	1.04	1.03	1.02	1.01
Ketones			1.08	1.07	1.06	1.06	1.05	1.04	1.04	1.03	1.02	1.01
Aldehydes	—	1.09	1.08	1.08	1.07	1.06	1.05	1.04	1.04	1.03	1.02	1.01
Nitrogen Compounds												
Primary amines	1.16	1.13	1.12	1.11	1.10	1.10	1.09	1.09	1.08	1.07	1.06	1.05[b]
Secondary amines		1.09	1.08	1.08	1.07	1.07	1.06	1.05	1.05	1.04	1.04	1.03[b]
Tertiary amines			1.01	1.01	1.01	1.01	1.01	1.01	1.01	1.01	1.01	1.01
Nitriles	—	1.05	1.07	1.06	1.06	1.05	1.05	1.04	1.04	1.03	1.02	1.01
Nitro compounds	1.07	1.07	1.07	1.06	1.06	1.05	1.05	1.04	1.04	1.03	1.02	1.01

(Continued)

TABLE 14-4 (Continued)

Compound Type	Number of carbon atoms (N) in compound, including carbon atoms of functional group											
	1	2	3	4	5	6	7	8	9	10	11	12-20
Sulfur Compounds												
Mercaptans	1.05	1.03	1.02	1.01	1.01	1.01	1.01	1.01	1.01	1.01	1.01	1.01
Sulfides		1.03	1.02	1.01	1.01	1.01	1.01	1.01	1.01	1.01	1.01	1.01
Alcohols												
Alcohols (single-OH group)	1.22	1.31	1.31	1.31	1.31	1.30	1.29	1.28	1.27	1.26	1.24	1.24[b]
Diols (glycols or condensed glycols)		1.33	1.33	1.33	1.33	1.33	1.33	1.33				
Triols (glycerol, etc.)			1.38	1.38	1.38							
Cyclohexanol, cyclohexyl methyl alcohol, etc.						1.20	1.20	1.21	1.24	1.26		
Miscellaneous Compounds												
Ethers (aliphatic only)		1.03	1.03	1.02	1.02	1.02	1.01	1.01	1.01	1.01	1.01	1.01
Oxides (cyclic ethers)		1.08	1.07	1.06	1.05	1.05	1.04	1.03	1.02	1.01	1.01	1.01

a. Carbocyclic or heterocyclic compounds having aliphatic properties.
b. For N = 12 only; no prediction is made for K_F where N > 12.

Notes:
1. Consider any phenyl group as a single carbon atom.
2. K_F factors are the same for all aliphatic isomers of a given compound. For example, $K_F = 1.31$ for n-butyl alcohol, i-butyl alcohol, t-butyl alcohol, and s-butyl alcohol.
3. In organometallic compounds, consider any metallic atom as a carbon atom.
4. For compounds not included in this table, assume $K_F = 1.06$.

Source: Fishtine [4]. *(Reprinted with permission from the American Chemical Society.)*

TABLE 14-5

Values of K_F for Aromatic Hydrogen Bonded Systems[a]

Compound Type	K_F
Phenols (single —OH)	1.15
Phenols (more than one —OH)	1.23
Anilines (single —NH_2)	1.09
Anilines (more than one —NH_2)	1.14
N-substituted anilines (C_6H_5NHR)	1.06
Naphthols (single —OH)	1.09
Naphthylamines (single —NH_2)	1.06
N-substituted naphthylamines	1.03

a. For mixed systems, K_F for the OH group takes precedence. Thus, K_F for p-aminophenol is 1.15.

Source: Fishtine [4]. (*Reprinted with permission from the American Chemical Society.*)

(4) Assume $\Delta Z_b = 0.97$ [8].

(5) Calculate C_2 using Eq. 14-15.

(6) Insert the above values into Eq. 14-14 and calculate ln P_{vp}.

(7) Take the antilog and multiply by 760 to obtain the vapor pressure in mm Hg.

Example 14-1 Estimate the vapor pressure of benzene at 20°C, given $T_b = 353.1$K.

(1) From Table 14-4, $K_F = 1.00$

(2) From Eq. 14-16

$$\Delta H_{vb}/T_b = 1.00 \times [8.75 + 1.987 \ln (353.1)]$$

$$= 20.41 \text{ cal/mol·K}$$

(3) From Eq. 14-15

$$C_2 = -18.00 + 0.19 (353.1)$$

$$= 49.09$$

(4)   From Eq. 14-14

$$\ln P_{vp} = \frac{20.41\ (353.1 - 49.09)^2}{0.97 \times 1.987 \times 353.1} \left[\frac{1}{(353.1 - 49.09)} - \frac{1}{(293 - 49.09)} \right]$$

$$= -2.24$$

$$P_{vp} = \text{antilog}\ (-2.24) \times 760 = 80.6\ \text{mm Hg}$$

The experimental value is 76 mm; hence, the deviation is 6.0%.

Example 14-2   Estimate the vapor pressure of 2-ethyl-butyraldehyde,

$$\overset{\displaystyle O}{\underset{\displaystyle \underset{C_2 H_5}{|}}{\overset{\displaystyle \|}{CH_3 CH_2 CHCH}}}, \text{at } 20^\circ C, \text{given } T_b = 390K.$$

(1)   From Table 14-4, $K_F = 1.06$

(2)   From Eq. 14-16

$$\frac{\Delta H_{vb}}{T_b} = 1.06\ [8.75 + 1.987 \ln (390)]$$

$$= 21.84$$

(3)   From Eq. 14-15

$$C_2 = -18 + 0.19\ (390)$$

$$= 56.1$$

(4)   From Eq. 14-14

$$\ln P_{vp} = \frac{21.84\ (390 - 56.1)^2}{0.97 \times 1.987 \times 390} \left[\frac{1}{(390 - 56.1)} - \frac{1}{(293 - 56.1)} \right]$$

$$= -3.972$$

$$P_{vp} = \text{antilog}\ (-3.972) \times 760 = 14.3\ \text{mm Hg}$$

The experimental value is 14 mm Hg, indicating a deviation of 2.3%.

14-4   METHOD 2

Derivation. As pointed out earlier, when Eq. 14-1 is integrated with the assumption that $\Delta H_v / \Delta Z$ is independent of temperature, the resulting expression yields inaccurate estimates of vapor pressure. However, if the temperature dependence of ΔH_v can be accurately expressed, the resulting integrated form of Eq. 14-1 should be more accurate. One correlation that the author has found to be satisfactory is a modification of the Watson correlation [17]:

$$\Delta H_v = \Delta H_{vb} \left(\frac{1 - T/T_c}{1 - T_b/T_c} \right)^m \tag{14-17}$$

where m is a constant. Equation 14-17 contains the critical temperature T_c. The approximation $T_c \approx 3\,T_b/2$ may be used, yielding

$$\Delta H_v \approx \Delta H_{vb}\,(3 - 2T_{\rho b})^m \tag{14-18}$$

where $T_{\rho b} = T/T_b$. For most organic materials the ratio T_c/T_b varies from 1.3 to 1.7; however, at temperatures below the boiling point, the maximum deviations in ΔH_v using Eq. 14-18 instead of Eq. 14-19 are $+ 2\%$ and -5% respectively. When Eq. 14-18 is substituted into Eq. 14-1 and the result is integrated twice, by parts, the result is

$$\ln P_{vp} \approx - \frac{\Delta H_{vb}}{\Delta Z_b RT_b} \left[\left(\frac{(3 - 2T_{\rho b})^m}{T_{\rho b}} + 2m\,(3 - 2T_{\rho b})^{m-1}\,\ln T_{\rho b} \right)_1^{T_{\rho b}} \right.$$
$$\left. - 4m\,(m - 1) \int_1^{T_{\rho b}} (3 - 2T_{\rho b})^{m-2}\,\ln T_{\rho b}\,dT_{\rho b} \right] \tag{14-19}$$

The integration may be carried out to as many terms as desired; however, sufficient accuracy is obtained by setting the integral in Eq. 14-19 equal to zero. The final result is

$$\ln P_{vp} \approx \frac{\Delta H_{vb}}{\Delta Z_b RT_b} \left[1 - \frac{(3 - 2T_{\rho b})^m}{T_{\rho b}} - 2m\,(3 - 2T_{\rho b})^{m-1}\,\ln T_{\rho b} \right] \tag{14-20}$$

The value of m in Eq. 14-20 depends upon the physical state at the temperature of interest. For all liquids, $m = 0.19$. For solids the following values are recommended:

$$T_{\rho b} > 0.6; \ m = 0.36$$

$$0.6 > T_{\rho b} > 0.5; \ m = 0.8$$

$$T_{\rho b} < 0.5; \ m = 1.19$$

Basic Steps

(1) Obtain T_b (K) from the literature. If it is unavailable, use one of the estimation methods outlined in Chapter 12. (Note that $T\rho_b$ in Eq. 14-20 is equal to T/T_b.)

(2) Obtain K_F from Table 14-4 or 14-5 as appropriate.

(3) Set m = 0.19 for liquids. For solids: if $T\rho_b > 0.6$, m = 0.36; if $0.6 > T\rho_b > 0.5$, m = 0.8; and if $T\rho_b < 0.5$, m = 1.19.

(4) Calculate $\Delta H_{vb}/T_b$ from Eq. 14-16. The value of R is 1.987 cal/mol·K.

(5) Assume $\Delta Z_b = 0.97$.

(6) Insert values obtained above into Eq. 14-20 and calculate ln P_{vp}.

(7) Take the antilog and multiply by 760 to obtain the vapor pressure in mm Hg.

Example 14-3   Estimate the vapor pressure of benzene at 20°C, given $T_b = 353.1$K.

(1)   From Table 14-4, $K_F = 1.00$.

(2)   Since benzene is a liquid at 20°C, m = 0.19.

(3) $T_{\rho b} = 293/353.1 = 0.830$

(4)   From Eq. 14-16

$$\frac{\Delta H_{vb}}{T_b} = 1.00 \times [\,8.75 + 1.987 \ln (353.1)\,]$$

$$= 20.4 \text{ cal/mol·K}$$

(5)   From Eq. 14-20

$$\ln P_{vp} = \frac{20.4}{0.97 \times 1.987} \left\{ 1 - \frac{[3-2(0.830)]^{0.19}}{0.830} \right.$$
$$\left. -2(0.19)\,[3-2(0.830)]^{-0.81} \ln (0.830) \right\}$$

$$= -2.3$$

$$P = \text{antilog}\,(-2.3) \times 760 = 76 \text{ mm Hg}$$

Since the experimental value is 76 mm, the deviation is 0%.

Example 14-4 Estimate the vapor pressure of DDT at 20°C.

(1) As DDT does not have a normal boiling point, the techniques of Chapter 12 are used to estimate it. T_b (est.) = 613K via Meissner's method (§12-3).

(2) From the listing for polyhalides in Table 14-4 (see also note 1), K_F = 1.05.

(3) $T_{\rho b}$ = 293/613 = 0.478

(4) As DDT is a solid at 20°C and $T_{\rho b} < 0.5$, m = 1.19.

(5) From Eq. 14-16

$$\frac{\Delta H_{vb}}{T_b} = 1.05 \times [8.75 + 1.987 \ln (613)]$$

$$= 22.6 \text{ cal/mol·K}$$

(6) From Eq. 14-20

$$\ln P_{vp} = \frac{22.6}{0.97 \times 1.987} \left\{ 1 - \frac{[3 - 2(0.478)]^{1.19}}{0.478} \right.$$

$$\left. - 2(1.19) [3 - 2(0.478)]^{0.19} \ln (0.478) \right\}$$

$$= -22.1$$

P = antilog (−22.1) × 760 = 1.9 × 10⁻⁷ mm Hg, which agrees with the experimental value of 2 × 10⁻⁷.

14-5 ESTIMATION FROM BOILING POINTS AT REDUCED PRESSURE

Quite often, data are available in which a boiling point at reduced pressure is given. As pointed out in §14-2, an estimate of the ratio $\Delta H_v/T$ must then be made. A simple but adequate approximation may be obtained by considering the origin of the Kistiakovskii equation (Eq. 14-16). At the normal boiling point, Eq. 14-16 may be written as

$$\Delta H_{vb} = K_F P_{vp} V_{vb} \ln V_{vb} \qquad (14\text{-}21)$$

where P_{vp} = pressure in atmospheres (= 1 atm at T_b)

V_{vb} = molar volume of the vapor at T_b (cm³).

If we use the ideal gas law, Eq. 14-16 is obtained.

It seems reasonable that an approximate value of $\Delta H_{v1}/T_1$ can be obtained at other temperatures by writing

$$\Delta H_{v1} = K_F P_1 V_{v1} \ln V_{v1} \qquad (14\text{-}22)$$

Again, using the ideal gas law, we obtain

$$\frac{\Delta H_{v1}}{T_1} \approx K_F R \ln\left(\frac{RT_1}{P_1}\right) \qquad (14\text{-}23)$$

where R, outside the parentheses, has the value 1.987 cal/mol·K and a value of 82.05 cm³·atm/K, inside the parentheses. Equation 14-23 can be rearranged to yield

$$\frac{\Delta H_{v1}}{T_1} \approx K_F \left[8.75 + R \left(\ln T_1 - \ln P_1\right)\right] \qquad (14\text{-}24)$$

With this modification, the reference temperature T_1 can be substituted directly for T_b in Eq. 14-14 (Method 1) to yield

$$\ln P_{vp} = \ln P_1 + \frac{\Delta H_{v1} (T_1 - C_2)^2}{\Delta Z_b RT_1^2} \left[\frac{1}{(T_1 - C_2)} - \frac{1}{(T - C_2)}\right] \qquad (14\text{-}25)$$

The same substitution is not strictly valid for Eq. 14-20 (Method 2). However, for practical purposes, little error is introduced if this is done. Thus

$$\ln P_{vp} = \ln P_1 + \frac{\Delta H_{v1}}{\Delta Z_b RT_1} \left\{ 1 - [3 - 2 (T/T_1)]^m \frac{T_1}{T} - \right.$$

$$\left. 2m [3 - 2 T/T_1)]^{m-1} \ln (T/T_1) \right\} \qquad (14\text{-}26)$$

Equation 14-25 is valid for liquids while Eq. 14-26 is applicable to liquids and solids.

Basic Steps. The basic steps are similar to those outlined in §14-3 (Method 1) or §14-4 (Method 2). The only differences are: (1) T_1 is substituted for T_b, and (2) Eqs. 14-24 and 14-25 or 14-26 are used to estimate ΔH_{v1} and P_{vp}, respectively. P_1 in Eqs. 14-24, -25 and -26 must be expressed in atmospheres.

Example 14-5   Estimate the vapor pressure of acetone cyanohydrin, $CH_3C(OH)(CN)CH_3$, at 20°C, given that $P_1 = 23$ mm and $T_1 = 354K$.

(1)   As the material is a liquid at 20°C, Eq. 14-25 may be used.

(2)   From Table 14-4, Note 4, $K_F = 1.06$.

(3) $P_1 = \dfrac{23}{760} = 3.03 \times 10^{-2}$ atm.

(4)   From Eq. 14-24

$$\frac{\Delta H_{v1}}{T_1} = 1.06 \left\{ 8.75 + 1.987 \left[\ln(354) - \ln(0.0303) \right] \right\}$$

$$= 29.0 \text{ cal/mol·K}$$

(5)   From Eq. 14-15

$$C_2 = -18 + 0.19\,(354)$$

$$= 49.3$$

(6)   From Eq. 14-25

$$\ln P_{vp} = \ln(0.0303) + \frac{29.0\,(354-49.3)^2}{0.97 \times 1.987 \times 354} \left[\frac{1}{(354-49.3)} - \frac{1}{(293-49.3)} \right]$$

$$= -6.74$$

$$P_{vp} = \text{antilog}\,(-6.74) \times 760 = 0.9 \text{ mm}$$

The experimental value is 0.8 mm, indicating a deviation of + 12.5%.

Example 14-6   Estimate the vapor pressure, at 298K, of methylene diphenyl isocyanate, $[(OCN)C_6H_4]_2CH_2$, given that $P_1 = 5$ mm and $T_1 = 463K$.

(1)   As the compound is a solid (m.p. = 41°C) at 298K (25°C), Eq. 14-26 will be used.

(2)   From Table 14-4, Note 4, $K_F = 1.06$

(3) $P_1 = \dfrac{5}{760} = 6.58 \times 10^{-3}$ atm

$T/T_1 = 298/463 = 0.644$

(4)   From Eq. 14-24

$$\frac{\Delta H_{v1}}{T_1} = 1.06 \left\{ 8.75 + 1.987 \left[\ln(463) - \ln(0.0066) \right] \right\}$$

$$= \quad 32.8 \text{ cal/mol·K}$$

(5) As we do not know the normal boiling point, a value of 0.8 is chosen for m. (If more confidence in the selected value of m is desired, a boiling point could be estimated via the methods in Chapter 12.)

(6) From Eq. 14-26

$$\ln P_{vp} = \ln (0.0066) + \frac{32.8}{0.97 \times 1.987} \left\{ 1 - \frac{[3-2(0.644)]^{0.8}}{0.644} - \right.$$

$$\left. 2 \times 0.8 \times [3-2(0.644)]^{-0.2} \times \ln(0.644) \right\}$$

$$= -17.9$$

$$P_{vp} = \text{antilog} (-17.9) \times 760 = 1.29 \times 10^{-5} \text{ mm}$$

The experimental value is 1.0×10^{-5} mm, indicating a deviation of 29%.

14-6 AVAILABLE DATA

Vapor pressure data for petroleum chemicals can be obtained from the general compilations listed in Appendix A. References 2, 5, and 14 list the vapor pressures of some other materials, such as pesticides.

14-7 SYMBOLS USED

A_1 = constant in Eq. 4-3

A_2 = constant in Eq. 14-12

B_1 = constant in Eq. 14-3

B_2 = constant in Eq. 14-12

C_2 = constant in Eq. 14-12

ΔH_v = heat of vaporization (cal/mol)

ΔH_{v1} = heat of vaporization at T_1 (cal/mol)

ΔH_{vb} = heat of vaporization at the normal boiling point (cal/mol) in Eq. 14-16

K_F = constant in Eq. 14-16

K_{ow} = octanol-water partition coefficient

m = exponent in Watson correlation (Eq.14-17)

P_1 = reference vapor pressure (mm Hg)

P_c = critical pressure (atm)

P_{vp} = vapor pressure (mm Hg)

R = gas constant $\equiv$ 1.987 cal/mol·K (= 82.057 cm³·atm/mol·K in Eq. 14-2)

ΔS_{vb} = entropy of vaporization at the normal boiling point (cal/mol·K) in Eq. 14-16

t   = temperature (°C)

T   = temperature (K)

T_1 = reference temperature (K)

T_b = temperature of the normal boiling point (K)

T_c = critical temperature (K)

T_m = melting point (K)

$T\rho_b$ = T/T_b

V_{vb} = molar volume of vapor at T_b (cm³)

V_{v1} = molar volume of vapor at T_1 (cm³)

ΔV = volume difference (cm³/mol) between vapor and liquid in Eq. 14-2

ΔZ = compressibility factor in Eq. 14-2

ΔZ_b = compressibility factor at the normal boiling point

Subscripts

b   = boiling point

c   = critical point

m   = melting point

ρ = ratio

v   = vapor, vaporization

vp   = vapor; used with P (P_{vp}) to denote vapor pressure

14-8   REFERENCES

1. Antoine, C., "Tensions des Vapeurs: Nouvelle Relation Entre les Tensions et les Températures," *Compt. Rend.,* **107**, 681 (1888).

2. Balson, E.W., "Studies in Vapor Pressure Measurement, Part III, An Effusion Manometer Sensitive to 5 x 10⁻⁶ mm of Mercury: Vapor Pressure of D.D.T. and Other Slightly Volatile Substances," *Trans. Faraday Soc.,* **43**, 54 (1947).

3. Chiou, C.T. and V.J. Freed, "Chemodynamic Studies on Bench Mark Industrial Chemicals," Annual Report to National Science Foundation on Contract AEN7617700, Report No. NSF/RA-770286, National Science Foundation, Washington, D.C. (1977). (Available from NTIS as PB 274263.)

4. Fishtine, S.H., "Reliable Latent Heats of Vaporization," *Ind. Eng. Chem.,* **55**, 47 (June 1963).

5. Goring, C.A.I. and J.W. Hanamaker (eds.), *Organic Chemicals in the Soil Environment,* Marcel Dekker, New York (1972).

6. Kistiakovskii, Vl. A., "Latent Heat of Vaporization," *J. Russ. Phys. Chem. Soc.,* **53**,I, 256-64 (1921).

7. Mackay, D., University of Toronto, Toronto, Canada, personal communication, June 30, 1980.

8. Miller, D.G., "Estimating Vapor Pressures — a Comparison of Equations." *Ind. Eng. Chem.,* **56**, 46 (March 1964).

9. Miller, D.G., "Derivation of Two Equations for the Estimation of Vapor Pressures," *J. Phys. Chem.,* **68**, 1399 (1964).

10. Partington, J.R., *An Advanced Treatise on Physical Chemistry,* Vol. 2, Longmans, Green and Co., London, 1951.

11. Perry, R.H. and C.H. Chilton (eds.), *Chemical Engineers' Handbook,,* 5th ed., McGraw-Hill Book Co., New York, p. 3-246 (1973).

12. Reid, R.C., J.M. Prausnitz and T.K. Sherwood, *The Properties of Gases and Liquids,* 3rd ed, McGraw-Hill Book Co., New York (1977).

13. Stull, D.R., "Vapor Pressure of Pure Substances: Organic Compounds," *Ind. Eng. Chem.,* **39**, 517 (1947).

14. Swan, R.L., P.J. McCall and S.M. Unger (Dow Chemical USA, Midland, MI), "Volatility of Pesticides from Soil Surfaces," unpublished manuscript (no date), rec. by ADL 11-79.

15. Thomson, G.W., in *Techniques of Organic Chemistry,* ed. by A. Weissberger, 3rd ed., Vol. I, Part 1, Interscience, New York, p. 473 (1959).

16. Yaws, C.L. (ed.), *Physical Properties: A Guide to the Physical, Thermodynamic and Transport Property Data of Industrially Important Chemical Compounds,* Chemical Engineering, McGraw-Hill, New York (1977).

17. Watson, K.M., "Thermodynamics of the Liquid State — Generalized Prediction of Properties," *Ind. Eng. Chem.,* **35**, 398 (1943).

15

VOLATILIZATION FROM WATER

Richard G. Thomas

15-1   INTRODUCTION

The vaporization of organic chemicals from water bodies is an important mass-transfer pathway from water to air. Knowledge of volatilization rates is necessary to determine the amount of chemical that enters the atmosphere and the change of pollutant concentrations in water bodies. The transfer process from the water to the atmosphere is dependent on the chemical and physical properties of the pollutant in question, the presence of other pollutants, and the physical properties (e.g., flow velocity, depth, and turbulence) of the water body and atmosphere above it. The factors that control volatilization are the solubility, molecular weight, and vapor pressure of the chemical and the nature of the air-water interface through which it must pass.

The mathematical modeling of volatilization involves the use of interphase exchange coefficients that depend on the properties mentioned above, some of which are difficult to measure or estimate in the actual environment. They can be measured under controlled laboratory conditions, but the results often cannot be extended with confidence to the varied and changeable conditions encountered in the environment. Part of this difficulty is due to the lack of environmental volatilization data against which laboratory-based hypotheses can be tested.

Thus, estimates of volatilization rates from surface waters on the basis of mathematical data and laboratory measurements are necessarily of unknown precision. No attempt is made here to give a quantitative estimate of the error implicit in the methods to be described; however, comparisons of experimental results with theoretical predictions indicate that these predictive techniques are generally in agreement with actual processes within about a factor of ten at the most and probably within a factor of two or three in most cases.

Volatilization rates from water vary over a large range. Some chemicals volatilize from well-mixed surface waters quite rapidly, with a half-life on the order of hours; others may remain in the water almost indefinitely unless they degrade or are removed by a different transfer mechanism. For example, trichloroethylene has a computed half-life of three to five hours for volatilization from a river; the pesticide dieldrin, on the other hand, volatilizes more slowly than water and its concentration would actually increase, at least in the short term, so that its half-life due to volatilization is on the order of a year or more.

This chapter describes the volatilization process, discusses some of the theoretical methods that have been developed to model the volatilization of chemicals from surface waters, and presents a method for estimating the rate at which this process takes place. Various organic chemicals are listed together with their properties related to volatilization, and basic steps and examples are presented to show how to compute the mass transfer coefficients and half-lives in water for organic chemicals.

15-2 MODELING VOLATILIZATION IN THE ENVIRONMENT

Many factors affect the volatilization process. Although these factors are known, they can change rapidly and over a wide range in a natural environment. This complicates the task of providing average or mean values for use in an analytical model. The processes are often nonlinearly interdependent and do not behave in simple, deterministic ways.

Specifically, the volatilization process depends on the thermodynamic or physical properties of a chemical, particularly its aqueous solubility, vapor pressure, Henry's law constant and diffusivity coefficient, and the presence of modifying materials such as adsorbents, organic films, electrolytes, and emulsions [10,11]. The true "dissolved" and "total" concentrations are highly dependent on the presence of these modifying materials [3]. The values of the rate-controlling factors also depend on the physical and chemical properties of the water body, such as its depth, flow rate, the presence of waves, sediment content, and the other pollutants present. Atmospheric conditions, particularly wind speed and stability, also affect the rate-controlling factors. Volatilization is, in general, relatively temperature-insensitive, since the principal effect of temperature is on the vapor pressure. The latter has little influence on volatility except for the few classes of chemicals whose volatilization is controlled by processes that occur in the vapor phase. (These are addressed later.)

The transport and transfer of a chemical may involve several sequential stages depending on the type of water body involved. Each of these stages has a characteristic rate, diffusion velocity, or resistance [12], and the slowest stage controls the overall volatilization rate. For a stratified lake, the stages may be: (1) release from the sorbed state on sediments; (2) diffusion through the hypolimnion; (3) diffusion through the thermocline; (4) diffusion through the epilimnion to the near surface (approximately one millimeter below the surface); (5) diffusion through the liquid surface "stagnant film"; (6) transfer across the water/air interface; and (7) diffusion through the atmospheric film to the bulk of the atmosphere. Transfer through the different layers in the water can occur only by bulk movement (as in the case of turbulent eddy motion) or by molecular diffusion [9,14]. If the water body is well mixed, as in a flowing river, most of the resistance to transport lies in the gas- and liquid-phase interfacial layers a few millimeters or centimeters above or below the surface [10,11,13]. The interface between the gas- and liquid-phase interfacial layers is believed to offer little or no resistance. Hence, a concentration gradient develops in the surface layers. For most substances and most water bodies, resistance in one phase tends to dominate [9].

Under given conditions of turbulence, layer thicknesses vary both spatially and temporally [9]. High turbulence in the liquid causes the liquid film or boundary layer to be thin; similarly, high turbulence in the gas causes the gas layer to be thin [18]. Wind also modifies surface hydrodynamics, affecting chemical mass-transfer coefficients.

Vertical transport to the surface of rivers and large lakes is controlled by currents, both direct and wind-induced. In rivers, diffusion from the bottom to the surface is accelerated by eddies caused by the interaction of the current with the bottom [12]. Since turbulence is generated mainly at the river bottom, the deeper rivers have more quiescent surfaces, and the transfer resistance tends to be higher [10]. In lakes, wind speed and fetch (the length of water over which the wind blows) are the controlling factors: turbulence is normally present in the atmosphere, but there is usually little in the water unless it is induced by wind-generated waves on the surface, subsurface springs, or thermally driven convective action. Regardless of the cause, turbulence can greatly increase the liquid-phase exchange coefficient.

Horizontal and vertical turbulent diffusion affect the rate and extent of mixing [10]. In water, horizontal turbulent diffusion is usually an order of magnitude faster than vertical turbulent diffusion; as a result, pollutants spread faster laterally than they do vertically. In the atmosphere, vertical diffusion is usually more rapid than in the water, and chemicals are transported from the interface quickly. During the stable conditions of temperature inversions (when temperature increases with height), however, vertical atmospheric diffusion decreases, the water surface is calmer, resistance to chemical transfer through the air and water boundary films increases, and volatilization rates decrease.

Sediments are important because they can act as permanent or temporary sinks for chemicals in the water [10,12]. The exchange between the water column and the sediment has a significant effect on the rate of removal from the water proper. The sediment/water partition (sorption) coefficient is affected by the characteristics of the sediment, e.g., type, spatial distribution, particle size and density, and organic matter. Chapter 4 of this handbook describes the adsorption phenomenon and methods for estimating adsorption coefficients.

15-3   APPROACHES TO ESTIMATION OF THE VOLATILIZATION RATE

The two-layer film or resistance concept of the interface — i.e., the theory that resistance to mass transport exists in both the gas- and liquid-phase interfacial layers — was first discussed in 1923 [23]. However, most of the analytical work has been done only in the last several years. Four basic approaches have been used; these are described below.

Method of Mackay and Wolkoff. These authors [14] analyzed the volatilization of chemicals from bodies of water on the basis of thermodynamic equilibrium considerations. This theory expresses the flux from a solution in water to the air above in terms of the ratio of contaminant mass in the vapor phase to the total vapor mass of the water plus chemical, expressed as a function of the chemical vapor pressure. The following assumptions are made:

(1) The contaminant concentration used is that which is truly in solution; there are no colloidal, suspended, ionic, complexed, or adsorbed forms of the contaminant.

(2) The concentration of the diffusing substance in the vapor adjacent to the interface is that which is in equilibrium with the concentration in the liquid at the interface.

(3) Diffusion or mixing in the liquid is sufficiently rapid that concentrations at the liquid side of the interface are equal to concentrations in the bulk of the liquid (which implies thorough mixing).

(4) The water evaporation rate is negligibly affected by the presence of the contaminant.

Ancillary to these assumptions are two others that affect the utility of the theory: (a) evaporation is the limiting process in the total volatilization process, and (b) there are no concentration gradients in the upper layers due to the evaporation. The latter follows from assumption (3) above, namely, that there is perfect mixing in the water phase and equilibration between the water and air. Because of these assumptions, this method will overestimate the volatilization rate if mixing or diffusion in the water body is slow, thus retarding the overall process.

This method is applicable only to a restricted class of compounds and is thus not recommended for general use. The significant conclusion reached by Mackay and Wolkoff was that volatilization may be significant, i.e., half-lives short, for compounds which have vapor pressures much lower than that of water, provided that the compound is sparingly soluble.

In a subsequent study, Mackay and Leinonen [11] extended the method to include consideration of the resistance due to diffusion in the liquid phase and estimated volatilization half-lives for all classes of compounds. Their work was an extension of methods previously developed by Liss and Slater [9].

Method of Liss and Slater. The volatilization process was analyzed on the basis of a two-layer film by Liss and Slater [9]. This has been shown [16] to be a more realistic approach than that originally developed by Mackay and Wolkoff [14]. The main water body is assumed to be well mixed, with a thin layer on the surface in which there is a concentration gradient. The air above is assumed to be well mixed (i.e., the background concentration is low), and a thin layer in contact with the surface contains another concentration gradient. (Thus, diffusion in the water body is not assumed to be a rate-limiting process.) At the interface between these two layers is a concentration discontinuity, and the ratio of concentrations across it (air to water) is assumed to equal the Henry's law constant.

Transfer through these films is by straightforward molecular diffusion. The molecules are assumed to diffuse through the layers at a rate dependent on the phase exchange coefficients found in the equations rather than to vaporize directly from solution along with the water vapor.

Environmental conditions leading to turbulence in either phase influence the thickness, diffusivity, resistance, and geometry of the layers. Since resistance to diffusion is dependent on layer geometry and composition, the molecular phase exchange coefficients used in the determination of the overall mass transfer are affected by environmental conditions. These coefficients are somewhat empirical, in that they cannot yet be readily computed using basic physical principles, but values for the gas-phase and liquid-phase exchange constants have been determined for the transfer of certain gases across the air/sea interface. These can be adjusted to apply to certain other classes of chemicals; however, the best method for doing this is not clear [10,15]. Schwarzenbach *et al.* [17] indicate that the coefficients based on open-ocean data may overestimate the transfer in lakes.

Method of Chiou and Freed. Chiou and Freed [1,2] present another method for estimating the volatilization of chemicals. It appears to be based on gas dynamic and thermodynamic considerations involving the mean free path of molecules and the vapor pressures of the chemical. A Langmuir-type equation is used to describe observed rates of volatilization from both single-component and multicomponent systems. No data have been offered to support the validity of this method, and its efficacy is not known. Little notice has been taken of it in recent studies of volatilization of chemicals from water.

Method of Smith et al. This approach is based on reaeration studies by Tsivoglou [21], who demonstrated that inert gases could be used as tracers for oxygen reaeration measurement. Since the transfer rates of oxygen and the inert gases are controlled by diffusion in the near surface film, these rates are similar; a correction is necessary only for the differing diffusivities, which are related primarily to molecular diameter. Smith *et al.* [18,19] applied this approach to other compounds and verified that the magnitude of the diffusivity correction term was as expected for compounds that experience the same liquid-phase resistance as oxygen. Compounds with lower Henry's law constants which also experience a gas-phase resistance volatilize more slowly, so this method of approximation is not applicable to such compounds.

For the class of compounds to which it is applicable, the advantage of this approach is that the volatilization rate can be related to the rate of oxygen reaeration. The latter is known for natural systems such as rivers and lakes. If one can obtain a laboratory-measured ratio of the reaeration rate of a chemical to that of oxygen, it can simply be multiplied by the environmental value of oxygen reaeration to yield an environmental value of the volatilization rate constant for the chemical. These investigators have also developed procedures for deriving the ratio when an experimental value cannot be found.

Smith *et al.* have demonstrated that their method is valid for the class of chemicals which have high volatility, high molecular weight, and low solubility (i.e., high Henry's law constants).

15-4 METHOD ERRORS

The environmental, physical, and chemical processes that control the volatilization rate have been described above. Their number and variety indicate the difficulties involved in devising an adequate model of the process. The fundamental mechanisms are reasonably well understood, however, and can be described by relatively simple mathematical expressions [10]; the results are often practically the same as those from more sophisticated models. The present inadequacies are primarily due to a lack of data for some of the parameters in the equations [10].

At the most basic level, some properties of chemicals may not be accurately known; these are needed to determine the Henry's law constant, which indicates the propensity of a chemical to volatilize. The gas- and liquid-phase exchange coefficients, on which the predictive techniques ultimately depend, are not known with certainty for diverse environmental conditions and a wide range of chemicals. Environmental and hydrodynamic factors that affect the movement of a chemical in water and its transfer into the air are difficult to quantify and relate to the volatilization rate; these factors may include wind speed, stratification, sediment content, and the presence of other pollutants. The range of values for these factors over a period of time, their variable nature, and their nonlinear interdependencies indicate the problems inherent in using time-averaged or mean values and reduce the validity of the calculations.

The overall environmental mass transfer coefficient for a particular chemical leaving a given water body, which is dependent on gas- and liquid-phase transfer coefficients and chemical properties, is difficult to predict on the basis of laboratory studies. This is especially true if average values are desired, since the phase exchange coefficients are sensitive to variable environmental and hydrodynamic factors [17]. The methods used for estimating the exchange coefficients have been verified for only a few chemicals in a few laboratory experiments using stirrers and fans to simulate environmental conditions [3-5,8,14,16,18,19]. The overall volatilization rates measured in these experiments are often quite similar to those computed by theoretical methods [20]. This agreement may be fortuitous, however, since the basic factors are so dissimilar: the computations use phase exchange coefficients based loosely on environmental data, while the experiments use measured rates of stirring, air speed, etc. Nevertheless, the differences between experiment and computation are well within an order of magnitude, lending credibility to the predictive techniques.

Schwarzenbach *et al.* [17], in a study of dichlorobenzene (DCB) and tetrachloroethylene in Lake Zurich, Switzerland, suggested that the average mass transfer coefficient computed from an overall mass balance is about a factor of ten larger than that derived from mass balance computations based on other measurements of DCB in the lake. However, they note that their observed value of the overall mass transfer coefficient for the lake compares well with those found in a similar study of small Canadian lakes by different investigators.

The laboratory-based volatilization rates (computed) are comparable to those found for the open ocean; this is not surprising, as the phase exchange coefficients used in the calculations are based on open-ocean data. It may be incorrect to apply open-ocean data to lakes and other smaller bodies of water, but this question has not been discussed in the literature.

In view of these observations and the difficulty of performing in-situ volatilization experiments, it is not possible to quantify the error in the calculated values of the volatilization rate constants. The lake example indicates that the error may be as large as a factor of ten, although laboratory data suggest that it could be much less. When one is applying the results of calculations to actual environmental situations, it would probably be advisable to assume that the values of volatilization rate may be high by a factor of ten at most and low by a smaller factor of possibly three.

15-5   METHODS OF ESTIMATION

Recommended General Method. The methods of estimation recommended in this chapter follow the two-film concept for estimating the flux of volatiles across the air-water interface. This was described by Liss and Slater [9] and extended by Mackay and others [3,10-15]. Additional refinements suggested by other investigators are introduced where useful. Figure 15-1 shows the basics of this concept.

The method is based on a finite difference approximation to Fick's law of diffusion, which can be written as

$$N = k \, \Delta C \tag{15-1}$$

where

N =   flux (g/cm^2·s)
k = D/z, a first-order exchange constant (cm/s)
D =   coefficient of molecular diffusion of chemical in the film (cm^2/s)
z =   film thickness (cm)
ΔC =   concentration difference across the film (g/cm^3)

In a steady-state process, Eq. 15-1 becomes

$$N = k_g \, (C_g - C_{sg}) = k_l \, (C_{sl} - C_l) \tag{15-2}$$

where

k_g =   gas-phase exchange coefficient (cm/s)
C_g =   concentration in gas phase at the outer edge of the film (g/cm^3)
C_{sg} =   concentration in gas phase at interface (g/cm^3)
k_l =   liquid-phase exchange coefficient (cm/s)
C_{sl} =   concentration in liquid phase at interface (g/cm^3)
C_l =   concentration in liquid phase at the outer edge of the film (g/cm^3)

The nondimensional Henry's law constant (H′) relates the concentration of a compound in the gas phase to its concentration in the liquid phase:

$$H' = C_{sg}/C_{sl} \tag{15-3}$$

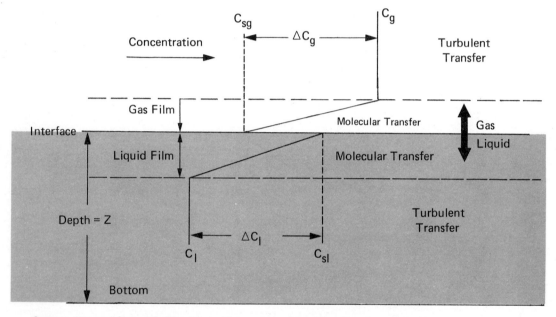

Source: Liss and Slater [9]. *(Reprinted with permission from Macmillan Journals Ltd.)*

FIGURE 15–1   Two-Layer Model of Gas-Liquid Interface

Equation 15-2 can then be written as

$$N = \frac{C_g - H'C_l}{1/k_g + H'/k_l} = \frac{C_g/H' - C_l}{1/k_l + 1/H'k_g} \qquad (15\text{-}4)$$

The overall mass transfer coefficients for the gas phase (K_G) and liquid phase (K_L) can be defined as follows:

$$1/K_G = 1/k_g + H'/k_l \qquad (15\text{-}5)$$

and

$$1/K_L = 1/k_l + 1/H'k_g \qquad (15\text{-}6)$$

By substitution in Eq. 15-4,

$$N = K_G (C_g - H'C_l) = K_L (C_g/H' - C_l) \qquad (15\text{-}7)$$

The Henry's law constant can also be written in the form:

$$H = P_{vp}/S \qquad (15\text{-}8)$$

where P_{vp} is in atm, S is in mol/m^3, and H is in atm-m^3/mol. Figure 15-2, a graphical representation of Eq. 15-8, shows values of H for water, air, and numerous organic compounds.

When H is calculated by Eq. 15-8, the data must be for the same temperature and applicable to the same physical state of the compound. For example, P_{vp} for a liquid should not be divided by S for the solid state. (This error can occur if P_{vp} is estimated by extrapolating data from higher pressures.) Furthermore, only data for the pure compound should be used, as the vapor pressures and solubilities of mixtures — e.g., polychlorinated biphenyl isomers — may be suspect.

Note that Eq. 15-8 is only approximate. If measured values of P_{vp} and S are not available, they can be estimated by methods given in Chapters 14 and 2, respectively. An alternate method of estimating S, via estimated activity coefficients, is provided in Chapters 3 and 11.

Mackay and Leinonen [11] give a slightly different but equivalent expression for $1/K_L$ to be used when H is computed according to Eq. 15-8:

$$1/K_L = 1/k_l + 1/(Hk_g/RT) \qquad (15\text{-}9)$$

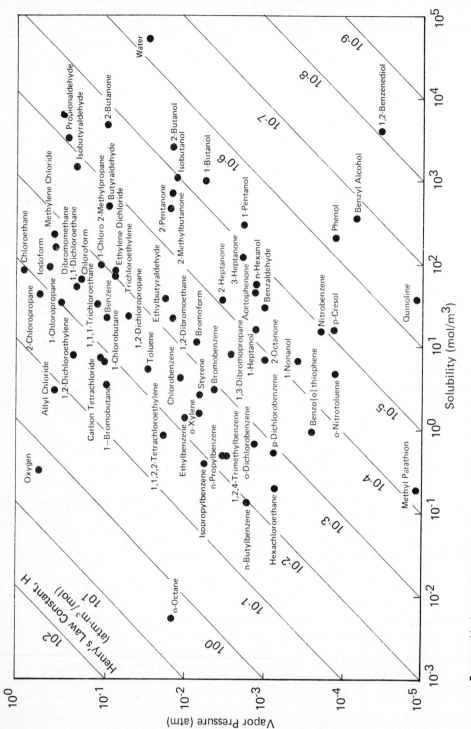

Source: Mackay and Yuen [15]. *(Reprinted with permission from the authors.)*

FIGURE 15-2   Solubility, Vapor Pressure and Henry's Law Constant for Selected Chemicals

where T is the absolute temperature (K) and R is the gas constant, 8.2×10^{-5} atm-m^3/mol-K. (At 20°C, RT = 2.4×10^{-2} atm-m^3/mol.) Equations 15-6 and 15-9 give identical values of $1/K_L$, since H/RT equals H'.

The same authors give the following equation for the flux:

$$N = K_L \left(C - \frac{P}{H} \right) \qquad (15\text{-}10)$$

P/H must be expressed in g/cm^3. If we assume the background atmospheric level to be negligible and integrate Eq. 15-10, the concentration at any time t can be expressed as

$$C = C_0 e^{-k_v t} = C_0 e^{-K_L \, t/Z} \qquad (15\text{-}11)$$

where

C_0 = initial concentration (g/cm^3)
k_v = K_L/Z = volatilization rate constant
Z = mean depth of the water body (cm)

The half-life can be written as

$$\tau_{1/2} = 0.69 \, Z/K_L = 0.69/k_v \qquad (15\text{-}12)$$

The term $1/K_L$ $(=R_L)$ can be thought of as the total resistance to flux [2]. It depends on the exchange constants of the individual phases and the value of the Henry's law constant. Similarly, the individual terms $1/k_l$ and RT/Hk_g (or $1/H'k_g$) can be thought of as liquid-phase resistance r_l and gas-phase resistance r_g respectively. The values of these resistances indicate the relative importance of the gas and liquid phases in the exchange of a compound.

In addition to the resistances offered by the gas and liquid phases, another resistance can be analyzed [15] — namely, the resistance r_w to transfer from the bulk of the water body to the interface. (This resistance can be added to r_l and r_g to give the total resistance, R_L.) The bulk water resistance can be expressed as

$$r_w = \tau_D /Z \qquad (15\text{-}13)$$

where τ_D is a characteristic time for the eddying motion of a turbulent water body to transport the volatile material to the surface. The value of τ_D indicates whether volatilization is limited by turbulent diffusion to the surface.

Associated with this transport is a turbulent diffusivity coefficient, which is a measure of the efficiency of macroscopic eddy motion in mixing the water. Because of stream geometry, the turbulent diffusivity takes on different values associated with different directions. Generally the value in the longitudinal direction, which is usually reported in the literature, is much larger than the values for the other directions because of larger eddy scale and intensity. Elder [6] relates the perpendicular turbulent diffusivity, D_z, to the longitudinal value, D_L, by

$$D_z \approx 0.039 \, D_L \qquad (15\text{-}14)$$

The time τ_D associated with movement a mean distance Z, which in this case is the mean depth of the water body, is

$$\tau_D = Z^2/1.3 D_z \qquad (15\text{-}15)$$

If this time is assumed to be equivalent to the half-life for the turbulent transfer process from depth Z to the surface, it can be compared with the volatilization half-life given in Eq. 15-12.

Typical values of the aquatic turbulent diffusivity are shown in Table 15-1. Values for streams, rivers, and estuaries were given as longitudinal diffusivities and converted to perpendicular values by Eq. 15-14.

TABLE 15-1

Typical Values of Aquatic Turbulent Diffusivities

Water Body	D_z (m^2/sec)	Source
Flumes and Small Streams	10^{-4} to 10^{-2}	Gloyna [7]
Large Rivers	10^{-2} to 1	
Estuaries	$1-20$	
Lakes		
Hypolimnion	10	Mackay [10]
Thermocline	1	
Epilimnion	3×10^5	

Surface active agents (surfactants) can reduce or inhibit volatilization. These agents form a layer one or more molecules thick on the surface of the water. The resistance of this layer is given by Smith *et al.* [19] as

$$r_s = 1/H'k_s = RT/Hk_s \qquad (15\text{-}16)$$

k_s is the mass transfer coefficient at the interface and is defined as

$$k_s = \alpha\sqrt{RT/2\pi M} \qquad (15\text{-}17)$$

where

α =   the accommodation coefficient, or the fraction of molecules striking the surface that condense on the surface
R   =   gas constant = 8.3×10^7 ergs/mol-K
T   =   temperature (K)
M   =   molecular weight (g/mol)

These authors give no values for α and no method for estimating it.

The values of H for different chemicals give some insight into the controlling rate processes. Figure 15-3 postulates certain ranges of H [10,15] and presents some generalizations regarding the volatility of chemicals that fall in these ranges. (As in Figure 15-2, H is computed by Eq. 15-8 with P_{vp} in atm and S in mol/m^3.)

- If H is less than about 3×10^{-7} atm-m^3/mol, the substance is less volatile than water and its concentration will increase as the water evaporates. Humidity in the air reduces the volatilization rate of water somewhat, so the lower limit can be set at about 10^{-7}. The substance could be considered essentially nonvolatile.
- In the range $10^{-7} < H < 10^{-5}$ atm-m^3/mol, the substance volatilizes slowly at a rate dependent on H. The gas-phase resistance dominates the liquid-phase resistance by a factor of ten at least. The rate is controlled by slow molecular diffusion through air.
- For H below about 2×10^{-5} atm-m^3/mol, the pollutant tends to partition into the liquid (i.e., it is quite soluble) and the transfer is gas-phase-controlled.

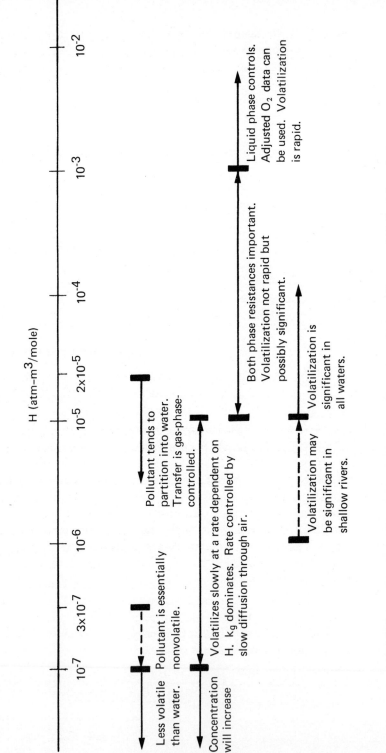

H (atm-m^3/mole)

Less volatile than water. Pollutant is essentially nonvolatile.

Concentration will increase

Pollutant tends to partition into water. Transfer is gas-phase-controlled.

Volatilizes slowly at a rate dependent on H. k_g dominates. Rate controlled by slow diffusion through air.

Both phase resistances important. Volatilization not rapid but possibly significant.

Liquid phase controls. Adjusted O$_2$ data can be used. Volatilization is rapid.

Volatilization may be significant in shallow rivers.

Volatilization is significant in all waters.

FIGURE 15—3   Volatility Characteristics Associated with Various Ranges of Henry's Law Constant

- In the range of $10^{-5} < H < 10^{-3}$ atm-m³/mol, liquid-phase and gas-phase resistances are both important. Volatilization for compounds in this range is less rapid than for compounds in a higher range of H but is still a significant transfer mechanism. Polycyclic aromatic hydrocarbons and halogenated aromatics lie in this range.

- Where H is high ($> 10^{-3}$ atm-m³/mol), the resistance of the water film dominates by a factor of at least ten. The transfer is liquid-phase-controlled. In this region $k_l \ll Hk_g/RT$ (or $r_l \gg r_g$), and Eq. 15-10 becomes

$$N = k_l \left(C - \frac{P}{H} \right)$$

(15-18)

where the flux N is in g/cm²·s and P/H is g/cm³. For most hydrocarbons that are only sparingly water-soluble (hydrophobic) and have relatively high values of the Henry's law constant, the resistance lies in the liquid phase.

If the atmospheric concentration (and thus P/H) is negligible, the transfer coefficient is independent of the value of the Henry's law constant, so the latter can be disregarded in the overall mass transfer rate equation (15-10).

Temperature affects volatilization mainly through its effect on H via its effect on vapor pressure, but it also influences k_l through its effect on diffusivity. Since in this case the volatilization rate is independent of H, the temperature effect is slight.

Method of Smith et al. For high-volatility compounds with $H > 10^{-3}$, Smith *et al.* [18,19] have developed a method for using the oxygen reaeration rate constant to determine the first-order volatilization rate constant. They show that[1]

$$k_v^c/k_v^o = K_L^c/K_L^o = d^o/d^c \approx D^c/D^o$$

(15-19)

where

k_v^c = overall liquid-phase exchange coefficient or first-order volatilization rate constant — chemical (hr^{-1})

---

1. The relation to D^c/D^o, while not exact, is useful for estimation purposes.

k_v^o = oxygen overall liquid-phase exchange coefficient or oxygen reaeration rate constant (hr⁻¹)

K_L^c = overall liquid-film mass transfer coefficient — chemical (hr⁻¹)

K_L^o = overall liquid-film mass transfer coefficient — oxygen (hr⁻¹)

D^c = diffusion coefficient in solution — chemical (cm²/s)

D^o = diffusion coefficient in solution — oxygen (cm²/s)

d^o = molecular diameter — oxygen (cm)

d^c = molecular diameter — chemical (cm)

If the oxygen reaeration rate constant is known for a given water body or type of water body, it is clear from Eq. 15-19 that the volatilization rate constant can be estimated from either the ratio of diffusivities or the ratio of molecular diameters:

$$k_v^c = k_v^o (d^o/d^c) \approx k_v^o (D^c/D^o) \qquad (15\text{-}20)$$

Table 15-2 compares measured and predicted values of these ratios.

TABLE 15-2

Measured Reaeration Coefficient Ratios for High-Volatility Compounds

Compound	H $\left(\dfrac{atm\text{-}m^3}{mole}\right)$	Measured k_v^c/k_v^o	Predicted d^o/d^c	Diffusion Coeff. Ratio D^c/D^o	Molecular Wt. Ratio $(M^o/M^c)^{0.5}$
Chloroform	3.8×10^{-3}	$\left\{\begin{array}{l}.57 \pm .02 \\ .66 \pm .11\end{array}\right\}$	.40	.47	.52
1,1-Dichloroethane	5.8×10^{-3}	$.71 \pm .11$	.44	.47	.57
Oxygen	7.2×10^{-2}	1.0	1.0		
Benzo[b]thiophene	2.7×10^{-4}	$.38 \pm .08$	.38		
Dibenzothiophene	4.4×10^{-4}	.14	.33		
Benzene	5.5×10^{-3}	$.57 \pm .02$		.45	.64
Carbon dioxide		$.89 \pm .03$		.84	.85
Carbon tetrachloride	2.3×10^{-2}	$.63 \pm .07$		.43	.47
Dicyclopentadiene		$.54 \pm .02$		.31	.49
Ethylene	8.6	$.87 \pm .02$		.70	1.06
Krypton		$.82 \pm .08$		.78	.62
Propane		$.72 \pm .01$		.53	.85
Radon		$.70 \pm .08$		.66	.38
Tetrachloroethylene	8.3×10^{-3}	$.52 \pm .09$		.40	.44
Trichloroethylene	1×10^{-2}	$.57 \pm .15$		.44	.49

Source: Smith *et al.* [18,19]. *(Reprinted with permission from the American Institute of Chemical Engineers.)*

To extend the utility of this method, Smith *et al.* [19] have measured the ratio k_v^c/k_v^o in the laboratory for several chemicals. The volatilization coefficient $(k_v^c)_{env}$ can be estimated from

$$(k_v^c)_{env} = (k_v^c/k_v^o)_{lab} \, (k_v^o)_{env} \qquad (15\text{-}21)$$

In principle, k_v^c is the same as K_L/Z, but since k_v^c has the depth and other water-body characteristics embedded within it due to the use of k_v^o, no adjustment is required to use it directly in Eq. 15-11 or 15-12.

The ratio $(k_v^c/k_v^o)_{lab}$ was found to be independent of turbulence conditions for high-volatility compounds with $H > 6.5 \times 10^{-3}$ atm-m^3/mol. It was also found to be independent of k_v^o over the range $0.05 < k_v^o < 15$ hr^{-1} and independent of temperature from 4°C to 50°C.

Equation 15-21 applies particularly to rivers. For lakes and ponds, the following equation may be more accurate:

$$(k_v^c)_{env} = (k_v^c/k_v^o)_{lab}^{1.6} \, (k_v^o)_{env} \qquad (15\text{-}22)$$

Alternatively, $(k_v^c)_{env}$ can be estimated for lakes and ponds by

$$(k_v^c)_{env} \approx (D^c/D^o) \, (k_v^o)_{lake} \qquad (15\text{-}23)$$

and for rivers by

$$(k_v^c)_{env} \approx (D^c/D^o) \, (k_v^o)_{river} \qquad (15\text{-}24)$$

Diffusion coefficients for compounds in water can be estimated by the following equation:[2]

$$D^c = \frac{14 \times 10^{-5}}{\mu_w{}^{1.1}/V_b^{0.6}} \qquad (15\text{-}25)$$

where μ_w is the viscosity of water (cp) and V_b is the molar volume of the chemical at its normal boiling point (cm^3/mol). The value of μ_w is about 1.0 cp at 20°C.

Typical values of k_v^o in the environment are given in Table 15-3 or can be computed from the equations below. If a $(k_v^c/k_v^o)_{lab}$ value is not known, one for a similar high-volatility chemical should be a reasonable

---

2. Additional estimation methods are given in Chapter 17.

TABLE 15-3

Oxygen Reaeration Coefficients, $(k_v^o)_{env}$, for Water Bodies

Water Body	Literature Values (hr^{-1})	Calculated Values[a] (hr^{-1})
Pond	0.0046 – 0.0096	0.008
River	0.008, 0.04 – 0.39	0.04
Lake	0.004 – 0.013	0.01

a. From Tsivoglou [21]

Source: Smith *et al.* [19]. *(Reprinted with permission from the American Chemical Society.)*

substitute. The values of k_v^o for ponds and lakes are speculative and depend on depth.

For predicting reaeration rates in rivers, Mackay and Yuen [15] present the equations listed below; these correlate k_v^o with river flow velocity, depth, and slope.

Tsivoglou-Wallace: $k_v^o = 638\ V_{curr} s\ hr^{-1}$ (15-26)

Parkhurst-Pomeroy: $k_v^o = 1.08\ (1 + 0.17\ F^2)\ (V_{curr} s)^{0.0375}\ hr^{-1}$ (15-27)

Churchill *et al.*: $k_v^o = 0.00102 V_{curr}^{2.695}\ Z^{-3.085}\ s^{-0.823}\ hr^{-1}$ (15-28)

If no slope data are available:

Isaacs-Gundy: $k_v^o = 0.223\ V_{curr}\ Z^{-1.5}\ hr^{-1}$ (15-29)

Langbein-Durum: $k_v^o = 0.241\ V_{curr}\ Z^{-1.33}\ hr^{-1}$ (15-30)

where

V_{curr} = river flow velocity (m/s)
s = river bed slope = m drop/m run (nondimensional)
Z = river depth (m)
F = Froude number = $V_{curr}/\sqrt{gZ}$ (dimensionless)
g = acceleration of gravity = 9.8 m/s²

Since none of the foregoing is clearly superior to the others, the best approach is probably to use all that are applicable and then average the results.[3] The values of k_v^o, D^c and D^o are then used in Eq. 15-19 to determine k_v^c.

For the range $10^{-5} < H < 10^{-3}$ atm-m^3/mol, Southworth [20] developed a method for estimating the volatilization rates of polycyclic aromatic hydrocarbons (PAHs). He derived equations for estimating the phase exchange coefficients k_g and k_l from laboratory data, which are used in computing the overall liquid-phase mass transfer coefficient,

$$K_L = \frac{H'k_g k_l}{H'k_g + k_l} \text{ cm/hr} \tag{15-31}$$

For the gas-phase exchange coefficient, Southworth's equation is

$$k_g = 1137.5 \ (V_{wind} + V_{curr})\sqrt{18/M} \text{ cm/hr} \tag{15-32}$$

where V_{wind} and V_{curr} are in m/s. The equation used for the liquid-phase exchange coefficient depends on the wind speed. For $V_{wind} < 1.9$ m/s,

$$k_l = 23.51 \left(\frac{V_{curr}^{0.969}}{Z^{0.673}}\right)\sqrt{32/M} \text{ cm/hr} \tag{15-33}$$

where Z is in meters. For $1.9 < V_{wind} < 5$ m/s,

$$k_l = 23.51 \left(\frac{V_{curr}^{0.969}}{Z^{0.673}}\right)\sqrt{32/M} \ e^{0.526(V_{wind}-1.9)} \text{ cm/hr} \tag{15-34}$$

Estimated values of k_l and k_g from these equations are plotted in Figures 15-4 and 15-5, respectively, for a range of molecular weights and environmental parameters. They are also listed in Table 15-4 for a variety of organic compounds.

If values for the phase exchange coefficients are not available, they can be roughly estimated. For k_l, Cohen *et al.* [3] have defined three regions:

(1) $V_{wind} < 3$ m/s
   - Water surface is relatively calm.
   - Flow is aerodynamically smooth.

---

3. The reaeration rate for a river 2 m deep and flowing at 1 m/s is about 0.042/hr.

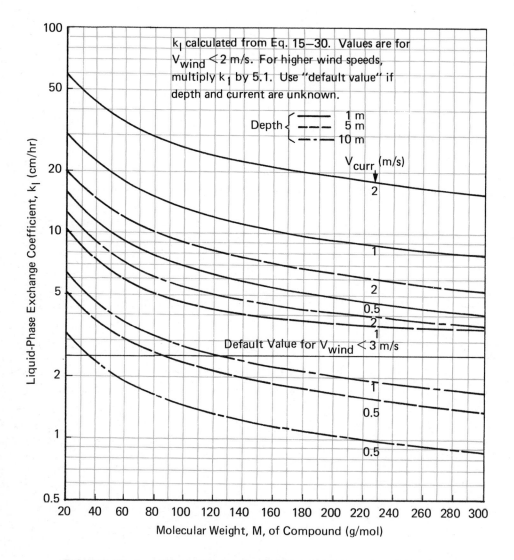

FIGURE 15—4   Effect of Molecular Weight and Environmental Characteristics
on Liquid-Phase Exchange Coefficient

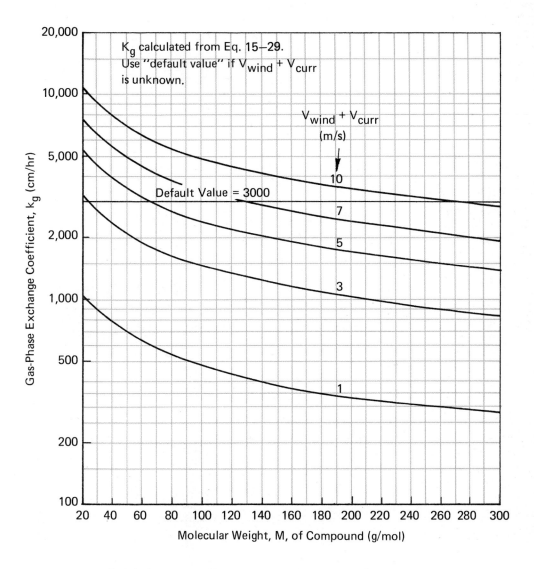

FIGURE 15—5   Effect of Molecular Weight, Wind Speed and Current on Gas-Phase Exchange Coefficient

TABLE 15-4

Volatilization Parameters for Selected Chemicals

Chemical	M (g/mol)	Solubility[a] mg/L	Solubility[a] mol/m³	Vapor Pressure at 20°C mm Hg	Vapor Pressure at 20°C atm	Henry's Law Const. H $\left(\frac{atm\text{-}m^3}{mol}\right)$	Henry's Law Const. H' (Non-dim.)	Phase Exchange Coeff.[b] (cm/hr) Liquid k_l	Phase Exchange Coeff.[b] (cm/hr) Gas k_g	Mass Transfer Coeff. K_L (cm/hr)	Half-life,[c] $\tau_{1/2}$ (hr)
Low Volatility ($H < 3\times10^{-7}$)											
3-Bromo-1-propanol	139	1.7×10^{-5}	1223	0.1	1.3×10^{-4}	1.1×10^{-7}	4.6×10^{-6}	16	1600	7.4×10^{-3}	9400
Dieldrin	381	0.25	6.6×10^{-4}	1×10^{-7}	1.3×10^{-10}	2×10^{-7}	8.9×10^{-6}	12	990	8.8×10^{-3}	7840
Middle Range ($3\times10^{-7} < H < 10^{-3}$)											
Lindane	290.9	7.3	2.5×10^{-2}	9.4×10^{-6}	1.2×10^{-8}	4.8×10^{-7}	2.2×10^{-5}	14	1130	0.025	2760
m-Bromonitrobenzene	170	10^4	58.8	0.07	9.2×10^{-5}	1.6×10^{-6}	7.4×10^{-5}	16	1500	0.11	626
Pentachlorophenol	266	14	5.3×10^{-2}	1.4×10^{-4}	1.8×10^{-7}	3.4×10^{-6}	1.5×10^{-4}	12	1150	0.17	406
4-t-Butylphenol	150	10^3	6.7	0.046	6.1×10^{-5}	9.1×10^{-6}	3.8×10^{-4}	17	1600	0.59	117
Triethylamine	101	7.3×10^4	723	7.0	9.2×10^{-3}	1.3×10^{-5}	5.4×10^{-4}	22	1900	0.98	7
Aldrin	365	0.2	5.5×10^{-4}	6×10^{-6}	7.9×10^{-9}	1.4×10^{-5}	6.1×10^{-4}	12	1810	1.0	68
Nitrobenzene	123	2×10^3	16.3	0.27	3.5×10^{-4}	2.2×10^{-5}	9.3×10^{-4}	20	1700	1.5	45
Epichlorohydrin	92.5	6.6×10^4	711	17.3	2.3×10^{-2}	3.2×10^{-5}	1.3×10^{-3}	24	2000	2.3	29
DDT	354.5	1.2×10^{-3}	3.4×10^{-6}	1×10^{-7}	1.3×10^{-10}	3.8×10^{-5}	1.7×10^{-3}	13	1025	1.5	45
Phenanthrene	178	1.29	7×10^{-3}	2.1×10^{-4}	2.8×10^{-7}	3.9×10^{-5}	1.7×10^{-3}	18	1450	2.2	31
Acenaphthene	154	3.9	2.5×10^{-2}	2.8×10^{-3}	3.7×10^{-6}	1.5×10^{-4}	6.2×10^{-3}	19	1560	6.4	11
Acetylene tetrabromide	344	650	1.9	0.3	3.9×10^{-4}	2.1×10^{-4}	8.9×10^{-3}	13	1000	5.3	13.1
Aroclor 1242	254	0.24	9.5×10^{-4}	4.1×10^{-4}	5.3×10^{-7}	5.6×10^{-4}	2.4×10^{-2}	15	1210	9.9	7
Ethylene dibromide	188	4.3×10^3	22.9	11.6	1.5×10^{-2}	6.6×10^{-4}	2.8×10^{-2}	16	1400	11.4	6.1

TABLE 15-4 (Continued)

Chemical	M (g/mol)	Solubility[a] mg/L	Solubility[a] mol/m³	Vapor Pressure at 20°C mm Hg	Vapor Pressure at 20°C atm	Henry's Law Const. H $\left(\frac{atm\text{-}m^3}{mol}\right)$	Henry's Law Const. H' (Non-dim.)	Phase Exchange Coeff.[b] Liquid k_l (cm/hr)	Phase Exchange Coeff.[b] Gas k_g (cm/hr)	Mass Transfer Coeff. K_L (cm/hr)	Half-life,[c] $\tau_{1/2}$ (hr)
High Volatility ($H > 10^{-3}$)											
Ethylene dichloride	99	8.0×10^3	80.8	67	0.09	1.1×10^{-3}	4×10^{-2}	22	1900	17.1	4
Naphthalene	128	33	0.26	0.23	3×10^{-4}	1.15×10^{-3}	4.9×10^{-2}	21	1765	16.9	4.1
Biphenyl	154	7.5	0.05	0.06	7.5×10^{-5}	1.5×10^{-3}	6.8×10^{-2}	19	1554	16.2	4.3
Aroclor 1254	326	1.2×10^{-2}	3.7×10^{-5}	7.7×10^{-5}	1×10^{-7}	2.7×10^{-3}	1.2×10^{-1}	13	1070	11.9	5.8
Methylene chloride	85	1.3×10^4	155	349	0.46	3×10^{-3}	1.3×10^{-1}	25	2000	22.8	3
Aroclor 1248	292	0.054	1.85×10^{-4}	4.9×10^{-4}	6.5×10^{-7}	3.5×10^{-3}	1.6×10^{-1}	14	1130	12.9	5.3
Chlorobenzene	113	472	4.2	11.8	1.6×10^{-2}	3.7×10^{-3}	1.65×10^{-1}	22	1820	15.1	4.6
Chloroform	119	8×10^3	67	246	0.32	4.8×10^{-3}	2.0×10^{-1}	20	1700	18.9	3.7
o-Xylene	106	175	1.7	6.6	8.7×10^{-3}	5.1×10^{-3}	2.2×10^{-1}	23	1870	22	3.2
Benzene	78	1780	22.8	95.2	1.25×10^{-1}	5.5×10^{-3}	2.4×10^{-1}	27	2180	26	2.7
Toluene	92	515	5.6	28.4	3.7×10^{-2}	6.6×10^{-3}	2.8×10^{-1}	25	2010	24	2.9
Aroclor 1260	361	2.7×10^{-3}	7.5×10^{-6}	4.1×10^{-5}	5.3×10^{-8}	7.1×10^{-3}	3.0×10^{-1}	13	1020	12.5	5.5
Perchloroethylene	166	400	2.4	14.3	2×10^{-2}	8.3×10^{-3}	3.4×10^{-1}	17	1450	16.4	4.2
Ethyl benzene	106	152	1.43	9.5	1.25×10^{-2}	8.7×10^{-3}	3.7×10^{-1}	23	1870	22.3	3.1
Trichloroethylene	131	1×10^3	7.6	60	8×10^{-2}	1×10^{-2}	4.2×10^{-1}	21	1700	20.4	3.4
Mercury	201	3×10^{-2}	1.5×10^{-4}	1.3×10^{-3}	1.7×10^{-6}	1.1×10^{-2}	4.8×10^{-1}	17	1360	16.6	4.2
Methyl bromide	95	1.3×10^4	137	1.4×10^3	1.8	1.3×10^{-2}	5.6×10^{-1}	23	2000	22.5	3.1
Cumene (isopropyl benzene)	120	50	0.416	4.6	6.1×10^{-3}	1.5×10^{-2}	6.2×10^{-1}	22	1760	21.6	3.2
1,1,1-Trichloroethane	133	950	7.1	100	0.13	1.8×10^{-2}	7.7×10^{-1}	19	1650	18.7	3.7
Carbon tetrachloride	154	800	5.2	91	0.12	2.3×10^{-2}	9.7×10^{-1}	19	1500	18.8	3.7
Methyl chloride	50.5	7.4×10^3	146	3.6×10^3	4.74	2.4×10^{-2}	3.6×10^{-1}	30	2600	29	2.4
Ethyl bromide	109	900	8.3	460	6.1×10^{-1}	7.3×10^{-2}	3.1	22	1800	22	3.1
Vinyl chloride	62.5	90	1.44	2580	3.4	2.4	99	28	2400	28	2.5
2,2,4-Trimethyl pentane	114	2.44	2.1×10^{-2}	49.3	6.5×10^{-2}	3.1	129	22	1810	22	3.1
n-Octane	114	0.66	5.8×10^{-3}	14.1	1.85×10^{-2}	3.2	136	22	1810	22	3.1
Fluorotrichloromethane	137		5.0			5.0		11.3	1090	11.3	6.1
Ethylene	28	131	4.7		>40	>8.6	~360	40	3700	40	1.7

a. Source: Refs. 11, 13, 14, 22

b. From Eqs. 15-32 to -34 and Figs. 15-4 and -5, using $V_{wind} = 3m/s$ and $V_{curr} = 1m/s$

c. From Eq. 15-12, with Z = 1m

- k_l values are typically 1-3 cm/hr and appear to be strongly influenced by mixing originating from within the water body.

- Wind velocity in this range has no apparent effect on the value of k_l.

- A suggested value for k_l is 2-3 cm/hr.

- The mass transfer rate is dominated by the underlying hydrodynamics, which are very site-specific and dependent on recent environmental conditions.

(2) $3 \text{ m/s} < V_{wind} < 10 \text{ m/s}$

- k_l increases from 3.5 to 30 cm/hr.

- In the range of 3-6 m/s, the increase in k_l is attributable to the onset of ripples and an increase in surface roughness.

- Above 6 m/s, wave growth is appreciable. Flow becomes completely rough, which increases the rate of mass transfer appreciably.

(3) $10 \text{ m/s} < V_{wind}$

- Waves may begin to break.

- k_l increases due to greater surface area, spray, bubble entrainment, and disintegration of wave crests.

- k_l values can reach 70 cm/hr.

Liss and Slater [9] give values of k_l for several gases and suggest that 20 cm/hr is appropriate for the sea surface. This value should be applicable for gases of $15 < M < 65$; outside this range, k_l can be adjusted by multiplying by the square root of the ratio of the molecular weight of CO_2 to that of the other gas, i.e.,

$$k_l = 20\sqrt{44/M} \text{ cm/hr} \tag{15-35}$$

The correction is not well established in extending data for low-molecular-weight gases to high-molecular-weight compounds such as PCBs.

The ratio k_g/k_l ranges from about 50 to 200 [10]. Liss and Slater [9] suggest a value of about 96 for oceans. For the transfer of water vapor from the ocean surface, they give a value of about 3000 cm/hr for k_g. The value of k_g for some other compound can be estimated by multiplying 3000 by the square root of the ratio of the molecular weights:

$$k_g = 3000\sqrt{18/M} \text{ cm/hr} \tag{15-36}$$

There is evidence that the phase exchange coefficients for transfer across the air-sea interface may be too high for lakes and other smaller bodies of water [17].

Basic Steps of Calculation. The following data are the minimum required for calculating rates of vaporization:

- Chemical properties — vapor pressure, aqueous solubility, molecular weight.[4]
- Environmental characteristics — wind speed, current speed, depth of water body.

(1) Find Henry's law constant (H) from Eq. 15-8 and/or Figure 15-2.

(2) If $H < 3 \times 10^{-7}$, volatilization can be considered unimportant as an inter-media transfer mechanism, and no further calculations are necessary.

(3) If $H > 3 \times 10^{-7}$, the chemical can be considered volatile. Determine the nondimensional Henry's law constant, $H' = H/RT$ (RT = 0.024 at 20°C).

(4) Compute the liquid-phase exchange coefficient (k_l). For a compound of low molecular weight (<65), use Eq. 15-35. If M>65, use Eq. 15-33 for $V_{wind} < 1.9$ m/s or Eq. 15-34 for $1.9 < V_{wind} < 5$ m/s. Alternatively, obtain the approximate value of k_l from Figure 15-4.

(5) Compute the gas-phase exchange coefficient (k_g). For a compound of low molecular weight (<65), use Eq. 15-36; if M> 65, use Eq. 15-32. Alternatively, read the approximate value of k_g from Figure 15-5.

(6) If the necessary data are available, compute the surfactant-resistance mass transfer coefficient (k_s), using Eq. 15-17.

(7) Compute overall liquid-phase mass transfer coefficient, K_L. If just the gas-phase and liquid-phase resistances are to be considered, use Eq. 15-31 or (if the dimensional Henry's law constant is used) the following equation, which is a rearrangement of Eq. 15-9:

$$K_L = \frac{(H/RT)k_g \, k_l}{(H/RT)k_g + k_l} \text{ cm/hr}$$

---

4. Estimation methods for aqueous solubility and vapor pressure are given in Chapters 2 and 14, respectively.

If resistances other than those of the gas and liquid phases must be considered, they can be included here (see Eqs. 15-13 and 15-16).

$$R_L = 1/K_L, r_1 = 1/k_1, r_g = 1/H'k_g = RT/Hk_g, r_w = \tau_D/Z = Z/1.3\,D_z$$

$$r_s = 1/H'k_s = RT/Hk_s$$

and

$$R_L = r_I + r_g + r_w + r_s$$

or

$$1/K_L = 1/k_1 + 1/H'k_g + Z/1.3\,D_z + 1/H'k_s$$

(8) Compute the half-life ($\tau_{1/2}$) from the above value of K_L and the depth of the water (Z), using Eq. 15-12.

Example 15-1: A High-Volatility Chemical Estimate the half-life of trichloro-ethylene at 20°C in a river 1 meter deep flowing at 1 m/s and with a wind velocity of 3 m/s. The vapor pressure of this compound is 0.08 atm, its molecular weight is 131 g/mol, and its solubility is 1.1 g/L (8.4 mol/m^3).

(1) Calculate the Henry's law constant from Eq. 15-8.

$$H = 0.08/8.4 = 0.01 \text{ atm-m}^3/\text{mol}$$

(2) Since $H > 10^{-3}$ atm-m^3/mol, trichloroethylene is highly volatile.

(3) Calculate the nondimensional Henry's law constant:

$$H' = 0.01/0.024 = 0.42 \text{ (at 20°C)}$$

(4) Compute the liquid-phase exchange coefficient, k_1. Since $M > 65$ and $1.9 < V_{wind} < 5$, Eq. 15-34 is used.

$$k_1 = 23.51(1^{0.969}/1^{0.673}) \sqrt{32/131} \; e^{0.526(3-1.9)}$$

$$= 21 \text{ cm/hr}$$

(5) Compute the gas-phase exchange coefficient, k_g, from Eq. 15-32.

$$k_g = 1137.5(3+1) \sqrt{18/131} = 1700 \text{ cm/hr}$$

(6) Use Eq. 15-31 to find the overall liquid-phase mass transfer coefficient, K_L. (It is assumed that the gas and liquid phases account for the only important resistance.)

$$K_L = \frac{0.42 \times 1700 \times 21}{(0.42)(1700) + 21} = 20.4 \text{ cm/hr}$$

(7) Use Eq. 15-12 to find the half-life.

$$\tau_{\frac{1}{2}} = 0.69(100/20.4) = 3.4 \text{ hr}$$

Alternatively, by the reaeration coefficient method:

(1) Calculate the oxygen reaeration rate constant by Eqs. 15-29 and -30. (Eqs. 15-26, -27, and -28 are not usable, because the slope of the river bed is unknown in this example.)

$$(k_v^o)_{env} = 0.223(1)(1)^{-1.5} = 0.223 \text{ hr}^{-1}$$

and

$$(k_v^o)_{env} = 0.241(1)(1)^{-1.33} = 0.241 \text{ hr}^{-1}$$

The average of these values is about 0.23 hr^{-1}. This is within the range listed in Table 15-3 ($0.008 - 0.39 \text{ hr}^{-1}$).

(2) The laboratory-measured value of k_v^c/k_v^o (see Table 15-2) is 0.57 ± 0.15. Therefore, by Eq. 15-21,

$$(k_v^c)_{env} = 0.57(0.23) = 0.13 \text{ hr}^{-1}$$

For comparison, the rate constant corresponding to the value of K_L found in the preceding method is $K_L/Z = 20/100 = 0.2 \text{ hr}^{-1}$.

(3) Using Eq. 15-12,

$$\tau_{\frac{1}{2}} = 0.69(Z/K_L) = 0.69(k_v^c)^{-1}$$

$$= 5.8 \text{ hr}$$

This agrees fairly well with the 3.4-hour half-life found by the preceding method.

Example 15-2: A Medium-Volatility Chemical   Estimate the half-life of acenaphthene under the same environmental conditions as in Example 15-1. Given: $P_{vp} = 3.72 \times 10^{-6}$ atm, M = 154 g/mol, S = 3.9 mg/L = 0.025 mol/m³.

(1) From Eq. 15-8,

$$H = 3.72 \times 10^{-6}/0.025 = 1.5 \times 10^{-4} \text{ atm-m}^3/\text{mol}$$

(2) Since $10^{-5} < H < 10^{-3}$ atm-m³/mol, acenaphthene is of medium volatility.

(3) Calculate the nondimensional Henry's law constant:

$$H' = H/RT = \frac{1.5 \times 10^{-4}}{0.024} = 6.2 \times 10^{-3} \text{ at } 20°C$$

(4)   Compute the liquid-phase exchange coefficient, k_l. Since $M > 65$ and $1.9 < V_{wind} < 5$, Eq. 15-34 is used.

$$k_l = 23.51 \, (1^{0.969}/1^{0.673}) \, \sqrt{32/154} \; e^{0.526 \,(3-1.9)}$$

$$= 19.1 \; cm/hr$$

(5)   Compute the gas-phase exchange coefficient, k_g, from Eq. 15-32.

$$k_g = 1137.5 \,(3+1) \, \sqrt{18/154} = 1560 \; cm/hr$$

(6)   Use Eq. 15-31 to find the overall liquid-phase mass transfer coefficient, K_L. (It is assumed that the gas and liquid phases account for the only important resistances.)

$$K_L = \frac{(6.2 \times 10^{-3}) \,(1560)\,(19.1)}{(6.2 \times 10^{-3})\,(1560) + 19.1} = \frac{185}{28.8} = 6.4 \; cm/hr$$

(7)   Use Eq. 15-12 to find the half-life.

$$\tau_{1/2} = 0.69 \,(100/6.4) = 11 \; hr$$

Basic Steps of Calculation via Reaeration Coefficient. For a high-volatility chemical, the volatilization rate constant can be estimated by the reaeration coefficient method. The following data are required:

- $(k_v^c/k_v^o)_{lab}$ *or* the ratio of diffusion coefficients D^c/D^o *or* the ratio of molecular diameters d^o/d^c;
- $(k_v^o)_{env}$ *or* stream flow parameters (velocity, stream bed slope, depth).

(1) Find the oxygen reaeration coefficient, $(k_v^o)_{env}$, from Table 15-3, or compute it from Eqs. 15-26 to 15-30. (It is recommended that k_v^o be computed from all the appropriate equations and an average be taken.)

(2) If $(k_v^c/k_v^o)_{lab}$ is known, calculate $(k_v^c)_{env}$ by Eq. 15-21 and use this value to find the half-life by Eq. 15-12.

(3) If molecular diameters or molecular diffusivities are known, calculate $(k_v^c)_{env}$ with Eq. 15-20 and then use Eq. 15-12 to find $\tau_{1/2}$.

(4) If neither (2) nor (3) is applicable, compute diffusivities for the chemical and oxygen via Eq. 15-25 and proceed as in step (3). (See Chapter 17 for other estimation methods.)

Example 15-3: A Low-Volatility Chemical Estimate the half-life of dieldrin in the same environment as in Example 15-1. Given: $P_{vp} = 1.3 \times 10^{-10}$ atm, M = 381 g/mol, S = 0.25 mg/L = 6.6×10^{-4} mol/m^3.

(1) From Eq. 15-8,

$$H = 1.3 \times 10^{-10}/6.6 \times 10^{-4} = 2.0 \times 10^{-7} \text{ atm-m}^3/\text{mol}$$

(2) Since the Henry's law constant is less than 3×10^{-7}, volatilization is unimportant as a transfer mechanism for dieldrin, and further calculations are unnecessary.

15-6 SYMBOLS USED[5]

C	=	concentration (M/L^3)
C_g	=	well-mixed concentration in gas phase (M/L^3)
C_l	=	well-mixed concentration in liquid phase (M/L^3)
C_o	=	initial concentration (M/L^3)
C_{sg}	=	concentration in gas phase at interface (M/L^3)
C_{sl}	=	concentration in liquid phase at interface (M/L^3)
ΔC	=	concentration difference (M/L^3)
D	=	diffusion coefficient (L^2/T)
D_L	=	longitudinal value of turbulent diffusivity in Eq. 15-14 (L^2/T)
D_z	=	perpendicular value of turbulent diffusivity in Eq. 15-14 (L^2/T)
d	=	molecular diameter (L)
F	=	Froude number
g	=	acceleration of gravity (L/T^2)
H	=	Henry's law constant (atm-m^3/mol)
H'	=	nondimensional Henry's law constant = C_{sg}/C_{sl} or H/RT
K_G	=	overall gas-phase mass transfer coefficient (L/T)
K_L	=	overall liquid-phase mass transfer coefficient (L/T)
k	=	first-order rate constant (T^{-1})
k_g	=	gas-phase exchange coefficient (L/T)
k_l	=	liquid-phase exchange coefficient (L/T)
k_s	=	mass transfer coefficient at interface for surfactants (L/T)

---

5. *M, L,* and *T* indicate mass, length, and time units respectively.

k_v =   volatilization rate constant (T^{-1})

M   =   molecular weight (g/mol)

N   =   flux (M/L^2T)

P   =   partial pressure (atm or mm Hg)

P_{vp} =   vapor pressure of compound (atm or mm Hg)

R   =   gas constant = 8.2×10^{-5} atm-m^3/mol-K or 8.3×10^7 ergs/mol-K

R_L =   overall liquid-phase resistance (T/L)

r_g =   gas-phase resistance (T/L)

r_l =   liquid-phase resistance (T/L)

r_s =   surfactant resistance (T/L)

r_w =   bulk water body resistance (T/L)

S   =   aqueous solubility or saturation concentration (mol/m^3 or M/L^3)

s   =   slope of river bed (L/L)

T   =   temperature (K or °C)

t   =   time (T)

V_b =   molar volume of chemical at normal boiling point $(L^3$/mol)

V_{curr} =   current speed (L/T)

V_{wind} =   wind speed (L/T)

Z   =   average or mean water body depth (L)

z   =   layer thickness (L)

Greek

α =   accommodation coefficient, or fraction of molecules striking the surface that condense on the surface

μ_w =   viscosity of water (centipoise) $\approx$ 1.0 cp at 20°C

$\tau_{1/2}$ =   half-life (T)

τ_D =   half-life for turbulent diffusion (T)

Subscripts

env   =   environmental value

lab   =   laboratory value

Superscripts

c   =   chemical

o   =   oxygen

15-7   REFERENCES

1. Chiou, C.T. and V.H. Freed, *Chemodynamic Studies on Benchmark Industrial Chemicals,* NSF/RA-770286 (1977). NTIS PB 274263

2. Chiou, C.T. and V.H. Freed, "Evaporation Rates from Single-Component and Multicomponent Systems," *Preprints of Papers Presented at the 177th National Meeting,* Honolulu, Hawaii, April 1-6, 1979, American Chemical Society, Division of Environmental Chemistry, **19,** No. 1 (1979).

3. Cohen, Y., W. Cocchio and D. Mackay, "Laboratory Study of Liquid-Phase Controlled Volatilization Rates in Presence of Wind Waves," *Environ. Sci. Technol.,* **12,** 553-58 (1978).

4. Dilling, W.L., "Interphase Transfer Processes. II. Evaporation Rates of Chloromethanes, Ethanes, Ethylenes, Propanes, and Propylenes from Dilute Aqueous Solutions. Comparisons with Theoretical Predictions," *Environ. Sci. Technol.,* **11,** 405-9 (1977).

5. Dilling, W.L., N.B. Tefertiller and G.J. Kallos, "Evaporation Rates and Reactivities of Methylene Chloride, Chloroform, 1,1,1-Trichloroethane, Trichloroethylene, and Other Chlorinated Compounds in Dilute Aqueous Solutions," *Environ. Sci. Technol.,* **9,** 833-38 (1975).

6. Elder, J., "The Dispersion of Marked Fluid in Turbulent Shear Flow," *J. Fluid Mech.,* **5,** 544 (1959).

7. Gloyna, G., "Prediction of Oxygen Depletion and Recent Developments in Stream Model Analyses," in *Stream Analyses and Thermal Pollution,* Vol. II, University of Texas, Austin TX, prepared for Poland Project 26, World Health Organization (1967).

8. Liss, P.S., "Processes of Gas Exchange Across an Air-Water Interface," *Deep-Sea Res.,* **20,** 231-38 (1973).

9. Liss, P.S. and P.G. Slater, "Flux of Gases Across the Air-Sea Interface," *Nature (London),* **247,** 181-84 (January 25, 1974).

10. Mackay, D., "Volatilization of Pollutants from Water," in "Aquatic Pollutants: Transformations and Biological Effects," edited by E.O. Hutzinger, I.H. Van Lelyveld, and B.C.J. Zoetemans, *Proceedings of the Second International Symposium on Aquatic Pollutants,* Amsterdam, The Netherlands, September 26-28, 1977.

11. Mackay, D. and P.J. Leinonen, "Rate of Evaporation of Low Solubility Contaminants from Water Bodies to Atmosphere," *Environ. Sci. Technol.,* **9,** 1178-80 (1975).

12. Mackay, D., W.Y. Shiu and R.J. Sutherland, "Volatilization of Hydrophobic Contaminants from Water," *Preprints of Papers Presented at the 176th National Meeting,* Miami, Florida, September 10-15, 1978, American Chemical Society, Division of Environmental Chemistry, **18,** No. 2 (Publication date unknown).

13. Mackay, D., W.Y. Shiu and R.P. Sutherland, "Determination of Air-Water Henry's Law Constants for Hydrophobic Pollutants," *Environ. Sci. Technol.,* **13,** 333-37 (1979).

14. Mackay, D. and A.W. Wolkoff, "Rate of Evaporation of Low Solubility Contaminants from Water Bodies to Atmosphere," *Environ. Sci. Technol.,* **7,** 611-14 (1973).

15. Mackay, D. and T.K. Yuen, "Volatilization Rates of Organic Contaminants from Rivers," *Proceedings of 14th Canadian Symposium, 1979: Water Pollution Research Canada* (Publication date unknown).

16. Neely, W.B. "Predicting the Flux of Organics Across the Air/Water Interface," in *Proceedings of the 1976 National Conference on the Control of Hazardous Material Spills,* New Orleans, LA, April 25-28, 1976.

17. Schwarzenbach, R.P., E. Molnar-Kubica, W. Giger and S. Wakeham, "Distribution, Residence Time, and Fluxes of Tetrachloroethylene and 1,4-Dichlorobenzene in Lake Zurich, Switzerland," *Environ. Sci. Technol.,* **13**, 1367-73 (1979).

18. Smith, J.H. and D.C. Bomberger, "Prediction of Volatilization Rates of Chemicals in Water," *Water: 1978,* AIChE Symposium Series, 190, **75**, 375-81 (1979).

19. Smith, J.H., D.C. Bomberger and D.L. Haynes, "Prediction of the Volatilization Rates of High Volatility Chemicals from Natural Water Bodies," *Environ. Sci. Technol.,* **14**, 1332-37 (1980).

20. Southworth, G.R., "The Role of Volatilization in Removing Polycyclic Aromatic Hydrocarbons from Aquatic Environments," *Bull. Environ. Contam. Toxicol.,* **21**, 507-14 (1979).

21. Tsivoglou, E.C., "Tracer Measurements of Stream Reaeration," Federal Water Pollution Control Administration, Washington, D.C. (1967). NTIS PB 229923/BA.

22. Verschueren, K., *Handbook of Environmental Data on Organic Chemicals,* Van Nostrand Reinhold Company, New York (1977).

23. Whitman, W.G., *Chem. Metall. Eng.,* **29**, 146 (1923).

16

VOLATILIZATION FROM SOIL

Richard G. Thomas

16-1   INTRODUCTION

Volatilization is the process by which a compound evaporates in the vapor phase to the atmosphere from another environmental compartment. Volatilization may be an important mechanism for the loss of chemicals from the soil and transfer to the air. In the case of spills or purposeful application, for example, it may be useful to know how long the compounds will persist in the soil, and volatilization is one of the factors on which persistence depends. This chapter presents methods to estimate the volatilization of a chemical from the soil to the air. Most of these methods are mathematical descriptions of the physical process of volatilization and do not generate a single number or parameter that represents a volatilization rate. These methods and descriptions will be referred to as the volatilization models.

To estimate concentrations in the soil column and/or flux of chemicals from the soil to the air, the models require as input data several chemical and environmental parameters, which must be obtained by the user. Some of the models are complex; therefore, a programmable calculator is useful for working with them.

The rate at which a chemical volatilizes from soil is affected by many factors, such as soil properties, chemical properties, and environmental conditions. These factors are discussed in the literature primarily in terms of their effect on the evaporation of pesticides, because most previous studies of volatilization rates have concentrated on these chemicals. Other than their use, there is apparently little that distinguishes pesticides from other organics; therefore, we assumed in this chapter that the observations based on pesticides are applicable to organic chemicals in general.

A rate constant for volatilization is not a feature of the models, except in the special case of volatilization of a surface-applied chemical. For cases in which the chemical is incorporated into the soil, methods are discussed which predict the concentration in the soil as a function of time or which predict the rate of evaporation.

The methods presented here do not explicitly address all the many factors influencing volatilization from soil and may exclude certain processes for the sake of analytical tractability. Thus, the results are not extremely accurate and should be considered only generally indicative of the actual environmental behavior. Measured values are given, where available, to indicate the accuracy of the different methods.

A discussion of the volatilization process and important factors affecting the volatilization of a chemical from the soil follows this introduction. Next, the available models developed to estimate concentrations of chemicals in the soil and the flux of chemicals from the soil to the air are presented and discussed. The final main section of this chapter considers the selection of the model most applicable to the user's needs.

16-2 FACTORS AFFECTING THE VOLATILIZATION PROCESS

Properties on which Volatilization is Dependent. Chemical, soil, and general environmental properties affect volatilization. Some of the properties of a chemical involved in volatilization are its vapor pressure, solubility in water, basic structural type, and the number, kind, and position of its basic functional groups [6]. Subtle differences in chemical structure can cause a large change in vapor pressure; they can also affect the charge on a molecule, causing it to sorb on soil more or less strongly than a similar compound [6]. This can determine the fraction of compound in the soil that will volatilize. Other factors affecting volatilization rate are: compound concentration in the soil, soil water content, airflow rate over the surface, humidity, temperature, sorptive and diffusion characteristics of the soil, and bulk properties of the soil such as organic matter content, porosity, density, and clay content. All of these factors affect the distribution of a compound between the soil, soil water, soil air, and atmosphere.

The above factors can be grouped into three categories: (1) those which affect movement away from the evaporating surface into the atmosphere, (2) those which affect the vapor density (concentration of the compound in air) of the chemical, and (3) those which control the rate of movement to the evaporating surface [19]. All of the processes and factors described below fall into one or another of these categories.

Compound Distribution and Equilibria. A compound in the soil may be partitioned between the soil water, soil air, and the soil constituents. Considered as a whole, the soil represents all three phases of matter rather than one, as does the atmosphere or water. The atmosphere constitutes another air compartment which is distinct from the soil air. The rate of volatilization of an organic molecule from a sorption site on the solid phase in the soil (or in solution in the soil water) to the vapor phase in soil air and then to the atmosphere is dependent on many physical and chemical properties of both the chemical and the soil and on the process of moving from one phase to another. (These properties and processes are discussed more fully below.) The three main distribution or transport processes involved are:

- Compound in soil   ↔   Compound in solution
- Compound in solution   ↔   Compound in vapor phase in soil air
- Compound in vapor phase in soil air   →   Compound in atmosphere

Partitioning of a chemical among the three phases can be estimated from either vapor-phase or solution-phase desorption isotherms [16].

Adsorption. Adsorption reduces the chemical activity below that of the pure compound and affects the vapor density and the volatilization rate [18]. Sorption may be a result of chemical adsorption (coulombic forces), physical adsorption (Van der Waals forces), or hydrogen bonding [6]. The concentration of the compound present in a desorbed state in solution in the soil water, along with other properties, controls the vapor density of the compound in soil air. The vapor density is directly related to the volatilization rate. The vapor density has also been shown to be inversely related to both the surface area of the soil particles and the organic matter content in the soil [6]. This follows, since sorption is usually directly related to the available surface area and organic matter content, and sorption reduces the amount of compound available to partition between the water and air.

Vapor Density. Vapor density is the concentration of a chemical in the air, the maximum concentration being a saturated vapor. The vapor density of a compound in the soil air ultimately determines the volatilization rate.

Vapor density and vapor pressure are affected by interaction with soils [16]. Some chemicals require only a low total soil concentration to have a saturated vapor in moist soil; thus, weakly adsorbed compounds may volatilize rapidly, especially if applied only to the soil surface. If the chemicals are incorporated into the soil, the concentration at the evaporating surface of the soil particles is reduced and the total volatilization rate decreases [19].

Spencer [16] found that the vapor density of dieldrin in several types of soils increased linearly with concentration but was inversely related to the organic matter content and was relatively independent of clay content when sufficient water was present to cover the surfaces. As mentioned above, even a low soil-water content increases the vapor density of the compound in the soil air by displacing it from soil surfaces. The effective vapor pressure may differ from the vapor pressure of the chemical itself because of adsorption on the soil or other matter, solution in the soil water, or retreat into deeper capillary spaces [19].

Water Content of the Soil. Water content also affects volatilization losses by competing for adsorption sites on the soil [6,16]. For non-polar and weakly polar compounds, water is preferentially sorbed onto soil particles and can displace the chemical. As most pesticides are relatively insoluble in water, a soil-water adsorption isotherm can be used to estimate concentrations in the soil/soil-water system [16]. The determination of these isotherms may be difficult, and not all compounds behave similarly; some compounds are strongly adsorbed as cations on soils, whereas others are weakly adsorbed on soils and are easily displaced by water. Between these two extremes the degree of adsorption is dependent on the polarity of the compound, which in turn may be dependent on pH and other factors [16]. For weakly polar or nonionic pesticides the fraction of soil organic matter is the most important soil factor in adsorption. Since most of the more volatile pesticides are nonionic or only weakly polar, their adsorption by soils is closely related to organic matter content [18]. Once the soil surfaces are saturated with just a molecular layer of water, the vapor density of a weakly polar compound in the soil air is greatly increased, and additional soil water does not influence the tendency of the compound to leave its sorbed site. The quantity of water required to achieve a monomolecular layer is about 2.8% in a Gila silt loam [17,19,21].

The concept of "co-distillation," in which a compound is assumed to evaporate along with the soil moisture in the same ratio as they are present in the soil/soil-water system, is not applicable. Chemicals in the soil do not behave in this manner. The loss of a compound in the presence of water is not due to co-distillation. When the water evaporates, the compound sorbs onto dry soil. Volatilization of the compound is enhanced by the *presence* of water, not its evaporation [16]. Compound loss becomes insignificant when water loss decreases, because the vapor density is reduced by adsorption on the dry soil. Conversely, the presence of water causes desorption of the compound, increasing vapor density.

As a consequence of this behavior, a compound may hold very strongly on a dry soil, thereby reducing its volatilization rate; but when the soil is wetted, the stronger affinity of the water displaces the compound, allowing volatilization to occur at a faster rate. Keeping the soil dry will reduce or stop the volatilization of some chemicals, since the soil needs some moisture for the displacement or solubilization of the chemical [6]. However, if the concentration of a compound in soil becomes high enough that its chemical activity approaches that of the pure compound, the presence or absence of water will not affect its volatilization rate.

Partitioning between Water, Soil, and Air. The concentration of a compound in the soil water can be determined from sorption and solution equilibria. These equilibria depend on the soil, the chemical, and the interaction with water [16]. The vapor density in the air in equilibrium with the soil solution can be determined from Henry's law regardless of solubility or volatility [16].

Desorption isotherms relating compound adsorbed by the soil to relative vapor density and to relative solution concentration (Figure 16-1) were all the same for lindane [17]. These relationships indicate that, for a compound that is relatively insoluble in water and weakly polar, the quantity that must be adsorbed to create a saturated solution is the same as that required for a saturated vapor; at higher soil concentrations the air is saturated with the vapor [17]. These observations would probably hold for most compounds regardless of their solubilities, although more soluble compounds may not exhibit linear vapor density/solution concentration isotherms over a wide concentration range [16].

The relationships between the equilibrium concentrations for each soil component[1] — soil solids (c_s), soil water (c_ℓ/S), and soil air

1. See §16-6 for symbol definitions.

(ρ/ρ_{max}) — will probably not change with varying water: soil ratios over the usual field soil water content range; as the water content decreases, however, the compound concentration in solution, and hence vapor density, increases. The equilibrium shifts to a higher amount of compound adsorbed, as shown in Figure 16-1.

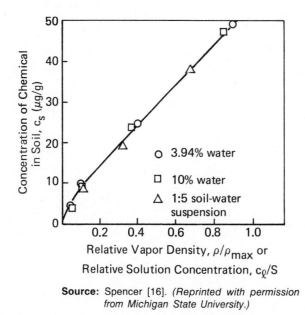

Source: Spencer [16]. *(Reprinted with permission from Michigan State University.)*

FIGURE 16-1   Desorption Isotherms for Lindane

The change in concentration in the soil solid phase with a change in soil water content is greater with more soluble compounds, since a greater proportion of the total compound in the soil is in solution at any time. As a general rule, for compounds that can be displaced by water, the higher the sorption partition coefficient (c_s/c_ℓ), the less is the effect of changing water content on solution concentrations and vapor densities within the "wet" soil moisture range above one monolayer of water [16].

Temperature. The effect of temperature on the volatilization process is unpredictable. The primary effect is on vapor density: an increase in temperature normally increases the equilibrium vapor density, which in turn increases the volatilization rate [6]. However, complicating factors can alter the expected result. Spencer [16] discusses experiments with lindane in which increasing temperature resulted in a lower relative vapor density (ρ/ρ_{max}). This ratio should have remained constant if changes in solubility or vapor density were the only operant effects. In this instance, adsorption reduced the rate of increase of ρ relative to ρ_{max}. Lindane adsorption appears to be a net endothermic process that results in a decrease in free energy and an increase in entropy, the opposite of the usual results. Thus, an increase in temperature will not necessarily lead to an increase in volatilization.

Low temperature may not eliminate volatilization entirely, since diffusion may continue even in frozen soil [7]. The likely explanation for this phenomenon is that water near clay surfaces is subject to attractive forces that prevent it from assuming the crystalline structure of ice, which would inhibit diffusion. This effect has not been demonstrated for many organic chemicals in soil; there is a general lack of data on the temperature effect for organic chemicals.

Atmospheric Conditions. Air flow over the soil surface plays an important role in the volatilization process. The evaporation rate is determined by the diffusion of vapor into the surrounding air. Close to the surface (i.e., in the boundary layer) there may be relatively little movement of air when the substance is transported only by molecular diffusion through more stagnant portions of the layer. (Transport across the boundary layer also depends on turbulence in the air in the boundary layer. The thickness of this layer, and the air velocity and turbulence within this layer, will depend on air velocity, surface geometry and distance from the surface.) Farther away from the surface there is more air movement, which can carry the vapor away more rapidly than molecular diffusion [8].

The rate of air flow can influence volatilization both directly, as discussed above, and indirectly [6]. If the relative humidity of the air is not 100%, an increase in air speed will hasten drying of the soil. This will reduce the soil water content and thereby ultimately reduce the volatilization of the chemical. Volatilization, originally rapid, will slowly decrease until the surface concentrations of water and/or compound are depleted. The rate will eventually become constant, and the loss of the compound will become a function of the movement of compound or moisture from deeper levels. The subsequent loss of chemicals for which water can compete for adsorption sites will be related to the time it takes to dry the soil sufficiently to reduce the vapor density to an insignificantly low value [19]. The soil water content at which vapor density begins to decrease will depend upon the soil and the sorption competition with water: the more strongly adsorbed the compound is, the higher the water content at which an appreciable decrease in vapor density will occur [19].

Diffusion. When little water is being lost due to evaporation, transport of the chemical through the soil occurs by diffusion, which then becomes the rate-controlling process. Diffusion can occur in vapor and nonvapor phases. In moist soils, chemicals evidently vaporize from the water-air interface, and the concentrations at the interface are maintained by diffusion of the compound through the soil water from its adsorption sites [17]. Diffusion through the soil or soil water is a slow process, however; the chemical on the surface will evaporate relatively quickly, but that which remains in the soil will have a long lifetime.

Liquid-phase diffusion is much slower than gaseous diffusion, proceeding at a rate of only a few centimeters per month, and is associated with a much smaller scale of movement in the soil [7]. The overall diffusion rate, as well as the proportion of the total diffusion in the vapor phase, is dependent on soil water content and other factors that influence the partitioning [16]. If the soil is dry and volatilization is then related to the moisture content at the surface, the loss rate would still be controlled by diffusion, but at a much slower rate, due to lowered vapor density and hence a lowered diffusion rate in the soil.

Diffusion can occur along four major pathways, one in the vapor phase and three in nonvapor phases: air-water interface, water-water pathway, and water-solid interface. The relative amount of diffusion via each pathway depends on temperature, bulk density, and other variables that affect the partitioning between air, water, and soil [15]. Only small quantities are in the vapor phase compared with the amount in solution or adsorbed on the soil. However, as coefficients of diffusion in air are several thousand times greater than those for water or surface diffusion, total mass transport by diffusion through the vapor phase can be approximately equal to that through nonvapor phases.

Since the vapor phase diffusion coefficient is approximately 10^4 larger than the solution phase diffusion coefficient, a partition coefficient for the distribution of a chemical between the soil water and the soil air of 10^4 (g/cm^3 in soil water $\div$ g/cm^3 in soil air) may be considered as a transition for determining when vapor or nonvapor phase diffusion becomes dominant. Chemicals with partition coefficients much smaller than 10^4 will diffuse mainly in the vapor phase, and those with partition coefficients much greater than 10^4 will diffuse primarily in the solution phase (Ref. 3 and Ref. 4 as cited in Ref. 11).

Ehlers *et al.* [1] show that vapor-phase diffusion is inversely proportional to ambient pressure, whereas nonvapor-phase diffusion is not. They give the following relationship:

$$D = D_s + D_v' \, (P_0/P) \qquad (16\text{-}1)$$

where[2]

D = actual diffusion coefficient in soil (L^2/T)
D_s = nonvapor-phase diffusion coefficient (L^2/T)
D_v' = apparent vapor-phase diffusion coefficient (L^2/T)
P_0 = reference pressure (standard atmospheric)
P = ambient pressure

In practice, P will probably not deviate significantly from P_0, so that $P_0/P \approx 1$.

The influence of vapor-phase diffusion becomes less important (the apparent vapor-phase diffusion coefficient decreases) as bulk density increases and air-filled porosity decreases, leaving fewer pathways by which vapor can diffuse. The apparent vapor-phase diffusion coefficient can be expressed as

$$D_v' = (D_v s^{10/3}/\rho_b s_t^2) \, \frac{d\rho}{dc} \qquad (16\text{-}2)$$

where

D_v = vapor-phase diffusion coefficient in air (L^2/T)
s = air-filled porosity
ρ_b = bulk density (M/L^3)
s_t = total porosity (air + water)
ρ = vapor density (M/L^3)
c = concentration in soil (M/L^3)

D_v' will increase with increasing $d\rho/dc$ (the change of vapor density with concentration), but above a certain critical concentration the vapor density becomes a constant and nonvapor movement controls the rate of diffusion. It has been shown for lindane [2] that the apparent nonvapor-phase diffusion coefficient increases as soil bulk density increases; there-fore, if an increase in soil bulk density (and a corresponding decrease in air-filled porosity) decreases vapor movement and reduces volatility, there will be a counteracting increase in nonvapor movement.

Another equation is given by Farmer *et al.* [3]. Based on an analysis in Ref. 13, an expression for the diffusion coefficient was derived that

---

2. Dimensions for parameters are identified as follows: L = length, T = time, M = mass, K = temperature in degrees Kelvin.

included soil porosity terms to account for the geometric effects of the soil on the apparent steady-state diffusion:

$$D_V' = D_V \, (s^{10/3}/s_t^2) \tag{16-3}$$

This equation is similar to Eq. 16-2 except that it omits bulk density and vapor density correction factors.

According to Hamaker [7], vapor-phase diffusion coefficients at different temperatures and pressures may be compared as

$$D_1/D_2 = P_2/P_1 \, (T_1/T_2)^m \tag{16-4}$$

where P = ambient total pressure (consistent units) and T = temperature (K); m is theoretically 1.5 and is measured as 1.75-2.0.

Diffusion coefficients of different chemicals can be related by their molecular weights:

$$D_1/D_2 = \sqrt{M_2/M_1} \tag{16-5}$$

When volatilization becomes a diffusion-controlled process, diffusion equations can be used to predict both concentrations in the soil and loss rates from the soil surface by volatilization if diffusion coefficients in the soil are known [19]. (Diffusion coefficients can also be estimated by methods given in Chapter 17.)

Mass Transport by the Wick Effect. Chemicals in the soil can volatilize through the action of a process called the wick effect or wick evaporation. The chemical is transported from the soil body to the surface by capillary action. Its rate of evaporation is enhanced by the evaporation of the water causing the capillary action. Hartley [8] discusses this transport mechanism in detail. The soil column acts as a wick; the water in the soil moves up the capillaries of the wick to replenish that lost at the top by evaporation. The diffusive escape in the vapor phase causes an actual mass flow in the wick. If the liquid contains a low concentration of a solute less volatile than water, the solute will increase in concentration near the evaporating surface and cannot diffuse back down the wick as quickly as it accumulates, since diffusion in the liquid phase competes poorly with this mass flow.

Hartley states that one or more of the following can result from this process:

- The solute can become so concentrated at the surface that water evaporation is greatly reduced;

- The concentration of the solute at the surface may increase until its rate of evaporation balances its rate of arrival; and/or

- The solute may form a saturated solution in the soil water, leading to an efflorescence of crystals or powder, such as that seen on the surface of a fertile soil after drying.

The second phenomenon noted above can greatly increase the evaporation of a moderately volatile, moderately water-soluble compound. The compound will tend to be carried along with the water in the ratio present in the undisturbed soil water, not because of any peculiarity in the evaporation process (such as codistillation), but because the bulk flow of the solution up the capillaries is too fast for diffusion in the soil water to compete. The evaporation rate from the surface will remain approximately constant and rapid as long as water continues to flow up and keep the surface moist. It is only during this period of rapid water loss that evaporation of water indirectly accelerates that of the chemical. The amount of chemical volatilized is related to the time needed to dry the soil sufficiently to reduce the vapor density of the chemical [19].

16-3 METHODS FOR ESTIMATING VOLATILIZATION OF CHEMICALS FROM SOIL

A comprehensive model of the volatilization process would have to consider not only all the factors mentioned above but also soil type, ground cover, terrain, weather, soil pH, organic matter content, type of input, method of incorporation into the soil, and other factors. The magnitude and complexity of these considerations indicate the shortcomings that can be expected in theoretical studies of soil volatilization. The approximations currently available do not explicitly address the complexities of the volatilization process. The methods therefore appear to have deficiencies, although their originators point out their usefulness when applied to the particular situations and chemicals for which they were developed.

Hartley Method. Hartley [8] presents two fairly simple equations to estimate the volatilization of chemicals from soil. The method is based on an analysis of the heat balance between the evaporating chemical (or water) and air. The flux (quantity volatilizing from soil to air) is expressed as

$$f = \frac{\rho_{max} \; (1-h)}{\delta} \bigg/ \left[\frac{1}{D_v} + \frac{\lambda_v^2 \; \rho_{max} \; M}{kR \; T^2} \right]$$

(16-6)

where

f	= flux of compound (M/L^2T)
ρ_{max}	= saturated vapor concentration at the temperature of the outer air (M/L^3)
h	= humidity of the outer air $(0 \le h \le 1)$
δ	= thickness of stagnant layer through which the chemical must pass (L)
D_v	= diffusion coefficient of vapor in the air (L^2/T)
λ_v	= latent heat of vaporization (cal/M)
M	= molecular weight (M/mol)
k	= thermal conductivity of air (cal/LK)
R	= gas constant (cal/mol)
T	= temperature (K)

The second term in the denominator is a thermal component of the resistance to volatilization. It is significant primarily for water or quite volatile compounds; for less volatile compounds, this term can be ignored, resulting in the following simplified form:

$$f = \frac{D_v \rho_{max} \; (1-h)}{\delta}$$

(16-7)

Hamaker Method. Hamaker [7] proposed a method for estimating the volatilization rate of chemicals from soil. The impregnated soil layer is assumed to be semi-infinite; i.e., the total depth is large in comparison with the depth to which the soil is significantly depleted of the chemical by diffusion and volatilization. The loss of chemical is expressed as

$$Q_t = 2c_o \sqrt{Dt/\pi}$$

(16-8)

where

Q_t	= total loss of chemical per unit area over some time t (M/L^2)
c_o	= initial concentration of chemical in soil (M/L^3)
D	= diffusion coefficient of vapor through the soil (L^2/T)
π	= 3.14159 . . .

This equation may give low values, since it ignores mass transfer due to the wick effect.

Hamaker presented a second method based on total water loss — i.e., the loss due to vapor diffusion as well as the mass transfer of soil solution. The approximation for the loss of a dissolved, volatile chemical is

$$Q_t = \frac{P_{vp}}{P_{H_2O}} \frac{D_v}{D_{H_2O}} (f_w)_V + c \, (f_w)_L \tag{16-9}$$

where

f_w	= loss of water per unit area (M/L^2)
P_{vp}	= vapor pressure of chemical
P_{H_2O}	= vapor pressure of water
D_v	= diffusion coefficient of chemical in air (L^2/T)
D_{H_2O}	= diffusion coefficient of water vapor in air (L^2/T)
sub V	= loss of vapor
sub L	= loss of liquid
c	= concentration of chemical in soil solution (M/M)

The use of this equation requires knowledge of the water flow in the soil as well as diffusion coefficients for water vapor and the chemical in question.

Mayer, Letey, and Farmer Method. Mayer *et al.* [12] applied the diffusion laws to the mathematical description of the movement of chemicals in soils under a concentration gradient. Diffusion is assumed to be the only mechanism supplying chemicals to the soil surface; therefore, the approach probably underestimates the volatilization rate. An analogy is drawn between the heat transfer equation (Fourier's law) and the transfer of matter under a concentration gradient (Fick's law). The equation is solved for various boundary conditions.

The one-dimensional diffusion equation, with a constant diffusion coefficient, D, is

$$\frac{\partial^2 c}{\partial z^2} - \frac{1}{D} \frac{\partial c}{\partial t} = 0 \tag{16-10}$$

where

c = concentration in the soil (M/L^3)

z = distance measured normal to the soil surface (down is positive, surface is zero) (L)

D = diffusion coefficient in soil (L^2/T)

t = time (T)

Five different solutions are presented, each applicable to a different set of boundary conditions. The models and their basic assumptions are described below. In addition to those defined above, the following variables are used:

L   = soil layer depth (L)
v   = air flow velocity (L/T)
c_a = concentration in air (M/L^3)
R_o = isotherm coefficient, ratio of concentration in air to concentration in soil. (In moist soils this ratio might be approximated by the ratio of Henry's Law constant to the soil adsorption coefficient.)

• *Model I.* In this model it is assumed that the compound volatilizes and is removed rapidly from the soil surface, maintaining a zero concentration at the surface. The flux at any time in the model depends only on the diffusion coefficient, which must remain constant over the time period of interest. No diffusion occurs across the lower boundary. It would be difficult to predict volatilization rates in the field using this model, because the boundary conditions would not necessarily conform to those in the model or supporting experiments. However, the assumed zero concentration at the surface would be approached even if wind speeds were only about 2 cm/s.

The boundary conditions are:

$$\left. \begin{array}{l} c = c_o \text{ at } t = 0 \quad 0 \leqslant z \leqslant L \\[10pt] c = 0 \text{ at } z = 0, t > 0 \\[10pt] \dfrac{\partial c}{\partial z} = 0 \text{ at } z = L \end{array} \right\} \qquad (16\text{-}11)$$

The solution of the diffusion equation with these boundary conditions is[3]

$$c(z,t) = \frac{4c_o}{\pi} \sum_{n=0}^{\infty} \frac{(-1)^n}{(2n+1)} \, e^{-D(2n+1)^2 \, \pi^2 t/4L^2} \, \cos \frac{(2n+1)\,\pi\,(L-z)}{2L} \qquad (16\text{-}12)$$

---

3. Eqs. 16-12, -13, -22, -24, -26, -29 and -31 all contain either a $\sum_{n=0}^{\infty}(...)$ or a $\sum_{n=1}^{\infty}(...)$ term. In practice the summation can be terminated after the terms converge for the second or third significant figure, and this will almost always be less than ten terms (i.e., $n \leq 10$). In some cases convergence at $n = 2$ will be found.

and the flux is given by

$$f = \frac{Dc_0}{\sqrt{\pi Dt}} \, [1 + 2\sum_{n=1}^{\infty}(-1)^n \, e^{-n^2 L^2/Dt}] \qquad (16\text{-}13)$$

• *Model II.* If the value of the summation term in Eq. 16-13 is negligible in comparison to 1, the equation can be simplified. The summation will be small if the expression in the exponential, n^2L^2/Dt, is large, say on the order of 10 or more. This simplified model is applicable until the concentration at the lower boundary drops by 1%, i.e. until time $t = L^2/14.4 \, D$.

As in model I, this model may not predict actual volatilization rates in the field, where the boundary conditions may not prevail.

With increasing L and decreasing D or t (which increases the value of the term in the exponential of Eq. 16-13), the flux becomes

$$f = \frac{Dc_0}{\sqrt{\pi Dt}} = c_0 \, \sqrt{D/\pi t} \qquad (16\text{-}14)$$

The total quantity volatilized equals the product of flux and time, giving total loss per unit area, $Q = c_0 \sqrt{Dt/\pi}$. As noted in Eq. 16-8, the value of Q derived by Hamaker [7] is twice as large. The reason for this discrepancy is not clear.

The concentration in the soil column for the same boundary conditions as in Model I can be expressed as

$$c\,(z,t) = c_0 \, \text{erf} \left(\frac{z}{2\sqrt{Dt}} \right) \qquad (16\text{-}15)$$

where "erf" is the error function, which may be evaluated by the methods described in §16-5. These equations are valid until time $t = L^2/14.4D$.

• *Model III.* This model is similar to Model I, except that it allows for diffusion downward across a boundary at $z = L$.

The boundary conditions are

$$\left. \begin{array}{l} c = c_0 \text{ at } t = 0, 0 \leqslant z \leqslant L \\[2mm] c = 0 \text{ at } t = 0, z > L \\[2mm] c = 0 \text{ at } t > 0, z = 0 \end{array} \right\} \qquad (16\text{-}16)$$

The concentration in the soil column is

$$c\,(z,t) = \frac{c_o}{2}\left[2\,\mathrm{erf}\left(\frac{z}{2\sqrt{Dt}}\right) - \mathrm{erf}\left(\frac{z-L}{2\sqrt{Dt}}\right) - \mathrm{erf}\left(\frac{z+L}{2\sqrt{Dt}}\right)\right] \qquad (16\text{-}17)$$

The flux is

$$f = D\,\frac{\partial c}{\partial z}\,\bigg|_{z=0} \qquad (16\text{-}18)$$

so that

$$f = \frac{Dc_o}{\sqrt{\pi Dt}}\,\left(1 - e^{-L^2/4Dt}\right) \qquad (16\text{-}19)$$

For large values of $L^2/4Dt$, less than 1% error will result in using

$$f = \frac{Dc_o}{\sqrt{\pi Dt}} \qquad (16\text{-}20)$$

if $e^{-L^2/4Dt} < 0.01$ or $t < L^2/18.4\ D$.

- *Model IV.* This model accounts for a concentration that varies with time at the soil surface instead of being always zero. The rate at which the chemical is removed by the air is then a limiting factor in the volatilization rate. The chemical is uniformly distributed to depth L, and there is no flux through the lower boundary. A continuous flow of fresh air at velocity v passes over the soil surface, so background concentration is low. The concentration in the air, c_a, after removal is uniform at a constant fraction of the concentration at the soil surface. This implies rapid mixing in the air, due either to turbulence or a large air diffusion coefficient. The model does not include diffusion in the transfer process, only convective mass transfer: if the air stops moving, the flux of chemical into the air becomes zero. Thus, the model is appropriate if the convective flow due to air movement is considerably greater than the flow due to a diffusion process. Again, this model may not accurately predict volatilization rates in the field, where the boundary conditions may not prevail.

The boundary conditions are:

$$c = c_0 \text{ at } t = 0, 0 \leqslant z \leqslant L$$

$$\frac{\partial c}{\partial z} = 0 \text{ at } z = L$$

$$f = vc_a \text{ at } t > 0, z = 0$$

$$(16\text{-}21)$$

The concentration in the soil is

$$c\,(z,t) = 2c_0 \sum_{n=1}^{\infty} \left\{ \frac{e^{-D\alpha_n^2 t}\;(h - R_0\,\alpha_n^2)\cos[\alpha_n\,(L\text{-}z)]}{[L(h - R_0\alpha_n^2)^2 + \alpha_n^2\,(L + R_0) + h]\cos\alpha_n L} \right\} \quad (16\text{-}22)$$

where $h = R_0 v/D$ and α_n are the roots of

$$\alpha_n \tan(\alpha_n L) = \frac{R_0 v}{D} - R_0 \alpha_n^2 \quad (16\text{-}23)$$

R_0 is an adsorption isotherm coefficient which relates concentration in air, c_a, over a solid onto which the compound is adsorbed to the concentration of the compound on the solid, c_s.

For most cases, $R_0 \alpha_n{}^2 \ll R_0 v/D$, so Eq. 16-22 can be simplified to

$$c\,(z,t) = 2c_0 \sum_{n=1}^{\infty} \left\{ \frac{e^{-D\alpha_n^2 t}\;(R_0\,v/D)\cos[\alpha_n\,(L\text{-}z)]}{[L\,(R_0 v/D)^2 + \alpha_n^2\,L + (R_0 v/D)]\cos\alpha_n\,L} \right\} \quad (16\text{-}24)$$

where α_n are the roots of

$$\alpha_n \tan(\alpha_n L) = \frac{R_0 v}{D} \quad (16\text{-}25)$$

Flux through the surface is

$$f = 2\,Dc_0 \sum_{n=1}^{\infty} \left[\frac{(R_0 v/D)^2\;e^{-D\alpha_n^2 t}}{L\,(R_0 v/D)^2 + \alpha_n^2\,L + (R_0 v/D)} \right] \quad (16\text{-}26)$$

The concentration in the air (c_a) is simply R_0 times the estimated concentration in the soil (c_s) at the surface, for the range of concentrations over which R_0 is valid.

• *Model V.* In this model, chemical at the surface diffuses into a stationary air layer. The diffusion coefficient of the chemical in the air is D_v and the thickness of the air layer is d. The concentration in the air at the soil surface is $R_o c$ (where c is concentration at the soil surface, c_s), so the flux from the soil is

$$f = D_v \, R_o c/d \qquad (16\text{-}27)$$

Initial and boundary conditions are

$$\left.\begin{array}{l} c = c_o \text{ at } t = 0, 0 \leqslant z \leqslant L \\[1em] \dfrac{\partial c}{\partial z} = 0 \text{ at } z = L \\[1em] f = (D_v \, R_o c_s)/d \text{ at } t > 0, x = 0 \\[1em] c = c_s \text{ at } t > 0, z = 0 \end{array}\right\} \qquad (16\text{-}28)$$

The concentration in the soil is

$$c\,(z,t) = \frac{2D_v \, R_o c_o}{Dd} \sum_{n=1}^{\infty} \left\{ \frac{e^{-D\alpha_n^2 \, t} \cos\,[\alpha_n \,(L\text{-}z)]}{[L\,(D_v \, R_o/Dd)^2 + L\alpha_n^2 + D_v \, R_o/Dd]\,\cos\,\alpha_n \, L} \right\} \qquad (16\text{-}29)$$

where α_n are the roots of

$$\alpha_n \, \tan\,(\alpha_n \, L) = \frac{D_v \, R_o}{Dd} \qquad (16\text{-}30)$$

The flux at the surface is

$$f = 2Dc_o \sum_{n=1}^{\infty} \left[\frac{e^{-D\alpha_n^2 \, t}(D_v \, R_o/Dd)^2}{L\,(D_v \, R_o/Dd)^2 + L\alpha_n^2 + D_v \, R_o \,/Dd} \right] \qquad (16\text{-}31)$$

• *Choosing the Appropriate Model.* Models I, II, and III appear to be valid when compared with the results of laboratory experiments, but predictions for field conditions would be subject to error because the boundary and environmental conditions such as wind velocity, incorporation depth, and water movement are not as well defined as they are in the laboratory. The models should make reasonable estimates of losses by diffusion, however.

For a chemical incorporated in a soil surface layer, if diffusion is the only transfer mechanism to the surface, use:

Models I,   If wind velocity is greater than about 2 cm/s.
  II or III

Model IV   If air is moving over the soil but is already enriched with chemical, due possibly to movement over contaminated soil upwind.

Model V   If a stationary air layer is in contact with the surface, such as when the air movement is restricted by a standing crop.

Mayer *et al.* [12] compared results from published laboratory experiments on the volatilization of dieldrin and lindane with predictions from these models. Figure 16-2 shows a comparison with model II. The conditions of the experiments were compatible with the requirements of the model; i.e., airflow over the surface maintained zero concentration at the air/soil interface, and the depth of the soil column could be assumed infinite for several hundred days. The model appears to offer good predictions under these conditions.

Figure 16-3 compares the results of models I, II and IV with data from a lindane volatilization experiment. Under the conditions of the experiment, the concentration at the lower boundary decreased by one percent quite rapidly, which accounts for the difference between the results of models I and II. Model IV predictions are better, because the air velocity across the experimental setup was not sufficient to maintain a zero concentration at the interface.

Figure 16-4 compares models II and IV with experimental data and illustrates the importance of choosing an appropriate model when airflow velocity affects chemical concentrations at the surface. Model II, which does not take air velocity (v) into account, is reasonably accurate at higher velocities. The model IV equations, however, contain v and adjust the calculated flux to reflect non-zero surface concentrations caused by low-velocity airflow.

Jury, Grover, Spencer, and Farmer Method. Jury *et al.* [10] have presented another method for predicting the flux of a chemical in both

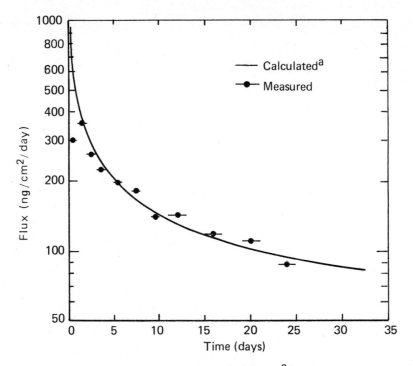

a. Diffusion coefficient assumed to be 2.3 mm^2/week.

Source: Mayer *et al.* [12]. *(Reprinted with permission from Soil Science Society of America.)*

FIGURE 16-2   Dieldrin Flux: Comparison of Measured Values with Those Calculated by Model II

the presence and absence of the wick effect. In their work with the pesticide triallate, they expressed the total concentration of a chemical in the soil as:

$$c_t = \rho_b c_s + \theta c_\ell + \eta c_g \qquad (16\text{-}32)$$

where

c_t = total concentration in soil (M/L^3 soil)
ρ_b = soil bulk density (M/L^3)
c_s = adsorbed concentration (M/M)
θ = volumetric soil water content (L^3/L^3)
η = soil air content (L^3/L^3)
c = concentration in the liquid phase (M/L^3 water)
c_g = concentration in the gas phase (M/L^3 air)

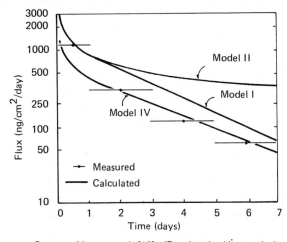

Source: Mayer *et al.* [12]. *(Reprinted with permission from Soil Science Society of America.)*

FIGURE 16-3 **Lindane Flux from Treated Gila Silt Loam: Comparison of Measured and Calculated Values**

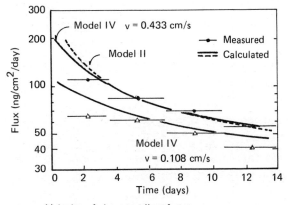

v = Velocity of air over soil surface

Source: Mayer *et al.* [12]. *(Reprinted with permission from Soil Science Society of America.)*

FIGURE 16-4 **Dieldrin Volatilization Flux from Uniformly Treated Gila Silt Loam at Two Air Velocities: Comparison of Measured and Calculated Values**

The flux is:

$$f = -D_g \frac{\partial c_g}{\partial z} - D_\varrho \frac{\partial c_\varrho}{\partial z} - f_w c_\varrho \qquad (16\text{-}33)$$

where

$\quad f \quad = $ chemical flux $(M/L^2 \cdot T)$
$\quad D_g \quad = $ gas diffusion coefficient (L^2/T)
$\quad D_\varrho \quad = $ liquid diffusion coefficient (L^2/T)
$\quad f_w \quad = $ water flux $(L^3/L^2 \cdot T)$

The transport terms in the gas phase, $D_g(\partial c_g/\partial z)$, in the liquid phase, $D_\varrho(\partial c_\varrho/\partial z)$, and along with the soil water, $f_w c_\varrho$, can be seen. The continuity equation is:

$$\frac{\partial c_t}{\partial t} + \frac{\partial f}{\partial z} = 0 \qquad (16\text{-}34)$$

These three equations can be solved to give concentrations in the soil column, along with the flux and total amount lost. The following assumptions have been made:

(1) c_ϱ and c_g are related by Henry's Law, with $K_H = c_\varrho/c_g$. (K_H will have units of L^3 air/L^3 water. Note that K_H is the inverse of the usual Henry's Law constant (H) used in other chapters.)

(2) Adsorption isotherms are linearized over the range of concentrations so that $c_s = \alpha c_\varrho + \beta$ (where α and β are parameters of the adsorption isotherm);

(3) Diffusion coefficients D_g and D_ϱ are constant (i.e., the soil is homogeneous);

(4) The water flux, f_w, is either zero or equal to a constant evaporation rate;

(5) The gas concentration c_g above the soil remains zero, indicating that the air carries the pollutant away rapidly; and

(6) The soil column is infinitely deep.

Assumptions (1) and (2) permit the elimination of c_s and c_ϱ from Eq. 16-32:

$$c_t = \rho_b \left[\alpha \left(K_H c_g \right) + \beta \right] + \theta \left(K_H c_g \right) + \eta c_g$$

$$= \left(\rho_b K_H \alpha + \theta K_H + \eta \right) c_g + \beta \rho_b \qquad (16\text{-}35)$$

$$c_t = \epsilon c_g + \gamma \qquad (16\text{-}36)$$

Similarly, through assumptions (1) and (3), Eq. 16-33 can be written as:

$$f = -D_g \frac{\partial c_g}{\partial z} - D_\varrho K_H \frac{\partial c_g}{\partial z} - f_w (K_H c_g)$$

$$= -(D_g + K_H D_\varrho) \frac{\partial c_g}{\partial z} - K_H f_w c_g \qquad (16\text{-}37)$$

or

$$f = -D_e \frac{\partial c_g}{\partial z} - V_e c_g \qquad (16\text{-}38)$$

where

D_e = effective diffusion coefficient $(D_g + K_H D_\varrho)$
V_e = effective convection velocity $(K_H f_w)$

As expressed by Eq. 16-38, all transport is combined into the gas phase.

The continuity equation (16-34) can be expressed solely in terms of the gas-phase concentration by the substitution of Eqs. 16-36 and -38:

$$\frac{\partial}{\partial t} (\epsilon c_g + \gamma) = -\frac{\partial}{\partial z} \left(-D_e \frac{\partial c_g}{\partial z} - V_e c_g \right)$$

$$\epsilon \frac{\partial c_g}{\partial t} = D_e \frac{\partial^2 c_g}{\partial z^2} + V_e \frac{\partial c_g}{\partial z} \qquad (16\text{-}39)$$

This reduces the number of parameters necessary to apply the equations. The boundary conditions are expressed as follows:

$$\left.\begin{array}{l} c_g = 0 \text{ at } z = 0, t \geqslant 0 \\[2mm] c_g = (c_{t_o} - \gamma)/\epsilon \text{ at } t = 0, z \geqslant 0 \\[2mm] c_g = (c_{t_o} - \gamma)/\epsilon \text{ at } t = 0, z = \infty \end{array}\right\} \qquad (16\text{-}40)$$

These solutions to Eq. 16-40 are the concentrations as a function of depth and time:

For $f_w = 0$,

$$c_g (z,t) = \frac{c_{t_o} - \gamma}{\epsilon} \; erf \left(\frac{z}{2} \sqrt{D_e t / \epsilon} \right) \tag{16-41}$$

For $f_w \neq 0$,

$$c_g (z,t) = \left(\frac{c_{t_o} - \gamma}{\epsilon} \right) \left[1 - \frac{1}{2} \; erfc \left(\frac{z + (V_e t / \epsilon)}{2 \sqrt{D_e t / \epsilon}} \right) \right.$$

$$\left. - \frac{1}{2} \; e^{-V_e z / D_e} \; erfc \left(\frac{z - (V_e t / \epsilon)}{2 \sqrt{D_e t / \epsilon}} \right) \right] \tag{16-42}$$

where:

$$erfc \, (x) = 1 - erf \, (x) \tag{16-43}$$

The flux at the surface for any time t, f (0,t), for the two cases of soil water flux $= 0$ and $\neq 0$ is:

For $f_w = 0$,

$$f \, (0,t) = - (c_{t_o} - \gamma) \sqrt{D_e / \epsilon \pi t} \tag{16-44}$$

For $f_w \neq 0$,

$$f \, (0,t) = - (c_{t_o} - \gamma) \sqrt{D_e / \pi \epsilon t} \; e^{-w^2} - V_e \left(\frac{c_{t_o} - \gamma}{2\epsilon} \right) [1 + erf \, (w)] \tag{16-45}$$

where

$$w^2 = V_e^2 \, t / 4 D_e \tag{16-46}$$

The total chemical loss when $f_w = 0$ can be found by integrating Eq. 16-44, which yields the following:

$$Q_t \, (t) = 2 \, (c_{t_o} - \gamma) \sqrt{D_e t / \pi \epsilon} \tag{16-47}$$

Error may result from these models because of the behavior of the diffusion coefficient. As soil becomes more saturated with water, the pore space is decreased. This, in turn, reduces the diffusion coefficient proportionally to soil air space, η, at high air content until intermediate satura-

and the rate constant for vola

$$k_v = \frac{0.693}{t_{1/2}}$$

The agreement between meas good (Table 16-2). Predictio pared with measured rates of chemicals and the compariso between laboratory data b appears to be within a factor ties such as soil moisture, soil not incorporated in this simp

16-4   SELECTION OF METI

The methods discussed quantity of chemical that vo applicable to a given situati method is selected and used. of the methods, and Figure model selection process. Refe given below.

Basic Steps

(1) If the chemical is n Dow method, which surface-applied chen the chemical is distr step (3).

(2) The Dow method chemical applied to include solubility adsorption coefficien 48 can be used to es the soil. Equation constant based on t tion of chemical on t from

tion is reached and then proportionally to some higher power of η. The use of a measured effective diffusion coefficient corresponding to the conditions of interest would lessen the potential for error. Some discrepancy may also be introduced by the use of linear adsorption isotherms; the adsorption isotherm may be curvilinear, with larger differences occurring at low concentrations.

In laboratory experiments with triallate and two soils, Jury *et al.* found reasonable agreement between the models and measured data. They noted a smaller discrepancy in the absence of water flux, when diffusion was the only transport mechanism. Results also indicated that flow occurred primarily by liquid convection below the surface layer. Depletion of the pesticide was high only in the top 1 centimeter or so of the soil column; 90% of the pesticide remained after about a month. Surface depletion caused a rapid decrease in volatilization with time, and rate of movement to the soil surface became the controlling factor in chemical loss. Most of the compound was sorbed and not free to move. The experiments indicated the relative importance of convective mass flow, diffusive transport, and soil properties. In the two soils used, a higher adsorptive capacity was partially mitigated by a lower bulk density, resulting in a higher diffusion coefficient. The two soils, though different in these respects, showed similar losses under diffusion-controlled volatilization.

These models generally predict the upper limit of the volatilization rate.

Dow Method. Scientists at Dow Chemical Company have conducted experiments to establish the relationships between vapor pressure, water solubility, and soil adsorption coefficient as they relate to the volatility from the soil of a chemical applied to the soil surface [20]. They studied nine chemicals (Table 16-1) and derived a correlation that could predict the volatilization of these chemicals from a soil surface under laboratory conditions. The volatilization rate was found to be proportional to the factor P_{vp}/SK_{oc} where P_{vp} is vapor pressure in mm Hg, S is water solubility in mg/L or ppm, and K_{oc} is the soil adsorption coefficient in $(\mu g/g)/(\mu g/mL)$. Half-life for depletion of the chemical from the surface was found to be:

$$t_{1/2} = 1.58 \times 10^{-8} \left(\frac{K_{oc}S}{P_{vp}} \right) \text{ days} \qquad (16\text{-}48)$$

TABLE 16-1

Chemical Properties and Volatilization Rate Constants for Chemicals Applied to the Soil Surface

Chemical	Vapor Pressure, P_{vp} (mm Hg)	Water Solubility (ppm)	Soil Adsorption Constant, K_{oc}	$\dfrac{P_{vp}}{S\,K_{oc}}$	Initial Rate Constant (hr^{-1})	Predicted Rate Constant (hr^{-1}) Eq. 16-49
Nitrapyrin	3×10^{-3}	4	600	1.2×10^{-6}	6.4×10^{-1}	2.2

that are computationally more complex. It is suggested that values of the flux be estimated by all methods if possible and compared before they are accepted.

- *Simple Method* — Hartley proposes the use of Eq. 16-6 to predict the volatilization flux of chemicals that are highly volatile and Eq. 16-7 for those that are less volatile. These equations require the chemical saturated vapor concentration (ρ_{max}), humidity of the air (h), the thickness of the stagnant boundary layer through which the chemical must pass (δ), and the vapor diffusion coefficient (D_v). For the more volatile compounds, one must also know the latent heat of vaporization (λ_v), molecular weight (M), thermal conductivity of air (k), and the gas constant (R).

  Hamaker suggests the use of Eq. 16-8 to predict the total amount of chemical volatilized as a function of time. Required data are the diffusion coefficient in soil (D) and initial soil concentration (c_o). If the diffusion coefficients are not known, they can be estimated by Eqs. 16-1 through 16-5, as appropriate; other methods are discussed in Chapter 17.

- *More Complex Method of Jury et al.* — The method presented by Jury *et al.* [10] can be used to predict concentrations and flux. The data required for these computations are the diffusion coefficient — either an effective diffusion coefficient (D_e) or gas-phase and liquid-phase diffusion coefficients (D_g and D_ℓ) — soil bulk density (ρ_b), Henry's law constant for the chemical (H), adsorption isotherms (α, β parameters), soil air content (η), soil water content (θ), and soil water flux (f_w). The equations applicable to the case of no water flux ($f_w = 0$) are 16-41 and 16-44. Equation 16-47 estimates the total amount lost.

- *More Complex Model of Mayer et al.* — The models proposed by Mayer *et al.* [12] can be used to estimate concentrations in the soil column. Five different models, applicable to different conditions, are available. The proper one should be selected according to its applicability to the environmental conditions of interest. Although these models are computationally complex, simplifications can be introduced to make them more tractable. Some of the equations involve a series summation requiring the addition of an infinite number of terms. In practice, the summation can be terminated after the terms begin to converge for the second or third significant figure. Ten terms should show the convergence, and the series can be terminated when the last term is less than ten percent of the sum. This summation will be a reasonable approximation of the whole value.

Equations 16-25 and -30 used in models IV and V require a solution for the roots of a transcendental equation. The user may employ an iterative procedure to find the roots or a programmable calculator or computer having a program to find values of the free parameter (α_n) that solve the equation.

Model I (see Eqs. 16-11 through 16-13) is used when the concentration of a chemical remains at zero at the surface due to a wind speed greater than about 2 cm/s. No diffusion occurs across the lower boundary of the soil column. Input data required for the use of this model are initial concentration in the soil column (c_o), depth of the soil column (L), and the effective diffusion coefficient of the chemical in the soil (D). Equation 16-12 can be used to estimate concentrations c(z,t) in the soil column as a function of time and depth. Equation 16-13 can be used to estimate the flux of the chemical.

Model II (see Eqs. 16-14 and -15) can be used for the same environmental conditions as model I. A simplification introduced in Eq. 16-13 in model I is also applicable to model II, as described in the text. Equation 16-14 can be used to estimate the flux. Equation 16-15 can be used to estimate concentrations in the soil column.

Model III (see Eqs. 16-16 through 16-20) is applicable under the same environmental conditions as for model I but allows for diffusion downward across the lower boundary. Equation 16-17 can be used to compute concentrations in the soil column, and Eqs. 16-19 and -20 give estimates of the flux.

Model IV (see Eqs. 16-21 through 16-26) accounts for a finite chemical concentration at the surface. The lower boundary is assumed to be impermeable. The chemical does not diffuse into the air; it is carried away only by convection. Consequently, if the air is not moving, the model will not accurately predict the flux. This model should be used if there is a finite initial concentration of chemical in the air through depletion from the soil surface upwind. Equation 16-24 estimates the concentration in the soil column, and Eq. 16-26 estimates the flux.

Model V (see Eqs. 16-27 through 16-31) is applicable if the air in contact with the soil surface is stagnant, such as might occur if there is a standing crop on the surface. Equation 16-29 estimates the concentration in the soil column, and Eq. 16-31 estimates the flux.

In the examples that follow, concentrations and fluxes are estimated by each of the principal methods described above. To facilitate comparison of the results, all examples refer to trichloroethylene (TCE) and the conditions are as similar as possible. The chemical and environmental data used in the calculations are listed in Table 16-5.

Example 16-1   Hartley Method, No Water Flux

(1)   Since TCE is quite volatile, Eq. 16-6 is appropriate. Table 16-5 lists all the required parameters. The diffusion coefficient of the vapor in air (D_v) was estimated by Eq. 16-5. Compounds whose vapor-phase diffusion coefficients at 293K are known include carbon disulfide $(M = 76, D = 0.102)$ and diethyl ether $(M = 74, D = 0.089)$. Substitute the values of M and D for each of these compounds in Eq. 16-5 and average the results.

$$\text{Using carbon disulfide: } D_v/0.102 \;=\; \sqrt{76/131.5}$$

$$D_v \;=\; 0.077 \text{ cm}^2/\text{s}$$

$$\text{Using diethyl ether: } \quad D_v/0.089 \;=\; \sqrt{74/131.5}$$

$$D_v \;=\; 0.067 \text{ cm}^2/\text{s}$$

Average $D_v = 0.072 \text{ cm}^2/\text{s}$

(2)   Substitute the above in Eq. 16-6 to find the flux, f.

$$f = \cfrac{\cfrac{(4.3 \times 10^{-4} \text{ g/cm}^3)\,(0.5)}{0.3 \text{ cm}}}{\cfrac{1}{0.072 \text{ cm}^2/\text{s}} + \cfrac{(63.2 \text{ cal/g})^2\,(4.3 \times 10^{-4} \text{ g/cm}^3)\,(131.5 \text{ g/mol})}{(61 \times 10^{-6} \text{ cal/s} \cdot \text{cm} \cdot \text{K})\,(1.987 \text{ cal/mol} \cdot \text{K})\,(293\text{K})^2}}$$

$$= 5.04 \times 10^{-5} \text{ g/cm}^2 \cdot \text{s}$$

TABLE 16-5

Chemical and Environmental Data for
Estimation of Trichloroethylene Volatilization[a]

Parameter	Symbol	Value
Characteristics of TCE at T = 293K		
Saturated vapor concentration	ρ_{max}	4.3×10^{-4} g/cm^3
Diffusion coefficient in air	D_v	0.072 cm^2/s
		(see Example 16-1)
Heat of vaporization	λ_v	63.2 cal/g
Thermal conductivity of air	k	61×10^{-6} cal/s-cm-K
Gas constant	R	1.987 cal/mole-K
Diffusion coefficient of vapor through soil	D	0.039 cm^2/s
		(see Example 16-2)
Vapor pressure of TCE	P_{vp}	60 mm Hg
Vapor pressure of water	P_{H_2O}	17.54 mm Hg
Diffusion coefficient of water vapor		
through air	D_{H_2O}	0.239 cm^2/s
Initial concentration in soil (assumed)	c_o	0.05 g/cm^3
	α	1
Adsorption coefficients (assumed)	β	0
	K_{oc}	360
Solubility	S	1100 mg/L
Molecular weight	M	131.5 g/mole
Ratio c_ℓ/c_g (= S/ρ_{max})	K_H	2.56 cm^3 air/cm^3 water
Environmental Characteristics (assumed)		
Humidity	h	0.5 (=50%)
Stagnant air layer thickness	δ	0.3 cm
Temperature	T	293K
Wind speed	v	100 cm/s
Soil solid density	ρ_{solid}	2.65 g/cm^3
Soil bulk density = $(1-\eta-\theta)\rho_{solid}$	ρ_b	1.32 g/cm^3
Volumetric soil water content	θ	0.2 cm^3/cm^3
Soil air content	η	0.3 cm^3/cm^3
Depth of soil column	L	20 cm
Water vapor flux per unit area	f_w	6.7×10^{-2} g/cm^2/day

a. Values for certain input parameters may be found in various chemistry and physics handbooks or in the published literature. This chapter and other chapters also provide estimation methods for some of the parameters.

(3) The total loss (Q_t) over a given time period is the product of the flux and the time (in seconds). The total loss per day (86,400 s) is

$$Q_t = (5.04 \times 10^{-5}) (86,400)$$

$$= 4.4 \text{ g/cm}^2$$

Example 16-2 Hamaker Method, No Water Flux

(1) The total loss of chemical per unit area is given by Eq. 16-8. This equation requires values for c_o and D. Table 16-5 lists the initial concentration, but the diffusion coefficient of TCE vapor through the soil (D) must be estimated. As in the preceding example, a reasonable estimate can be obtained by substituting known diffusion coefficients of other compounds in Eq. 16-5 and averaging the results. Table 16-6 lists vapor diffusion coefficients for various compounds at or near the desired temperature. (The diffusion coefficients used in the table were measured for conditions different from those assumed here; however, in the absence of other information we will assume that they are at least generally indicative of the desired value.) The final column lists the corresponding values of D for TCE as calculated from Eq. 16-5 and indicates an average of 0.039 cm^2/s.

(2) Substituting the above in Eq. 16-8, the total loss of chemical per day (86,400 s) is

$$Q_t = 2(0.05 \text{ g/cm}^3) \sqrt{(0.039 \text{ cm}^2/\text{s}) (86,400 \text{ s})/\pi}$$

$$= 3.3 \text{ g/cm}^2$$

Example 16-3 Hamaker Method, Water Flux

(1) In the case of water flux in the soil column, Hamaker's equation (Eq. 16-9) permits estimation of the TCE flux. The loss of liquid water will be neglected, so the $c\,(f_w)_L$ term is zero.

(2) Substituting the values from Table 16-5 in Eq. 16-9,

$$Q_t = \left(\frac{60 \text{ mm Hg}}{17.54 \text{ mm Hg}} \right) \left(\frac{0.072 \text{ cm}^2/\text{s}}{0.239 \text{ cm}^2/\text{s}} \right) \quad (6.7 \times 10^{-2} \text{ g/cm}^2) + 0$$

$$= 6.9 \times 10^{-2} \text{ g/cm}^2$$

Example 16-4 Mayer/Letey/Farmer Method, No Water Flux (Model IV is used here to illustrate the five models in this method.)

(1) An adsorption isotherm coefficient, R_o, relating c_{air} to c_{soil} must be known to use this model. R_o for TCE is apparently not available in the literature, but the parameter is known for lindane (3×10^{-5}) and dieldrin (2.4×10^{-6}). Using bounds of an order of magnitude either way will indicate the sensitivity of the estimated flux to these values.

TABLE 16-6

TCE Vapor Diffusion Coefficients Through Soil, Calculated from Known Properties of Other Compounds Using Eq. 16-5
(See Example 16-2)

Chemical	M	D (cm²/s)	T (K)	Soil	Soil Porosity	D_{TCE} (cm²/s)
Ethylene Dibromide	187.9	0.015	293	Garden Soil	0.389	0.018
Ethylene Dibromide	187.9	0.005	293	Ashhurst	0.199	0.006
Ethylene Dibromide	187.9	0.011	293	Ashhurst	0.303	0.013
CO_2	44	0.043	296	Dry	—	0.025
CS_2	76	0.193	288.6	Dry Sand	0.374	0.147
Ethanol	46	0.0415	294.5	Dry Quartz Sand	0.415	0.025
O_2	32	0.105	298	32.7% moist	0.38	0.052
O_2	32	0.056	298	45.3% moist	0.27	0.028
					Avg.	0.039

Source: Hamaker [7]. (Reprinted with permission from Marcel Dekker, Inc.)

(2)   This model requires that the roots of Eq. 16-25 be found. Substituting the values of L, v, and D from Table 16-5, Eq. 16-25 becomes

$$\alpha_n \tan (20\alpha_n) = 100 \, R_o/0.039$$

Insert values of R_o ranging from 3×10^{-4} to 2.4×10^{-7} and find the roots, α_n. A programmable calculator is helpful in this task. Table 16-7 lists the first 10 roots for four values of R_o. It is apparent that only a few roots may need to be evaluated, since the terms in the summation rapidly approach zero.

(3)   Equation 16-24 must now be solved to find the concentration in the soil. For the conditions of this example (Table 16-5), Eq. 16-24 becomes, for $z = 0$ (surface) and $t = 1$ day (86,400 s),

$$c \, (0,1 \text{ day}) = 2 \, (0.05) \sum_{n=1}^{\infty} \frac{e^{-0.039 \, \alpha_n^2 \, (86,400)} \, R_o \left(\dfrac{100}{0.039}\right) \cos \, [\alpha_n \, (20 - 0)]}{[20 \, (R_o/0.039)^2 + \alpha_n^2 \, (20) + R_o \left(\dfrac{100}{0.039}\right)] \, \cos \, 20\alpha_n}$$

$$= 0.1 \sum_{n=1}^{\infty} \frac{128 \, R_o}{e^{3370 \, \alpha_n^2} \, (657 \, R_o^2 + 128 \, R_o + \alpha_n^2)}$$

Substitution of the values of R_o and corresponding values of α_n in this equation result in the values for each term n in the summation listed in the s_n columns of Table 16-7. These terms are summed and multiplied by 0.1 to find the concentration in the soil for each R_o, as shown beneath the table.

(4)   In a similar manner, Eq. 16-26 is used to find the flux of TCE that volatilizes from the soil after one day. The following values result:

R_o	$f \, (g/cm^2 \cdot s)$
3×10^{-4}	1.94×10^{-12}
3×10^{-5}	2.28×10^{-8}
2.4×10^{-6}	4.3×10^{-6}
2.4×10^{-7}	1.1×10^{-6}

This model is obviously quite sensitive to the value of R_o. At the upper limit of R_o (3×10^{-4}), the calculations show very small values of c and f after one day, indicating that most of the TCE has volatilized. At the lower limit of R_o, on the other hand, the surface concentration drops only 6% below the initial concentration in one day. Therefore, the value of R_o should be carefully chosen when this model is used. If this value is not available, model results are uncertain.

TABLE 16-7

Sample Calculations of Soil Concentration by Eq. 16-24
(See Example 16-4)

n	$R_o = 3 \times 10^{-4}$ α_n	s_n^*	$R_o = 3 \times 10^{-5}$ α_n	s_n^*	$R_o = 2.4 \times 10^{-6}$ α_n	s_n^*	$R_o = 2.4 \times 10^{-7}$ α_n	s_n^*
1	0.074	5.87×10^{-10}	0.0496	7.64×10^{-5}	0.0172	0.1771	0.00531	0.47
2	0.0784	6.1×10^{-11}	0.0784	2.39×10^{-10}	0.0784	4.79×10^{-11}	0.07844	4.9×10^{-12}
3	0.2216	$1.9 \times 10^{-7.3}$	0.1778	5.08×10^{-48}	0.1591	1.09×10^{-39}	0.1572	8.5×10^{-40}
4	0.2353	$8.5 \times 10^{-8.3}$	0.2350	9×10^{-83}	0.2353	5.23×10^{-84}	0.2353	5.3×10^{-85}
5	0.3703	0	0.3259	0	0.3153	0	0.3141	0
6	0.3928	0	0.3298	0	0.3928	0	0.3928	0
7	0.5203	0	0.479	0	0.4722	0	0.4716	0
8	0.5497	0	0.5497	0	0.5497	0	0.5497	0
9	0.6709	0	0.634	0	0.6291	0	0.6284	0
10	0.7066	0	0.7066	0	0.7066	0	0.7066	0
Σs_n		6.48×10^{-10}		7.6×10^{-5}		0.18		0.47
c (0,1 day) g/cm³		6.48×10^{-11}		7.6×10^{-6}		0.018		0.047

$^*s_n$ are terms n of the summation in Eq. 16-24 (see step 3, Example 16-4).

Source: Roots α_n found with TI-59 calculator Master Library program ML-08, "Zeros of Functions."

Example 16-5 Jury/Grover/Spencer/Farmer Method, No Water Flux

(1) Although they are not needed for the flux calculation, the concentrations of TCE in the soil, soil water, and soil air may be of interest. From Eq. 16-32,

$$0.05 = 1.32\ c_s + 0.2\ c_\varrho + 0.3\ c_g$$

As stated in the assumptions made by Jury *et al*, $K_H = c_\varrho/c_g$ and $c_s = \alpha c_\varrho + \beta$. Thus,

$$c_\varrho = 2.56\ c_g$$

and

$$c_s = 1\ c_\varrho$$

Solving the above equations for c_s, c_ϱ, and c_g yields

$$c_s = 0.030\ \text{g/g}$$
$$c_\varrho = 0.030\ \text{g/cm}^3$$
$$c_g = 0.012\ \text{g/cm}^3$$

(2) The applicable flux equation is 16-41. The constant γ is defined as $\beta \rho_b$. Since β is 0, $\gamma = 0$.

(3) The constant ϵ is defined as $\rho_b K_H \alpha + \theta K_H + \eta$. From Table 16-5,

$$\epsilon = (1.32)(2.56)(1) + (0.2)(2.56) + 0.3$$

$$= 4.19$$

(4) In this model, c_g at $z = 0$ is always zero. Therefore, a value of $z = 1$ cm will be used to estimate gas-phase soil concentration.

(5) D_e, the effective diffusion coefficient, is defined as $D_g + K_H D_\varrho$. However, as D_g is usually orders of magnitude larger than D_ϱ, $D_e = D_g$. This value was estimated earlier (Example 16-2) as 0.039 cm²/s.

(6) Substituting the above in Eq. 16-41, the gas-phase soil concentration at $t = 1$ day (86,400 s) and $z = 1$ cm is

$$c_g(1, 86400) = \frac{0.05 - 0}{4.19}\ \text{erf}\left(\tfrac{1}{2} \sqrt{(0.039)(86,400)/4.19}\ \right)$$

$$= 0.012\ \text{erf}(14.2)$$

$$= 0.012\ \text{g/cm}^3$$

(Evaluation of the error function is described in §16-5.) Since this value of c_g is higher than the saturated vapor concentration, ρ_{max}, of 4.3×10^{-4} g/cm³ (Table 16-5), this method of estimation may not be valid for the conditions of this example.

(7)   The flux is estimated by Eq. 16-44:

$$f(0,t) \; = \; -(0.05-0) \sqrt{\frac{0.039}{4.19\,\pi\,(86,400)}}$$

$$= \; -9.3 \times 10^{-6} \; g/cm \cdot s$$

The negative sign indicates that the flux is out of the soil column.

(8)   The total amount of TCE lost per day can be estimated from Eq. 16-47:

$$Q_t \, (1 \; day) \; = \; 2 \,(0.05) \sqrt{\frac{0.039 \cdot 86,400}{\pi \cdot 4.19}}$$

$$= \; 1.6 \; g/cm^2$$

This is a considerable amount, since in a 1-cm^2 soil column, TCE at 0.05 g/cm^3 would have to be removed entirely to a depth of 32 cm in one day. This value for total loss again casts doubt on the validity of this model for chemicals other than those for which it was developed.

For comparison, c_g and $f(0,t)$ will be calculated by the water flux model (Eqs. 16-42 and -45). The only additional parameter needed is V_e.

(9)   From Table 16-5,

$$f_w \; = \; 6.7 \times 10^{-2} \; g/cm^2/day$$

$$= \; 7.8 \times 10^{-7} \; g/cm^2/s$$

(10) V_e, the effective convection velocity, is defined as the product of f_w and K_H.

$$V_e \; = \; (7.8 \times 10^{-7})(2.56)$$

$$= \; 2.0 \times 10^{-6} \; cm/s$$

(11)   By Eq. 16-42,

$$c_g\,(1,86400) = \frac{0.05-0}{4.19} \left\{ 1 - \tfrac{1}{2} \, \mathrm{erfc} \left[\frac{1+ \dfrac{(2.0 \times 10^{-6})\,(86,400)}{4.19}}{2\sqrt{(0.039)\,(86,400)/4.19}} \right] \right. $$

$$\left. - \tfrac{1}{2}\, e^{-(2.0 \times 10^{-6})\,(1)/0.039} \; \mathrm{erfc} \left[\frac{1 - \dfrac{(2.0 \times 10^{-6})\,(86,400)}{4.19}}{2\sqrt{(0.039)\,(86,400)/4.19}} \right] \right\}$$

$$= 0.012 \left\{ 1 - \tfrac{1}{2}\,[1 - \mathrm{erf}\,(0.0184)] \right.$$

$$\left. - \tfrac{1}{2}\, e^{-(5.1 \times 10^{-5})}\,[1 - \mathrm{erf}\,(0.0173)] \right\}$$

$$= 1.3 \times 10^{-4} \text{ g/cm}^3$$

(Evaluation of the error function is described in §16-5.)

(12)   By Eq. 16-46,

$$w^2 = \frac{(2.0 \times 10^{-6})^2 \ (86{,}400)}{4 \ (0.039)} = 2.2 \times 10^{-6} \approx 0$$

(13)   Then, by Eq. 16-45,

$$f\,(0,t) = -(0.05 - 0)\ \sqrt{0.039/(\pi)\,(4.19)\,(86{,}400)}\ e^{o}$$

$$- 2.0 \times 10^{-6} \left[\frac{0.05 - 0}{2\,(4.19)} \right]\ [1 + \text{erf}\,(1.49 \times 10^{-3})]$$

$$= -9.3 \times 10^{-6} - (1.2 \times 10^{-8})\,[1 + \text{erf}\,(1.49 \times 10^{-3})]$$

$$\approx -9.3 \times 10^{-6}$$

Thus, both models give about the same results.

Example 16-6   Dow Method

(1)   For a surface-applied chemical, the half-life on the surface can be estimated by Eq. 16-48.

$$t_{\frac{1}{2}} = 1.58 \times 10^{-8} \left(\frac{360 \cdot 1100}{60} \right)$$

$$= 1.0 \times 10^{-4} \text{ day}$$

$$\approx 9 \text{ seconds}$$

This half-life indicates that the chemical would probably volatilize soon after it was applied.

(2)   The volatilization rate constant is found by Eq. 16-49:

$$k_v = 0.693/1.0 \times 10^{-4}$$

$$= 6.9 \times 10^3 \text{ day}^{-1}$$

With k_v, the concentration of the chemical at any time can be found from the equation:

$$c = c_o\, e^{-k_v t}$$

16-5 EVALUATING THE ERROR FUNCTION

The solution of some of the diffusion equations presented in this chapter requires the evaluation of the error function, erf (x). The error function is defined as

$$\text{erf}(x) = \frac{2}{\sqrt{\pi}} \int_{o}^{x} e^{-y^2} \, dy \qquad (16\text{-}51)$$

Since the error function involves an integral of the Gaussian or normal distribution, further information is usually found in standard reference texts on probability and statistics. Some of these texts contain tabulated values of erf (x) for various values of x.

Four methods are presented below for evaluating erf (x) without recourse to other works. They are:

Method 1 — Tabulated values for x $\leq$ 2.2 (Table 16-8)
Method 2 — Graphical interpolation (Figure 16-6)
Method 3 — Series expansion
Method 4 — Approximations for: (a) large x (x $\gtrsim$ 2)
                                        (b) intermediate x (values near 1)
                                        (c) small x (x $\lesssim$ 0.1)

It should be noted that erf $(-x) = -$erf (x), and that values of erf (x) range from 0 to +1 (0 to -1 for erf $(-x)$).

In particular:

$$\text{erf}(0) = 0 \qquad (16\text{-}52)$$

and

$$\lim_{x \to \infty}(\text{erf}(x)) = 1 \text{ and } \lim_{x \to \infty}(\text{erf}(-x)) = -1 \qquad (16\text{-}53)$$

Method 1. Table 16-8 provides values of erf (x) for values of x up to 2.2. Values of erf (x) are given for each 0.01 increment up to 1.5.

Method 2. Figure 16-6 is a plot of erf (x) vs x for values of x up to 2.0. For values of x above 2, one can assume erf (x) $\simeq$ 1.0.

TABLE 16-8

Values of the Error Function for x ≤ 2.2

x	0	1	2	3	4	5	6	7	8	9
0.0	0.0000	0.0113	0.0226	0.0338	0.0451	0.0564	0.0676	0.0789	0.0901	0.1013
0.1	0.1125	0.1236	0.1348	0.1459	0.1569	0.1680	0.1790	0.1900	0.2009	0.2118
0.2	0.2227	0.2335	0.2443	0.2550	0.2657	0.2763	0.2869	0.2974	0.3079	0.3183
0.3	0.3286	0.3389	0.3491	0.3593	0.3694	0.3794	0.3893	0.3992	0.4090	0.4187
0.4	0.4284	0.4380	0.4475	0.4569	0.4662	0.4755	0.4847	0.4937	0.5027	0.5117
0.5	0.5205	0.5292	0.5379	0.5465	0.5549	0.5633	0.5716	0.5798	0.5879	0.5959
0.6	0.6039	0.6117	0.6194	0.6270	0.6346	0.6420	0.6494	0.6566	0.6638	0.6708
0.7	0.6778	0.6847	0.6914	0.6981	0.7047	0.7112	0.7175	0.7238	0.7300	0.7361
0.8	0.7421	0.7480	0.7538	0.7595	0.7651	0.7707	0.7761	0.7814	0.7867	0.7918
0.9	0.7969	0.8019	0.8068	0.8116	0.8163	0.8209	0.8254	0.8299	0.8342	0.8385
1.0	0.8427	0.8468	0.8508	0.8548	0.8586	0.8624	0.8661	0.8698	0.8733	0.8768
1.1	0.8802	0.8835	0.8868	0.8900	0.8931	0.8961	0.8991	0.9020	0.9048	0.9076
1.2	0.9103	0.9130	0.9155	0.9181	0.9205	0.9229	0.9252	0.9275	0.9297	0.9319
1.3	0.9340	0.9361	0.9381	0.9400	0.9419	0.9438	0.9456	0.9473	0.9490	0.9507
1.4	0.9523	0.9539	0.9554	0.9569	0.9583	0.9597	0.9611	0.9624	0.9637	0.9649
1.5	0.9661	0.9672	0.9684	0.9695	0.9706	0.9716	0.9726	0.9736	0.9746	0.9755
1.6	0.9763					0.9804				
1.7	0.9838					0.9867				
1.8	0.9891									
1.9	0.9928									
2.0	0.9953									
2.1	0.9970									
2.2	0.9981									

Source: Jost [9]. *(Reprinted with permission from Academic Press, Inc.)*

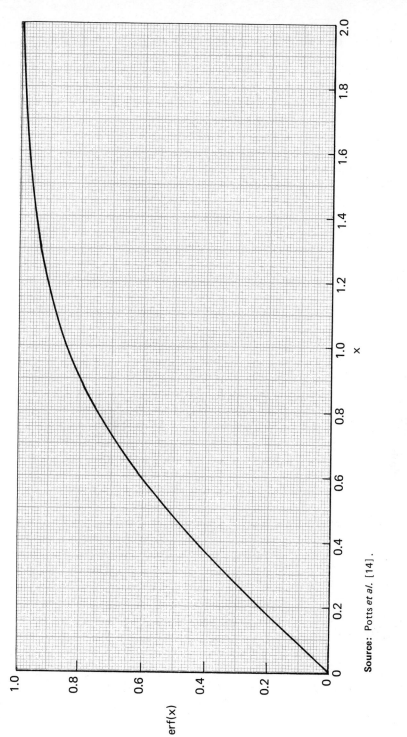

Source: Potts *et al.* [14].

FIGURE 16–6   Error Function

Method 3. The series expansion given in Eq. 16-54 may be used to calculate erf (x) [5].

$$\text{erf}(x) = \frac{2}{\sqrt{\pi}} \sum_{n=0}^{\infty} \frac{(-1)^n x^{2n+1}}{n!\,(2n+1)} \qquad (16\text{-}54)$$

In practice, only a few terms in the series must be added to obtain just two or three significant figures for erf (x). For example, summing the first five terms of Eq. 16-54 gives a value of erf (1) = 0.8434. The correct value of erf (1) is 0.8427.

Method 4. This method is based upon a more detailed description given in Ref. 2.

(a) For values of $x \gtrsim 2$ use:

$$\text{erf}(x) \simeq 1 - \frac{e^{-x^2}}{x\sqrt{\pi}} \qquad (16\text{-}55)$$

where e = 2.71828 and π = 3.14159.

(b) For values of x close to 1, set $x = 1 + \nu$ ($\nu \ll 1$) and then use:

$$\text{erf}(x) \simeq \text{erf}(1) + \frac{2\nu}{e\sqrt{\pi}} = 0.8427 + \frac{2\nu}{e\sqrt{\pi}} \qquad (16\text{-}56)$$

(c) For values of $x \leq 0.1$ use:

$$\text{erf}(x) \simeq \frac{2x}{\sqrt{\pi}} \qquad (16\text{-}57)$$

16-6   SYMBOLS USED[4]

c	=	concentration (M/L^3 or M/M)
c_a	=	concentration in air (M/L^3)

---

4. Since most of the parameters are not restricted to particular units, only dimensions are given (M = mass, L = length, T = time).

c_g =   concentration in gas phase (in soil air) (M/L^3)

c_l =   concentration in liquid phase (M/L^3)

c_o =   initial concentration (M/L^3)

c_s =   concentration in soil (adsorbed) $(M/L^3$ or $M/M)$

c_t =   total concentration in soil (M/L^3)

c_{t_o} =   initial total concentration in soil (M/L^3)

d   =   thickness of non-moving air layer in Eq. 16-27 (L)

D   =   diffusion coefficient in soil in Eq. 16-1 (L^2/T)

D_e =   effective diffusion coefficient = D_g + HDd in Eq. 16-38 (L^2/T)

D_g =   diffusion coefficient in gas phase (soil air) in Eq. 16-33 (L^2/T)

D   =   diffusion coefficient in liquid phase (soil water) in Eq. 16-33 (L^2/T)

D_s =   nonvapor-phase diffusion coefficient in Eq. 16-1 (L^2/T)

D_v =   vapor-phase diffusion coefficient in air in Eq. 16-2 (L^2/T)

D'_v =   apparent vapor-phase diffusion coefficient in Eq. 16-1 (L^2/T)

erf(x)   = error function of value x in Eq. 16-15 (see Table 16-8)

erfc(x) = complementary error function of value x = $1-$erf(x) in Eq. 16-42 (see   16-5)

f   =   flux of chemical in Eq. 16-6 $(M/L^2 \cdot T)$

f_w =   flux of water in Eqs. 16-9 and -33 $(M/L^2 \cdot T)$

h   =   humidity of outer air in Eq. 16-6

H   =   Henry's law constant

k   =   thermal conductivity of air in Eq. 16-6 $(cal/L \cdot K)$

k_v =   rate constant for volatilization in Eq. 16-49 (T^{-1})

K_H =   constant for relating c_l and c_g, based on Henry's Law, = c_l/c_g $(cm^3$ air/cm^3 water)

K_{oc} =   soil adsorption coefficient (soil/water) in Eq. 16-48 $(M/M)/(M/L^3)$

L   =   depth of soil column in Eq. 16-11 (L)

m   =   exponent in Eq. 16-4

M   =   molecular weight in Eq. 16-5 (M/mol)

n   =   index in Eqs. 16-12 and -54

P   =   ambient pressure in Eq. 16-1

P_o	=	reference pressure in Eq. 16-1
P_{vp}	=	vapor pressure of chemical in Eq. 16-9
Q_t	=	total loss of chemical/unit area in Eq. 16-9 (M/L^2)
R	=	universal gas constant
R_o	=	isotherm coefficient in Eq. 16-22
s	=	air-filled porosity in Eq. 16-2
s_t	=	total porosity in Eq. 16-2
S	=	solubility in Eq. 16-48 (M/L^3)
t	=	time (T)
$t_{1/2}$	=	half-life time in Eq. 16-48 (T)
T	=	temperature in Eq. 16-4
v	=	air velocity in Eq. 16-21 (L/T)
V	=	mean velocity of molecules in Eq. 16-4 (L/T)
V_e	=	effective convection velocity = $K_H f_w$ in Eq. 16-38
w	=	parameter in Eqs. 16-45, -46
x,y	=	parameters in Eq. 16-51
z	=	vertical (depth) coordinate in soil column (L)

Greek

α	=	parameter of adsorption isotherm in Eq. 16-35
α_n	=	root of Eq. 16-23
β	=	parameter of adsorption isotherm in Eq. 16-35
γ	=	$\beta \rho_b$ in Eq. 16-36
δ	=	thickness of stagnant layer in Eq. 16-6 (L)
ϵ	=	$(\rho_b K_H \alpha + \theta K_H + \eta)$ in Eq. 16-36
η	=	soil air content in Eq. 16-32 (L^3/L^3)
θ	=	volumetric soil water content in Eq. 16-32 (L^3/L^3)
λ_v	=	latent heat of vaporization in Eq. 16-6 (cal/M)
ν	=	parameter in Eq. 16-56
π	=	3.14159
ρ	=	vapor density in Eq. 16-2 (M/L^3)
ρ_b	=	soil bulk density in Eq. 16-2 (M/L^3)
ρ_{max}	=	saturated vapor concentration at the temperature of the outside air in Eq. 16-6 (M/L^3)

16-7   REFERENCES

1. Ehlers, W., J. Letey, W.F. Spencer and W.J. Farmer, "Lindane Diffusion in Soils: I. Theoretical Considerations and Mechanism of Movement," *Soil Sci. Soc. Am. Proc.* **33,** 501-4 (1969).

2. Farmer, W.J., K. Igue and W.F. Spencer, "Effects of Bulk Density on the Diffusion and Volatilization of Dieldrin from Soil," *J. Environ. Qual.,* **2,** 107-09 (1973).

3. Farmer, W.J., M.S. Yang, J. Letey and W.F. Spencer, "Hexachlorobenzene; Its Vapor Pressure and Vapor Phase Diffusion in Soil," unpublished, undated manuscript received as personal communication from W.J. Farmer, 4 December 1979. [For further discussion, see the report by these authors titled "Land Disposal of Hexachlorobenzene Wastes: Controlling Vapor Movement in Soil," EPA-600/2-80-119, Office of Research and Development, U.S. Environmental Protection Agency, Cincinnati, Ohio (August 1980).]

4. Goring, C.A.I., "Theory and Principles of Soil Fumigation," *Advan. Pest. Control Res.,* **5,** 47-84 (1972).

5. Gradshteyn, I.S. and I.M. Ryzhik, *Table of Integrals, Series and Products,* 4th ed., prepared by Yu. V. Geronimus and M. Yu. Tseytlin, Academic Press, New York (1965).

6. Guenzi, W.D. and W.E. Beard, "Volatilization of Pesticides," Chap. 6 in *Pesticides in Soil and Water,* ed. by W.D. Guenzi, Soil Science Society of America, Madison, Wis. (1974).

7. Hamaker, J.W., "Diffusion and Volatilization," Chap. 5 in *Organic Chemicals in the Soil Environment,* Vol. 1, ed. by C.A.I. Goring and J.W. Hamaker, Marcel Dekker, New York (1972).

8. Hartley, G.S., "Evaporation of Pesticides," Chap. II in "Pesticidal Formulations Research, Physical and Colloidal Chemical Aspects," *Advances in Chemistry Series,* **86,** American Chemical Society, Washington, D.C. (1969).

9. Jost. W., "Diffusion in Solids, Liquids, Gases," in *Physical Chemistry — A Series of Monographs,* ed. by E. Hutchinson, Academic Press, New York (1952).

10. Jury, W.A., Grover, R., Spencer, W.F. and Farmer, W.J., "Modeling Vapor Losses of Soil-Incorporated Triallate," *Soil Sci. Soc. Am. J.,* **44,** 445-50 (May-June 1980).

11. Letey, J. and W.J. Farmer, "Movement of Pesticides in Soil," Chap. 4 in *Pesticides in Soil and Water,* ed. by W.D. Guenzi, Soil Science Society of America, Madison, Wis. (1974).

12. Mayer, R., J. Letey and W.J. Farmer, "Models for Predicting Volatilization of Soil-Incorporated Pesticides," *Soil Sci. Soc. Am. Proc.,* **38,** 563-68 (1974).

13. Millington, R.J. and J.M. Quirk, "Permeability of Porous Solids," *Trans. Faraday Soc.,* **57,** 1200-7 (1961).

14. Potts, R.G., J.H. Hagopian, C.R. Woodruff and P.P. Raj, "Mathematical Development of the Spill Assessment Model (SAM) for Hydrazine and Similar Acting Materials in Water Bodies," Report No. ESL-TR-80-07 prepared for the Department of the Air Force, Headquarters Air Force Engineering and Service Center, Tyndall Air Force Base, Florida (1980).

15. Shearer, R.C., J. Letey, W.J. Farmer and A. Klute, "Lindane Diffusion in Soil," *Soil Sci. Soc. Am. Proc.,* **37,** 189-93 (1973).

16. Spencer, W.F., "Distribution of Pesticides Between Soil, Water and Air," in *Pesticides in the Soil: Ecology, Degradation and Movement,* Michigan State University, East Lansing, Mich., pp. 120-28 (1970).

17. Spencer, W.F. and M.M. Cliath, "Desorption of Lindane from Soil as Related to Vapor Density," *Soil Sci. Soc. Am. Proc.,* **34,** 574-78 (1970).

18. Spencer, W.F. and M.M. Cliath, "Vaporization of Chemicals," in *Environmental Dynamics of Pesticides,* ed. by R. Haque and V.H. Freed, Plenum Press, New York (1975).

19. Spencer, W.F., W.J. Farmer, and M.M. Cliath, "Pesticide Volatilization," *Residue Rev.,* **49,** 1-47 (1973).

20. Swann, R.L., P.J. McCall and S.M. Unger, "Volatility of Pesticides from Soil Surfaces," unpublished, undated manuscript received as personal communication from Dow Chemical USA, Midland, Mich., 16 November 1979.

21. Tinsley, I.J., *Chemical Concepts in Pollutant Behavior,* John Wiley and Sons, New York, p. 51 (1979).

17

DIFFUSION COEFFICIENTS IN
AIR AND WATER

William A. Tucker and Leslie H. Nelken

17-1   INTRODUCTION

Molecular diffusion is the net transport of a molecule in a liquid or gas medium and is a result of intermolecular collisions rather than turbulence or bulk transport. The process is promoted by gradients, such as pressure, temperature, and concentration, the last of which is considered in this chapter. The rate of diffusion is a function of the properties of two compounds; i.e., it depends not only on the nature of the compound in question but also on the medium through which the compound moves. The *diffusion coefficient*, or diffusivity, is defined as

$$\mathcal{D}_{BA} = \frac{J_B}{\nabla X_B} \qquad (17\text{-}1)$$

where $\mathcal{D}_{BA}$ = diffusion coefficient of compound B in compound A (cm²/s)

J_B = net molal flux of B across a hypothetical plane (mol/cm²·s)

∇X_B = concentration gradient of B at the hypothetical plane (mol/cm³·cm)

A cursive $\mathcal{D}$ is used to distinguish this property from the *apparent* diffusion coefficient, which is typically represented by D. The components that comprise the binary system are indicated by subscripts to the symbol; in this chapter, one of the components is always air (A) or water (W).

It should be emphasized that the methods described in this chapter are *not* useful for estimating the dispersion[1] of contaminants in either the atmosphere or surface water bodies. Most environmental fluid media — the atmosphere, rivers, lakes and oceans — are turbulent, and dispersion in these media is therefore controlled by the intensity of turbulent mixing rather than molecular diffusion. For example, in meteorological or oceanographic literature, the apparent dispersion coefficient, turbulent diffusion coefficient, diffusivity, or dispersivity (denoted by D_s in this chapter) all refer to the ratio of the contaminant flux to its gradient; in turbulent flow, this coefficient is always much greater than the molecular diffusion coefficient, $\mathcal{D}_{BA}$.

Even in highly stable layers of the atmosphere, such as inversions or in thermoclines of oceans and lakes, actual transfer rates are significantly higher than they would be if caused solely by molecular diffusion. Although these stabilized layers are usually laminar flow regions, intermittent bursts of turbulence caused by flow instabilities account for most of the dispersion through them [9]. Published coefficients of dispersion, turbulent diffusion, or eddy diffusion in the atmosphere or surface waters are generally several orders of magnitude larger than the diffusion coefficients derivable by the methods explained in this chapter.

There are, however, several special environmental situations in which molecular diffusion is a significant, or even controlling, factor in determining chemical fluxes. These include air-water interfaces, the interstitial waters of sediments, and groundwater (saturated or unsaturated soils).

Air-Water Interfaces. The air-water interface is probably the most important of the fluid interfaces considered in environmental analyses. Its thickness, which varies according to the rate of turbulent mixing in the free fluids, has a direct effect on the transfer rate for a given chemical.

Even when the rate of turbulent mixing is constant and the thickness of the interface does not change, however, the transfer rate varies from one chemical to another. This is because of the action of the laminar boundary layers on both sides of the interface; in these boundary layers,

---

1. The combined effect of molecular diffusion and contaminant migration associated with fluid motion.

turbulence is suppressed, and the flux of chemicals across the fluid-fluid (or fluid-solid) interface is largely a process of molecular diffusion. Thus, chemicals with different molecular diffusion coefficients exhibit different interfacial fluxes.

This relationship has prompted the method of Smith *et al.* [22,23], which relates the fluxes of gases across the surface of rivers and lakes to the reaeration rate constant (oxygen mass transfer coefficient). The flux of a chemical is approximately equal to the oxygen transfer rate times the ratio of the chemical's diffusion coefficient to that of oxygen. (See Eq. 15-20.) The formulation established by Smith *et al.* is a special case of the more general relationship presented by Kraus [8], who derived a gas-phase exchange coefficient, k_g, that characterizes the rate of diffusion in cm/s across the molecular boundary layers.[2] For the air side,

$$k_g = \frac{\mathcal{D}_{AB} U}{75 \, \nu_A} \tag{17-2}$$

where ν_A is the kinematic viscosity of air at 20°C (0.15 cm²/s), and U is the wind speed (cm/s) at a height of 10 meters. The water-phase exchange coefficient is given by:

$$k_l = \frac{\mathcal{D}_{WB} \, U}{75 \, \nu_W} \sqrt{\frac{\rho_A}{\rho_W}} \tag{17-3}$$

where ν_W is the kinematic viscosity of water (0.01 cm²/s) and ρ_A and ρ_W are the density of air and water (at 20°C), respectively. This water-phase exchange coefficient is valid only when momentum is transferred from air to water, as in lakes, oceans, and slowly flowing streams. For rapidly flowing water, the water current becomes a factor.

Liss and Slater [12] and Liss [11] have further developed this concept, defining the overall interfacial resistance to transfer as

$$R_T = \frac{1}{k_l} + \frac{1}{H' k_g} \tag{17-4}$$

where H' is the Henry's law constant. This approach is described in more detail in Chapter 15. Equation 17-4 implies that transfer is limited by water phase exchange when $k_l << H' k_g$.

---

2. Kraus used the term "diffusion velocity" instead of "phase exchange coefficient." The latter is used here to keep the terminology consistent with that of Chapter 15.

Interstitial Waters. Other environmental media where molecular diffusion can be important include groundwaters and the interstitial waters of sediments, where the movement of water is not turbulent. However, the flow paths of molecules through porous media are not parallel as in typical laminar flow because of the tortuous path taken through the pore spaces; these meandering flow paths through the porous medium are analogous to the eddy motion of a turbulent fluid, causing the observed dispersivities to exceed molecular diffusion rates. In many instances, particularly when the water moves slowly, the molecular diffusion coefficient affects the overall dispersion process. In a recent review of exchange at the sediment-water interface, Lerman [10] discusses data of Manheim [14] which indicate that the ratio of the apparent diffusion coefficient to the molecular diffusion coefficient varies as the sediment porosity raised to the nth power, where $1.2 < n < 2.8$.

The vertical distribution of anthropogenic contaminants in undisturbed lake, estuarine, or ocean sediments reflects historical contaminant loadings to the water body. The analysis of potential contaminant migration in the interstitial waters after deposition requires an analysis of molecular diffusion in the aqueous phase of the sediments.

Groundwater. In saturated groundwater (aquifers) the effective dispersivity is generally much greater than the molecular diffusion coefficient. Scheidigger [20] postulated that the apparent dispersivity is proportional to the flow velocity and that molecular diffusion may therefore be significant in aquifer systems with extremely slow flow rates. Rifai *et al.* (1956) (cited in Ref. 7) determined that molecular diffusion is important at pore water velocities of less than 2.5×10^{-4} cm/s, in saturated flow. Harleman and Rumer (1962) (also cited in Ref. 7) proposed the following relationship:

$$D_s = f_t \mathcal{D}_{BW} + \alpha u^m \qquad (17\text{-}5)$$

where D_s = apparent dispersion coefficient (cm²/s)
   f_t = the tortuosity factor, a soil property (0.01-0.5)
   u = pore water velocity (cm/s)
   α, m = empirically determined soil constants

Scheidigger's theoretical analysis suggests $m=1$, while empirically determined values range from 1 to 1.4 [3]. Kirda *et al.* [7] found that the ratio of D_s to $\mathcal{D}_{BW}$ is greater than 100 for $u > 0.002$ cm/s. Therefore, molecular diffusion can probably be ignored at pore water velocities exceeding 0.002 cm/s.

The analysis is more complicated in unsaturated soils, since these represent a three-phase system consisting of soil particles, water, and air. Liquid-phase and gas-phase diffusion, as well as diffusion along the water-air and water-solid interfaces, contribute to dispersion. (Air-solid interfaces are of lesser importance, because any available water will exist as a coating on soil particles and minimize their contact with the air.) Two equations (17-6 or 17-8) have been developed for describing dispersion in these cases, assuming there is no net infiltration of water into the soil. The more rigorous, by Shearer *et al.* [21], yields the following expression for the effective diffusion coefficient (D_{EF}):

$$D_{EF} = \frac{\mathcal{D}_{AB}\, p^{7/3}}{p_T^2\,(S+1)} + \left(\frac{S}{1+S}\right)\left(\frac{D_S + D_H K'\beta + \beta D_I S'}{\beta K' + \theta + \beta S'}\right) \qquad (17\text{-}6)$$

where $\mathcal{D}_{AB}$ = gas diffusion coefficient in air (cm²/s)

p = air-filled porosity of the soil (0-0.7 cm³/cm³)

p_T = total porosity of the soil (0-0.7 cm³/cm³)

S = equilibrium coefficient of proportionality between vapor mass density and total concentration of pesticide in soil (g/cm³)/(g/cm³)

D_S = apparent solution phase diffusion coefficient (cm²/s)

D_H = apparent diffusion coefficient of molecules adsorbed at the solution-solid interface (cm²/s)

K' = adsorption coefficient (cm³/g)

β = soil bulk density (g/cm³)

D_I = apparent diffusion coefficient of molecules adsorbed at the air-solution interface (cm²/s)

S' = coefficient of proportionality between solution concentration and vapor concentration at the air-solution interface (cm³/g)

θ = fractional volumetric water content (cm³/cm³)

Furthermore, it was shown in Ref. 19 that

$$D_S = (\theta/p_T)^2 \theta^{4/3}\, \mathcal{D}_{BW} \qquad (17\text{-}7)$$

but the authors did not derive similar theoretical expressions for D_H and D_I.

Walker and Crawford [25] developed the following equation for use when liquid-phase diffusion is assumed to predominate:

$$D_{EF} = \frac{\mathcal{D}_{BW}\, \theta\, f_t}{\beta K' + \theta} \qquad (17\text{-}8)$$

For this case, where $\mathcal{D}_{BA} = D_H = D_I = S' = 0$, Shearer's expression (Eq. 17-6) reduces to

$$D_{EF} = \frac{S\left[(\theta/p_T)^2\, \theta^{4/3}\, \mathcal{D}_{BW}\right]}{(1+S)(\beta K' + \theta)} \qquad (17\text{-}9)$$

Clearly, by either analysis, the effective diffusion coefficient is proportional to the molecular diffusion coefficient and inversely proportional to the term $(\beta K' + \theta)$. Available information does not permit a comparative evaluation of these different expressions relating the effective diffusion coefficient to the molecular diffusion coefficient.

17-2   DIFFUSIVITY IN AIR

Methods for estimating the diffusion coefficient of a binary gas system have foundations in the theoretical equation derived independently by Chapman and Enskog (cited in Ref. 17) for dilute gases at low pressures. They found that the diffusion of gases by intermolecular collision is a function of Boltzmann's constant (k), the molecular weight as described by M_r, the collision integral (Ω), and the characteristic length (σ_{AB}) of molecule A interacting with molecule B, according to the following equation:

$$\mathcal{D}_{AB} = 1.858 \times 10^{-3} \left(\frac{T^{3/2}\sqrt{M_r}}{P\sigma_{AB}^2\, \Omega} \right) \qquad (17\text{-}10)$$

where $M_r = (M_A + M_B)/M_A M_B$
$\quad\quad\ \ P\ \ = $ pressure (atm)
$\quad\quad\ \ T\ \ = $ temperature (K)

This equation was derived for non-polar, monatomic spherical molecules and has since been tested on a number of organic gases over a wide temperature range. Values of Ω and σ_{AB} are functions of temperature and depend upon the intermolecular potential function selected. For this chapter, the Lennard-Jones 12-6 potential is used [17].

Empirical equations for estimating the diffusion coefficient reflect much of the Chapman-Enskog equation form. The equation of Wilke and

Lee [28], for example, differs only in the constant (1.858×10^{-3}), which is expressed as a function of the molecular weight.

The gaseous diffusion coefficient is a function of density, pressure, and temperature. It is inversely related to density and pressure; the product of $\mathscr{D}_{BA}\rho$ decreases with increasing density, while the product $\mathscr{D}_{BA}P$ is nearly constant at low pressures [17].

The gaseous diffusion coefficient is theoretically related to temperature (T) in the following manner:

$$\mathscr{D}_{BA} \propto \frac{T^{3/2}}{\Omega(T)} \qquad (17\text{-}11)$$

The exponential coefficient for temperature varies from 1.5 to 2 over a wide temperature range. Among recommended methods, the Wilke and Lee (WL) method uses 1.5; Fuller, Schettler and Giddings (FSG) uses 1.75, which Barr and Watts [1] found to give the best value for the gaseous diffusion coefficient. The WL method retains the temperature-dependent collision integral function and therefore allows for prediction of $\mathscr{D}_{BA}$ with good accuracy over a wide temperature range. The FSG method does not incorporate any temperature-dependent functions to compensate for the fixed exponential coefficient associated with T, so the diffusion coefficient calculated by this method is accurate only over a limited temperature range; however, as the temperature range encountered in environmentally important systems is not very wide, the accuracy is normally adequate.

17-3 AVAILABLE METHODS OF ESTIMATING DIFFUSION COEFFICIENTS IN AIR

A number of methods exist for estimating the gaseous diffusion coefficient. Jarvis and Lugg [6,13] compared the results of nine estimation techniques with measured diffusion coefficients for approximately 150 compounds in air. For the most part, these methods require few data inputs. Only two of the methods listed in Table 17-1 require critical properties; the others use readily available or easily estimated data. The method errors shown in the table, except for that of Venezian [24], were calculated from the findings of Jarvis and Lugg. Although the reported absolute average errors do not differ greatly, a few of the methods listed are applicable only to a limited number of chemical groups. The two recommended methods are described in the following section.

TABLE 17-1

Methods for Estimating Gaseous Diffusion Coefficients

Method	Inputs (excluding air parameters)	Applicability	Compounds Tested (n)	Absolute Average Error (%)	% n ≤ ±5% Error	Comments
Fuller, Schettler and Giddings[a,b]	M_B, V_B	Applicable to nonpolar gases at low to moderate temperatures	128	7.6	42	Does not distinguish isomers
Wilke and Lee[a]	M_B, σ_{AB}, Ω	Applicable to a wide range of compounds and temperatures	137	4.3	69	See notes c and d. Does not distinguish isomers
Chen and Othmer[e]	M_B, T_{C_B}, V_{C_B}	Unsuitable for amides and amines	66	4.0	76	Requires critical properties
Gilliland	M_B, V_B	Applicable only to restricted range of compounds	151	10	26	Does not distinguish isomers
Arnold	M_B, V_B		151	8.9	54	
Hirschfelder, Bird and Spotz	M_B, σ_{AB}, Ω	Greatest accuracy for aliphatic esters, acids, alcohols	151	9.3	16	See notes c and d. Accuracy increases with increasing molecular weight
Andrusson	M_B, V_B, P	Accurate only for sub-stituted aromatics and glycols	151	9.1	33	Does not distinguish isomers. Deviations observed were most erratic of methods using molar volume
Othmer and Chen	M_B, V_{C_B}		66	8.3	25	Requires critical parameters
Venezian [24]	M_B, n_D		28	9.1		

a. Recommended method.
b. Calculations used experimental atomic diffusion data.

c. $\sigma_{AB} = f(V'_B)$.
d. $\Omega = f(T, \epsilon_{AB})$.

e. From Ref. 2 except for error analysis, which is from Ref. 6.

Source: Jarvis and Lugg [6] except where noted.

17-4   SELECTED METHODS OF ESTIMATING GASEOUS DIFFUSION COEFFICIENTS OF ORGANICS IN AIR

The criteria for choosing a specific estimation technique include (1) ease of use, (2) availability of input data, and (3) accuracy of results for a general chemical population. Of those listed in Table 17-1, only the first two fulfill all of these requirements. Both methods estimate the gaseous diffusion coefficient from the structure of the chemical, although neither can distinguish between isomers.

The FSG method is much easier and less time-consuming than the WL method and is reportedly applicable to non-polar gases at low to moderate temperatures. Jarvis and Lugg [6] state that the method is most accurate for chlorinated aliphatics and that aromatics, alkanes, and ketones deviate ±5% from the measured value. It does not give accurate estimates for esters and alcohols higher than C_5 and C_7, respectively. The reported average absolute error of 7.6% is based on values of the atomic diffusion volume, a parameter which is derived from measured diffusion coefficient values.

The WL method, as mentioned earlier, is very similar to the theoretical expression derived by Chapman and Enskog and is applicable to many kinds of compounds over a fairly wide temperature range. It is a much more tedious computation, since parameters such as the collision integral and characteristic length must be calculated by additional approximations. However, as shown in Table 17-2, the average errors obtained with this method are considerably less than those of the FSG method for some classes of chemicals, particularly nitriles, glycols, and aromatic esters.

FSG Method. The method of Fuller, Schettler and Giddings [4] is most accurate for non-polar gases at low to moderate temperatures. Its accuracy is poorest with the polar acids and glycols (Table 17-2); minimal error is associated with the aliphatics and aromatics. The method is based on the following correlation:

$$\mathcal{D}_{BA} = \frac{10^{-3} \, T^{1.75} \sqrt{M_r}}{P \, (V_A^{1/3} + V_B^{1/3})^2} \tag{17-12}$$

TABLE 17-2

Comparison of Absolute Average Errors by Chemical Class

	Number of Compounds Tested	WL Method (%)	FSG Method (%)
Aliphatics	3	1.8	3.9
Aromatics	13	3.9	4.2
Substituted Aromatics	9	4.2	9.9
Alcohols	17	5.8	9.6
Ethers	5	6.1	7.5
Ketones	5	1.4	3.0
Acids	8	8.7	12.4
Aliphatic Esters	36	3.9	8.5
Aromatic Esters	4	1.8	9.8
Halogen Hydrocarbons	22	3.4	1.7
Glycols	7	1.0	11.5
Amines and Amides	6	7.6	4.3
Nitriles	2	1.9	7.8

Source: Jarvis and Lugg [6]

where T and M_r are as previously defined, P is the pressure (atm), and V_A and V_B are the molar volumes for air and the gas in question, respectively. Values for V_A and other properties of air are listed in Table 17-3. V_B can be estimated from the chemical structure of the molecule using the increments listed in Table 17-4. For a chemical containing atoms that are not listed in Table 17-4, the diffusion volume can be estimated as 85-90% of the LeBas volume, V_B' [17], i.e.,

$$V_B \approx (0.85 \text{ to } 0.90)V_B' \qquad (17\text{-}13)$$

Increments for calculating LeBas volumes are listed in Table 17-5. (See description of the WL method below.) The molal volumes of some chemicals, notably organophosphorus compounds, cannot be estimated from either Table 17-4 or 17-5 because the related atoms are not listed.

Basic Steps

(1) Find M_B, the molecular weight of gas B.

(2) Use Table 17-3 to find M_A and V_A.

(3) Calculate M_r from $(M_A + M_B)/M_A M_B$.

TABLE 17-3

Some Physical Properties of Air

M_A	28.97 g/mol
V_A	20.1 cm^3/mol
ϵ/k	78.6 K
σ_A	3.711 Å

Source: Jarvis and Lugg [6]. Values of ϵ/k and σ_A obtained from Lennard-Jones 12-6 potential function.

TABLE 17-4

Atomic and Structural Diffusion Volume Increments[a] (cm^3/mol)

	ΔV_B
C	16.5
H	1.98
O	5.48
N	(5.69)
Cl	(19.5)
S	(17.0)
Aromatic and heterocyclic rings	−20.2

a. Values in parentheses are based upon few data points.

Source: Fuller, Schettler, and Giddings [4]. *(Reprinted with permission from the American Chemical Society.)*

TABLE 17-5

Additive Volume Increments for Calculating LeBas Molar Volume, V_B'

Atom	Increment (cm^3/mol)	Atom	Increment (cm^3/mol)
C	14.8	Br	27.0
H	3.7	Cl	24.6
O (except as noted below)	7.4	F	8.7
In methyl esters and ethers	9.1	I	37.0
In ethyl esters and ethers	9.9	S	25.6
In higher esters and ethers	11.0		
In acids	12.0	Ring	
Joined to S, P, N	8.3	3-Membered	−6.0
N		4-Membered	−8.5
		5-Membered	−11.5
Double bonded	15.6	6-Membered	−15.0
In primary amines	10.5	Naphthalene	−30.0
In secondary amines	12.0	Anthracene	−47.5

Source: Reid, *et al.* [17]. *(Reprinted with permission from McGraw-Hill Book Co.)*

(4) Calculate V_B by adding the incremental values for the atoms listed in Table 17-4; for molecules containing atoms not listed in this table (except P), calculate the LeBas volume, V_B', from Table 17-5 and use 87.5% of this value as the atomic diffusion volume.

(5) Calculate $\mathcal{D}_{BA}$ using Eq. 17-12 for the temperature and pressure (P) of interest. P will usually be 1 atm.

Example 17-1   Calculate the diffusion coefficient of *m*-chlorotoluene (C_7H_7Cl) into air at 25°C and 1 atm, given M_B = 126.59 g/mol.

(1)   From Table 17-3,

$$M_A = 28.97 \text{ g/mol and } V_A = 20.1 \text{ cm}^3/\text{mol}$$

(2)   Calculate M_r

$$M_r = (28.97 + 126.59)/(28.97)(126.59) = 0.0424$$

(3)   Calculate V_B for *m*-chlorotoluene from Table 17-4:

$$
\begin{aligned}
7(C) &= 7(16.5) \\
7(H) &= 7(1.98) \\
1(Cl) &= 1(19.5) \\
1(Ring) &= 1(-20.2) \\
V_B &= 128.7 \text{ cm}^3/\text{mol}
\end{aligned}
$$

(4)   Using Eq. 17-12,

$$\mathcal{D}_{BA} = \frac{10^{-3} (298)^{1.75} \sqrt{0.0424}}{1 \cdot \left[(20.1)^{1/3} + (128.7)^{1/3} \right]^2} = 7.31 \times 10^{-2} \text{ cm}^2/\text{s}$$

The deviation is +13% from the literature value, 6.45×10^{-2} cm^2/s [6].

Example 17-2   Calculate the diffusion coefficient of isopropyl iodide (($CH_3)_2$CHI) into air at 25°C and 1 atm, given M_B = 169.9 g/mol.

(1)   Since M_A = 28.97 g/mol (Table 17-3), M_r is:

$$M_r = (28.97 + 169.9)/(28.97)(169.9) = 0.0404$$

(2) V_B of $(CH_3)_2$CHI cannot be determined by using Table 17-4. The LeBas volume (Table 17-5) is:

$$
\begin{aligned}
3(C) &= 3(14.8) \\
7(H) &= 7(3.7)
\end{aligned}
$$

$$1(I) = 1(37.0)$$

$$V'_B = 107.3 \text{ cm}^3/\text{mol}$$

Thus, $V_B \simeq 0.875 \times 107.3 = 93.9 \text{ cm}^3/\text{mol}$.

(3) From Eq. 17-14

$$\mathcal{D}_{BA} = \frac{10^{-3} (298)^{1.75} \sqrt{0.0404}}{1 \cdot \left[(20.1)^{1/3} + (93.9)^{1/3} \right]^2} = 8.15 \times 10^{-2} \text{ cm}^2/\text{s}$$

The literature value is $8.78 \times 10^{-2} \text{ cm}^2/\text{s}$ [6], indicating a deviation of -7.2%.

WL Method. The method of Wilke and Lee [28] for estimating gaseous diffusion coefficients is reported to be usable for a wider range of compounds and temperatures than is the FSG method. According to Jarvis and Lugg [6], the absolute average error for about 150 compounds tested was 4.3%; all classes of compounds had average errors of less than 8%, except for acids. (See Table 17-2.) This method is significantly more accurate than the FSG method for nitriles, aromatic esters, and glycols.

The correlation used is very similar to that developed by Chapman and Enskog (see Eq. 17-10):

$$\mathcal{D}_{BA} = \frac{B'T^{3/2} \sqrt{M_r}}{P\sigma_{AB}^2 \Omega} \tag{17-14}$$

where

$$B' = 0.00217 - 0.00050 \sqrt{\frac{1}{M_A} + \frac{1}{M_B}} \tag{17-15}$$

The other parameters are as defined for Eq. 17-10. The greater accuracy is obtained at the cost of more tedious computation, involving calculation of the parameters Ω and σ_{AB}.

The collision integral, Ω, is a function of the molecular energy of attraction, ϵ, and the Boltzmann constant, k, as shown in Eqs. 17-16 and -17 [17].

$$\Omega = \frac{a}{(T^*)^b} + \frac{c}{e^{T^*d}} + \frac{e}{e^{T^*f}} + \frac{g}{e^{T^*h}} \tag{17-16}$$

where the values of a-h are as follows:

$$a = 1.06036 \qquad c = 0.19300 \qquad e = 1.03587^3 \qquad g = 1.76474$$
$$b = 0.15610 \qquad d = 0.47635 \qquad f = 1.52996 \qquad h = 3.89411$$

and

$$T^* = \frac{T}{(\epsilon/k)_{AB}} \qquad (17\text{-}17)$$

The denominator of Eq. 17-17 is defined as

$$(\epsilon/k)_{BA} = \sqrt{(\epsilon/k)_A \, (\epsilon/k)_B} \qquad (17\text{-}18)$$

where $(\epsilon/k)_A = 78.6K$ (Table 17-3) and, from Ref. 28,

$$(\epsilon/k)_B = 1.15 \, T_b \ (K) \qquad (17\text{-}19)$$

The characteristic length, σ_{AB}, is a function of the molal volume at the boiling point:

$$\sigma_{AB} = \frac{\sigma_A + \sigma_B}{2} \qquad (17\text{-}20)$$

where $\sigma_A = 3.711$ Å (Table 17-3) and, from Ref. 28,

$$\sigma_B = 1.18 \, (V'_B)^{1/3} \qquad (17\text{-}21)$$

Thus, calculation of Ω and σ_{AB} ultimately requires knowledge of the molecular weight, boiling point, and molal volume of compound B. If the boiling point is unknown, it can be estimated by methods presented in Chapter 12, and the LeBas method (Table 17-5) provides a means of estimating molal volume.

Basic Steps

(1) Retrieve the molecular weight and boiling point (K) of the compound from the literature.

(2) From Table 17-3, obtain M_A, $(\epsilon/k)_A$, and σ_A.

(3) Calculate the LeBas molal volume (V'_B) for the compound of interest using Table 17-5.

(4) Substitute V'_B in Eq. 17-21 to find σ_B.

---

3. The constant e is used only once in Eq. 17-16. The e's printed in bold face are the base of natural logarithms (2.718+).

(5) Calculate σ_{AB} from Eq. 17-20.

(6) Use Eq. 17-19 to determine $(\epsilon/k)_B$ from T_b.

(7) Calculate T^* for the desired temperature by using Eq. 17-18 and 17-17 in that order.

(8) Insert T^* and constants a-h in Eq. 17-16 for evaluation of Ω.

(9) Find B' from Eq. 17-15.

(10) Calculate $(M_A + M_B)/M_A M_B$ to obtain M_r.

(11) Calculate $\mathcal{D}_{BA}$ from Eq. 17-14, using the known values of T and P and the derived values of B', M_r, σ_{AB}, and Ω.

Example 17-3   Calculate the diffusion coefficient of m-chlorotoluene (C_7H_7Cl) in air at 25°C (298K) and 1 atmosphere.

(1)   From the literature, $M_B = 126.59$ g/mol, and $T_b = 435$K.

(2)   From Table 17-3,

$$M_A = 28.97 \text{ g/mol}$$
$$\sigma_A = 3.711 \text{ Å}$$
$$(\epsilon/k)_A = 78.6 \text{K}$$

(3)   From Table 17-5, the increments of the LeBas molal volume are:

$$7(C) = 7(14.8) = 103.6$$
$$7(H) = 7(3.7) = 25.9$$
$$1(Cl) = 24.6$$
$$6\text{-membered ring} = \underline{-15.0}$$
$$V'_B = 139.1 \text{ cm}^3/\text{mol}$$

(4)   From Eq. 17-21,

$$\sigma_B = 1.18(139.1)^{1/3} = 6.11 \text{ Å}$$

(5)   From Eq. 17-20,

$$\sigma_{AB} = (3.711 + 6.11)/2 = 4.91 \text{ Å}$$

(6)   From Eq. 17-19,

$$(\epsilon/k)_B = 1.15(435) = 500\text{K}$$

(7)   From Eq. 17-18,

$$(\epsilon/k)_{AB} = \sqrt{(78.6)(500)} = 198\text{K}$$

which yields the following (Eq. 17-17):

$$T^* = 298/198 = 1.51$$

(8) From Eq. 17-16,

$$\Omega = \frac{1.06}{1.51^{0.156}} + \frac{0.193}{e^{(1.51)(0.476)}} + \frac{1.04}{e^{(1.51)(1.53)}} + \frac{1.76}{e^{(1.51)(3.89)}}$$

$$= 1.20$$

(9) From Eq. 17-15,

$$B' = 0.00217 - 0.00050 \sqrt{(1/28.97)+(1/126.59)}$$

$$= 2.07 \times 10^{-3}$$

(10) $M_r = (28.97 + 126.59)/(28.97)(126.59)$

$$= 0.0424$$

(11) From Eq. 17-14,

$$\mathcal{D}_{BA} = \frac{(2.07 \times 10^3)(298)^{3/2} \sqrt{0.0424}}{(1)(4.91)^2(1.20)}$$

$$= 7.58 \times 10^{-2} \text{ cm}^2/\text{s}$$

This value deviates +17.5% from the literature value of 6.45×10^{-2} cm^2/s [6].

Example 17-4 Calculate the diffusion coefficient of isopropyl iodide, $(CH_3)_2 CHI$, in air at 298K and 1 atmosphere.

(1) From the literature, $M_B = 169.9$ g/mol and $T_b = 362.45$K.

(2) The values of M_A, σ_A and $(\epsilon/k)_A$ are as given in Example 17-3.

(3) From Table 17-5, the increments of the LeBas molal volume are:

$$3(C) = 3(14.8) = 44.4$$
$$7(H) = 7(3.7) = 25.9$$
$$1(I) \qquad\quad = \underline{\ 37\ }$$
$$V'_B = 107.3 \text{ cm}^3/\text{mol}$$

(4) $\sigma_B = 1.18(107.3)^{1/3} = 5.61$ Å (Eq. 17-21)

(5) $\sigma_{AB} = (3.71 + 5.61)/2 = 4.66$ Å (Eq. 17-20)

(6) $(\epsilon/k)_B = 1.15(362.45) = 417$K (Eq. 17-19)

(7) $(\epsilon/k)_{AB} = \sqrt{(417)(78.6)} = 181.0$K (Eq. 17-18)

Therefore,

T* = 298/181 = 1.65 (Eq. 17-17)

(8) Substituting the above values in Eq. 17-16, Ω = 1.15

(9) B' = $0.00217 - 0.00050 \sqrt{(1/28.97) + (1/169.9)}$

  = 2.07×10^{-3} (Eq. 17-15)

(10) M_r = (28.97 + 169.9)/(28.97)(169.9)

  = 0.0404

(11) $\mathcal{D}_{BA} = \dfrac{(2.07 \times 10^{-3})(298)^{3/2} \sqrt{0.0404}}{(1)(4.66)^2 (1.15)}$

  = 8.57×10^{-2} cm^2/s

This deviates −2.4% from the literature value of 8.78×10^{-2} cm^2/s [6].

17-5 DIFFUSIVITY IN WATER

Equation 17-22, the theoretically derived Stokes-Einstein equation, is the foundation of existing methods for estimating the liquid diffusion coefficient.

$$\mathcal{D}_{BW} = \frac{RT}{6 \, \eta_W r_B} \qquad (17\text{-}22)$$

where η_W = viscosity of water (cp)
  r_B = radius of molecule B (cm)

This equation applies to large, spherical molecules diffusing in a continuous solution (i.e., one consisting of small molecules) [17]. The correlation equations that have been developed for $\mathcal{D}_{BW}$ are generally functions of solute size, temperature, and solution viscosity. A number of researchers [17] have found that $\mathcal{D}_{BW}$ is not simply proportional to the inverse of viscosity, as suggested by Eq. 17-22; rather, it is more closely correlated to η raised to a power between −0.45 and −0.66. The deviation from the Stokes-Einstein equation is caused by variations in molecular shape and size.

The dependence of $\mathcal{D}_{BW}$ upon temperature over a wide range has not been sufficiently studied for firm conclusions to be reached [17]. The correlation of viscosity, with temperature suggests that $\mathcal{D}_{BW}$ is similarly dependent, since $\mathcal{D}_{BW}\eta/T$ remains constant when T changes; over the small temperature range typically encountered in the environment, however, a nonlinear temperature dependence of $\mathscr{D}_{BW}$ should be insignificant.

The methods presented in this chapter assume that diffusion of the organic compound occurs in an infinitely dilute solution, or at least where the concentration of solute is less than 0.05 molar. The actual diffusion coefficient, D_{BW} (i.e., where the solute concentration >0.05 molar) is observed to be a function of the activity and mole fraction of the solute. Reid *et al.* [17] provide a detailed explanation of the methods proposed to compensate for the effects of concentration on $\mathscr{D}_{BW}$. They conclude by recommending the following equation suggested by Vignes:

$$D_{BW} = (\mathscr{D}_{WB})^{X_B} (\mathscr{D}_{BW})^{X_W} \qquad (17\text{-}23)$$

where X_W and X_B are the mole fractions of water and compound B. A plot of log D_{BW} vs. mole fraction should be linear and give data that can be used for concentrated binary liquid systems.

17-6   AVAILABLE METHODS FOR ESTIMATING DIFFUSION COEFFICIENTS IN WATER

Table 17-6 summarizes six methods for estimating the liquid diffusion coefficient. The Wilke-Chang method [27] incorporates a solution association constant (ϕ), which is a function of the solution polarity. Wilke and Chang assigned a value of 2.6 for aqueous solutions, fitting experimental data published prior to 1950. They calculated an average error of 10% for an unspecified number of compounds. Using a data set of more recently measured values, Hayduk and Laudie [5] found that an association parameter of 2.26 reduced the average error to 5.8%.

The method of Scheibel [19] requires the molal volumes of both the solute and solvent. Scheibel theorized that the mechanism of diffusion is a function of the relative sizes of the molecules comprising a liquid binary system; unlike the Wilke-Chang equation, Scheibel's equation does not take solution polarity into account. Nevertheless, Hayduk and Laudie determined that Scheibel's method results in an absolute average error slightly smaller than that of Wilke-Chang.

The last two methods listed in Table 17-6 are not recommended here because of (1) their numerous input parameters, (2) lack of general applicability to aqueous solutions, and/or (3) relatively low accuracy.

TABLE 17-6

Available Methods for Estimating Diffusivity into Water

Method [Ref.]	Formula	Inputs (excluding water parameters)	Absolute Average Error (%) [5]
Hayduk and Laudie [5] [a]	$\mathcal{D}_{BW} = \dfrac{13.26 \times 10^{-5}}{\eta_W^{1.14}\, V_B^{0.589}}$	V_B	5.8 (87 solutes)
Wilke-Chang [27]	$\mathcal{D}_{BW} = \dfrac{7.4 \times 10^{-8}\,(\phi_W M_W)^{1/2}\, T}{\eta_W\, V_B^{0.6}}$	V_B	8.8 (87 solutes)
Scheibel [19]	$\mathcal{D}_{BW} = \dfrac{8.2 \times 10^{-8}\, T}{\eta_W\, V_B^{1/3}}\left[1 + \left(\dfrac{3V_W}{V_B}\right)^{2/3}\right]$	V_B	6.7 (87 solutes)
Othmer and Thakar [5]	$\mathcal{D}_{BW} = \dfrac{1.4 \times 10^{-5}}{\eta_W^{1.1}\, V_B^{0.6}}$	V_B	5.9 (87 solutes)
Reddy and Doraiswamy [17] [b]	$\mathcal{D}_{BW} = \dfrac{M_W^{1/2}\, T K'}{\eta_W\,(V_W V_B)^{1/3}}$	V_B	< 20 (96 solutes)
Venezian [24]	$\mathcal{D}_{BW} = \dfrac{6 \times 10^{-10}\, T}{\eta_W\,(R_M - 0.855)}$ where $R_M = \left(\dfrac{n_D^2 - 1}{n_D^2 + 2}\right)\left(\dfrac{M_B}{\rho_B}\right)^{1/3}$	$n_D, \rho_B,$ M_B	

a. Recommended method. b. $K' = 10 \times 10^{-8}$ for $V_W/V_B \leq 1.5$ and 8.5×10^{-8} for $V_W/V_B > 1.5$.

The following section describes the computation method of Hayduk and Laudie, which is a modification of the Othmer-Thakar equation. Further discussion of Scheibel's and Wilke and Chang's methods is not warranted, since they do not require unique input parameters and are not quite as accurate as the Hayduk and Laudie method.

17-7 RECOMMENDED METHOD OF ESTIMATING $\mathcal{D}_{BW}$ FOR ORGANIC LIQUIDS AND VAPORS

The Hayduk and Laudie method for estimating the diffusion coefficient of organic compounds in water is comparable to the Wilke-Chang and Scheibel methods in terms of input parameters, accuracy, and general applicability to a wide range of compounds. This method has been recommended by Reid *et al.* [17] because its computation is slightly easier than that required by the latter two methods, and because it has been validated by a more recently compiled data base.

The Hayduk and Laudie method is based on the following equation:

$$\mathcal{D}_{BW} = \frac{13.26 \times 10^{-5}}{\eta_W^{1.14} \ V_B^{\prime \, 0.589}} \tag{17-24}$$

Unlike the previously mentioned correlations, Eq. 17-24 does not include a temperature term, since temperature dependence is incorporated in the viscosity term [5]. Hayduk and Laudie report an average absolute error of 5.8% for 87 solutes diffusing into water, using the LeBas method for molal volume, as is recommended below.

Basic Steps

(1) From Table 17-7 obtain the viscosity of water, η_W, at the desired temperature.

(2) Calculate $V_B^\prime$ by the LeBas method, summing the applicable incremental values listed in Table 17-5.

(3) Substitute the above values of η_W and $V_B^\prime$ in Eq. 17-24 and compute the diffusion coefficient.

Example 17-5 Compute the diffusion coefficient of aniline ($C_6H_5NH_2$) in water at 25°C.

(1) From Table 17-7, η_W at 25°C = 0.8904 cp

TABLE 17-7

Viscosity of Water at Various Temperatures

°C	η_W (cp)	°C	η_W (cp)	°C	η_W (cp)
0	1.787	11	1.271	21	0.9779
1	1.728	12	1.235	22	0.9548
2	1.671	13	1.202	23	0.9325
3	1.618	14	1.169	24	0.9111
4	1.567	15	1.139	25	0.8904
5	1.519	16	1.109	26	0.8705
6	1.472	17	1.081	27	0.8513
7	1.428	18	1.053	28	0.8327
8	1.386	19	1.027	29	0.8148
9	1.346	20	1.002	30	0.7975
10	1.307				

Source: Weast and Astle [26].

(2) From Table 17-5,

$$
\begin{aligned}
6(C) = 6(14.8) &= 88.8 \\
7(H) = 7(3.7) &= 25.9 \\
1(\text{N-primary amine}) &= 10.5 \\
1(\text{6-membered ring}) &= \underline{-15.0} \\
V'_B &= 110.2 \text{ cm}^3/\text{mol}
\end{aligned}
$$

(3) From Eq. 17-24,

$$
\mathscr{D}_{BW} = \frac{13.26 \times 10^{-5}}{(0.8904)^{1.14} \, (110.2)^{0.589}} = 0.95 \times 10^{-5} \text{ cm}^2/\text{s}
$$

This estimate deviates −10% from the literature value of 1.05×10^{-5} cm^2/s [5].

Example 17-6 Compute the diffusion coefficient of ethyl acetate ($CH_3CO_2CH_2CH_3$) in water at 25°C.

(1) From Table 17-7, η_W at 25°C = 0.8904 cp

(2) From Table 17-5,

$$
\begin{aligned}
4(C) = 4(14.8) &= 59.2 \\
8(H) = 8(3.7) &= 29.6 \\
2(0) = 2(9.9) &= \underline{19.8} \\
V'_B &= 108.6 \text{ cm}^3/\text{mol}
\end{aligned}
$$

(3)   From Eq. 17-24,

$$\mathscr{D}_{BW} = \frac{13.26 \times 10^{-5}}{(0.8904)^{1.14}\,(108.6)^{0.589}} = 0.96 \times 10^{-5}\ cm^2/s$$

This deviates -14% from the measured value of $1.12 \times 10^{-5}\ cm^2/s$ [5].

17-8   AVAILABLE DATA

Measured values of diffusion coefficients may be found in Refs. 5, 13, and 15. Values of M and T_b required as inputs for the recommended methods are readily available from sources such as those listed in Appendix A.

17-9   SYMBOLS USED

a through h = constants in Eq. 17-16

B' = constant in Eqs. 17-14, -15

    = theoretical diffusion coefficient (cm²/s)

D = apparent diffusion coefficient (cm²/s)

D_s = apparent dispersion coefficient in Eq. 17-5 (cm²/s)

D_S = apparent solution phase diffusion coefficient in Eq. 17-6 (cm²/s)

f_t = tortuosity factors in Eqs. 17-5 and -8

H = Henry's law constant in Eq. 17-4

J = flux in Eq. 17-1 (mol/cm²-s)

k = Boltzmann's constant = 1.38062×10^{-16} ergs/K

k_g = gas-phase exchange coefficient in Eq. 17-2 (cm/s)

k_1 = liquid-phase exchange coefficient in Eq. 17-3 (cm/s)

K' = adsorption coefficient in Eq. 17-6

M = molecular weight (g/mol)

M_r = $(M_A + M_B)/M_A M_B$. (Note that the M_r used here is the reciprocal of what is usually called the reduced mass.)

m = soil constant in Eq. 17-5

n_D = refractive index in method of Venezian (Table 17-4)

p = air-filled porosity in Eq. 17-6 (cm³/cm³)

p_T = total porosity in Eq. 17-6 (cm³/cm³)

P = pressure (atm)

r = molecular radius in Eq. 17-22 (cm)

R   = gas constant

R_M = constant in method of Venezian (Table 17-6)

R_T = overall interfacial resistance to transfer in Eq. 17-4

S   = proportionality constant of vapor density to solution concentration in Eq. 17-6

S'   = equilibrium constant at air/solution interface in Eq. 17-6

T   = temperature (K)

T*   = constant in Eqs. 17-16, -17

u   = pore water velocity in Eq. 17-5 (cm/s)

U   = wind velocity in Eqs. 17-2, -3 (cm/s)

v   = atomic diffusion volume (cm³/mol)

V   = molar volume (Schroeder method) (cm³/mol)

V   = molar volume (LeBas method)

X   = mole fraction

∇X = concentration gradient in Eq. 17-1 (mol/cm³ · cm)

Greek

α = soil constant in Eq. 17-5

β = soil bulk density in Eq. 17-6

ϵ = molecular energy of attraction (ergs)

η = viscosity (cp)

θ = volumetric water content in Eq. 17-6

ν = kinematic viscosity in Eq. 17-2 (cm²/s)

ρ = density (g/cm³)

σ = characteristic length of molecule (Å)

ϕ = solution association constant in Wilke-Chang equation

Ω = collision integral in Eq. 17-10

Subscripts

A   = air

B   = compound B

c   = critical property

EF   = effective (diffusion) in Eq. 17-6

I   = air/solution interface in Eq. 17-6

H   = solution/solid interface in Eq. 17-6

T   = total porosity

W   = water

17-10 REFERENCES

1. Barr, R.F. and H.F. Watts, "Diffusion of Some Organic and Inorganic Compounds in Air," *J. Chem. Eng. Data,* **17,** 45-46 (1972).

2. Chen, N.H. and D.F. Othmer, "Net Generalized Equation for Gas Diffusion Coefficient," *J. Chem. Eng. Data,* **7,** 37-41 (1962).

3. Fried, J.J. and M.A. Cambarnons, "Dispersion in Porous Media," *Advances in Hydroscience,* **7,** 169-82 (1971).

4. Fuller, E.N., P.D. Schettler and J.C. Giddings, "A New Method for Prediction of Binary Gas-Phase Diffusion Coefficients," *Ind. Eng. Chem.,* **58,** 19-27 (1966).

5. Hayduk, W. and H. Laudie, "Prediction of Diffusion Coefficients for Non-electrolysis in Dilute Aqueous Solutions," *AIChE J.,* **20,** 611-15 (1974).

6. Jarvis, M.W. and G.A. Lugg, "The Calculation of Diffusion Coefficients of Vapors of Liquids into Air," Report 318, Australian Defense Scientific Service, Alexandria, Virginia, p. 33 (July 19, 1968).

7. Kirda, C., D.R. Nielsen, and J.W. Biggar, "Simultaneous Transport of Chloride and Water During Infiltration," *Soil Sci. Soc. Am. Proc.,* **37,** 339-45 (1973).

8. Kraus, E.B., *Atmospheric-Ocean Interaction,* Oxford Monographs on Meteorology, Oxford University Press, New York (1972).

9. Kullenberg, G., "Investigation of Small-scale Vertical Mixing — Relation to the Temperature Structure in Stably Stratified Waters," in *Adv. Geophys.,* **18A,** International Symposium on Turbulent Diffusion in Environmental Pollution, ed. by F.N. Frenkiel and R.E. Munn, Academic Press, New York (1974).

10. Lerman, A., "Chemical Exchange Across Sediment Water Interface," *Annu. Rev. Earth Planet. Sci.,* **6,** 281-303 (1978).

11. Liss, P.S., "The Exchange of Gases Across Lake Surfaces," *J. Great Lakes Res.,* International Association for Great Lakes Research, State University College at Buffalo, **2,** 88-99 (1976).

12. Liss, P.S. and P.G. Slater, "Flux of Bases Across the Air-Sea Interfaces," *Nature (London),* **247,** 181-84 (1974).

13. Lugg, G.A., "Diffusion Coefficients of Some Organic and Other Vapors in Air," *Anal. Chem.,* **40,** 1072-77 (1968).

14. Manheim, F.J., "The Diffusion of Ions in Unconsolidated Sediments," *Earth Planet Sci. Lett.,* **9,** 307-9 (1970).

15. Marrero, T.R. and E.A. Mason, "Gaseous Diffusion Coefficients," *J. Phys. Chem. Ref. Data,* **1,** 3-118 (1972).

16. Perry, R.H. and C.H. Chilton (eds.) *Chemical Engineers' Handbook,* 5th ed., McGraw-Hill Book Co., New York, pp. 3-230-235 (1973).

17. Reid, R.C., J.M. Prausnitz and T.K. Sherwood, *The Properties of Gases and Liquids,* 3rd ed., McGraw-Hill Book Co., New York, pp. 544-601 (1977).

18. Saksena, M.P. and S.C. Saxena, "Calculation of Diffusion Coefficients of Binary Gas Mixtures," *Indian J. Pure Appl. Phys.*, **4**, 109-16 (1966).

19. Scheibel, E.G., "Liquid Diffusivities," *Ind. Eng. Chem.*, **46**, 2007-8 (1954).

20. Scheidigger, A.E., *The Physics of Flow through Porous Media*, 3rd ed., University of Toronto Press, Toronto (1974).

21. Shearer, R.C., J. Letey, W.J. Farmer and A. Klute, "Lindane Diffusion in Soil," *Soil Sci. Soc. Am. Proc.*, **37**, 189-93 (1973).

22. Smith, J.H. and D.C. Bomberger, "Prediction of Volatilization Rates of Chemicals in Water," in *Water: 1978*, AIChE Symposium Series, 190, **75**, 375-81 (1979).

23. Smith, J.H., D.C. Bomberger and D.C. Haynes, "Prediction of the Volatilization Rates of High Volatility Chemicals from Natural Water Bodies," submitted in August 1979 for publication in *Environ. Sci. Technol.*

24. Venezian, E.C., *Pre-Screening for Environmental Hazards — A System for Selecting and Prioritizing Chemicals*, Report to U.S. EPA by Arthur D. Little, Inc. (June 1976).

25. Walker, A. and D.V. Crawford, "Diffusion Coefficients for Two Triazine Herbicides in Six Soils," *Weed Res.*, **10**, 126-32 (1970).

26. Weast, R.C. and M.J. Astle (eds.), *CRC Handbook of Chemistry and Physics*, 59th ed., CRC Press, Inc., West Palm Beach, FL (1978).

27. Wilke, C.R. and P. Chang, "Correlation of Diffusion Coefficients in Dilute Solutions," *AIChE J.*, **1**, 264-70 (1955).

28. Wilke, C.R. and C.Y. Lee, "Estimation of Diffusion Coefficients for Gases and Vapors," *Ind. Eng. Chem.*, **47**, 1253-57 (1955).

18

FLASH POINTS OF PURE SUBSTANCES

John H. Hagopian

18-1 INTRODUCTION

The flash point of a substance is commonly defined [8] as the minimum temperature at which it emits sufficient vapor to form an ignitable mixture with air near its surface or within the chamber of a testing apparatus. An "ignitable mixture" is in turn defined as a fuel-air mixture within the explosive range (i.e., with a gaseous fuel concentration in air between the lower and upper flammability limits of the fuel) that is capable of propagating flame away from a source of ignition.

Although flash points are normally associated with flammable or combustible liquids, they are also useful for characterizing solids that sublime, as they indicate the relative ease with which substances can be ignited at a given temperature. Typical measured values range from $-36°F$ ($-38°C$) for acetaldehyde to $450°F$ ($232°C$) for diisooctyl phthalate.[1]

In theory, the flash point of a pure substance in equilibrium with its vapors is the temperature at which the vapor concentration equals the lower flammable limit (LFL) or lower explosive limit (LEL) concentration at standard atmospheric pressure. Several estimation methods take advantage of this fact to provide "ideal" or equilibrium flash points.

---

1. These are closed-cup flash points, cited simply as examples. Flash points measured in other ways and for other substances may have higher or lower values.

Most flash points listed in the literature have been experimentally determined in one or more of several types of testing apparatus. Among these are the Tag Closed Tester, the Pensky-Martens Closed Tester, the Setaflash Closed Tester, the Cleveland Open Tester, and the Tag Open Tester. Standard specifications for these units are published by the American Society for Testing and Materials (ASTM). Results from the first three types are referred to as "closed-cup" flash points, while those from the latter two types are called "open-cup" flash points.

Because of differences in equipment design and testing procedure, the numerical value of the closed-cup flash point is typically some 5-10°F lower than that of the open-cup flash point for the same liquid. The difference can be greater or smaller in particular cases, however, depending on the particular apparatus used, the purity of the substance, and the degree of equilibrium achieved. It is not surprising, therefore, that the literature often reports a number of different closed-cup and open-cup flash points for the same substance, all of which may differ from the "ideal" or equilibrium flash point.

All of the estimated methods presented in this chapter take advantage of a general relationship between the flash points and normal boiling points of a homologous series of liquids. Affens [1] used vapor pressure and lower flammable limit data for n-alkanes to define equilibrium flash points and, ultimately, an equation for estimating flash point as a function of boiling point. Prugh [12] generalized and simplified the approach, at some loss in accuracy, to cover a wide variety of organic substances and developed a nomograph. Butler *et al.* [5] simply used experimentally derived closed-cup flash points and normal boiling points to develop a relationship for pure hydrocarbons, and Hagopian [6] extended this work, developing similar relationships for aldehydes, amines, ketones, and alcohols.

A different method, by Shimy [13], is not presented here, primarily because of its limited applicability. It requires an intermediate calculation of the autoignition temperature,[2] using the total number of carbon atoms in the molecule and the number of carbon atoms in branches. The flash point is then calculated with an expression that contains the autoignition temperature and the number of hydrogen atoms in radicals. The method applies only to paraffinic hydrocarbons, olefins, the benzene series, and alcohols.

---

2. The autoignition temperature, sometimes simply called the "ignition temperature," is the minimum temperature at which a material will ignite without a spark or flame being present. The ASTM provides a standard method for its measurement.

18-2   SELECTION OF APPROPRIATE METHOD

Table 18-1 lists recommended methods of estimation, indicates the kinds of substances to which they apply, and comments upon their bases and overall accuracy. Further details are given in the following sections.

It is difficult, if not impossible, to assess the accuracy of any of the methods. Since flash point data in the literature are derived experimentally and often span a range of temperatures for any given substance, and since the results of the estimation methods can be compared only with these values, it is entirely possible that a given method is more or less accurate than noted in Table 18-1 or the following text. Indeed, comments on the overall accuracy of any given method are intended only to give a rough approximation for the general case.

With many substances, more than one method can be used. For example, both Prugh's nomograph and the equation of Butler *et al.* can be used for pure hydrocarbons, and the nomograph can be used for all types of substances addressed by this author. In such cases, it is recommended that all pertinent methods be applied. If the results are comparable, they are probably fairly accurate; if they are not comparable, the estimates can be compared with literature values for similar substances with slightly higher or lower boiling points. Since higher boiling points generally suggest higher flash points, a significantly erroneous estimate should be immediately evident.

18-3   AFFENS' METHOD

Principles of Use. Affens [1] utilized actual curves of vapor pressure versus temperature together with experimentally derived lower flammability limits to estimate the equilibrium flash points for n-alkanes with 1 to 14 carbon atoms. Using boiling point data, he then plotted boiling points versus flash points and found the line that best fit these points. The line corresponded to the equation:

$$t_f = 0.6946 \, t_b - 122.9 \qquad (18\text{-}1)$$

where t_f is the flash point and t_b is the normal boiling point, both in degrees Fahrenheit. The equation does not work well for methane; for other n-alkanes in the series, however, it predicts flash points generally within $9°F$ ($5°C$) of an average of flash points reported in the literature.

What is most useful here is not the equation, since the flash points of the n-alkanes are fairly well defined in the literature, but the basic

TABLE 18-1

Overview of Flash Point Estimation Methods

Method	Applicability	Basis	Accuracy
Affens [1]	Provides equilibrium flash points for n-alkanes.	Uses actual vapor pressure and LFL data to determine flash points graphically. Basic methodology can be applied to other types of substances.	Generally within 9°F (5°C) of literature value and within 4°F (2°C) of precisely computed equilibrium flash points for n-alkanes with 2 to 14 carbon atoms.
Prugh [12]	Provides closed-cup flash points for pure organic compounds containing carbon, hydrogen, oxygen, sulfur, and halogens.	Nomograph based on estimates of LFL and vapor pressure data, along with actual data for 200 chemicals.	Author claims nomograph yields predictions within 20°F (11°C).
Butler [5]	Provides closed-cup flash points for pure hydrocarbons. Normal paraffins, cyclo-paraffins, aromatics, and sharply separated fractions all fit the given equation.	Based on relationship between actual flash-point and boiling-point data for similar substances. Basic methodology can be applied to other types of substances.	Not specifically expressed. Of 29 points on graph presented, 24 are within 14°F (8°C) of equation curve. Maximum error is 40°F (22°C) for tetralin.
Hagopian [6]	Provides both open- and closed-cup flash points for alcohols, aldehydes, amines, and ketones.	Same as Butler *et al.* [5].	Average errors range from 5.2°F (2.9°C) to 17.0°F (9.4°C), depending on combination of flash point and material type.

methodology for estimating equilibrium flash points for pure substances.[3]
This is described below.

Basic Steps

(1) Plot vapor pressure versus temperature for the substance,
using literature data or one of the estimation methods given
in Chapter 14. The result will correspond to curve A in
Figure 18-1.

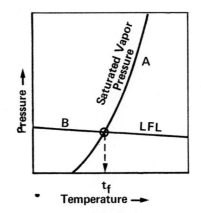

**FIGURE 18–1   Graphical Determination
of Equilibrium Flash Point**

(2) Find the lower flammable limit (LFL) or lower explosive
limit (LEL) of the substance in the literature. (Some
sources are listed in §18-7.) These data are usually given in
units of volume percent. It is preferable to obtain sufficient
data to plot a curve of LFL or LEL as a function of temperature, if such data are available.

(3) Convert the LFL or LEL data to the same units as the vapor
pressure data. The equation for converting from volume
percent to mm Hg is:

$$\text{LFL (mm Hg)} = \frac{760}{100} \times \text{LFL (vol \%)} \qquad (18\text{-}2)$$

(4) Plot the LFL or LEL data in units of pressure on the same
graph with the vapor pressure data, as in curve B of Figure
18-1. The intersection of the two curves indicates the equilibrium flash point, t_f, of the substance. If only a single
value of LFL or LEL is obtainable, as will frequently be the

---

3. This and subsequent methods generally apply when the flash point of the substance
is below the temperature at which it begins to decompose.

case, find the temperature that corresponds to it on the vapor pressure curve. Since LFL and LEL are generally not strong functions of temperature, the loss in accuracy should not be great.

Example 18-1 Find the equilibrium flash point of n-hexane, given data on its saturated vapor pressure and lower flammability limit from Ref. 1 as functions of temperature.

(1) Plot the vapor pressure data. The result is shown in curve A of Figure 18-2.

(2) Plot the LFL data after converting from volume percent to mm Hg with Eq. 18-2. The result is shown as curve B.

(3) The intersection of curves A and B indicates that the equilibrium flash point of n-hexane (t_f) is approximately -14°F. In comparison, Ref. 16 lists a closed-cup value of -7°F, and Ref. 1 gives -9.4°F as an average of literature values.

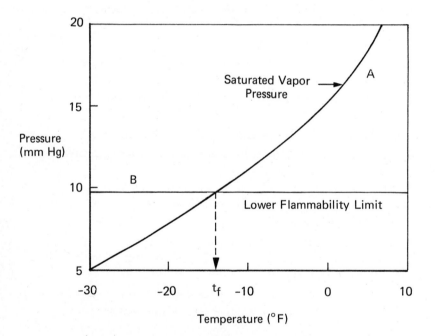

FIGURE 18–2   Determination of Equilibrium Flash Point of n-Hexane by Affens' Method

18-4   PRUGH'S METHOD

Principles of Use. Prugh [12] made a number of generalizations that permit prediction of the vapor pressure versus temperature curves and lower flammability or explosive limits of organic compounds solely on the basis of their normal boiling points and chemical structures. Subsequently, he utilized the definition of equilibrium flash point, together with actual data for 200 chemicals, to develop a nomograph and a procedure for its use.

Prugh's nomograph is presented in Figure 18-3. The procedure that he recommended for its use is as follows:

Basic Steps

(1) Calculate the stoichiometric concentration X_{st} of the vapor in air from the equation:

$$X_{st} = \frac{83.8}{4(C) + 4(S) + (H) - (X) - 2(O) + 0.84} \text{ vol \%} \qquad (18\text{-}3)$$

where C, S, H, X, and O are respectively the number of carbon, sulfur, hydrogen, halogen, and oxygen atoms in the molecule of fuel.

(2) Find the point on the appropriate line for "alcohols" or "most other organic materials" on Figure 18-3 that corresponds to the stoichiometric concentration computed above.

(3) Mark the "boiling-point/flash-point" scale (the vertical axis of the graph) at the location corresponding to this point.

(4) Mark the boiling-point scale (i.e., the left-most scale) at the appropriate boiling point.

(5) Connect the points marked in steps (3) and (4).

(6) Read the approximate flash point where the connecting line intersects the flash-point scale between the boiling-point scale and the left axis of the graph.[4]

Prugh states that the nomograph will yield flash points within approximately 20°F (11°C) of the correct value. To check the validity of the nomograph further, the author of this chapter extracted values for 14 compounds at random from a handbook and compared them with predictions. The overall average error was 7.5°F (4.2°C). Seven of the predictions (50%) were within 3°F (1.7°C) of the literature value, four (29%) were within 5-10°F (2.8-5.6°C), two were within 15-20°F (8.3-16.7°C),

---

4. Although not explicitly stated, this scale probably indicates closed-cup flash points, since the data in the referenced source are mostly of this kind.

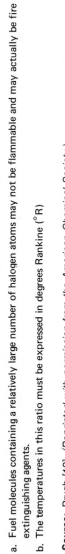

FIGURE 18—3 Nomograph for Estimation of Flash Point

a. Fuel molecules containing a relatively large number of halogen atoms may not be flammable and may actually be fire extinguishing agents.

b. The temperatures in this ratio must be expressed in degrees Rankine (°R)

Source: Prugh [12]. *(Reprinted with permission from the American Chemical Society.)*

and one was 25°F (13.9°C) in error. Since results were found to be sensitive to slight variations in the position of marks and connecting lines, the user is advised to be as precise as possible.

Example 18-2   Estimate the flash point of methyl butyrate, given a boiling point of 215°F (101.7°C) and a chemical formula of $CH_3OOCCH_2CH_2CH_3$.

(1)   The molecule contains 5 carbon atoms, 10 hydrogen atoms, and 2 oxygen atoms. From Eq. 18-3.

$$X_{st} = \frac{83.8}{(4 \times 5) + 0 + 10 - 0 - (2 \times 2) + 0.84} = 3.12 \text{ vol } \%$$

(2)   Find the value of X_{st} on the bottom axis of the graph in Figure 18-3 and proceed upward vertically to the line labeled "Most Other Organic Materials."

(3)   Proceed horizontally to the left axis of the graph and mark the "boiling point/flash point" scale at the appropriate value. For this example, the appropriate value is 1.343.

(4)   Draw a line between this point and 215°F on the boiling-point scale.

(5)   The intersection of the drawn line with the flash-point scale provides a flash-point estimate in the range of 50-55°F. For comparison, the handbook value is 57°F.

18-5   BUTLER'S METHOD

Principles of Use. Butler *et al.* [5] plotted the flash points of pure hydrocarbons (values obtained by means of the Pensky-Martens closed-cup flash point tester) against their respective normal boiling points and then fit a curve through the points. The equation corresponding to this curve was:

$$t_f = 0.683 \, t_b - 119 \tag{18-4}$$

where t_f is the flash point and t_b is the normal boiling point, both in degrees Fahrenheit. The boiling points considered ranged from 150°F (66°C) to 550°F (288°C).

The available data did not indicate any great variation among hydrocarbon types. Normal paraffins, cycloparaffins, aromatics, and sharply defined fractions all fit the above equation fairly well. Of 29 points on the graph presented, 24 were within approximately 14°F (8°C) of the curve; the maximum deviation was approximately 40°F (22°C).

Hagopian [6] used the same method to develop relationships for the open- and closed-cup flash points of assorted alcohols, aldehydes, amines, and ketones with varying degrees of success. Table 18-2 presents the resulting equations, together with ancillary data describing degree of error, number of data sets used, temperature range addressed, and other details.

As with Affens' method, the technique for developing the equations is of greater significance than the equations themselves. The basic steps are outlined below.

Basic Steps

(1) From the literature, find the normal boiling points and flash points of substances similar to that for which a flash-point prediction is desired.

(2) Plot the flash points versus the associated boiling points on regular graph paper.

(3) Fit the best curve through the points.

(4) Use the normal boiling point of the substance in question to find its flash point from the curve.

In general, there appear to be no great variations between the flash-point characteristics of any specific homologous series of substances and other series within the same basic chemical family. This was indicated by Butler's findings, as well as by those of Hagopian. Nevertheless, it is probably advisable to use data only for substances within exactly the same homologous series, if sufficient data are available.

The curve through the points on the resulting graph can either be "eyeballed" or fitted by means of numerical regression. While the latter approach is more complicated, it obviously provides a more precise fit with the data.

Example 18-3   Find the closed-cup flash point of acetaldehyde ($t_b = 68.7°F$).

(1)   The appropriate equation in Table 18-2 is:

$$t_f = 0.6901\ t_b - 92.3$$

(2)   Substitution of the boiling point in this equation gives:

$$t_f = 0.6901\ (68.7) - 92.3 = -44.9°F$$

In comparison, the handbook value of the closed-cup flash point is $-36°F$.

TABLE 18-2

Flash Point Estimation Equations

Chemical Family	Type of Measurement	Number of Data Points Used	Boiling Point Range (°F)	Equation[a]	Average Deviation (°F)	Maximum Deviation (°F)
Alcohols	Closed cup	31	180[b] to 491	$t_f = 0.7056\ t_b - 77.1$	6.7	14.2
	Open cup	25	180[b] to 506	$t_f = 0.6875\ t_b - 66.4$	7.4	18.5
Aldehydes	Closed cup	8	69 to 329	$t_f = 0.6901\ t_b - 92.3$	6.9	13.9
	Open cup	17	69 to 469	$t_f = 0.7491\ t_b - 107.8$	5.8	18.9
Amines	Closed cup	19	120 to 531	$t_f = 0.7962\ t_b - 120.9$	17.0	38.6
	Open cup	22	90 to 532	$t_f = 0.7988\ t_b - 114.2$	13.3	41.1
Ketones	Closed cup	18	133 to 433	$t_f = 0.6717\ t_b - 95.2$	8.0	20.4
	Open cup	17	133 to 425	$t_f = 0.6917\ t_b - 89.7$	5.2	16.1

a. Both t_f and t_b must be expressed in degrees Fahrenheit.
b. Extrapolation below this temperature will produce erroneous results.

Source: Hagopian [6].

Example 18-4   Find the flash point of a substance, given the flash points and normal boiling points for compounds within the same chemical family.

(1)   Plot the flash points against associated boiling points as in Figure 18-3.

(2)   Draw a curve through the points. (The best fit is not always a straight line.)

(3)   Find t_f for the substance in question by using its normal boiling point and the curve drawn. The dashed lines on Figure 18-4 demonstrate the procedure.

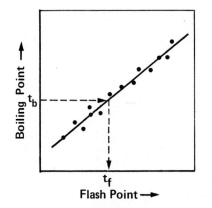

FIGURE 18—4   Graphical Determination of Open- or Closed-Cup Flash Point

18-6   FLASH POINTS OF MIXTURES

It is frequently necessary to estimate the flash point of a mixture of substances. Some references that address this subject are cited below.

Thiele [14] provides a method for estimating flash points for blends of lubricating oils. Butler [5] does the same for middle distillates, claiming that methods given by Nelson [11] and the Factory Mutual Fire Insurance Companies [3] for hydrocarbon fractions are less accurate. In a more recent paper, Hu and Burns [7] address distillate-fuel blends and reference a method by Wickey and Chittenden [19]. Other sources of information on hydrocarbon fractions are Burger [4] and Lenoir [8].

Wu and Finkelman [20] discuss a method for predicting the closed-cup flash points of solvent blends. They cite an earlier method by Walsham [17] for predicting open-cup flash points. Walsham [18] discusses this topic in a later paper also.

18-7 AVAILABLE DATA

There are a number of compilations of flash-point data that can be useful in applying the approaches of Butler and of others. The following are examples:

Fire Protection Handbook, NFPA [10] — Contains flash point, boiling point, LFL and other data for hundreds of compounds. Flash points are mostly of the closed-cup type.

Fire Protection Guide on Hazardous Materials, NFPA [9] — Contains the same tables as the previous source in a less expensive volume.

CHRIS Hazardous Chemical Data, USCG [16] — Contains individual data sheets for 900 compounds. Items include both closed-cup and open-cup flash points, when available, and boiling points, LFL's, etc. Earlier version for 400 substances also had vapor pressure vs. temperature curves.

Chemical Data Guide, USCG, [15] — Contains flash point, boiling point, LFL and other data for 279 common compounds. Flash points are mostly of the open-cup type.

"Aldrich Catalog Handbook of Fine Chemicals" [2] — Updated frequently, this catalog contains listings for over 9,000 chemicals. Pertinent data include flash points determined by a Setaflash Closed-Cup Tester, and boiling points.

18-8 SYMBOLS USED

C	= number of carbon atoms in the fuel molecule
χ_{st}	= stoichiometric concentration of fuel vapor in air (vol %)
H	= number of hydrogen atoms in the fuel molecule
LEL	= lower explosive limit concentration of fuel in air (vol %)
LFL	= lower flammable limit concentration of fuel in air (vol %)
O	= number of oxygen atoms in the fuel molecule
S	= number of sulfur atoms in the fuel molecule
X	= number of halogen atoms in the fuel molecule
t_f	= flash point temperature
t_b	= normal boiling point temperature

18-9   REFERENCES

1. Affens, W.A., "Flammablility Properties of Hydrocarbon Fuels," *J. Chem. Eng. Data,* 11, 197-202 (1966).

2. Aldrich Chemical Co., Inc. *Aldrich Catalog Handbook of Fine Chemicals,* Milwaukee, Wisconsin (1978).

3. Associated Factory Mutual Insurance Cos., "Properties of Flammable Liquids, Gases, and Solids," *Ind. Eng. Chem.,* 32, 880 (1940).

4. Burger, L.L., U.S. Atomic Energy Commission Publication 4W-35579 (1955).

5. Butler, R.M., G.M. Cooke, G.G. Lukk and B.G. Jameson, "Prediction of Flash Points of Middle Distillates," *Ind. Eng. Chem.,* 48, 808-12 (1956).

6. Hagopian, J.H., unpublished data (1979).

7. Hu, J. and A.M. Burns, "New Method Predicts Cloud, Pour, Flash Points of Distillate Blends," *Hydrocarbon Process.,* 49, 213-16 (November 1970).

8. Lenoir, J.M., "Predict Flash Points Accurately," *Hydrocarbon Process.,* 54, 95-99 (January 1975).

9. National Fire Protection Association, *Fire Protection Guide on Hazardous Materials,* 6th ed., Boston, Massachusetts (1975).

10. National Fire Protection Association, *Fire Protection Handbook,* 14th ed., Boston, Massachusetts (1976).

11. Nelson, W.L., *Petroleum Refinery Engineering,* 3rd ed., McGraw-Hill Book Co., New York, p. 210 (1949).

12. Prugh, R.W., "Estimation of Flash Point Temperature," *J. Chem. Educ.,* 50, A85-A89 (1973).

13. Shimy, A.A., "Calculating Flammability Characteristics of Hydrocarbons and Alcohols," *Fire Technol.,* 6, 135-39 (1970).

14. Thiele, E.W., "Prediction of Flash Points of Blends of Lubricating Oils," *Ind. Eng. Chem.,* 19, 259-62 (1927).

15. U.S. Coast Guard, *Chemical Data Guide for Bulk Shipment by Water,* Publ. No. CG-388, U.S. Government Printing Office, Washington, D.C. (1976).

16. U.S. Coast Guard, *CHRIS Hazardous Chemical Data,* Publ. No. CG-446-2, U.S. Government Printing Office, Washington, D.C. (1978).

17. Walsham, J.G., ACS Washington Meeting Preprints, 31, 405 (1971).

18. Walsham, J.G., "Prediction of Flash Points for Solvent Mixtures," *Adv. in Chem. Ser.,* Publ. 73, Ser. 124, American Chemical Society, Washington. D.C., 56-69 (1973).

19. Wickey, R.O. and D.H. Chittenden, "Flash Points of Blends Correlated," *Hydrocarbon Process. Pet. Refiner,* 42, 157 (June 1963).

20. Wu, D.T. and R. Finkelman, "A Mathematical Model for the Prediction of Closed-Cup Flash Points," ACS Preprints V38, 175th National Meeting, Anaheim, California, 61-67 (1977).

19

DENSITIES OF VAPORS, LIQUIDS AND SOLIDS

Leslie H. Nelken

19-1 INTRODUCTION

The density of a substance, ρ, is the ratio of its mass to its volume. The property varies not only with molecular weight but also with molecular interaction and structure: for instance, although primary and tertiary butyl alcohol have the same molecular weight, their respective densities at 20°C are 0.8098 and 0.7887 g/mL. In environmental analysis, the primary reason for estimating the density of a substance is to determine whether gases are heavier or lighter than air, or whether liquids and solids will float or sink in water. Specific applications for density include chemical spill models for substances such as oil or toxic gases. In addition, density is often required for the estimation of other chemical properties, such as molar refraction and viscosity.

Units of density express a mass to volume ratio. For liquids the units are g/mL or mol/mL, and for solids, g/cm^3. Liquid densities range from 0.6 to 2.9 g/mL, and the range for solids is about 0.97 to 2.7 g/cm^3. Vapor densities are in units of g/L or mg/m^3; about 0.5 to 3.0 g/L is the typical range for vapor densities at normal temperatures and pressures.

Specific gravity (SG) is a dimensionless parameter derived from density. It is defined as the ratio of the weight of a given volume of substance at a specified temperature to the weight of the same volume of water at a given temperature. In the notation $SG = 0.8076_4^{20}$, the superscript refers to the temperature of the substance at the time of measurement, and the subscript is the water temperature. At 4°C, the density of water is 1.0000 g/mL, so the specific gravity of the liquid is identical to the liquid density relative to the density of water at 4°C.

Methods for estimating the density of the vapor, liquid, and solid phases are discussed separately. A brief overview of the recommended methods is given below.

•   *Vapor density* (§19-2) is estimated by use of the ideal gas law. This is an easy and sufficiently accurate method at the pressures and temperatures encountered in environmental problems.

•   Two methods are recommended to estimate *liquid density*. The first (§19-5) requires the boiling point of the compound and knowledge of the molecular structure. The average error obtained by this method on a group of about 30 compounds was near 3.4%. The second method (§19-6) is more rigorous but slightly more accurate; inputs include the acentric factor (or boiling point in the absence of the acentric factor), the critical temperature, and the critical pressure. With the exception of the acentric factor, these properties may be estimated by methods detailed in Chapters 12 and 13.

•   *Solid density* (§19-8) can be estimated by a method based upon fragment constants. Only the structure of the chemical is required for this method. The following atoms and ions are included in this method: H, C, O, N, S, F, Cl, Br, I, Na+, K+, Rb+, F−, Cl−, Br−, and I−. The densities of hydrated solids and various hydrogen-bonded solids can also be estimated by this approach.

19-2   VAPOR DENSITY ESTIMATION

Numerous equations of state have been formulated to account for the nonideal behavior of gases at high temperatures and pressures. These equations have been reviewed by Reid *et al.* [14] and Tarakad [17]. However, for the purposes of environmental analysis, where temperatures are in the range of 0-50°C and pressures approximate 1 atmosphere, the ideal gas law provides a simple, accurate estimation method. For the few applications in which the temperature or pressure is outside the normal range, the Redlich-Kwong equation of state [13,14] is an accurate, though more complicated, technique for estimating the vapor density.

The ideal gas law is expressed as:

$$PV = nRT \qquad (19\text{-}1)$$

where

P = pressure (atmospheres)
V = volume (liters)
n = the number of moles of gas
R = ideal gas constant (0.082 atm-L/mol-K)
T = temperature (K)

Density is defined as weight per volume, in this case by n/V (moles/liter) or, preferably, by nM/V (g/liter) where M is the molecular weight. Rearrangement of Eq. 19-1 results in an expression for the vapor density, ρ_v:

$$\rho_v = \frac{PM}{RT} \qquad (19\text{-}2)$$

Table 19-1 lists the percent error obtained by estimating the vapor density of 15 gases using the ideal gas law. For purposes of comparison, the error obtained by the more complicated Redlich-Kwong method (which requires an estimate of critical temperature and pressure) is also included for some of the compounds. A calculator capable of solving cubic-order equations is helpful, though not necessary.

Basic Steps

(1) Obtain the molecular weight of the compound, M.

(2) Assign values to the temperature T (K) and pressure P (atm) for the system of interest. Note that R = 0.082 atm-L/mol-K.

(3) Calculate ρ_v (g/L) using Eq. 19-2.

Example 19-1   Calculate the density of methyl chloride at 0°C and 1 atm, given M = 50.49 g/mol.

By Eq. 19-2, ρ_v = (1) (50.49)/(0.082) (273) = 2.26 g/L. The vapor density of methyl chloride given in the literature is 2.31 g/L [14], indicating an error of −2.6%.

19-3   AVAILABLE METHODS FOR ESTIMATING LIQUID DENSITY

Most methods for estimating liquid density are based upon the law of corresponding states. According to this law, properties that are dependent upon intermolecular forces are related to the critical properties in the same way for all compounds. Thus, the estimation of density

TABLE 19-1

Accuracy of the Ideal Gas Law and Redlich-Kwong Methods
for Estimation of Vapor Density

	ρ_v (Literature Value, g/L at 0°C)[a]	Percent Deviation of Estimated Value	
		Ideal Gas Law	Redlich-Kwong
Hydrocarbons			
Methane	0.7168	−0.1	+0.3
Ethane	1.357	−1.0	+0.7
Propene	1.937	−2.9	−1.4
Acetylene	1.173	−0.8	0.0
Allene	1.787	+0.1	+2.4
Halogenated Hydrocarbons			
Methyl chloride	2.310	−2.6	−0.8
Ethyl fluoride	2.198	−2.3	−0.8
Bromoacetylene	4.684	0.0	
O-containing Compounds			
Methyl ether	2.091	−1.6	+0.6
Ethylene oxide	1.965	+0.1	−0.4
Carbon dioxide	1.977	−0.5	
Carbon monoxide	1.250	0.0	
N-containing Compounds			
Cyanogen	2.355	−0.4	+1.1
Ammonia	0.7710	−1.3	
Other			
COS	2.721	−1.4	
Absolute Average Error		1.0	0.8

a. **Source:** Hodgman *et al.* [7]

generally requires a knowledge of certain critical properties — namely, temperature (T_c), pressure (P_c), and sometimes also the compressibility factor (Z_c). Many of the methods developed for estimating liquid density are described and discussed in Refs. 1,2,6,12,14-17,19 and 20.

The method of Kier *et al.* [10] is an exception to the above, in that it relies only upon the molecular structure of a chemical. The method is based on a topological index called the *connectivity function,* χ, which is correlated with liquid density by a regression equation developed for the applicable chemical class. The density of aliphatic hydrocarbons, mono-alcohols, and acids at 20°C can presently be estimated by substituting χ in the appropriate regression equations. The method is not included in this handbook, however, because of the limited number of chemical classes for which it is applicable and the labor involved in calculating χ for a complex molecule.[1]

There are approximately twenty methods for estimating liquid density. Table 19-2 summarizes the characteristics of a few of the more common ones. From the equations, it is obvious that the density of a liquid is a function of temperature. It usually decreases as temperature increases, although there are anomalies (the most familiar being that of water between 0° and 4°C). The density of a liquid is also affected by the polarity of the molecule and by molecular interactions. For the majority of estimation methods, accuracy decreases as the polarity of the compound increases. The critical compressibility factor and acentric factor account for some additional deviation from theoretical behavior. Liquid density increases with pressure, but only slightly; for example, a liquid with a density of 0.4 g/mL at 1 atmosphere increases in density by only 0.01 g/mL at 34 atmospheres in an isothermal system [12].

19-4   SELECTION OF APPROPRIATE METHOD FOR LIQUID DENSITY

The two methods described here have been chosen on the basis of ease of use, minimum amount of input data, and accuracy. The method of Grain [3] is a modification of the Goldhammer equation (cited in Ref. 12) and is invaluable because it requires no critical properties, only molecular weight (M), boiling point (T_b), and molar volume (V_b) at the boiling point.[2] V_b can be estimated from chemical structure. The second method, that of Bhirud [1], requires more input data, namely, the critical temperature and pressure and the acentric factor. These values are

---

1. A computer program has been developed for calculating χ [5]. Thus, the method of Kier *et al.* could have significant future potential when regression equations for other chemical classes are developed.

2. Estimation methods for T_b, T_c and P_c are described in Chapters 12 and 13 of this book. Reid *et al.* [14] also provides estimation techniques for T_c and P_c; molar volume can be estimated by the method given in §19-5.

TABLE 19-2

Some Available Methods for Estimating Liquid Density

Method	Formula[a]	Inputs[b]	Applicable Temperature Range[c]	Remarks
Grain [3][d] (see § 19-5)	$\rho_L = M\rho_{Lb}\,[3 - 2\,(T/T_b)]^n$	M, T_b, V_b	$T < T_b$	Technique available for estimating V_b within 3-4%. [14]
Bhirud [1][d] (see § 19-6)	$\rho_L = MP_c/RTe^{a} + \omega b$	M, T_c, P_c, ω	$T_r < 1.0$	Average absolute error = 0.76% (max 2.23%) for 24 polar and nonpolar compounds [1].
Gunn and Yamada [4]	$\rho_L = M\,[V_{sc}\,V_r^{(0)}\,(1-\omega\Gamma)]^{-1}$	M, T_b, T_c, P_c	$0.2 \leqslant T_r \leqslant 0.99$	Average absolute error = 0.22% (max = 4.90%) for 32 nonpolar and slightly polar compounds. [4]
Yen and Woods [22]	$\rho_L = M\rho_c\,[1 + \sum_{i=1}^{4} K_i\,(1-T_r)^{i/3}]$	$M, T_b, T_c,$ P_c, Z_c, ρ_c	$T_r < 1.0$	Average absolute error = 2.1% for 200 nonpolar and slightly polar compounds. [6]
Guggenheim [12]	$\rho_L = M\rho_c\,[1 + \dfrac{7}{4}\,(1-T_r)^{1/3} + \dfrac{3}{4}\,(1-T_r)]$	T_c, M, ρ_c	$T_r < 1.0$	Average error = 6%; requires only 1 critical property. [12]
Modified Rackett Eq. [16]	$1/\rho_L = \dfrac{MRT_c}{P_c}\,Z_{RA}^{\,1+(1-T_r)^{2/7}}$	T_c, P_c, M	Wide	Z_{RA} must be determined experimentally. Average absolute error = 0.6% for 100 compounds. [16]

a. See § 19-10 for explanation of symbols.   b. Z_c and ρ_c can be estimated.   c. T_r = reduced temperature (T/T_c).   d. Recommended method.

unlikely to be available in the literature for a compound whose density is not reported. Methods for estimating the first and second parameters are given in Chapters 12 and 13, respectively; the acentric factor, ω, may be estimated by a method described in §19-6 of this chapter.

Table 19-3 compares the accuracy of the above methods. Grain's method is more accurate for each of the five chemical groups, achieving average and maximum errors of 3.1% and 8.6% respectively. Bhirud's method averages a 7.2% error for the compounds sampled, partly because of the estimated value of the acentric factor used in these calculations.

Grain's method [3] has not been tested on a large number of chemicals, but it appears to work quite well for the compounds in the sample group. It differs from the original Goldhammer equation in that the critical temperature is replaced by the term $1.5\,T_b$. The critical temperature actually ranges from 1.2 to 1.7 times the boiling point, but the error introduced by the 1.5 approximation is minimal because it is raised to a fractional power (0.25-0.31). Table 19-4 shows the accuracy of Grain's method in comparison with the results of Goldhammer's method. The additional error caused by the approximation averages 3.7% for 11 sample points; the maximum error is 11%.

Bhirud's method [1] estimates the saturated liquid density of polar and nonpolar organic liquids over a wide temperature range.[3] The saturated liquid density is measured when the liquid is in equilibrium with the vapor of the pure compound. Since the pressure exerted by the pure vapor is different from atmosphere pressure, it can slightly affect precise measurements of liquid density; however, as indicated earlier, the effect of a change in pressure of even several atmospheres can be ignored for the applications addressed here [12]. Bhirud's method is reported [1] to achieve an overall average error of 0.76% and maximum error of 2.23% for 24 compounds. Table 19-3 indicates, however, that when one of the required parameters (acentric factor) must be estimated, the error increases about tenfold. Should the critical parameters also require estimation, even greater errors can be expected.

19-5   GRAIN'S METHOD (LIQUID DENSITY)

Grain's method [3] is a modification of a density estimation technique derived by Goldhammer, as reviewed by Reid *et al.* [14] and

---

3. It should be noted that Bhirud's objective was to develop a method just for normal (nonpolar or slightly polar) fluids [1].

TABLE 19-3

Accuracy of Two Methods for Estimating Liquid Density

Chemical	Literature Value (g/mL) at 20°C Unless Indicated[a]	Percent Deviation of Estimated Value	
		Grain	Bhirud
Hydrocarbons			
Naphthalene, 1-methyl	1.0202	−4.5	−10.9
Benzene, 1-isopropyl, 3-methyl	0.8610	−0.2	− 6.1
Cyclopentene	0.7720	+1.1	+ 1.9
2-Hexadecyne	0.7965	+0.8	b
Halogenated Hydrocarbons			
Fluorobenzene	1.0225	+1.9	− 1.2
Hexanoic chloride, 3-methyl	0.967	+0.7	b
Dibromoethylene	2.1792	−5.9	b
2-Butanone, 3-chloro, 3-methyl	1.0083	−0.7	b
Allyl chloride	0.937	−2.3	− 8.8
p-Dichlorobenzene	1.248 (55°C)	+1.1	− 7.8
O-containing Compounds			
3-p-Cymenol	0.990	−4.7	+14.8
Methoxybenzene	0.9961	−2.6	+ 3.5
Isopropyl ether	0.7214	−8.6	+ 6.4
Butanoic acid, ethyl ester	0.8785	+2.5	+ 5.3
Diethyl ketone	0.8138	+5.7	+ 3.0
p-Dioxane	1.0337	+4.0	+ 1.3
Malonic acid, acetyl, diethyl ester	1.0834 (26°C)	+3.5	b
2-Butanone, 3-chloro, 3-methyl	1.0083	−0.7	b
Hexanoic chloride, 3-methyl	0.967	+0.7	b
Tetrahydrofuran	0.889	+3.5	− 0.9
Methyl isobutyl ketone	0.801	+1.2	+ 0.9
Octadecanol	0.812 (59°C)	−1.8	+43.5
N-containing Compounds			
Aniline, N,N-dimethyl	0.9557	−2.1	+ 1.1
Piperidine	0.8606	−0.8	− 2.0
n-Propylamine	0.7173	−4.8	+ 0.7
Propionitrile	0.7818	+1.9	−29.0
Benzonitrile	1.0102 (15°C)	+2.2	− 8.5
Ethylenediamine	0.896	−9.5	− 6.6

TABLE 19-3 (Continued)

Chemical	Literature Value (g/mL) at 20°C Unless Indicated[a]	Percent Deviation of Estimated Value	
		Grain	Bhirud
S-containing Compounds			
Thiophene	1.0649	+3.4	− 0.8
Methylmercaptan	0.866	−4.0	− 1.0

SUMMARY OF ERRORS

	Grain		Bhirud	
	n	%E	n	%E
Absolute Average Error	28	3.1	23	7.2
Hydrocarbons	4	1.6	3	6.3
Halogenated Hydrocarbons	6	2.1	3	5.9
O-containing	12	3.2	9	8.8
N-containing	6	3.5	6	8.0
S-containing	2	3.7	2	0.9

a. Values from Refs. 7 and 14.
b. Critical points not available for these compounds in the literature.

Gambill [2]. The Goldhammer method requires input of the molar density at the normal boiling point (ρ_{Lb}) and is expressed by the following equation:

$$\rho_L = M\rho_{Lb} \left(\frac{1 - T_r}{1 - T_{br}} \right)^n \qquad (19\text{-}3)$$

where M is the molecular weight, $T_r = T/T_c$ and $T_{br} = T_b/T_c$. The value of the exponent n varies, depending on the chemical class:

	n
Alcohols	0.25
Hydrocarbons	0.29
Other organics	0.31

Grain modified Eq. 19-3 by using the following approximation:

$$T_c \approx \frac{3}{2} T_b \qquad (19\text{-}4)$$

TABLE 19-4

Percent Difference Obtained by Grain's Equation
Using the Approximation $T_c = 1.5\, T_b$

Compound	T_b (K)	T_c (K)	T_c/T_b	Δ a	Δ b	% Difference $\frac{(a-b)}{a}$ 100
Acetone	329.7	509.1	1.54			
200K				1.23	1.25	− 1.6
273K				1.11	1.12	− 0.9
Isoamyl propionate	433.4	611.4	1.41			
200K				1.375	1.320	4.0
273K				1.277	1.234	3.4
383K				1.099	1.083	1.5
Methane	111.7	190.7	1.71			
95K				1.075	1.105	− 2.8
110K				1.007	1.011	− 0.4
n-Octadecane	589.3	756	1.28			
323K				1.44	1.28	11.1
450K				1.26	1.16	8.0
550K				1.08	1.05	2.8
Dimethyl ether	249.5	400.1	1.61			
100K				1.30	1.35	− 3.8

a. Goldhammer's equation

$$\Delta = \left(\frac{1 - T/T_c}{1 - T_b/T_c} \right)^{0.38}$$

b. Grain's equation

$$\Delta = [3 - 2\,(T/T_b)]^{0.38}$$

(The exponent 0.38 was chosen to represent a worst-case example.)

T_r is then equal to $2T/3T_b$, $T_{br} = \frac{2}{3}$, and Eq. 19-3 becomes:

$$\rho_L = M\rho_{Lb}\,[3-2\,(T/T_b)]^n \tag{19-5}$$

The density at the boiling point is easily determined by estimating the inverse property, the molar volume at the boiling point. Methods for doing this are reviewed by Reid *et al.* [14]; the one used here (that of Schroeder) is a simple, additive method applicable to a large number of compounds, with an average error of 3.4%. Table 19-5 lists the incremental values used in this calculation.

TABLE 19-5

Incremental Values for Estimating Molar Volume, V_b, by Schroeder's Method

V_b is calculated by summing the incremental values for every atom, chemical structure, or bond listed; e.g., cyclopentene $(C_5 H_8) = 5(C) + 8(H) + 1$ (Ring) $+ 1$ (C=C) $= 91$ cm^3/g-mol.

Molecular Feature	Increment (cm^3/g-mol)	Molecular Feature	Increment (cm^3/g-mol)
Atoms		I	38.5
C	7	S	21
H	7		
N	7	Rings	− 7
O	7		
Br	31.5	Bonds	
Cl	24.5	Single	0
F	10.5	Double	7
		Triple	14

Source: Reid, *et al.* [14]. *(Reprinted with permission from McGraw-Hill Book Co.)*

Basic Steps

(1) Look up the boiling point, T_b (K), and calculate the molecular weight, M, of the chemical. (T_b may be estimated by the methods given in Chapter 12.)

(2) Calculate V_b from the molecular structure, using Table 19-5. Calculate $\rho_{Lb} = 1/V_b$.

(3) Assign the appropriate value to n for the chemical class of the compound, as listed above.

(4) Apply these values to Eq. 19-5 to estimate ρ_L at the desired temperature, T (K).

Example 19-2   Calculate the density of fluorobenzene (C_6H_5F) at 20°C (293K), given M = 96.11 g/mol and T_b =85.1°C (358K).

(1)   Determine V_b from Table 19-5:

$$
\begin{array}{lll}
\text{6 carbons} & = 6(7) = & 42 \\
\text{5 hydrogens} & = 5(7) = & 35 \\
\text{1 fluorine} & = & 10.5 \\
\text{1 ring} & = & -7 \\
\text{3 double bonds} & = 3(7) = & \underline{21} \\
& & 101.5
\end{array}
$$

$\rho_{Lb} = 1/101.5 = 9.852 \times 10^{-3}$ g/cm³

(2)   Assign fluorobenzene to the class "other organics."

Therefore, n = 0.31

(3)   Substituting into Eq. 19-5,

$$\rho_L = (96.11)\,(9.852 \times 10^{-3})\,[3 - 2\,(293/358)]^{0.31}$$

$$= 1.04 \text{ g/cm}^3$$

Since the literature value is 1.0225, the error is +2%.

19-6   BHIRUD'S METHOD (LIQUID DENSITY)

Bhirud [1] developed this method of liquid density estimation to suit a wide range of chemical compounds over a large temperature span. The equation is based on the law of corresponding states and requires the acentric factor (ω), critical temperature (T_c), and critical pressure (P_c) as inputs. The following equation is written in terms of parameters a and b, which, in turn, are functions of the reduced temperature, T_r:

$$\ln\,(P_c V/RT) = a + \omega b \qquad (19\text{-}6)$$

where

$$a = 1.39644 - 24.076 T_r + 102.615 T_r^2 - 255.719 T_r^3$$

$$+ 355.805 T_r^4 - 256.671 T_r^5 + 75.1088 T_r^6 \qquad (19\text{-}7)$$

$$b = 13.4412 - 135.7437 T_r + 533.380 T_r^2 - 1091.453 T_r^3$$

$$+ 1231.43 T_r^4 - 728.227 T_r^5 + 176.737 T_r^6 \qquad (19\text{-}8)$$

Rearrangement of Eq. 19-6 and incorporation of molecular weight allows for direct calculation of the liquid density:

$$\rho_L = MP_c/RTe^{a + \omega b} \tag{19-9}$$

In this equation, the gas constant, R, is expressed as 82.04 cm³-atm/mol-K. The acentric factor, ω, is a measure of the sphericity of the molecule and is affected by the molecular weight and the polarity. It can be calculated by Eq. 19-10, which was developed by Edmister (cited in Ref. 14).[4]

$$\omega = \frac{3}{7}\left(\frac{T_{br}}{1 - T_{br}}\right)\log P_c - 1 \tag{19-10}$$

Basic Steps

(1) Obtain the values of M, T_c, P_c and ω from the literature. If ω cannot be found, obtain the value of T_b. (See Chapters 12 and 13 for estimation of T_b, T_c and P_c. T_c may also be approximated by Eq. 19-4.)

(2) If ω was not found, estimate it by Eq. 19-10. Note that $T_{br} = T_b/T_c$.

(3) Calculate T_r (= T/T_c) and evaluate Eqs. 19-7 and 19-8.

(4) Insert values determined above into Eq. 19-9 and calculate ρ_L.

Example 19-3   Calculate the density of fluorobenzene at 20°C, given M = 96.11 g/mol, T_b = 85.1°C (358.3K), T_c = 286.95°C (560.1K), P_c = 44.6 atm, and R = 82.04 cm³-atm/mol-K.

(1) T_{br} = 358.3/560.1 = 0.6397

(2) $\omega = \left(\dfrac{3}{7}\right)\left(\dfrac{0.6397}{1 - 0.6397}\right)(\log 44.6) - 1 = 0.255$

(3) T_r = 293/560.1 = 0.5231

(4)   Input into Eqs. 19-7 and 19-8 yields

   a = −1.5951

   b = −0.5410

---

4. Although Eq. 19-10 contains neither molecular weight nor polarity *per se*, both of these properties are represented by the reduced boiling point (T_{br}), because it is affected by them.

(5)   Substituting the above into Eq. 19-9,

$$\rho_L = \frac{(96.11)\,(44.6)}{(82.04)\,(293)\,e^{-\,1.5951\,+\,(0.255)\,(-0.5410)}}$$

$$= 1.01 \text{ g/mL}$$

This corresponds well with the literature value, 1.0225; the deviation is −1.2%.

19-7   AVAILABLE METHODS FOR ESTIMATING SOLID DENSITY

In molecular crystals, the actual volume (crystal volume) occupied by a molecule consists of the molecular volume in addition to the empty spaces in the crystal. The ratio of molecular volume to crystal volume expresses the "packing coefficient" of the crystal and ranges from 0.65 to 0.77 [9]. Cady (cited in Ref. 8) incorporates these ideas, which were developed by the Russian crystallographer Kitaigorodsky, into a method to estimate solid densities. He calculates the volume of the molecule as the sum of atomic spheres of Van der Waals radii, corrected for the overlap created by covalently bonded atoms. He fit a large number of explosive-type compounds to a linear function relating the packing coefficient to the mole fraction of hydrogen atoms bonded to carbon in the molecule.

Tarver [18] used an empirical approach to solid density in which atomic volumes are added together. He compiled values for 74 molecular groupings, so that the special volume effects of a common molecular group can be considered. Table 19-6 summarizes the errors associated with 173 compounds, primarily explosives. The error is less than 1% for 40% of these compounds. The largest errors are encountered with compounds containing several polar and bulky constituents, and with those whose substituent molar volumes are based upon only a few data points. The major disadvantages of this approach are (1) the large data base required to provide a sufficient number of examples to test the accuracy of the method and (2) the narrow category of compounds to which the method applies.

Nielsen [11] has developed a somewhat similar approach based on 43 fragment constants, which he has tested on various solids and liquids, again predominantly explosives. As shown in Table 19-7, the percentage error is larger than with the method of Tarver. In his paper, Nielsen discusses the effect of steric factors on density measurements and suggests a qualitative method of correcting for these factors.

TABLE 19-6

Percent Error in Calculated Density Using the Method of Tarver

Class	No. of Compounds	Avg. Error (%)	Number of Compounds Within Error Range		
			0-1%	1-2%	2-3%
Solid Aromatic Explosives Containing NO_2, OH, CH_3 and NH_2	25	1.29	11	9	5
Aliphatic Compounds Containing NO_2, OH, F, CO_2H, and ONO_2	75	1.51	35	24	8
Amines, Nitramine, and Heterocyclic Compounds	73	1.68	24	24	18
Overall	173	1.55	70	57	31

Source: Tarver [18]

TABLE 19-7

Absolute Average Error Calculated for Density Using the Method of Nielsen

Compound Class	No. of Compounds	Average Error (%)
Amines	16	2.0
Hydrocarbons (Aromatic and Acyclic)	28	4.3
Explosives	16	9.7
Miscellaneous (Halogenated, Ethers, Ketones, Alcohols)	22	6.7
Overall	82	5.5

Source: Nielsen [11]

A third method of estimation, that of Immirzi and Perini [9], is recommended here because it was developed from a data base that was not limited to explosive-type compounds but, rather, 500 organic crystalline compounds ranging in molecular weight from 50 to 1000 g/mole. To compensate for the empty space in an organic crystal, which is accomplished in the Cady and Kitaigorodsky methods by the packing coefficient, Immirzi and Perini assign atomic volumes based upon the number of other atoms bonded to the specific atom. Carbon, for instance, may bond to 2, 3, or 4 other atoms, oxygen to 1 or 2, and nitrogen (in this method) to a maximum of 3 and a minimum of 1.

The following restrictions apply to the method of Immirzi and Perini: (1) crystals that have a structural disorder or are not solids at room temperature are excluded; (2) with the exception of water, crystals may not contain molecules of the solvent; (3) only the elements H, C, O, N, S, F, Cl, Br, I, Na, K, and Rb are considered; and (4) cyclic compounds are limited to derivatives of benzene and naphthalene.

Immirzi and Perini analyzed the method errors by selecting a random sample of 53 crystalline organic compounds whose composition was compatible with the data set used for their regression analysis. Table 19-8 lists the chemical formulas of these compounds and the errors in the estimated molecular volumes. The data indicate an average absolute error of 2.0% in the estimates of crystal density. Approximately 40% of the estimates were within 1% of the measured values, and 57% of the estimates were within 2%. The maximum error was +6.9%.

19-8 METHOD OF IMMIRZI AND PERINI

The additivity method of Immirzi and Perini is based upon the following equation:

$$V_s = \sum_i m_i v_i \qquad (19\text{-}11)$$

where

V_s = calculated crystal volume for a single molecule (Å^3/molecule)
m_i = relative stoichiometric multiplicities
v_i = unit volumes of atomic elements (Å^3)

Values for v_i are listed in Table 19-9 in units of cubic angstroms; these values were selected to minimize the quantity $(V_e - V_s)^2$, where V_e is the experimentally derived crystal volume.

TABLE 19-8

Errors in Estimates of Crystal Volume for 53 Compounds, by Method of Immirzi and Perini

Compound	Experimental Crystal Volume, V_e (nm^3/molecule)	Deviation of Estimate[a] (%)	Compound	Experimental Crystal Volume, V_e (nm^3/molecule)	Deviation of Estimate[a] (%)
$C_{20}H_{26}O_2$	0.4400	0.95	$C_6H_{10}O_7 \cdot H_2O$	0.2201	0.84
$C_6H_6O_2 \cdot CON_2H_4$	0.2075	4.32	$C_{17}H_{10}N_2Cl_2$	0.3750	0.43
$C_{20}H_{12}O_5 \cdot C_3H_6O$	0.4676	0.11	$C_{23}H_{35}O_4N_5S$	0.6194	1.59
$C_8H_2O_4N_2 \cdot 6H_2O$	0.3175	1.40	$(C_6H_6N)_2 \cdot C_6H_6N_3O_6 \cdot \frac{1}{2}C_6H_6$	0.5159	1.76
$C_6H_2O_4F_2$	0.1574	0.99	$C_{16}H_{10}N_3O_3Cl$	0.3540	2.49
$C_6H_3N_2O_5Cl$	0.2068	2.20	$C_{22}H_{31}O_2N$	0.4697	4.28
$C_6H_4NO_3^- \cdot K^+ \cdot \frac{1}{2}H_2O$	0.1859	2.20	$C_{24}H_{31}O_6F$	0.5714	1.30
$C_5H_6O_3$	0.1295	3.10	$C_6H_{12}N_2O_4S_2 \cdot 2HBr \cdot 2H_2O$	0.3850	0.65
$C_{17}H_{16}$	0.3190	0.56	$C_6H_{12}Cl_2$	0.3333	2.30
$C_{21}H_{24}O_2$	0.4132	4.78	$C_{18}H_{21}N_4O_2Cl$	0.4361	0.16
$C_{13}H_{12}N_4O_2 \cdot 2C_{10}H_8O_2 \cdot 3H_2O$	0.7533	0.07	$C_{20}H_{14}O_4$	0.4044	0.25
$C_{10}H_{20}N_2O_4S_2 \cdot 2HCl$	0.4234	0.70	$C_{20}H_{14}O_4$	0.4129	2.31
$C_8H_{15}N_2O_2$	0.2442	4.60	$C_7H_{15}NO$	0.2151	6.90
$C_8H_{15}NO_6$	0.2536	0.24	$C_9H_{19}NO$	0.2629	5.10
$C_6H_6N_4O_2 \cdot HCl \cdot H_2O$	0.2280	2.19	$C_{10}H_{20}N_2O_3S_2$	0.3370	0.88
$C_{18}H_{22}O_2$	0.3641	2.64	$C_8H_{15}NO_7$	0.2710	0.00
$C_{17}H_{22}N_2O_3$	0.3835	0.75	$C_7H_{13}NO_2$	0.1959	1.40
$C_{27}H_{46}$	0.5576	6.55	$C_4H_{13}NO_2^+ \cdot C_8H_{11}N_2O_3^-$	0.3853	5.67
$C_{29}H_{50}$	0.6141	4.76	$C_4H_{12}NO^+ \cdot C_8H_{11}N_2O_3^-$	0.3644	1.65
$C_{16}H_{20}N_4O_6$	0.4475	6.48	$C_4H_{12}NS_2^+ \cdot Br^- \cdot H_2O$	0.2510	1.54
$C_{16}H_{27}O_5N$	0.4319	1.51	$C_{10}H_{15}BrO$	0.2561	0.22
$C_{16}H_{16}N_4$	0.3334	2.10	$C_{10}H_{14}Br_2O$	0.2898	2.81
$C_6H_{13}NO_2$	0.1828	0.40	$C_{14}H_{20}NCl$	0.3229	2.02
$C_{21}H_{30}O_2 \cdot C_6H_6O_2$	0.5906	0.02	$C_{15}H_{10}O_5$	0.2990	3.44
$C_{30}H_{22}O_2N_2Br_2$	0.6091	0.33	$C_{15}H_{13}O_2F$	0.3153	2.15
$C_9H_{17}N_3O_4 \cdot \frac{1}{2}H_2O$	0.3021	1.21	$C_{23}H_{29}ClO_4$	0.5333	0.31
$C_{16}H_{22}O_3N_2$	0.3954	0.76		Avg. absolute error = 2.0	

a. Estimated values included corrections for aromatic and non-aromatic 5- or 6-membered rings and for hydrogen bonds in $-CONH_2$, $-CONH$, and $-COOH$ groups.

Source: Immirzi and Perini [9], based on data from *Acta Crystallogr. Sect. B, 31* (1975). *(Reprinted with permission from the International Union of Crystallography.)*

TABLE 19-9

Volume Increments (v_i) for Common Elements and Ions[a]

Element or Ion	v_i (Å^3)	Std. Error, σ	No. of Contributors
—H	6.9	0.4	5228
=C= −C≡	15.3	0.7	74
—C⪜	13.7	0.6	453
>C<	11.0	0.9	1165
=O	14.0	0.5	649
—O—	9.2	0.5	468
N≡	16.0	1.3	30
N⪜	12.8	0.8	68
N⪝	7.2	0.8	354
S	23.8	0.9	92
—F	12.8	1.5	14
—Cl	26.7	0.5	134
—Br	33.0	0.5	120
—I	45.0	1.3	26
Cl^-	28.9	1.5	39
Br^-	39.3	1.5	20
I^-	56.6	2.5	11
Na^+	13.6	2.2	16
K^+	27.3	1.6	32
Rb^+	34.1	2.2	15
H_2O	21.5	0.8	68
Benzene frame (carbons only)	75.2	2.5	443
O—H $\cdots$ O hydrogen-bond	− 2.6	0.7	206
N—H $\cdots$ O hydrogen-bond	− 2.8	0.5	152
N—H $\cdots$ N hydrogen-bond	− 0.3	1.7	11
Non-aromatic rings (rough est.)	− 3.0		
Naphthalene frame (carbons only)	123.7		

a. Different coordination numbers are considered for C,N, and O.

Source: Immirzi and Perini [9]. *(Reprinted with permission from the International Union of Crystallography.)*

Once V_s is determined, it is substituted in Eq. 19-12 to calculate the solid density, ρ_s. In this equation, M is the molecular weight of the compound.

$$\rho_s = \frac{1.660M}{V_s} \qquad (19\text{-}12)$$

Basic Steps

(1) Diagram the structure of the chemical, being careful to identify sites of intramolecular hydrogen bonding.

(2) Find in Table 19-9 the volume increment (v_i) for each atom, ion, or structure in the molecule. Sum these values in proportion to their frequency (m_i) in the molecule, as indicated in Eq. 19-11. The resulting sum is the estimated crystal volume, V_s, in Å³/molecule.

(3) Obtain the crystal density, ρ_s, in g/cm³ by substituting V_s and the molecular weight, M, in Eq. 19-12.

Example 19-4   Estimate the crystal density of morphine ($C_{17}H_{19}NO_3 \cdot H_2O$). The molecular weight (M) is 303.35 g/mol.

(1)   The structure is:

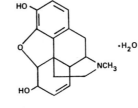

(2)   The volume increments from Table 19-9 are:

		$m_i \, (v_i)$
Benzene frame carbons	:	1 (75.2)
$\rangle C \langle$	:	9 (11.0)
$\rangle C =$	:	2 (13.7)
$-H$	:	19 (6.9)
$-O-$	:	3 (9.2)
$N \langle$	:	1 (7.2)
H_2O	:	1 (21.5)

Non-aromatic rings   : $\underline{4\ (-3.0)}$

$$V_s = \sum_i m_i\, v_i = 377.0 \text{ Å}^3/\text{molecule}$$

(3)   Substituting V_s and M in Eq. 19-12:

$$\rho_s = 1.660\ (303.35)/377.0 = 1.336 \text{ g/cm}^3$$

The observed value is 1.317 g/cm^3 [19], indicating an error of +1.4%.

Example 19-5   Estimate the crystal density of maleic acid, *cis*-(COOH) CHCH (COOH). The molecular weight, M, is 116.07 g/mol.

(1)   The structure is:

Note the presence of intramolecular hydrogen bonding of the O–H $\cdots$ O type.

(2)   The volume increments from Table 19-9 are:

		$m_i\ (v_i)$
$\rangle C =$	:	4 (13.7)
$= O$	:	2 (14.0)
$- O -$	:	2 (9.2)
$- H$	:	4 (6.9)
O–H $\cdots$ O hydrogen bond	:	$\underline{1\ (-2.6)}$

$$V_s = \sum_i m_i\, v_i = 126.2 \text{ Å}^3/\text{molecule}$$

(3)   Substituting V_s and M into Eq. 19-12,

$$\rho_s = 1.660\ (116.07)/126.2 = 1.527 \text{ g/cm}^3$$

The observed value is 1.590 g/cm^3 ($20°/4°$) [19], indicating an error of –4.0%.

Example 19-6   Estimate the crystal density of 1,2,4,5-tetrabromobenzene, $C_6 H_2 Br_4$. The molecular weight, M, is 393.74 g/mol.

(1) The structure is:

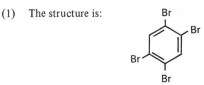

(2) The volume increments from Table 19-9 are:

$$m_i \; (v_i)$$

Benzene frame carbons	:	1 (75.2)
$-$ Br	:	4 (33.0)
$-$ H	:	2 (6.9)

$$V_s = \sum_i m_i \, v_i = 221.0 \text{ Å}^3/\text{molecule}$$

(3) Substituting V and M in Eq. 19-12:

$$\rho_s = 1.660 \, (393.74)/221.0 = 2.958 \text{ g/cm}^3$$

The observed value is 3.072 g/cm³ (20°/4°) [19], indicating an error of −3.7%.

19-9   AVAILABLE DATA

Standard handbooks of chemical properties usually include density. Appendix A lists some general handbooks. For measured or calculated crystal densities, a good source is the following reference, which lists values of 7,500 organic and organo-metallic chemicals:

Donnay, J.D.H. and H.M. Ondik, *Organic Compounds,* Vol. I of *Crystal Data Determinative Tables,* 3rd ed., U.S. Dept. of Commerce, National Bureau of Standards, and the Joint Committee on Powder Diffraction Standards (1972).

Volume II of this series includes principally inorganic compounds, although cyanides and cyanates are also listed.

19-10   SYMBOLS USED

a   = constant in Bhirud's method (Eq. 19-6)
b   = constant in Bhirud's method (Eq. 19-6)
m_i = relative stoichiometric multiplicities (Eq. 19-11)
M   = molecular weight (g/mol)

n   = number of moles in Eq. 19-1; also, coefficient in Eq. 19-3

P   = pressure (atm)

P_c = critical pressure (atm)

R   = gas constant = 82.04 cm^3-atm/mol-K = 0.082 L-atm/ mol-K

SG = specific gravity

T   = temperature (K)

T_b = temperature at the boiling point (K)

T_{br} = T_b/T_c

T_c = critical temperature (K)

T_r = T/T_c

U   = parameter in Eq. 19-6

V   = volume (L)

V_b = molar volume at the boiling point (Å^3/molecule)

V_c = critical volume (L)

V_e = experimental crystal volume in Table 19-8

v_i = atomic volume in Eq. 19-11 (Å^3)

V_s = calculated crystal volume of a single molecule in Eq. 19-11 (Å^3/molecule)

V_{sc} = scaling volume (Table 19-2)

$V_r^{(o)}$ = volume parameter (Table 19-2)

Z_c = critical-compressibility factor in Table 19-2

Z_{RA} = parameter in Table 19-2

Greek

ω = Pitzer acentric factor (unitless)

ρ = density

ρ_c = critical density (mol/mL)

ρ_L = liquid density (g/mL)

ρ_{Lb} = molar density at the boiling point (mol/mL)

ρ_s = solid density (g/cm^3)

ρ_v = vapor density (g/L)

χ = connectivity function

19-11 REFERENCES

1. Bhirud, V.L., "Saturated Liquid Densities of Normal Fluids," *AIChE J.,* **24**, 1127-31 (1978).

2. Gambill, W.R., "How to Estimate Liquid Densities," *Chem. Eng. (NY),* **66**, 193-94 (1959).

3. Grain, C.F., Arthur D. Little, Inc., Cambridge, Mass., personal communication, November 1979.

4. Gunn, R.D. and T. Yamada, "A Corresponding States Correlation of Saturated Liquid Volumes," *AIChE J.,* **17**, 1341-45 (1971).

5. Hall, L.H., Eastern Nazarene College, Quincy, Mass., personal communication, October 16, 1979.

6. Hankinson, R.W. and G.H. Thomson, "A New Correlation for Saturated Densities of Liquids and Their Mixtures," *AIChE J.,* **25**, 653-63 (1979).

7. Hodgman, C.D. (ed.), *Handbook of Chemistry and Physics,* 35th ed., Chemical Rubber Publishing Company, Cleveland, Ohio, 1953.

8. Holden, J.R. and C. Dickson, "Prediction of Crystal Density Through Molecular Packing Analysis," Naval Service Weapons Center, Report NSWC MP 79-185, Silver Spring, Md. (Sept. 27, 1979).

9. Immirzi, A. and B. Perini, "Prediction of Density in Organic Crystals," *Acta Crystallogr. Sect. A,* **33**, 216-18 (1977).

10. Kier, L.B., W.J. Murray, M. Randie and L.H. Hall, "Molecular Connectivity: V. Connectivity Series Concept Applied to Density," *J. Pharm. Sci.,* **65**, 1226-30 (1976).

11. Nielsen, A.T., "Calculation of Densities of Fuels and Explosives from Molar Volume Additive Increments," Naval Weapons Center, TP 5452, China Lake, CA (1973).

12. Perry, R.H. and C.H. Chilton (eds.), *Chemical Engineers' Handbook,* 5th ed., McGraw-Hill Book Co., New York, 1973, p. 3-231.

13. Redlich, O. and J.N.S. Kwong, "On the Thermodynamics of Solutions: V. An Equation of State. Fugacities of Gaseous Solutions," *Chem. Rev.,* **44**, 233-44 (1949).

14. Reid, R.C., J.M. Prausnitz and T.K. Sherwood, *The Properties of Gases and Liquids,* 3rd ed., McGraw-Hill Book Co., New York, 1977 (see esp. pp. 16-21).

15. Spencer, C.F. and S.B. Adler, "A Critical Review of Equations for Predicting Saturated Liquid Density," *J. Chem. Eng. Data,* **23**, 82-89 (1978).

16. Spencer, C.F. and R.P. Danner, "Improved Equation for Prediction of Saturated Liquid Density," *J. Chem. Eng. Data,* **17**, 236-41 (1972).

17. Tarakad, R.R., C.F. Spencer and S.B. Adler, "A Comparison of Eight Equations of State to Predict Gas-Phase Density and Fugacity," *Ind. Eng. Chem. Process Des. Dev.,* **18**, 726-39 (1979).

18. Tarver, C.M., "Density Estimations for Explosives and Related Compounds Using the Group Additivity Approach," preprint UCRL-81192, University of California, Lawrence Livermore Laboratory, CA (1978). [See Tarver, C.M., *J. Chem. Eng. Data,* **24,** 136-45 (1979) for the published version of this report.]

19. Weast, R.C. (ed.), *CRC Handbook of Chemistry and Physics,* 58th ed., CRC Press, Inc., Cleveland, Ohio, 1978-1979.

20. Weber, J.H., "Predicting Volumes of Pure Liquids," *Chem. Eng. (NY),* **86,** 173-77 (March 26, 1979).

21. Weber, J.H., "Predicting Volumes of Saturated Liquid Mixtures," *Chem. Eng. (NY),* **86,** 95-99 (May 7, 1979).

22. Yen, L.C. and J.J. Woods, "A Generalized Equation for Computer Calculation of Liquid Densities," *AIChE J.,* **12,** 95-99 (1966).

20

SURFACE TENSION

Clark F. Grain

20-1 INTRODUCTION

Surface tension affects the extent of spreading when a liquid chemical is spilled on land or water. It is also important with respect to the emulsification of liquids that are mixed with water and in their adsorption on solid surfaces.

The molecules at the surface of a liquid are subjected to unequal forces. In general, we consider only the interface between the surface of the liquid and a reference gas (air or the vapor of the liquid). If the density of the gas or vapor is low, the molecules at the liquid surface are attracted toward the bulk liquid; the gas or vapor phase exerts considerably less attraction. As a result, the surface is in *tension*, which causes it to contract to a minimum area consistent with the mass of material and the container walls. This force is a characteristic property of a given liquid at a given temperature.

The surface tension is defined as the force per unit length (usually dynes/cm) in the plane of the surface and is represented by the symbol σ. It may also be expressed as energy per unit area (ergs/cm²). Surface tensions for most organic liquids are between 25 and 40 dynes/cm at room temperature, although polyhydric alcohols range up to 65 dynes/cm. For comparison, the surface tension of water at 25°C is 72 dynes/cm.

The two methods for estimating σ that are recommended in this chapter require, as inputs, the parachor, density (or molar volume), boiling point, and values of characteristic constants. Method 1 requires only the parachor and measured density values; more parameters are needed with method 2, but all of them can be estimated.

By either method, errors typically average 5%, but individual values can deviate as much as 30% from measured values.

20-2   AVAILABLE METHODS

The most general estimation method is that of MacLeod and Sugden. MacLeod [5] suggested an empirical relationship between surface tension and liquid density. Sugden [9] later showed that the empirical constant appearing in MacLeod's formulation was identical to a constitutive constant called the parachor. The equation introduced by Sugden is:

$$\sigma^{1/4} = P\frac{(\rho_L - \rho_v)}{M} \tag{20-1}$$

where

$$
\begin{array}{lll}
\sigma & = & \text{surface tension (dynes/cm)} \\
P & = & \text{parachor} \\
\rho_L & = & \text{liquid density (g/cm}^3\text{)} \\
\rho_v & = & \text{vapor density (g/cm}^3\text{)} \\
M & = & \text{molecular weight (g/mol)}
\end{array}
$$

The parachor can be visualized as a comparative volume between two liquids and is reasonably independent of temperature. Quayle [7] presented a comprehensive review of methods for estimating the parachor and devised one that is suitable for a wide variety of organic compounds.

Equation 20-2, a more familiar form of Eq. 20-1, can be used to calculate σ if the temperature dependency of ρ_L and ρ_v are known.

$$\sigma = \left[P\frac{(\rho_L - \rho_v)}{M}\right]^4 \tag{20-2}$$

For most purposes ρ_v is so much smaller than ρ_L that it can be ignored.

Equation 20-2 then becomes

$$\sigma = \left(\frac{P\,\rho_L}{M}\right)^4 \tag{20-3}$$

Equation 20-3 is the basic equation for method 1. In method 2 the equation is altered to permit calculation of surface tension when a measured value of ρ_L is unavailable. One of the methods discussed in Chapter 19 is used to estimate ρ_L. Details of the derivation are given in §20-4.

Several other methods are available that use as inputs experimental or estimated values of the heat of vaporization, viscosity, critical temperature, and critical pressure. These have been reviewed by Gambill [1,2] and Reid *et al.* [8]. Table 20-1 lists a representative sample of available methods and comments on their applicability.

The surface tension of mixtures of organic liquids can be treated, in a first approximation, as a linear function of the mole fractions of the pure components. The estimates obtained are very approximate; further refinements are necessary for more accurate estimates. The available methods for these mixtures are discussed by Reid *et al.* [8].

Mixtures of organic liquids and water are even more complicated, and the methods for mixtures of organic liquids only cannot be used. Discussions of methods for obtaining the mixture surface tension of organic liquids and water are given in Chapter 21 and in Ref. 8.

Methods 1 and 2 were chosen because of their generality: they can be applied to a wide variety of materials, including polar and associated organic liquids. The other methods are restricted to hydrocarbon and polar (but nonassociated) liquids.

20-3   METHOD 1 (MacLEOD-SUGDEN)

If measured liquid densities are available as a function of temperature, Eq. 20-3 provides a quite simple and surprisingly accurate method for estimating surface tension. Parachor values are estimated from the table devised by Quayle [7], which is reproduced here as Table 20-2.

Reid *et al.* [8] tested Eq. 20-3 for 28 compounds, including alcohols. They found an average error of 4.5% and a maximum error of −20% (Table 20-3).

TABLE 20-1

Some Methods for Estimating Liquid Surface Tension

Method	Formula[b]	Inputs[b]	Range (dynes/cm)	Comments
Macleod-Sugden [5,9] [a]	$\sigma = (P\rho_L/M)^4$	P, ρ_L, M	15-65	P can be estimated; measured ρ_L must be available; versatile. Average error ~5%.
Grain[a]	$\sigma = [P(1+k)/V_b \cdot (3-2T/T_b)^n]^4$	P, n, k, T_b, V_b	15-65	All inputs can be estimated; versatile. Average error ~5%.
Walden [10]	$\sigma = \Delta H_{vb}/3.64V_b$	$\Delta H_{vb}, V_b$	20-40	Inputs can be estimated; hydrocarbons only. Average error ~5%.
Mayer [4]	$\sigma = \dfrac{(1-A)\,\Delta H_v}{(36\pi \cdot N_L \cdot V^2)^{1/3}}$	$A, \Delta H_v, N_L, V$	20-40	$N_L = 2.687 \times 10^{19}/cm^3$; other parameters can be estimated; not applicable to alcohols. Average error ~4%.

a. Recommended method.
b. See §20-5 for explanation of symbols.

TABLE 20-2

Parachor Increments

Group	Increment	Group	Increment
CH_2 in $-(CH_2)_n$		Single bond	0.0
n < 12	40.0	Semipolar bond	0.0
n > 12	40.3	Triple bond	40.6
C	9.0	Carbonyl bond in ketones:	
H	15.5	R + R' = 2	22.3
in OH	10.0	3	20.0
in HN[a]	12.5	4	18.5
O	19.8	5	17.3
O_2 in esters	54.8	6	17.3
N	17.5	7	15.1
S	49.1	8	14.1
P	40.5	9	13.0
F	26.1	10	12.6
Cl	55.2	Alkyl groups	
Br	68.0	1-Methylethyl	133.3
I	90.3	1-Methylpropyl	171.9
Se	63	1-Methylbutyl	211.7
Si	31	2-Methylpropyl	173.3
Al	55	1-Ethylpropyl	209.5
Sn	64.5	1,1-Dimethylethyl	170.4
As	54	1,1-Dimethylpropyl	207.5
Singlet linkage[b]	− 9.5	1,2-Dimethylpropyl	207.9
Hydrogen bridge	−14.4	1,1,2-Trimethylpropyl	243.5
Ethylenic bond		Position differences in benzene:[c]	
Terminal	19.1	Ortho-meta	1.8-3.4
2,3-Position	17.7	Meta-para	0.2-0.5
3,4-Position	16.3	Ortho-para	2.0-3.8
Chain branching, per branch	− 3.7	Ring size	
Secondary-secondary adjacency	− 1.6	3-membered ring	12.5
Secondary-tertiary adjacency	− 2.0	4-membered ring	6.0
Tertiary-tertiary adjacency	− 4.5	5-membered ring	3.0
		6-membered ring	0.8
		7-membered ring	4.0

a. For second H in primary amines use increment of 15.5.
b. Bond containing an unpaired electron.
c. Small differences are seen in the Parachor values of di-substituted benzenes. Ortho compounds have the smallest values followed, in turn, by meta and para compounds. The "increments" given are the range of differences noted for the indicated compound pairs.

Source: Quayle [7]. *(Reprinted with permission of the Williams & Wilkins Co., Baltimore.)*

TABLE 20-3

Error in Estimating Surface Tension of Pure Liquids by Macleod-Sugden Method

Compound	t (°C)	Measured σ (dynes/cm)	Error (%)
Acetic acid	20	27.59	−4.6
	60	23.62	−3.2
Acetone	25	24.02	−5.4
	35	22.34	−3.9
	45	21.22	−4.5
Aniline	20	42.67	−3.2
	40	40.50	−6.0
	60	38.33	−7.1
	80	36.15	−8.8
Benzene	20	28.88	−5.1
	40	26.25	−5.6
	60	23.67	−5.0
	80	21.20	−3.9
Benzonitrile	20	39.37	−2.0
	50	35.89	−3.4
	90	31.26	−4.2
Bromobenzene	20	35.82	−0.7
	50	32.34	−1.4
	100	26.54	−0.7
n-Butane	− 70	23.31	11
	− 40	19.69	5.2
	20	12.46	1.5
Carbon disulfide	20	32.32	3.8
	40	29.35	3.8
Carbon tetrachloride	15	27.65	−1.1
	35	25.21	−1.2
	55	22.76	−1.0
	75	20.31	0.1
	95	17.86	2.5
Chlorobenzene	20	33.59	−0.6
	50	30.01	0.7
	100	24.06	5.8
p-Cresol	40	34.88	0.5
	100	29.32	−0.3

TABLE 20-3 (Continued)

Compound	t (°C)	Measured σ (dynes/cm)	Error (%)
Cyclohexane	20	25.24	−3.9
	40	22.87	−3.5
	60	20.49	−2.2
Cyclopentane	20	22.61	−5.6
	40	19.68	−2.4
Diethyl ether	15	17.56	0
	30	16.20	0.4
2,3-Dimethylbutane	20	17.38	−0.9
	40	15.38	0.6
Ethyl acetate	20	23.97	−4.6
	40	21.65	−4.8
	60	19.32	−4.3
	80	17.00	−2.7
	100	14.68	0.5
Ethyl benzoate	20	35.04	−1.9
	40	32.92	−2.7
	60	30.81	−3.1
Ethyl bromide	10	25.36	−5.3
	30	23.04	−6.1
Ethyl mercaptan	15	23.87	−6.7
	30	22.68	−9.1
Formamide	25	57.02	−8.8
	65	53.66	−15
	100	50.71	−20
n-Heptane	20	20.14	−0.6
	40	18.18	0.7
	60	16.22	3.1
	80	14.26	6.8
Isobutyric acid	20	25.04	1.2
	40	23.20	0.5
	60	21.36	−1.2
	90	18.60	−3.5
Methyl formate	20	24.62	−7.6
	50	20.05	−7.2
	100	12.90	−7.4
	150	6.30	−8.8

TABLE 20-3 (Continued)

Compound	t (°C)	Measured σ (dynes/cm)	Error (%)
Methyl alcohol	20	22.56	−13
	40	20.96	−15
	60	19.41	−17
Phenol	40	39.27	−6.7
	60	37.13	−7.3
	100	32.86	−7.8
n-Propyl alcohol	20	23.71	−0.6
	40	22.15	−1.9
	60	20.60	−3.3
	90	18.27	−4.0
n-Propyl benzene	20	29.98	0.2
	40	26.83	0.8
	60	24.68	2.1
	80	22.53	3.9
	100	20.38	6.6
Pyridine	20	37.21	−2.8
	40	34.60	−3.6
	60	31.98	−4.1

Source: Reid, Prausnitz, and Sherwood, [8]. *(Reprinted with permission from McGraw-Hill Book Co.)*

Basic Steps

(1) Draw the structure of the compound.

(2) Estimate the parachor (P) using Table 20-2. Sum the atomic, group, and structural fragment constants in this table in direct proportion to their occurrence in the molecule.

(3) Obtain the measured density of the liquid, ρ_L, from the literature. (See §20-6.)

(4) Insert the above values along with the molecular weight into Eq. 20-3 and calculate the surface tension, σ, in dynes/cm.

Example 20-1   Calculate the surface tension of aniline, $C_6H_5NH_2$, at 20°C. The measured value of σ is 42.7 dynes/cm [8].

The structure is

(2)   From Table 20-2,

$$
\begin{aligned}
1\text{N} &= 1 \times 17.5 & &= 17.5 \\
\text{H in HN} &= 1 \times 12.5 & &= 12.5 \\
6\text{H} &= 6 \times 15.5 & &= 93.0 \\
6\text{C} &= 6 \times 9.0 & &= 54.0 \\
\text{Dbl. bond} &= 3 \times 19.1 & &= 57.3 \\
\text{Ring closure (6-membered)} & & &= \underline{0.8} \\
& & \text{P} &= 235.1
\end{aligned}
$$

(3)   From the literature (CRC handbook [11]):

$$\rho_L = 1.022 \ \text{g/cm}^3 \ \text{at } 20^\circ\text{C}$$
$$M = 93.12 \ \text{g/mole}$$

(4)   Substituting in Eq. 20-3,

$$\sigma = \left(\frac{235.1 \times 1.022}{93.12} \right)^4$$

$$= (2.580)^4$$

$$= 44.3 \ \text{dynes/cm, which deviates} + 3.7\% \ \text{from the measured value.}[1]$$

Example 20-2   Estimate the surface tension of ethyl acetate. (Measured $\sigma = 24.0$ dynes/cm at 20°C [8].)

(1)   The structure is $CH_3 \overset{\overset{\displaystyle O}{\|}}{C} - O - CH_2 CH_3$

(2)   From Table 20-2,

$$
\begin{aligned}
4\text{C} &= 4 \times 9.0 & &= 36.0 \\
8\text{H} &= 8 \times 15.5 & &= 124.0 \\
\text{O}_2 \text{ in ester} & & &= \underline{54.8} \\
& & \text{P} &= 214.8
\end{aligned}
$$

(3)   From the literature (CRC handbook [11]):

$$\rho_L = 0.901 \ \text{g/cm}^3 \ \text{at } 20^\circ\text{C}$$
$$M = 88.1 \ \text{g}$$

(4)   Substituting in Eq. 20-3,

$$\sigma = \left(\frac{214.8 \times 0.901}{88.1} \right)^4$$

---

1. Table 20-3 shows a different estimated value by Reid *et al.* [8]; the difference in estimated values is presumably due to different assumptions on the use of the parachor increments and/or a different density value.

$$= (2.197)^4$$

$$= 23.3 \text{ dynes/cm}$$

The deviation from the measured value is –2.9%. (See footnote 1, p. 20-9)

20-4   METHOD 2 (GRAIN)

If the density of the liquid is unknown, one of the methods presented in Chapter 19 may be used. The one shown in Eq. 20-4 is particularly useful. (Details are given in §19-4.)

$$\rho_L = \rho_{Lb} \, (3-2 \, T/T_b)^n \tag{20-4}$$

where

ρ_{Lb} = liquid density at T_b (g/cm³)
T_b = boiling point (K)
n = constant for class of compound

The density at T_b can be estimated from the molar volume, V_b, using the additive schemes of either Schroeder (in Ref. 6) or LeBas [3]. The volume increments for these methods were listed earlier in Tables 19-4 and 12-12 respectively.

Substitution of Eq. 20-4 into Eq. 20-3 and replacement of ρ_{Lb} with its equivalent, M/V_b, yields the following:

$$\sigma = \left[\frac{P}{V_b} \, (3-2 \, T/T_b)^n \right]^4 \tag{20-5}$$

Values of n for various classes of compounds are shown in Table 20-4.

When applied to primary amines and amides, Eq. 20-5 severely underestimates the surface tension if V_b is calculated by the Schroeder method. The use of LeBas volumes, on the other hand, leads to overestimates for these compounds if they contain less than six carbon atoms. A weighted average value of V_b from both methods could be used; however, the values of the LeBas volumes approach those obtained by the Schroeder method as the number of carbon atoms increases, so the surface tension would still be underestimated. The degree of underestimation is even greater for diols and triols but considerably less for alcohols and phenols. The surface tension of completely halogenated hydrocarbons, in contrast, tends to be overestimated.

TABLE 20-4

Values of Constants k and n

Compound Class	k	n
Alcohols and Phenols	0.020	0.25
Amides	0.065	0.25
Amines (primary)	0.065	0.25
Amines (secondary and tertiary)	0	0.25
Completely Halogenated Hydrocarbons	−0.028	0.29
Diols, Triols	0.10	0.25
Hydrocarbons	0	0.29
All Other Compounds	0	0.31

Sources:

k: Calculated, by the author, from experimental data on one member of the homologous series. The same value of k can be used on other members of the series.

n: Reid, Prausnitz, and Sherwood, [8] . *(Reprinted with permission from McGraw-Hill Book Co.)*

The obvious solution is to modify the value of V_b to suit the class of compound. This is accomplished by Eq. 20-6, where the value of k is chosen from those listed in Table 20-4.

$$V_b' = V_b/(1+k) \qquad (20\text{-}6)$$

Incorporation of the correction factor in Eq. 20-5 yields the following:

$$\sigma = \left[\frac{P(1+k)}{V_b} (3-2\,T/T_b)^n \right]^4 \qquad (20\text{-}7)$$

Equation 20-7 can be applied to a wide variety of liquids. However, difficulties may be encountered with perhalogenated compounds such as tetrafluorodichloroethane: a k factor can be estimated, but it will not be applicable to all perhalogenated materials. Surface tensions estimated from Eq. 20-7 are compared with measured values for 32 common organic compounds in Table 20-5.

To determine how well Eq. 20-7 accounts for the influence of temperature, the surface tensions of three compounds were estimated for various temperatures and compared with measured values (Table 20-6). Except

TABLE 20-5

Comparison of Measured Surface Tensions with Values Calculated by Method 2

Compound	t (°C)	Surface Tension (dynes/cm)		Error (%)
		Measured[a]	Estimated	
Acetamide	85	39.0	39.1	0.3
Acetic acid	20	27.6	29.1	5.4
Acetone	25	24.0	24.0	0
Acetonitrile	56	29.1	31.0	6.5
Aniline	20	42.7	42.9	+0.9
Benzaldehyde	30	37.4	39.9	6.7
Benzamide	130	38.1	40.4	6.0
Benzene	20	28.8	27.0	−6.3
Benzonitrile	20	39.4	42.9	8.9
Benzylamine	20	39.1	38.7	−0.1
Bromobenzene	20	35.8	34.3	4.1
n-Butane	−40	19.7	19.4	−1.5
Carbon disulfide	20	32.3	34.3	6.2
Carbon tetrachloride	35	25.2	25.9	2.8
Chlorobenzene	20	33.6	34.9	3.9
Diethylamine	25	16.3	15.7	−3.9
Ethyl acetate	20	24.0	25.0	4.2
Ethyl alcohol	20	22.4	23.9	6.7
Ethylamine	25	19.2	18.3	−4.7
Formamide	0	59.6	55.1	−7.6
Glycerol	20	63.4	64.8	2.2
Heptaldehyde	20	26.6	29.2	9.7
Heptane	20	20.1	20.9	4.0
2-Hexanol	25	24.3	25.5	4.9
Methanol	20	22.6	23.3	3.0
Methylamine	−20	23.0	22.6	−1.7
Phenol	40	39.3	35.8	−8.9
n-Propyl alcohol	20	23.7	24.6	3.8
Propylamine	19	22.0	22.2	0.9
Tetrachlorodifluoroethane	30	22.7	28.7	25.9
Tetrachloroethylene	20	31.7	31.8	+0.3
Trimethylamine	−32	20.0	21.0	5.0

Average error = 5.1%
Maximum error = 25.9%

a. From Quayle [7] and Reid, Prausnitz and Sherwood [8]. *(Reprinted with permission from McGraw-Hill Book Co.)*

TABLE 20-6

Temperature Dependence of Measured and Calculated Surface Tensions

Compound	t (°C)	σ_t (dynes/cm)		Error (%)
		Measured	Calculated	
Acetone	25	24.0	24.0	0
	35	22.3	22.5	0.8
	45	21.2	21.0	−0.9
	56.6 (t_b)	19.5	19.3	−1.0
Glycerol	20	63.4	64.8	2.2
	90	58.6	56.6	−3.4
	150	51.9	49.5	−4.6
	291 (t_b)	40.0	33.1	−17.3
Carbon tetrachloride	15	27.7	29.1	5.0
	35	25.2	25.9	2.8
	55	22.8	23.5	3.1
	75	20.3	20.7	2.0
	95	17.9	18.0	0.6

for glycerol, the correlation is satisfactory for temperatures up to the normal boiling point; even with glycerol, reasonable results are obtained over at least a range of 100°C. This equation is not recommended if T is above the normal boiling temperature.

Basic Steps

(1) Draw the structure of the compound.

(2) Calculate the parachor (P) as in method 1.

(3) Estimate the molar volume (V_b) using Table 19-5 or 12-12 (Schroeder method, Table 19-5, preferred).

(4) Obtain k and n from Table 20-4.

(5) Obtain the boiling point, T_b (K), from the literature (see Appendix A) or estimate via one of the methods in Chapter 12.

(6) Calculate T/T_b. (T = system temperature in K.)

(7) Substitute these values in Eq. 20-7 and calculate the surface tension, σ, in dynes/cm.

Example 20-3   Calculate the surface tension of aniline at 20°C. (The measured value [8] is 42.7 dynes/cm.)

(1)   The structure is

(2)   From Table 20-2, P = 235.1 (see Example 20-1).

(3)   Determine V_b from Table 19-4:

$$
\begin{array}{llr}
6C & = 6(7) & = \ \ 42 \\
7H & = 7(7) & = \ \ 49 \\
1N & & = \ \ \ \ 7 \\
1\ \text{6-membered ring} & & = - 7 \\
3\ \text{double bonds} = 3(7) & & = \ \ \underline{21} \\
& V_b & = 112\ \text{cm}^3/\text{mol}
\end{array}
$$

(4)   From Table 20-4,

$$k = 0.065$$
$$n = 0.25$$

(5)   From the literature,

$$T_b = 457\text{K}$$

(6) $T/T_b = (20+273)/457 = 0.641$

(7)   Substitute the above values in Eq. 20-7:

$$
\sigma = \left\{ \frac{235.1(1+0.065)}{112} \times [3-(2 \times 0.641)]^{0.25} \right\}^4
$$

$$= (2.559)^4$$

$$= 42.9\ \text{dynes/cm, which deviates } +0.5\% \text{ from the measured value.}$$

Example 20-4   Estimate the surface tension of ethyl acetate at 20°C. (Measured $\sigma = 24.0$ dynes/cm [8].)

(1)   The structure is $CH_3 \overset{\overset{\displaystyle O}{\|}}{C}-O-CH_2 CH_3$

(2)   From Table 20-2,

$$P = 214.8 \text{ (see Example 20-2).}$$

(3)   Determine V_b from Table 19-4:

$$
\begin{array}{llr}
4C = 4(7) & = 28 \\
8H = 8(7) & = 56
\end{array}
$$

$$2O = 2(7) \quad = \quad 14$$
$$1 \text{ double bond} = \quad \underline{\quad 7 \quad}$$
$$V_b \quad = 105 \text{ cm}^3/\text{mol}$$

(4) From Table 20-4,

$$k = 0$$
$$n = 0.31$$

(5) From the literature,

$$T_b = 350K$$

(6) $T/T_b = 293/350 = 0.837$

(7) Substituting in Eq. 20-7:

$$\sigma = \left\{ \frac{214.8\,(1+0)}{105} \, [3-(2\times0.837)]^{0.31} \right\}^4$$

$$= (2.233)^4$$

$$= 24.9 \text{ dynes/cm, which deviates } +3.8\% \text{ from the measured value.}$$

20-5 SYMBOLS USED

A   = constant in equation for Mayer's method (Table 20-1) = 0.43 at 25°C

ΔH_v = heat of vaporization (cal/mol)

k   = constant in Eq. 20-7

M   = molecular weight (g/mol)

N_L = Loschmidt's number ($2.687 \times 10^{19}/\text{cm}^3$) in equation for Mayer's method (Table 20-1)

n   = constant in Eq. 20-7

P   = parachor

T   = temperature (K)

T_b = normal boiling temperature (K)

t   = temperature (°C)

V   = molar volume (cm³/mol)

V_b = molar volume at the normal boiling temperature (cm³/mol)

V_b' = adjusted molar volume at the normal boiling temperature in Eq. 20-6 (cm³/mol)

Greek

ρ_L = liquid density (g/cm³)
ρ_v = vapor density (g/cm³)
σ = surface tension (dynes/cm)

Subscripts

b   = boiling point
v   = vapor, vaporization

20-6   AVAILABLE DATA

Surface tension data can be obtained from the general compilations listed in Appendix A. Data on many materials are also available in Refs. 7 and 8.

20-7   REFERENCES

1. Gambill, W.R., "Surface Tension for Pure Liquids," *Chem. Eng. (NY)*, **65**(7), 146-50 (1958).

2. Gambill, W.R., "How Temperature and Composition Affect Surface and Interfacial Tensions," *Chem. Eng. (NY)*, **65**(9), 143 (1958).

3. LeBas, G., *The Molecular Volumes of Liquid Chemical Compounds*, Longmans, Green, New York (1915).

4. Mayer, U., "Relationship Between Vapor Pressure, Enthalpy of Vaporization and Surface Tension of Aprotic Solvents," *Monatsh. Chem.*, **110**, 191-99 (1979).

5. MacLeod, D.B., "On a Relation Between Surface Tension and Density," *Trans. Faraday Soc.*, **19**, 38-42 (1923).

6. Partington, J.R., *An Advanced Treatise on Physical Chemistry*, Vol. II, John Wiley & Sons, New York (1951).

7. Quayle, O.R., "The Parachors of Organic Compounds," *Chem. Rev.*, **53**, 439-589 (1953).

8. Reid, R.C., J.M. Prausnitz and T.K. Sherwood, *The Properties of Gases and Liquids*, 3rd ed., McGraw-Hill Book Co., New York (1977).

9. Sugden, S., "The Influence of the Orientation of Surface Molecules on the Surface Tension of Pure Liquids," *J. Chem. Soc.*, **125**, 1167-89 (1924).

10. Walden, P., "Coefficient of Expansion, Specific Cohesion and Molecular Complexity of Solvents," *Z. physik. Chem.*, **65**, 129-225 (1909).

11. Weast, R.C. and M.J. Astle (eds.), *Handbook of Chemistry and Physics*, 60th ed., CRC Press, Boca Raton, Fla. (1979).

21

INTERFACIAL TENSION WITH WATER

Clark F. Grain

21-1 INTRODUCTION

The interfacial tension between an organic liquid and water affects such processes as the formation of stable emulsions, the resistance to flow through orifices, and the dispersion of droplets. A measured or estimated value of the interfacial tension may be important when one attempts to determine the fate of a chemical of environmental concern or desires to remove a hazardous liquid from an aqueous environment. For example, in liquid-liquid extraction processes a solvent with a high interfacial tension is required to obtain good phase separation after mixing.

When two immiscible or partially miscible liquids are brought into contact, the interface thus formed possesses free surface energy. This surface energy is numerically equal to the interfacial tension. The magnitude of the interfacial tension is less than the larger of the surface tension values for the pure liquids, because the mutual attraction of unlike molecules at the interface reduces the large unbalance of forces present.

The units commonly used to express interfacial tension are the same as for surface tension, namely, dynes/cm. Interfacial tensions of organic liquids with water range from zero (for completely miscible liquids) up to the surface tension of water, which is 72 dynes/cm at 25°C.

21-2 AVAILABLE METHODS

Table 21-1 summarizes some basic information on two recommended methods for estimating this property. Method 1 requires as input information the mutual solubilities of the organic liquid and water. Method 2 requires the mutual solubility data and also the surface tensions of the pure components. The surface tension of pure water is well documented over an extensive temperature range; values of this parameter are provided in Table 21-4. The other input data can be obtained from the literature or estimated using the methods outlined in this handbook.

TABLE 21-1

Overview of Recommended Estimation Methods

Method	Information Required[a]	Average Error[b]
1 — Donahue and Bartell [2]	X_o, X_w	15.4%
2 — Antonov [1]	X_o^w, X_w^w,	11.8%
	σ_o, σ_w,	
	V_o, V_w	

a. X_o = solubility of organic phase in water (mole fraction); X_w = solubility of
water in organic phase (mole fraction); X_o^w = mole fraction of organic in water
= X_o; X_w^w = mole fraction of water remaining in water phase = $1 - X_o^w \neq X_w$;
σ = surface tension of organic (o) and water (w); V = molar volume of organic (o)
and water (w).

b. From Table 21-2.

Method 1 has an average error of 15.4% and a maximum error of 48.5%; method 2 has an average error of 11.8% and a maximum error of 38.6%.

Method 1, the simplest method, is based on a correlation first introduced by Donahue and Bartell [3]. These authors observed a linear relationship between the logarithm of the sum of the solubilities of a mutually saturated liquid with water and the interfacial tension which exists when these phases are in contact. Their correlation can be expressed analytically as

$$\sigma_{ow} = a - b \ln (X_o + X_w) \qquad (21\text{-}1)$$

where

$$
\begin{array}{rcl}
\sigma_{ow} &=& \text{interfacial tension (dynes/cm)} \\
a &=& -3.33 \\
b &=& 7.21 \\
X_o &=& \text{mole fraction of organic phase in water} \\
X_w &=& \text{mole fraction of water in organic phase}
\end{array}
$$

The correlation is strictly valid at 25°C only. However, for moderate departures from this temperature (and if solubility data are available), Eq. 21-1 should be reasonably valid without revised values of a and b.

The second method is based upon an empirical correlation reported by Antonov [1] and known as Antonov's rule:

$$\sigma_{ow} = \left| \sigma_{os} - \sigma_{ws} \right| \qquad (21\text{-}2)$$

where σ_{os} and σ_{ws} are the mixture surface tensions of the mutually saturated water and organic phases, respectively. In general, the surface tensions of pure components should not be used. A method of estimating the saturated phase surface tension is described in § 21-4.

Equation 21-2 has been criticized by Donahue and Bartell [3] and by Good [7]. However, it is generally applicable and reasonably accurate as long as saturated phase surface tensions are used.

The estimation method with the strongest theoretical base has been described by Girifolco and Good [6] and, more recently, by Good [7]. It is based upon some of the ideas put forth in Hildebrand's [8] treatment of the solubility of nonelectrolytes and the Berthelot [2] "geometric mean" hypothesis for the attractive constant in the Van der Waals equation. The method is expressed by the following equation:

$$\sigma_{ow} = \sigma_o + \sigma_w - 2\phi \sqrt{\sigma_o \cdot \sigma_w} \qquad (21\text{-}3)$$

where the parameter ϕ is related to the similarity of the cohesive forces across the liquid interface. Values of ϕ approaching or exceeding 1 suggest similar forces, and values less than 1 suggest increasingly disparate forces.

Equation 21-3 was originally proposed for estimating interfacial tensions using pure component surface tensions only. Unfortunately, reliable values of ϕ have been calculated only for hydrocarbon-water systems; when applied to hydrogen-bonded systems such as alcohol-water, calculated values of ϕ do not give acceptably accurate results. Even when the effect of mutual solubility is taken into account, errors in the estimated interfacial tension are too high. Therefore, the method of Eq. 21-3 is not recommended despite its sound theoretical base, and the rest of this chapter will be devoted to the two methods described earlier.

Further information with respect to calculating and/or estimating interfacial tension is given in reviews by Fowkes [4] and Gambill [5].

21-3 METHOD 1

If experimental mutual solubility data are available, Eq. 21-1 represents the simplest and fastest method for estimating interfacial tension. The same calculation can be done graphically with Figure 21-1 if desired. Table 21-2 lists the results obtained with 19 systems representing a diverse group of chemical classes.

Every effort should be made to obtain reliable solubility data. If no experimental values are available, the methods of Chapters 2, 3, and 11 can be used to estimate them.

Basic Steps

(1) Obtain the mutual solubilities, in mole fraction, of the organic phase in water (X_o) and of water in the organic phase (X_w). If experimental data are not available, the methods of Chapters 2, 3, and 11 may be used to obtain an estimate.

(2) Calculate $\ln (X_o + X_w)$.

(3) Calculate the interfacial tension (σ_{ow}) by Eq. 21-1, or read off the value from Fig. 21-1. The result is in dynes/cm.

Example 21-1 Estimate the interfacial tension of the system methyl ethyl ketone (MEK) and water at 25°C.

(1) From Table 21-2, the solubility of MEK in water at 25°C is 5.6×10^{-2} mole fraction. The solubility of water in MEK at 25°C is 0.35 mole fraction.

(2) $X_o + X_w \quad = 0.056 + 0.35 = 0.41$

   $\ln (X_o + X_w) = \ln (0.41) = -0.89$

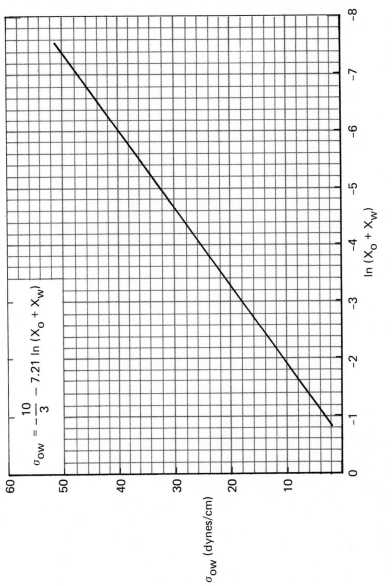

$$\sigma_{ow} = -\frac{10}{3} - 7.21 \ln (X_o + X_w)$$

σ_{ow} (dynes/cm)

$\ln (X_o + X_w)$

FIGURE 21—1   Graphical Estimation of Interfacial Tension (Method 1)

TABLE 21-2

Estimated vs Experimental Interfacial Tensions with Water

Compound	Temperature (°C)	Solubility (mole fraction) X_o	X_w	σ_o (dynes/cm)	Calculated σ_{ow} Method 1	Method 2	Experimental σ_{ow}
Aniline	20	7.3×10^{-3}	2.2×10^{-1}	42.9	7.5	3.7	5.8
Benzene	25	4.9×10^{-4}	4.4×10^{-3}	28.9	34.6	34.0	33.9
Bromobenzene	25	5.3×10^{-5}	2.9×10^{-3}	35.2	38.4	36.5	36.5
n-Butanol	25	1.7×10^{-2}	5.0×10^{-1}	25.1	1.4	2.3	1.9
Butyl acetate	20	1.0×10^{-3}	7.5×10^{-2}	25.2	15.2	8.9	14.5
Carbon tetrachloride	20	9.1×10^{-5}	8.3×10^{-4}	26.2	47.1	46.3	45.0
Cyclohexanol	25	5.0×10^{-3}	4.1×10^{-1}	33.6	3.2	3.9	3.9
Diethyl ether	20	1.8×10^{-2}	5.3×10^{-2}	17.0	15.7	9.7	10.7
Diisopropyl ether	20	1.9×10^{-3}	3.3×10^{-2}	17.3	21.9	12.8	17.9
Ethyl acetate	20	1.6×10^{-2}	1.4×10^{-1}	23.9	10.1	7.1	6.8
Heptaldehyde	20	3.4×10^{-3}	—	26.6	—	10.1	13.7
Heptylic acid	20	3.8×10^{-4}	2.7×10^{-1}	28.0	6.1	7.9	7.7
Hexane	25	2.9×10^{-5}	6.0×10^{-4}	19.1	50.1	50.3	50.0
Methyl n-butyl ketone	20	5.0×10^{-3}	1.8×10^{-1}	25.0	9.1	8.9	9.7
Methyl ethyl ketone	25	5.6×10^{-2}	3.5×10^{-1}	24.8	3.1	2.2	3.0
n-Octane	20	3.0×10^{-6}	9.0×10^{-4}	21.8	47.2	50.5	50.8
n-Octanol	20	8.9×10^{-5}	2.5×10^{-1}	27.5	6.7	8.3	8.5
Toluene	20	1.0×10^{-4}	2.5×10^{-3}	27.6	39.6	34.9	36.1
Trichloroethane	20	2.1×10^{-3}	1.9×10^{-3}	34.1	37.0	37.6	37.4
					Avg. error = 15.4%	11.8%	
					Max. error = 48.5%	38.6%	

Source: Experimental values from Refs. 3, 6, and 7.

(3)   From Eq. 21-1:

$$\sigma_{ow} = -3.33 - 7.21\,(-0.89)$$
$$= 3.1 \text{ dynes/cm}$$

Within the limits of graphical accuracy, the same value is obtained from Figure 21-1. The experimental value is 3.0; hence, the error is +3.3%.

Example 21-2   Estimate the interfacial tension of benzene and water, assuming that experimental mutual solubility data are not available.

(1)   From Chapter 11, the calculated infinite dilution activity coefficient, γ^{∞}, for benzene in water is 2094; for water in benzene, $\gamma^{\infty} = 329$. From Chapter 3 (Eq. 3-11), the solubility of benzene in water is given by $X_o = 1/\gamma^{\infty} = 1/2046 = 4.8 \times 10^{-4}$. From Figure 3-3 of Chapter 3, the solubility of water in benzene is estimated as $X_w = 3.3 \times 10^{-3}$.

(2) $X_o + X_w \quad = 4.8 \times 10^{-4} + 3.3 \times 10^{-3} = 3.8 \times 10^{-3}$

$\ln\,(X_o + X_w) = -5.6$

(3)   From Eq. 21-1:

$$\sigma_{ow} = -10/3 - 7.21\,(-5.6)$$
$$= 37 \text{ dynes/cm}$$

A similar value is obtained from Figure 21-1. The experimental value is 33.9 dynes/cm, indicating an error of +8.8%. (The use of measured values of X_o and X_w would have yielded an estimated σ_w of 34.6 dynes/cm, as shown in Table 21-2.)

21-4   METHOD 2

The second method utilizes Eq. 21-2. Proper interpretation of Antonov's rule requires that only saturated phase surface tensions be used. This requirement can be relaxed somewhat with respect to the organic phase saturated with water. Figure 21-2 is a typical plot of saturated phase or mixture surface tension as a function of concentration for the system *n*-butanol-water. Notice that an X_o of only 0.01 mole fraction reduces the surface tension of the solution by a factor of ≈ 2. However, for $X_o > 0.03$ ($X_w < 0.97$), the mixture surface tension is nearly as low as that of pure *n*-butanol. Thus, the substitution of σ_o for σ_{ws} in Eq. 21-2 should result in relatively minor errors.

Methods of estimating mixture surface tensions have been reviewed in Ref. 9. The most versatile method was first proposed by Tamura *et*

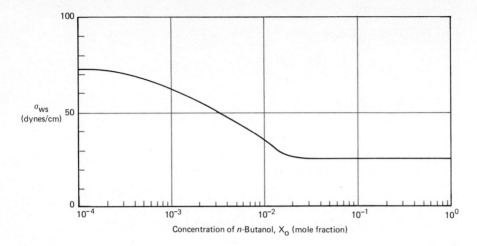

Concentration of n-Butanol, X_o (mole fraction)

Source: Meissner, H.P. and A.S. Michaels, "Surface Tensions of Pure Liquids and Liquid Mixtures," *Ind. Eng. Chem.*, **41**, 2782-87 (1949).

FIGURE 21—2   Mixture Surface Tension of n-Butanol — Water

al. [10]. The basic equation is the Macleod-Sugden correlation[1] as applied to mixtures. For a binary mixture and neglecting vapor densities,

$$\sigma_m^{1/4} = \frac{\rho_{Lm}}{M_m} (P_o X_o^w + P_w X_w^w) \tag{21-4}$$

where

σ_m =   surface tension of mixture
ρ_{Lm} =   density of mixture
M_m =   molecular weight of mixture
P_o, P_w =   parachors for organic component and water
X_o^w, X_w^w =   mole fractions of organic component and water

The significant concentrations and densities were taken to be those characteristic of the surface layer. Thus, the mixture density is replaced by a hypothetical liquid molar volume $(V^\sigma)^{-1}$. V^σ is estimated from

$$V^\sigma = X_o^\sigma V_o + X_w^\sigma V_w \tag{21-5}$$

where

X_o^σ, X_w^σ =   mole fractions of the organic component and water in the surface layer

---

1. The Macleod-Sugden correlation is discussed in some detail in Chapter 20.

V_o, V_w =   pure component molar volumes

Using the definition of the parachor, Tamura *et al.* rewrote Eq. 21-4 as

$$\sigma_m^{1/4} = \psi_w^\sigma \, \sigma_w^{1/4} + \psi_o^\sigma \, \sigma_o^{1/4} \tag{21-6}$$

where $\psi_w^\sigma = X_w^\sigma V_w / V^\sigma$ is the hypothetical volume fraction of water in the surface layer. Similarly, for the organic phase, $\psi_o^\sigma = X_o^\sigma V_o / V^\sigma$. By assuming an equilibrium between the surface and bulk phases, the authors were able to estimate ψ_w^σ and ψ_o^σ. The final equation is

$$\log \frac{\left(\psi_w^\sigma\right)^p}{\psi_o^\sigma} = \log \frac{\left(\psi_w\right)^p}{\psi_o} + \frac{0.441 \, q}{T} \left(\frac{\sigma_o \, V_o^{2/3}}{q} - \sigma_w \, V_w^{2/3}\right) \tag{21-7}$$

where:

ψ_w = bulk volume fraction of water = $X_w^w V_w / (X_o^w V_o + X_w^w V_w)$
ψ_o = bulk volume fraction of organic material = $X_o^w V_o /$
          $(X_o^w V_o + X_w^w V_w)$
σ_o, σ_w = pure component surface tensions of organic phase and
          water
T = temperature (K)
p, q = constants, characteristic of the organic compound

and

$$\psi_o^\sigma + \psi_w^\sigma = 1 \tag{21-8}$$

$$X_o^w + X_w^w = 1 \tag{21-9}$$

When Eq. 21-7 is solved for ψ_w^σ and ψ_o^σ and these values are substituted in Eq. 21-6, the resulting estimates of σ_m (the mixture surface tension) are quite good for hydrocarbons but very poor for polar and hydrogen-bonded materials. For the compounds listed in Table 21-2, interfacial tensions estimated from these values of σ_m deviate from measured values by an average of only 2% for the first group but 260% for the second. This problem is related to the value assigned to p, which was identical to q in Tamura's original paper. We have found, however, that if p is arbitrarily set equal to 1.0, good estimates of σ_m are obtained for all classes of materials, and the subsequent estimates of interfacial tension have average errors of 11.8%.

Table 21-3 lists values of q for various compound classes. Once σ_m for the organic component in water is determined, it can be substituted directly for σ_{os} in Eq. 21-2. Also, as shown earlier, σ_{ws} is very nearly equal to σ_o. Table 21-2, which compares calculated and experimental values of σ_{ow}, shows that method 2 is substantially more accurate than method 1.

TABLE 21-3

Values of q for Equation 21-7

Compound Class	$q^{a,b}$
Acids, alcohols, esters, primary amines	4N/5
Hydrocarbons (aliphatic and aromatic), ethers	2N/3
Ketones and aldehydes	1.833 + N/3
Perhalogenated aliphatic hydro-carbons	N
Monohalogenated aromatics	2N/3

a. N = number of carbon atoms.

b. The values of q listed were determined by the author of this chapter and differ from those given by Tamura *et al.* [10]. However, these values appear to fit the experimental data and also extend the method to more classes of materials.

Basic Steps

(1) Draw structure of organic component. This will assist in the determination of q in Step 5.

(2) Obtain the solubility of the organic component in water (X_o^w) in mole fraction. If solubility data are not available, they may be estimated using methods outlined in Chapters 2, 3, and 11.

(3) Obtain the pure component surface tension of the organic phase (σ_o). (If a measured value is not available, see estimation methods in Chapter 20.) Also obtain the surface tension of water (σ_w), which is listed as a function of temperature in numerous handbooks. An abbreviated list is presented in Table 21-4.

TABLE 21-4

Surface Tension of Water at Various Temperatures

Temperature (°C)	σ_w (dynes/cm)
5	74.9
10	74.2
15	73.5
20	72.8
25	72.0
30	71.2
40	69.6

Source: Weast [11]. *(Reprinted with permission from the Chemical Rubber Co., CRC Press, Inc.)*

(4) Obtain values of the molar volume (V_o and V_w) in cm^3. This can be easily computed from available density data.[2] Again, data for water are available for a range of temperatures. Over the temperature range of 5-40°C, however, the density of water varies by only 0.008 g/cm^3 (1.00 g/cm^3 to 0.992 g/cm^3). If density or volume data for the organic component are not available, they may be estimated using the methods of Chapter 19.

(5) Obtain q from Table 21-3.

(6) Substitute these values in Eq. 21-7. From the definition of ψ_w and ψ_o, $\psi_w/\psi_o = X_w^w V_w/X_o^w V_o$. Note that $X_w^w = 1 - X_o^w$.

(7) Calculate ψ_w^σ from Eq. 21-7, with p = 1. Note that $\psi_o^\sigma = 1 - \psi_w^\sigma$.

(8) Calculate σ_m from Eq. 21-6 and set equal to σ_{os} in Eq. 21-2. Assume that $\sigma_{ws} = \sigma_o$.

(9) Calculate σ_{ow} using Eq. 21-2.

Example 21-3   Estimate the interfacial tension between methyl ethyl ketone (MEK) and water at 25°C. The molecular weights are 72.10 and 18.02, respectively.

(1)   The structure of MEK is $CH_3-\overset{\overset{O}{\|}}{C}-C_2H_5$; hence N = 4.

(2)   The solubility of MEK in water (X_o^w) is 5.6 × 10^{-2} mole fraction [7].

---

2. V (cm^3) = molecular weight (g)/density (g/cm^3).

(3) At 25°C, the surface tension of MEK (σ_o) is 24.8 dynes/cm [7], and the surface tension of water (σ_w) is 72.0 dynes/cm.

(4) At 25°C, the density of water (ρ_w) is 0.997 g/cm^3; hence, its molar volume (V_w) is 18.1 cm^3. Similarly, for MEK, $\rho_o = 0.81$ g/cm^3 [11] and $V_o = 89.0$ cm^3.

(5) From Table 21-3, q = 1.833 + (4/3) = 3.17

(6) $$\frac{\psi_w}{\psi_o} = \frac{X_w^w V_w}{X_o^w V_o} = \frac{0.944 \times 18.1}{0.056 \times 89.0} = 3.43$$

Substituting in Eq. 21-7,

$$\log \frac{\psi_w^\sigma}{\psi_o^\sigma} = \log (3.43) + \frac{0.441 \times 3.17}{298} \left[\frac{24.8\,(89.0)^{2/3}}{3.17} - 72\,(18.1)^{2/3} \right]$$

$$= 0.535 - 1.595$$

$$= -1.06$$

(7) $$\frac{\psi_w^\sigma}{\psi_o^\sigma} = \frac{\psi_w^\sigma}{1 - \psi_w^\sigma} = \text{antilog}\,(-1.06) = 0.087$$

Solving for ψ_w^σ,

$$\psi_w^\sigma = 0.080$$

$$\psi_o^\sigma = 1 - \psi_w^\sigma = 0.920$$

(8) From Eq. 21-6,

$$\sigma_m^{1/4} = 0.080\,(72.0)^{1/4} + 0.920\,(24.8)^{1/4}$$

$$= 2.28$$

$$\sigma_m = (2.28)^4 = 27.0 = \sigma_{os}$$

$$\sigma_{ws} = \sigma_o = 24.8$$

(9) From Eq. 21-2,

$$\sigma_{ow} = \left| 27.0 - 24.8 \right| = 2.2$$

The experimental value is 3.0; hence, the error is −27%.

Example 21-4 Estimate the interfacial tension between benzene and water at 25°C using estimated values of the input parameters for the organic component.

(1)   The structure of benzene is: ⬡

(2)   The solubility of benzene in water (X_o^w) was estimated in Example 21-2 as 4.8×10^{-4} mole fraction.

(3)   The estimated surface tension of benzene (σ_o) is 27 dynes/cm (see Chap. 20, method 2). For water, $\sigma_w = 72$ dynes/cm.

(4)   Using method 2 of Chap. 20, the molar volume at the normal boiling point, V_b, is estimated at 98 cm^3. By the Grain method outlined in Chapter 19, the molar volume of benzene (V_o) at 298K is 90.8 cm^3. V_w is, as in Example 21-3, 18.1 cm^3.

(5)   From Table 21-3, q = 2(6)/3 = 4.

(6) $X_w^w = 1 - X_o^w = 0.9995$. Therefore,

$$\frac{\psi_w}{\psi_o} = \frac{0.9995 \times 18.1}{4.8 \times 10^{-4} \times 90.8} = 415$$

Substituting in Eq. 21-7,

$$\log \frac{\psi_w^\sigma}{\psi_o^\sigma} = \log(415) + \frac{0.441\,(4)}{298} \left[\frac{27\,(90.8)^{2/3}}{4} - 72\,(18.1)^{2/3} \right]$$

$$= 2.62 - 2.13$$

$$= 0.49$$

(7) $\psi_o^\sigma = 1 - \psi_w^\sigma$. Thus,

$$\frac{\psi_w^\sigma}{1 - \psi_w^\sigma} = \text{antilog}\,(0.49) = 3.1$$

Solving for ψ_w^σ,

$$\psi_w^\sigma = 0.76$$

$$\psi_o^\sigma = 1 - 0.76 = 0.24$$

(8)   From Eq. 21-6,

$$\sigma_m^{1/4} = 0.76\,(72)^{1/4} + 0.24\,(27)^{1/4}$$

$$= 2.8$$

$$\sigma_m = (2.8)^4 = 61 = \sigma_{os}$$

$$\sigma_{ws} = \sigma_o = 27$$

(9)   From Eq. 21-2,

$$\sigma_{ow} = \left| 61 - 27 \right| = 34$$

This agrees with the experimental value of 33.9 within the limits of accuracy of the calculation.

21-5   AVAILABLE DATA

Limited data on the interfacial tension between organic liquids and water can be obtained from the references listed in Appendix A. Somewhat more extensive compilations are given in Refs. 3, 6, and 7.

21-6   SYMBOLS USED

a	=	constant in Eq. 21-1
b	=	constant in Eq. 21-1
M	=	molecular weight (g/mol)
N	=	number of carbon atoms (Table 21-3)
P	=	parachor
p	=	parameter in Eq. 21-7
q	=	parameter in Eq. 21-7
T	=	temperature (K)
V	=	molar volume (cm^3)
V_b	=	molar volume at the normal boiling point (cm^3)
V^σ	=	hypothetical surface molar volume
X_o	=	solubility of organic phase in water (mole fraction)
X_w	=	solubility of water in the organic phase (mole fraction)
X_o^σ	=	mole fraction of the organic phase in the surface layer
X_w^σ	=	mole fraction of water in the surface layer
X_o^w	=	mole fraction of organic component in the water phase
X_w^w	=	mole fraction of water component in the water phase = $1 - X_o^w$

Greek

ϕ	=	parameter in Eq. 21-3
ψ	=	bulk volume fraction
ψ^σ	=	volume fraction in surface layer

ρ_{Lm} = mixture liquid density (g/cm³)

σ_m = surface tension of mixture (dynes/cm)

σ_o = pure component surface tension of organic phase (dynes/cm)

σ_{os} = surface tension of water saturated with organic phase (dynes/cm)

σ_{ow} = interfacial tension (dynes/cm)

σ_w = pure component surface tension of water (dynes/cm)

σ_{ws} = surface tension of organic phase saturated with water (dynes/cm)

Subscripts

m = mixture
o = organic phase
w = water

Superscripts

σ = surface layer
w = water phase

21-7   REFERENCES

1. Antonov, G.N., "Tension at the Limit of Two Layers," *J. Russ. Phys. Chem. Soc.*, **93**, 342 (1907).

2. Berthelot, D., *Compt. Rend.*, **126**, 1703, 1857 (1898).

3. Donahue, J.D. and F.E. Bartell, "The Boundary Tension at Water-Organic Liquid Interfaces," *J. Phys. Chem.*, **56**, 480 (1952).

4. Fowkes, F.M., "Attractive Forces at Interfaces," *Ind. Eng. Chem.*, **56**, 40 (1964).

5. Gambill, W.R., "How Temperature and Composition Affect Surface and Interfacial Tensions," *Chem. Eng. (NY)*, **64**, 143 (1958).

6. Girifolco, L.A. and R.J. Good, "A Theory for the Estimation of Surface and Interfacial Energies: I. Derivation and Applications to Interfacial Tension," *J. Phys. Chem.*, **61**, 904 (1957).

7. Good, R.J., "Generalization of Theory for Estimation of Interfacial Energies," *Ind. Eng. Chem.*, **62**, 55 (1970).

8. Hildebrand, J.H. and R.L. Scott, *Solubility of Nonelectrolytes*, 3rd ed., Rheinhold, New York (1950).

9. Reid, R.C., J.M. Prausnitz and T.K. Sherwood, *The Properties of Gases and Liquids*, 3rd ed., McGraw-Hill Book Co., New York (1977).

10. Tamura, M., M. Kurata and H. Odani, "Surface Tension of Solutions of Chain Molecules: II. A Practical Method for Estimating the Surface Tension," *Busseiron Kenkyu* (Researches on Chem. Phys.), **57**, 1 (1952).

11. Weast, R.C. (ed.), *CRC Handbook of Chemistry and Physics,* 59th ed., CRC Press, West Palm Beach, Fla. (1978-79).

22

LIQUID VISCOSITY

Clark F. Grain

22-1 INTRODUCTION

The viscosity of a liquid is a measure of the forces that work against movement or flow when a shearing stress is applied. It has an important bearing on several problems relating to the transfer or movement of bulk quantities of the liquid. For example, a knowledge of the viscosity is required in formulas relating to the pumpability of a liquid, the rate of flow (e.g., from a tank), or spreading (e.g., on water) of a chemical spill. However, as the solutions to these formulas are not very sensitive to variations in viscosity, the focus of this chapter is on estimation methods that yield reasonable accuracy from a minimum of input data. More accurate methods have been developed, but the input data they generally require are available for relatively few chemicals.

Viscosity is commonly reported in units of centipoise (cp)[1] and is represented by the symbol η (subscript L for liquids). Values of η_L for organic liquids generally range from 0.3 to 20 cp at ambient temperatures. Water has a viscosity of 1 cp at 20°C.

Gas viscosity is rather well understood: the viscous force is related to the transfer of momentum as the gas molecules collide and thus is based on the kinetic theory. The theory of liquid viscosity, on the other hand, is

_____

1. 1 cp = 0.01 poise = 0.01 g/s·cm.

not yet firmly established. The molecules in a liquid are held together much more strongly than in a gas. Here, viscosity may be thought of as a measure of the force needed to overcome the mutual attraction of the molecules so that they can be displaced relative to each other. The more strongly the molecules are held together, the smaller the flow for a given shearing stress. With increasing temperature, the random kinetic energy of the molecules helps to overcome the intermolecular forces and reduces the viscosity.

The theories that have developed around this main theme include the following:

• Van der Waals [20] and Andrade [1] assumed that the viscous force is due to transfer of momentum, which is strongly influenced by molecular forces;

• Prandtl [16] and Taylor [18] assumed the viscous force to be molecular in nature but computed it from a mechanical analogy;

• Eyring [5] also considered the force to be molecular in nature but computed it on the basis of chemical kinetics.

Thorough reviews of the various theories have been given by Ge-Mant [9] and Brush [3]. In all of the theories mentioned, the temperature dependence of liquid viscosity, η_L, is described by an equation of the form

$$\eta_L = Ae^{B/T} \tag{22-1}$$

This general form has been verified for a wide variety of liquids and forms the basis for most estimation techniques [17].

In recent years Hildebrand [10] has argued that liquid viscosity is not a direct function of temperature and instead is dependent upon the molar volume. He modified an equation due to Bachinski [2] such that

$$\frac{1}{\eta_L} = \phi = B_2 \left(\frac{V - V_o}{V_o} \right) \tag{22-2}$$

where ϕ is the fluidity, V the molar volume, V_o the molar volume when $\phi = 0$, and B_2 a constant. Equation 22-2 has been tested with several nonpolar liquids and has been found to be adequate [4,11].

For estimation purposes, however, Eq. 22-2 is somewhat limited in that it applies only to nonpolar liquids and requires measured or estimated data on molar volume as a function of temperature. Furthermore, the constants B_2 and V_o can be reliably estimated for straight-chain hydrocarbons only.

22-2 AVAILABLE ESTIMATION METHODS

This chapter is limited to a consideration of estimation methods for pure substances below their boiling points. Estimation methods for liquid mixtures are discussed by Reid *et al.* [17]. In considering estimation methods, we have excluded those that require extensive experimental data. Most of these are mere curve-fitting schemes designed to reproduce the viscosity/temperature relationship accurately. Quite often, extrapolation of the analytical expression outside the range of the experimental data base leads to erroneous results. Furthermore, these empirical equations can only be used for a limited number of liquids. Partington [15] and Brush [3] list several of these equations.

Where adjustable parameters are required, we have featured the methods that allow one to estimate the parameters either from structure alone or from some readily available physical property.

Table 22-1 lists the methods considered and brief comments about their limitations. In the following section the recommended methods are analyzed with respect to ease of use and reliability.

22-3 SELECTION OF APPROPRIATE METHOD

A good estimation method should be easy to use, require a minimum amount of input data, and be reasonably accurate. The three methods recommended in this section satisfy these criteria. The first is strictly graphical and requires a knowledge of the viscosity, η_L, at some temperature, T. A means of estimating viscosity at the normal boiling point, T_b, is given.

The other two methods are simply ways of estimating the constants in Eq. 22-1 using either a knowledge of the molecular structure or a knowledge of T_b and the heat of vaporization at that temperature.

Many other methods exist and have been reviewed by Gambill [7] and by Reid, Prausnitz and Sherwood [17]. However, all require a knowledge of the variation of density with temperature and/or the critical constants, T_c and V_c.

TABLE 22-1

Methods for Estimating Liquid Viscosity

Method	Formula[b]	Input[b]	Range	Remarks
Method 1 Lewis and Squires [12] [a]	Graphical	Viscosity at one temp.	0-2 cp	Approximate only; limited range
Method 2 van Velzen et al. [21] [a]	$\log \eta_L = B_3 \left(\dfrac{1}{T} - \dfrac{1}{T_o} \right)$	Structure	0-15 cp	S compounds cannot be treated; B_3 and T_o estimated from structure
Method 3 Grain[a]	$\ln \left(\dfrac{\eta_L}{\eta_{Lb}} \right) = \dfrac{(\Delta H_{vb} - RT)}{n} \left(\dfrac{1}{T} - \dfrac{1}{T_b} \right)$	T_b	0-15 cp	η_{Lb}, ΔH_{vb}, and n estimated from structure
Orrick and Erbar [14]	$\ln \left(\dfrac{\eta_L}{\rho_L M} \right) = A_1 + B_1/T$	ρ_L, A_1, B_1	0-15 cp	N, S compounds cannot be treated; A_1 and B_1 estimated by group contributions
Thomas [19]	$\log \left(8.569 \, \dfrac{\eta_L}{\rho_L^{1/2}} \right) = \theta \left(\dfrac{1}{T_r} - 1 \right)$	ρ_L, θ, T_c	0-15 cp	N, S compounds cannot be treated; θ estimated by group contributions
Hildebrand [10]	$\dfrac{1}{\eta_L} = B_2 \left(\dfrac{V - V_o}{V_o} \right)$	V, V_o, B_2	0-15 cp	Limited to nonpolar compounds, V_o and B_2 estimated from structure
Marris [13]	$\log \left(\dfrac{\eta_L}{\eta_+} \right) = J \left(\dfrac{1}{T_r} - 1 \right)$	T_c, η_+, J	0-15 cp	η_+ and J estimated from group contributions; S compounds cannot be treated

a. Methods recommended in this chapter.
b. See § 22-7 for definition of symbols.

The recommended methods are described in the three sections that follow. The first and the third are the most general and can be applied to almost any organic liquid. While both require information on the normal boiling point and the viscosity at that temperature, these properties can be estimated if they are not known.

Method 2 requires no experimental data, only a knowledge of the molecular structure; thus, it would appear to be the most attractive of the three. However, it is not applicable to sulfur or phosphorus compounds and nitriles. Also, as noted by van Velzen *et al.* [21], the method is unsatisfactory for the first members of a homologous series.

In Table 22-2 the viscosity estimated by each of the methods is compared with experimental values for a representative list of liquids. It is apparent that method 1 is appropriate only for viscosities up to about 2 cp, a rather limited range. Estimates made outside this range are generally much too high; however, for viscosities within the range, the method offers a rapid and simple means of estimation. This table also shows the average and maximum deviations, in percent, for each of the recommended methods. Values obtained outside the range of validity of the method were excluded.

22-4   METHOD 1

Derivation. The method described here was first proposed by Lewis and Squires [12] and has been found to be reasonably accurate as long as one datum point is known. It is based on the exponential relationship between viscosity and temperature set forth in Eq. 22-1. This equation implies that a curve of the kind shown in Figure 22-1 can be used to estimate the viscosity of any chemical at any temperature, given its viscosity at some other temperature. Thus, for example, if the viscosity at 0°C is 0.7 cp, its value at 100°C is 0.2 cp. The problem, however, is obtaining the single datum; a convenient point is the viscosity at the normal boiling point, η_{Lb}. Approximate values of η_{Lb} for a wide variety of liquids (estimated by the author) are listed in Table 22-3. If the normal boiling point is not known, one of the estimation techniques described in Chapter 12 of this handbook may be used.

TABLE 22-2

Comparison of Calculated and Experimental Liquid Viscosities

Compound	Temperature (°C)	Viscosity, η_L (cp)			
		Exp.[a]	Method 1	Method 2	Method 3
Hexane	-60	0.89	1.1	0.93	0.85
Hexadecane	20	3.34	40.0	2.95	2.02
Ethanol	0	1.77	1.8	2.26	1.68
n-Butanol	40	1.77	1.8	1.77	1.65
Ethylene glycol	20	19.9	25.0	4.1	12.9
Diethylamine	25	0.35	0.3	0.35	0.31
Acetone	-90	2.10	1.7	1.67	2.64
	0	0.39	0.4	0.38	0.39
Valeric acid	20	2.30	2.2	2.20	2.30
Stearic acid	70	11.6	300.0	12.3	10.9
Chloroform	20	0.56	0.4	0.60	0.65
Butyl acetate	0	1.00	1.1	0.95	1.36
Acetonitrile	25	0.35	0.4	––	0.43
Carbon disulfide	20	0.36	0.3	––	0.26
Benzene	5	0.83	0.64	0.64	0.56
Toluene	20	0.59	0.6	0.41	0.64
Phenol	50	3.0	5.0	2.3	3.1
Aniline	- 5	13.4	30.0	6.4	11.8
	20	4.4	9.0	3.4	5.9
Benzophenone	55	4.67	20.0	5.1	2.7
Bromobenzene	15	1.20	1.5	1.04	1.46
Nitrobenzene	20	2.03	4.0	1.52	2.63
Average Error (%)			19[b]	22	19
Maximum Error (%)			49[b]	79	49

a. From Ref. 17, 21 and 22.
b. Calculated values for aniline, benzophenone, hexadecane and stearic acid were not included in these computations. Values of η_L for these compounds are outside of the applicable range for Method 1. If ethylene glycol is also excluded, the average error is still ∿ 19%.

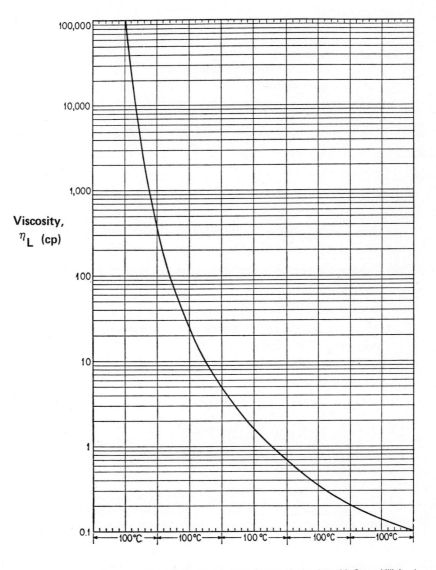

Viscosity, η_L (cp)

FIGURE 22-1 Variation of Viscosity with Temperature

TABLE 22-3

Values of η_{Lb} to be Used with Figure 22-1

Substance	η_{Lb} (cp)
Alcohols (aliphatic and aromatic)[a]	0.45
Primary amines (aliphatic and aromatic)	0.45
All other organic liquids[b]	0.2

a. Not applicable to triols.
b. Exceptions are benzene (η_{Lb} = 0.3) and cyclohexane (η_{Lb} = 0.4).

Basic Steps

(1) Estimate η_{Lb} using Table 22-3 if a measured value is not available.

(2) Determine temperature difference $\Delta t = t_b - t$, where t_b is the boiling point in °C.

(3) Locate η_{Lb} in Figure 22-1.

(4) Move Δt degrees to the left along the abscissa and locate the corresponding value of η_L.

Example 22-1   Estimate the viscosity of n-butanol at 0°C and 40°C.

(1)   Table 22-3 lists η_{Lb} = 0.45 at the boiling point of n-butanol (117.71°C).

(2)   Locate the point corresponding to a viscosity of 0.45 cp on the curve in Figure 22-1.

(3)   For 0°C, Δt = 118 − 0 = 118°C. Find the point on the curve that is 118° to the left of the point determined in (2) above. The new point corresponds to a viscosity of η_L = 4 cp. (The measured value of this viscosity is 5.1 cp, indicating a deviation of −22% for the estimated value.)

(4)   Similarly, at 40°C, Δt = 78°C and η_L = 1.8 cp. (Since the measured viscosity listed in Table 22-2 is 1.77 cp, the deviation of this estimated value is only 1.7%.)

Example 22-2   Estimate the viscosity of n-hexane at −60°C and at 0°C.

(1)   From Table 22-3, $\eta_{Lb} = 0.2$ cp.

(2)   Locate the point corresponding to a viscosity of 0.2 cp on the curve in Figure 22-1.

(3)   Since the normal boiling point of n-hexane is 69°C, $\Delta t = 129°$ for $t = -60°$C.

(4)   Find the point on the curve that is 129° to the left of the point determined in (2) above. The new point corresponds to a viscosity of $\eta_L = 1.1$ cp. (This is a deviation of +18% from the measured value of 0.89 cp listed in Table 22-2.)

(5)   For 0°C, $\Delta t = 69°$. The corresponding viscosity from Figure 22-1 is 0.45 cp. (Since the measured viscosity is 0.38 cp, the deviation is again +18%.)

22-5   METHOD 2

Derivation. The estimation method described here was introduced by van Velzen *et al.* [21]. It is based on the use of Eq. 22-1 at a temperature (T_o) at which $\eta_L = 1$, so that the equation can be rewritten as $1 = Ae^{B/T_o}$, or

$$\log \eta_L = 0 = A_3 + (B_3/T_o) \,\Big|\, \eta_L = 1 \qquad (22\text{-}3)$$

whence

$$A_3 = -B_3/T_o$$

At any other temperature T,

$$\log \eta_L = B_3 \left(\frac{1}{T} - \frac{1}{T_o} \right) \qquad (22\text{-}4)$$

Note that T and T_o are absolute temperatures.

For straight-chain hydrocarbons containing N carbon atoms, the authors found that the values of B_3 and T_o could be described by the following equations:

$$N \leqslant 20 \begin{cases} B_3 = 24.79 + 66.885\,N - 1.3173\,N^2 - 0.00377\,N^3 & (22\text{-}5) \\[2mm] T_o = 28.86 + 37.439\,N - 1.3547\,N^2 + .02076\,N^3 & (22\text{-}6) \end{cases}$$

$$N > 20 \begin{cases} B_3 = 530.59 + 13.740\,N & (22\text{-}7) \\[2mm] T_o = 8.164\,N + 238.59 & (22\text{-}8) \end{cases}$$

For molecules other than straight-chain hydrocarbons, N is replaced by the "equivalent chain length," NE; this is the chain length of a hypothetical n-alkane with viscosity equal to 1 cp at the temperature at which the viscosity of the compound in question is also 1 cp. NE is calculated as the sum of N (the total number of carbon atoms) and one or more structural and/or configurational factors:

$$NE = N + \sum_i \Delta N_i \qquad (22\text{-}9)$$

Similarly, B_3 is calculated as

$$B_3 = B_a + \Delta B \qquad (22\text{-}10)$$

where B_a is the value of B for the hypothetical alkane with equivalent chain length NE and ΔB is a correction factor that varies with the structure of the compound.

If the compound contains two or more functional groups, the ΔN correction must be multiplied by the number of groups. The ΔB correction is applied only once, regardless of the number of functional groups.

Once NE is known, B_a can be calculated from Eq. 22-5 or 22-7 and T_o is found via either Eq. 22-6 or 22-8. Table 22-4 lists the functions for ΔN and ΔB for various functional groups and structural configurations.

Basic Steps

(1) Count the number of carbon atoms (N) in the molecular formula.

(2) Determine the ΔN correction using Table 22-4.

(3) Sum N and ΔN to obtain NE.

(4) Calculate B_a by substituting NE in Eq. 22-5 or 22-7 as appropriate.

(5) Determine the ΔB correction from Table 22-4 and add this to B_a to obtain B_3.

(6) Calculate T_o by substituting the value of NE in Eq. 22-6 or 22-8 as appropriate.

(7) Use the calculated values of T_o and B_3 in Eq. 22-4 to obtain log η_L at temperature T. Taking the antilog yields η_L.

TABLE 22-4

Functions for ΔN and ΔB

FUNCTIONAL GROUPS	ΔN	ΔB
n-Alkanes		
Alkene	$-0.152 - 0.042N$	$-44.94 + 5.410NE$
Acid $3 \leqslant N \leqslant 10$	$6.795 + 0.365N$	$-249.12 + 22.449NE$
$N > 10$	10.71	$-249.12 + 22.449NE$
Ester	$4.337 - 0.230N$	$-149.13 + 18.695NE$
Primary alcohol	$10.606 - 0.276N$	$-589.44 + 70.519NE$
Secondary alcohol	$11.200 - 0.605N$	497.58
Tertiary alcohol	$11.200 - 0.605N$	928.83
Diol	Alcohol correction + configurational factor	557.77
Ketone	$3.265 - 0.122N$	$-117.21 + 15.781NE$
Ether	$0.298 + 0.209N$	$-9.39 + 2.848NE$
Primary amine	$3.581 + 0.325N$	$25.39 + 8.744NE$
Secondary amine	$1.390 + 0.461N$	$25.39 + 8.744NE$
Tertiary amine[a]	3.27	$25.39 + 8.744NE$
Fluoride	1.43	5.75
Chloride	3.21	-17.03
Bromide	4.39	$-101.97 + 5.954NE$
Iodide	5.76	-85.32
Aromatic and 1-nitro	$7.812 - 0.236N$	$-213.14 + 18.330NE$
2-Nitro	5.84	$-213.14 + 18.330NE$
3-Nitro	5.56	$-338.01 + 25.086NE$
4-,5-Nitro	5.36	$-338.01 + 25.086NE$

CONFIGURATIONAL FACTORS
Correction for Aromatic Nucleus

	ΔN	ΔB
Alkyl-, halogen-, nitrobenzenes, secondary and tertiary amines $8 \leqslant N \leqslant 15$	0.60	$-140.04 + 13.869NE$
$N > 15$	$3.055 - 0.161N$	$-140.04 + 13.869NE$
Acids	4.81	$-188.40 + 9.558NE$
Esters	$-1.174 + 0.376N$	$-140.04 + 13.869NE$
Alcohols: OH attached to nucleus: take for all phenolic compounds		
$NE = 16.17$[b]		213.68
Alcohols: OH in side chain	-0.16	213.68
Ketones	2.70	$-760.65 + 50.478NE$
Ethers: take for all aromatic ethers		
$NE = 11.50$[b]		$-140.04 + 13.869NE$
Primary amines: NH_2 attached to nucleus; take for all anilinic compounds		
$NE = 15.04$[b]		

(Continued)

TABLE 22-4 (Continued)

	ΔN	ΔB
Correction for Aromatic Nucleus (Cont'd.)		
Primary amines: NH_2 in side chain	− 0.16	
Polyphenyls	− 5.340 + 0.815N	−188.40 + 9.558NE
Ortho configuration: OH group present	0.51	−571.94
Without OH		54.84
Meta configuration	0.11	27.25
Para configuration	− 0.04	− 17.57
Cyclopentane $7 \leqslant N \leqslant 15$	0.205 + 0.069N	− 45.96 + 2.224NE
$N > 15$	3.971 − 0.172N	−339.67 + 23.135NE
Cyclohexane $8 \leqslant N \leqslant 16$	1.48	−272.85 + 25.041NE
$N > 16$	6.517 − 0.311N	−272.85 + 25.041NE
Iso Configuration		
Alkanes	1.389 − 0.238N	15.51
Double iso in alkanes (extra correction)	0.93	
Alkenes	1.389 − 0.238N	8.93
Alcohols	0.24	94.23
Esters, alkylbenzenes, halogenides, ketones	− 0.24	8.93
Acids	− 0.24	
Ethers, amines	− 0.50	8.93
Various		
$C(Cl)_x$ configuration	1.91 − 1.459X	− 26.38
-CCl-CCl-	0.96	
-C(Br)$_x$	0.50	81.34 − 86.850X
-CBr-CBr-	1.60	− 57.73
CF_3- (in alcohols)	− 3.93	341.68
(other compounds)	− 3.93	25.55
Diols	− 2.50 + N	See alcohols

a. Includes pyridine.

b. Other substituents, such as Cl, CH_3, and NO_2, are neglected for the determination of NE. For the calculation of B_3, they have to be taken into account.

Source: van Velzen et al. [21]. *(Reprinted with permission from the American Chemical Society.)*

Example 22-3   Find the viscosity of valeric acid at 20°C (293K).

(1)   The molecular formula of valeric acid, C_4H_9COOH, indicates that $N = 5$.

(2)   Table 22-4 lists the ΔN correction for acids with less than 10 carbon atoms as $6.795 + 0.365N$. For $N = 5$, $\Delta N = 8.62$.

(3) $NE = N + \Delta N = 13.62$.

(4)   From Eq. 22-5,
$$B_a = 24.79 + 66.885(13.62) - 1.3173(13.62)^2 - 0.00377(13.62)^3$$
$$= 681.86$$

(5)   From Table 22-4, $\Delta B = -249.12 + 22.449NE$
$$= 56.64$$

(6) $B_3 = B_a + \Delta B = 681.86 + 56.64 = 738.50$

(7)   From Eq. 22-6,
$$T_o = 28.86 + 37.439(13.62) - 1.3547(13.62)^2 + 0.02076(13.62)^3$$
$$= 339.93K$$

(8)   Substituting the above values of B_3, T, and T_o in Eq. 22-4,
$$\log \eta_L = 738.50(1/293 - 1/339.93)$$
$$= 0.347$$

Therefore, η_L = antilog (0.347) = 2.22 cp. (The experimentally measured viscosity of valeric acid at 20°C is listed in Table 22-2 as 2.30 cp. The value estimated above thus has a deviation of −3.5%.)

Example 22-4   Find the viscosity of chloroform at 20°C.

(1)   Since the chloroform molecule contains 1 carbon atom, $N = 1$.

(2)   Table 22-4 lists two ΔN corrections for $CHCl_3$:
- Each chloride functional group requires a correction of 3.21
- The $C(Cl)_x$ configuration requires a correction of $1.91 - 1.495(x) = -2.47$

(3) $NE = N + \Delta N = 1 + 3(3.21) - 2.47 = 8.16$

(4)   From Eq. 22-5, $B_a = 480.81$

(5)   Two ΔB corrections are also given in Table 22-4:
- The chloride functional groups require a correction of −17.03·
- The $C(Cl)_x$ configuration requires a correction of −26.38

(6) $B_3 = B_a + \Delta B = 480.81 + (-17.03 - 26.38) = 437.40$

(7)   From Eq. 22-6, T_o = 255.44K

(8)   Using Eq. 22-4,

$$\log \eta_L = 437.40(1/293 - 1/255.44)$$
$$= -0.2195$$

$$\eta_L = \text{antilog } (-0.2195)$$
$$= 0.603 \text{ cp}$$

Since Table 22-2 lists an experimental value of 0.56 cp, the estimated value has a deviation of +7.1%.

Example 22-5   Find the viscosity of benzophenone at 55°C.

(1)   For $(C_6H_5)_2 CO$, N = 13.

(2)   The ΔN corrections (Table 22-4) are:
   - 3.265 − 0.122(13) for the ketone functional group
   - 2.70(2) for the two aromatic nuclei

(3)   NE = 13 + 3.265 − 1.586 + 5.40
       = 20.08

(4)   From Eq. 22-7, B_a = 530.59 + 13.740(20.08)
                      = 806.49

(5)   The ΔB corrections (Table 22-4) are:
   - −117.21 + 15.781(20.08) for the ketone functional group
   - −760.65 + 50.478(20.08)

(6) $B_3 = B_a + \Delta B$ = 806.49 − 117.21 + 316.88 − 760.65 + 1013.60
       = 1259.11

(7)   From Eq. 22-8, T_o = 402.5K

(8)   Using Eq. 22-4,

$$\log \eta_L = 1259.11(1/328 - 1/402.5)$$
$$= 0.7105$$

$$\eta_L = \text{antilog } (0.7105)$$
$$= 5.13 \text{ cp}$$

(Table 22-2 lists an experimental value of 4.67 cp; thus, the estimated viscosity has a deviation of +10.0%.)

22-6   METHOD 3

Derivation. The third method can be considered a variation of method 2 or even an analytical extension of method 1. The method is simply based on applying Eq. 22-3 at the normal boiling point, T_b. Thus,

$$\ln \eta_{Lb} = A_4 + (B_4/T_b) \tag{22-11}$$

and

$$A_4 = \ln \eta_{Lb} - (B_4/T_b) \tag{22-12}$$

Hence, at any other temperature T,

$$\ln \eta_L = \ln \eta_{Lb} + B_4 \left(\frac{1}{T} - \frac{1}{T_b} \right) \tag{22-13}$$

The viscosity at T_b is found from Table 22-3. Following the ideas of Eyring [5], we equate the parameter B_4 to the energy of vaporization such that

$$B_4 = \Delta E_v/n \tag{22-14}$$

where n is an integer. For an ideal gas, the energy of vaporization is related to the heat of vaporization as follows:

$$\Delta E_v = \Delta H_v - RT \tag{22-15}$$

Thus

$$B_4 = \frac{1}{n} (\Delta H_v - RT) \tag{22-16}$$

As a first approximation, the heat of vaporization at the normal boiling point, ΔH_{vb}, can be substituted for ΔH_v. We can then use the method developed by Fishtine [6], in which ΔH_{vb} is given by

$$\Delta H_{vb} = K_F \cdot T_b (8.75 + R \ln T_b) \tag{22-17}$$

where R is the gas constant (1.987 cal/mol-degree) and K_F is a constant. From a consideration of the dipole moments of several compounds, Fishtine derived a table (presented earlier as Table 14-4) that relates the value of K_F to structure.

Combining Eqs. 22-17 and 22-16, we obtain

$$B_4 = \frac{1}{n} [K_F T_b (8.75 + R \ln T_b) - RT] \qquad (22\text{-}18)$$

Appropriate values of n are given in Table 22-5.

TABLE 22-5

Values of n (Equation 22-18)

Compound Type	n
Aliphatic hydrocarbons	8
Ketones (aliphatic and aromatic)	7
All other organic compounds	5

Basic Steps

(1) Obtain η_{Lb} from Table 22-3.

(2) Obtain K_F from Table 14-4.

(3) Obtain n from Table 22-5.

(4) Obtain the normal boiling point from either the CRC handbook [22] or from Ref. 17. If T_b is unavailable, it may be estimated by one of the techniques given in Chapter 12.

(5) Calculate B_4 by using Eq. 22-18.

(6) Calculate $\ln \eta_L$ at temperature T (K) using Eq. 22-13. The antilog yields η_L.

Example 22-6   Find the viscosity of hexane at −60°C (213K).

(1)   From Table 22-3, $\eta_{Lb} = 0.2$ cp

(2)   From Table 14-4, $K_F = 1.0$

(3)   From Table 22-5, n = 8

(4)   From the CRC Handbook [22], $T_b = 342K$

(5)   Substituting the above values in Eq. 22-18,

$$B_4 = \frac{1}{8} [1.0 \times 342 (8.75 + 1.987 \ln (342)) - (1.987 \times 213)]$$

$$= 817$$

(6)   From Eq. 22-13,

$$\ln \eta_L = \ln (0.2) + 817 \left(\frac{1}{213} - \frac{1}{342} \right)$$

$$= -0.16$$

$$\eta_L = 0.85 \text{ cp}$$

The experimental value is 0.89 cp. Hence, the deviation is −4.4%.

Example 22-7   Find the viscosity of aniline at −5°C (268K).

(1) $\eta_{Lb} = 0.45$ cp (Table 22-3)

(2) $K_F = 1.16$ (Table 14-4)

(3)   n = 5 (Table 22-5)

(4) $T_b = 457.4$K [22]

(5)   Then, using Eq. 22-18,

$$B_4 = \frac{1}{5} \; [1.16 \times 457.4 \, (8.75 + 1.987 \ln (457.4)) - (1.987 \times 268)]$$

$$= 2114$$

(6)   From Eq. 22-13,

$$\ln \eta_L = \ln (0.45) + 2114 \left(\frac{1}{268} - \frac{1}{457.4} \right)$$

$$= 2.47$$

$$\eta_L = 11.8 \text{ cp}$$

The experimental value is 13.4 cp. Hence, the deviation is −11.9%.

22-7   AVAILABLE DATA

The following sources were used in obtaining experimental liquid viscosity data:

Weast, R.C. (ed.), *CRC Handbook of Chemistry and Physics*, 59th ed., CRC Press, Inc., West Palm Beach, Fla., 1978-1979.

Reid, Prausnitz and Sherwood [17].

Additional sources of experimental data are listed in Appendix A.

22-8 SYMBOLS USED

A_1 = constant in Table 22-1

A_3 = constant in Eq. 22-3

A_4 = constant in Eq. 22-11

B_a = constant in Eq. 22-10

B_1 = constant in Table 22-1

B_2 = constant in Eq. 22-2

B_3 = constant in Eq. 22-4

B_4 = constant in Eq. 22-11

ΔB = correction factor in Eq. 22-10

ΔE_v = energy of vaporization in Eq. 22-14 (cal/mol)

ΔH_v = heat of vaporization in Eq. 22-15 (cal/mol)

ΔH_{vb} = heat of vaporization at the normal boiling point in Eq. 22-17 (cal/mol)

J = viscosity constant in Table 22-1

K_F = constant in Eq. 22-17

M = molecular weight (g)

N = number of carbon atoms

n = integer in Eq. 22-14

ΔN = correction factor in Eq. 22-9

NE = equivalent chain length in Eq. 22-9

R = gas constant (cal/mol·K)

t = temperature (°C)

T = temperature (K)

T_b = temperature at the normal boiling point (K)

T_c = critical temperature (K)

T_o = temperature at which $\eta_L = 1$ in Eq. 22-3 (K)

T_r = ratio in Table 22-1 equal to T/T_c

V = molar volume in Eq. 22-2 (cm³/mol)

V_c = critical volume (cm³/mol)

V_o = molar volume when $\phi = 0$ in Eq. 22-2 (cm³/mol)

Greek

η_+ = pseudo critical viscosity in Table 22-1 (cp)

η_L = liquid viscosity (cp)

η_{Lb} = liquid viscosity at the normal boiling point (cp)

ϕ = fluidity = $1/\eta_L$ in Eq. 22-2 (cp^{-1})

ρ_L = liquid density in Table 22-1 (g/cm^3)

θ = constant in Table 22-1

22-9 REFERENCES

1. Andrade, E.N. DaC., "Viscosity of Liquids," *Proc. R. Soc. London, Ser A,* **215A**, 36 (1952).

2. Bachinski, A.I., "Inner Friction of Liquids," *Z. Phys. Chem. (Leipzig),* 84, 643 (1913).

3. Brush, S.G., "Theories of Liquid Viscosity," *Chem. Rev.,* **62**, 513 (1962).

4. Ertl, H. and F.A.L. Dullien, "Hildebrand's Equation for Viscosity and Diffusivity," *J. Phys. Chem.,* **77**, 3007 (1973).

5. Eyring, H. and R.P. Marchi, "Significant Structure Theory of Liquids," *J. Chem. Ed.,* **40**, 563 (1963).

6. Fishtine, S.H., "Reliable Latent Heats of Vaporization," *Ind. Eng. Chem.,* **55**, 47 (June 1963).

7. Gambill, W.R., "How to Calculate Liquid Viscosity Without Experimental Data," *Chem. Eng. (NY),* 127 (Jan. 12, 1959).

8. Gambill, W.R., "How P and T Change Liquid Viscosity," *Chem. Eng. (NY),* 123 (Feb. 9, 1959).

9. GeMant, A., "Frictional Phenomena V, B. Liquids," *J. Apply. Phys.,* **12**, 827 (1941).

10. Hildebrand, J.H., "Motions of Molecules in Liquids: Viscosity and Diffusivity," *Science,* **174**, 490 (1971).

11. Hildebrand, J.H. and R.H. Lamoreaux, "Fluidity and Liquid Structure," *J. Phys. Chem.,* **77**, 1471 (1973).

12. Lewis, W.K. and L. Squires, "The Structure of Liquids and the Mechanism of Viscosity," *Refiner Nat. Gasoline Manuf.,* **13**, 448 (1934).

13. Morris, P.S., M.S. Thesis, Polytechnic Institute of Brooklyn, Brooklyn, N.Y., 1964.

14. Orrick, C. and J.H. Erbar, private communication, reported in Ref. 17.

15. Partington, J.R., *Properties of Liquids,* Vol. 2 of *An Advanced Treatise on Physical Chemistry,* Longmans, Green and Co., London (1951).

16. Prandtl, L., *Essentials of Fluid Dynamics,* Hafner Publishing Co., New York (1952).

17. Reid, R.C., J.M. Prausnitz and T.K. Sherwood, *The Properties of Gases and Liquids,* Ch. 9, McGraw-Hill, New York, p. 391 (1977).

18. Taylor, G.I., "Volume Viscosity of Water Containing Air Bubbles," *Proc. R. Soc. London, Ser. A,* **A226**, 39 (1954).

19. Thomas, L.H., "The Dependence of the Viscosities of Liquids on Reduced Temperature and the Relation Between Viscosity, Density and Chemical Constitution," *J. Chem. Soc.,* **1946**, 573 (1946).

20. van der Waals, J.D., Jr., "Theory of the Brownian Movement," *Proc. Acad. Sci. Amsterdam,* **21**, 1057 (1919).

21. van Velzen, D., R. Lopes Cardozo and H. Lagenkamp, "A Liquid Viscosity-Temperature-Chemical Constitution Relation for Organic Compounds," *Ind. Eng. Chem. Fundam.,* **11**, 20 (1972).

22. Weast, R.C. (ed.), *CRC Handbook of Chemistry and Physics,* 59th ed., CRC Press, Inc., West Palm Beach, Fla. (1978-1979).

23

HEAT CAPACITY

James D. Birkett

23-1 INTRODUCTION

Heat capacity, C, is defined as the ratio of the heat absorbed by a system to the resulting temperature increase. It is commonly expressed in calories per gram per degree Celsius (cal/g·°C) or calories per mole per degree Celsius (cal/mol·°C).[1] Because C is also a function of temperature, it is the limiting value of the ratio as ΔT approaches zero, and is specified at a given temperature. For gases, two heat capacities may be defined: one, C_{pg}, where heat is absorbed with the system held at constant pressure, and the other, C_{vg}, where heat is absorbed with the system held at constant volume. Thermodynamically, it can be shown [5] that, for ideal gases, the difference in molar heat capacities $(C_{pg}-C_{vg})$ is equal to the gas constant, R (1.987 cal/mol·°C). Heat capacities typically range from 25 to 55 cal/mol·°C for organic liquids and 15 to 50 cal/mol·°C for organic gases at constant pressure.

It can further be shown [5] that $C_{vg} = 3/2R$ and $C_{pg} = 5/2R$ for a monatomic gas whose energy is solely translational. These relationships are borne out experimentally with helium, argon, neon, mercury, and sodium. For polyatomic molecules, however, rotational and vibrational components of energy must also be considered; predictions based only on theory require accurate measurement of spectroscopic parame-

---

1. Many publications use the notation cal/g-mol°C, which is equivalent to cal/mol·°C, the notation used in this chapter. Note also that cal/mol·°C and cal/mol·K are equivalent.

ters (moments of inertia, vibrational frequencies) and lengthy calculations more suited to computerization.

Heat capacities are required for calculating many thermochemical and engineering parameters. The values are obtained either by direct reference to experimental data, which are not always available, or by various empirical estimation methods. Not surprisingly, most of the methods are based on structural considerations, such as the contribution of the component bonds and/or functional groups to the total heat capacity of the molecule. Values contributed by these groups are derived from experimentally determined heat capacities of numerous compounds.

This chapter presents methods for estimating the heat capacity at constant pressure for gases (C_{pg}) and liquids (C_{pl}), as these are most often required when dealing with an organic chemical that has been introduced into the environment. Of particular concern are spills of bulk quantities of organic chemicals. A number of mathematical models have been derived to predict the fate of chemicals after such spills. Heat capacity is a required parameter in models predicting venting rates and the heating and/or cooling of bulk quantities of the chemical under various conditions, e.g., contact with water and exposure to thermal radiation from a fire.

The estimation methods described in this chapter are summarized in Table 23-1. All are group-contribution methods in which the sum of individual group contributions directly yields a value of C. *Thus, the only input information required is the chemical structure.*

All of the recommended methods allow the estimation of heat capacities at ambient temperatures ($\sim$25-30°C). Values of C usually increase with temperature, but for purposes of hazard assessment (chemical spill modeling), changes in C may be considered negligible over the normal range of ambient temperatures.

23-2 ESTIMATION METHODS FOR GASES

There are several methods for estimating C_{pg}, including those of Dobratz and Meghreblian [4,8] and Thinh *et al.* [12]. This chapter recommends two methods — one derived by Rihani and Doraiswamy [11] and the other by Benson [1,2] — which provide good compromises among ease of use, accuracy, and applicability to a wide range of compounds. The first method is based on adding the contribution to C_{pg} of each of the component groups within the molecule; Rihani and Doraiswamy have tabulated values of these contributions for 41 common groups. These

TABLE 23-1

Recommended Methods for Estimating Heat Capacity

Method	Basis	Comments
Gases		
Rihani and Doraiswamy [11]	Group contributions	Errors typically 2-10%
		Can handle most aromatics, ring systems, conjugated double bonds, triple bonds
Benson, Cruickshank, *et al.* [2]	Group contributions	Errors typically < 5%
		Numerous group contributions evaluated; more complex to use than that of Rihani and Doraiswamy
Liquids		
Johnson and Huang [6]	Group contributions	Errors usually < 10%
		Limited number of group contributions available
Chueh and Swanson [3]	Group contributions	Errors usually < 3%
		More group contributions available than for Johnson and Huang

contributions are given as the coefficients of a polynomial in temperature, T:

$$C_{pg} = a + bT + cT^2 + dT^3 \qquad (23\text{-}1)$$

This method is easy to use, if the contributions for all groups present are listed; it generally predicts C_{pg} to within 5% and is often within 2% of measured values. Rihani and Doraiswamy state [11], based on their own calculations including a test set of 36 compounds, that at 300K the average method error is about 3%, while in the temperature range of 400 to 1500K it is less than 2%. In addition, the error for hydrocarbons (1.5%) is less than that for nonhydrocarbons (4%). The procedure is described in §23-3.

Benson's method, applicable only at 300 K, assigns a value to each divalent or polyvalent atom based on the groups or other atoms to which it is bonded. Values have been compiled for most of the groups likely to be encountered. While the method is a bit cumbersome for large and complex molecules, patience is generally rewarded by estimation within 1% of the observed values.

Table 23-2 gives a sampling of C_{pg} values calculated by both of the above methods and compares the results with measured values. The more complex method of Benson, Cruickshank, *et al.* is generally more accurate, but in many cases the added accuracy may not warrant the extra effort involved.

TABLE 23-2

Measured vs. Estimated Heat Capacities for Gases at 300K
(cal/mol·K)

| Compound | Literature Value [13] | Estimated C_{pg} | | | |
		Rihani and Doraiswamy [11]	% Error	Benson, Cruickshank, et al. [2]	% Error
2,2-Dimethylpropane	29.21	29.29	+ 0.27	29.13	−0.27
2-Methylbut-1-ene	26.41	27.33	+ 3.5	27.08	+2.5
3-Methyl-1,2-butadiene	25.3	25.38	+ 0.32	24.95	- 1.38
cis 1,3-Pentadiene	22.7	21.57	- 5.0	22.51	-0.84
1,2,3,5-Tetramethylbenzene	44.57	46.94	+ 5.3	44.16	- 0.92
o-Methylstyrene	34.9	32.91	- 5.7	35.17	+0.77
Biphenyl	39.05	38.69	- 0.92	39.06	+0.03
p-Cresol	29.8	30.31	+ 1.7	30.02	+0.74
Trimethylene oxide	14.3	15.49	+ 8.3	14.28	-0.14
Methyl ethyl ketone	24.7	24.39	- 1.3	23.47	-5.0
Nitromethane	15.36	13.76	-10.4	15.29	-0.45
1,2-Diiodopropane	24.1	25.10	+ 4.1	24.59	+2.0
1,1-Difluoro-2-iodoethene	19.3	17.49	- 9.4	18.5	-4.1
Avg. Error			4.3%		1.5%

23-3   METHOD OF RIHANI AND DORAISWAMY

Principles of Use. The method of Rihani and Doraiswamy [11] is the quickest way to estimate C_{pg} for organic gases and is applicable to several broad classes of compounds, including many aromatics, acyclic ring systems, conjugated double bonds, and acetylenes. Table 23-3 lists the bond contributions calculated by Rihani and Doraiswamy.

Basic Steps

(1) Draw the structure of the compound.

(2) Find the individual group contributions to the coefficients of Eq. 23-1 in Table 23-3 and add them in proportion to the

TABLE 23-3

Group Contributions for Eq. 23-1

Group	a	$b \times 10^2$	$c \times 10^4$	$d \times 10^6$
Aliphatic Hydrocarbon Groups				
$-CH_3$	0.6087	2.1433	− 0.0852	0.001135
$-\overset{\mid}{C}H_2$	0.3945	2.1363	− 0.1197	0.002596
$=CH_2$	0.5266	1.8357	− 0.0954	0.001950
$-\overset{\mid}{\underset{\mid}{C}}-H$	− 3.5232	3.4158	− 0.2816	0.008015
$-\overset{\mid}{\underset{\mid}{C}}-$	− 5.8307	4.4541	− 0.4208	0.012630
$\overset{H}{\underset{/}{\diagdown}}C=CH_2$	0.2773	3.4580	− 0.1918	0.004130
$\overset{\diagdown}{\underset{/}{}}C=CH_2$	− 0.4173	3.8857	− 0.2783	0.007364
$\overset{H}{\diagdown}C=C\overset{H}{\diagup}$	− 3.1210	3.8060	− 0.2359	0.005504
$\overset{H}{\diagdown}C=C\underset{\diagdown H}{\diagup}$	0.9377	2.9904	− 0.1749	0.003918
$\overset{\diagdown}{\diagup}C=C\overset{\diagup H}{\diagdown}$	− 1.4714	3.3842	− 0.2371	0.006063
$\overset{\diagdown}{\diagup}C=C\overset{\diagup}{\diagdown}$	0.4736	3.5183	− 0.3150	0.009205
$\overset{H}{\diagdown}C=C=CH_2$	2.2400	4.2896	− 0.2566	0.005908
$\overset{\diagdown}{\diagup}C=C=CH_2$	2.6308	4.1658	− 0.2845	0.007277

(continued)

TABLE 23-3 (Continued)

Group	a	b × 10²	c × 10⁴	d × 10⁶
Aliphatic Hydrocarbon Groups (cont'd.)				
$\begin{array}{c}H \qquad H \\ \backslash \qquad \\ C{=}C{=}C \\ / \qquad \backslash \end{array}$	− 3.1249	6.6843	− 0.5766	0.017430
≡CH	2.8443	1.0172	− 0.0690	0.001866
−C≡	− 4.2315	7.8689	− 0.2973	0.00993
Aromatic Hydrocarbon Groups				
HC $<$	− 1.4572	1.9147	− 0.1233	0.002985
−C $<$	− 1.3883	1.5159	− 0.1069	0.002659
⟶C $<$	0.1219	1.2170	− 0.0855	0.002122
Oxygen-Containing Groups				
−OH	6.5128	− 0.1347	0.0414	− 0.001623
−O−	2.8461	− 0.0100	0.0454	− 0.002728
$\begin{array}{c}H \\ | \\ -C{=}O\end{array}$	3.5184	0.9437	0.0614	− 0.006978
$\begin{array}{c}\backslash \\ C{=}O \\ /\end{array}$	1.0016	2.0763	− 0.1636	0.004494
$\begin{array}{c}O \\ \| \\ -C-O-H\end{array}$	1.4055	3.4632	− 0.2557	0.006886
$\begin{array}{c}O \\ \!\!\!\!\nearrow \\ -C \\ \!\!\!\!\searrow \\ O-\end{array}$	2.7350	1.0751	0.0667	− 0.009230
O $<$	− 3.7344	1.3727	− 0.1265	0.003789
Nitrogen-Containing Groups				
−C≡N	4.5104	0.5461	0.0269	− 0.003790
−N≡C	5.0860	0.3492	0.0259	− 0.002436
−NH₂	4.1783	0.7378	0.0679	− 0.007310

TABLE 23-3 (Continued)

Group	a	b × 10²	c × 10⁴	d × 10⁶
Nitrogen-Containing Groups (cont'd.)				
＼NH／	− 1.2530	2.1932	− 0.1604	0.004237
＼N−／	− 3.4677	2.9433	− 0.2673	0.007828
N≼	2.4458	0.3436	0.0171	− 0.002719
−NO₂	1.0898	2.6401	− 0.1871	0.004750
Sulfur-Containing Groups				
−SH	2.5597	1.3347	− 0.1189	0.003820
−S−	4.2256	0.1127	− 0.0026	− 0.000072
S≼	4.0824	− 0.0301	0.0731	− 0.006081
−SO₃H	6.9218	2.4735	0.1776	− 0.022445
Halogen-Containing Groups				
−F	1.4382	0.3452	− 0.0106	− 0.000034
−Cl	3.0660	0.2122	− 0.0128	0.000276
−Br	2.7605	0.4731	− 0.0455	0.001420
−I	3.2651	0.4901	− 0.0539	0.001782
Contributions Due to Ring Formation				
3-membered ring	− 3.5320	− 0.0300	0.0747	− 0.005514
4-membered ring	− 8.6550	1.0780	0.0425	− 0.000250
5-membered ring				
Pentane	− 12.2850	1.8609	− 0.1037	0.002145
Pentene	− 6.8813	0.7818	− 0.0345	0.000591
6-membered ring				
Hexane	− 13.3923	2.1392	− 0.0429	− 0.001865
Hexene	− 8.0238	2.2239	− 0.1915	0.005473

Source: Rihani and Doraiswamy [11]. *(Reprinted with permission from the American Chemical Society.)*

number of such groups in the molecule. Include ring contributions if necessary.

(3) Substitute the coefficients in Eq. 23-1 and calculate C_{pg} using the appropriate value of T (K).

(4) The total is in units of cal/mol · K (25°C). To convert to units of cal/g · K (same as cal/g·°C), divide by the molecular weight.

Example 23-1   Calculate C_{pg} for 3-methyl-1,2-butadiene at 300K. (Although this compound is a liquid at room temperature, C_{pg} for the vapors over the liquid can still be calculated.)

(1)   The molecular structure is:

$$H_2C=C=C-CH_3$$
$$|$$
$$CH_3$$

(2)   The group contributions from Table 23-3 are:

	a	b × 10²	c × 10⁴	d × 10⁶
2[−CH₃] :	2(0.6087)	2(2.1433)	2(−0.0852)	2(0.001135)
[C=C=CH₂] :	2.6308	4.1658	−0.2845	0.007277
	3.8482	8.4524	−0.4549	0.009547

(3)   Substituting in Eq. 23-1, with T = 300K:

$$C_{pg} = 3.8482 + 8.4524 \times 10^{-2} \, (300) - 0.4549 \times 10^{-4} \, (300)^2$$
$$+ .009547 \times 10^{-6} \, (300)^3$$
$$= 25.38 \text{ cal/mol·°C}$$

The literature value of C_{pg}^{300} is 25.3 cal/mol·°C [13], and thus the calculated value is within 0.3%.

Example 23-2   Calculate C_{pg} for 1,2-diiodopropane at 300K.

(1)   The molecular structure is:

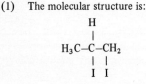

(2)   The bond contributions from Table 23-3 are:

	a	$b \times 10^2$	$c \times 10^4$	$d \times 10^6$
[—CH$_3$]:	0.6087	2.1433	−0.0852	0.001135
[—CH]: $\mid$ $\mid$	−3.5232	3.4158	−0.2816	0.008015
[—CH$_2$]: $\mid$	0.3945	2.1363	−0.1197	0.002596
2[—I]:	2(3.2651)	2(0.4901)	2(−0.0539)	2(0.001782)
	4.0102	8.6756	−0.5943	0.1531

(3)   Substituting in Eq. 23-1:

$$C_{pg} = 4.0102 + 8.6756 \times 10^{-2}\,(300) - 0.5943 \times 10^{-4}\,(300)^2$$
$$+ 0.01531 \times 10^{-6}\,(300)^3$$
$$= 25.10 \text{ cal/mol} \cdot {}^\circ\text{C}$$

The literature value of C_{pg}^{300} is 24.1 cal/mol·°C [13], indicating an error of 4.1%.

23-4   METHOD OF BENSON, CRUICKSHANK, et al.

Principles of Use. The method of Benson, Cruickshank, *et al.* [2] is considerably more complex than the preceding method, but its greater accuracy and utility become apparent with a little practice. Instead of focusing on each group within the molecule, as specified by the atoms connected therein, it deals with each atom, as differentiated by the number and type of bonds to neighboring atoms. The contributions of each such atom are then combined to give the total C_{pg}. Corrections are also made for the presence of certain overall structures, such as rings.

Table 23-4 lists the contributions at 300K (27°C) and various correction terms. Use of the table requires not only knowledge of the molecular structure but also familiarity with the notational system employed. (See footnotes to the table.) For molecules with both *trans* and *cis* isomers, contributions are tabulated for the *trans* form, which is generally the more stable. However, a *cis* correction is given, as well as one for *ortho* positioning of functional groups on aromatic compounds. Corrections are also given for some ring compounds, due to the strain imposed by such configurations on "normal" bond angles.

TABLE 23-4
Benson Group Contributions to Heat Capacities at 300K

A. Basic Hydrocarbon Groups[a,b]

Group	Contribution	Group	Contribution	Group	Contribution
$C-(C)(H)_3$	6.19	$C_d-(C_t)(C)$	4.40	$C-(C_B)(C)_3$	(4.37)
$C-(C)_2(H)_2$	5.50	$C-(C_d)(H)_3$	6.19	$C-(C_B)_2(C)(H)$	3.74
$C-(C)_3(H)$	4.54	$C-(C_d)_2(H)_2$	(4.7)	$C-(C_B)_2(C)_2$	3.57
$C-(C)_4$	4.37	$C-(C_d)_2(C)_2$	3.57	$C-(C_B)(C_d)(H)_2$	(4.7)
		$C-(C_d)(C)_3$	(3.99)	$C_t-(H)$	5.27
$C_d-(H)_2$	5.10	$C-(C_d)(C)(H)_2$	5.12	$C_t-(C)$	3.13
$C_d-(C)(H)$	4.16	$C-(C_d)(C)_2(H)$	(4.16)	$C_t-(C_d)$	(2.57)
$C_d-(C)_2$	4.10	$C-(C_d)_2(C)(H)$	3.74	$C_t-(C_B)$	2.57
$C_d-(C_d)(H)$	4.46	$C-(C_t)(H)_3$	6.19		
$C_d-(C_d)(C)$	(4.40)	$C-(C_t)(C)(H)_2$	4.95	$C_B-(H)$	3.24
$C_d-(C_B)(H)$	4.46	$C-(C_t)(C)_2(H)$	(3.99)	$C_B-(C)$	2.67
$C_d-(C_B)(C)$	(4.40)			$C_B-(C_d)$	3.59
$C_d-(C_t)(H)$	4.46	$C-(C_B)(H)_3$	6.19	$C_B-(C_t)$	3.59
		$C-(C_B)(C)(H)_2$	5.84	$C_B-(C_B)$	3.33
		$C-(C_B)(C)_2(H)$	(4.88)		
				C_a	3.9

Correction for Next-Nearest Neighbor

Cis	−1.34
Ortho	1.12

Corrections for Ring Compounds

Cyclopropane	−3.05
Cyclobutane	−4.61
Cyclobutene	−2.53
Cyclopentane	−6.5
Cyclopentene	−5.98
Cyclopentadiene	−4.3
Cyclohexane	−5.8
Cyclohexene	−4.28

B. Oxygen-containing Compounds[a,b]

Group	Contribution	Group	Contribution	Group	Contribution
$CO-(CO)(H)$	6.72	$CO-(C_B)(C)$	5.68	$O-(CO)(C_d)$	1.44
$CO-(CO)(C)$	5.46	$CO-(C_B)(H)$	6.40	$O-(CO)(C)$	3.90
$CO-(O)(C_d)$	5.97	$CO-(C)_2$	5.59	$O-(CO)(H)$	3.81
$CO-(O)(C_B)$	2.18	$CO-(C)(H)$	7.03	$O-(O)(C)$	(3.7)
$CO-(O)(C)$	5.97	$CO-(H)_2$	8.47	$O-(O)_2$	(3.7)
$CO-(O)(H)$	7.03	$O-(C_B)(CO)$	2.06	$O-(O)(H)$	5.17
$CO-(C_d)(H)$	7.03	$O-(CO)_2$	−0.41	$O-(C_d)_2$	3.4
$CO-(C_B)_2$	5.26	$O-(CO)(O)$	3.7	$O-(C_d)(C)$	3.4

(continued)

TABLE 23–4 (Continued)

B. Oxygen-containing Compounds (Continued)

Group	Contribution	Group	Contribution	Group	Contribution
$O-(C_B)_2$	1.09	$C_d-(O)(H)$	4.16	$C-(O)_2(C)(H)$	5.06
$O-(C_B)(C)$	3.4	$C_B-(CO)$	2.67	$C-(O)_2(H)_2$	2.83
$O-(C_B)(H)$	4.3	$C_B-(O)$	3.9	$C-(O)(C_B)(H)_2$	3.71
$O-(C)_2$	3.4	$C-(CO)_2(H)_2$	5.60	$C-(O)(C_B)(C)(H)$	5.14
$O-(C)(H)$	4.33	$C-(CO)(C)_2(H)$	6.21	$C-(O)(C_d)(H)_2$	4.66
$C_d-(CO)(O)$	5.59	$C-(CO)(C)(H)_2$	6.2	$C-(O)(C)_3$	4.33
$C_d-(CO)(C)$	3.73	$C-(CO)(C)_3$	5.07	$C-(O)(C)_2(H)$	4.80
$C_d-(CO)(H)$	3.79	$C-(CO)(H)_3$	6.19	$C-(O)(C)(H)_2$	4.99
$C_d-(O)(C_d)$	(4.40)	$C-(O)_2(C)_2$	1.59	$C-(O)(H)_3$	6.19
$C_d-(O)(C)$	4.10				

Strain or Ring Corrections

Ether oxygen, gauche		−0.10
Ditertiary ethers		−3.94

Ethylene oxide   −2.0

Trimethylene oxide   −4.6

Tetrahydrofuran   −4.25

Tetrahydropyran   −4.28

1,3-Dioxane   −2.51

1,4-Dioxane   −4.16

1,3,5-Trioxane   1.79

Furan   −4.19

Dihydropyran   −4.44

Cyclopentanone   −8.53

Cyclohexanone   −8.10

Succinic anhydride   −7.90

Glutaric anhydride   −7.93

Maleic anhydride   −5.12

(continued)

TABLE 23—4 (Continued)

C. Nitrogen-containing Compounds[a,c]

Group	Contribution	Group	Contribution	Group	Contribution
C—(N)(H)$_3$	6.19	N—(C$_B$)(H)$_2$	5.72	C—(CN)(C)(H)$_2$	11.10
C—(N)(C)(H)$_2$	5.25	N—(C$_B$)(C)(H)	3.82	C—(CN)(C)$_2$(H)	11.00
C—(N)(C)$_2$(H)	4.67	N—(C$_B$)(C)$_2$	0.62	C—(CN)(C)$_3$	8.65
C—(N)(C)$_3$	4.35	N—(C$_B$)$_2$(H)	2.16	C—(CN)$_2$(C)$_2$	14.72
N—(C)(H)$_2$	5.72	C$_B$—(N)	3.95	C$_d$—(CN)(H)	9.80
N—(C)$_2$(H)	4.20	N$_A$—(N)	2.12	C$_d$—(CN)(C)	9.74
N—(C)$_3$	3.48	CO—(N)(H)	7.03	C$_d$—(CN)$_2$	13.60
N—(N)(H)$_2$	6.10	CO—(N)(C)	5.37	C$_d$—(NO$_2$)(H)	12.3
N—(N)(C)(H)	4.82	N—(CO)(H)$_2$	4.07	C$_B$—(CN)	9.8
N—(N)(C)$_2$	1.56	N—(CO)(C)(H)	3.87	C$_t$—(CN)	10.30
N—(N)(C$_B$)(H)	3.28	N—(CO)(C)$_2$	1.83	C—(NO$_2$)(C)(H)$_2$	12.59
N$_I$—(H)	2.95	N—(CO)(C$_B$)(H)	3.03	C—(NO$_2$)(C)$_2$(H)	11.99
N$_I$—(C)	2.48	N—(CO)$_2$(H)	3.59	C—(NO$_2$)(C)$_3$	9.89
N$_I$—(C$_B$)	2.60	N—(CO)$_2$(C)	1.07	C—(NO$_2$)$_2$(C)(H)	17.32
N$_A$—(H)	4.38	N—(CO)$_2$(C$_B$)	0.98	O—(NO)(C)	9.10
N$_A$—(C)	2.70			O—(NO$_2$)(C)	9.54

Ring Corrections

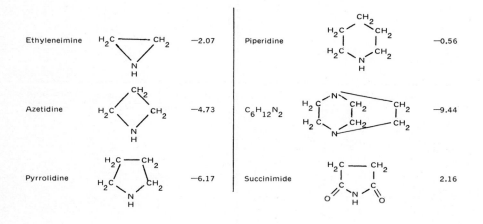

Ethyleneimine		−2.07	Piperidine	−0.56
Azetidine		−4.73	C$_6$H$_{12}$N$_2$	−9.44
Pyrrolidine		−6.17	Succinimide	2.16

TABLE 23—4 (Continued)

D. Halogen Groups[a]

Group	Contribution	Group	Contribution	Group	Contribution
C—(F)$_3$(C)	12.7	C—(I)(H)(C)$_2$	9.2	C$_d$—(I)(H)	8.8
C—(F)$_2$(H)(C)	9.9	C—(I)(C)(C$_d$)(H)	8.13	C$_d$—(C)(Cl)	8.0
C—(F)(H)$_2$(C)	8.1	C—(I)(C$_d$)(H)$_2$	8.82	C$_d$—(C)(I)	8.9
C—(F)$_2$(C)$_2$	9.9	C—(I)(C)$_3$	9.83	C$_d$—(C$_d$)(Cl)	8.3
C—(F)(H)(C)$_2$	7.30	C—(Cl)(Br)(H)(C)	12.4	C$_d$—(C$_d$)(I)	9.2
C—(F)(C)$_3$	6.80	N—(F)$_2$(C)	8.25	C$_t$—(Cl)	7.9
C—(F)$_2$(Cl)(C)	13.7	C—(Cl)(C)(O)(H)	9.85	C$_t$—(Br)	8.3
C—(Cl)$_3$(C)	16.3	C—(I)$_2$(C)(H)	12.69	C$_t$—(I)	8.4
C—(Cl)$_2$(H)(C)	12.1	C—(I)(O)(H)$_2$	8.22	C$_B$—(F)	6.3
C—(Cl)(H)$_2$(C)	8.9	C$_d$—(F)$_2$	9.7	C$_B$—(Cl)	7.4
C—(Cl)$_2$(C)$_2$	12.2	C$_d$—(Cl)$_2$	11.4	C$_B$—(Br)	7.8
C—(Cl)(H)(C)$_2$	9.0	C$_d$—(Br)$_2$	12.3	C$_B$—(I)	8.0
C—(Cl)(C)$_3$	9.3	C$_d$—(F)(Cl)	10.3	C—(C$_B$)(F)$_3$	12.5
C—(Br)$_3$(C)	16.7	C$_d$—(F)(Br)	10.8	C—(C$_B$)(Br)(H)$_2$	9.29
C—(Br)(H)$_2$(C)	9.1	C$_d$—(Cl)(Br)	12.1	C—(C$_B$)(I)(H)$_2$	9.78
C—(Br)(H)(C)$_2$	8.93	C$_d$—(F)(H)	6.8	C—(Cl)$_2$(CO)(H)	12.8
C—(Br)(C)$_3$	9.3	C$_d$—(Cl)(H)	7.9	C—(Cl)$_3$(CO)	17.0
C—(I)(H)$_2$(C)	9.2	C$_d$—(Br)(H)	8.1	CO—(Cl)(C)	8.87

Corrections for Next-nearest Neighbors

Ortho (F)(F)	0
Ortho (Cl)(Cl)	—0.50
Ortho (alkane) (halogen)	0.42
Cis (halogen) (halogen)	—0.19
Cis (halogen) (alkane)	—0.97

E. Organosulfur Groups[a]

Group	Contribution	Group	Contribution	Group	Contribution
C—(H)$_3$(S)	6.19	S—(S)$_2$	4.7	SO$_2$—(C$_d$)(C$_B$)	9.89
C—(C)(H)$_2$(S)	5.38	C—(SO)(H)$_3$	6.19	SO$_2$—(C$_d$)$_2$	11.52
C—(C)$_2$(H)(S)	4.85	C—(C)(SO)(H)$_2$	4.55	SO$_2$—(C)$_2$	10.18
C—(C)$_3$(S)	4.57	C—(C)$_3$(SO)	3.06	SO$_2$—(C)(C$_B$)	9.94
C—(C$_B$)(H)$_2$(S)	4.11	C—(C$_d$)(SO)(H)$_2$	4.40	SO$_2$—(C$_B$)$_2$	8.36
C—(C$_d$)(H)$_2$(S)	5.00	C$_B$—(SO)	2.67	SO$_2$—(SO$_2$)(C$_B$)	9.81
C$_B$—(S)	3.90	SO—(C)$_2$	8.88	CO—(S)(C)	5.59
C$_d$—(H)(S)	4.16	SO—(C$_B$)$_2$	5.72	S—(H)(CO)	7.63
C$_d$—(C)(S)	3.50	C—(SO$_2$)(H)$_3$	6.19	C—(S)(F)$_3$	9.88
S—(C)(H)	5.86	C—(C)(SO$_2$)(H)$_2$	5.38	CS—(N)$_2$	5.59
S—(C$_B$)(H)	5.12	C—(C)$_2$(SO$_2$)(H)	4.42	N—(CS)(H)$_2$	6.07
S—(C)$_2$	4.99	C—(C)$_3$(SO$_2$)	2.32	S—(S)(N)	3.7
S—(C)(C$_d$)	4.22	C—(C$_d$)(SO$_2$)(H)$_2$	5.00	N—(S)(C)$_2$	3.97
S—(C$_d$)$_2$	4.79	C—(C$_B$)(SO$_2$)(H)$_2$	3.71	SO—(N)$_2$	5.59
S—(C$_B$)(C)	3.02	C$_B$—(SO$_2$)	2.67	N—(SO)(C)$_2$	4.20
S—(C$_B$)$_2$	2.00	C$_d$—(H)(SO$_2$)	3.04	SO$_2$—(N)$_2$	5.59
S—(S)(C)	5.23	C$_d$—(C)(SO$_2$)	1.85	N—(SO$_2$)(C)$_2$	6.02
S—(S)(C$_B$)	2.89				

(continued)

TABLE 23-4 (Continued)

Ring Corrections

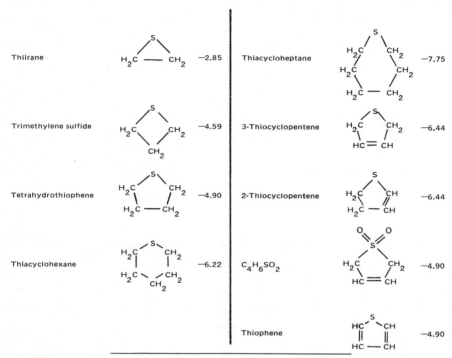

Thiirane		−2.85
Thiacycloheptane		−7.75
Trimethylene sulfide		−4.59
3-Thiocyclopentene		−6.44
Tetrahydrothiophene		−4.90
2-Thiocyclopentene		−6.44
Thiacyclohexane		−6.22
$C_4H_6SO_2$		−4.90
Thiophene		−4.90

Notes:

a. C_d represents a carbon atom that is joined to another carbon atom by a double bond. It is considered divalent. Example: 2-pentene has the groups $C-(C_d)$ $(H)_3$, $C_d-(C)(H)$ twice, $C-(C_d)(H)_2$, and $C-(C)(H)_3$.

C_t represents a carbon atom that is joined to another carbon atom by a triple bond. It is considered monovalent. Example: propyne has the groups $(C_t-(H)$, $C_t-(C)$, and $C-(C_t)(H)_3$.

C_B represents a carbon atom in an aromatic ring. It is considered monovalent. Example: p-ethyl toluene has the groups $C-(C)(H)_3$, $C-(C_B)(C)(H)_2$, $C-(C_B)(H)_3$, $C_B-(C)$ twice, and $C_B-(H)$ four times.

C_a represents the allene group, $>C=C=C<$; the end carbons are treated as normal C atoms. Example: 1,2-butadiene has the groups C_a, $C_d-(H)_2$, $C_d-(C)(H)$, and $C-(C_d)(H)_3$.

b.   Numbers in parentheses are estimated or derived from other than direct experimental data.

c. N_1 represents a double-bonded nitrogen in imines. $N_1-(C_B)$ represents a pyridine nitrogen. N_A represents a double-bond nitrogen in azo compounds.

Source: Benson et al. [2] and Reid et al. [10]. (Reprinted with permission from the American Chemical Society.)

Basic Steps

(1) Draw the structure of the compound.

(2) Find the individual group contributions to C_{pg} in Table 23-4 and add them in proportion to the number of such groups in the molecule.

(3) Add (or subtract, as applicable) any necessary corrections for next-nearest neighbors and ring structures (other than benzene rings).

(4) The resulting value is in units of cal/mol · °C and applies to a temperature of 300K (27°C). To convert to units of cal/g·°C (same as cal/g · K), divide by the molecular weight.

Example 23-3   Calculate the C_{pg} of 3-methyl-1,2-butadiene.

(1)   The molecular structure is:

$$H_2C{=}C{=}C{-}CH_3$$
$$|$$
$$CH_3$$

(2)   The group contributions from Table 23-4 are:

$$C_d{-}(H)_2 \qquad 5.10$$
$$C_a \qquad\qquad 3.90$$
$$C{-}(C_d)(C)_2 \quad 3.57$$
$$2[C{-}(C)(H)_3] \quad \underline{2\,(6.19)}$$
$$C_{pg}^{300} = 24.95 \text{ cal/mol·°C}$$

The literature value is 25.3 cal/mol·°C [13], indicating an error of -1.4%.

Example 23-4   Calculate the C_{pg} of p-cresol.

(1)   The molecular structure is:

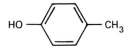

(2)   The group contributions from Table 23-4 are:

$$4[C_B{-}(H)] \qquad 4\,(3.24)$$
$$C_B{-}(C) \qquad\qquad 2.67$$

$$C_B-(O) \qquad 3.9$$

$$C-(C_B)(H)_3 \qquad 6.19$$

$$O-(C_B)(H) \qquad \underline{4.3}$$

$$C_{pg}^{300} = 30.02 \text{ cal/mol} \cdot {}^\circ C$$

Since the literature value is 29.8 cal/mol·°C [13], the error is 0.74%.

23-5 ESTIMATION METHODS FOR LIQUIDS

Heat capacities of liquids, like those of gases, are estimated from the aggregate contribution of bonds or groups. Included here are the simplified group method of Johnson and Huang [6] and the more complex method of Chueh and Swanson [3].

As noted previously, liquid heat capacity (C_{pl}) usually increases with temperature. However, it is a very weak function of temperature except in the region just below the critical temperature. At ambient temperatures, it can be assumed to be constant for most purposes.

The method of Johnson and Huang generally allows estimation of C_{pl} within 10%, while that of Chueh and Swanson usually involves errors of less than 3%. Table 23-5 compares values of C_{pl} for several organic compounds with those estimated by both methods.

23-6 METHOD OF JOHNSON AND HUANG

Principles of Use. The method of Johnson and Huang [6] is based on molecular structure and requires the identification of various molecular groups and the summation of their contributions to the overall value of C_{pl}. Table 23-6 lists such group contributions at 20°C.

Basic Steps

(1) Draw the structure of the compound.

(2) Find the individual group contributions to C_{pl} in Table 23-6 and add them in proportion to the number of such groups in the molecule.

(3) The value obtained is in units of cal/mol·°C at 20°C. To convert to units of cal/g·°C (same as cal/g·K), divide by the molecular weight.

TABLE 23-5

Measured vs. Estimated Heat Capacities for Organic Liquids
(cal/mol·K)

Compound	Literature Value 298K [13]	Estimated C_{pl} at 293K			
		Johnson & Huang [6]	% Error	Chueh & Swanson [3]	% Error
3- Ethylpentane	52.5	54.0	+ 2.8	53.18	+1.3
2,3-Dimethylbutene-2	41.7	38.0	- 8.9	42.8	+2.6
1,2,4-Trimethylbenzene	51.6	53.0	+ 2.7	51.0	- 1.2
Furan	27.4	30.0	+ 9.5	29.6	+8.0
1,1-Dimethylhydrazine	39.2	43.0	+ 9.7	39.1	-0.3
trans-1,2-Dimethylcyclopentane	45.1	49.5	+ 9.8	45.0	-0.2
1,2-Dibromoethane	32.2	20.0	-37.8	32.5	+0.9
Cyclopentylmercaptan	39.5	44.8	+13.4	39.9	+1.0
Ethyl acetate	40.6	40.6	0	39.42	-2.9
sec-Butylmercaptan	40.9	44.7	+ 9.3	40.56	-0.8
Avg. Error			10.4%		1.9%

Example 23-5   Calculate C_{pl} for 1,2,4-trimethylbenzene.

(1)   The molecular structure is:

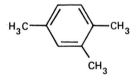

(2)   The group contributions from Table 23-6 are:

$$3\,(CH_3-) \quad 3\,(9.9)$$
$$C_6H_5- \quad 30.5$$
$$-2(-H) \quad -2(3.6)$$
$$C_{pl} = 53.0 \text{ cal/mol·°C}$$

Note that because the ring contribution is for the C_6H_5- group, and the compound has only three hydrogens directly on the ring, the value of two H- groups was *subtracted* from the total. Comparing the result with the measured value of 51.6 at 25°C [13], we find an error of 2.7%. Part of this is attributable to the 5° temperature difference.

TABLE 23-6

Group Heat Capacities for Organic Liquids

Group	Contribution[a] (cal/mol·°C)
CH$_3$—	9.9
—CH$_2$—	6.3
—C—H	5.4
—COOH	19.1
—COO—(esters)	14.5
>CO (ketones)	14.7
—CN	13.9
—OH	11.0
—NH$_2$ (amines)	15.2
—Cl	8.6
—Br	3.7
—NO$_2$	15.3
—O—	8.4
—S—	10.6
C$_6$H$_5$—	30.5
H—(formic acid, formates)	3.55

a. At 20°C only. For certain atomic configurations, it may be necessary to use values for C alone and for H alone; values of 1.76 and 3.6, respectively, appear adequate.

Source: Johnson and Huang [6].

Example 23-6   Calculate C$_{pl}$ for cyclopentylmercaptan.

(1)   The molecular structure is:

 — SH

(2)   The group contributions from Table 23-6 are:

4 (—CH$_2$—)	4 (6.3)
—CH—	5.4
—S—	10.6
H—	3.6
	44.8 cal/mol·°C at 20°C

The literature value is 39.5 [13], indicating an error of about 13%.

23-7 METHOD OF CHUEH AND SWANSON

Principles of Use. The method of Chueh and Swanson [3], like that of Johnson and Huang, is based on the principle of aggregating group contributions. While somewhat more complicated, it is generally more accurate: the results are usually within 3% of the experimental values. Contributions at 20°C are shown in Table 23-7. The footnotes and exceptions are important and should be studied carefully. While the method is accurate only at reduced temperatures[2] below about 0.74, this is rarely a drawback in actual practice.

Basic Steps

(1) Draw the structure of the compound.

(2) Find the individual group contributions to C_{pl} in Table 23-7 and add them in proportion to the number of such groups in the molecule.

(3) The value obtained is in units of cal/mol·°C at 20°C. To convert to units of cal/g·°C (same as cal/g·K), divide by the molecular weight.

Example 23-7   Calculate C_{pl} for 1,2,4-trimethylbenzene.

(1)   The molecular structure is:

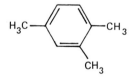

(2)   The group contributions from Table 23-7 are:

$$3(\text{ring–CH=})\quad 3(5.3)$$
$$3(\text{ring–C=})\quad 3(2.9)$$
$$\underline{3(\text{–CH}_3)\quad\quad 3(8.8)}$$
$$C_{pl}= 51.0 \text{ cal/mol·°C at 20°C}$$

Since the literature value is 51.6 cal/mol·°C at 25°C [13], the error is only −1.2%.

---

2. Reduced temperature = (system temperature)/(critical temperature).

TABLE 23-7

Group Contributions for Molar Liquid Heat Capacity at 20°C

Group	Value	Group	Value	Group	Value
Alkane					
$-CH_3$	8.80	$-C=O$ $\vert$ H	12.66	$-N=$ (in a ring)	4.5
$-CH_2-$	7.26			$-C\equiv N$	13.9
$\vert$ $-CH-$ $\vert$	5.00	O $\parallel$ $-C-OH$	19.1	**Sulfur** $-SH$	10.7
$\vert$ $-C-$ $\vert$	1.76	O $\parallel$		$-S-$	8.0
Olefin[a]		$\parallel$ $-C-O-$	14.5	**Halogen** $-Cl$ (first or second	8.6
$=CH_2$	5.20	$-CH_2OH$	17.5	on a carbon)	
$\vert$ $=C-H$	5.10	$\vert$ $-CHOH$ $\vert$	18.2	$-Cl$ (third or fourth on a carbon)	6.0
$\vert$ $=C-$	3.80	$-COH$ $\vert$	26.6	$-Br$	9.0
Alkyne[a]				$-F$	4.0
$CH\equiv$	5.90	$-OH$	10.7	$-I$	8.6
$-C\equiv$	5.90	$-ONO_2$	28.5	**Hydrogen** H$-$ (for formic acid,	3.5
In a Ring		**Nitrogen**		formates, hydrogen	
$\vert$ $-CH-$ $\vert$	4.4	H $\vert$ $H-N-$	14.0	cyanide, etc.)	
$\vert$ $-C=$ or $-C-$ $\vert$	2.9	H $\vert$ $-N-$	10.5		
$-CH\equiv$	5.3	$\vert$ $-N-$	7.5		
$-CH_2-$	6.2				
Oxygen					
$-O-$	8.4				
$\diagdown C=O$ $\diagup$	12.66				

a. Add 4.5 for any carbon group which is joined by a single bond to a second carbon group connected by a double or triple bond to a third carbon group. If a carbon group can meet this criterion in more ways than one, add 4.5 for each time it can do so, with the following exceptions:

● Do not add an extra 4.5 for $-CH_3$ groups or for a carbon group in a ring.
● For a $-CH_2-$ group fulfilling the 4.5 addition criterion, add 2.5 instead of 4.5. However, when the $-CH_2-$ group fulfills the addition criterion in more ways than one, the addition should be 2.5 the first time and 4.5 for each subsequent addition.

For example:

$\vert$ $-C-$ $\vert$ $\vert$ $\vert$ $\vert$ $-C-CH-C=C-$ $\vert$	$\vert$ $\vert$ $\vert$ $-C-CH_2-C=C-$ $\vert$	$\vert$ $\vert$ $CH_3-C=C-$
Add 4.5	Add 2.5	No addition

Source: Chueh and Swanson [3]. *(Reprinted with permission from the American Institute of Chemical Engineers and the Canadian Society for Chemical Engineering.)*

Example 23-8   Calculate C_{pl} for cyclopentylmercaptan.

(1)   The molecular structure is:

(2)   The group contributions from Table 23-7 are:

4($-CH_2-$)	4(6.2)
$-CH-$	4.4
$-SH$	10.7

$$C_{pl} = 39.9 \text{ cal/mol·°C at } 20°C$$

This is within 1% of the literature value of 39.5 cal/mol·°C at 25° [13].

23-8   AVAILABLE DATA

Heat capacity data for both liquids and gases are found in most conventional handbooks; a listing of such handbooks is provided in Appendix A. In addition, the extensive article in *Chemical Reviews* by Benson, Cruickshank *et al.* [2] contains many examples of experimental data.

23-9   SYMBOLS USED

a,b,c,d = parameters in Eq. 23-1
C   = heat capacity (cal/g·°C or cal/mol·°C)
R   = gas constant = 1.987 cal/mol·°C
T   = temperature (K)

Subscripts

pg = gas at constant pressure
vg = gas at constant volume
pl   = liquid at constant pressure

23-10   REFERENCES

1. Benson, S.W., *Thermochemical Kinetics,* 2nd ed., John Wiley & Sons, New York (1978).
2. Benson, S.W., F.R. Cruickshank, D.M. Golden, G.R. Haugen, H.E. O'Neal, A.S. Rogers, R. Shaw and P. Walsh, "Additivity Rules for the Estimation of Thermochemical Properties," *Chem. Rev.,* **69,** 279-324 (1969).
3. Chueh, C.F. and A.C. Swanson: (a) "Heat Transfer: Estimating Liquid Heat Capacity," *Chem. Eng. Prog.,* **69** (7), 83-85 (1973); (b) "Estimation of Liquid Heat Capacity," *Can. J. Chem. Eng.,* **51,** 596-600 (1973).

4. Dobratz, C.J., "Heat Capacities of Organic Vapors," *Ind. Eng. Chem.*, **33**, 759-62 (1941).

5. Glasstone, S., *Thermodynamics for Chemists*, D. Van Nostrand, New York (1958).

6. Johnson, A.I. and C.-J. Huang, "Estimation of the Heat Capacities of Organic Liquids," *Can. J. Technol.*, **33**, 421-25 (1955).

7. Luria, M. and S.W. Benson, "Heat Capacities of Liquid Hydrocarbons: Estimation of Heat Capacities at Constant Pressure as a Temperature Function, Using Additivity Rules," *J. Chem. Eng. Data*, **22**, 90-99 (1977).

8. Meghreblian, R.V. "Approximate Calculations of Specific Heats for Polyatomic Gases," *J. Am. Rocket Soc.*, **21**, 127-28 (1951).

9. Perry, R.H. and C.H. Chilton (eds.), *Chemical Engineers' Handbook*, 5th ed., McGraw-Hill Book Co., New York (1973).

10. Reid, R.C., J.M. Prausnitz and T.K. Sherwood, *The Properties of Gases and Liquids*, 3rd ed., McGraw-Hill Book Co., New York (1977).

11. Rihani, D.N. and L.K. Doraiswamy, "Estimation of Heat Capacity of Organic Compounds from Group Contributions," *Ind. Eng. Chem. Fundam.*, **4**, 17-21 (1965).

12. Thinh, T.-P., J.-L. Duran and R.S. Ramlho, "Estimation of Ideal Gas Heat Capacities of Hydrocarbons from Group Contribution Techniques," *Ind. Eng. Chem. Process Des. Dev.*, **10**, 576 (1971).

13. Weast, R.C. (ed.), *Handbook of Chemistry and Physics*, 53rd ed., Chemical Rubber Co., Cleveland, Ohio (1972).

24

THERMAL CONDUCTIVITY

James D. Birkett

24-1 INTRODUCTION

Thermal conductivity, λ, is defined as the quantity of heat which will traverse a medium of unit thickness and cross-sectional area per unit time, under the influence of an applied temperature gradient. In scientific notation, it is expressed as

$$\lambda = \frac{\text{calories} \cdot \text{centimeters of thickness}}{\text{square centimeters of area} \cdot \text{seconds} \cdot \text{K (temperature gradient)}} \quad (24\text{-}1)$$

which reduces to

$$\lambda = \text{cal}/\text{cm} \cdot \text{s} \cdot \text{K} \quad (24\text{-}2)$$

However, the literature abounds in data given in English, engineering, and mixed units; readers will frequently encounter Btu, feet, inches, °F, etc., in various reference works and should be prepared to make the appropriate conversions.

This chapter addresses the estimation of thermal conductivities of organic liquids, λ_ℓ, as this parameter is often required for calculating rates of evaporation of large quantities of spilled liquids. Values of λ_ℓ are usually in the range of 250-400 $\times$ 10^{-6} cal/cm·s·K, but some liquids with a high degree of association, such as may occur with hydrogen bonding, have higher conductivities.

The thermal conductivity of organic liquids is generally estimated by means of equations that utilize other known properties of the material and, to a lesser extent, structural considerations (as with heat capacity in the preceding chapter). Two methods are frequently employed, depending upon the required accuracy (Table 24-1).

TABLE 24-1

Recommended Methods for Estimating Thermal Conductivity

Method	Inputs Required	Method Error[a]	Comments
Sato and Riedel [9] (§ 24-3)	M, T_b, T_c	5.5% [9] 7.0% (Table 24-3)	Convenient to use
Robbins and Kingrea [11] (§ 24-4)	$\Delta H_{vb}, C_{p\ell}, \rho_\ell,$ $T_b, T_c,$ structural parameters	4.8% [9] 12% (Table 24-3)	Not recommended for routine use because of excessive input requirements

a. Average absolute errors for selected test sets. See § 24-2 for details.

Combining the methods of Sato [5] and Riedel [10], Reid *et al.* [9] developed the equation

$$\lambda_\ell = \frac{2.64 \times 10^{-3}}{M^{1/2}} \left[\frac{3 + 20(1 - T_r)^{2/3}}{3 + 20(1 - T_{rb})^{2/3}} \right] \qquad (24\text{-}3)$$

in which only the reduced temperature, T_r (= T/T_c), reduced boiling point, T_{rb} (= T_b/T_c), and molecular weight, M, are needed. While errors may be as much as ± 20%, Eq. 24-3 may be adequate for many purposes and is rapidly calculated.

A more elaborate method has been reported by Robbins and Kingrea [11], who propose the following equation:

$$\lambda_\ell = \frac{(88.0 - 4.94H)\,10^{-3}}{\Delta S^*} \left(\frac{0.55}{T_r} \right)^N C_{p\ell}\rho_\ell^{4/3} \qquad (24\text{-}4)$$

where

H   =   structural factor determined from Table 24-2

TABLE 24-2

H Factors for Method of Robbins and Kingrea

Functional Group	Number of Groups	H^a
Unbranched hydrocarbons:		
Paraffins		0
Olefins		0
Rings		0
F substitutions	1	1
	2	2
Cl substitutions	1	1
	2	2
	3 or 4	3
Br substitutions	1	4
	2	6
I substitutions	1	5
OH substitutions	1 (iso)	1
	1 (normal)	−1
	2	0
	1 (tertiary)	5
CH_3 branches	1	1
	2	2
	3	3
C_2H_5 branches	1	2
i-C_3H_7 branches	1	2
C_4H_9 branches	1	2
Oxygen substitutions:		
$-\overset{\mid}{C}=O$ (ketones, aldehydes)		0
$-\overset{\overset{\textstyle O}{\|}}{C}-O-$ (acids, esters)		0
$-O-$ (ethers)		2
NH_2 substitutions	1	1

a. For compounds containing multiple functional groups, the H-factor contributions are additive.

Source: Robbins and Kingrea [11]. *(Reprinted with permission from McGraw-Hill, Inc., and Gulf Publishing Co.)*

ΔS^* = entropy term equal to $\Delta H_{vb}/T_b + R\ln(273/T_b)$
ΔH_{vb} = molar heat of vaporization at normal boiling point
ρ_ℓ = liquid density
$C_{p\ell}$ = molar liquid heat capacity
R = universal gas constant
N = $\begin{cases} 1 \text{ for liquids with densities of } <1\text{g/cm}^3 \\ 0 \text{ for liquids of greater density} \end{cases}$

Deviations from experimental values range from a few percent to perhaps 25%. (However, disagreement between separately measured values of λ_ℓ are often this great.)

Several other estimation methods are available; some are discussed by Reid *et al.* [9], and others are described in Refs. 1, 2, 6, 7 and 8.

Thermal conductivity is only weakly a function of temperature, usually decreasing as temperature increases. At ambient conditions, any temperature correction to λ would likely be less than other uncertainties in the calculation.

24-2   METHOD ERRORS

The combined method of Sato and Riedel for the estimation of thermal conductivity requires very little effort and produces results which are generally within a few percent of the literature value. As mentioned earlier, considerable disagreement may be found in the measured literature values, so these may be acceptable deviations. Reid *et al.* [9] provide measured and estimated values using this method for 31 compounds at one or more temperatures. For the 64 data points covered, the errors range from < 0.1% to 27%, and the average absolute error is 5.5%. An average absolute error of 7.0% was found for the 14 compounds listed in Table 24-3; the errors ranged up to 30%. The method requires only the molecular weight, the boiling point (at normal pressure), and the critical temperature of the compound.

The method developed by Robbins and Kingrea reportedly yields λ_ℓ within 5% of the literature value, with maximum deviations up to 15% [11]. This claim appears justified from the comparison of measured and estimated values (using this method) for 66 data points (32 compounds) given by Reid *et al.* [9].[1] They show errors ranging from < 0.1% to 28% and an average absolute error of 4.8%. However, for the compounds tested in Table 24-3, this method was not outstandingly accurate

---

1. The 66 data points included all of those used to test the Sato and Riedel method plus two additional points for ethylene glycol.

TABLE 24-3

Measured vs. Estimated Thermal Conductivities of Organic Liquids
$(10^{-6}$ cal/cm·s·K)

Compound	Literature Value [12]	Estimated λ_ℓ			
		Sato & Riedel	Error (%)	Robbins & Kingrea	Error (%)
Acetaldehyde	393	399	+ 1.5	313	-20
Acetone	385	377	- 2.1	365	- 5.2
Allyl alcohol	430	418	- 2.8		
Aniline	424	369	-13	382	- 9.9
Bromobenzene	266	268	+ 0.8	286	+ 7.5
n-Butyl acetate	327	226	-31		
Carbon tetrachloride	261	251	- 3.8	192	-26
m-Cresol	358	342	- 4.7		
Ethyl ether	328	337	+ 2.7	263	-20
Iodobenzene	242	240	- 0.8	201	-17
N-Methylaniline	442	325	-26		
Toluene	338	345	+ 2.1	324	- 4.1
Triethylamine	289	313	+ 8.3	283	- 2.1
p-Xylene	325	326	+ 0.3		
Average Absolute Error			7.1%		12%

(average error was 12%). The calculation itself is not difficult, but a substantial amount of user time can be spent in obtaining values for the required inputs — the critical and boiling point temperatures, molecular weight, heat of vaporization, liquid heat capacity, molar and gram densities, and a calculated structural value, H. The latter calculation is subject to error, and other properties must often be estimated because of a lack of measured values. For instance, literature values of the liquid heat capacity were readily available for only 9 of the 14 compounds listed in Table 24-3. Among these nine, the error ranged from 2.1% to 26%. In contrast, the conductivities of all 14 could be estimated by the Sato-Riedel method. For this reason, we do not recommend the method of Robbins and Kingrea for routine use.

24-3 METHOD OF SATO AND RIEDEL

The use of this method and the derived equation [9] is very rapid, requiring only the molecular weight, M, and the reduced temperature and boiling point:

$$\lambda_\ell = \frac{2.64 \times 10^{-3}}{M^{1/2}} \left[\frac{3 + 20(1 - T_r)^{2/3}}{3 + 20(1 - T_{rb})^{2/3}} \right] \qquad (24\text{-}3)$$

Basic Steps

(1) Obtain molecular weight, M.

(2) Look up or estimate critical temperature, T_c, and boiling point, T_b, of the compound. (Methods for estimating T_c and T_b are provided in Chapter 12.)

(3) Calculate T_r and T_{rb} from the simple relationships $T_r = T/T_c$, $T_{rb} = T_b/T_c$.

(4) Insert these parameters into Eq. 24-3 and compute λ_ℓ in cal/cm·s·K.

Example 24-1   Calculate λ_ℓ for acetaldehyde at 293K.

(1)   The molecular weight is 44.05.

(2) T_c = 461K, T_b = 294K [12]

(3) T_r = 293/461 = 0.636, T_{rb} = 294/461 = 0.638

(4) $\lambda_\ell = \dfrac{2.64 \times 10^{-3}}{\sqrt{44.05}} \left[\dfrac{3 + 20\,(1 - 0.636)^{2/3}}{3 + 20\,(1 - 0.638)^{2/3}} \right] = 3.99 \times 10^{-4}$ cal/cm·s·K

The literature value is 3.93×10^{-4} cal·cm·s·K [12]; thus, the error is 1.5%.

Example 24-2   Calculate λ_ℓ for *m*-cresol at 300K.

(1)   The molecular weight is 108.

(2) T_c = 705K, T_b = 475K [12]

(3) T_r = 300/705 = 0.426, T_{rb} = 475/705 = 0.674

(4) $\lambda_\ell = \dfrac{2.64 \times 10^{-3}}{\sqrt{108}} \left[\dfrac{3 + 20\,(1 - 0.426)^{2/3}}{3 + 20\,(1 - 0.674)^{2/3}} \right] = 3.42 \times 10^{-4}$ cal/cm·s·K

The literature value is 3.58×10^{-4} cal·cm·s·K [12]; thus, the error is −4.7%.

24-4   METHOD OF ROBBINS AND KINGREA

This method entails use of the following equation:

$$\lambda_\ell = \frac{(88.0 - 4.94\text{H}) \ 10^{-3}}{\Delta S^*} \left(\frac{0.55}{T_r}\right)^N C_{p\ell}\rho_\ell^{4/3} \qquad (24\text{-}4)$$

Measured or estimated values must be obtained for $C_{p\ell}$, ρ_ℓ, T_c, T_b and ΔH_{vb}. T_r is determined by the relationship $T_r = T/T_c$, and ΔS^* is calculated from the equation

$$\Delta S^* = \Delta H_{vb}/T_b + R \ln (273/T_b) \qquad (24\text{-}5)$$

H is obtained by adding the contributions from Table 24-2, and N is set at 1 or 0, depending on whether ρ_ℓ is less than or greater than unity, respectively.

Basic Steps

(1) Obtain the following input parameters:
    Liquid heat capacity, $C_{p\ell}$ (cal/mol·K)
    Molar liquid density, ρ_ℓ (mol/cm³)
    Critical temperature, T_c (K)
    Boiling point, T_b (K)
    Heat of vaporization at T_b, ΔH_{vb} (cal/mol)

    (Estimation methods for these inputs are available elsewhere in this handbook: $C_{p\ell}$, Chapter 23; ρ_ℓ, Chapter 19; T_c, Chapter 12 (§12-4); T_b, Chapter 12; and ΔH_{vb}, Chapter 13.)

(2) For the temperature of interest, T(K), calculate the reduced temperature, T_r ($= T/T_c$).

(3) Use Eq. 24-5 to obtain ΔS^*. For the gas constant, R, use 1.987 cal/mol·K.

(4) Find the appropriate value of H from Table 24-2.

(5) Set the value of N equal to 1 if $\rho_\ell < 1.0$ g/cm³ (at 20°C) and equal to 0 if $\rho_\ell > 1.0$ g/cm³ (at 20°C).

(6) Substitute all of these input parameters into Eq. 24-4 to obtain λ_ℓ in cal/cm·s·K.

Example 24-3   Calculate λ_ℓ for toluene at 273K.

(1)   From Ref. 12,

$$C_{p\ell} = 35.6 \ \text{cal/mol·K}, \rho_\ell = 0.00941 \ \text{mol/cm}^3, T_c = 594\text{K},$$

$$T_b = 384\text{K, and } \Delta H_{vb} = 8581 \ \text{cal/mol}.$$

(2) $T_r = 273/594 = 0.460$

(3)   From Eq. 24-5,

$$\Delta S^* = 8581/384 + 1.987 \ln (273/384)$$
$$= 21.7$$

(4)   Table 24-2 lists a value of 1 for H, since toluene has one CH_3 branch.

(5) $\rho_\ell < 1 g/cm^3$. Therefore, N=1.

(6)   Substituting these values into Eq. 24-4,

$$\lambda_\ell = \frac{(88.0 - 4.94)\ 10^{-3}}{21.7} \left(\frac{0.55}{0.46}\right)^1 35.6\ (0.00941)^{4/3}$$

$$= 0.324 \times 10^{-3}\ cal/cm \cdot s \cdot K$$

Since the literature value [12] of λ_ℓ is 0.338×10^{-3} cal/cm·s·K, the error with this method is −4.1%.

24-5   AVAILABLE DATA

Thermal conductivity data are not as readily available as most other physical properties. Conventional handbooks such as Weast [12] list some values, but there are no major literature compilations other than Jamieson [2,3]. Manufacturers' data sheets and similar commercial publications should be consulted when possible. Appendix A lists a few additional sources.

24-6   SYMBOLS USED

$C_{p\ell}$	=	molar liquid heat capacity (cal/mol·K)
H	=	structural factor from Table 24-2
ΔH_{vb}	=	molar heat of vaporization at normal boiling point (cal/mol)
M	=	molecular weight (g/mol)
N	=	exponent in Eq. 24-4, (1 for liquid density <1g/cm, 0 for density >1g/cm)
R	=	gas constant (1.987 cal/mol·K)
ΔS^*	=	entropy term in Eq. 24-4; evaluated with Eq. 24-5
T	=	temperature (K)
T_b	=	boiling point (K)
T_c	=	critical temperature (K)
T_r	=	reduced temperature = T/T_c
T_{rb}	=	reduced boiling point = T_b/T_c

Greek

ρ_ℓ =   molar liquid density (mol/cm³)
λ =   thermal conductivity (cal/cm·s·K)
λ_ℓ =   liquid thermal conductivity (cal/cm·s·K)

24-7   REFERENCES

1. Christensen, P.L., and A.A Fredenslund, "A Corresponding States Model for the Thermal Conductivity of Gases and Liquids," *Chem. Eng. Sci.,* **35,** 871 (1980).

2. Jamieson, D.T., "Thermal Conductivities of Liquids," *J. Chem. Eng. Data,* **24,** 244 (1979).

3. Jamieson, D.T., J.B. Irving and J.S. Tudhope, *Liquid Thermal Conductivity: A Data Survey to 1973,* H.M. Stationery Office, Edinburgh, 1975.

4. Jamieson, D.T. and J.S. Tudhope, Report 137, National Engineering Laboratory, Glasgow (March 1964).

5. Maejima, T., Private communication to R. Reid, 1973, offering suggestions of Professor K. Sato of Tokyo Institute of Technology. (Cited in Ref. 9, p. 541.)

6. Mathur, V.K., J.D. Singh, and W.M. Fitzgerald, "Estimation of Thermal Conductivity of Liquid Hydrocarbons," *J. Chem. Eng. (Japan),* **11,** 67 (1978).

7. Miller, J.W., J.J. McGinley and C.L. Yaws, "Thermal Conductivities of Liquids," *Chem. Eng. (NY),* 133 (Oct. 25, 1976).

8. Missenard, F.A., "A Propos de la Conductivité Thermique des Séries de Liquides Organiques," *Rev. Gén. Therm. Fr.,* **141,** 751-59 (September 1973).

9. Reid, R.C., J.M. Prausnitz and T.K. Sherwood, *The Properties of Gases and Liquids,* 3rd ed., McGraw-Hill Book Co., New York (1977).

10. Riedel, L., *Chem. Ing. Tech.,* **21,** 349 (1949) and **23,** pp. 59, 321, 465 (1951). (Cited in Ref. 9, p. 542.)

11. Robbins, L.A. and C.L. Kingrea: (a) *Hydrocarbon Process. Pet. Refiner,* **41** (5), 133 (1962); (b) preprint of paper presented at the Sess. Chem. Eng. 27th Midyear Meet. Am. Pet. Inst., Div. Refining, San Francisco, May 14, 1962. (Cited in Ref. 9, p. 542.)

12. Weast, R.C. and M.J. Astle (eds.), *Handbook of Chemistry and Physics,* 60th ed., The Chemical Rubber Co., Cleveland (1979).

25

DIPOLE MOMENT

Leslie H. Nelken and James D. Birkett

25-1   INTRODUCTION

Dipole moments result from electronegativity differences of the atoms within a molecule, electronegativity being defined by Pauling as the "power of an atom in a molecule to attract electrons to itself" [3]. A molecule becomes dipolar if the electrons forming the bond accumulate toward more electronegative atoms, leaving less electronegative atoms with a slight positive charge. The dipole moment, μ, can be defined as the vector in the direction of negative to positive charge of magnitude Qr, where Q is the charge and r is the charge separation distance. The unit of the dipole moment is the debye (D). One debye is equivalent to 10^{-18} electrostatic units-cm (esu-cm). For example, one positive and negative charge (4.8×10^{-10} esu/charge), separated by a distance of 0.1 nm (or 10^{-8} cm) would equal 4.8×10^{-18} esu-cm, or 4.8 debyes. Typically, dipole moments of organic molecules fall in the range of 0 to 5 debyes. Large molecules, such as dyes, may have a moment of up to 17 debyes [6].

Property correlations for polar compounds often require this parameter. For example, the boiling point of a chemical may be related to its heat of vaporization with a parameter (K_F) that may be derived from the dipole moment (see §14-3). This particular relationship is used in this handbook in one of the estimation methods for vapor pressure (see §14-3) and one for liquid viscosity (see §22-6). The dipole moment is also used in some pressure-volume-temperature relationships and in some estimation methods for the diffusion coefficients of polar gases.

The dipole moment is a fairly significant parameter used to aid in the determination of molecular structure, bond angles, and resonance. For instance, the fact that carbon dioxide has no dipole moment leads one to conclude the molecule is linear; water, on the other hand, cannot be linear, because it has a measured dipole moment. Both of these conclusions are in agreement with the structures predicted from molecular orbital considerations. The basis of light scattering in Raman spectroscopy is the dipole moment induced in the molecule when hit by the incident light beam [3]. Dipole moments are also responsible for the interaction of the molecules with radiation in the infrared and microwave regions, giving rise to their characteristic vibrational-rotational spectra. Lastly, the dipole moment is useful in ascertaining the potential for a molecule to interact with its surrounding medium via hydrogen bonding, Van der Waals forces, and dipole-dipole attractions [1]. (Van der Waals force differs from dipole-dipole attraction in that the former is an attraction between small, transient charges in molecules which are normally considered non-polar, whereas the latter involves permanent molecular dipoles [2].)

Dipole moments may be experimentally determined by a number of methods, including microwave spectroscopy, indices of refraction, infrared vibration/rotation spectra, and electron resonance. The most common method is based upon Debye's equation relating the dipole moment to the dielectric constant [1] and temperature [3,8]:

$$\frac{(\epsilon-1)M}{(\epsilon+2)\rho} = \frac{4\pi\, N\alpha}{9\epsilon_0 kT} + \frac{4\pi\, N\mu^2}{9\epsilon_0 kT} \qquad (25\text{-}1)$$

where

ϵ	=	dielectric constant (dimensionless)
ϵ_0	=	dielectric constant of vacuum (dimensionless)
M	=	molecular weight (g/mol)
ρ	=	density (g/cm^3)
N	=	Avogadro's number, 6.02×10^{23} molecules/mole
α	=	polarizability factor, or dipole induced by the electric field
k	=	Boltzmann's constant, 1.3806×10^{-23} J/K
T	=	temperature (K)

---

1. A dielectric is an insulating substance, and ϵ is the factor by which the dielectric increases the capacitance above that measured in a vacuum.

The dielectric constant, ϵ, can be calculated from the capacitance of a parallel-plate capacitor in which the dielectric is the vapor or solution between the plates [3]. A plot of the left-hand side of Eq. 25-1 versus $1/T$ over a wide temperature range will provide a slope (b) from which μ in the vapor or liquid state can be determined. LeFevre [2] has shown that when the constants are inserted into Eq. 25-1, the following equation is obtained:

$$\mu = 0.012816 \ \sqrt{b} \times 10^{-18} \qquad (25\text{-}2)$$

There are two kinds of dipole moments — induced and permanent. The *induced* dipole, reflected by α in Eq. 25-1, is created by an electric field. Its value is temperature-independent, since the molecule will reorient itself in the direction of the field after it is perturbed by a thermally agitated molecule. The *permanent* dipole is caused by the electronegativity differences of the atoms within the molecule and is temperature-dependent. At higher temperatures, the random movement of molecules opposes their tendency to become oriented in the direction of the electric field.

Equation 25-1 applies most accurately to gases when the pressure is low enough so that the gas molecules have little opportunity to interact. However, measurement of μ (or ϵ) in the gaseous state is a difficult procedure, so the measurement is often made with the molecule in the liquid state. Measurement of ϵ in the liquid state is affected by the forces exerted by the closely packed molecules upon each other, and by the solvent interacting with the molecules of interest. The effects of solvation may be minimized by measuring ϵ in a dilute solution in which the solvent is a non-polar organic liquid such as benzene or carbon tetrachloride.

25-2   AVAILABLE ESTIMATION METHODS

The most accurate methods for estimating dipole moments require a knowledge of the bond angles between atoms. Only a few such methods have been developed, and their complexity makes them unsuitable for inclusion in this handbook. Readers who desire further details should refer to Smyth's comprehensive dissertation on this subject [6].

The dipole moment may also be calculated with a computer program, CAMSEQ, which performs a variety of calculations related to molecular mechanics and molecular structure [4]. The heart of this program involves the calculation of conformational potential energy for selected molecular conformations until a structure of minimum energy is found. Various properties of the molecule are then calculated for the preferred structure. A significant amount of computer time is required to carry out this analysis.

Dipole moments of certain molecular structures can be closely estimated without resorting to large-scale computations. Fishtine [1] developed a fairly easy method for calculating the dipole moment of substituted benzenes, naphthalene derivatives, and heterocyclics containing nitrogen, oxygen, and sulfur. Unfortunately, similar methods are not available for aliphatic and alicyclic compounds. Table 25-1 lists the average values (x) of experimentally determined dipole moments for about 20 chemical classes of aliphatic and alicyclic liquids and vapors and also shows the standard deviations (σ) from the average. Most dipole moments fall in the range of 0 to 5 debye units.

25-3   ESTIMATION OF DIPOLE MOMENTS FOR AROMATIC COMPOUNDS

Fishtine [1] describes methods for calculating the dipole moment of substituted benzenes, naphthalenes, and heterocyclic compounds such as pyridine, furan, and thiophene. Excluded from this group are substituents that participate in hydrogen bonding, such as phenols and anilines. The steps in the calculation differ for each class of compound. There are three sub-categories of substituted benzenes to reflect variations in the symmetry of the substituents.[2]

Table 25-2 lists the dipole moment contributions (compiled by Smyth [6]) of about 30 substituents to benzene rings, including halogens, cyanides, sulfide groups, methyl esters, and cyanates. These are grouped according to the direction in which the dipole points (i.e., the location of the negative end of the dipole). Asymmetrical substituents are designated by an "x" in the third column.

The dipole moment was calculated for a sample group of aromatic compounds (Table 25-3) to find the absolute error associated with this method. Most errors were in the 2-30% range. From the small group of compounds tested, it is not possible to determine the chemical classes for which the method will yield poor results and why.

---

2. "Symmetry" in this case refers to that of the atoms that comprise the substituent rather than the arrangement of substituents around the benzene ring.

TABLE 25-1

Typical Values of Experimentally Determined
Dipole Moments of Aliphatic and Alicyclic Chemicals

Chemical Class	Liquids			Vapors		
	n	$\bar{x}$	σ	n	$\bar{x}$	σ
Paraffins	4	0	0.0	7	0	0.0
Saturated cyclic hydrocarbons	8	0	0.0	4	0	0.0
Mono/di-alkenes				15	0.4	0.3
Alkyl acetylenes				7	0.6	0.4
Alkyl halides				20	2.0	0.1
Polyhalogenated alkanes	15	1.35	0.7	26	1.1	0.8
Halogenated cyclic alkanes	28	1.68	1.6			
Halogenated alkenes	21	1.1	0.7	12	1.6	0.5
Halogenated alkynes	9	0.8	0.4			
Cyanogens, HCN, substituted cyanides	6	2.9	0.7	3	1.9	1.6
Alkyl, cyclic, and unsaturated cyanides	11	3.7	0.2	8	4.0	0.2
Isocyanides, isocyanates, thiocyanates, isothiocyanates	7	3.3	0.3			
Nitroalkanes, chloronitroalkanes	9	2.7	1.1	11	3.0	1.1
Aldehydes	15	2.6	0.5	7	2.8	0.4
Ketones	42	2.7	0.7			
Ethers	11	1.3	0.3	5	1.2	0.1
Alicyclic oxides	21	1.6	0.6			
Alcohols	8	1.6	0.0	6	1.7	0.0
Sulfides/mercaptans	13	1.4	0.3	6	1.3	0.4
Monocarboxylic acids, esters	34	1.9	0.2			
Amines	10	1.2	0.2			
Fulvenes and azulene	10	1.2	0.3			

n = no. of compounds; $\bar{x}$ = avg. value (debyes); σ = std. deviation (debyes)

Source: Smyth [6]. *(Reprinted with permission from McGraw-Hill Book Co.)*

Calculation of the dipole moment by this method entails the addition of vectors lying in the same plane. Each substituent dipole moment (μ) can be considered to have a vertical component (μ_v) and a horizontal component (μ_h):

$$\mu_v = \mu \sin \theta \qquad (25\text{-}3)$$

$$\mu_h = \mu \cos \theta \qquad (25\text{-}4)$$

TABLE 25-2

Magnitude and Direction of
Electrical Moments of Benzene Derivatives

Group A in C_6H_5-A	Dipole Moment of C_6H_5-A	Substituent is Asymmetric	Vector Points:
–H	0.0		
–CH=CH$_2$	a		
–CH$_3$	0.4		
–C$_2$H$_5$	0.4	x	
–CH(CH$_3$)$_2$	0.6	x	Toward Ring
–C(CH$_3$)$_3$	0.5		
–N(CH$_3$)$_2$	1.57	x	
–F	1.46		
–Cl	1.58		
–Br	1.54		
–I	1.30		
–CH$_2$Cl	1.85	x	
–CHCl$_2$	2.04	x	
–CCl$_3$	2.11		
–CH$_2$Br	1.86	x	
–CN	3.90		
–CH$_2$CN	3.5	x	
–NC	3.5		
–NCO	2.32		Away from Ring
–SCN thiocyanate	3.59		
–NCS isothiocyanate	2.9		
–NO$_2$	3.98		
–CH$_2$NO$_2$	3.3	x	
–NO	3.14		
–CHO	2.76	x	
–COCH$_3$	2.89	x	
–OCH$_3$	1.25	x	
–SCH$_3$	1.3	x	
–OH	1.6	x	
–SH	1.3	x	
–COOCH$_3$	1.83	x	

a. Small, but greater than zero.

Source: Fishtine [1]. *(Reprinted with permission from the American Chemical Society.)*

TABLE 25-3

Comparison of Observed and Estimated Dipole Moments

| Compound | Dipole Moment, μ (debyes) | | Error |
	Observed[a]	Estimated[b]	(%)
p-Bromophenoxy-*p*-toluene	2.45	2.30	− 6
p-Nitrophenoxybenzene	4.54	3.5	−23
m-Nitromethoxybenzene	4.55	3.5	−23
o-Fluorotoluene	1.86	1.7	− 9
o-Chloro (trifluoromethyl) benzene	3.46	3.2	− 7
m-Difluorobenzene	1.58	1.5	− 5
2,3,4,5-Tetrabromothiophene	0.73	0.5	−26
2-Mercaptobenzothiazole	4.0	2.4	−40
4-Methoxypyridine	2.94	1.9	−34
2,7-Dichloronaphthalene	1.53	1.2	−21
1-Bromo-5-nitronaphthalene	2.49	2.4	− 2
p-Nitrobiphenyl	4.22	4.0	− 5
1,2,3,4-Tetramethyldinitrobenzene	6.86	7.6	+10
1,2-Dimethyl-3,4,5-trichlorobenzene	2.46	3.7	+50
m-Chloronitrobenzene	3.40	3.5	+ 2
p-Bromobenzaldehyde	2.20	2.4	+ 9
p-Benzenedicarboxaldehyde	2.35	2.8	+17

Average Absolute Error   17%

Standard Deviation   14.1%

a. As reported by Smyth [6]. *(Reprinted with permission from McGraw-Hill Book Co.)*
b. Calculated from substituent moments listed in Table 25-2.

The magnitude of these components can be calculated by Eqs. 25-3 and -4 from μ and the angle (θ) of the substituent vector from horizontal. Just as a point may be represented in Cartesian coordinates as (x,y), we may represent the substituent vector as (μ_h, μ_v). The sum of two vectors (1 and 2) is, in this manner, represented by the sum of their individual horizontal and vertical components, i.e., ($\mu_{h_1} + \mu_{h_2}$, $\mu_{v_1} + \mu_{v_2}$). Vector components that point in the positive direction of the selected coordinates are positive, and those that point in the negative direction are negative. In Cartesian coordinates, positive signs are normally associated with the "up" direction (μ values above the abscissa) and with the "right" direction (x values to the right of the ordinate). Thus, in the diagram above, both μ_v and μ_h would be positive; if μ had pointed in the opposite direction, both μ_v and μ_h would be negative.

The length of any vector expressed by its vertical and horizontal components, μ_v and μ_h, is

$$\mu = \sqrt{\mu_v^2 + \mu_h^2} \qquad (25\text{-}5)$$

When this vector is the sum of several vectors,

$$\mu = \left[\left(\sum_{i=1}^{n} \mu_{v_i} \right)^2 + \left(\sum_{i=1}^{n} \mu_{h_i} \right)^2 \right]^{\frac{1}{2}} \qquad (25\text{-}6)$$

where the individual values of μ_{v_i} and μ_{h_i} for the n substituents on the benzene ring are the values obtained from Eqs. 25-3 and -4, respectively.

The angles of the benzene-substituent bonds are individually separated by 60°, as shown below.

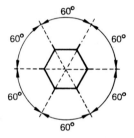

Thus, if one substituent position is set at $\theta = 0°$, the θ for any other substituent must be some multiple of 60°. The signs (+ or —) of the μ_v and μ_h values are automatically calculated if θ is always measured from the same reference line (e.g., the positive direction of the abscissa) and in the same direction (e.g., counterclockwise) around to the direction in which the substituent dipole is pointing. These directions are either directly toward or away from the center of the benzene ring (see Table 25-2).

An easy way to obtain the correct θ for each substituent is to draw the basic benzene structure with arrows at each substituent location pointing in the correct direction. Then transfer the arrows to a diagram where the tails of all arrows start at a common point. Finally, measure the θ angles by moving counterclockwise from the positive abscissa direction. Examples are given in Figure 25-1.

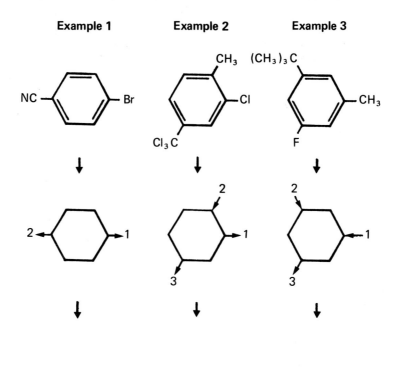

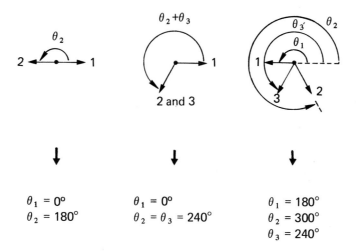

$\theta_1 = 0°$	$\theta_1 = 0°$	$\theta_1 = 180°$
$\theta_2 = 180°$	$\theta_2 = \theta_3 = 240°$	$\theta_2 = 300°$
		$\theta_3 = 240°$

FIGURE 25-1   Examples of the Determination of θ

The method is described in more detail below for four different classes of compounds:

(1) Substituted benzenes with no hydrogen bonding
  • Symmetrical functional groups
  • One asymmetrical functional group
  • More than one asymmetrical functional group
(2) Naphthalene derivatives
(3) Heterocyclics
(4) Miscellaneous compounds

Substituted Benzenes with No Hydrogen Bonding

• *Symmetrical Functional Groups.* Addition of symmetrical functional groups on the benzene ring is a case of straightforward vector addition.

Basic Steps

(1) Draw the chemical structure.

(2) Obtain the dipole moment of each substituent (μ) from Table 25-2. Note the direction of the dipole moment as indicated by the direction of the arrow in the right-hand column of the table.

(3) Draw a benzene ring with an arrow at each substituent position pointing in the direction indicated by Step 2. (See Figure 25-1 for examples.) Then determine the angle, θ, for each substituent vector (arrow).

(4) Calculate the vertical and horizontal components of each substituent dipole (μ_v and μ_h) using Eqs. 25-3 and -4.

(5) Calculate the resulting dipole for the whole molecule, μ, from Eq. 25-6. The answer is in units of debyes.

Example 25-1   Estimate the dipole moment of p-nitrotoluene.

(1)   The structure is

(2)   From Table 25-2, the substituent dipoles (μ) for –CH_3 and O_2N– are 0.4 and 3.98, respectively. Table 25-2 indicates an arrow pointing toward the ring for –CH_3 and away from the ring for O_2N–.

(3)   The vectors are represented as:

θ is 180° for both vectors 1 and 2

(4)   The vertical components (μ_v) of both vectors are 0, since sin (180°) = 0. The horizontal components (Eq. 25-4) are

$$\mu_h \ (O_2N-) = 3.98 \cos (180°) = 3.98 \ (-1) = -3.98$$

$$\mu_h \ (-CH_3) = 0.4 \cos (180°) = 0.4 \ (-1) = -0.4$$

(5)   The resulting dipole for the molecule, from Eq. 25-6, is

$$\mu = \sqrt{(-0.4 - 3.98)^2} = 4.4 \text{ D}$$

The measured value is also 4.4 D [1].

Example 25-2   Estimate the dipole moment of 1,2,3-trichlorobenzene

(1)   The structure is

(2)   From Table 25-2, the substituent dipole (μ) for −Cl is 1.58 and the vector points away from the ring.

(3)   The benzene vector diagram is:

θ for vector 1 is 360° or 0°:

θ for vector 2 is 60°

θ for vector 3 is 120°

(4)   The vertical component of each vector is:

$$\mu_v(1) \; = \; 1.58 \sin (0°) = 0$$
$$\mu_v(2) \; = \; 1.58 \sin (60°) = 1.37$$
$$\mu_v(3) \; = \; 1.58 \sin (120°) = 1.37$$

The horizontal component of each vector is:

$$\mu_h(1) \; = \; 1.58 \cos (0°) = 1.58$$
$$\mu_h(2) \; = \; 1.58 \cos (60°) = 0.79$$
$$\mu_h(3) \; = \; 1.58 \cos (120°) = -0.79$$

(5)   The resulting dipole for the molecule, from Eq. 25-6, is

$$\mu = \sqrt{[\, 0 + 2\,(1.37)]^2 + (1.58 + 0.79 - 0.79)^2} = 3.16 \text{ D}$$

The measured value is 2.31 [1], giving a deviation of 37%.

• *One Asymmetrical Functional Group*

Basic Steps.   The calculation of μ is the same as in the case of symmetrical functional groups, except that if the asymmetrical substituent is in the para position, its horizontal and vertical components should be calculated as though it were in the meta position.

Example 25-3   Estimate the dipole moment of p-chlorobenzaldehyde.

(1)   The structure is

(2)   From Table 25-2, the substituent dipole (μ) is 1.58 for $-$Cl and 2.76 for $-$CHO. The table shows that $-$CHO is asymmetrical and that both dipoles point away from the ring.

(3)   The benzene vector diagram is as follows (the asymmetrical $-$CHO group is treated as though it were in a position meta to the symmetrical Cl group):

θ for Cl is $0°$:

θ for $-$ CHO is $120°$:

(4) The vertical component of each vector is:

$$\mu_v(-Cl) \quad = \quad 1.58 \sin (0°) = 0$$

$$\mu_v(-CHO) = 2.76 \sin (120°) = 2.39$$

The horizontal components are:

$$\mu_h (-Cl) \quad = \quad 1.58 \cos (0°) = 1.58$$

$$\mu_h(-CHO) = 2.76 \cos (120°) = -1.38$$

(5) From Eq. 25-6, the resulting dipole for the molecule is

$$\mu = \sqrt{(2.39)^2 + (1.58 - 1.38)^2} = 2.40$$

The experimental value is 2.03 [1]; the deviation is +18%.

• *More Than One Asymmetrical Functional Group*

Basic Steps. If the asymmetrical groups are ortho or meta to each other, the calculation of μ is the same as for symmetrical functional groups. If they are para to each other, perform the calculation as though one group were shifted to the meta position, such that the maximum moment for the compound is achieved.

Example 25-4 Estimate the dipole moment of p-acetylacetophenone.

(1) The structure is:

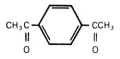

(2) From Table 25-2, the substituent dipole value for $-COCH_3$ is 2.89. The substituent is asymmetrical, and its vector points away from the ring.

(3) The benzene vector diagram is drawn as though one $-COCH_3$ is moved to a position meta to the other:

θ for vector 1 is $0°$:

θ for vector 2 is $120°$:

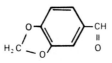

(4)   The vertical component of each vector is:

$$\mu_v (CH_3CO)_1 = 2.89 \sin (0°) = 0$$

$$\mu_v (CH_3CO)_2 = 2.89 \sin (120°) = 2.50$$

The horizontal component is:

$$\mu_h (CH_3CO)_1 = 2.89 \cos (0°) = 2.89$$

$$\mu_h (CH_3CO)_2 = 2.89 \cos (120°) = -1.44$$

(5)   From Eq. 25-6, the resulting dipole for the molecule is:

$$\mu = \sqrt{(2.50)^2 + (2.89 - 1.44)^2} = 2.89$$

The experimental value is 2.71 [1], indicating a + 7% error.

Example 25-5   Estimate the dipole moment of piperonal.

(1)   The structure is:

(2)   Since the μ value of the cyclic ether substituent is not listed in Table 25-2, replace it with two separate substituents with known dipole moments. The $-OCH_3$ group is the most appropriate candidate.

(3)   The molecule is now assumed to have one $-CHO$ substituent ($\mu = 2.76$) and two $-OCH_3$ substituents ($\mu = 1.25$). All of these are asymmetrical, and their vectors point away from the ring.

(4)   Since two asymmetrical groups lie para to each other, the benzene vector diagram is modified as follows:

The vector angles are $180°$, $240°$, and $300°$ from the horizontal.

The other option would be to move the p-OCH$_3$ group to a position meta to the CHO group. This option is discounted since it will not result in a maximum dipole for the compound.

(5) The vertical component of each vector is:

$$\mu_v\,(-CHO) \;\;\;= \;2.76\sin(300°) = -2.39$$

$$\mu_v\,(-OCH_3)_1 = \;1.25\sin(240°) = -1.08$$

$$\mu_v\,(-OCH_3)_2 = \;1.25\sin(180°) = 0$$

The horizontal component of each vector is:

$$\mu_h\,(-CHO) \;\;\;= \;2.76\cos(300°) = -1.38$$

$$\mu_h\,(-OCH_3)_1 = \;1.25\cos(240°) = -0.625$$

$$\mu_h\,(-OCH_3)_2 = \;1.25\cos(180°) = -1.25$$

(6) From Eq. 25-6, the resulting dipole for the molecule is:

$$\mu = \sqrt{(-2.39 - 1.08)^2 + (1.38 - 0.625 - 1.25)^2} = 3.50$$

No value for piperonal was found in the literature.

Naphthalene Derivatives. For estimation purposes, assume the naphthalene molecule to be octagonal, making the angle between two adjacent substituents $45°$. Use Table 25-2 for the moments of substituent groups and apply the rules for symmetrical substituents given in the preceding section.

Basic Steps

(1) Draw the chemical structure.

(2) Obtain the dipole moment of each substituent, μ, from Table 25-2. Note the direction of the dipole moment indicated in the right-hand column.

(3) Draw the naphthalene structure with vectors at each substituent position in the direction found in Step 2. Then transform the structure into an octagon (representing the

eight possible points of attachment in naphthalene), with the x-axis passing through the middle of the vertical sides. Determine the angle, θ, for each substituent by moving counterclockwise from $0°$.

(4) Calculate the vertical and horizontal components of each vector (μ_v and μ_h) using Eqs. 25-3 and -4.

(5) Calculate the resulting dipole for the whole molecule, μ, from Eq. 25-6.

Example 25-6   Calculate the dipole moment of 1,4-dichloronaphthalene.

(1)   The structure is

(2)   From Table 25-2, the substituent dipole value of $-Cl$ is 1.58. The substituent is symmetrical, and its vector points away from the ring.

(3)   The naphthalene vector diagram develops as follows:

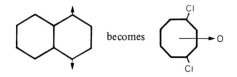

becomes

The vector angles are $67.5°$ and $292.5°$ from the horizontal:

(4)   The vertical components of the vectors are:

$$\mu_v \,(Cl_1) = 1.58 \sin (67.5°) = 1.46$$

$$\mu_v \,(Cl_2) = 1.58 \sin (292.5°) = -1.46$$

The horizontal component of the vector is:

$$\mu_h \,(Cl_1) = 1.58 \cos (67.5°) = 0.60$$

$$\mu_h \,(Cl_2) = 1.58 \cos (292.5°) = 0.60$$

(5)   The resultant dipole moment (Eq. 25-6) is

$$\mu = \sqrt{(1.46 - 1.46)^2 + (0.60 + 0.60)^2} = 1.2$$

The measured value of the dipole is 0.48 [6], indicating a large error of 152%. The measured dipole of 1,5-dichloronaphthalene is 1.78. Obviously, the angle approximation of 45° in the molecule contributes to the error.

Example 25-7   Estimate the dipole moment for 1,5-bromonitronaphthalene.

(1)   The structure is

(2)   The substituent values for $-Br$ and $-NO_2$ from Table 25-2 are 1.54 and 3.98, respectively, and both dipoles point away from the ring.

(3)   The naphthalene vector diagram appears as follows:

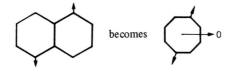

becomes   0

The vector angle for $-NO_2$ is 67.5°

The vector angle for $-Br$ (247.5°) appears as

(4)   The vertical component of the vector is:

$$\mu_v\,(-NO_2) = 3.98 \sin\,(67.5°) = 3.68$$
$$\mu_v\,(-Br)\ \ = 1.54 \sin\,(247.5°) = -1.42$$

The horizontal component is

$$\mu_h\,(-NO_2) = 3.98 \cos\,(67.5°) = 1.52$$
$$\mu_h\,(-Br)\ \ = 1.54 \cos\,(247.5°) = -0.59$$

(5)   From Eq. 25-6, the calculated dipole moment is

$$\mu = \sqrt{(3.68 - 1.42)^2 + (1.52 - 0.59)^2} = 2.44$$

Smythe [6] reports a measured value of 2.49, indicating an error of -2.0%.

Heterocyclics. The dipole vector diagrams for pyridine, furan, and thiophene are illustrated below. The arrow indicates that for all three heterocyclics, the positive end of the dipole is at the ring. For five-membered rings, assume that the bond angles are 60°, as for six-membered rings [1].

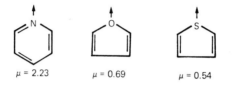

$\mu = 2.23$ $\mu = 0.69$ $\mu = 0.54$

Basic Steps. Follow the sequence of steps outlined for symmetrical functional groups on benzene. The only difference is that the parent molecule possesses a dipole moment.

Example 25-8   Estimate the dipole moment of 4-methylpyridine.

(1)   The structure is:

(2)   Table 25-2 indicates the substituent value of $-CH_3$ is 0.4, pointing toward the ring. The dipole value of pyridine is 2.23, pointing away from the ring.

(3)   The pyridine vector diagram is

θ is 180° for both vectors.

(4)   The vertical component, μ_v, of both vectors is 0, since sin 180° = 0. The horizontal components, from Eq. 25-4, are:

$$\mu_h (-CH_3) = 0.4 \cos (180°) = -0.4$$

$$\mu_h (\text{pyridine}) = 2.23 \cos (180°) = -2.23$$

(5)   The resulting dipole for the molecule, from Eq. 25-6, is

$$\mu = \sqrt{(0)^2 + (-0.4 - 2.23)^2} = 2.63$$

The observed value is 2.57 [6], indicating a +2.3% error.

Example 25-9   Estimate the dipole moment of 2,5-dichlorothiophene.

(1)   The structure is:

(2)   Table 25-2 lists the substituent value of $-Cl$ as 1.58. All dipole vectors point away from the ring.

(3)   The thiophene diagram is:

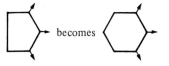

becomes

(4)   The vector angles are 60° and 300° for the $-Cl$ substituents and 0° for thiophene.

The vertical components of the vectors are:

$$\mu_v (-Cl)_1 = 1.58 \sin 60° = 1.37$$

$$\mu_v (\text{thiophene}) = 0.54 \sin 0° = 0$$

$$\mu_v (-Cl)_2 = 1.58 \sin 300° = -1.37$$

The horizontal components are:

$$\mu_h (-Cl)_1 = 1.58 \cos 60° = 0.79$$

$$\mu_h (\text{thiophene}) = 0.54 \cos 0° = 0.54$$

$$\mu_h (-Cl)_2 = 1.58 \cos 300° = 0.79$$

(5)   The resulting dipole moment from Eq. 25-6 is

$$\mu = \sqrt{(1.37 + 0 - 1.37)^2 + (0.79 + 0.54 + 0.79)^2} = 2.12$$

The value reported in the literature is 1.12 [6], indicating a large (89%) error.

Miscellaneous Compounds. Calculation of μ by this method requires judgment on the part of the user. Fishtine [1] presents a few cases (e.g., Example 25-5) where the procedure was not strictly circumscribed. Another example is salicylaldehyde (o-hydroxybenzaldehyde): typically, this molecule would be excluded from the estimation technique, since the $-OH$ substituent could participate in hydrogen bonding. Knowledge of the chemistry of this molecule, however, would permit evaluation of μ, since the $-OH$ participates in *intra*molecular hydrogen bonding and thus cannot interact in *inter*molecular hydrogen bonding.

A third example requiring judgment on the part of the user is estimation of μ for phthalic anhydrides:

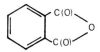

Fishtine suggests converting the molecule to

for which the information necessary to calculate μ exists.

25-4   AVAILABLE DATA

Values for dipole moments are given by Smyth [6] and in the following two books referenced by Reid *et al.* [5]:

McClellan, A.L., *Tables of Experimental Dipole Moments,* Freeman, San Francisco (1963).

Minkin, V.I., O.A. Osipov and Y.A. Zhdanov, *Dipole Moments in Organic Chemistry,* translated from Russian by B.J. Hazard, Plenum Press, New York (1970).

Another source is:

Gordon, A.J. and R.A. Ford, *The Chemist's Companion. A Handbook of Practical Data, Techniques, and References,* John Wiley and Sons, New York (1972).

The dipole moments of a few compounds are listed in Ref. 7 and in other sources listed in Appendix A.

25-5   SYMBOLS USED

b	=	parameter in Eq. 25-2
k	=	Boltzmann's constant in Eq. 25-1
M	=	molecular weight in Eq. 25-1 (g/mol)
N	=	Avogadro's number in Eq. 25-1
n	=	number of samples
Q	=	charge on one electron (= 4.8×10^{-10} esu)
r	=	distance of charge separation (cm)
T	=	temperature in Eq. 25-1 (K)
$\bar{x}$	=	average of n values
x	=	abscissa component of vector
y	=	ordinate component of vector

Greek

α	=	polarity in Eq. 25-1
ϵ	=	dielectric constant in Eq. 25-1
μ	=	dipole moment in Eq. 25-1 (debye)
ρ	=	density in Eq. 25-1 (g/cm^3)
σ	=	standard deviation
θ	=	angle of substituent vector from horizontal (see Fig. 25-1)

Subscripts

h	=	horizontal component of vector
v	=	vertical component of vector

25-6   REFERENCES

1. Fishtine, S.H., "Reliable Latent Heats of Vaporization," *Ind. Eng. Chem.,* **55** (6), 47-56 (1963).

2. LeFevre, R.J.W., *Dipole Moments, Their Measurement and Application in Chemistry,* John Wiley and Sons, New York (1953).

3. Moore, W.J., *Physical Chemistry,* 4th ed., Prentice-Hall, Englewood Cliffs, N.J. (1972).

4. Potenzone, R., Jr., E. Cavicchi, H.J.R. Weintraub and A.J. Hopfinger, "Molecular Mechanics and the CAMSEQ Processor,"*Comput. Chem.,* **1,** 187-94 (1977).

5. Reid, R.C., J.M. Prausnitz, and T.K. Sherwood, *The Properties of Gases and Liquids,* 3rd. ed., McGraw-Hill Book Co., New York (1977).

6. Smyth, C.P., *Dielectric Behavior and Structure,* McGraw-Hill Book Co., New York (1955).

7. Weast, R.C. (ed.), *Handbook of Chemistry and Physics,* 59th ed., Chemical Rubber Co., Cleveland, Ohio (1979-80).

8. Wheatley, P.J., *The Determination of Molecular Structure,* Oxford University Press, London (1959).

26

INDEX OF REFRACTION

Leslie H. Nelken

26-1   INTRODUCTION

The index of refraction (n) and molar refraction (R_D) are indications of the manner in which a molecule interacts with light. The index of refraction is the ratio of the speed of light in a vacuum (c) to the speed of light in the medium (v):

$$n = c/v \qquad\qquad (26\text{-}1)$$

This is a dimensionless parameter which ranges between 1.3 and 1.5 for organic liquids. (See, for example, Ref. 15.)

The refractive index is measured using a beam of monochromatic light — typically, the yellow light of the sodium D line (wavelength λ = 589.3 nm). Thus, n_D^{20} indicates the wavelength used, D, and the temperature, 20°C. Other wavelengths used are the C and F lines of hydrogen (λ = 656.3 nm and 486.1 nm, respectively) and the G line of mercury (λ = 435.8 nm) [12].

Molar refraction, R_D, is a function of the density, ρ, of the medium. The Lorentz-Lorenz equation (26-2) expresses the relationship between R_D, ρ, and n, based upon electromagnetic theory:

$$R_D = \left(\frac{n^2 - 1}{n^2 + 2} \right) \frac{M}{\rho} \qquad\qquad (26\text{-}2)$$

where M is molecular weight and R_D has units of volume. (A term related to R_D is the specific refraction, which equals R_D divided by M.) Rearrangement of Eq. 26-2 allows evaluation of n:

$$n = \sqrt{\frac{M + 2\rho R_D}{M - \rho R_D}} \tag{26-3}$$

Molar refraction and refractive indices have many uses. They are often required in confirming the identity and purity of a compound. Determination of molecular structure and weight is often aided by these parameters [6]. R_D is also used in other estimation schemes, such as in critical properties [9], surface tension [11], and the solubility parameter, which is a measure of intermolecular forces [8].

The refractive index, n, is affected by changes in temperature, pressure, and wavelength of radiation. (R_D remains nearly constant with changes of temperature and pressure by virtue of the density factor, which is a function of temperature and pressure and, thus, offsets these effects.) Methods for extrapolating values of n from one temperature to another are provided in § 26-7.

The refractive index increases as pressure increases, due to the resulting increase in density. This effect is not as significant with liquids as with gases [12]. Lastly, the refractive index decreases as the wavelength increases. For this reason, one cannot compare indices measured at different wavelengths of light [12].

26-2   AVAILABLE ESTIMATION METHODS

Group Contribution Methods. Only two types of methods are available for estimating n or R_D. The first of these is additive in nature. The earliest method of this kind was developed by Eisenlohr [1] in 1910 and subsequently published by Ward and Kurtz [15], Meissner [9], Reid and Sherwood [11], and Gold and Ogle [2]. The method relies upon atomic and structural contributions to estimate R_D. The density and molecular weight are then required to calculate a value of n. The error in R_D is usually less than ±5% but can be as much as 10% for two specific classes of compounds [2]:

- The calculated R_D for a compound containing conjugated double bonds is always *less* than the observed value. For polynuclear aromatics, the discrepancy can be large.

- The calculated R_D for furan-type compounds is always *greater* than the observed value.

Vogel developed two group contribution methods. By one of these [13], the molar refractivity and index of refraction for the sodium D line at 20°C can be calculated by adding atomic, structural, and group contributions; by the other [14], the parameters are found by adding bond refractions at 20°C. Vogel's group contribution method may handle a more diverse set of chemical classes: unlike that of Eisenlohr, it can be used for phosphates, sulfites, sulfates, nitro compounds, nitrates, and carbonates.

Hansch *et al.* [4] applied the group contributions for R_D tabulated by Vogel to an even larger variety of aromatic substituents. These values were collected to study the quantitative relationships between structure and activity in biochemical/biomedical systems. For very complex aromatics, use of the group contributions collected by Hansch *et al.* should provide a good estimate of the molar refraction.

Connectivity Method. The other general method for computing molar refraction is also based upon molecular structure but involves the use of a connectivity function, χ, developed by Kier and Hall [5]. This function can be correlated with R_D by regression equations, which presently exist for the following groups of chemical classes: alcohols, ethers, amines, halogen-containing saturated compounds, alkanes, substituted alkyl benzenes, and mono-olefinic alkenes. The correlation coefficient is generally near 0.99 [5]. Although this method is quite accurate, hand computation of χ is time-consuming. A computer program has been developed by Kier [3] for calculating χ, which greatly improves the likelihood that the connectivity method will be applied to more chemical classes and even adapted to many estimation methods besides that for R_D. For the present, however, the additive methods provide a simple and accurate alternative.

26-3   SELECTION OF APPROPRIATE METHOD

Recommended Methods. Table 26-1 summarizes the estimation methods recommended in this chapter. The methods of Eisenlohr and Vogel require only structural information to estimate R_D, and both require the molecular weight and density for the subsequent calculation of n. Furthermore, both methods use atomic and fragment constants to

TABLE 26-1

Overview of Recommended Methods

Method	Information Required	Applicability	Method Error[a]
Eisenlohr [1] (§26-4)	Structure only for R_D. ρ needed to calculate n.	Compounds with C, H,O,N,S,F, Cl, Br, I; double or triple bonds; aromatics	~ 1.6%[b]
Vogel [13, 14] (§26-5)	Structure only for R_D (ρ needed to calculate n from R_D). n (at 20°C) may also be calculated directly from structural information.	As above, plus phosphates, sulfites, sulfates, nitrates and carbonates	~ 1.1%
Hansch, et al. [4] (§26-6)	Structure only for R_D. ρ needed to calculate n.	Substituted benzenes	~ 0.7%

a. Percentage errors apply to R_D and are the averages from the test sets in Tables 26-2 and 26-3. A 1% error in R_D typically translates into an error of ~0.3% in n. A 5% error in R_D (about the maximum observed) translates into a 1-2% error in n.

b. See §26-4 for additional information.

calculate R_D and appear to involve about the same method error. Vogel's method is applicable to a somewhat larger number of chemicals. The third method, that of Hansch *et al.*, is only for substituted benzene compounds.

Method Errors. Table 26-2 compares the values and errors resulting from estimating molar refraction by Eisenlohr's and Vogel's group contribution methods. The average absolute error achieved by Eisenlohr's group contributions is 1.6%; for Vogel's method, the error is 1.1%. Both methods achieved the same approximation range or error for the separate chemical classes. Additional information, given in §26-4 below, indicates that a method error of 0.2-0.3% may be expected for some chemical classes with Eisenlohr's method.

Table 26-3 similarly indicates the error resulting from estimating R_D by Hansch's method. The error attributable to this method was not evident in the literature, and the value generated in Table 26-3 is not

TABLE 26-2

Comparison of Observed Molar Refraction with Values Estimated by Methods of Eisenlohr and Vogel

	Observed[b] R_D (cm^3)	Eisenlohr		Vogel	
		Estimated R_D (cm^3)	Error (%)	Estimated R_D (cm^3)	Error (%)
Hydrocarbons					
4-Methylheptane	39.12	39.14	+0.1	39.12	0
2,4-Dimethyl-3-isopropylpentane	47.91	48.38	+1.0	48.39	+1.0
3,3-Diethylpentane	43.12	43.76	+1.5	43.79	+1.5
1-Methyl-3-propylbenzene	45.36	44.78	−1.3	44.80	−1.2
1-Isopropyl-2-methylbenzene	45.09	44.78	−0.7	43.20	−4.2
sec-Butylbenzene	45.04	44.78	−0.6	44.78	−0.6
O-Containing Compounds					
1,1-Dimethyl-3-propanol	26.70	28.81	+7.9	26.76	+0.2
1,1,3,3-Tetramethyl-2-propanol	35.67	36.05	+1.0	36.05	+1.1
4-Ethyl-4-heptanol	44.92	45.29	+0.8	45.30	+0.8
Triethylene glycol monobutyl ether	54.85	54.83	0	55.31	+0.8
1,2-Dimethoxyethane	24.17	23.96	−0.9	24.13	−0.1
Methylpropyl ether	22.05	22.31	+1.2	22.36	+1.4
Methyl-*n*-butyl ether	27.02	26.93	−0.3	27.01	0
3-Chlorobenzaldehyde[a]	36.89	35.80	−2.9	35.65	−3.4
1-Methoxy-2-nitrobenzene[a]	36.89	38.16	+3.4	38.31	+3.8
1,3-Dichloro-2-butanone[a]	29.92	30.42	+1.7	29.18	−2.5
Diethylcamphoric acid ester	67.35	67.72	+0.5	67.86	+0.7
t-Butylphenylketone	50.27	49.41	−1.7	49.36	−1.8
N-Containing Compounds					
1-Aminobutane	24.08	24.09	0	24.02	−0.2
Triethylamine	33.79	33.85	+0.2	33.64	−0.4
Dimethylpentylamine	38.28	38.47	+0.5	38.30	0
Cyclopropylcyanide	19.09	20.69	+8.3	18.97	−0.6
Chlorotrinitromethane[a]	29.72	28.11	−5.4	28.57	−3.8
2,4-Dimethylquinoline	51.20	50.20	−1.9	51.58	+0.7
1-Methoxy-2-nitrobenzene[a]	36.89	38.16	+3.4	38.31	+3.8
Halide-Containing Compounds					
3-Chloropentane	30.16	30.16	0.0	30.09	+0.2
1-Bromobutane	28.35	28.44	+0.3	28.32	−0.1
1-Iodohexane	42.89	42.71	−0.4	42.81	−0.1
3-Chlorobenzaldehyde[a]	36.89	35.80	−2.9	35.65	−3.4
1,3-Dichloro-2-butanone[a]	29.92	30.42	+1.7	29.18	−2.5
Chlorotrinitromethane[a]	29.72	28.11	−5.4	28.57	−3.8
3-Chloro-1,3-octadiene	43.83	43.08	−1.7	43.05	−1.7
Decachloro-1,5-hexadiene	76.73	77.64	−1.2	77.14	+0.5
Average Absolute Error (N = 29)			1.6		1.1

a. Compounds listed twice because they fit two categories.
b. **Source:** Refs. 3 and 13.

TABLE 26-3

Observed Molar Refraction versus Values Estimated by
Hansch's Method

Compound	Observed[a] R_D (cm³)	Estimated R_D (cm³)	Error (%)
2-Methoxyethylbenzoic acid, ethyl ester	49.44	49.67	+0.5
2-Bromo-4-methylphenol	40.08	40.68	+1.5
4-Allylnaphthalene	57.33	57.32	0.0
1,6-Naphthylenediamine	53.77	53.67	−0.2
1,8-Naphthalenedicarboxylic acid, diethyl ester	77.31	77.76	+0.5
1-Naphthonitrile	50.31	49.16	−2.2
Nicotine	49.49	49.67	+0.4
3-Hydroxystyrene	38.94	38.17	−2.0
1-Naphthyl azide	53.00	53.03	0.0
1-Dichlorophosphino-4-isopropylbenzene	60.66	60.73	+0.1
Average absolute error (N = 10)			0.7

a. Calculated by Eq. 26-2, using measured values of n and ρ from Ref. 16.

highly reliable because of the small sample set. However, the method
does appear to be a very accurate means of estimating R_D for aromatic
compounds; verification of this finding must await further testing.

The method errors given above are for the estimation of R_D, which
must be combined with ρ and M in Eq. 26-3 to calculate n. A 1% error in
R_D will typically translate into an error of ~0.3% in n, while a 5% error
in R_D will typically translate into an error of ~1-2% in n. The method
error for a particular estimated value of n should be calculated using the
appropriate values of ρ and M. Details of methods for calculating propagated errors are given in Appendix C.

26-4 EISENLOHR'S METHOD

The method of Eisenlohr [1] allows calculation of R_D by simple
addition of atomic and structural group contributions. The values of the
increments are listed in Table 26-4. As stated earlier, estimates by this
method are generally within ±5% of observed values. Gold and Ogle [2]

list the following average errors and 95% confidence limits of R_D for N number of compounds:

	Average Error (%)	95% Confidence Limit
All organics (N=177)	−0.26	± 4.53 cm³
Polar organics (N=99)	−0.34	± 3.43 cm³
Nonpolar organics (N=18)	+0.19	±10.55 cm³

For the sample of 29 selected compounds listed in Table 26-2, the average absolute error in R_D is 1.3%.

TABLE 26-4

Atomic and Structural Contributions to Molar Refraction by Additive Method of Eisenlohr

Group	Contribution (cm³)	Group	Contribution (cm³)
$-CH_2-$	4.618	O(hydroxyl)	1.525
C	2.418	O(ether)	1.643
H	1.100	O(carbonyl)	2.211[b]
S as SH	7.69	OO(ester)	3.736[b]
S as RSR	7.97	N(pri-amine)	2.322
S as RCNS	7.91	N(sec-amine)	2.502
S as RSSR	8.11	N(tert-amine)	2.840
F	0.95[a]	N(nitrile)	5.516[c]
Cl	5.967	Double bond	1.733
Br	8.865	Triple bond	2.398
I	13.900		

a. Only for one F attached to C. (1.1 for each F in polyfluorides.)
b. Includes allowance for double bond.
c. Includes allowance for triple bond.

Source: Adapted from Gold and Ogle [2]. *(Reprinted with permission from McGraw-Hill, Inc.)*

Basic Steps

(1) Calculate the molecular weight (M) of the compound and find the density (ρ) at the temperature of interest. (Methods for estimating density are detailed in Chapter 19.)

(2) Draw the structure of the compound.

(3) Total the incremental values from Table 26-4, accounting for all bonds and atoms. The sum is R_D, in cm³.

(4) Using Eq. 26-3, determine n.

Example 26-1   Estimate n_D^{20} for *m*-chlorobenzaldehyde, C_7H_5OCl.

(1)   M = 140.57 g/mol

ρ = 1.2410 g/cm³ at 20°C [16]

(2)   The structure is:

(3)   From Table 26-4,

7 C = 7 (2.418)	=	16.926
5 H = 5 (1.100)	=	5.500
1 Cl	=	5.967
1 carbonyl O	=	2.211
3 dbl bonds = 3 (1.733)	=	5.199

$$R_D = 35.803 \text{ cm}^3$$

(4)   Substitute the values of R_D, ρ, and M into Eq. 26-3:

$$n = \sqrt{\frac{140.57 + 2\,(1.2410)\,(35.803)}{140.57 - (1.2410)\,(35.803)}} = 1.545$$

The observed value of n_D^{20} is 1.565 [13], indicating an error of −1.3%.

Example 26-2   Estimate n_D^{20} for 2,4-dimethylquinoline, $C_{11}H_{11}N$.

(1)   M = 157.22 g/mol

ρ = 1.0611 g/cm³ at 20°C [16]

(2)   The structure is:

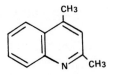

(3)   From Table 26-4,

$$
\begin{array}{lll}
11\ C = 11\ (2.418) & = & 26.598 \\
11\ H = 11\ (1.100) & = & 12.100 \\
1\ t\text{-amine N} & = & 2.840 \\
5\ \text{dbl bonds} = 5\ (1.733) & = & \underline{8.665} \\
& R_D = & 50.203\ cm^3
\end{array}
$$

(4)   Substitute the values of R_D, ρ, and M into Eq. 26-3:

$$
n = \sqrt{\dfrac{157.22 + 2\ (1.0611)\ (50.203)}{157.22 - (1.0611)\ (50.203)}} = 1.593
$$

This underestimates the observed value, 1.6075 [13], by 0.9%.

Example 26-3   Estimate n_D^{20} for the ethyl ester of 2-thiophenecarboxylic acid, $C_7H_8O_2S$.

(1)   M = 156.22 g/mol

$\rho = 1.1623$ g/cm^3 at 16°C [16]

(2)   The structure is:

(3)   From Table 26-4,

$$
\begin{array}{lll}
7\ C = 7\ (2.418) & = & 16.926 \\
8\ H = 8\ (1.100) & = & 8.800 \\
1\ OO\ (\text{ester}) & = & 3.736 \\
1\ S\ \text{as RSR} & = & 7.97 \\
2\ \text{dbl bonds} = 2\ (1.733) & = & \underline{3.466} \\
& R_D = & 40.90\ cm^3
\end{array}
$$

(4)   Substitute the values of R_D, ρ, and M into Eq. 26-3:

$$
n = \sqrt{\dfrac{156.22 + 2\ (1.1623)\ (40.90)}{156.22 - (1.1623)\ (40.90)}} = 1.520\ (\text{at }16°C)
$$

The measured value of n cited in the literature is 1.5248 at 20°C [13]. Using the approximation that n decreases 5×10^{-4} per Celsius degree rise in temperature (§ 26-7), the estimated value of n adjusted to 20°C is $1.520 - (20 - 16)$ $(5 \times 10^{-4}) = 1.518$. The error in the temperature-adjusted value is −0.45%.

26-5 VOGEL'S METHODS

The methods of Vogel [13,14] allow the calculation of molar refractions for the C and F lines of hydrogen and G line of mercury as well as the sodium D line. Direct computation of n_D^{20} is also possible, whereas Eisenlohr's method requires its computation from R_D and the density of the compound. The incremental values for the group contribution method are listed in Table 26-5. Vogel's method based on bond increments [14] is not discussed here but is probably equally accurate. The accuracy of the group contribution method is not stated in the literature; for a sample group of the compounds tested in Table 26-5, however, the average absolute error (1.1%) was slightly less than that of Eisenlohr's procedure.

Basic Steps

(1) Calculate the molecular weight (M) of the compound. For estimation of R_D or n_D at temperatures other than 20°C, find the density of the compound at the temperature of interest.

(2) Draw the structure of the compound.

(3) Total the incremental values of R_D or Mn_D^{20} from Table 26-5 for all bonds, atoms, and molecular groups. The sum is either $R_D(cm^3)$ or Mn_D^{20}.

(4) To determine n_D^{20}, divide Mn_D^{20} by the molecular weight of the compound. If R_D was calculated, use Eq. 26-3 to obtain n_D^{20}.

Example 26-4   Estimate n_D^{20} for *m*-chlorobenzaldehyde, C_7H_5OCl

(1)   M = 140.57 g/mol

(2)   The structure is:

(3)   The increments of Mn_D^{20} from Table 26-5 are:

1 C_6H_5	=	122.03
−1 H	=	+2.56
1 Cl	=	50.41
1 CO (carbonyl)	=	42.41
1 H	=	−2.56
	Mn_D^{20} =	214.85 g/mol

TABLE 26-5

Atomic, Structural, and Group Increments for Vogel's Method

Group[a]	R_C	R_D	R_F	$R_{G'}$	Mn_D^{20}
CH_2	4.624	4.647	4.695	4.735	20.59
H (in CH_2)	1.026	1.028	1.043	1.040	−2.56
C (in CH_2)	2.572	2.591	2.601	2.655	25.71
O (in ethers)	1.753	1.764	1.786	1.805	22.74
O (in acetals)	1.603	1.607	1.618	1.627	22.41
CO (in carbonyls)	4.579	4.601	4.654	4.702	42.41
CO (in methyl ketones)	4.730	4.758	4.814	4.874	42.42
COO (in esters)	6.173	6.200	6.261	6.315	64.14
OH (in alcohols)	2.536	2.546	2.570	2.588	23.94
$CO_2 H$	7.191	7.226	7.308	7.368	63.98
Cl	5.821	5.844	5.918	5.973	50.41
Br	8.681	8.741	8.892	9.011	118.07
I	13.825	13.954	14.310	14.620	196.27
F	0.81	0.81	0.79	0.78	21.84
NH_2 (in primary aliphatic amines)	4.414	4.438	4.507	4.570	22.64
NH (in secondary aliphatic amines)	3.572	3.610	3.667	3.732	23.34
NH (in secondary aromatic amines)	4.548	4.678	5.000	5.273	29.52
N (in tertiary aliphatic amines)	2.698	2.744	2.820	2.914	24.37
N (in tertiary aromatic amines)	4.085	4.243	4.675	5.155	30.23
NO (nitroso)	5.130	5.200	5.397	5.577	43.14
O·NO (nitrite)	7.187	7.237	7.377	7.507	62.27
NO_2 (nitro)	6.662	6.713	6.823	6.918	65.61
N·NO (nitrosoamine)	7.748	7.850	8.100	8.358	69.67
S (in sulfides)	7.852	7.921	8.081	8.233	52.86
S_2 (in disulfides)	15.914	16.054	16.410	16.702	106.52
SH (in thiols)	8.691	8.757	8.919	9.057	50.20
CS (in xanthates)	12.84	13.07	13.67	14.22	77.20
SCN (in thiocyanates)	13.313	13.400	13.603	13.808	88.90
NCS (in *iso*thiocyanates)	15.445	15.615	15.980	16.300	93.11
Carbon-carbon double bond	1.545	1.575	1.672	1.720	−6.07
Carbon-carbon triple bond, terminal	1.959	1.977	2.061	2.084	−12.56
CN (in nitriles)	5.431	5.459	5.513	5.561	36.46
Three-carbon ring	0.592	0.614	0.656	0.646	−4.72
Four-carbon ring	0.303	0.317	0.332	0.322	−4.67
Five-carbon ring	−0.19	−0.19	−0.19	−0.22	−4.56
Six-carbon ring	−0.15	−0.15	−0.16	−0.17	−3.53
CO_3 (carbonates)	7.662	7.696	7.754	7.818	86.35

TABLE 26-5 (Continued)

Group[a]	R_C	R_D	R_F	$R_{G'}$	Mn_D^{20}
SO_3 (sulfites)	11.273	11.338	11.468	11.550	118.09
NO_3 (nitrates)	8.973	9.030	9.170	9.293	87.59
SO_4 (sulfates)	11.050	11.090	11.153	11.225	138.86
PO_4 (orthophosphates)	10.733	10.769	10.821	10.905	139.74
CH_3	5.636	5.653	5.719	5.746	18.13
C_2H_5	10.260	10.300	10.414	10.481	38.72
$C_3H_7{}^n$	14.895	14.965	15.125	15.235	59.25
$C_3H_7{}^i$	14.905	14.975	15.145	15.255	58.95
$C_4H_9{}^n$	19.500	19.585	19.800	19.950	79.81
$C_4H_9{}^i$	19.530	19.620	19.840	19.990	79.54
$C_4H_9{}^s$	19.330	19.420	19.625	19.775	80.21
$C_5H_{11}{}^n$	24.140	24.250	24.515	24.700	100.46
$C_5H_{11}{}^i$ (from the synthetic alcohol)	24.095	24.195	24.460	24.650	100.30
$C_5H_{11}{}^i$ (from fermentation alcohol)	24.170	24.280	24.540	24.720	100.21
$C_6H_{13}{}^n$	28.725	28.855	29.160	29.385	121.10
$C_7H_{15}{}^n$	33.395	33.550	33.905	34.170	141.75
$C_8H_{17}{}^n$	37.960	38.135	38.535	38.830	162.43
C_3H_5 (allyl)	14.425	14.520	14.745	14.920	57.60
$C_6H_5{}^b$	25.136	25.359	25.906	26.356	122.03

a. Superscripts: n = straight chain; i = iso-alkyl; s = sec-alkyl.

b. Benzene ring missing one hydrogen. This value is useful for calculating R_D and Mn_D for mono-substituted benzenes. No ring correction should be applied.

Source: Vogel [13]. *(Reprinted with permission from the Royal Society of Chemistry.)*

(4)   Divide Mn_D^{20} by M:

$$214.85/140.57 = 1.528$$

The literature value of n_D^{20} is 1.565 [16], giving an error of −2.3%.

Example 26-5   Estimate n_D^{20} for 2,4-dimethylquinoline, $C_{11}H_{11}N$.

(1)   M = 157.22 g/mol

(2)   The structure is:

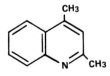

(3)   The increments of Mn_D^{20} from Table 26-5 are:

$$
\begin{array}{llll}
9\,C & = & 9\,(25.71) & = & 231.39 \\
5\,H & = & 5\,(-2.56) & = & -12.80 \\
5\ \text{dbl bonds} & = & 5\,(-6.07) & = & -30.35 \\
2\,CH_3 & = & 2\,(18.13) & = & 36.26 \\
1\,N\ \text{(tertiary} & & & & \\
\quad \text{aromatic amine)} & & & = & 30.23 \\
2\ \text{6-membered rings}^1 & = & 2\,(-3.53) & = & \underline{-7.06} \\
\end{array}
$$

$$Mn_D^{20} = \ 247.67\ \text{g/mol}$$

(4)   Divide this answer by M:

$$247.67/157.22 = 1.5753$$

The value of n_D^{20} reported in the literature is 1.6075 [16]; the error is thus −2.0%.

Example 26-6   Estimate R_D for the ethyl ester of 2-thiophenecarboxylic acid, $C_7H_8O_2S$, at 16°C.

(1)   M = 156.22 g/mol

ρ = 1.1623 g/cm³ at 16°C [16]

(2)   The structure is:

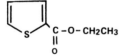

(3)   The increments of R_D from Table 26-5 are:

$$
\begin{array}{lll}
1\ \text{5-membered ring}^2 & = & -0.19 \\
1\,S\ \text{(sulfide)} & = & 7.921 \\
1\,COO & = & 6.200 \\
1\,C_2H_5 & = & 10.300 \\
2\ \text{dbl bonds} = 2\,(1.575) & = & 3.150 \\
4\,C = 4\,(2.591) & = & 10.364 \\
3\,H = 3\,(1.028) & = & \underline{3.084} \\
\end{array}
$$

$$R_D = \ 40.83\ \text{cm}^3$$

The value predicted from the Lorentz-Lorenz equation using ρ = 1.1623 at 16°C is 41.18 cm³, indicating a −0.8% error.

---

1. The table indicates *6-carbon* rings; it is assumed this applies to *6-membered* rings.

2. The table indicates a *5-carbon* ring; it is assumed this value applies to *5-membered* rings.

26-6   HANSCH'S METHOD

The method of Hansch et al. [4] is, like the previous recommended methods, a simple approach to the calculation of R_D using atomic and group fragment constants. The method is limited to substituted benzene compounds. The incremental values of R_D for approximately 240 aromatic substituents are given in Table 26-6. The arrangement of substituents in the table follows alphabetical order, beginning with the atom attached to the benzene ring. Within each group, C and H are listed first, with the remaining atoms sorted alphabetically. If there is no C, then H precedes the alphabetical arrangement, and if C and H are not in the substituent group, the group follows alphabetically.

The substituent R_D values in Table 26-6 may also be used for substituted benzene compounds with more than one substituent. However, interactions between ortho substituents may affect the R_D values in an unknown way; thus, methods errors for ortho-substituted benzenes may be larger than for those that contain only meta or para substituents. When the compound has more than one substituent, the R_D for C_6H_5 is diminished by 1.03 (the value of one H atom) for every substituent after the first.

The method error has not been established. For mono-substituted benzenes, one would expect very high accuracy, since the substituent R_D values were presumably derived from measured values of n and ρ for that particular compound. For compounds with two or more substituents, the error is likely to vary with the nature of the substituents and their relative positions on the ring. For the small sample set in Table 26-3, the error ranged from 0 to 2.2% and averaged 0.7%.

Basic Steps

(1) Calculate the molecular weight (M) of the compound, and find the density (ρ) at the temperature of interest. (Methods for estimating density are provided in Chap. 19.)

(2) Draw the molecular structure.

(3) For mono-substituted benzenes, find R_D for the substituent in Table 26-6 and add it to the R_D for C_6H_5 (25.36). If the compound has two or more substituents on the benzene ring, subtract 1.03 (the value for one H atom) from the total for every substituent after the first.

(4) Calculate n_D with Eq. 26-3.

TABLE 26-6

Hansch's Group Contributions to R_D of Aromatic Compounds

Group	Contribution	Group	Contribution
$B(OH)_2$	11.04	$CH=CHCN$	16.23
Br	8.88	$C\equiv CCH_3$	14.14
CBr_3	28.81	$CH=CHCHO$	16.88
CCl_3	20.12	$CH=CHCOOH$	17.91
CF_3	5.02	$CH_2CH=CH_2$	14.49
CN	6.33	Cyclopropyl	13.53
COO^-	6.05	CH_2COCH_3	15.06
CHO	6.88	$CO_2C_2H_5$	17.47
COOH	6.93	$CH_2OC=O(CH_3)$	16.48
CH_2Br	13.39	$CH_2CH_2CO_2H$	16.52
CH_2Cl	10.49	$3,4\text{-}(CH_2CH_2CH_2)$	13.94
CH_2I	18.60	C_3H_7	14.96
$CONH_2$	9.81	$CH(CH_3)_2$	14.98
$CH=NOH$	10.28	$CH_2N(CH_3)_2$	18.74
$C=O(NHOH)$	11.22	$CF_2CF_2CF_2CF_3$	17.65
CH_3	5.65	2-Thienyl	24.04
CH_2OH	7.19	$3,4\text{-}(CH=CHCH=CH)$	17.47
CH_2NH_2	9.09	$CH=CHCOCH_3$	21.10
$C=O(CF_3)$	11.17	Cyclobutyl	17.88
$3,4\text{-}(CF_2OCF_2)$	10.19	$3,4\text{-}(CH_2)_4$	18.59
$C\equiv CH$	9.55	C_4H_9	19.59
CH_2SCF_3	17.59	$C(CH_3)_3$	19.62
$CH_2SO_2CF_3$	17.51	$CH_2Si(CH_3)_3$	29.61
CH_2CN	10.11	4-Pyridyl	23.03
$CH=CHNO_2$ (trans)	16.42	$CH=CHCO_2C_2H$	27.21
$CH=CH_2$	10.99	Cyclopentyl	22.02
$COCH_3$	11.18	C_5H_{11}	24.25
CO_2CH_3	12.87	$(CH_2)_3N(CH_3)_2$	28.04
CH_2COOH	11.88	C_6Cl_5	49.53
$C=O(NHCH_3)$	14.57	C_6F_5	23.98
CH_2CONH_2	14.41	$C_6H_2[2,4,6\text{-}(NO_2)_3]$	42.21
$C=S(NHCH_3)$	22.33	C_6H_5	25.36
C_2H_5	10.30		
$C\equiv CCF_3$	14.13		24.80
$CF(CF_3)_2$	13.44		
$C(OH)(CF_3)_2$	15.18		
$CH=CHCF_3$ (trans)	15.57		
$CH=CHCF_3$ (cis)	15.57	Cyclohexyl	29.69

(continued)

TABLE 26-6 (Continued)

Group	Contribution	Group	Contribution
2-Benzoxazolyl	32.74	I	13.94
		IO	39.06
		IO_2	63.51
		NO	5.2
2-Benzthiazolyl	38.88	NO_2	7.36
		NNN	10.2
		NH_2	5.42
		NHOH	7.22
		$NHNH_2$	8.44
		$NHSO_2 NHSO_2\text{-}NH_2$	28.40
$C=O(C_6H_5)$	30.33	5-Cl-1-tetrazolyl	23.16
$CH=NC_6H_5$	33.01	$N=CCl_2$	18.35
$CH_2C_6H_5$	30.01	$N=C=O$	8.82
$CH(OH)C_6H_5$	31.52	$N=C=S$	17.24
		5-Azido-1-tetrazolyl	26.85
	29.44	NHCN	10.14
		1-Tetrazolyl	18.33
		5-OH-1-tetrazolyl	19.77
		5-SH-1-tetrazolyl	26.06
		NHCHO	10.31
$C{\equiv}CC_6H_5$	33.21	$NHCONH_2$	13.72
$CH=CHC_6H_5$	34.17	$NHCSNH_2$	22.19
$CH_2CH_2C_6H_5$	34.65	$NHCH_3$	10.33
$CH=CHCOC_6H_4\text{-}(4\text{-}NO_2)$	45.68	$NHSO_2CH_3$	18.17
$CH=CHCOC_6H_5$	40.25	$N(CF_3)_2$	14.28
Ferrocenyl	48.24	$NHCOCF_3$	14.30
Adamantyl	40.63		
1-Phenyl-2-benzimidazolyl	59.08		49.17
		$NHCOCH_2Cl$	19.77
$CO_2CH(C_6H_5)_2$	60.37	$NHCOCH_3$	14.93
Cl	6.03	$NHCSCH_3$	23.40
F	0.92	NHC_2H_5	14.98
$GeBr_3$	36.35	$N(CH_3)_2$	15.55
$GeCl_3$	25.85	$N(SO_2CH_3)_2$	31.22
GeF_3	6.95	$N=NN(CH_3)_2$	20.88
H	1.03	$NHCOC_2H_5$	19.58
$HgCH_3$	19.43	$NHCO_2C_2H_5$	21.18

TABLE 26-6 (continued)

Group	Contribution	Group	Contribution
$NHCONHC_2H_5$	23.19	PH_2	12.19
$NHCSNHC_2H_5$	31.66	$P(Cl)N(CH_3)_2$	27.01
$NHCOCH(CH_3)_2$	24.25	$PO(CH_3)_2$	19.93
$NHCH_2CO_2C_2H_5$	25.82	$PO(OCH_3)_2$	21.87
NHC_4H_9	24.26	$P(CH_3)_2$	21.19
$N=NC_6H_5$	31.31	$P(OC_2H_5)_2$	32.42
NHC_6H_5	30.04	$PO(OC_2H_5)_2$	31.16
$NHSO_2C_6H_5$	37.88	$PO(Cl)C_6H_4$-3-F	39.49
$N=CHC_6H_5$	33.01	$P(Cl)C_6H_4$-3-F	40.75
$NHCOC_6H_5$	34.64	$PS(Cl)C_6H_4$-3-F	47.62
$N=NC_6H_3$-(2-OH)(5-CH_3)	37.45	$P(Cl)C_6H_5$	40.99
$N=CHC_6H_4$-(4-OCH_3)	39.29	$P(H)C_6H_4$-3-F	36.14
$NHCOC_6H_4$-(4-OCH_3)	41.03	$PO(OC_3H_7)_2$	40.46
$N(C_6H_5)_2$	54.96	$P(OCH_3)C_6H_4$-3-F	41.68
OH	2.85	$PO(CH_3)C_6H_4$-3-F	39.37
3,4-(OCF_2O)	8.95	$P(CH_3)C_6H_4$-3-F	40.63
OCF_3	7.86	$PO(C_4H_9)_2$	47.81
$OCHF_2$	7.86	$PO(C_6H_5)_2$	59.29
$OCONH_2$	11.28	$P(C_6H_5)_2$	60.55
3,4-(OCH_2O)	8.96	$PS(C_6H_5)_2$	67.42
OCH_3	7.87	$SO_2(F)$	8.65
OSO_2CH_3	16.99	SF_5	9.89
OCF_2CHFCl	17.30	SH	9.22
$OCOCH_3$	12.47	$SO_2(NH_2)$	12.28
OCH_2COOH	13.99	$SCCl_3$	28.34
OEt	12.47	$S=O(CF_3)$	13.07
$OPO(OCH_3)_2$	22.02	$SO_2(CF_3)$	12.86
$OCH(CH_3)_2$	17.06	SCF_3	13.81
OC_3H_7	17.06	SCN	13.40
OC_4H_9	21.66	$SCHF_2$	13.81
OC_5H_{11}	26.26	$SOCHF_2$	13.28
OC_6H_5	27.68	SO_2CHF_2	13.08
$OSO_2C_6H_5$	36.70	$SOCH_3$	13.70
O-β-glucose	36.53	SO_2CH_3	13.49
$OCOC_6H_5$	32.33	SCH_3	13.82
$POCl_2$	20.16	SCF_2CHF_2	18.40
PCl_2	21.42	$SCOCH_3$	18.42
POF_2	9.58	SC_2H_5	18.42
PF_2	11.02	$SO_2(C_6H_5)$	33.20
$PSCl_2$	28.29	SC_6H_5	34.29

TABLE 26-6 (Continued)

Group	Contribution	Group	Contribution
$SeCF_3$	16.32	SiF_3	7.62
SeCN	16.82	$Si(CH_3)_3$	24.96
$SeCH_3$	17.03	$Si(CH_3)_2[OSi-(CH_3)_3]$	43.64
$SiBr_3$	32.76	$Si(CH_3)[OSi-(CH_3)_3]_2$	62.32
$SiCl_3$	23.85	$Si[OSi(CH_3)_3]_3$	80.99

Source: Hansch *et al.* [4]. *(Reprinted with permission from the American Chemical Society.)*

Example 26-7   Estimate n_D^{20} for *m*-chlorobenzaldehyde.

(1)   M = 140.57 g/mol

   ρ = 1.2410 g/cm³ (20°C)

(2)   The structure of the molecule is:

(3)   Total the incremental values in Table 26-6:

   C_6H_5 = 25.36
   Cl   = 6.03
   CHO   = 6.88
   —H   = -1.03
   ────────
      R_D = 37.24 cm³

(4)   Substitute the above in Eq. 26-3:

$$n_D^{20} = \sqrt{\frac{140.57 + 2\,(1.2410)\,(37.24)}{140.57 - (1.2410)\,(37.24)}} = 1.571$$

The observed value is 1.565 [13], indicating an error of +0.4%.

Example 26-8   Estimate n_D^{20} for 1,8-naphthalene-dicarboxylic acid, diethyl ester.

(1)   M = 273.3 g/mol

   ρ = 1.1399 g/cm³ (at 70°C)

(2)   The structure of the molecule is

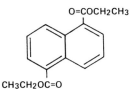

$O=COCH_2CH_3$

$CH_3CH_2OC=O$

(3)   Total the incremental values from Table 26-6:

C_6H_5		$= 25.36$
3,4-(CH=CH–CH=CH)		$= 17.47$
$2\ CO_2\ C_2\ H_5$	$= \quad 2(17.47)$	$= 34.94$
$-3H$	$= \quad -3\ (1.03)$	$= -3.09$
	R_D	$= \quad 74.68\ cm^3$

(4)   Substitute the above in Eq. 26-3:

$$n_D^{70} = \sqrt{\frac{273.3 + 2(1.1399)\,(74.68)}{273.3 - (1.1399)\,(74.68)}} = 1.535$$

The literature value is 1.559 [13], indicating an error of 1.5%.

26-7   EFFECTS OF TEMPERATURE ON n

The Lorentz-Lorenz equation (Eq. 26-2) reflects the theoretical relationship of R_D to n and ρ (which is temperature-dependent). Table 26-7 includes three other equations of empirical derivation which provide more accurate means to calculate n from R_D over a wide temperature range [7]. Of the listed equations, that by Eykman appears to be the most accurate. This equation has not received the exposure of the Lorentz-Lorenz equation, which appears (see Table 26-7) to give a positive bias in the error. Generally, the refractive index (n) decreases about 4×10^{-4} to 6×10^{-4} for each Celsius degree increase in temperature [12].

26-8   AVAILABLE DATA

Ward and Kurtz [15] list refractive indices for a large number of compounds, arranged by chemical class. References [6] and [8] also tabulate some values. Standard references such as [13] include values of n_D^{20}. Additional sources are listed in Appendix A.

26-9   SYMBOLS USED

c	$=$	speed of light in a vacuum $= 3 \times 10^8$ cm/s in Eq. 26-1
M	$=$	molecular weight (g/mol)

TABLE 26-7

Methods of Deriving n from R_D

Method of	Formula[a]	$\left(\dfrac{\text{Deviation per °C of}}{\text{Temperature Change}^b}\right) \times 10^6$	
		Liquid Hydrocarbons (N = 14)	Liquid Non-Hydrocarbons (N = 34)[c]
Eykman	$\dfrac{n^2 - 1}{n + 0.4} = R_D\, \rho$	±15	±11
Lorentz-Lorenz	$\dfrac{n^2 - 1}{n^2 + 2} = R_D\, \rho$	+47[d]	+60[d]
Gladstone-Dale	$n - 1 = R_D\, \rho$	+28	−16
Ward-Kurtz	$\Delta n = 0.6\, \Delta\rho$	±24	±56

a. The dependence of n on temperature is expressed here by equations containing density (ρ), which is a function of temperature.

b. Range of temperature change $\cong$ 16-79°C.

c. Average of 34 liquid non-hydrocarbons, including alcohols, saturated acids and esters, unsaturated acids, derivatives of phenols, aromatic esters, and aromatic ketones.

d. Deviation indicates positive bias for both chemical classes.

Source: Kurtz, Amon, and Sankin [7]. *(Reprinted in part with permission from the American Chemical Society.)*

N	=	number of compounds in sample
n	=	refractive index (unitless)
n_D	=	refractive index measured using light from sodium D line
R_D	=	molar refraction (cm³)
T	=	temperature (°C)
v	=	velocity of light in medium in Eq. 26-1 (cm/s)

Greek

λ	=	wavelength (nm)
ρ	=	density (g/cm³)
χ	=	connectivity parameter

26-10 REFERENCES

1. Eisenlohr, F., "A New Calculation of Atomic Refractions," *Z. Physik Chem.*, **75**, 585-607 (1910).

2. Gold, P.I. and G.J. Ogle, "Estimating Thermophysical Properties of Liquids — Part II: Parachor, Others," *Chem. Eng. (NY)*, **76**, 97-100 (August 11, 1969).

3. Hall, L.H., Eastern Nazarene College, Quincy, Mass., personal communication, October 16, 1979.

4. Hansch, C., A. Leo, S.H. Unger, K.H. Kum, D. Nikatiani and E.J. Lien, "Aromatic Substituent Constants for Structure-Activity Correlations," *J. Med. Chem.*, **16**, 1207-16 (1973).

5. Kier, L.B. and L.H. Hall, "Molecular Connectivity in Chemistry and Drug Research," Academic Press, New York (1976).

6. *Kirk-Othmer Encyclopedia of Chemical Technology,* 2nd ed., Vol. 17, John Wiley and Sons, New York, pp.218-21 (1968).

7. Kurtz, S.S., Jr., S. Amon and A. Sankin, "Effect of Temperature on Density and Refractive Index," *Ind. Eng. Chem.*, **42**, 174-76 (1950).

8. Lawson, D.D. and J.D. Ingham, "Estimation of Solubility Parameters from Refractive Index Data," *Nature (London)*, **223**, 614-15 (1969).

9. Meissner, H.P., "Critical Constants from Parachor and Molar Refraction," *Chem. Eng. Prog.*, **45**, 149-52 (1949).

10. Perry, R.H. and C.H. Chilton (eds.), *Chemical Engineers' Handbook,* 5th ed., McGraw-Hill Book Company, New York, pp. 3-240, 3-301 (1973).

11. Reid, R.C. and T.K. Sherwood, *The Properties of Gases and Liquids,* 2nd ed., McGraw-Hill Book Company, New York, p. 376 (1966).

12. Skoog, D.A. and D.M. West, *Principles of Instrumental Analysis,* Holt, Rinehart & Winston, New York (1971).

13. Vogel, A.I., "Physical Properties and Chemical Constitution, Part XXIII. Miscellaneous Compounds, Investigation of the So-Called Co-ordinate or Dative Link in Esters of Oxy-acids and in Nitroparaffins by Molecular Refractivity Determinations. Atomic, Structure and Group Parachors and Refractivities," *J. Chem. Soc.*, 1833-54 (1948).

14. Vogel, A.I., W.T. Cresswell, G.H. Jeffrey and J. Leicester, "Physical Properties and Chemical Constitution, Part XXIV. Aliphatic Aldoximes, Ketoximes and Ketoxime O-Alkyl Ethers, N,N-Dialkylhydrazines, Aliphatic Ketozines, Mono and Dialkyl Aminopropionitriles, Alkoxypropionitriles, Dialkyl Azodiformates and Dialkyl Carbonates. Bond Parachors, Bond Refractions, and Bond Refraction Coefficients," *J. Chem. Soc.*, 514-49 (1952).

15. Ward, A.L. and S.S. Kurtz, "Refraction, Dispersion and Related Properties of Pure Hydrocarbons," *Ind. Eng. Chem.*, **10**, 559-76 (1938).

16. Weast, R.C. (ed.), *Handbook of Chemistry and Physics,* 53rd ed., The Chemical Rubber Company, Cleveland (1972).

A

BIBLIOGRAPHY OF STANDARD CHEMICAL PROPERTY DATA SOURCES

This Appendix provides a listing of selected reference books which contain compilations of standard physical and chemical properties of organic compounds. It is meant to serve as a quick reference to important handbooks for standard parameters; it excludes environmental parameters such as the bioconcentration factor, process rate constants, and the octanol/water partition coefficient, which are adequately covered in the chapters covering these properties (see sections on "Available Data"). The listing is only for works covering organic compounds (although some may contain data on inorganics as well), and should be considered as supplemental to the listings given in the "Available Data" sections of each individual chapter.

Table A-1 indicates a number of published sources which may be consulted for data on the properties of interest to the users of this handbook. The numbers at the top of each column refer to the reference numbers in the bibliography. Although the properties listed do not distinguish between the gaseous and liquid state, most of the handbooks contain data for both phases or contain an identifier in the title of the reference. The user should be aware that some of the data in these sources have not been critically reviewed and that the referenced sources may contain scanty information or be specific to certain classes of chemicals. However, in all cases, we have cited the most recent edition of the publication so that the most up-to-date data will be available to the user.

TABLE A-1

Compilations of Data for Common Properties of Organic Chemicals

Property	Source (Ref. No.)																				Other Ref.
	1	2	3	4	5	6	7	8	9	10	11	12	13	14	15	16	17	18	19	20	
Solubility in Water	•	•		•	•	•	•		•	•		•		•			•	•			21
Acid/Base Dissociation	•	•		•		•	•										•	•			21
Boiling Point	•	•	•	•		•	•	•		•	•	•	•	•	•	•	•	•	•	•	22
Heat of Vaporization	•	•	•	•	•	•	•		•	•	•	•	•		•	•	•	•	•		23
Vapor Pressure	•	•	•			•				•	•	•	•		•	•	•	•	•	•	24
Diffusion Coefficient	•											•									
Flash Point			•				•	•	•	•	•	•	•	•		•	•	•	•	•	
Density	•	•	•	•	•	•	•	•	•	•	•	•	•	•	•	•	•	•	•	•	22, 25
Surface Tension	•	•	•	•		•	•			•	•	•	•								
Interfacial Tension	•	•	•																	•	22
Viscosity	•	•	•	•	•	•	•		•	•	•	•	•		•				•	•	21, 25, 26
Heat Capacity	•		•	•	•	•	•		•	•	•	•	•								26
Thermal Conductivity	•												•						•		
Dipole Moment	•			•														•			
Refractive Index	•	•	•	•	•	•	•	•	•	•				•	•	•	•			•	22
Melting Point	•	•	•	•	•	•		•	•	•	•	•	•	•	•	•	•	•	•		
Heat of Fusion	•	•	•	•											•		•		•	•	26
Critical Properties															•					•	25-28
Temperature	•	•	•	•	•	•	•			•	•	•	•								
Pressure	•	•	•	•	•	•	•			•	•	•	•								
Volume			•	•		•							•								
Critical Compressibility													•								

A wealth of information on chemical data is accessible through the National Bureau of Standards (NBS), Office of Standard Reference Data in Washington, D.C. This office administers a program of data compilations called the National Standard Reference Data System (NSRDS), which provides data in the areas of energy, environment, industry, materials and physical sciences. The industrial process data include information on thermodynamic, transport, and physical properties of industrially important chemicals. The physical science data include basic data and fundamental physical constants.

The data evaluated by the NSRDS program are brought to public attention in the following publications and journals:

- Journal of Physical and Chemical Reference Data — a quarterly review published for the NBS by the American Chemical Society and the American Institute of Physics.
- NSRDS-NBS Series — distributed through the Government Printing Office.
- Miscellaneous sources, such as technical society publications and on-line computer retrieval systems.

The NSRDS collects its data from a number of data centers which specialize in specific areas of chemical research. The addresses of data centers pertinent to this handbook are:

- Chemical Kinetics Information Center
  Center for Thermodynamics and Molecular Science
  Chemistry Building
  NBS, Washington, D.C. 20234
  Telephone: 301-921-2565
- Chemical Thermodynamics Data Center
  Center for Thermodynamics and Molecular Science
  Chemistry Building
  NBS, Washington, D.C. 20234
  Telephone: 301-921-2773
- Fundamental Constants Data Center
  Center for Absolute Physical Quantities
  Metrology Building
  NBS, Washington, D.C. 20234
  Telephone: 301-921-2701

- Thermodynamics Research Center
  Texas A&M University
  Department of Chemistry
  College Station, Texas 77843
  Telephone: 713-846-8765
  or   713-845-4971

The types of reports available through the NBS often encompass a small class of chemicals, or at times, just one chemical. NSRDS data compilations have been indexed by property for the years 1964-1972 (Report No. NSRDS-NBS-55). The Journal of Physical and Chemical Reference Data updates its index once a year. The properties indexed include a broad spectrum of parameters including activity coefficient, diffusion coefficient, dipole moment, rate constants and surface tension.

An important publication of the Thermodynamics Research Center (one of the NSRDS data centers) is their *Selected Values of Properties of Hydrocarbons and Related Compounds* [29]. Category A of this publication (8 volumes) contained, as of September 1979, tables of physical and thermodynamic properties of hydrocarbons on 3,194 data sheets. Supplements of new and revised loose-leaf data sheets are issued irregularly to keep the work current.

Another publication series that may be of interest is available from Engineering Sciences Data Units, Ltd. (ESDU). A sampling of reports issued in 1979 included:

1) Viscosity of Liquid Aliphatic Hydrocarbons: Alkanes
2) Thermal Conductivity of Liquid Carboxylic Acids
3) Heat Capacity and Enthalpy of Liquids: Aliphatic Alcohols
4) Vapor Pressures and Critical Points of Liquids. XIV: Aliphatic Oxygen-Nitrogen Compounds

Details on the availability of these and other reports in their Physical Data/Chemical Engineering series may be obtained from ESDU, 251-259 Regent Street, London W1R 7AD, England.

A very useful compilation of data sources for both inorganic and organic substances has been prepared by Armstrong and Goldberg [30] for the National Bureau of Standards.

BIBLIOGRAPHY

1. Weast, R.C. and M.J. Astle (eds.), *Handbook of Chemistry and Physics,* 60th ed., CRC Press, Inc., West Palm Beach, FL (1979).

2. Perry, R.H. and C.H. Chilton (eds.), *Chemical Engineers' Handbook,* 5th ed., McGraw-Hill Book Co., New York (1973).

3. *CHRIS Hazardous Chemical Data,* Report No. CG-446-2, U.S. Coast Guard, Washington, D.C. (1974).

4. Dean, J.A. (ed.), *Lange's Handbook of Chemistry,* 12th ed., McGraw-Hill Book Co., New York (1979).

5. Windholz, M. (ed.), *The Merck Index,* 9th ed., Merck & Co., Inc., Rahway, NJ (1976).

6. The Office of Critical Tables, *Consolidated Index of Selected Property Values, Physical Properties and Thermodynamics,* Publication 976, National Academy of Sciences, National Research Council, Washington, D.C. (1962).

7. Grayson, M. and D. Eckroth (eds.), *Kirk-Othmer Encyclopedia of Chemical Technology,* 3rd ed., Wiley-Interscience, New York (1978).

8. Coffey, S. (ed.), *Rodd's Chemistry of Carbon Compounds,* 2nd ed., Elsevier Publishing Company, New York (1967).

9. National Research Council, *International Critical Tables,* E.W. Washburn (ed.), McGraw-Hill Book Co., New York (1926).

10. Dreisbach, R.R., *Physical Properties of Chemical Compounds,* Vols. I, II, III, American Chemical Society, Washington, D.C. (1955-61).

    Vol. I — Cyclic Compounds
    Vol. II — Acyclic Compounds
    Vol. III — Aliphatic Compounds

11. Gallant, R.W. (ed.), *Physical Properties of Hydrocarbons,* Vols. I and II, Gulf Publishing Company, Houston, TX (1968).

    Vol. I — paraffinic, halogenated, oxygenated hydrocarbons (alcohols, oxides, glycols)

    Vol. II — organic acids, ketones, aldehydes, ethers, esters, nitrogen compounds, aromatics, cyclic hydrocarbons, sulfur compounds.

12. American Petroleum Institute, *Technical Data Book — Petroleum Refining,* Vols. I and II, 3rd ed., Washington D.C. (1976).

13. Reid, R.C., J.M. Prausnitz and T.K. Sherwood, *The Properties of Gases and Liquids,* 3rd ed., McGraw-Hill Book Co., New York (1977).

14. Verschueren, K., *Handbook of Environmental Data on Organic Chemicals,* Van Nostrand Reinhold Co., New York (1977).

15. Wilhoit, R.C. and B.J. Zwolinksi, "Physical and Thermodynamic Properties of Aliphatic Alcohols," *J. Phys. Chem. Ref. Data,* **2**, Supplement 1 (1973).

16. Sax, N.I., *Dangerous Properties of Industrial Materials,* 5th ed., Van Nostrand Reinhold Co., New York (1979).

17. Lowenheim, F.A. and M.K. Moran (eds.), *Faith, Keyes, and Clark, Industrial Chemicals,* 4th ed., John Wiley and Sons, New York (1975).

18. Gordon, A.J. and R.A. Ford, *The Chemist's Companion, A Handbook of Practical Data, Techniques, and References,* John Wiley and Sons, New York (1972).

19. Bondi, A., *Physical Properties of Molecular Crystals, Liquids, and Glasses,* John Wiley and Sons, New York (1968).

20. Timmermans, J., *Physico-Chemical Constants of Pure Organic Compounds,* (2 vols.), Elsevier Publishing Co., Amsterdam (1950 and 1965).

21. Sober, H.A. (ed.), *Handbook of Biochemistry — Selected Data for Molecular Biology,* The Chemical Rubber Co., Cleveland (1968 and 1970).

22. *Properties of Hydrocarbons of High Molecular Weight, 1967,* API Publication No. 951, available from University Microfilms International, Books and Collections, Ann Arbor, MI (1967).

23. Cox, J.D. and G. Pilcher, *Thermochemistry of Organic and Organometallic Compounds,* Academic Press, New York (1970).

24. Jordan, T.E., *Vapor Pressure of Organic Compounds,* Interscience Publishers, New York (1954).

25. Natural Gas Processors Association, *Engineering Data Handbook,* 9th ed., Natural Gas Processors Suppliers Assoc., Tulsa, OK (1972).

26. Rossini, F.D., D.D. Wagman, W.H. Evans, S. Levine and I. Jaffe, *Selected Values of Chemical Thermodynamic Properties,* Parts I and II, NBS Circular 500, National Bureau of Standards, Washington, D.C. (1952).

27. Kudchadkar, A.P., G.H. Alani and B.J. Zwolinski, "The Critical Constants of Organic Substances," *Chem. Rev.,* **68,** 659 (1968).

28. Passnt, C.A. and R.P. Danner, "Acentric Factor: A Valuable Corresponding Parameter for the Properties of Hydrocarbons," *Ind. Eng. Chem. Proc. Des. Dev.,* **12,** 365 (1973).

29. *Selected Values of Properties of Hydrocarbons and Related Compounds, 1977,* Thermodynamics Research Center, Texas A&M Research Foundation, College Station, TX (1977).

30. Armstrong, G.T. and R.N. Goldberg, "An Annotated Bibliography of Compiled Thermodynamic Data Sources for Biochemical and Aqueous Systems (1930 to 1975). Equilibrium, Enthalpy, Heat Capacity and Entropy Data," Special Publication 454, National Bureau of Standards, Washington, D.C. (1976).

B

SIMPLE LINEAR REGRESSION

Cathy Campbell

B-1 INTRODUCTION

Many research projects involve the measurement of an independent, predictor or control variable (X) and a response or dependent variable (Y) for different values of X. The values of Y that are observed vary because of (a) the value of the control variable and (b) random error. When a researcher believes that the variation is basically linear, this can be expressed by assuming that the data follow a simple linear model:

$$Y = A + BX + \text{error} \qquad \text{(B-1)}$$

The errors in the model may be positive or negative but on the average will be equal to 0. Figure B-1 shows a plot of 25 (X, Y) observations that appear to follow the simple linear model. The original data are in Table B-1.

The quantities A, B, and $\sigma_{Y \cdot X}$ are the parameters (unknown constants) that describe the linear relationship and are defined as follows:

A	=	intercept of regression line
B	=	slope of regression line
$\sigma_{Y \cdot X}$	=	standard deviation of the errors, also called standard error of the estimate

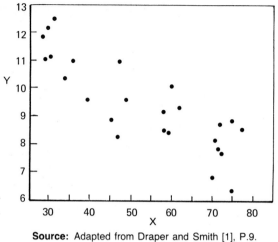

Source: Adapted from Draper and Smith [1], P.9.
(Reprinted with permission from John
Wiley & Sons, Inc.)

FIGURE B-1   Scatter Plot of Twenty-five
Observations of Two
Variables

TABLE B-1

Twenty-five Observations of Two Variables

Observation Number	X	Y	Observation Number	X	Y
1	35.3	10.98	14	39.1	9.57
2	29.7	11.13	15	46.8	10.94
3	30.8	12.51	16	48.5	9.58
4	58.8	8.40	17	59.3	10.09
5	61.4	9.27	18	70.0	8.11
6	71.3	8.73	19	70.0	6.83
7	74.4	6.36	20	74.5	8.88
8	76.7	8.50	21	72.1	7.68
9	70.7	7.82	22	58.1	8.47
10	57.5	9.14	23	44.6	8.86
11	46.4	8.24	24	33.4	10.36
12	28.9	12.19	25	28.6	11.08
13	28.1	11.88			

Source: Draper and Smith [1], p.8. (Reprinted with permission from John Wiley &
Sons, Inc.)

Linear regression analysis is the area of statistics which encompasses methods for:

- estimating the parameters of the simple linear model;
- determining if there is a "significant" linear relationship between X and Y;
- predicting future values of Y, given values of the control variable X.

The procedures used in a simple linear regression analysis are described in the following sections of this appendix:

B-2   Examining the Data
B-3   Parameter Estimation
B-4   Evaluating the Regression
B-5   Predicting Future Observations
B-6   Symbols Used

The necessary data for performing a regression analysis consist of n pairs of observations, $\{(X_i, Y_i): i = 1, ..., n\}$ which are usually arranged as in Table B-1. These data will be used throughout for demonstration purposes. The measurements were taken periodically on a steam plant, where

X   =   average atmospheric temperature (°F) for a month
Y   =   steam consumption (lb/mo)

B-2   EXAMINING THE DATA

Observed data for a linear regression problem represent a sample of an entire population of values that might have been observed. For the methods of linear regression to be valid, certain assumptions about the underlying population must be satisfied:

1. X and Y are linearly (not curvilinearly) related;

2. The long-run average of the errors is 0;

3. The errors are statistically independent; i.e., knowing the error for one observation gives no information about the sign or magnitude of the error for any other observation;

4. The errors are approximately symmetric around 0 and do not deviate dramatically from a normal distribution;

5. The average magnitude of the errors is constant for all values of X, called homoscedastic errors.

Before conducting a regression analysis, the analyst should examine the data to determine if these assumptions are (approximately) satisfied. For a given sample of data, none will be satisfied exactly, because the assumptions apply to the population that generates the data. Unless the observed data depart dramatically from these assumptions, simple linear regression is an appropriate analysis. Alternative analysis methods should be used if any of the assumptions appear to be severely violated.

Making a scatterplot, as in Figure B-1, is the best way to examine the data when only two variables are observed and should be a routine first step of any regression analysis. Since the plot in Figure B-1 does not show anything unusual, linear regression analysis would be appropriate for these data.

Figure B-2 contains scatterplots that illustrate violation of assumptions (1), (3) (4), and (5). Violation of assumption (2) cannot be detected from a sample.

Assumption (3), independence of errors, may be in doubt when the observations are taken sequentially or when different people or instruments are used for separate parts of the data collection. (See Section B-4.)

Assumption (4) often fails because of the presence of outliers — observations which depart from the general pattern of the data; they show up very clearly in the scatterplot. Outliers are troublesome in regression analysis, because one or two unusual observations can unduly influence the results of an analysis. When outliers appear in the scatterplot, the following steps should be taken:

(1) First determine if a simple recording error was made. Make the necessary correction, or delete the observation if correction is impossible.

(2) If no error occurred, see if the observation was made under unusual but observable circumstances that would explain the strange value. If a valid cause can be found, delete the observation from the analysis.

(3) When no cause for the outlier(s) can be found, perform two analyses — one that includes the observation(s) and one

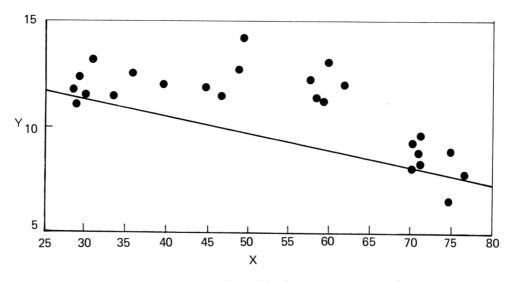

a. Plot Is Nonlinear

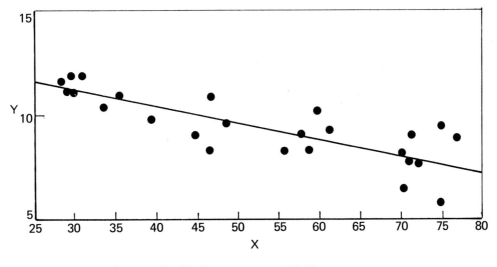

b. Variance Increases with X

FIGURE B-2   Violation of Regression Assumptions

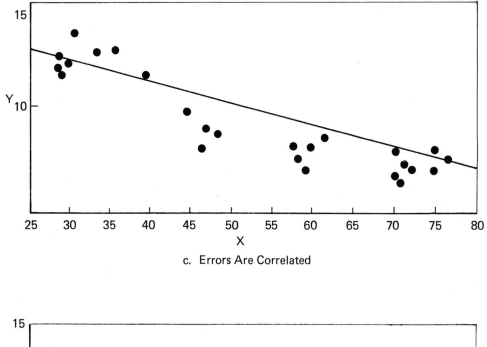

c. Errors Are Correlated

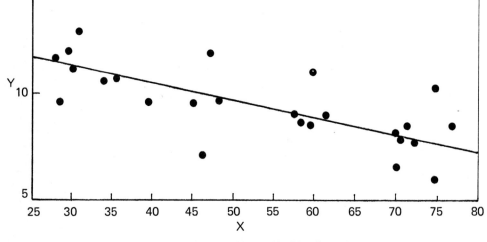

d. Errors Are Not Normally Distributed

FIGURE B-2   Violation of Regression Assumptions (continued)

that excludes them. If the two analyses essentially agree, report the full analysis. If they disagree, report both and explain why they differ.

After making the scatterplot and determining that simple linear regression is appropriate, one can begin formal analysis of the data.

B-3   PARAMETER ESTIMATION

With limited data, one cannot determine exactly the parameters (A, B and $\sigma_{Y \cdot X}$) of the regression equation (Eq. B-1). One must be satisfied with point estimates (single best values) and estimated standard errors that tell how far away from the "true" value the point estimate may be. Because of this uncertainty about the "true" values, it is recommended that parameter estimates be presented in the form of confidence intervals that realistically reflect the researcher's knowledge about the parameter value.

A confidence interval for a parameter B consists of two numbers, B_{lower} and B_{upper} (B_L and B_U), calculated from the sample data so that the researcher can say, "I am 95% (or 90%, 99%, etc.) confident that the true value of B lies between B_L and B_U." If the statement is made with a high degree, say 99%, of confidence, the interval will be larger than if a low degree of confidence, say 70%, is chosen.

The principle of least squares is used in calculating the point estimates of A and B. Estimating A and B is equivalent to finding the slope and intercept of the best-fitting line through the sample data. The equation of the sample regression line is:

$$\hat{Y} = a + bX \qquad \text{(B-2)}$$

where

$$
\begin{aligned}
a &= \text{estimate of A} \\
b &= \text{estimate of B} \\
\hat{Y} &= \text{predicted value of Y for a given X.}
\end{aligned}
$$

Estimates a and b are chosen so that the sum of the squared distances of the observed Y's from the Y's predicted by the sample regression line is as small as possible, i.e., so that

$$\sum_{i=1}^{n} (Y_i - a - bX_i)^2$$

is minimized. The quantity $Y_i - \hat{Y}_i = Y_i - a - bX_i$ is called the residual for the ith observation. See Figure B-3 for a display of these quantities. (The derivation of the regression equation is explained later in this section.)

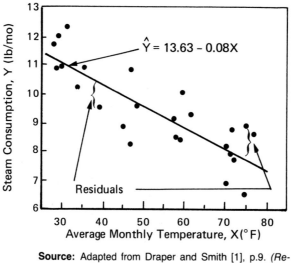

Source: Adapted from Draper and Smith [1], p.9. *(Reprinted with permission from John Wiley & Sons, Inc.)*

FIGURE B-3   Least-squares Regression Line

The least-squares estimate of B is:

$$b = \frac{\displaystyle\sum_{i=1}^{n} X_i Y_i - n\,\overline{X}\,\overline{Y}}{\displaystyle\sum_{i=1}^{n} X_i^2 - n\,\overline{X}^2} \tag{B-3}$$

where

$$\overline{X} = \frac{1}{n}\sum_{i=1}^{n} X_i \text{ and } \overline{Y} = \frac{1}{n}\sum_{i=1}^{n} Y_i \tag{B-4, -5}$$

are the sample means of X and Y. The estimate of A is

$$a = \overline{Y} - b\overline{X}. \tag{B-6}$$

The estimate of $\sigma_{Y \cdot X}$ is

$$s_{Y \cdot X} = \sqrt{\frac{n-1}{n-2} (s_Y^2 - b^2 s_X^2)} \qquad \text{(B-7)}$$

where s_Y^2 and s_X^2 are the sample variances of X and Y:

$$s_X^2 = \frac{1}{n-1}\left(\sum_{i=1}^{n} X_i^2 - n\,\overline{X}^2\right) \qquad \text{(B-8)}$$

$$s_Y^2 = \frac{1}{n-1}\left(\sum_{i=1}^{n} Y_i^2 - n\,\overline{Y}^2\right) \qquad \text{(B-9)}$$

The sample standard deviation $s_{Y \cdot X}$ depends on the magnitude of the sum of squared residuals; it can also be expressed as

$$s_{Y \cdot X} = \sqrt{\frac{1}{n-2} \sum_{i=1}^{n} (Y_i - \hat{Y}_i)^2} \qquad \text{(B-10)}$$

but this is not a useful computing form.

To find confidence intervals for A and B, estimates of the standard errors (s.e.) of a and b must be calculated as follows:

$$\text{s.e. (a)} = s_{Y \cdot X}\sqrt{\frac{1}{n} + \frac{\overline{X}^2}{(n-1)\,s_X^2}} \qquad \text{(B-11)}$$

$$\text{s.e. (b)} = \sqrt{\frac{s_{Y \cdot X}^2}{(n-1)\,s_X^2}} \qquad \text{(B-12)}$$

The magnitude of the estimated standard errors determines the width of the confidence intervals.

The actual form of (B_L, B_U) and (A_L, A_U) is given by

$$B_L = b - t_{1-\alpha/2}\ (n-2)\ \text{s.e. (b)} \qquad \text{(B-13)}$$
$$B_U = b + t_{1-\alpha/2}\ (n-2)\ \text{s.e. (b)}$$

and

$$A_L = a - t_{1-\alpha/2}\ (n-2)\ \text{s.e. (a)} \qquad \text{(B-14)}$$
$$A_U = a + t_{1-\alpha/2}\ (n-2)\ \text{s.e. (a)}$$

where 100 $(1 - \alpha)$ % is the desired confidence level and $t_{1-\alpha/2}$ $(n - 2)$ is the 100 $(1 - \alpha/2)$ percentile point of Student's t-distribution with n-2 degrees of freedom. The value of t can be read from tables found in any introductory statistics text; for convenience, a t-table is reproduced here as Table B-2.

Example B-1 In this sample, all of the preceding calculations are demonstrated and interpreted for the data in Table B-1.

(1) The sample size, n, is 25. The mean (Eq. B-4) and standard deviation (Eq. B-8) of X are

$$\bar{X} = \frac{1}{n} \sum_{i=1}^{n} X_i = \frac{1315}{25} = 52.60$$

$$s_X = \sqrt{\frac{1}{n-1} \sum_{i=1}^{n} X_i^2 - n\bar{X}^2} = \sqrt{\frac{1}{24} [76323 - 25(52.60^2)]}$$

$$= 17.27$$

(2) Performing the same calculations for Y gives

$$\bar{Y} = 9.424$$

$$s_Y = 1.63$$

(3) The first step in calculating b is to multiply the X and Y values for each observation and total the products.

$$\sum_{i=1}^{n} X_i Y_i = 11821$$

(4) Substituting in Eq. B-3,

$$b = \frac{\sum_{i=1}^{n} X_i Y_i - n\bar{X}\bar{Y}}{\sum_{i=1}^{n} X_i^2 - n\bar{X}^2} = \frac{11821 - (25)(52.60)(9.424)}{76323 - (25)(52.60^2)}$$

$$= -0.080$$

(5) Since $\quad a = \bar{Y} - b\bar{X}$ (Eq. B-6),

$$a = 9.424 - (-.080)(52.60)$$

$$= 13.63$$

TABLE B-2

Upper Percentage Points of the t Distribution

Degrees of Freedom	$1-\alpha$						
	0.75	0.90	0.95	0.975	0.99	0.995	0.9995
1	1.000	3.078	6.314	12.706	31.821	63.657	636.619
2	0.816	1.886	2.920	4.303	6.965	9.925	31.598
3	0.765	1.638	2.353	3.182	4.541	5.841	12.941
4	0.741	1.533	2.132	2.776	3.747	4.604	8.610
5	0.727	1.476	2.015	2.571	3.365	4.032	6.859
6	0.718	1.440	1.943	2.447	3.143	3.707	5.959
7	0.711	1.415	1.895	2.365	2.998	3.499	5.405
8	0.706	1.397	1.860	2.306	2.896	3.355	5.041
9	0.703	1.383	1.833	2.262	2.821	3.250	4.781
10	0.700	1.372	1.812	2.228	2.764	3.169	4.587
11	0.697	1.363	1.796	2.201	2.718	3.106	4.437
12	0.695	1.356	1.782	2.179	2.681	3.055	4.318
13	0.694	1.350	1.771	2.160	2.650	3.012	4.221
14	0.692	1.345	1.761	2.145	2.624	2.977	4.140
15	0.691	1.341	1.753	2.131	2.602	2.947	4.073
16	0.690	1.337	1.746	2.120	2.583	2.921	4.015
17	0.689	1.333	1.740	2.110	2.567	2.898	3.965
18	0.688	1.330	1.734	2.101	2.552	2.878	3.922
19	0.688	1.328	1.729	2.093	2.539	2.861	3.883
20	0.687	1.325	1.725	2.086	2.528	2.845	3.850
21	0.686	1.323	1.721	2.080	2.518	2.831	3.819
22	0.686	1.321	1.717	2.074	2.508	2.819	3.792
23	0.685	1.319	1.714	2.069	2.500	2.807	3.767
24	0.685	1.318	1.711	2.064	2.492	2.797	3.745
25	0.684	1.316	1.708	2.060	2.485	2.787	3.725
26	0.684	1.315	1.706	2.056	2.479	2.779	3.707
27	0.684	1.314	1.703	2.052	2.473	2.771	3.690
28	0.683	1.313	1.701	2.048	2.467	2.763	3.674
29	0.683	1.311	1.699	2.045	2.462	2.756	3.659
30	0.683	1.310	1.697	2.042	2.457	2.750	3.646
40	0.681	1.303	1.684	2.021	2.423	2.704	3.551
60	0.679	1.296	1.671	2.000	2.390	2.660	3.460
120	0.677	1.289	1.658	1.980	2.358	2.617	3.373
∞	0.674	1.282	1.645	1.960	2.326	2.576	3.291

Source: Adapted from Morrison [2]. *(Reprinted with permission from McGraw-Hill Book Co.)*

(6)   The equation of the sample regression line is therefore

$$\hat{Y} = 13.63 - 0.08X$$

The value of b indicates that a $1°F$ increase in average monthly temperature has been associated with an average decrease of 0.08 pound of steam used. A plot of the regression line appears in Figure B-3. It is not meaningful to interpret the value of the sample intercept (a) for these data, since no X's were observed near 0. The predicted amount of steam used in a month with average temperature, X, of $50°F$ is given by

$$\hat{Y} = 13.63 - 0.08(50)$$
$$= 9.63 \text{ lb/mo}$$

(7)   The residual for the first observation in Table B-1 is

$$Y_1 - \hat{Y}_1 = Y_1 - a - bX_1$$
$$= 10.98 - 13.63 - (-.08)(35.3)$$
$$= 0.17$$

(8)   From Eq. B-7,

$$s_{Y \cdot X} = \sqrt{\frac{n-1}{n-2}(s_Y^2 - b^2 s_X^2)}$$

$$= \sqrt{\frac{24}{23}[1.63^2 - (.08^2)(17.27^2)]}$$

$$= 0.88$$

When the basic assumptions for a regression analysis are satisfied, two thirds of the observed Y values should lie within $\pm s_{Y \cdot X}$ of the sample regression line.

(9)   The estimated standard errors of a and b (Eqs. B-11 and B-12) are:

$$\text{s.e. (a)} = .88 \sqrt{\frac{1}{25} + \frac{52.6^2}{24(17.27^2)}}$$

$$= 0.57$$

and

$$\text{s.e. (b)} = \sqrt{\frac{0.88^2}{(24)(17.27^2)}}$$

$$= 0.010$$

(10)   Both 90% and 95% confidence intervals for B will be calculated. The only difference in the calculations is the value of t that is used in Eq. B-13. For

a 90% confidence interval, $\alpha = 0.10$ and $(1 - \alpha/2) = 0.95$. From Table B-2, the 95th percentile of the t-distribution with $n - 2 = 23$ degrees of freedom is found to be 1.714. Therefore,

$$B_L = b - t_{.95}(23) \text{ s.e. (b)}$$
$$= -0.080 - 1.714 \,(0.010)$$
$$= -0.097$$

(11) Similarly,

$$B_U = -0.080 + 1.714 \,(0.010)$$
$$= -0.063$$

(12) Thus, with 90% confidence, one can assert that B is between -0.097 and -0.063, or that a 90% confidence interval for B is $(-0.097, -0.063)$. While the exact value of B is not known and may fall outside of the interval, 90% of all confidence intervals that are calculated using this method will contain B.

(13) To calculate a 95% confidence interval, use the 97.5th percentile of t with 23 degrees of freedom. From Table B-2, this number is 2.069. Using the same procedure as above, the 95% confidence interval for B is $(-0.101, -.059)$. This interval is, of course, larger than the 90% confidence interval because of the increased confidence that it will contain B.

(14) A confidence interval for A is not of substantive interest for this example. To demonstrate the calculation, however, the 95% confidence interval will be found. From Eq. B-14, the limits are:

$$A_L = 13.63 - 2.069 \,(0.57)$$
$$= 12.4$$

and

$$A_U = 14.8$$

The 95% confidence interval for A is therefore $(12.4, 14.8)$.

The value of A can be interpreted as the average response (Y) for units with $X = 0$. If the X's are all far from 0, as they are here, the value of A is of no practical importance.

B-4 EVALUATING THE REGRESSION

After finding the point estimates and possibly confidence intervals for A and B, it is often instructive to evaluate formally how well the proposed linear relationship describes the data at hand. This evaluation may consist of different steps, depending on individual needs:

(1) investigating again the appropriateness of the assumptions in Section B-2;

(2) calculating a summary statistic (R^2) which tells what proportion of the variation in Y has been explained by the sample regression;

(3) testing formally whether there is indeed a linear relationship between X and Y, i.e., is b significantly different from 0?

Steps to ascertain that the assumptions for this analysis are valid should always be done. The steps outlined here depend on the sample residuals ($Y_i - \hat{Y}_i$); this seems reasonable, since most of the assumptions deal with properties of the errors. Again the use of scatterplots is recommended, since most problems are easily seen this way. Several plots are listed along with a description of the desired appearance and problems that may be seen. As many of these plots as are pertinent and possible should be made.

1. *Histogram of residuals.* Should be fairly symmetrical and centered at 0. Outliers or other signs of non-normality may show up. (Example provided in Figure B-4.)

2. *Probability plot of residuals.* Should be approximately a straight line. Curves or straggling ends indicate a non-normal distribution of errors. (Example provided in Figure B-5.)

3. *Residuals versus $\hat{Y}_i$, called residuals versus predicteds,* should look random with no pattern. Curves or other trends indicate a nonlinear relationship or nonindependent errors. A megaphone shape (as shown below)

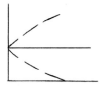

or other uneven spread indicates heterogeneous variance. (Example of residuals versus predicteds plot given in Figure B-6a.)

4. *Residuals versus order of observation* should be random. Any pattern or trend indicates systematic variation in Y across time and leads to nonindependent errors. (Example in Figure B-6b.)

5. *Residuals versus operator, temperature, day, etc.* (any quantities which varied as the data were observed); trends indicate nonindependent errors or the need for other variables

in the regression. Uneven spread indicates heteroscedastic variances.

Remedies for these problems do exist, but a detailed discussion is beyond the scope of this discussion. Multiple linear regression allows use of additional predictor variables. Transformations of X and/or polynomial regression are helpful for nonlinearity. Transformations of Y may solve non-normality or heteroscedasticity problems. Weighted least-squares may be done to account for heteroscedastic variances. These topics are discussed in Draper and Smith [1] and other statistics books. It may be necessary to consult with a statistician.

Steps (2) and (3) in the evaluation both depend on an analysis of variance of the sample data. The rationale behind the analysis of variance is that the total variation in Y,

$$\sum_{i=1}^{n} (Y_i - \overline{Y})^2$$

can be split into two components: the variation explained by the regression,

$$\sum_{i=1}^{n} (\hat{Y}_i - \overline{Y})^2$$

and the residual or unexplained variation, which is

$$\sum_{i=1}^{n} (Y_i - \hat{Y}_i)^2$$

Each of the quantities is called a "sum of squares," and the relationship

$$
\begin{array}{c}
\text{Total (corrected) sum of squares} = \\
\text{(SST)} \\
\text{Sum of squares due to regression} + \text{Error sum of squares} \\
\text{(SSR)} \qquad\qquad\qquad \text{(SSE)}
\end{array}
\qquad \text{(B-15)}
$$

will always hold.

Results of an analysis of variance are usually displayed in an analysis of variance (ANOVA) table as shown in Table B-3. Computing formulas for the various quantities are also given. Each mean square is the sum of squares divided by the corresponding degrees of freedom.

TABLE B-3

Analysis of Variance (ANOVA) Table

Source	Sum of Squares	Degrees of Freedom	Mean Square
Regression	$SSR = (n-1)\, b^2 s_X^2$	1	$MSR = SSR$
Error	$SSE = (n-2)\, s_{Y\cdot X}^2$	$n-2$	$MSE = s_{Y\cdot X}^2$
Total	$SST = (n-1)\, s_Y^2$	$n-1$	s_Y^2

From the ANOVA table, one can complete steps (2) and (3) of the evaluation. The squared multiple correlation coefficient (R^2) is the proportion of the total sum of squares that has been explained by the sample regression. From the ANOVA table we see that

$$R^2 = \frac{SSR}{SST} = \frac{b^2 s_X^2}{s_Y^2} \qquad \text{(B-16)}$$

Obviously, R^2 lies between 0 and 1. The value of R^2 may be distorted by the presence of outliers or by small sample size; its significance should be interpreted cautiously in these cases. Otherwise, a large value of R^2 indicates that simple linear regression does a good job of explaining the observed variability in Y. R^2 is the square of r, the simple correlation coefficient between X and Y. The correlation coefficient

$$r = \frac{b s_X}{s_Y} \qquad \text{(B-17)}$$

lies between -1 and $+1$, depending on the sign of b and the degree of linearity in the relationship.

At times, it is important to decide formally whether the data support the existence of a linear relationship between X and Y. In statistical jargon, one wishes to test the null hypothesis H_o: $B = 0$ versus the alternative hypothesis H_A: $B \neq 0$. The rationale guiding this procedure stems from the fact that even if $B = 0$, the computed value of b will seldom, if ever, be zero. How far away from 0 must b lie before one has confidence that $B \neq 0$? The answer to this question depends on the sample size, n, the variability of the X's as measured by $(n-1)s_X^2$, and the risk one is willing to take of erroneously deciding $B \neq 0$ when it really is. The risk level, or level of significance, is usually set at 10%, 5%, or 1%, depending on the seriousness of an error.

The procedure for testing H_o: $B = 0$ is given below.

(1) Choose the significance level (α).

(2) Calculate the test statistic

$$F = \frac{MSR}{MSE} \qquad \text{(B-18)}$$

using values of MSR and MSE from Table B-3.

(3) Locate the comparison value of the F-distribution in Table B-4. The appropriate value depends on the degrees of freedom for the numerator (m) and the denominator (k) of the F statistic (Eq. B-18). These degrees of freedom are found in the ANOVA table (Table B-3) as $m = 1$ and $k = n - 2$. Therefore, the appropriate comparison value is $F_{1-\alpha}$ (1, $n - 2$).

(4) Compare F with $F_{1-\alpha}$ (1, $n - 2$):

- If $F \geq F_{1-\alpha}$ (1, $n - 2$), reject H_o; conclude $B \neq 0$.
- If $F < F_{1-\alpha}$ (1, $n - 2$), do not reject H_o; observed b is not significantly different from 0.

An alternative but equivalent test is based on the t-distribution:

(1) Choose α.

(2) Calculate

$$t = \left| \frac{b}{s.e. (b)} \right| \qquad \text{(B-19)}$$

using Eq. B-12 for s.e.(b).

(3) Find $t_{1-\alpha/2}$ ($n - 2$) in Table B-2.

(4) If $t \geq t_{1-\alpha/2}$ ($n - 2$), reject H_o.

If $t < t_{1-\alpha/2}$ ($n - 2$), do not reject H_o.

Some researchers choose to report the results of their hypothesis tests in a different manner. Rather than saying they have rejected or failed to reject H_o: $B = 0$, they report a p-value, which is a measure of the plausibility of the null hypothesis. A p-value is the smallest significance level for which the null hypothesis would be rejected. To find a p-value, find in Table B-2 or B-4 the largest F(1, $n - 2$) or t($n - 2$) which is

TABLE B-4

Upper Percentage Points of the F Distribution

$1 - \alpha$	k	1	2	3	4	5
0.90		39.9	49.5	53.6	55.8	57.2
0.95		161	200	216	225	230
0.975	1	648	800	864	900	922
0.99		4,050	5,000	5,400	5,620	5,760
0.995		16,200	20,000	21,600	22,500	23,100
0.90		8.53	9.00	9.16	9.24	9.29
0.95		18.5	19.0	19.2	19.2	19.3
0.975	2	38.5	39.0	39.2	39.2	39.3
0.99		98.5	99.0	99.2	99.2	99.3
0.995		199	199	199	199	199
0.90		5.54	5.46	5.39	5.34	5.31
0.95		10.1	9.55	9.28	9.12	9.01
0.975	3	17.4	16.0	15.4	15.1	14.9
0.99		34.1	30.8	29.5	28.7	28.2
0.995		55.6	49.8	47.5	46.2	45.4
0.90		4.54	4.32	4.19	4.11	4.05
0.95		7.71	6.94	6.59	6.39	6.26
0.975	4	12.2	10.6	9.98	9.60	9.36
0.99		21.2	18.0	16.7	16.0	15.5
0.995		31.3	26.3	24.3	23.2	22.5
0.90		4.06	3.78	3.62	3.52	3.45
0.95		6.61	5.79	5.41	5.19	5.05
0.975	5	10.0	8.43	7.76	7.39	7.15
0.99		16.3	13.3	12.1	11.4	11.0
0.995		22.8	18.3	16.5	15.6	14.9
0.90		3.78	3.46	3.29	3.18	3.11
0.95		5.99	5.14	4.76	4.53	4.39
0.975	6	8.81	7.26	6.60	6.23	5.99
0.99		13.7	10.9	9.78	9.15	8.75
0.995		18.6	14.5	12.9	12.0	11.5
0.90		3.59	3.26	3.07	2.96	2.88
0.95		5.59	4.74	4.35	4.12	3.97
0.975	7	8.07	6.54	5.89	5.52	5.29
0.99		12.2	9.55	8.45	7.85	7.46
0.995		16.2	12.4	10.9	10.1	9.52
0.90		3.46	3.11	2.92	2.81	2.73
0.95		5.32	4.46	4.07	3.84	3.69
0.975	8	7.57	6.06	5.42	5.05	4.82
0.99		11.3	8.65	7.59	7.01	6.63
0.995		14.7	11.0	9.60	8.81	8.30
0.90		3.36	3.01	2.81	2.69	2.61
0.95		5.12	4.26	3.86	3.63	3.48
0.975	9	7.21	5.71	5.08	4.72	4.48
0.99		10.6	8.02	6.99	6.42	6.06
0.995		13.6	10.1	8.72	7.96	7.47

TABLE B-4 (Continued)

$1-\alpha$	k	1	2	3	4	5
0.90		3.29	2.92	2.73	2.61	2.52
0.95		4.96	4.10	3.71	3.48	3.33
0.975	10	6.94	5.46	4.83	4.47	4.24
0.99		10.0	7.56	6.55	5.99	5.64
0.995		12.8	9.43	8.08	7.34	6.87
0.90		3.18	2.81	2.61	2.48	2.39
0.95		4.75	3.89	3.49	3.26	3.11
0.975	12	6.55	5.10	4.47	4.12	3.89
0.99		9.33	6.93	5.95	5.41	5.06
0.995		11.8	8.51	7.23	6.52	6.07
0.90		3.07	2.70	2.49	2.36	2.27
0.95		4.54	3.68	3.29	3.06	2.90
0.975	15	6.20	4.77	4.15	3.80	3.58
0.99		8.68	6.36	5.42	4.89	4.56
0.995		10.8	7.70	6.48	5.80	5.37
0.90		2.97	2.59	2.38	2.25	2.16
0.95		4.35	3.49	3.10	2.87	2.71
0.975	20	5.87	4.46	3.86	3.51	3.29
0.99		8.10	5.85	4.94	4.43	4.10
0.995		9.94	6.99	5.82	5.17	4.76
0.90		2.88	2.49	2.28	2.14	2.05
0.95		4.17	3.32	2.92	2.69	2.53
0.975	30	5.57	4.18	3.59	3.25	3.03
0.99		7.56	5.39	4.51	4.02	3.70
0.995		9.18	6.35	5.24	4.62	4.23
0.90		2.79	2.39	2.18	2.04	1.95
0.95		4.00	3.15	2.76	2.53	2.37
0.975	60	5.29	3.93	3.34	3.01	2.79
0.99		7.08	4.98	4.13	3.65	3.34
0.995		8.49	5.80	4.73	4.14	3.76
0.90		2.75	2.35	2.13	1.99	1.90
0.95		3.92	3.07	2.68	2.45	2.29
0.975	120	5.15	3.80	3.23	2.89	2.67
0.99		6.85	4.79	3.95	3.48	3.17
0.995		8.18	5.54	4.50	3.92	3.55
0.90		2.71	2.30	2.08	1.94	1.85
0.95		3.84	3.00	2.60	2.37	2.21
0.975	∞	5.02	3.69	3.12	2.79	2.57
0.99		6.63	4.61	3.78	3.32	3.02
0.995		7.88	5.30	4.28	3.72	3.35

Source: Adapted from Morrison [2]

smaller than the calculated t or F and report the observed significance level. (More details are given in the example below.) All the procedures described in this section are now demonstrated using the data from Table B-1.

Example B-2   Evaluating the regression.

(1)   First the residuals, $Y_i - \hat{Y}_i$, are examined. Table B-5 contains the X value, predicted value ($\hat{Y}_i$), and residual for each observation. A histogram of the residuals is given in Figure B-4.

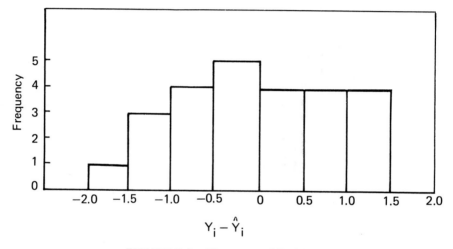

FIGURE B-4   Histogram of Residuals

The histogram does not show any obvious outliers; it does, however, indicate too many large residuals. With only 25 observations, a sample histogram may deviate strongly from a normal shape before giving firm evidence that the underlying distribution is non-normal. A normal probability plot of the residuals, in Figure B-5, shows some curvature, but the basic pattern is linear. This same pattern, with a larger sample, would cause one to consider corrective measures, perhaps a transformation of Y.

(2)   Figure B-6 shows two other residual plots which should be routinely done, if possible. Neither of these plots displays any obvious pattern. If values of other variables were available for these units, residual plots should be made for these variables also.

(3)   Since no alarming deviations were found in the previous steps, it is appropriate to continue. The analysis of variance for these data is in Table B-6.

TABLE B-5

Predicted Values and Residuals

Observation Number	X	$\hat{Y}$	$Y - \hat{Y}$
1	35.3	10.81	0.17
2	29.7	11.25	−0.12
3	30.8	11.17	1.34
4	58.8	8.93	−0.53
5	61.4	8.72	0.55
6	71.3	7.93	0.80
7	74.4	7.68	−1.32
8	76.7	7.50	1.00
9	70.7	7.98	−0.16
10	57.5	9.03	0.11
11	46.4	9.92	−1.68
12	28.9	11.32	0.87
13	28.1	11.38	0.50
14	39.1	10.50	−0.93
15	46.8	9.89	1.05
16	48.5	9.75	−0.17
17	59.3	8.89	1.20
18	70.0	8.04	0.07
19	70.0	8.04	−1.21
20	74.5	7.68	1.20
21	72.1	7.87	−0.19
22	58.1	8.98	−0.51
23	44.6	10.06	−1.20
24	33.4	10.96	−0.60
25	28.6	11.34	−0.26

Source: Adapted from Draper and Smith [1], p.12. *(Reprinted with permission from John Wiley & Sons, Inc.)*

(4)   The squared multiple correlation is calculated with Eq. B-16 as

$$R^2 = \frac{45.81}{63.76} = 0.718$$

Thus, the average monthly temperature explains about 72% of the observed variability in pounds of steam used. The other 28% of the variability cannot be explained with this linear regression.

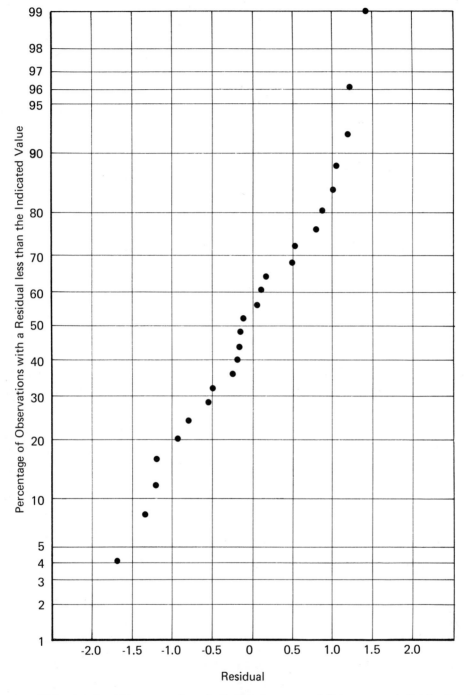

(This plot should be approximately linear when errors follow a normal distribution.)

FIGURE B-5   Normal Probability Plot of Residuals

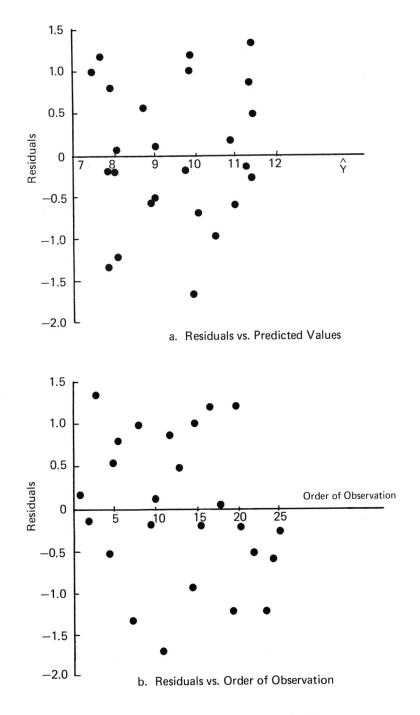

a. Residuals vs. Predicted Values

b. Residuals vs. Order of Observation

FIGURE B-6   Residual Plots

TABLE B-6

Analysis of Variance

Source	Sum of Squares	Degrees of Freedom	Mean Square
Regression	45.81	1	45.81
Error	17.95	23	.781
Total	63.76	24	2.657

(5)   The procedure for testing $H_0 : B = 0$ vs. $H_A : B \neq 0$ will be demonstrated. The significance level $\alpha = .05$ will be used. This means that if the average temperature (X) and pounds of steam used (Y) are actually unrelated, there is a 5% chance of erroneously concluding that they are related. The F-statistic (Eq. B-18) is

$$F = \frac{MSR}{MSE} = \frac{45.81}{0.781} = 58.7$$

It must be compared with $F_{.95} (1, 23)$ from Table B-4. Let m=1, n = 23, $1 - \alpha = .95$ and find $F_{.95} (1, 23) \cong 4.29$. (As only n=20 and n = 30 are given in the table, the approximate value is found using linear interpolation.) Since F = 58.7 is much larger than 4.29, the null hypothesis is rejected. One concludes that $B \neq 0$; there is a linear relationship between X and Y.

(6)   To test the same hypothesis using the t-distribution, evaluate Eq. B-19:

$$t = \left| \frac{b}{s.e. (b)} \right| = \left| \frac{-0.08}{0.010} \right| = 8.0$$

Compare this with $t_{.975} (23) = 2.069$ from Table B-2. As $8.0 > 2.069$, the same conclusion is reached.

(7)   To find a p-value for this hypothesis, it is necessary to compare the calculated t with several values. From Table B-2, one finds

$t_{0.975}(23) \quad = 2.069 \ (\alpha = 0.05)$

$t_{0.99}(23) \quad = 2.500 \ (\alpha = 0.02)$

$t_{0.995}(23) \quad = 2.807 \ (\alpha = 0.01)$

$t_{0.9995}(23) \quad = 3.767 \ (\alpha = 0.001)$

Since the calculated t = 8.0 is larger than all of these, the approximate p-value is $p < .001$. This is very convincing evidence that $B \neq 0$. Had t = 2.71 been observed, one would report $.01 < p < .02$, as 2.71 falls between 2.500 and 2.807.

B-5   PREDICTING FUTURE VALUES OF Y

A common use of the results of a regression analysis is predicting future Y values for known values of X. Since the prediction cannot be exact, it is recommended that one report a prediction interval, analogous to a confidence interval, in addition to the point prediction. Let X_o be the known value of X for which a prediction is desired. Based on the sample data, the best point prediction for Y at X_o is the value of the sample regression line:

$$\hat{Y}_{X_o} = a + b \, X_o \tag{B-20}$$

The next value of Y observed for $X = X_o$ will not be equal to $\hat{Y}_{X_o}$ for two primary reasons:

(1) $Y = A + BX$ is the true regression line; $\hat{Y} = a + bX$ is an estimate which is subject to error, because $a \neq A$ and $b \neq B$.

(2) Observed Y's do not fall exactly on the line $A + BX$ because of random error. The likely magnitude of this error is given by $\sigma_{Y \cdot X}$. If $\sigma_{Y \cdot X}$ is large, accurate point predictions cannot be made.

To account for these errors it is best to report a prediction interval. A prediction interval consists of two numbers P_L and P_U calculated so that the probability that the next Y at X_o falls between P_U and P_L is $1 - \alpha$. By choosing a small α, one can be quite certain that the value will fall in the prediction interval.

As a preliminary step to calculating the prediction interval, one needs s.e. $(\hat{Y}_{X_o})$, the estimated standard error of the prediction. This is given by

$$\text{s.e. } (\hat{Y}_{X_o}) = s_{Y \cdot X} \sqrt{1 + \frac{1}{n} + \frac{(X_o - \bar{X})^2}{(n-1) \, s_X^{\,2}}} \tag{B-21}$$

From Eq. B-21, one sees that predictions will be more accurate for X's near $\bar{X}$ than for those far away. The prediction interval is given by

$$P_L = \hat{Y}_{X_0} - t_{1-\alpha/2} \; (n-2) \; \text{s.e.} \; (\hat{Y}_{X_0})$$

and

$$P_U = \hat{Y}_{X_0} + t_{1-\alpha/2} \; (n-2) \; \text{s.e.} \; (\hat{Y}_{X_0})$$

(B-22)

The t-values may be found in a t-table as described in Section B-3. It is often useful to calculate prediction intervals for several values of X and summarize the results in a graph, as shown in Figure B-7.

A word of warning about the accuracy of prediction intervals: do not make predictions for X's outside the range of the original data. Extrapolation can lead to serious errors. The predictions are based on an assumed linear relationship between X and Y; it is known to be linear only for the X's in the original data set. For larger or smaller X's, the relationship may curve. Figure B-8 displays this difficulty.

Example B-3   Predicting future values of Y.

(1)   It is desired to predict pounds of steam used (Y) for a month with average temperature $X_0 = 43°F$. The point prediction (from Eq. B-20) is:

$$\hat{Y}_{43} = a + b \; (43)$$
$$= 13.63 - 0.08 \; (43)$$
$$= 10.2$$

(2)   The estimated standard error of the prediction (Eq. B-21) is:

$$\text{s.e.} \; (\hat{Y}_{43}) = s_{Y \cdot X} \sqrt{1 + \frac{1}{n} + \frac{(X_0 - \bar{X})^2}{(n-1) \, s_X^{\,2}}}$$

$$= 0.8835 \sqrt{1 + \frac{1}{25} + \frac{(43 - 52.6)^2}{24 \, (17.27^2)}}$$

$$= 0.9066$$

(3)   A 95% prediction interval is then obtained from Eq. B-22:

$$P_L = 10.2 - 2.069 \; (.9066)$$
$$= 8.3$$

and

$$P_U = 10.2 + 2.069 \; (.9066)$$
$$= 12.1$$

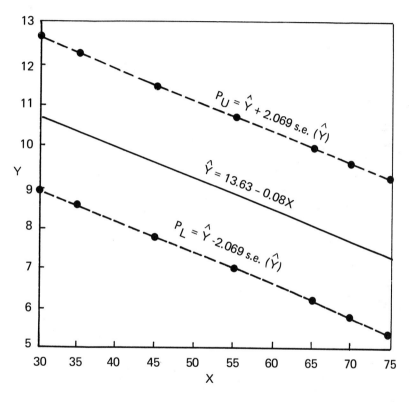

FIGURE B-7   95% Prediction Intervals

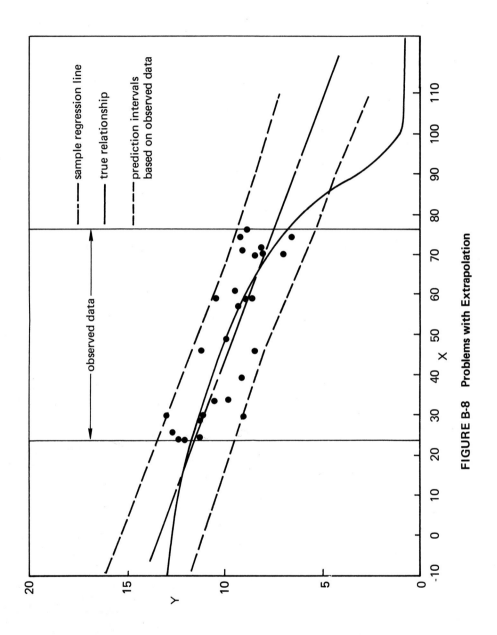

FIGURE B-8 Problems with Extrapolation

Thus, with probability of 95%, the pounds of steam used in a month with average temperature of 43° ($\hat{Y}_{43}$) will be between 8.3 and 12.1. Ninety-five percent prediction intervals are given in Table B-7 for several X's. The results were previously shown in Figure B-7.

TABLE B-7

95% Prediction Intervals

X_o	$\hat{Y}$	P_L	P_U
30	10.83	8.90	12.76
35	10.43	8.53	12.33
45	9.63	7.76	11.50
55	8.83	6.97	10.69
65	8.03	6.15	9.91
70	7.63	5.73	9.53
75	7.23	5.30	9.16
80	6.83	4.87	8.78

B-6 SYMBOLS USED

A	=	intercept of regression line in Eq. B-1
A_L	=	lower confidence interval for A, Eq. B-14
A_U	=	upper confidence interval for A, Eq. B-14
a	=	estimate of A in Eqs. B-2 and -6
B	=	slope of regression line in Eq. B-1
B_L	=	lower confidence interval for B, Eq. B-13
B_U	=	upper confidence interval for B, Eq. B-13
b	=	estimate of B in Eqs. B-2 and -3
F	=	test statistic in Eq. B-18
H_O	=	null hypothesis
H_A	=	alternative hypothesis
k	=	degrees of freedom for denominator of F statistic
m	=	degrees of freedom for numerator of F statistic
MSR	=	mean square of regression
MSE	=	mean square of error ($= s_{Y \cdot x}^2$)
n	=	number of data points in set
P_L	=	lower prediction limit in Eq. B-22
P_U	=	upper prediction limit in Eq. B-22
p	=	("p-value") the smallest significance level for which the null hypothesis would be rejected
R	=	correlation coefficient in Eq. B-16

r	=	simple correlation coefficient between X and Y in Eq. B-17
s_Y^2	=	sample variance of Y in Eq. B-9
s_X^2	=	sample variance of X in Eq. B-8
$s_{Y \cdot X}$	=	estimate of $\sigma_{Y \cdot X}$ in Eqs. B-7 and -10
s.e. (a)	=	estimate of the standard error of a in Eq. B-11
s.e. (b)	=	estimate of the standard error of b in Eq. B-12
s.e. $(\hat{Y}_{x_o})$	=	estimated standard error of the prediction in Eq. B-21
SSE	=	sum of squares due to error in Eq. B-15
SSR	=	sum of squares due to regression in Eq. B-15
SST	=	total (corrected) sum of squares in Eq. B-15
t	=	observed value of test statistic in Eq. B-19
X	=	predictor or control variable in Eq. B-1
X̄	=	simple mean of X values in Eq. B-4
Y	=	response or dependent variable in Eq. B-1
Ȳ	=	simple mean of Y values in Eq. B-5
Ŷ	=	predicted value of Y for a given X in Eq. B-2

Greek

α	=	significance level for testing hypotheses or forming confidence intervals
$\sigma_{Y \cdot X}$	=	standard deviation of the errors

B-7 REFERENCES

1. Draper, N.R. and H. Smith, *Applied Regression Analysis,* John Wiley & Sons, New York (1966).

2. Morrison, D.F., *Multivariate Statistical Methods,* 2nd ed., McGraw-Hill Book Co., New York (1976).

C

EVALUATING PROPAGATED AND TOTAL ERROR IN CHEMICAL PROPERTY ESTIMATES

Cathy Campbell

C-1   INTRODUCTION

Using the estimation procedures described in this handbook involves a two-step procedure. First, one obtains chemical-specific input values (and sometimes other parameters and constants) required by the estimation method. Second, these input values are used in an estimation process that yields an estimate of the unknown property of interest.

Assessing the likely error in an estimate is a key component in determining the applicability and usefulness of the estimate. All of the estimation methods are accompanied by available information on the magnitude of the error introduced by using that estimation method, by assuming no error in the input values. In many cases, however, the true error may be larger than the method error because the input values themselves have been estimated or otherwise imprecisely determined.

In this appendix a general procedure for assessing the total error in estimates of chemical properties is presented. This method applies to all estimates that are calculated using estimated or imprecise input values. Section C-2 contains a general description of the components of error in an estimated value which is followed, in Section C-3, with some theoretical concepts. Section C-4 contains three methods for estimating total error, one (Method 1) that applies when the estimation procedure uses a single estimated input and two (Methods 2 and 3) that apply for estimating total error when the procedure uses more than one estimated input. Method 2 is more exact, but also more complicated, than Method 3. Each method is demonstrated with an example, and further examples are given in Section C-5.

C-2   COMPONENTS OF ERROR

Concepts and procedures for estimating total error are introduced via a specific example that will be used to demonstrate the procedures later. Consider Method 3 for estimating the viscosity of a liquid, η_L, at a given temperature T. The procedure follows from three (or two) equations, earlier given as Eqs. 22-13, 22-16, and 22-17:

$$\hat{\eta}_L^* = \eta_{Lb} \exp\left\{B_4 \left(\frac{1}{T} - \frac{1}{T_b}\right)\right\} \tag{C-1}$$

$$B_4 = \frac{1}{n} (\Delta H_v - RT) \tag{C-2}$$

$$\Delta H_v = K_F T_b (8.75 + R \ln T_b) \tag{C-3}$$

where

η_{Lb}	=	viscosity of the liquid at its boiling temperature, T_b,
ΔH_v	=	heat of vaporization,
R	=	gas constant, and
n and K_F	=	constants which describe the class of chemicals.

When a procedure such as this is used to estimate a property, responsible scientists will want to assess and report the likely error in these results. Error in $\hat{\eta}_L$, or other estimates, may rise from two sources.

(1) *Method error* occurs because of approximations or inaccuracies in the prediction equation. The estimates are subject to this error even when the chemical-specific input properties, in this case η_{Lb}, T_b, and K_F, are known quite accurately. Table 22-2 gives some assessment of method error for the procedure described here; the average relative error is 19% and the maximum error is 49%.

---

*A " ˆ " above a property indicates the value is estimated.

(2) *Propagated error* is introduced because the input values have been estimated or imprecisely determined. For example, if T_b is estimated, then uncertainty about the value of T_b creates additional uncertainty in the estimate of η_L.

To demonstrate the effect of propagated error, data from Example 22-6 will be used. Suppose that η_{Lb} and B_4 are known quite precisely with η_{Lb} = 0.2 cp and B_4 = 817. To estimate η_L when T = −60°C = 213 K, Eq. (C-1) becomes

$$\hat{\eta}_L = 0.2 \, \exp\left\{3.84 \left(1 - \frac{213}{T_b}\right)\right\} \tag{C-4}$$

For different values of T_b, the relationship between $\hat{\eta}_L$ and T_b can be plotted as shown in Figure C-1. If, as in the example, T_b = 342 K, then η_L is estimated as 0.85 cp. However, if the value of T_b is known to be subject to a 5% error, then T_b could be as small as 325 K or as large as 359 K. This means, as shown in the figure, that $\hat{\eta}_L$ could be as small as 0.75 cp or as large as 0.95 cp. This uncertainty about $\hat{\eta}_L$ comes from the expected error in T_b and must be considered in addition to the method error. The estimation error that derives solely from errors in the input values will be referred to as propagated error.

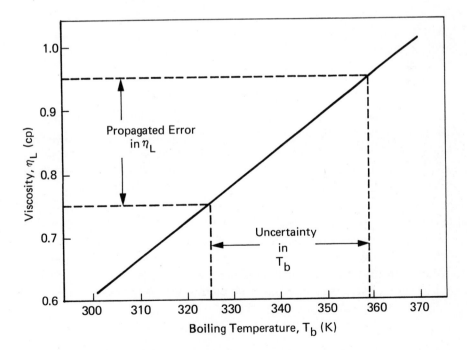

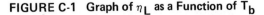

FIGURE C-1 Graph of η_L as a Function of T_b

To realistically assess the total error in $\hat{\eta}_L$, it is necessary to consider both sources of error simultaneously. The procedure given in this chapter leads to evaluating

$$\text{Total Error} = \sqrt{(\text{Method Error})^2 + (\text{Propagated Error})^2} \qquad \text{(C-5)}$$

For the example given here, if the average method error of 19% is used, and one observes that the propagated error of ±0.10 is 11.8% of the estimated value, then the total error in the estimate of η_L is approximately 22%.

In this section the concepts of method error, propagated error and total error have been informally introduced for an estimation process that has one input value subject to known error. In the following section, these ideas are described more formally, and the theory of estimating propagated error is briefly presented. The procedure will be extended to estimation methods which have error in more than one input value. Following that, simple and very general procedures for estimating propagated error will be given and demonstrated for a series of examples.

C-3   THEORETICAL BACKGROUND

Some theory for evaluating propagated error and total error is given in this section. First, estimation methods with a single estimated input will be considered. Then procedures for multiple inputs will be developed.

Any of the estimation methods in this handbook give rules for using input characteristics to determine an estimated value of the chemical property. Sometimes, the "rule" can be written down explicitly as an equation or series of equations. The "rule" may consist of a graphical method, such as finding the intersection of two lines or of looking up values in a nomogram. In any case the rule may be expressed as

$$\hat{C} = f(x_o) \qquad \text{(C-6)}$$

where $\hat{C}$ is the obtained estimate, and x_o is the estimated input value. The function $f(\cdot)$ is the rule linking the input value x_o to the estimated value $\hat{C}$. In the situation considered here, x_o is an estimate of X_o, the true input value. For instance, X_o may be boiling temperature of a liquid and x_o is the estimated boiling temperature.

The estimate $\hat{C}$ may differ from the true value C because of method error and propagated error. If X_o is known, then one would calculate $\hat{C} = f(X_o)$ and there would be no propagated error. The difference between $f(X_o)$ and $f(x_o)$ is the propagated error, while the difference between $f(X_o)$ and C, the true value, is method error.

To see the effect of method error and propagated error the estimate may be expressed as

$$\hat{C} = f(x_o) = C + [f(X_o) - C] + [f(x_o) - f(X_o)] \tag{C-7a}$$

or

$$\text{Estimate} = \text{True Value} + \text{Method Error} + \text{Propagated Error} \tag{C-7b}$$

For a particular estimate, the exact values of the method and propagated errors are not known, but using available information about the method error and the likely error in x_o, one can obtain an average or expected value for the total error.

In the development that follows, both error components are assumed to have an expected value of 0. That is, the differences $f(X_o) - C$ and $f(x_o) - f(X_o)$ may be either positive or negative, but in the long run will average to 0. This assumption will not be satisfied if:

(1) the method is biased,
(2) the estimate of X_o is biased, or
(3) the rule $f(\cdot)$ has a large second derivative, i.e., the graph of $\hat{C}$ vs. X is curving asymmetrically close to $X = X_o$.

If one also assumes that the magnitude of the method error is independent of the error in x_o, then one obtains

$$\text{avg. } [(\hat{C} - C)^2] = \text{avg. } [(f(X_o) - C)^2] + \text{avg. } [(f(x_o) - f(X_o))^2] \tag{C-8}$$

or

$$\text{Total Error} = \sqrt{\text{avg. (Method Error)}^2 + \text{avg. (Propagated Error)}^2}, \tag{C-9}$$

where "avg." represents average or expected value. Method, propagated, and total errors may be expressed either in absolute or relative units (error/estimate) so long as consistency is maintained. To evaluate Eq. C-9, one obtains an estimate of error from the literature in the chapters of this handbook and then uses the procedure given below to estimate propagated error. Theory of propagation of errors has been reviewed by Ku [1]; an outline will be given here.

In Figure C-1, one sees that the change in $\hat{\eta}_L$ when the value of T_b changes is related to the steepness of the graph at the point where T_b changes. Readers who have studied calculus realize that the steepness of a graph at a given point is measured by its first derivative. Although the succeeding development depends on some notation from calculus, the methods for estimating propagated error will not require knowledge of calculus.

If the estimation procedure depends only on a single input which is subject to error, then the estimated error is

$$\text{Propagated Error} = \sqrt{\left[\frac{df}{dX}\right]_{x_o}^2 s_{x_o}^2} \qquad \text{(C-10)}$$

where $\left[\dfrac{df}{dX}\right]_{x_o}$ is the first derivative of f evaluated at $X = x_o$, and s_{x_o} is the estimated total error in x_o.

When the estimation rule can be written explicitly and the derivative easily obtained, then use of Eq. C-10 is the easiest way to evaluate the propagated error. For the example in § C-2, Eq. C-4 gives the rule for estimating η_L for a given value of T_b. [Note that $\hat{T}_b$ corresponds to x_o and $\eta_L(T_b)$ corresponds to f(X).] The derivative of this function is

$$\frac{d\eta_L (T_b)}{dT_b} = 0.2 \left[\exp\left(3.84 \left(1 - \frac{213}{T_b}\right)\right)\right]\left(\frac{3.84\ (213)}{T_b^2}\right) \qquad \text{(C-11)}$$

When this is evaluated at $\hat{T}_b = 342$, one obtains

$$\left[\frac{d\eta_L}{dT_b}\right]_{\hat{T}_b}^2 = (0.00594)^2 \qquad \text{(C-12)}$$

It was reported that $\hat{T}_b$ was subject to a 5% error. In absolute units, $s_{\hat{T}_b} = (0.05)\ (342) = 17.1$. Substituting these values into Eq. C-10 gives progagated error = 0.10, as was found earlier by inspecting the graph.

Other cases, where the graph is more curved, will not necessarily give perfect agreement.

If the estimation rule is a function of p input variables, all of which are subject to error, the progagated error must be evaluated separately for each input variable. The assumption is made that the input values come from difference sources and that the errors in them are independent of each other. Let $\hat{C} = f(X_1, \ldots, X_p)$ be the estimation rule, and suppose the rule will be evaluated for numbers $(x_1, \ldots, x_p)$ which are estimates of $(X_{1o}, \ldots, X_{po})$, the "true" values of the input properties. The propagated error is the extent to which $f(x_1, \ldots, x_p)$ is expected to differ from $f(X_{1o}, \ldots, X_{po})$. Let $(s_{x_1}, \ldots, s_{x_p})$ be the estimated errors in $(x_1, \ldots, x_p)$. Then the estimated propagated error is given by

$$\text{Propagated Error} = \sqrt{\sum_{i=1}^{p} \left[\frac{\partial f}{\partial X_i}\right]^2_{(x_i)} s^2_{x_i}} \tag{C-13}$$

where $\left[\dfrac{\partial f}{\partial X_i}\right]_{x_i}$ is partial derivative of f with respect to X_i evaluated at the estimates $(x_1, \ldots, x_p)$.

The following section gives methods for evaluating propagated error that do not depend on having explicit expressions for the derivatives. This approach is somewhat analogous to the graphical approach used in § C-2.

C-4   EVALUATING PROPAGATED ERROR

Single Estimated Input (Method 1). Suppose $C_x = f(X)$ is the estimation rule, and that $\hat{C}_o = f(X_o)$ is the desired estimate. When X_o is not known, an estimate x_o will be used instead. Let s_{x_o} be the estimated error in x_o. In other words, the estimate of X_o is $x_o \pm s_{x_o}$. The estimate of C, then, is $f(x_o)$. The method for evaluating the propagated error consists of using different values in the estimation rule and observing how much the resulting estimate changes.

Basic Steps
(1) Find $\hat{C}_1 = f(x_o - s_{x_o})$ by using $x_o - s_{x_o}$ instead of x_o in the estimation rule.
(2) Find $\hat{C}_2 = f(x_o + s_{x_o})$ by using $x_o + s_{x_o}$ in the estimation rule.

(3) Estimate the propagated error as

$$\text{Propagated Error} = \sqrt{\frac{(\hat{C}_1 - \hat{C}_2)^2}{4}} \qquad\qquad \text{(C-14)}$$

(4) To estimate the total error in the estimate, a fourth step is to calculate

$$\text{Total Error} = \sqrt{(\text{Method Error})^2 + (\text{Propagated Error})^2} \qquad \text{(C-5)}$$

Example C-1 In § C-2, Eqs. C-1, -2, and -3 give a rule for estimating liquid viscosity. In the simple example there, it was assumed that B_4 was known. In practice, one estimates ΔH_v from T_b and then estimates B_4. Thus, T_b is involved in a more complicated fashion than is indicated by Eq. C-4. Again, following Example 22-6, suppose $\eta_{Lb} = 0.2$ cp is known quite precisely, and that $K_F = 1.0$, $R = 1.988$ and $n = 8$. Let $\hat{T}_b = 342K$ be subject to a 5% error so that $s_{\hat{T}_b} = 17.1K$. The viscosity estimate is desired for $T = 213K$.

(1) Evaluate the rule at $\hat{T}_b - s_{\hat{T}_b} = 324.9K$.

$$\Delta H_v = (1.0)(324.9)[8.75 + 1.988 \ln (324.9)] = 6578$$

$$B_4 = \frac{1}{8}[6578 - (1.988)(213)] = 769.4$$

$$\hat{C}_1 = 0.2 \exp\left\{769.4\left(\frac{1}{213} - \frac{1}{324.9}\right)\right\} = 0.694 \text{ cp}$$

(2) Evaluate the rule at $\hat{T}_b + s_{\hat{T}_b} = 359.1K$

$$\Delta H_v = (1.0)(359.1)[8.75 + 1.988 \ln (359.1)] = 7342$$

$$B_4 = \frac{1}{8}[7342 - (1.988)(213)] = 864.9$$

$$\hat{C}_2 = 0.2 \exp\left\{864.9\left(\frac{1}{213} - \frac{1}{359.1}\right)\right\} = 1.043 \text{ cp}$$

(3) Propagated Error $= \sqrt{\dfrac{(0.694 - 1.043)^2}{4}} = 0.175$ cp

(4) Recalling that the average method error is 19% of the true value, the estimated error, in absolute units, is obtained by multiplying the relative error by the estimated viscosity $\hat{\eta}_L = 0.85$.

Method Error $= 0.19 (0.85 \text{ cp}) = 0.162$ cp

Total Error $= \sqrt{(0.163)^2 + (0.175)^2} = 0.238$ cp

In relative terms, estimated total error is 28% of the estimated value.

Multiple Input Values (Method 2). If the estimation rule depends on more than one estimated input, then propagated error must be estimated separately for each estimated input variable. The separate estimates are combined as shown in Eq. C-15. Suppose that $\hat{C} = f(x_1, \ldots, x_p)$ is the estimate with $(x_1, \ldots, x_p)$ being the available estimates of $(X_{1o}, \ldots, X_{po})$. Let $(s_{x_1}, \ldots, s_{x_p})$ be the estimate of total error in $(x_1, \ldots, x_p)$. The estimate of propagated error is given by

$$\text{Propagated Error} = \sqrt{\sum_{i=1}^{p} (\text{Propagated Error for Input } i)^2} \qquad \text{(C-15)}$$

Total Error is found from Eq. C-5 as before.

To find the propagated error for the first input,

(1) Find $\hat{C}_{11} = f(x_1 - s_{x_1}, x_2, \ldots, x_p)$ by using $x_1 - s_{x_1}$ in the estimation rule, but leaving all the other quantities at their original values.

(2) Find $\hat{C}_{12} = f(x_1 + s_{x_1}, x_2, \ldots, x_p)$ by using $x_1 + s_{x_1}$ in the rule with the other values unchanged.

(3) Propagated Error for Input 1 $= \sqrt{\dfrac{(\hat{C}_{11} - \hat{C}_{12})^2}{4}}$ \qquad (C-16)

These three steps are repeated for each input. For the last variable the procedure is

(1) Find $\hat{C}_{p1} = f(x_1, \ldots, x_{p-1}, x_p - s_{x_p})$.

(2) Find $\hat{C}_{p2} = f(x_1, \ldots, x_{p-1}, x_p + s_{x_p})$.

(3) Propagated Error for Input p $= \sqrt{\dfrac{(\hat{C}_{p1} - \hat{C}_{p2})^2}{4}}$ \qquad (C-17)

Finally, Eqs. C-15 and -5 can be used to find the propagated and total error. A summary of the steps is given below.

Basic Steps

(1) Evaluate the propagated error for each input as:

$$\text{Propagated Error for Input } i = \sqrt{\dfrac{(\hat{C}_{i1} - \hat{C}_{i2})^2}{4}} \qquad \text{(C-18)}$$

(2) Use Eq. C-15 to find the overall propagated error.

(3) Use Eq. C-5 to find total error.

Continuation of Example C-1   Suppose now that η_{Lb} has been estimated as 0.2 cp with a 10% error, so, $s_{\hat{\eta}_{Lb}} = 0.02$ cp. The estimate $\hat{\eta}_L = f(\hat{T}_b, \hat{\eta}_{Lb})$ now depends on two estimated inputs.

(1)   To find the propagated error for input 1 $(\hat{T}_b)$, one finds $\hat{C}_{11} = f(\hat{T}_b - s_{\hat{T}_b}, \hat{\eta}_{Lb})$ and $\hat{C}_{12} = f(\hat{T}_b + s_{\hat{T}_b}, \hat{\eta}_{Lb})$. That is, for $\hat{C}_{11}$, substitute $T_b = 324.9$ and $\eta_{Lb} = 0.20$ in Eqs. C-1 to -3. For $\hat{C}_{12}$, use $T_b = 359.1$ and $\eta_{Lb} = 0.20$. These have already been found in steps (1) and (2) of the previous example which evaluated the error propagated by errors in T_b. The results were

$$\hat{C}_{11} = 0.694 \text{ cp and}$$

$$\hat{C}_{12} = 1.043 \text{ cp,}$$

so,

$$\text{Propagated Error for Input 1} = \sqrt{\frac{(0.694 - 1.043)^2}{4}} = 0.175 \text{ cp}$$

(2)   To find the propagated error for input 2 $(\hat{\eta}_{Lb})$, one needs $\hat{C}_{21} = f(\hat{T}_b, \hat{\eta}_{Lb} - s_{\hat{\eta}})$ and $\hat{C}_{22} = f(\hat{T}_b, \hat{\eta}_{Lb} + s_{\hat{\eta}})$. To find $\hat{C}_{21}$, evaluate the estimation rule with $T_b = 342K$ and $\eta_{Lb} = 0.18$ cp:

$$\Delta H_v = (1.0)(342)[8.75 + 1.988 \ln(342)] = 6960$$

$$B_4 = \frac{1}{8}[6960 - 1.988(213)] = 817.0$$

$$\hat{C}_{21} = 0.18 \exp\left\{817.0\left(\frac{1}{213} - \frac{1}{342}\right)\right\} = 0.765$$

Since η_{Lb} does not enter the exponent, $\hat{C}_{22}$ can be found by reevaluating Eq. C-1 with $\eta_{Lb} = 0.22$. Thus,

$$\hat{C}_{22} = 0.22 \exp\left\{817.0\left(\frac{1}{213} - \frac{1}{342}\right)\right\} = 0.935$$

which is found by using $\hat{\eta}_{Lb} + s_{\hat{\eta}}$ in the estimation rule. Now, using Eq. C-18:

$$\text{Propagated Error for Input 2} = \sqrt{\frac{(0.765 - 0.935)^2}{4}} = 0.085 \text{ cp}$$

(3)   Using Eq. C-15,

$$\text{Propagated Error} = \sqrt{0.175^2 + 0.085^2} = 0.194 \text{ cp}$$

(4) From Step (4) of the previous example, method error was estimated to be 0.162 cp, therefore:

$$\text{Total Error} = \sqrt{0.162^2 + 0.194^2} = 0.253 \text{ cp}$$

The total error is 30% of the estimated value of 0.85 cp.

Simplified Method for Multiple Inputs (Method 3). When the estimation process is difficult, it may be quite burdensome to estimate propagated error. The above examples show, however, that the method error may be a substantial understatement of the total error in the estimate. In the previous example, use of method error alone leads to a 35% understatement of the total error. Therefore some effort to estimate total error seems justified.

The process for estimating propagated error can be simplified somewhat, with a concurrent loss in the accuracy of the estimated error. The simplification may be worthwhile, however, when there are several input values, or if the estimation process is difficult. The simplified process requires that only one extra estimate, rather than two, be calculated for each input. In the previous method, propagated error for an input was estimated by looking at the difference between two estimates that were obtained by changing the value of that input — once by increasing it and once by decreasing it. Thus,

$$\text{Propagated Error for Input } i = \sqrt{\frac{(\hat{C}_{i1} - \hat{C}_{i2})^2}{4}} \qquad \text{(C-18)}$$

The effect of changing the input can also be estimated by changing the input only once and seeing how far the estimate moves from its original value $\hat{C}$. Either of the quantities given below can also be used as an estimate of propagated error:

$$\text{Propagated Error for Input } i = \left| \hat{C}_{i1} - \hat{C} \right| \qquad \text{(C-19)}$$

or

$$\text{Propagated Error for Input } i = \left| \hat{C}_{i2} - \hat{C} \right| \qquad \text{(C-20)}$$

For the examples here, $|\hat{C}_{i1} - \hat{C}|$ will be used.

Basic Steps

(1) Find $\hat{C}_{11} = f(x_1 - s_{x_1}, x_2, \ldots, x_p)$ by using $x_1 - s_{x_1}$ in the estimation rule, but leaving all the other quantities at their original values.

(2) Calculate Propagated Error for Input 1 using Eq. C-19.

(3) Repeat steps (1) and (2) for each input (2, . . ., p), each time using $x_i - s_{x_i}$ while leaving the other quantities at their original estimated value.

(4) Evaluate Propagated Error using Eq. C-15.

(5) Evaluate Total Error using Eq. C-5.

Example of Simplified Method. (Continuation of Example C-1.) In this example $\hat{C} = 0.85$ cp was the estimate actually obtained for η_L. From the above, it was found that

$$\hat{C}_{11} = 0.694 \text{ cp } [f(\hat{T}_b - s\hat{T}_b, \hat{\eta}_{Lb})]$$

and

$$\hat{C}_{21} = 0.765 \text{ cp } [f(\hat{T}_b, \hat{\eta}_{Lb} - s\hat{\eta}_{Lb})]$$

Thus,

Propagated Error for Input 1 $= |0.694 - 0.85| = 0.156$ cp

Propagated Error for Input 2 $= |0.765 - 0.85| = 0.085$ cp

Propagated Error $= \sqrt{0.156^2 + 0.085^2} = 0.178$ cp

This compares with a value of 0.194 cp obtained using the longer method. The Estimated Total Error is $\sqrt{0.162^2 + 0.178^2} = 0.240$ cp which compares quite favorably with the 0.253 cp obtained before. It cannot be identified in advance whether the simplified method will lead to a larger or smaller estimate of the propagated error.

C-5   ADDITIONAL EXAMPLES

Example C-2   Estimating K_{oc} from K_{ow} for 1-chloro-1-bromo-2,2,2-trifluoroethane.

Estimation of the soil adsorption coefficient, K_{oc}, is described in detail in Chapter 4. A variety of empirically-derived regression equations with log K_{oc} as the dependent variable are given in that chapter. Several of these use log K_{ow}, the logarithm of the octanol-water partition coefficient, as the predictor variable. Thus, the estimating equation is

$$\log K_{oc} = A_1 \log K_{ow} + A_2 \tag{C-21}$$

where A_1 and A_2 depend upon the data set used which may or may not represent a special class of chemicals. Estimation methods for K_{ow} are given in Chapter 1.

To estimate the adsorption coefficient for 1-chloro-1-bromo-2,2,2-trifluoroethane either Eq. 4-7 or 4-8 appears to be appropriate. Equation 4-8 is used for this example and gives:

$$\log \hat{K}_{oc} = 0.544 \log K_{ow} + 1.377 \tag{C-22}$$

Example 1-27 of Chapter 1 demonstrates the estimation of log K_{ow} for 1-chloro-1-bromo-2,2,2-trifluoroethane. Using the method of fragment constants,

$$\log \hat{K}_{ow} = 2.46 \pm 0.14$$

(The estimate of method error, ±0.14 log units, was obtained by averaging errors from 76 chemicals with known values.)

From Eq. C-22,

$$\log \hat{K}_{oc} = 0.544\,(2.46) + 1.377 = 2.72$$

$$\hat{K}_{oc} = 519$$

Error in $\hat{K}_{oc}$ depends on method error due to using Eq. C-22 and propagated error because of uncertainty in the value of log K_{ow}. Table 4-9 contains information on estimation errors for a number of chemicals. The ratio, Estimated K_{oc}/Measured K_{oc}, varies from 18 to nearly 0 for Eq. 4-8, giving relative errors up to 1700%. The average for those chemicals with K_{oc} between 100 and 1000 is 120%. That value is used for this example.

Estimation of propagated error follows Method 1 in § C-4. The estimation rule is:

$$\hat{K}_{oc} = 10^{[0.544 \log (K_{ow}) + 1.377]}$$

The single estimated input is log $\hat{K}_{ow}$. Following the Method 1 steps, we obtain:

(1) $\hat{C}_1 = 10^{[0.544\,(2.46\,-\,.14) + 1.377]} = 435.6$

(2) $\hat{C}_2 = 10^{[0.544\,(2.46\,+\,.14) + 1.377)]} = 618.6$

(3) Propagated Error $= \sqrt{\dfrac{(435.6 - 618.6)^2}{4}} = 91.5$

Expressing the method error in K_{oc} units gives:

Method Error $= 1.20\,(519.1) = 623$

(4) Total Error $= \sqrt{623^2 + 91.5^2} = 630$

An alternate (and, perhaps, preferable) approach would be to calculate the total error in terms of log K_{oc}; this avoids having the final range of K_{oc} values include negative numbers. Using this alternate approach, the results of steps 1 to 3 above are 2.64, 2.79, and 0.075, respectively (all in log K_{oc} units). The method error $= \log 623 = 2.79$ and the total error (from step 4) is also 2.79, again in log K_{oc} units. For this example, where the input value is estimated quite precisely, total error is almost all due to method error.

Example C-3   Estimating P_{vp} for aniline from ΔH_{vb} and T_b.

Assessing error in the estimated vapor pressure of aniline is the next example. The estimation rule (from Eq. 14-20) is:

$$\hat{P}_{vp} = 760 \exp \left\{ \frac{\Delta H_{vb}}{\Delta Z_b R T_b} \left[1 - \frac{(3-2T_{rb})^m}{T_{rb}} - 2m(3-2T_{rb})^{m-1} \ln T_{rb} \right] \right\} \quad \text{(C-23)}$$

where

$\hat{P}_{vp}$ = estimated vapor pressure (mm Hg)

ΔH_{vb} = heat of vaporization at the normal boiling point (cal/mol)

T_b = normal boiling point (K)

T = temperature (K)

T_{rb} = T/T_b

m, R and ΔZ_b are constants

The constants R = 1.987, ΔZ_b = 0.97, and m = 0.19 are considered error-free. However ΔH_{vb} and T_b must be estimated if measured values are not available. Note that $T_{rb} = T/T_b$, where T is the temperature for which the vapor pressure is desired.

Chapter 12 gives methods for estimating T_b. For aniline, the Lydersen, Forman, Thodos method gives $\hat{T}_b$ = 441.5K with an average method error of ±6.8%, or ±30K for aniline. The heat of vaporization of aniline can be estimated using methods from Chapter 13 as 11,090 cal/mole with an average method error of ±2% (±222 cal/mole for aniline). The vapor pressure for aniline will be estimated for T = 20°C (293K).

In this setting, estimated vapor pressure depends on *two* estimated input values, $\hat{T}_b$ and $\Delta \hat{H}_{vb}$. Methods 2 and 3 in §C-4 are applicable here. For completeness, both the full method and simplified method will be used. To set the notation, let

$$\hat{P}_{vp} = f(T_b, \Delta H_{vb}) \quad \text{(C-24)}$$

where $f(\cdot, \cdot)$ is the right-hand side of Eq. C-23. The estimate of P_{vp} is

$$\hat{C} = \hat{P}_{vp} = f(\hat{T}_b, \Delta \hat{H}_{vb}) \quad \text{(C-25)}$$

$$\hat{P}_{vp} = 760 \exp \left\{ \frac{11,090}{(0.97)(1.987)(441.5)} \left[1 - \frac{(3-2(293)/441.5)^{0.19}}{293/441.5} \right.\right.$$

$$\left.\left. - 2(0.19)(3-2(293)/441.5)^{0.19-1} \ln \left(\frac{293}{441.5}\right) \right] \right\}$$

$$= 0.52 \text{ mm Hg}$$

Using Method 2

To assess propagated error with the full method, one must calculate $\hat{C}_{11}, \hat{C}_{12}, \hat{C}_{21},$ and $\hat{C}_{22}$.

$$\hat{C}_{11} = f(441.5 - 30, 11{,}090) = 1.58 \text{ mm Hg}$$

$$\hat{C}_{12} = f(441.5 + 30, 11{,}090) = 0.20 \text{ mm Hg}$$

$$\hat{C}_{21} = f(441.5, 11{,}090 - 222) = 0.60 \text{ mm Hg}$$

$$\hat{C}_{22} = f(441.5, 11{,}090 + 222) = 0.45 \text{ mm Hg}$$

$$\text{Propagated Error for Input 1} = \sqrt{\frac{(1.58 - 0.20)^2}{4}} = 0.69 \text{ mm Hg}$$

$$\text{Propagated Error for Input 2} = \sqrt{\frac{(0.60 - 0.45)^2}{4}} = 0.075 \text{ mm Hg}$$

$$\text{Propagated Error} = \sqrt{0.69^2 + 0.075^2} = 0.694 \text{ mm Hg}$$

Table 14-1 shows that method error for this P_{vp} range is approximately 28% or 0.146 mm Hg. Thus,

$$\text{Total Error} = \sqrt{0.146^2 + 0.694^2} = 0.71 \text{ mm Hg}$$

Using Method 3

This method of estimating propagated error depends only on the values $\hat{C}_{11}$ and $\hat{C}_{21}$ which are calculated by setting $\hat{T}_b$ and $\Delta\hat{H}_{vb}$ respectively at one standard error below their estimated values.

$$\text{Propagated Error for Input 1} = \left| \hat{C}_{11} - \hat{C} \right| = \left| 1.58 - 0.52 \right| = 1.06 \text{ mm Hg}$$

$$\text{Propagated Error for Input 2} = \left| \hat{C}_{21} - \hat{C} \right| = \left| 0.60 - 0.52 \right| = 0.08 \text{ mm Hg}$$

$$\text{Propagated Error} = \sqrt{1.06^2 + 0.08^2} = 1.06 \text{ mm Hg}$$

$$\text{Total Error} = \sqrt{0.146^2 + 1.06^2} = 1.07 \text{ mm Hg}$$

Estimated total error from the simplified method is not qualitatively different from the estimate obtained using the full method (0.71 mm Hg). As in Example C-2, the ranges associated with the estimated errors imply (wrongly) the possibility of negative numbers. Thus an alternate approach of calculating the propagated and total error in $\ln P_{vp}$ might be preferable.

C-6   SYMBOLS USED[1]

A_1, A_2	=	parameters in Eq. C-21
B_4	=	parameter in Eqs. C-1, -2
C	=	surrogate symbol for any dependent variable, Eq. C-6

---

1. A " ˆ " above any symbol indicates the value is estimated.

C_1 = value of C obtained from $f(x_o - s_{x_o})$.
See § C-4, Method 1.

C_2 = value of C obtained from $f(x_o + s_{x_o})$.
See § C-4, Method 1.

C_{11} = value of C obtained from $f(x_1 - s_{x_1}, x_2, \ldots, x_p)$.
See § C-4, Method 2.

C_{12} = value of C obtained from $f(x_1 + s_{x_1}, x_2, \ldots, x_p)$.
See § C-4, Method 2.

C_{i1}, C_{i2} = values of C when the ith input variable is
$x_i - s_{x_i}$ and $x_i + s_{x_i}$, respectively.
See § C-4, Method 2.

$f(\cdot)$ = rule, or equation(s), linking the input variable(s)
to the dependent variable, C. See Eq. C-6.

$[df/dX]_{x_o}$ = first derivative of $f(\cdot)$ evaluated at $X = x_o$, Eq. C-10.

ΔH_v = heat of vaporization, Eqs. C-2, -3 (cal/mol)

ΔH_{vb} = heat of vaporization at the normal boiling point,
Eq. C-23 (cal/mol)

K_F = parameter in Eq. C-3

K_{oc} = soil (or sediment) adsorption coefficient based upon
organic carbon content, Eq. C-21

K_{ow} = octanol/water partition coefficient, Eq. C-21

m = parameter in Eq. C-23

n = parameter in Eq. C-2

p = number of input variables (subject to error) in $f(\cdot)$

P_{vp} = vapor pressure, Eq. C-23 (mm Hg)

R = gas constant, Eqs. C-3, -23 (=1.987)

s = estimated standard error

s_{x_o} = estimated standard error for the estimated input parameter X_o

T = temperature (K)

T_b = normal boiling point, Eqs. C-1, -3, -23 (K)

T_{rb} = T/T_b, Eq. C-23

X = surrogate symbol for any input variable in $f(\cdot)$

X_o = true value of X for some chemical

x_o = estimated value of X_o

X_i, x_i = true and estimated values for the ith input variable in
$f(\cdot)$

ΔZ_b = parameter in Eq. C-23

Greek

η_L = liquid viscosity, Eq. C-1 (cp)

η_{Lb} = liquid viscosity at the normal boiling point,
Eq. C-1 (cp)

C-7 REFERENCE

1. Ku, H.H., "Notes on the Use of Propagation of Error Formulas," *J. Res. Nat. Bur. Stand.,* Section C, **70**, 263-73 (1966).

INDEX

Absorption spectrum, 8-11
  wavelength of maximum absorption,
    8-11,-13ff
Acclimation in biodegradation, 9-17ff
Acentric factor, **19**-5,-7
  estimation of, **19**-13
  use in estimating liquid density, **19**-
    12,-13
Acid-base reactions, **7**-1
Acid dissociation constant:
  definition of, **6**-1
  effect on adsorption, 4-6
  estimation of, **6**-6
    Hammett correlation for aromatic
      acids, **6**-9
    method errors, **6**-22
    Taft correlation for aliphatic acids,
      6-20
  measurement of, **6**-5
  "mixed" or Bjerrum, **6**-3
  "true" or thermodynamic, **6**-3
  use in estimating rate of hydrolysis, **7**-
    19,-21,-32
Activation energy, **7**-11,-14ff,-21,
  -37ff; **9**-28
Activity of a chemical, 3-6ff; **11**-2

Activity coefficient:
  at infinite dilution:
    estimation of, *see* Chapter 11
    use in estimating solubility, *see*
      Chapter **3**; **2**-3ff,-13
  estimation of, **11**-1
    available methods, **11**-2
    recommended methods, **11**-8
    method errors, **11**-7
    Method 1 (Pierotti *et al.*), **11**-10
    Method 2 — UNIFAC, **11**-20
  relationship to dissociation constant,
    6-2
  relationship to octanol-water
    partition coefficient, **11**-2
  use in estimating:
    octanol-water partition coefficient,
      1-4ff,-47ff; **2**-24ff
    solubility, *see* Chapter **3**; **2**-3ff,-24ff
Addition reactions, **7**-1,-2
Adsorption, effect on:
  biodegradation, 9-32
  rate of hydrolysis, **7**-17
  rate of photolysis, 8-36
  volatilization from soil, **16**-3
  volatilization from water, **15**-3,-4,-7

Adsorption coefficient for soils and
    sediments:
  definition of, **4**-1
  estimation of, **4**-1
    available methods, **4**-8
    inputs required, **4**-3
    method errors, **4**-23
  factors influencing value of, **4**-4ff
  measurement of, **4**-2
  use in estimating:
    bioconcentration factor, **5**-3,-13
    water solubility, **2**-3ff
Affens' method for estimating
    flashpoint, **18**-3
Air, properties of, **17**-11
Air-water interface, **15**-3ff,-9,-10; **17**-2,-5
Alternative hypothesis, **B**-16
Analysis of variance, **B**-15, -16
ANOVA table, **B**-16
  example, **B**-24
Antoine's constant, **13**-13,-14
Antoine's equation, **13**-12; **14**-5,-7
Antonov's method for estimating
    interfacial tension with water, **21**-7
Antonov's rule, **21**-3,-7
Apparent diffusion coefficient, **17**-1
Aqueous photolysis, **8**-1
  absorption of light, **8**-16
    calculation of, **8**-16
    rate constant for, **8**-17
  deactivation process, **8**-14
  energy transfer: sensitization and
      quenching, **8**-7
  examples, **8**-36
  excitation processes, **8**-2
  half-life for, **8**-23,-38
  photochemical reactions in, **8**-29
  rate constant for, **8**-18
  Zepp *et al.* model for, **8**-17ff
    compound-specific inputs, **8**-19
    ecosystem-specific inputs, **8**-22
Arrhenius equation, **7**-14; **9**-28; **10**-26
ASOG, **11**-7,-21
Atmospheric residence time:
  estimates for chemicals, **10**-8ff
  estimation of, **10**-1
    method errors, **10**-4ff
    method selection, **10**-3ff
    methods:
      inputs required, **10**-5ff
      non-steady-state, one-
          compartment model, **10**-16

Atmospheric residence time,
    estimation of, methods *(Cont.)*:
      non-steady-state, two-
          compartment model, **10**-18
      steady-state model, **10**-13
      correlation with mean standard
          deviation (Junge's
          correlation), **10**-27
      use of chemical reactivity data,
          10-21
Atomic diffusion volume, **17**-11,-12
Atomic numbers:
  table of, **12**-43
  use in estimating boiling point, **12**-42
Atomic volume, **19**-14,-16,-18
Attenuation coefficient for absorption of
    light by water, **8**-22
Autodissociation constant of water, **6**-4
Autoignition temperature, **18**-2
Average transit time (in atmosphere),
    10-1
Avogadro's number, **25**-2

Base dissociation constant, **6**-4
Beer-Lambert Law, **8**-4,-19
Benson, Cruickshank *et al.*, method for
    estimating heat capacity, **23**-9
Bhirud's method for estimating liquid
    density, **19**-12
Bioaccumulation, **5**-2
Bioconcentration factor for aquatic life:
  definition of,**5**-2
  estimation of, **5**-1
    from adsorption coefficient, **5**-13
    from octanol-water partition
        coefficient, **5**-4
    from solubility, **5**-10
    inputs required, **5**-3,-4
    method errors, **5**-2,-17 to -20
  laboratory vs. field, **5**-21
  use in estimation of:
    adsorption coefficients, **4**-3,-8ff,-20
    water solubility, **2**-3ff
  variables affecting values of, **5**-17 to
      -22
Biodegradation:
  definitions of, **9**-2
  derivation of rate constants, **9**-47
  examples for organic compounds, **9**-51
  habitats of important
      microorganisms, **9**-8

Biodegradation *(Cont.):*
  important variables in, **9**-21
  environment-related, **9**-27
    organism-related, **9**-21
    substrate-related, **9**-21
  laboratory vs. field, **9**-52
  microorganisms responsible for, **9**-3,-15
    habitats of, **9**-8
    population densities, **9**-19
  rate of:
    estimation of, **9**-1,-57
      relationship to BOD/COD ratio,
        9-62
      relationship to rate of hydrolysis,
        9-69
      relationship to solubility, **9**-62
      rules of thumb, **9**-57
    measurement of, **9**-33
  reaction types, **9**-33ff
  role of acclimation, **9**-17ff
  role of cometabolism, **9**-19
Biomagnification, **5**-2
BOD/COD ratio, **9**-62
Boiling point:
  definition of, **12**-1
  effect of impurities on, **12**-50ff
  estimation of, **12**-2
    inputs required, **12**-3
    method errors, **12**-4 to -8
    by method of:
      Kinney, **12**-44
      Lyderson-Forman-Thodos, **12**-16
      Meissner, **12**-8
      Miller, **12**-33
      Ogata and Tsuchida, **12**-39
      Somayajulu and Palit, **12**-42
      Stiel and Thodos, **12**-47
  measurement of, **12**-1
  relationship to critical temperature,
    12-16
  use in estimation of:
    flash point, **18**-7,-9ff
    heat of vaporization, **13**-7ff,-13,-17,
      -24
    liquid density, **19**-9ff
    liquid viscosity, **22**-5ff,-15ff
    surface tension, **20**-10
    thermal conductivity, **24**-2
    vapor pressure, **14**-3,-5,-7ff,-12ff
    water solubility, **2**-4ff
Boiling point number, **12**-44,-45
Bond dissociation energies, **8**-2,-3

Boltzmann's constant, **7**-14; **17**-6,-13,-14;
  25-2
Butler's method for estimating flash
  point, **18**-9

Carbonium ion, **7**-7
Chen's method for estimating heat of
  vaporization, **13**-8,-9
Chiou and Freed's method for
  estimating volatilization, **15**-6
Chromophores, **8**-9,-11,-12,-13
Chueh and Swanson's method for
  estimating heat capacity, **23**-19
Clapeyron equation, **13**-2,-4
Clausius-Clapeyron equation, **12**-50; **14**-
  1,-2
Co-distillation, **16**-5
Collision integral, **17**-6,-11,-13,-14
Cometabolism in biodegradation, **9**-19
Compressibility factor, **12**-23; **13**-7,-8,
  -12,-14,-15; **14**-2ff,-8; **19**-4,-5
Confidence interval, **B**-7,-9
Confidence level, **B**-10
Conjugate acid, **6**-4,-5; **7**-1
Conjugate base, **6**-4,-5; **7**-1
Connectivity parameter, relationship to:
  boiling point, **12**-2
  density, **19**-5
  heat of vaporization, **13**-3 (Ref. 13)
  molar refraction, **26**-3
  octanol-water partition coefficient, **1**-6,
    -7,-9
  water solubility, **2**-4ff
Correlation coefficient, **B**-14,-16
Critical compressibility factor, **12**-33
Critical pressure:
  estimation of, **12**-34; **13**-9,-10
  use in estimating:
    boiling point, **12**-33
    heat of vaporization, **13**-7ff,-12,-13
    liquid density, **19**-12ff
    vapor pressure, **14**-1ff
Critical temperature:
  approximation for, **14**-13; **19**-19
  estimation of, **12**-17
  use in estimating:
    boiling point, **12**-16ff,-33ff
    heat of vaporization, **13**-7ff,-12,-13,-17,
      -23
    liquid density, **19**-9ff
    thermal conductivity, **24**-2

Critical temperature,
  use in estimating *(Cont.)*:
    vapor pressure, 14-1ff
Critical volume:
  estimation of, 12-34
  use in estimating boiling point, 12-33ff
Crystal volume, 19-14,-16,-19
Cyclohexane-water partition coefficient,
    1-42

Davies' equation, 6-5,-6
Debye's equation, 25-2
Density:
  of liquids, 19-1,-2
    at normal boiling point, 19-9ff
    effect of pressure on, 19-5
    effect of temperature on, 19-5
    estimation of, 19-3
      available methods, 19-3
      Bhirud's method, 19-12
      Grain's method, 19-7
      inputs required, 19-5,-6,-7
      method errors, 19-6 to -10
      method selection, 19-5
    use in estimating:
      liquid viscosity, 20-2,-10
      molar refraction, 26-1ff
      thermal conductivity, 24-2
  of solids, 19-1,-2
    estimation of, 19-14
      available methods, 19-14
      method errors, 19-14 to -17
  of vapors, 19-1,-2
    estimation of, 19-2
      inputs required, 19-3
      method errors, 19-2,-3,-4
    use in estimation of surface tension,
        20-2
Desorption coefficient, 4-26,-27
Dielectric constant, relationship to
    dipole moment, 25-2
Diffusion:
  eddy, 17-2
  molecular, 17-1ff,-6
  turbulent, 17-2
Diffusion coefficient:
  apparent, 16-9ff; 17-1,-4,-5
  definition of, 17-1
  effective, 17-5,-6
  in air, 16-9ff; 17-1
    effects of temperature on, 17-7

Diffusion coefficient,
  in air *(Cont.)*:
    estimation of, 17-6
      available methods, 17-7,-8
      FSG method, 17-9
      inputs required, 17-7,-8
      method errors, 17-8,-9,-10
      WL method, 17-13
  in water, 17-1
    effects of concentration on, 17-18
    effects of temperature on, 17-18
    estimation of, 17-17
      available methods, 17-18,-19
      inputs required, 17-19
      method errors, 17-19,-20
      recommended methods, 17-20
  in soil, 16-9
  use in estimating rate of volatilization
      from water, 15-17ff
Diffusion velocity, 17-3
Diffusion volume increments, 17-10,-11
Diffusivity:
  definition of, 17-1,-2
  turbulent, 15-14
Dipole moment:
  definition of, 25-1
  estimation of:
    available methods, 25-3
    general methods, 25-4
    method errors, 25-4,-7
    method for:
      heterocyclics, 25-18
      miscellaneous compounds, 25-20
      naphthalene derivatives, 25-10
      substituted benzenes, 25-10
    typical values for aliphatic and
      alicyclic compounds, 25-5
  use in estimating heat of vaporization,
      14-8
  values for mono-substituted benzene
      compounds, 25-6
Direct photolysis, 8-1,-2
Dispersion, 17-2,-5
Dispersivity, 17-4
Dissociation constant:
  acid, 6-1 (*see* Acid dissociation
      constant)
  base, 6-4
  water, 6-4
Donahue and Bartell's method for
    estimating interfacial tension with
    water, 21-3

Ecological magnification, **5**-2,-23ff
Eddy diffusion, **17**-2
Eisenlohr's method for estimating
    refractive index, **26**-6
Electrophile, **7**-4
Elimination reactions, **7**-1,-2,-12
Energy of mixing, **11**-3
Energy of vaporization, **22**-15
Enthalpic equations, **11**-4ff
Enthalpy of activation, **7**-14ff
Enthalpy of vaporization, *see* Heat of
    vaporization
Entropy of activation, **7**-14ff
Entropy of mixing, **11**-4
Entropy of vaporization, **13**-2
Error:
    method:
        definition of, **C**-2,-5
        for a particular property, *see* named
            property
    propagated:
        definitions, **C**-2,-5
        equations for, **C**-6,-7
        Method 1 (single estimated input),
            C-7
        Method 2 (multiple input values),
            C-9
        Method 3 (simplified method for
            multiple input values), **C**-11
    total, **C**-4,-5
Error function, evaluation of, **16**-43
Estimation rule, **C**-4
Excess Gibbs free energy, **11**-3
Exchange coefficient:
    gas phase, **15**-9ff; **17**-3
        effect of molecular weight and wind
            speed on, **15**-23
        values of, **15**-24,-25
    liquid phase, **15**-9ff; **17**-3
        effect of molecular weight and wind
            speed on, **15**-22
        values of, **15**-24,-25
Exchange rate, interhemispheric, **10**-18ff
Extinction coefficient, **8**-4
Eyring reaction rate theory, **7**-14

Factors for estimating octanol-water
    partition coefficient, **1**-22,-23,-25ff
F-distribution, **B**-17
    table of values, **B**-18
Fick's law of diffusion, **15**-9

Flammability limits, **18**-1
Flash point:
    definition of, **18**-1
    estimation of, **18**-1
        available methods, **18**-2,-3,-4
            Affens' method, **18**-3
            Butler's method, **18**-9
            Prugh's method, **18**-6
        inputs required, **18**-4
        method errors, **18**-3,-4,-7,-8,-9
    measurement of, **18**-2
    of mixtures, **18**-12
Flory-Huggins equation, **11**-21
Fluorescence, **8**-5,-6,-7
Fragment or substituent constants for
    estimating:
    acid dissociation constant, **6**-11ff,-21,
        -22
    activity coefficient, **11**-4,-20
    boiling point:
        B values for Meissner's equation,
            12-12
        ΔT values for Lydersen's equation
            for θ, **12**-18
        ΔY values, **12**-45
    critical pressure (ΔP values), **12**-35;
        13-10
    critical volume (ΔV values), **12**-35
    density of solids, **19**-18
    dipole moment, **25**-5,-6
    heat capacity, **23**-5,-10,-18,-20
    heat of vaporization (H_{vo} values),
        13-18
    molar refraction, **12**-10; **26**-7,-11,
        -15
    molecular connectivity, *see*
        Connectivity parameter (and
        cited references in text)
    octanol-water partition coefficient, **1**-3ff,
        -10ff,-16ff,-17
    parachor, **12**-10,-12; **20**-5
    solubility in water, **2**-40,-41
    thermal conductivity (H factors), **24**-3
    van der Waal's constants, **12**-21,-22,-25,
        -27
Free energy of solution:
    use in estimating octanol-water
        partition coefficient, **1**-6,-7,-9
Freundlich equation, **4**-2,-26,-27
Froude number, **15**-20
FSG, *see* Fuller, Schettler, and
    Giddings

Fuller, Schettler and Giddings (FSG)
     method, 17-9

Gas:
    density, 19-1,-2ff
    exchange coefficient, *see* Exchange
         coefficient
    heat capacity, 23-2ff
    solubility, 2-1,-39; 3-2,-5
    vapor pressure, 14-5
Giacalone method for estimating heat of
     vaporization, 13-7,-8
Gibbs free energy of mixing, 3-1,-6ff; 11-2ff
Goldhammer equation, 19-5,-10
Grain's equation, 19-10
Grain's method for estimating:
    liquid density, 19-7
    liquid viscosity, 22-15
    surface tension, 20-10
Grotthus-Draper Law, 8-2

Haggenmacher's method for estimating
     heat of vaporization, 13-12
Half-life for:
    aqueous photolysis, 8-23,-38
    atmospheric residence time, 10-1
    biodegradation, 9-47,-52ff
    hydrolysis, 7-6,-8
    volatilization from soil, 16-25,-27,-28
    volatilization from water, 15-13,-24,-25
Hamaker's method for estimating
     volatilization from soil, 16-12
Hammett:
    correlation, 7-19,-20,-22
        for aromatic acids, 6-9,-10
    equation, 7-20,-26,-27
    reaction constants, 6-9ff; 7-20,-22,-23
    relationship to adsorption
         coefficients, 4-4
    substituent constants, 6-9ff; 7-20
Hansch's method for estimating index of
     refraction, 26-14
Hartley's method for estimating
     volatilization from soil, 16-11
Hayduk and Laudie's method for
     estimating diffusion coefficient in
     water, 17-20
Heat capacity:
    definition of, 23-1
    estimation of, 23-1

Heat capacity, estimation of *(Cont.)*:
    available methods, 23-3
    for gases, 23-2
        method errors, 23-3,-4
        method of Benson *et al.*, 23-9
        method of Rihani and
             Doraiswamy, 23-4
    for liquids, 23-16
        method errors, 23-16,-17
        method of Chueh and Swanson,
             23-19
        method of Johnson and Huang,
             23-16
    inputs required, 23-2
    method errors, 23-2
Heat of condensation, 13-1
Heat of evaporation, *see* Heat of
     vaporization
Heat of fusion:
    approximation for, 2-25; 3-18
    use in estimating solubility, 2-4ff; 3-17ff
Heat of vaporization:
    definition of, 13-1
    effect of temperature on, 13-1
    estimation of, 13-1
        at temperatures other than the
             boiling point, 13-23
        available methods, 13-2
        from critical constants, 13-4
            Chen's method, 13-8,-9
            Giacalone's method, 13-7,-8
            Klein's method, 13-7
            Riedel's method, 13-8
        from compound structure, 13-16
        from K_F factors, 14-8
        from vapor pressure data, 13-12
        inputs required, 13-4,-5
        method errors, 13-4,-5,-6
        method selection, 13-4
    measurement of, 13-1
    use in estimating:
        liquid viscosity, 22-15
        thermal conductivity, 24-2,-4
        vapor pressure, 14-1ff
        volatilization from soil, 16-12
Henry's Law constant, 2-1; 3-6; 10-15;
     11-2,-7,-16,-17; 16-22; 17-3
    equations for, 15-9,-11
    nondimensional, 15-9
    relationship to volatility, 15-15ff
    values for selected chemicals, 15-12,-18,
    -24,-25

Hydration reactions, 7-1,-2
Hydrolysis:
  estimation of rate of, **7**-1
    available methods, **7**-18ff
    Hammett correlation for k_H, **7**-22
    Hammett equation for k_0, **7**-26
    Hammett equation for k_{OH}, **7**-28
    k_{OH} from acid dissociation constant, **7**-32
    method errors, **7**-22
    Taft correlation for k_H, **7**-25
    Taft equation for k_{OH}, **7**-31
  examples of, **7**-3,-4
  half-lives, range of, **7**-6
  measurement of rate, **7**-11
  rate law, **7**-7,-8
  rate constant, **7**-8ff
    effect of reaction medium on, **7**-16
    effect of temperature on, **7**-14,-21
    for acid-catalyzed hydrolysis, **7**-8ff
    for base-catalyzed hydrolysis, **7**-8ff
    for neutral hydrolysis, **7**-8ff
    measured values, **7**-37ff
Hydroxyl radical:
  concentration in atmosphere, **10**-27
  reaction rate constants, **10**-23,-24
Hyperbolic rate law, **9**-47,-49
Hysteresis in adsorption, **4**-7

Ideal gas law, **19**-2,-3
Immirzi and Perini's method for estimating solid density, **19**-16
Index of refraction:
  effects of temperature on, **26**-19,-20
  estimation of, **26**-1
    available methods, **26**-2
    method errors, **26**-2
    recommended methods, **26**-3
      Eisenlohr's method, **26**-6
      Hansch's method, **26**-14
      inputs required, **26**-3,-4
      method errors, **26**-4
      Vogel's method, **26**-10
Infinite dilution activity coefficient, *see* Activity coefficient
Insolation, **8**-9,-10
  seasonal variation, **8**-11,-23ff
  attenuation by natural waters, **8**-12
Interfacial tension with water:
  estimation of, **21**-1
    available methods, **21**-2

Interfacial tension with water, estimation of *(Cont.):*
    inputs required, **21**-2
    method error, **21**-2,-6,-9
    Method 1 (Donahue and Bartell), **21**-4
    Method 2 (Antonov), **21**-7
Internal conversion, **8**-4ff
Intersystem crossing, **8**-4ff
Irmann's method for estimating water solubility, **2**-39
Isolating carbon, **1**-16

Johnson and Huang method for estimating liquid heat capacity, **23**-16
Junge's correlation, **10**-27ff
Jury *et al.* method for estimating volatilization from soil, **16**-19

K_F factors, **14**-8ff
  relationship to dipole moment, **14**-8
Kinney's method for estimating boiling point, **12**-44
Kistiakovskii equation, **14**-8
Klein constant, **13**-7,-8
Klein method for estimating heat of vaporization, **13**-7

Law of corresponding states, **19**-3,-12
Least-squares analysis, **B**-7,-8
Leaving group (in hydrolysis), **7**-4,-5,-7, -8,-17,-20
LeBas volume, **17**-10,-11,-14,-20
Lennard-Jones potential, **17**-6,-11
Leo's method for estimating octanol-water partition coefficient, **1**-10
Lewis and Squires' method for estimating liquid viscosity, **22**-5
LFER, *see* Linear free energy relationships
Linear free energy relationships, use in estimating:
  acid dissociation constant, **6**-6ff
  rate of hydrolysis, **7**-18ff
Linear regression, **B**-1
Liquid density, *see* Density, liquid
Liquid viscosity:
  at normal boiling point, **22**-5ff,-15ff
  values of, **22**-8

Liquid viscosity *(Cont.):*
definition of, **22**-1
effect of temperature on, **22**-2,-5ff
estimation of, **22**-1
available methods, **22**-3
inputs required, **22**-4,-5
method errors, **22**-5,-6
Method 1 (Lewis and Squires), **22**-5
Method 2 (van Velzen *et al.*), **22**-9
Method 3 (Grain), **22**-15
Liss and Slater method for estimating
volatilization from water, **15**-5
Lorentz-Lorenz equation, **26**-1
Lower explosive limit (LEL), **18**-1,-5,-6
Lower flammable limit (LFL), **18**-1,-5,-6
Lydersen-Forman-Thodos method for
estimating boiling point, **12**-16
Lydersen's increments, **12**-35,-37
Lydersen's method for estimating
critical pressure, **13**-9,-10
Lyoparachor, **13**-3

Mackay and Wolkoff method for
estimating volatilization from
water, **15**-4
MacLeod-Sugden:
correlation, **21**-8
method for estimating surface
tension, **20**-3
Margules equation, **3**-7,-8,-19; **11**-4,-5
Mayer, Letey and Farmer method for
estimating volatilization from soil,
16-13
McGowan's method for estimating
parachor, **12**-9,-12
Mean of samples, **B**-8
Meissner's method for estimating
boiling point, **12**-8
Melting point, use in estimating
solubility of solids, **2**-4ff,-15,-25,-34,
-39; **3**-18ff
Method error, *see* Error, method
Miller's method for estimating boiling
point, **12**-33
Mixture surface tension, **21**-3,7ff
Model ecosystems, use for
bioconcentration factor
measurements and estimates,
5-23ff
Molal volume, **13**-4,-7
at boiling point, **17**-14,-18,-19

Molar absorptivity, **8**-4,-5,-11,-13,-14ff,
-19,-20ff
Molar refraction:
estimation of, **12**-8,-10; **26**-1
relationship to index of refraction, **26**-1ff
use in estimating boiling point, **12**-8, -10
Molar volume, **12**-33; **17**-10; **19**-5,-11,-14
estimation of (at boiling point), **19**-11
hypothetical liquid, **21**-8
use in estimating:
interfacial tension, **21**-2ff
liquid viscosity, **22**-2
surface tension, **20**-10
water solubility, **2**-4ff
Molecular connectivity, *see* Connectivity
parameter
Molecular diameter, use in estimating
volatilization rates, **15**-17ff
Molecular diffusion, **17**-1,-2
Molecular volume, **19**-14
Mole fraction, **2**-38; **3**-27
Monod kinetics in biodegradation, **9**-49ff
Multiple correlation coefficient, **B**-16

National Standard Reference Data
System, **A**-3
Nielsen's method for estimating solid
density, **19**-14,-15
Non-random two liquid (NRTL) equation,
11-5
Normal probability plot, **B**-20,-22
Nucleophile, **7**-4,-5,-9
Null hypothesis, **B**-16,-17

Octanol/water partition coefficient:
definition of, **1**-1
estimation of, **1**-1
available methods, **1**-5
from activity coefficients, **1**-47
from solvent regression equations, **1**-39
inputs required, **1**-39
method errors, **1**-42
Leo's method, **1**-10
inputs required, **1**-10
method error, **1**-12
recommended methods, **1**-3,-4
measurement of, **1**-1,-2
use in estimating:
adsorption coefficient, **4**-3,-8ff,-20

Octanol/water partition coefficient,
    use in estimating *(Cont.)*:
    bioconcentration factor, **5**-3,-4,-20
    vapor pressure, **14**-4
    relationship to solvent solubilities, **3**-4
Ogata and Tsuchida's method for
    estimating boiling point, **12**-39
Organic carbon, relationship to organic
    matter in soils, **4**-3
Organic content of soils, **4**-2,-5
Organic matter, relationship to organic
    carbon in soils, **4**-3
Outliers, **B**-4,-16
Overall mass transfer coefficient, **15**-11ff
    values of, **15**-24,-25
Ozone:
    concentration in atmosphere, **10**-27
    reaction rate constants, **10**-25

Packing coefficient, **19**-14,-16
Parachor, **13**-3; **21**-8,-9
    estimation of, **12**-9,-10,-12; **20**-5
    use in estimating:
        boiling point, **12**-8
        soil adsorption coefficients, **4**-3,-8ff,
            -19,-20
        surface tension, **20**-2
        water solubility, **2**-4ff
Partition coefficient, *see* Octanol/water
    or Solvent/water partition coefficient
Phase instability, **3**-7,-8
Phosphorescence, **8**-6,-7
Photolysis, *see* Aqueous photolysis
Photophysical deactivation, **8**-5
pH-rate profiles for hydrolysis, **7**-9ff
Pi (π) constants, **1**-5
Pierotti *et al.*, method for estimating
    activity coefficients, **11**-10
Pitzer's correlation, **13**-8
Planck's constant, **7**-14
Polarizability factor, relationship to di-
    pole moment, **25**-2
Porosity of soil, **16**-9ff
Power rate law for biodegradation, **9**-47
Prediction interval, **B**-25,-26
Probability plot, **B**-14,-20,-22
Propagated error, *see* Error, propagated
Prugh's method for estimating flash
    point, **18**-6
Prugh's nomograph, **18**-8
p-Value, **B**-17

Quantum yield, **8**-2,-5,-6,-9,-18ff,-29,-37
    dependence on wavelength, **8**-6,-18,-25
Quenching in photolysis, **8**-7ff

Rackett equation, **12**-33
Rainout, **10**-14
Raoult's Law, **12**-50,51
Rate of aqueous photolysis, *see* Aqueous
    photolysis
Rate of biodegradation, *see*
    Biodegradation
Rate of hydrolysis, *see* Hydrolysis
Reaction constants for hydrolysis, **7**-20ff
Reaction parameter, **6**-27
Reaction rate constants:
    with hydroxyl radicals in atmosphere,
        10-23,-24
    with ozone in atmosphere, **10**-25
Reaeration rate constant, **15**-17ff; **17**-3
Redlich-Kwong equation of state, **19**-2,-3,
    -4
Refractive index, *see* Index of refraction
Refractory index of organic compounds,
    9-63,-69,-70
Regression equations, **B**-1ff
Residuals, **B**-8,-14
    histograms of, **B**-14,-20
    probability plot of, **B**-14,-22
    vs order of observation, **B**-14,-23
    vs predicteds, **B**-14,-21,-23
Retention factor (R_F), correlation with
    adsorption coefficient, **4**-4
Retention time (on HPLC), correlation
    with octanol-water partition
    coefficient, **1**-2,-6,-8,-10
Riedel method for estimating heat of va-
    porization, **13**-8
Rihani and Doraiswamy's method for es-
    timating heat capacity, **23**-4
Risk level, **B**-16
Robbins and Kingrea method for esti-
    mating thermal conductivity, **24**-6

Sample mean, **B**-8
Sample variance, **B**-9
Sastri's method for estimating heat of
    vaporization, **13**-16
Sato and Riedel's method for estimating
    thermal conductivity, **24**-5
Saturated phase surface tension, **21**-3,-7

Scatter plot, **B**-2,-4,-5,-6
Schroeder's method for estimating
    molar volume, **19**-11
Sediment adsorption coefficient, *see*
    Adsorption coefficient
Sensitized photolysis, **8**-16,-25
Significance level, **B**-16,-25
Simple correlation coefficient, **B**-16
Simple linear regression, **B**-1ff
S_N1 and S_N2 processes, **7**-5,-7ff
Snow scavenging, **10**-15
Soil adsorption coefficient, *see*
    Adsorption coefficient
Solar radiation intensity, **8**-23ff
    (*See also* Insolation)
Solid density, *see* Density
Solubility in solvents:
    effect of temperature on, **3**-6ff,
        -17ff
    estimation of, **3**-1
        inputs required, **3**-1,-2
        liquid-liquid systems, **3**-6
            method errors, **3**-10
        solid-liquid systems, **3**-17
            inputs required, **3**-18
            method errors, **3**-20
    of gases, **3**-2,-5
Solubility in water:
    definition of, **2**-1
    effects of environmental variables on,
        2-11ff
    estimation of, **2**-1
        available methods, **2**-4
        from activity coefficients, *see*
            Chapter **3**
        from octanol-water partition
            coefficient, **2**-13
            equation selection, **2**-33ff
            method errors, **2**-25ff
        from structure (Irmann's method),
            2-39
            method errors, **2**-39
        inputs required, **2**-3
    of gases, **2**-1,-39
    liquids vs solids, **2**-10,-13,-25,-33,-34,
        -39
    measurement, **2**-11
    use in estimating:
        adsorption coefficient, **4**-4,-8ff,-20
        bioconcentration factor, **5**-3,-10
        Henry's law constant, **15**-11
        interfacial tension, **21**-2ff

Solubility in water,
    use in estimating (*Cont.*):
        octanol-water partition coefficient,
            1-3,-4
        relationship to rate of biodegradation,
            9-62
Solubility parameter, **3**-4,-20
Solution association parameter, **17**-18
Solvent regression equations:
    use in estimating octanol-water
        partition coefficient, **1**-3ff,-39ff
    table of, **1**-41
Solvent-water partition coefficient:
    potential use in estimating solubility,
        3-3
    use in estimating octanol-water
        partition coefficient, **1**-39
Somayajulu and Palit's method for
    estimating boiling point, **12**-42
Specific acid and base catalysis, **7**-8ff,-17
Specific gravity, **19**-1,-2
Specific refraction, **26**-2
Standard deviation, **B**-1,-9
Standard error, **B**-2,-4,-5,-6
Stark-Einstein-Bodenstein Law, **8**-5
Steric substituent constant, **7**-20,-25
Stiel and Thodos' method for estimating
    boiling point, **12**-47
Stokes-Einstein equation, **17**-7
Student's t-distribution, **B**-10,-17
    table, **B**-11
Substituent constants, *see* Fragment or
    substituent constants
Sugden's equation, **12**-9
Sugden's method for estimating
    parachor, **12**-9,-10
Sum of squares, **B**-15,-16
Surface area of molecule, use in
    estimating solubility, **2**-4ff
Surface tension: definition of, **20**-1
    estimation of, **20**-1
        available methods, **20**-2
        inputs required, **20**-4
        method errors, **20**-2,-3,-4,-6,-7,-8,-10,
            -12,-13
        Method 1 (MacLeod-Sugden), **20**-3
        Method 2 (Grain), **20**-10
    of water, **21**-11
    mixture, **21**-3,-7ff
    saturated phase, **21**-3,-7
    use in estimation of interfacial
        tension, **21**-2ff

Taft:
  constants, relationship to adsorption coefficient, **4**-4
  correlation, **7**-19,-20,-25
    for aliphatic acids, **6**-20
  equation, **6**-20; **7**-20
  substituent parameter, **7**-20
Tarver's method for estimating density of solids, **19**-14,-15
Test statistic, **B**-17
Thermal conductivity:
  estimation of, **24**-1
    inputs required, **24**-2
    method errors, **24**-2,-4,-5
    method of Robbins and Kingrea, **24**-6
    method of Sato and Riedel, **24**-5
Thiessen's correlation, **13**-23
Thiessen's equation, **13**-3
Thomson's rule, **14**-8
Time scales for atmospheric phenomena, **10**-4
Topological indices, use in estimating solubility, **2**-4ff
Tortuosity factor, **17**-5
Total error, *see* Error
Triplet energies, **8**-7,-8
Troposphere, **10**-2
Trouton's rule, **13**-2
t-Table, **B**-11
Turbulent diffusion, **17**-2
Turbulent diffusivity, **15**-14
Turn-over time, **10**-1

UNIFAC, **11**-7
  method for estimating activity coefficient, **11**-20
UNIQUAC, **11**-4
  equation, **11**-5,-7

van der Waal's constants:
  for estimating critical pressure, **13**-9
  for estimating critical temperature, **12**-17
van Laar equation, **11**-4,-5,-7,-17
van Velzen *et al.* method for estimating liquid viscosity, **22**-9
Vapor density: concentration in soils, **16**-3,-4
  (*See also* Density)

Vapor-liquid equilibria, calculation of, **3**-10
Vapor pressure:
  estimation of, **14**-1
    from boiling points at reduced pressure, **14**-15
    inputs required, **14**-5
    method errors, **14**-5,-6,-7
    Method 1, **14**-7
    Method 2, **14**-12
    problems with solids, **14**-4
    recommended methods, **14**-5
  relationship to octanol-water partition coefficient, **14**-4
  use in estimating:
    flash point, **18**-3ff
    heat of vaporization, **13**-12
    Henry's law constant, **15**-11
Variance:
  analysis of, **B**-15
  sample, **B**-9
Vavilov's rule, **8**-6
Viscosity, *see* Liquid viscosity
Vogel's method for estimating index of refraction, **26**-10
Volatilization from soil:
  estimation of, **16**-1
    Dow method, **16**-25
    Hartley method, **16**-11
    Haymaker method, **16**-12
    inputs required, **16**-27ff
    Jury *et al.* method, **16**-19
    Mayer, Letey and Farmer method, **16**-13
    method selection, **16**-27
  factors affecting process, **16**-2ff
Volatilization from water:
  environmental effects on, **15**-2ff
  estimation of, **15**-1
    available methods, **15**-4
    inputs required, **15**-27,-30
    method errors, **15**-2,-7,-8
    method of Chiou and Freed, **15**-6
    method of Liss and Slater, **15**-5
    method of Mackay and Wolkoff, **15**-4
    method of Smith *et al.* **15**-6,-17
    recommended general method, **15**-9
  half-life for, **15**-13
    values of, **15**-24,-25
  rate constant for, **15**-31,-17ff
  stages of, **15**-3

Water:
  density, **19**-2
  dissociation constant, **6**-4
  interfacial tension with, *see* Inter-
    facial tension
  surface tension, **21**-11
  viscosity, **17**-21
Watson correlation (modified), **14**-13
Weiner number, **12**-20,-22,-47,-48
Wick effect in volatilization from soil,
  16-10

Wilson equation, **11**-4,-5,-21
Winstein-Grunwald relation, **7**-18
Wilke and Lee (WL) method for estimat-
    ing diffusion coefficient in air, **17**-13
Woodward rules, **8**-13

Zepp *et al.* model for aqueous photolysis,
  8-17ff

LIST OF MATERIAL PROTECTED BY COPYRIGHTED SOURCES

CHAPTER 1

Table 1-3, p. 1-11, and Table 1-4, pp. 1-13 to 1-14, from T. J. Chou and P. C. Jurs, "Computer Assisted Computation of Partition Coefficients from Molecular Structures Using Fragment Constants," *J. Chem. Inf. Comput. Sci.,* **19,** 172–178 (1979). Reprinted with permission from the American Chemical Society.

Table 1-5, pp. 1-17 to 1-21, from C. Hansch and A. J. Leo, *Substituent Constants for Correlation Analysis in Chemistry and Biology,* John Wiley, New York (1979), except as noted. Reprinted with permission from John Wiley & Sons, Inc.

Table 1-6, p. 1-22, Table 1-7, p. 1-23, and Examples 1-1 to 1-43, pp. 1-30 to 1-38, from C. Hansch and A. J. Leo, *Substituent Constants for Correlation Analysis in Chemistry and Biology,* John Wiley, New York (1979). Reprinted with permission from John Wiley & Sons, Inc.

Table 1-8, p. 1-40, from A. Leo, C. Hansch, and D. Elkins, "Partition Coefficients and Their Uses," *Chem. Rev.,* **71,** 525–621 (1971). Reprinted with permission from the American Chemical Society.

Table 1-9, p. 1-41, adapted from A. Leo, C. Hansch, and D. Elkins, "Partition Coefficients and Their Uses," *Chem. Rev.,* **71,** 525–621 (1971), Table VIII. Reprinted with permission from the American Chemical Society.

Table 1-10, p. 1-43, from P. Seiler, "Interconversion of Lipophilicities from Hydrocarbon/Water Systems into the Octanol/Water System," *Eur. J. Med. Chem.–Chem. Ther.,* **9,** 473–479 (1974). Reprinted with permission from the Societe d'Etudes de Chimie Therapeutique.

Table 1-11, pp. 1-44 to 1-45, from A. Leo, C. Hansch, and D. Elkins, "Partition Coefficients and Their Uses," *Chem. Rev.,* **71,** 525–621 (1971). Reprinted with permission from the American Chemical Society.

CHAPTER 2

Figure 2-2, p. 2-27, from G. Dec, S. Banerjee, H. C. Sikka, and E. J. Pack, Jr., "Water Solubility and Octanol/Water Partition Coefficients of Organics: Limitations of the Solubility-Partition Coefficient Correlation," preprint submitted to *Environ. Sci. Technol.* Reproduced with permission from the American Chemical Society.

Table 2-16, pp. 2-40 to 2-41, Table 2-17, p. 2-42, and Table 2-18, p. 2-42, from F. Irmann, "A Simple Correlation Between Water Solubility and Structure of Hydrocarbons and Halohydrocarbons," *Chem. Ing. Tech.,* **37,** 789–798 (1965). Translation available from the National Translation Center, The John Crerar Library, 35 West 33rd St., Chicago, IL 60616. Reproduced with permission from Verlag Chemie International, Inc.

CHAPTER 3

Table 3-2, p. 3-21, Table 3-3, p. 3-22, Table 3-4, p. 3-23, and Table 3-5, p. 3-24, from J. G. Gmehling, T. F. Anderson, and J. M. Prausnitz, "Solid-Liquid Equilibria Using UNIFAC," *Ind. Eng. Chem. Fundam.,* **17,** 269–273 (1978). Reprinted with permission from the American Chemical Society.

CHAPTER 4

Table 4-3, pp. 4-11 to 4-13, from E. E. Kenaga and C. A. I. Goring, "Relationship Between Water Solubility, Soil-Sorption, Octanol-Water Partitioning, and Bioconcentration of Chemicals in Biota," prepublication copy of paper dated October 13, 1978, given at the American Society for Testing and Materials, Third Aquatic Toxicology Symposium, October 17–18, 1978, New Orleans, La. [Symposium papers published by ASTM, Philadelphia, Pa., as Special Technical Publication (STP) 707 in March 1980.] Reproduced with permission from the American Society for Testing and Materials.

Table 4-8, p. 4-15, from G. G. Briggs, "A Simple Relationship Between Soil Adsorption of Organic Chemicals and Their Octanol/Water Partition Coefficients," *Proc. 7th British Insecticide and Fungicide Conf.*, Vol. 1, The Boots Company Ltd., Nottingham, G.B. (1973). Reprinted with permission from Macmillan Journals Ltd.

Table 4-10, p. 4-27, from J. W. Hamaker and J. M. Thompson, "Adsorption," in *Organic Chemicals in the Soil Environment*, Vol. 1, C. A. I. Goring and J. W. Hamaker (eds.), Marcel Dekker, Inc., New York (1972). Reprinted with permission from Marcel Dekker, Inc.

CHAPTER 5

Table 5-3, pp. 5-7 to 5-9, from G. D. Veith, K. J. Macek, S. R. Petrocelli, and J. Carroll, "An Evaluation of Using Partition Coefficients and Water Solubility to Estimate Bioconcentration Factors for Organic Chemicals in Fish," *J. Fish. Res. Board Can.* (1980), prepublication copy. Reprinted with permission from the Canadian Department of Fisheries and Oceans, Ottawa, Ontario.

Table 5-4, pp. 5-11 to 5-12, from E. E. Kenaga and C. A. I. Goring, "Relationship Between Water Solubility, Soil-Sorption, Octanol-Water Partitioning, and Bioconcentration of Chemicals in Biota," prepublication copy of paper dated Oct. 13, 1978, given at the Third Aquatic Toxicology Symposium, American Society for Testing and Materials, New Orleans, La., October 17–18, 1978. [Symposium papers to be published by ASTM, Philadelphia, Pa., as Special Technical Publication (STP) 707 in 1980.] Reproduced with permission from the American Society for Testing and Materials.

Table 5-7, p. 5-16, from G. R. Southworth, J. J. Beauchamp, and P. K. Schmieder, "Bioaccumulation Potential and Acute Toxicity of Synthetic Fuels Effluents in Fresh Water Biota: Asaarenes," *Environ. Sci. Technol.*, **12**, 1062–1066 (1978). Reprinted in part with permission from the American Chemical Society.

Table 5-8, p. 5-16, from G. R. Southworth, J. J. Beauchamp, and P. K. Schmieder, "Bioaccumulation Potential of Polycyclic Aromatic Hydrocarbons in *Daphnia pulex*," *Water Res.*, **12**, 973–977 (1978). Reprinted with permission from Pergamon Press, Ltd.

Table 5-9, p. 5-16, from C. T. Chiou, V. H. Freed, D. W. Schmedding, and R. L. Kohnert, "Partition Coefficient and Bioaccumulation of Selected Organic Chemicals," *Environ. Sci. Technol.*, **11**, 475–478 (1977). Reprinted with permission of the American Chemical Society.

Figure 5-1, p. 5-20, adapted from G. D. Veith, K. J. Macek, S. R. Petrocelli, and J. Carroll, "An Evaluation of Using Partition Coefficients and Water Solubility to Estimate Bioconcentration Factors for Organic Chemicals in Fish." *J. Fish. Res. Board Can.* (1980), prepublication copy. Reprinted with permission from the Canadian Department of Fisheries and Oceans, Ottawa, Ontario.

CHAPTER 6

Table 6-1, p. 6-5, from J. B. Hendrickson, D. J. Cram, and G. S. Hammond, *Organic Chemistry*, 3rd ed., McGraw-Hill, New York, pp. 303–307 (1970).

Reprinted with permission from McGraw-Hill Book Co.

Table 6-4, pp. 6-13 to 6-14, and Table 6-6, pp. 6-17 to 6-18, from H. H. Jaffe, "A Reexamination of the Hammett Equation," *Chem. Rev.,* **53,** 191–261 (1953). Reprinted with permission from the Williams & Wilkins Co.

Table 6-8, p. 6-22, from P. R. Wells, *Linear Free Energy Relationships,* Academic Press, New York (1968).

CHAPTER 7

Figure 7-2, p. 7-10, from W. Mabey and T. Mill, "Critical Review of Hydrolysis of Organic Compounds in Water Under Environmental Conditions," *J. Phys. Chem. Ref. Data,* **7,** 383–415 (1978). Printed in part with permission from the American Chemical Society.

Table 7-7, p. 7-27, and Table 7-8, pp. 7-29 to 7-30, from P. R. Wells, *Linear Free Energy Relationships,* Academic Press, New York (1968), except as noted. Reprinted with permission from Academic Press, Inc.

Figure 7-3, p. 7-35, from N. L. Wolfe, "Organophosphate and Organophosphorothioate Esters: Application of Linear Free Energy Relationships to Estimate Hydrolysis Rate Constants for Use in Environmental Fate Assessment," *Chemosphere,* **9,** 571–579 (1980). Reprinted with permission from Pergamon Press, Ltd.

Table 7-10, pp. 7-37 to 7-42, from W. Mabey and T. Mill, "Critical Review of Hydrolysis of Organic Compounds in Water Under Environmental Conditions," *J. Phys. Chem. Ref. Data,* **7,** 383–415 (1978), except as noted. Reprinted in part with permission from the American Chemical Society.

CHAPTER 8

Table 8-3, p. 8-8, from J. G. Calvert and J. N. Pitts, Jr., *Photochemistry,* John Wiley & Sons, New York (1966). Reprinted with the permission of the authors.

Figure 8-1, p. 8-10, from E. P. Odum, *Fundamentals of Ecology,* 3rd ed., W. B. Saunders Co., Philadelphia, pp. 40–43 (1971). Reprinted with permission from Holt, Rinehart and Winston.

Figure 8-2, p. 8-11, adapted from data of W. E. Reifsnyder and H. W. Lull, "Radiant Energy in Relation to Forests," Technical Bulletin No. 1344, U.S. Department of Agriculture Forest Service (1965), as cited in E. P. Odum, *Fundamentals of Ecology,* 3rd ed., W. B. Saunders Co., Philadelphia (1971). Reprinted with permission from Holt, Rinehart and Winston.

Table 8-4, p. 8-13, from J. G. Calvert and J. N. Pitts, Jr., *Photochemistry,* John Wiley & Sons, New York (1966). Reprinted with the permission of the authors.

Table 8-5, pp. 8-14 to 8-16, from J. G. Calvert and J. N. Pitts, Jr., *Photochemistry,* John Wiley & Sons, New York (1966), and from R. M. Silverstein and G. C. Bassler, *Spectrometric Identification of Organic Compounds,* John Wiley & Sons, New York, pp. 99–103 (1963), except as noted in table footnotes. Reprinted with permission of John Wiley & Sons, Inc., J. G. Calvert, and J. N. Pitts, Jr.

Table 8-6, p. 8-17, from N. J. Turro, *Molecular Photochemistry,* W. A. Benjamin, New York (1965), p. 46. Reprinted with permission from Benjamin/Cummings, Inc.

Table 8-8, pp. 8-24 to 8-25, and Table 8-9, pp. 8-26 to 8-27, from R. G. Zepp and D. M. Cline, "Rates of Direct Photolysis in the Aqueous Environment," *Environ. Sci. Technol.,* **11,** 359–366 (1977). Reprinted with permission from the American Chemical Society.

Figure 8-5, p. 8-30, Figure 8-6, p. 8-30, and Figure 8-7, p. 8-31, from R. G. Zepp and D. M. Cline, "Rates of Direct Photolysis in the Aqueous Environment." *Environ. Sci. Technol.,* **11,** 359–366 (1977). Reprinted with permission from the American Chemical Society.

CHAPTER 9

Figure 9-1, p. 9-3, from R. W. Meikle, "Decomposition: Qualitative Relationships," in *Organic Chemicals in the Soil Environment,* Vol. 1, ed. by C. A. I. Goring and J. W. Hamaker, Marcel Dekker, New York (1972). Reprinted with the permission of Marcel Dekker, Inc.

Figure 9-5, p. 9-15, from S. E. Manahan, *Environmental Chemistry,* Willard Grant Press, Inc., Boston (1969). Reprinted with the permission of Prindle, Weber & Schmidt.

Figure 9-11, p. 9-29, from R. J. Larson, "Role of Biodegradation Kinetics in Predicting Environmental Fate," in *Biotransformation and Fate of Chemicals in the Aquatic Environment,* ed. by A. Maki, K. Dickson, and J. Cairns, Jr., American Society for Microbiology, Washington, D.C. (1980). Reprinted with permission from the American Society for Microbiology.

Figure 9-12, p. 9-30, from C. E. Collins, C. P. L. Grady, Jr., and F. P. Incropera, "The Effects of Temperature on Biological Wastewater Treatment Processes," Purdue University Water Resources Center Technical Report No. 98, West Lafayette, Ind. (March 1978). Reprinted with permission from the Purdue University Press.

Table 9-11, p. 9-37, from R. D. Swisher, *Surfactant Biodegradation,* Marcel Dekker, New York (1970). Reprinted with permission from Marcel Dekker, Inc.

Figure 9-14, p. 9-51, from R. J. Larson, "Role of Biodegradation Kinetics in Predicting Environmental Fate," in *Biotransformation and Fate of Chemicals in the Aquatic Environment,* ed. by A. Maki, K. Dickson, and J. Cairns, Jr., American Society for Microbiology, Washington, D.C. (1980). Reprinted by permission from the American Society for Microbiology.

Table 9-22, pp. 9-66 to 9-69, from P. Pitter, "Determination of Biological Degradability of Organic Substances," *Water Res.,* **10,** 231 (1976). Reprinted with permission from Pergamon Press, Ltd.

Figure 9-15, p. 9-71, from N. L. Wolfe, D. F. Paris, W. C. Steen, and G. L. Baughman, "Correlation of Microbial Degradation Rates with Chemical Structure," *Environ. Sci. Technol.,* **14,** 1143 (1980). Reprinted with permission from the American Chemical Society.

CHAPTER 10

Table 10-8, p. 10-28, from C. E. Junge, "Residence Time and Variability of Tropospheric Trace Gases," *Tellus,* **26,** 477–488 (1974). Reprinted with permission from the International Union of Crystallography.

CHAPTER 11

Table 11-1, pp. 11-5 to 11-6, from R. C. Reid, J. M. Prausnitz, and J. K. Sherwood, *The Properties of Gases and Liquids,* 3rd ed., McGraw-Hill Book Co., New York (1977). Reprinted with permission from McGraw-Hill Book Co.

Table 11-3, pp. 11-9 to 11-10, from A. Fredenslund, J. Gmehling, and P. Rasmussen, *Vapor-Liquid Equilibrium Using UNIFAC,* Elsevier Scientific Publishing Co., Amsterdam (1977). Reprinted with permission from Elsevier Scientific Publishing Co.

Table 11-4, pp. 11-12 to 11-14, from G. Pierotti, C. Deal, and E. Derr, "Activity Coefficients and Molecular Structure," *Ind. Eng. Chem.,* **51,** 95 (1959), as modified by R. C. Reid, J. M. Prausnitz, and T. K. Sherwood, *The Properties of Gases and Liquids,* 3rd ed., McGraw-Hill Book Co., New York (1977), and by Clark F. Grain, the author of this chapter. Reprinted with permission from the American Chemical Society and McGraw-Hill Book Co.

Table 11-5, p. 11-15, from C. Tsonopoulos and J. Prausnitz, "Activity Coefficients of Aromatic Solutes in Dilute Aqueous Solutions," *Ind. Eng. Chem. Fundam.*, **10,** 593 (1971). Reprinted with permission from the American Chemical Society and McGraw-Hill Book Co.

Table 11-6, pp. 11-23 to 11-24, and Table 11-7, pp. 11-26 to 11-32, from J. Gmehling, P. Rasmussen, and A. Fredenslund, "Vapor-Liquid Equilibria by UNIFAC Group Contribution. Revision and Extension II," *Ind. Eng. Chem. Process Des. Dev.* (in press 1981). Reprinted with permission from the American Chemical Society.

CHAPTER 12

Table 12-3, pp. 12-10 to 12-11, from R. H. Perry and C. H. Chilton (eds.), *Chemical Engineers' Handbook,* 5th ed., McGraw-Hill Book Co., New York, 1973. Reprinted with permission from McGraw-Hill Book Company.

Table 12-4, p. 12-12, from J. C. McGowan, "Molecular Volumes and the Periodic Table," *Chem. Ind.,* London, **1952,** 495–496.

Table 12-5, p. 12-12, from H. P. Meissner, "Critical Constants from Parachor and Molar Refraction," *Chem. Eng. Prog.,* **45,** 149–153. Reprinted with permission from the American Institute of Chemical Engineers.

Table 12-6, p. 12-18, from R. C. Reid and T. K. Sherwood, *The Properties of Gases and Liquids—Their Estimation and Correlation,* 2nd ed., McGraw-Hill Book Co., New York (1966). Reprinted with permission from McGraw-Hill Book Company.

Table 12-7, p. 12-21, and Table 12-9, p. 12-25, from J. C. Forman and G. Thodos, "Critical Temperatures and Pressures of Hydrocarbons," *AIChE J.,* **4,** 356–361 (1958). Reprinted with permission from the American Institute of Chemical Engineers.

Table 12-11, p. 12-29, from J. C. Forman and G. Thodos, "Critical Temperatures and Pressures of Organic Compounds," *AIChE J.,* **6,** 206–209 (1960). Reprinted with permission from the American Institute of Chemical Engineers.

Table 12-12, pp. 12-35 to 12-37, from R. C. Reid and T. K. Sherwood, *The Properties of Gases and Liquids—Their Estimation and Correlation,* 2nd ed., McGraw-Hill Book Co., New York (1966). Reprinted with permission from McGraw-Hill Book Co.

Table 12-13, p. 12-40, and Table 12-14, p. 12-41, from Y. Ogata and M. Tsuchida, "Linear Boiling Point Relationships," *Ind. Eng. Chem.,* **49,** 415–417 (1957). Reprinted with permission from the American Chemical Society.

Table 12-16, p. 12-43, from G. R. Somayajulu and S. R. Palit, "Boiling Points of Homologous Liquids," *J. Chem. Soc.,* **1957,** 2540–2544. Reprinted with permission from the Royal Society of Chemistry, London.

Table 12-17, p. 12-45, from C. R. Kinney, "A System Correlating Molecular Structure of Organic Compounds with Their Boiling Points. I: Aliphatic Boiling Point Numbers," *J. Am. Chem. Soc.,* **60,** 3032–3039 (1938), and "Calculation of Boiling Points of Aliphatic Hydrocarbons," *Ind. Eng. Chem.,* **32,** 559–562 (1940). Reprinted with permission from the American Chemical Society.

CHAPTER 13

Table 13-3, p. 13-8, from S. H. Fishtine, "Reliable Latent Heats of Vaporization, Pt. 1," *Ind. Eng. Chem.,* **55** (4) 20–28 (1963). Reprinted with permission from the American Chemical Society.

Table 13-4, p. 13-10, from A. L. Lydersen, "Estimation of Critical Properties of Organic Compounds," College of Engineering, Univ. of Wisconsin, Engineering Experiment Report 3, Madison, Wis. (April 1955); also tabulated in R. C. Reid and T. K. Sherwood, *The Properties of Gases and Liquids—Their Estimation and Correlation,* 2nd ed., McGraw-Hill Book Co., New York (1966). Reprinted with permission from McGraw-Hill Book Co.

Table 13-5, p. 13-13, and Table 13-6, p. 13-15, from S. H. Fishtine, "Reliable Latent Heats of Vaporization, Pt. 2," *Ind. Eng. Chem.,* **55** (5) 49–54 (1963). Reprinted with permission from the American Chemical Society.

Table 13-8, pp. 13-18 to 13-19, Table 13-9, p. 13-20, and Table 13-10, p. 13-21, from S. R. S. Sastri, M. V. Ramana Rao, K. A. Reddy, and L. K. Doraiswamy, "A Generalized Method for Estimating the Latent Heat of Vaporization of Organic Compounds," *Br. Chem. Eng.,* **14,** 959–963 (1969). Reprinted with permission from IPC Industrial Press, Ltd.

CHAPTER 14

Table 14-4, pp. 14-9 to 14-10, and Table 14-5, p. 14-11, from S. H. Fishtine, "Reliable Latent Heats of Vaporization," *Ind. Eng. Chem.,* **55,** 47 (June 1963). Reprinted with permission from the American Chemical Society.

CHAPTER 15

Figure 15-1, p. 15-10, from P. S. Liss and P. G. Slater, "Flux of Gases Across the Air-Sea Interface," *Nature (London),* **247,** 181–184 (January 25, 1974).

Figure 15-2, p. 15-12, from D. Mackay and T. K. Yuen, "Volatilization Rates of Organic Contaminants from Rivers," *Proceedings of 14th Canadian Symposium, 1979: Water Pollution Research Canada* (publication date unknown). Reprinted with permission from the authors.

Table 15-2, p. 15-18, from J. H. Smith and D. C. Bomberger, "Prediction of Volatilization Rates of Chemicals in Water," *Water: 1978,* AIChE Symposium Series, 190, **75,** 375–381 (1979), and J. H. Smith, D. C. Bomberger, and D. L. Haynes, "Prediction of the Volatilization Rates of High Volatility Chemicals from Natural Water Bodies," *Environ. Sci. Technol.,* **14,** 1332–1337 (1980). Reprinted with permission from the American Institute of Chemical Engineers.

Table 15-3, p. 15-20, from J. H. Smith, D. C. Bomberger, and D. L. Haynes, "Prediction of the Volatilization Rates of High Volatility Chemicals from Natural Water Bodies," *Environ. Sci. Technol.,* **14,** 1332–1337 (1980). Reprinted with permission from the American Chemical Society.

CHAPTER 16

Figure 16-1, p. 16-6, from W. F. Spencer, "Distribution of Pesticides Between Soil, Water and Air," in *Pesticides in the Soil: Ecology, Degradation and Movement,* Michigan State University, East Lansing, Mich., pp. 120–128 (1970). Reprinted with permission from Michigan State University.

Figure 16-2, p. 16-20, Figure 16-3, p. 16-21, and Figure 16-4, p. 16-21, from R. Meyer, J. Letey, and W. J. Farmer, "Models for Predicting Volatilization of Soil-Incorporated Pesticides," *Soil Sci. Soc. Am. Proc.,* **38,** 563–568 (1974). Reprinted with permission from Soil Science Society of America.

Table 16-6, p. 16-37, from J. W. Hamaker, "Diffusion and Volatilization," Chap. 5 in *Organic Chemicals in the Soil Environment,* Vol. 1, ed. by C. A. I. Goring and J. W. Hamaker, Marcel Dekker, New York (1972). Reprinted with permission from Marcel Dekker, Inc.

Table 16-8, p. 16-44, from W. Jost, "Diffusion in Solids, Liquids, Gases," in *Physical Chemistry—A Series of Monographs,* ed. by E. Hutchinson, Academic Press, New York (1952). Reprinted with permission from Academic Press, Inc.

CHAPTER 17

Table 17-4, p. 17-11, from E. N. Fuller, P. D. Schettler, and J. C. Giddings, "A New Method for Prediction of Binary Gas-Phase Diffusion Coefficients," *Ind. Eng. Chem.*, **58,** 19–27 (1966). Reprinted with permission from the American Chemical Society.

Table 17-5, p. 17-11, from R. C. Reid, J. M. Prausnitz, and T. K. Sherwood, *The Properties of Gases and Liquids,* 3rd ed., McGraw-Hill Book Co., New York, pp. 544–601 (1977). Reprinted with permission from McGraw-Hill Book Co.

CHAPTER 18

Figure 18-3, p. 18-8, from R. W. Prugh, "Estimation of Flash Point Temperature," *J. Chem. Educ.,* **50,** A85–A89 (1973). Reprinted with permission from the American Chemical Society.

CHAPTER 19

Table 19-5, p. 19-11, from R. C. Reid, J. M. Prausnitz, and T. K. Sherwood, *The Properties of Gases and Liquids,* 3rd ed., McGraw-Hill Book Co., New York (1977). Reprinted with permission from McGraw-Hill Book Co.

Figure 19-8, p. 19-17, from A. Immirzi and B. Perini, "Prediction of Density in Organic Crystals," *Acta Crystallogr. Sect. A,* **33,** 216–218 (1977), based on data from *Acta Crystallogr. Sect. B,* **31** (1975). Reprinted with permission from the International Union of Crystallography.

Table 19-9, p. 19-18, from A. Immirzi and B. Perini, "Prediction of Density in Organic Crystals." *Acta Crystallogr. Sect. A,* **33,** 216–218 (1977). Reprinted with permission from the International Union of Crystallography.

CHAPTER 20

Table 20-2, p. 20-5, from O. R. Quayle, "The Parachors of Organic Compounds," *Chem. Rev.,* **53,** 439–589 (1953). Reprinted with permission of the Williams & Wilkins Co., Baltimore.

Table 20-3, pp. 20-6 to 20-8, and Table 20-4, p. 20-11, from R. C. Reid, J. M. Prausnitz, and T. K. Sherwood, *The Properties of Gases and Liquids,* 3rd ed., McGraw-Hill Book Co., New York (1977). Reprinted with permission from McGraw-Hill Book Co.

Table 20-5, p. 20-12, from O. R. Quayle, "The Parachors of Organic Compounds," *Chem. Rev.,* **53,** 439–589 (1953), R. C. Reid, J. M. Prausnitz, and T. K. Sherwood, *The Properties of Gases and Liquids.* 3rd ed., McGraw-Hill Book Co., New York (1977). Reprinted with permission of the Williams & Wilkins Co. and permission from McGraw-Hill Book Co.

CHAPTER 21

Table 21-4, p. 21-11, from R. C. Weast (ed.), *CRC Handbook of Chemistry and Physics,* 59th ed., CRC Press, West Palm Beach, Fla. (1978–1979). Reprinted with permission from the Chemical Rubber Co., CRC Press, Inc.

CHAPTER 22

Figure 22-1, p. 22-7, from W. R. Gambill, "How P and T Change Liquid Viscosity," *Chem. Eng.* (NY), 123 (Feb. 9, 1959). Reprinted with permission from McGraw-Hill, Inc.

Table 22-4, pp. 22-11 to 22-12, from D. van Velzen, R. Lopes Cardozo, and H.

Lagenkamp, "A Liquid Viscosity-Temperature-Chemical Constitution Relation for Organic Compounds," *Ind. Eng. Chem. Fundam.*, **11**, 20 (1972). Reprinted with permission from the American Chemical Society.

CHAPTER 23

Table 23-3, pp. 23-5 to 23-7, from D. N. Rihani and L. K. Doraiswamy, "Estimation of Heat Capacity of Organic Compounds from Group Contributions," *Ind. Eng. Chem. Fundam.*, **4**, 17–21 (1965). Reprinted with permission from the American Chemical Society.

Table 23-4, pp. 23-10 to 23-14, from S. W. Benson, F. R. Cruickshank, D. M. Golden, G. R. Haugen, H. E. O'Neal, A. S. Rogers, R. Shaw, and P. Walsh, "Additivity Rules for the Estimation of Thermochemical Properties," *Chem. Rev.*, **69**, 279–324 (1969), and R. C. Reid, J. M. Prausnitz, and T. K. Sherwood, *The Properties of Gases and Liquids*, 3rd ed., McGraw-Hill Book Co., New York (1977). Reprinted with permission from the American Chemical Society and from McGraw-Hill Book Co.

Table 23-6, p. 23-18, from A. I. Johnson and C. J. Huang, "Estimation of the Heat Capacities of Organic Liquids," *Can. J. Technol.*, **33**, 421–425 (1955).

Table 23-7, p. 23-20, from C. F. Chueh and A. C. Swanson: (a) "Heat Transfer: Estimating Liquid Heat Capacity," *Chem. Eng. Prog.*, **69** (7) 83–85 (1973); (b) "Estimation of Liquid Heat Capacity," *Can. J. Chem. Eng.*, **51**, 596–600 (1973). Reprinted with permission from the American Institute of Chemical Engineers and the Canadian Society for Chemical Engineering.

CHAPTER 24

Table 24-2, p. 24-3, from L. A. Robbins and C. L. Kingrea: (a) *Hydrocarbon Process. Pet. Refiner*, **41** (5), 133 (1962); (b) preprint of paper presented at the Sess. Chem. Eng. 27th Midyear Meet. Am. Pet. Inst., Div. Refining, San Francisco, May 14, 1962. Reprinted with permission from McGraw-Hill, Inc., and Gulf Publishing Co.

CHAPTER 25

Table 25-1, p. 25-5, from C. P. Smyth, *Dielectric Behavior and Structure,* McGraw-Hill Book Co., New York (1955). Reprinted with permission from McGraw-Hill Book Co.

Table 25-2, p. 25-6, from S. H. Fishtine, "Reliable Latent Heats of Vaporization," *Ind. Eng. Chem.*, **55** (6) 47–56 (1963). Reprinted with permission from the American Chemical Society.

Observed dipole moments in Table 25-3, p. 25-7, as reported by C. P. Smyth, *Dielectric Behavior and Structure,* McGraw-Hill Book Co., New York (1955). Reprinted with permission from McGraw-Hill Book Co.

CHAPTER 26

Table 26-4, p. 26-7, from P. I. Gold and G. J. Ogle, "Estimating Thermophysical Properties of Liquids—Part II: Parachor, Others," *Chem. Eng.* (NY), **76**, 97–100 (August 11, 1969). Reprinted with permission from McGraw-Hill, Inc.

Table 26-5, pp. 26-11 to 26-12, from A. I. Vogel, "Physical Properties and Chemical Constitution, Part XXIII. Miscellaneous Compounds, Investigation of the So-Called Coordinate or Dative Link in Esters of Oxy-acids and in Nitroparaffins by Molecular Refractivity Determinations. Atomic, Structure and Group Parachors and Refractivities," *J. Chem. Soc.*, 1833–1854 (1948). Reprinted with permission from the Royal Society of Chemistry.

Table 26-6, pp. 26-15 to 26-18, from C. Hansch, A. Leo, S. H. Unger, K. H. Kum, D. Nikatiani, and E. J. Lien, "Aromatic Substituent Constants for Structure-Activity Correlations," *J. Med. Chem.*, **16**, 1206–1216 (1973). Reprinted with permission from the American Chemical Society.

Table 26-7, p. 26-20, from S. S. Kurtz, Jr., S. Amon, and A. Sankin, "Effect of Temperature on Density and Refractive Index," *Ind. Eng. Chem.*, **42**, 174–176 (1950). Reprinted in part with permission from the American Chemical Society.

APPENDIX B

Figure B-1, p. B-2, adapted from N. R. Draper and H. Smith, *Applied Regression Analysis,* John Wiley & Sons, New York (1966), p. 9. Reprinted with permission from John Wiley & Sons, Inc.

Table B-1, p. B-2, from N. R. Draper and H. Smith, *Applied Regression Analysis,* John Wiley & Sons, New York (1966), p. 8. Reprinted with permission from John Wiley & Sons, Inc.

Figure B-3, p. B-8, adapted from N. R. Draper and H. Smith, *Applied Regression Analysis,* John Wiley & Sons, New York (1966), p. 9. Reprinted with permission from John Wiley & Sons, Inc.

Table B-2, p. B-11, and Table B-4, pp. B-18 to B-19, adapted from D. F. Morrison, *Multivariate Statistical Methods,* 2nd ed., McGraw-Hill Book Co., New York (1976). Reprinted with permission from McGraw-Hill Book Co.

Table B-5, p. B-21, adapted from N. R. Draper and H. Smith, *Applied Regression Analysis,* John Wiley & Sons, New York (1966), p. 12. Reprinted with permission from John Wiley & Sons, Inc.